Optimization of Large Structural Systems

NATO ASI Series

Advanced Science Institutes Series

A Series presenting the results of activities sponsored by the NATO Science Committee, which aims at the dissemination of advanced scientific and technological knowledge, with a view to strengthening links between scientific communities.

The Series is published by an international board of publishers in conjunction with the NATO Scientific Affairs Division

A Life Sciences
B Physics
Plenum Publishing Corporation
London and New York

C Mathematical and Physical Sciences
D Behavioural and Social Sciences
E Applied Sciences
Kluwer Academic Publishers
Dordrecht, Boston and London

F Computer and Systems Sciences
G Ecological Sciences
H Cell Biology
I Global Environmental Change
Springer-Verlag
Berlin, Heidelberg, New York, London, Paris and Tokyo

NATO-PCO-DATA BASE

The electronic index to the NATO ASI Series provides full bibliographical references (with keywords and/or abstracts) to more than 30000 contributions from international scientists published in all sections of the NATO ASI Series.
Access to the NATO-PCO-DATA BASE is possible in two ways:

– via online FILE 128 (NATO-PCO-DATA BASE) hosted by ESRIN, Via Galileo Galilei, I-00044 Frascati, Italy.

– via CD-ROM "NATO-PCO-DATA BASE" with user-friendly retrieval software in English, French and German (© WTV GmbH and DATAWARE Technologies Inc. 1989).

The CD-ROM can be ordered through any member of the Board of Publishers or through NATO-PCO, Overijse, Belgium.

Series E: Applied Sciences - Vol. 231–I

Optimization of Large Structural Systems

Volume I

edited by

G. I. N. Rozvany
Professor of Structural Design,
Essen University,
Germany

Springer-Science+Business Media, B.V.

Proceedings of the NATO/DFG Advanced Study Institute on
Optimization of Large Structural Systems
Berchtesgaden, Germany
23 September – 4 October 1991

Library of Congress Cataloging-in-Publication Data

NATO/DFG Advanced Study Institute (1991 : Berchtesgaden, Germany)
Optimization of large structural systems : proceedings of the NATO/DFG Advanced Study Institute, Berchtesgaden, Germany, 23 Sept.-4 Oct. 1991 / edited by G.I.N. Rozvany.
p. cm. -- (NATO ASI series. Series E, Applied sciences ; vol. 231)
Includes index.
ISBN 978-94-010-9579-2 ISBN 978-94-010-9577-8 (eBook)
DOI 10.1007/978-94-010-9577-8
1. Structural optimization--Congresses. I. Rozvany, G. I. N. II. North Atlantic Treaty Organization. III. Deutsche Forschungsgemeinschaft. IV. Title. V. Series: NATO ASI series. Series E, Applied sciences ; no. 231.
TA658.8.N39 1991
624.1'771--dc20 92-43799

ISBN 978-94-010-9579-2

Printed on acid-free paper

TABLE OF CONTENTS

VOLUME I

Mathematical Programming and Global Optima

VOLUME II

Composite, Anisotropic and Nonlinear Materials

Neural Networks, Parallel Processing, Multicriteria and Control Problems

Applications

Miscellaneous Topics

NOTE: Short contributions to this NATO/DFG ASI whose text does not appear in full in these proceedings will be considered for publication in the international journal **"Structural Optimization"** (Springer-Verlag).

PREFACE

G.I.N. Rozvany
ASI Director,
Professor of Structural Design,
FB 10, Essen University, Essen, Germany

Structural optimization deals with the optimal design of all systems that consist, at least partially, of solids and are subject to stresses and deformations. This integrated discipline plays an increasingly important role in all branches of technology, including aerospace, structural, mechanical, civil and chemical engineering as well as energy generation and building technology. In fact, the design of most man-made objects, ranging from space-ships and long-span bridges to tennis rackets and artificial organs, can be improved considerably if human intuition is enhanced by means of computer-aided, systematic decisions.

In analysing highly complex structural systems in practice, discretization is unavoidable because closed-form analytical solutions are only available for relatively simple, idealized problems. To keep discretization errors to a minimum, it is desirable to use a relatively large number of elements. Modern computer technology enables us to analyse systems with many thousand degrees of freedom. In the optimization of structural systems, however, most currently available methods are restricted to at most a few hundred variables or a few hundred active constraints. To eliminate this discrepancy between analysis capability and optimization capability, radically new or greatly improved methods are necessary for the optimal design of large and highly complex structural systems. The aim of this NATO ASI was to review critically such new methods, including novel optimality criteria techniques, systematic decomposition of engineering systems with an emphasis on the multi-disciplinary approach, improved methods of sensitivity analysis, topology optimization, approximation concepts, advances in mathematical programming, new composite materials, artificial neural networks and genetic algorithms, multicriteria problems, passive and active control as well as applications of all the above methods to various classes of practical problems.

It is hoped that in fulfilling the above objectives, the body of knowledge compiled at this meeting will result not only in substantial financial savings in all branches of technology but also, through reductions in material consumption, in energy savings, improved environment protection, better public safety and a diminished exploitation of mineral resources.

On the basis of extensive discussions with members of the organizing committee, lecturers and other participants, we can conclude that the achievements of this NATO ASI included the following positive aspects.

- In terms of the number of participants, it has been the biggest international meeting devoted exclusively to structural optimization.

- The number of lecturers was unusually large (40) for a NATO ASI and included most leading experts in this important field.
- In accordance with the guidelines of the NATO Scientific Division, the lecturers gave a comprehensive state-of-the-art report on various fields that have an impact on the optimization of large structural systems.
- The International Society of Structural and Multidisciplinary Optimization (ISSMO) was founded during this NATO ASI.
- This was the first NATO ASI on structural optimization which included Eastern European participants subsidized by the NATO.
- Following NATO ASI guidelines, frequent social programs, ranging from coffee breaks and after-session get-togethers with free catering as well as an extensive excursion and sight-seeing program ensured that scientists and practitioners engaged in structural optimization get to know each other and discuss their technical problems at an informal level during such a relatively long (two-week) meeting.

Finally, it should be mentioned that two important factors contributed to the obvious popularity of this ASI.

First, as even Alexander von Humboldt, a leading German scientist and philantropist, remarked some 200 years ago, the picturesque scenery around Berchtesgaden can compete with the most beautiful ones in the world. This, and the reliably warm and sunny weather in late autumn, provided a very pleasant setting for the meeting and, in particular, for all non-scientific programs, such as walks in the national park with breathtaking views, jovial "Hüttenabende" at various mountain tops or sightseeing trips to places of cultural or historic interest, including the neighbouring Salzburg.

Second, the large number of subsidised participants, and the additional benefits offered to them, were made possible by keeping a very tight control on finances, with a lot of voluntary labour by the staff of my university department, who produced very economically not only publicity material, programs and lecture notes but also meal tickets and carrying bags for this meeting. The participants also benefitted from the very reasonable accommodation, catering and congress centre charges in Berchtesgaden.

I wish to express my most sincere gratitude to the NATO Scientific Division and the German Research Foundation (DFG) for their financial support; to all speakers for the careful preparation of their presentations and lecture notes; to all other participants for their high level of interest and stimulating discussions; to the Director of the Berchtesgaden congress centre, Renate Rauscher for her outstanding organisational support; to the staff of my department, in particular to Anne Fischer, Sabine Kramer, Wilfried Gollub, Ming Zhou, Dirk Gerdes, Torben Birker and Ole Sigmund for their help; and last but not least, to the organizing secretary, Susann Rozvany for her untiring efforts in preparing and running the meeting, as well as for putting together these proceedings.

LIST OF PARTICIPANTS

DIRECTOR
Prof. Dr. G. Rozvany, Essen University, Germany

DEPUTY DIRECTOR
Prof. U. Kirsch, Technion – Israel Institute of Technology, Haifa, Israel

ORGANIZING COMMITTEE
Prof. E.J. Haug, The University of Iowa, USA
Dr. G. Lecina, Dassault, Suresnes, France
Prof. C.A. Mota Soares, CEMUL, Lisbon, Portugal
Prof. G.I.N. Rozvany, Essen University, Germany
Dr. J. Sobieski, NASA Langley Research Center, Hampton, VA, USA
Prof. B.H.V. Topping, Heriot-Watt University, Edinburgh, UK

LECTURERS
Dr. H. Baier, Dornier, Friedrichshafen, Germany
Prof. N. Banichuk, Institute of Mechanical Problems, Moscow, USSR
Prof. M.P. Bendsøe, Technical University of Denmark
Dr. L. Berke, NASA Lewis Research Center, Cleveland, Ohio, USA
Prof. V. Braibant, SAMTECH, Liège, Belgium
Prof. G. Cheng, Dalian Institute of Technology, P.R. China
Prof. K.K. Choi, The University of Iowa, USA
Prof. C. Cinquini, University of Pavia, Italy
Prof. K. Dems, Lódź Technical University, Poland
Prof. H. Eschenauer, Siegen University, Germany
Dr. B. Esping, ALFGAM, Stockholm, Sweden
Prof. C. Fleury, University of Liège, Belgium
Prof. M.B. Fuchs, Tel-Aviv University, Israel
Prof. D.E. Grierson, University of Waterloo, Ontario, Canada
Prof. W. Gutkowski, Polish Academy of Sciences, Warsaw, Poland
Prof. R.T. Haftka, Virginia Polytechnic Institute and State University, USA
Prof. J. Herskovits, Federal University of Rio de Janeiro, Brazil
Prof. B.L. Karihaloo, The University of Sydney, Australia
Prof. N. Kikuchi, The University of Michigan, USA
Prof. U. Kirsch, Technion – Israel Institute of Technology, Haifa, Israel
Dr. N.S. Khot, Wright Patterson Air Force Base, Dayton, Ohio, USA
Dr. V. Kobelev, USSR Academy of Sciences, Moscow, USSR
Prof. J. Koski, Tampere University of Technology, Finland
Dr. T. Lekszycki, Polish Academy of Sciences, Warsaw, Poland
Prof. C.A. Mota Soares, CEMUL, Lisbon, Portugal

Prof. Z. Mróz, Polish Academy of Sciences, Warsaw, Poland
Prof. N. Olhoff, Aalborg University, Denmark
Prof. P. Pedersen, Technical University of Denmark
Mr. H. Rapp, MBB, München, Germany
Dr. U.T. Ringertz, The Aeronautical Research Institute of Sweden
Dr. M. Rönnqvist, Linköping University, Sweden
Prof. B. Rousselet, University of Nice, France
Prof. G.I.N. Rozvany, Essen University, Germany
Prof. E. Schnack, Karlsruhe University, Germany
Dr. J. Sobieski, NASA Langley Research Center, Hampton, Virginia, USA
Prof. K. Svanberg, The Royal Institute of Technology, Stockholm, Sweden
Prof. J.E. Taylor, The University of Michigan, USA
Prof. B.H.V. Topping, Heriot-Watt University, Edinburgh, UK
Dr. V.B. Venkayya, Wright-Patterson Air Force Base, Dayton, Ohio, USA
Prof. Yamakawa, Waseda University, Tokyo, Japan

OTHER PARTICIPANTS

Belgium
Mr. P. Duysinx, University of Liège
Mr. P. Morelle, SAMTECH, Liège

Canada
Mr. C.M. Chan, University of Waterloo, Ontario
Prof. M.Z. Cohn, University of Waterloo, Ontario
Mr. L. Xu, University of Waterloo, Ontario

P.R. China
Dr. Y. Gu, Dalian Institute of Technology

Denmark
Mr. M. Hansen, Aalborg University
Mr. F. Jensen, Aalborg University
Mr. O. Jørgensen, The Technical University of Denmark
Mr. E. Lund, Aalborg University
Mr. T. Petersen, The Technical University of Denmark
Mr. O. Sigmund, The Technical University of Denmark

Estonia
Prof. U. Lepik, Tartu University

France
Prof. D. Chenais, University of Nice, France

Germany
Mr. W. Achtziger, University of Bayreuth
Mr. T. Birker, Essen University
Mr. D. Gerdes, Essen University
Dr. W. Gollub, Essen University
Mr. H. Hallmann, Essen University
Prof. D. Hartmann, Bochum University
Mr. O. Iancu, Daimler Benz, Stuttgart

Mr. S. Kimmich, T-Programme, Reutlingen
Ms. S. Kramer, Essen University
Dr. K.R. Leimbach, Bochum University
Prof. K. Leśniak, TH Darmstadt
Prof. O. Mahrenholtz, TU Hamburg-Harburg
Mr. W. Marb, MacNeal and Schwendler, München
Dr. H.P. Mlejnek, Stuttgart University
Mr. Nottebaum, Aachen University
Mr. R. Reitinger, Stuttgart University
Prof. C. Richter, TH Köthen
Prof. W. Schäfer, TH Leipzig
Mr. G. Schumacher, Siegen University
Mr. F. Spengemann, Essen University
Dr. G. Trippler, TH Leipzig
Dr. P.K. Umesha, Bochum University
Mr. C.T. Weber, TH Köthen
Mr. K. Wieghardt, Bochum University
Mr. B. Yuan, University of Karlsruhe
Mr. M. Zhou, Essen University
Prof. J. Zowe, University of Bayreuth

Greece
Dr. C.C. Baniotopoulos, Aristotle University, Thessaloniki
Mr. P. Georgiou, National Technical University of Athens
Mr. V.K. Koumousis, National Technical University of Athens

Hungary
Dr. K. Jarmai, Technical University of Miskolc
Dr. J. Lógó, Technical University of Budapest
Mr. T. Havady, Technical University of Budapest
Dr. A. Vásárhelyi, Technical University of Budapest

Israel
Dr. R. Levy, Technion – Israel Institute of Technology, Haifa

Italy
Mr. M. Lombardi, University of Pavia
Mr. Strona, Centri Recherchi Fiat, Orbassani

Japan
Mr. K. Akahori, Fujitsu America, Inc., USA
Dr. J. Fukushima, Toyota Technical Center, USA

Korea
Dr. Y.M. Yoo, Korea Automotive Tech. Inst., Seoul

The Netherlands
Mr. J. Mendoza, ESA, Noordwijk
Prof. D.H. van Campen, Technical University of Eindhoven
Dr. A.J.G. Schoofs, Technical University of Eindhoven

Poland
Dr. T. Burczynski, Silesian Technical University
Dr. P. Fedelinski, Silesian Technical University
Dr. S. Jendo, Polish Academy of Sciences, Warsaw
Dr. S. Imielowski, Polish Academy of Sciences, Warsaw
Dr. T. Lewinski, Warsaw University of Technology
Prof. A. Osyczka, Technical University of Cracow
Dr. K. Szuwalski, Technical University of Cracow

Portugal
Prof. A. Adao-da-Fonseca, University of Porto
Mr. J.I. Barbosa, Cemul, Lisbon
Prof. R. Leal, University of Coimbra
Prof. C.M. Mota Soares, Cemul, Lisbon
Prof. H. Rodrigues, Cemul, Lisbon
Prof. J.L.T. Santos, Instituto Superior Tecnico, Lisbon
Prof. L.M.C. Simoes, University of Coimbra
Prof. A. Soeiro, FEUP, Porto

Slovenia
Mr. A. Mihelic, University of Ljubljana

South Africa
Prof. J.A. Snyman, University of Pretoria
Prof. N. Stander, University of Pretoria

Spain
Dr. S. Hernandez, University of Zaragoza
Dr. P. Marti-Montrull, Universidad Politecnica, Valencia
Prof. J.A. Tarrago, Escuela Tecnico Superior de Ingenieros, Bilbao

Republic of China
Prof. Chen, National Cheng Kung Univ., Tainan

Turkey
Dr. N. Akkas, Middle East Technical University, Ankara
Dr. M. Utku, Middle East Technical University, Ankara

UK
Ms. O. Abraham, University of Wales, Cardiff
Dr. J. Blachut, The University of Liverpool
Dr. R. Butler, University of Bath
Mr. P.D. Gosling, Warwick University, Coventry
Mr. A.I. Khan, Heriot-Watt University, Edinburgh
Dr. W. McKenzie, Napier Polytechnic
Dr. M. Philip, Brighton Polytechnic
Mr. N.V. Ramana Rao, University College of Swansea
Mr. J. Sienz, University College of Swansea
Mr. C. York, University of Wales, Cardiff

USA
Prof. J.S. Arora, Iowa University, Iowa

Dr. J.-F. Barthelemy, NASA Langley Research Center, Virginia
Prof. A.D. Belegundu, Pennsylvania State University, Pennsylvania
Dr. C. Bloebaum NASA Langley Research Center, Virginia
Prof. A. Chattopadhyay, Arizona State University, Arizona
Dr. Y.W. Chun, Villanova University, Pennsylvania
Prof. A. Diaz, Michigan State University, Michigan
Dr. M.E.M. El-Sayed, University of Missouri-Columbia, Missouri
Prof. F.E. Fagundo, University of Florida, Florida
Prof. R.V. Grandhi, Wright State University, Ohio
Prof. Z. Gürdal, Virginia Polytechnic Institute and State University, Virginia
Prof. P. Hajela, Rensselaer Polytechnic, Troy, N.Y.
Prof. J.S. Lamancusa, Pennsylvania State University
Prof. E. Nikolaidis, Virginia Polytechnic and State University, Virginia
Dr. H.L. Thomas, VMA Engineering, California
Prof. D.A. Tortorelli, University of Illinois at Urbana-Champaign
Prof. B.P. Wang, The University of Texas at Arlington, Texas

USSR

Mr. A.D. Larichev, TsNIISK, Moscow
Prof. Malkov, Nizhny Novgorod University
Dr. S.V. Selyugin, TsAGI, Moscow
Prof. A.P. Seyranian, Institute of Mechanical Problems, Moscow
Dr. V. Toropov, Nizhny Novgorod University

Bird's-eye view of Berchtesgaden with Watzmann in the background
(Photo: Baumann-Schicht, courtesy of Kurdirektion)

OPENING ADDRESS

L. Berke
Chief Scientist, Structures Division,
NASA Lewis Research Center,
Cleveland, Ohio, USA

It is a very great pleasure indeed to have this opportunity to present a few opening remarks at this NATO Advanced Study Institute on the Optimization of Large Structural Systems, held in these beautiful surroundings. This geographical area is something very special for me, I am the third time among these mountains, but this occasion is by far the most pleasant one. The first time I came to Salzburg, on the other side of these picturesque mountains, was after escaping from a Russian prison camp in Austria in August 1945. I ended up washing ten thousand pots and pans for a year at an American military kitchen in Salzburg, while learning to speak what since then became the international language of science, broken English. The second time I ended up in this neighbourhood was in 1956 escaping the Russians again, this time during the Hungarian revolution. My wife and I ended up temporarily in a Salzburg refugee camp. So you can see that this NATO ASI is a much more pleasant occasion for me.

Let me turn now to the reason for our gathering here among these beautiful mountains. For the past two decades every NATO ASI on optimization was a landmark event. The collection of the proceedings of these NATO Advanced Study Institutes is a library of structural optimization that contains all the important advancements one needs to know. This ASI will follow in the same tradition. I would like to point out now a few aspects that are already influencing the state of affairs in structural optimization.

The acceptance of a new technology requires a certain degree of change in our technical culture. The slide rule culture was replaced by the computer culture. The culture of expert approximations by great structural engineers with decades of experience was replaced by the culture of automated finite element analysis, etc. These cultural changes are brought about by the synergism of new technical generations on one hand and new needs and new opportunities of technological advancements on the other.

Currently the new needs are evolving from such extreme projects as the European and the American single and two-stage-to-orbit hypersonic/orbital vehicles, and from the very strong competition in the commerical transport industry for the development of more efficient vehicles in the subsonic and now also in the supersonic regimes. Recently, the U.S. formally started a High Speed Civil Transport Program for a 200-300 passenger Mach 2.4 aircraft with operating efficiencies that would keep the air fare close to current subsonic air fares. For the airframes and engines of such vehicles new material systems and new structural concepts are being developed. Human design experience does not exist in these areas and, therefore, formal optimization needs to be applied both *to design* the new material systems and *to design with* the new material systems. The design has to be optimal for its total life. As opposed to previous approaches not only its performance but also its manufacturing and operating considerations have to enter the design optimization in a simultaneous, or as it is now called, in a "concurrent engineering" mode. Probability considerations and risk assessment are important aspects of machinery operating in extreme environments. All these methods are to operate in a highly automated CAD/CAM world.

At the same time computer science is in a state of fermentation developing new computer architectures and concepts to complement the traditional von Neumann machines. Parallel machines from transputer based research systems to massively parallel fine grain machines of various kinds and connectivities are being offered and are being further developed. The famous Japanese 5th generation project ends in mid 1992 after producing a series of interference machines. The goal was a massively parallel knowledge based non-numeric interference machine with spoken communication and language to language translation capabilities reminding one of the conversational machine HAL in the science fiction movie: *2001, A Space Odessey.* Except for the speech capabilities the goals were met, even if on a smaller scale than originally stated. Now a sixth generation project might get under way with number crunching capabilities and biologically motivated neural nets added to the inferencing capabilities. The US has the High Performance Computer Initiative on its way towards a terraflop number cruncher for engineering supercomputing. The work station world is experiencing rapid evolution both in computing power and in networking.

The software environment is also evolving out of the traditional FORTRAN world and the C and lately the oriented C^{++} environment is here. The latest FEM and multidisciplinary framework codes we wrote in C^{++} in my organization at NASA Lewis Research Center.

One could go on and on and the question we now ask is what does all that mean if examined through the eyes of the structural optimization methods developer. I think what it means is very exciting times ahead with much increased need in the changing design environment for formal optimization methods, that is, for the fruits of our labour. We recently had two NASA-Propulsion Industry Workshops discussing Computational Structures Technology needs and opportunities. For the first time leading technical people in the propulsion industry asked for automated design methods based on optimization and artificial intelligence techniques. It should feel good to be needed more than ever, or at least "finally". With that let's go to work and let's have an excellent ASI here in beautiful Berchtesgaden.

CLOSING ADDRESS

B.L. Karihaloo
School of Civil and Mining Engineering, The University of Sydney

Having had at times to perform the somewhat unpleasant duty of keeping the speakers from straying outside the allocated time for their presentations, I have at last gained poetic justice by having to perform a very pleasant task. Let me say straightaway what that is. It is to thank George for having organised a very successful ASI. The ingredients of any successful conference are the quality of speakers, the quality of organisation and the quality of servicing mundane human needs. I am sure you will all agree with me that on all these criteria this NATO ASI attained the global optimum. To attain the global optimum required all of George's organisational talents and his taste for all things, places and human beings that are the most beautiful.

But with apologies to our female friends and colleagues, I believe the optimum strategy proved successful because behind George stood a very dedicated and talented woman. I mean Susann Rozvany, who has been responsible in no uncertain terms for George's moment of glory.

This ASI has set new standards that will be hard to emulate let alone surpass, in the future. All of us no doubt gained some new insights into optimization which will help us in our future research pursuits. In difficult moments during these pursuits I have no doubt that the fond memories of this NATO ASI will tide us over.
Thank you very much Sue and George.

Church of Ramsau outside Berchtesgaden
(Photo: Baumann-Schicht, courtesy of Kurdirektion)

Idyllic setting in the outskirts of Berchtesgaden with Watzmann
(Photo: Baumann-Schicht, courtesy of Kurdirektion)

Königssee with St. Bartholomä Church and Watzmann
(Photo: Baumann-Schicht, courtesy of Kurdirektion)

CONTINUUM-BASED OPTIMALITY CRITERIA (COC) METHODS: AN INTRODUCTION

G.I.N. Rozvany and M. Zhou
FB 10, Essen University, D-4300 Essen 1, Germany

Abstract. After reviewing briefly problems and methods of structural optimization, continuum-based optimality criteria (COC) are presented for elastic design with stress and deflection constraints and then the criteria for optimal plastic design are derived as a special case of these general optimality conditions. The general proof of the proposed optimality criteria is formulated for a relaxed problem which requires only statical admissibility of the real and virtual stresses. Then it is shown by means of variational principles and an energy theorem that kinematic admissibility is ensured by an optimality condition. Alternate proofs on the basis of (a) the stationary mutual energy principle and (b) variational calculus (special case of Bernoulli-beams) is also given. Some of the criteria derived are interpreted in terms of a fictitious "adjoint" structure which has significant computational advantages in iterative procedures. Finally, reasons for non-fully-stressed solutions in stress design are explained through two analytical examples and a preview of the achievements of the iterative COC method is given.

1. Problems and Methods of Structural Optimization

In this introductory lecture, we explain

- the importance of the proposed (COC) methods, and
- the theoretical background of this approach.

The *aim of structural optimization* is

- the minimization (or maximization) of an *objective function* (e.g. cost of materials and labour, structural weight, storage capacity etc.),
- subject to
  (i) *geometrical constraints* (e.g. restriction on height, prescribed variation of the cross-sections over given "segments"), and
  (ii) *behavioural constraints* (e.g. restrictions on stresses, displacements, buckling load, natural frequency, etc.).

  Behavioural constraints can be
  (a) *local constraints*, in which only stresses or stress resultants for a given cross-section are involved, or
  (b) *global constraints*, which contain integrals of stresses or stress resultants for the entire structure. System instability (buckling), natural frequency and deflection constraints are global ones.

G. I. N. Rozvany (ed.), Optimization of Large Structural Systems, Vol. I, 1–26.

Depending on the relative proportions of their dimensions, structures may be idealised as

- *one-dimensional* continua (e.g. bars, beams, arches, rings, frames, trusses, beam-grids or grillages, shell grids, cable nets);
- *two-dimensional* continua (plates, disks, structures subject to plane stress or plane strain, shells, folded plates, "truss-like continua" (Prager, see Section 18), "grillage-like continua", "shell-grid-like continua", etc.);
- *three-dimensional* continua (stressed systems for which the above idealizations are not possible).

Problems of structural optimization may be classified as

- optimization of the *cross-sectional dimensions* (or "sizing") of one- or two-dimensional structures, for which the cross-sectional geometry is partially prescribed, so that the cross-section can be fully described by a finite number of variables;
- *shape optimization* (e.g. the shape of the centroidal axis of bars and the middle surface of shells, boundaries of continua or interfaces between different materials in composites);
- *layout optimization* which consists of three simultaneous operations:
  (a) topological optimization (spatial sequence or configuration of members and joints),
  (b) geometrical optimization (location of joints and shape of member axes), and
  (c) optimization of the cross-sections.

In the design of complex, real structural systems, *discretization* and the use of numerical methods is unavoidable. However, in order to achieve a reasonable accuracy it is necessary to use a *very large number of elements.*

Whereas the *analysis capability* of modern finite element software at present is between *ten thousand* and *hundred thousand* degrees of freedom, the *optimization capability* is restricted to a *few hundred variables* or a *few hundred active constraints*, if conventional methods (e.g. primal or dual mathematical programming, respectively) are used. For cross-sectional optimization problems without variable linking, this results in a *discrepancy between analysis capability and optimization capability*, which was pointed out repeatedly by Berke and Khot (e.g. [1]).

Methods of structural optimization fall into two major categories:

- *direct minimization techniques* (e.g. mathematical programming, MP), and
- *indirect methods* (e.g. optimality criteria, OC, methods).

In most of the so-called *"primal" mathematical programming* (MP) methods each iteration consists of two basic steps:

(1) calculation of the value of the *objective function* and *its gradients* with respect to all design variables for a solution; and
(2) calculation of a *change* of the design variables resulting in a cost reduction.

Steps (1) and (2) are repeated until a local minimum of the objective function is reached.

The *main advantage of MP methods* is their *robustness* which means that they are readily applicable to most problems within and outside the field of structural optimization. However, all MP methods require sensitivities, whose efficient calculation can be highly problem-dependent.

The *main disadvantage of primal MP methods* is their very limited optimization capability in terms of the number of variables.

Optimality criteria are necessary (and sometimes sufficient) conditions for minimality of the objective function.

Applications of optimality criteria methods include

- *small, idealized systems*, for which closed-form analytical solutions are obtained either
  (i) by hand, or
  (ii) by special analytically-based computer programs,
  with a view to determining
  (a) fundamental features of optimal solutions,
  (b) the range of validity and applicability of various numerical methods, and
  (c) the relative economy of realistic designs (basis of comparison); and
- *large, real systems*, where OC methods can be used for
  (a) checking the validity of solutions determined by other methods, and
  (b) developing efficient iterative re-sizing strategies.

Optimality criteria (OC) methods fall into two distinct categories, i.e. discretized and continuum-based OC-methods.

Discretized optimality criteria (DOC) methods have been developed since the late sixties by aero-space scientists (Berke, Khot, Venkayya) who used the Kuhn-Tucker minimality condition in *finite dimensional* design space, expressed in terms of *nodal forces*. Earlier OC approaches included such *intuitive methods* as fully stressed design, stress ratio methods and constant mutual energy design.

Continuum-based optimality criteria (COC) methods employ Euler-Lagrange type minimality conditions in *infinite dimensional design spaces*, using calculus of variations, control-theory and energy theorems. The optimality conditions, some of which are interpreted in terms of an *"adjoint structure"*, are differential equations in terms of *generalized stresses* (stresses or stress resultants), e.g. bending moments or shear forces.

It will be seen in Sections 15-17, however, that COC can also be formulated directly for discretized systems (termed DCOC). Therefore, *the most important difference between DOC and COC* methods is that in DOC the only variables considered in the formulation of the problem are the design variables (e.g. cross-sectional dimensions), whereas in the COC formulation the variables are the design variables as well as the real and virtual stress resultants (generalized stresses). This, and other differences between DOC and COC are summarized in Table 0 below.

In *iterative COC-methods*, each iteration consists of two basic steps:
(a) the analysis of the real and adjoint systems, and
(b) updating of the cross-sectional dimensions.

In various methods of structural optimization, the storage capacity and computer time requirement largely depend on the quantity indicated in Table 1, in which N is the number of design variables and m is the number of *active* behavioural constraints, out of which m_g is the number of global (e.g. deflection) constraints and m_ℓ is the number of local (e.g. stress) constraints.

	DOC	COC
Variables	Design Variables	Design Variables, Real and Virtual Forces
Formulation of Stress Contraints	through Equivalent Displacements	Directly in Terms of Real Forces
Equilibrium	Implicit	Explicit Equality Constraints
Compatibility	Implicit	Not Included, Implied by Optimality

Table 0. Main differences between the DOC and COC methods.

primal MP	dual MP, DOC	COC
N	$m = m_g + m_\ell$	m_g

Table 1. Quantities influencing the optimization capability of various methods.

For *large structural systems* the number of active global constraints (m_g) is usually much smaller than the number of active local constraints (m_ℓ), which results in a *very high efficiency of the COC method* for such systems. Therefore, *the main advantage of the COC method* is that, in general, it can highly efficiently optimize systems with an extremely large number of variables and active (local) constraints.

Disadvantages of the COC method are that (i) for each type of structure and for each type of design condition, a lengthy analytical derivation of the relevant optimality criteria is necessary, and (ii) in its present form it is limited to problems of structural optimization.

One of the sources of difficulties in the communication between analytical and numerical schools of structural optimization is the difference in the notation used, which will be illustrated by a simple example.

Considering a plane frame with flexural and axial deformations, the *analytical (continuum) notation* for curvature and axial strain, respectively, is κ and ε, and for the bending moment and axial force, respectively, M and N. The relation between these two sets of quantities is then given by

$$\begin{bmatrix} \kappa \\ \varepsilon \end{bmatrix} = \begin{bmatrix} \frac{1}{EI} & 0 \\ 0 & \frac{1}{EA} \end{bmatrix} \begin{bmatrix} M \\ N \end{bmatrix}, \tag{1}$$

where E is Young's modulus, I is the moment of inertia and A is the cross-sectional area.

In Prager's notation, the relation (1) is represented in the general form

$$\mathbf{q} = [\mathbf{F}]\mathbf{Q}, \tag{2}$$

where $\mathbf{q} = (\kappa, \varepsilon)$ are termed the ***generalized strains*** (referring to an entire cross-section), $\mathbf{Q} = (M, N)$ the ***generalized stresses*** (stress resultants) and

$$[\mathbf{F}] = \begin{bmatrix} \frac{1}{EI} & 0 \\ 0 & \frac{1}{EA} \end{bmatrix}$$

is the ***generalized flexibility matrix.*** All the above quantities vary continuously along the frame and, therefore, they depend on the longitudinal coordinate x: $\mathbf{q}(x)$, $\mathbf{Q}(x)$, $[\mathbf{F}](x)$. The advantage of Prager's notation is that theorems and formulae can be derived not only for three-dimensional continua but also for one- or two-dimensional structures (e.g. beams, frames, plates and shells).

In the case of the formulation for ***discretized systems***, the general form of flexibility relations for a structural element is usually expressed as

$$\left\{\mathbf{u}_f^e\right\} = \left[\mathbf{f}_f^e\right]\left\{\mathbf{F}_f^e\right\}, \tag{3}$$

where $\left\{\mathbf{u}_f^e\right\}$ is the element relative displacement vector, $\left[\mathbf{f}_f^e\right]$ is the element flexibility matrix and $\left\{\mathbf{F}_f^e\right\}$ is the element nodal force vector. The subscript f signifies the fact that in the flexibility formulation the element is supported in a stable, statically determinate manner. For a ***plane frame element*** with flexural and axial deformations, the support conditions can be chosen as those for a cantilever fixed at end A and then the symbols in (3) take on the following meaning:

$$\left\{\mathbf{u}_f^e\right\} = \left\{ u_{fB}^e \quad v_{fB}^e \quad \theta_{fB}^e \right\}, \quad \left\{\mathbf{F}_f^e\right\} = \left\{ F_{xB}^e \quad F_{yB}^e \quad M_B^e \right\}, \tag{4}$$

where B denotes the other end of the element, F_x amd F_y are horizontal and vertical forces, M is the end moment, and u, v and θ refer to horizontal and vertical displacements and rotation. Moreover

$$\left[\mathbf{f}_f^e\right] = \frac{1}{E}\begin{bmatrix} \frac{L}{A} & 0 & 0 \\ 0 & \frac{L^3}{3I} & \frac{L^2}{2I} \\ 0 & \frac{L^2}{2I} & \frac{L}{I} \end{bmatrix}. \tag{5}$$

The ***difference between the two formulations*** is that whilst the continuum formulation is expressed in terms of generalized strains and stresses at each cross-section, in the discretized formulation the strains are integrated over the element to give relative displacements.

The term "continuum-based" optimality criteria is sometimes misinterpreted by assuming that it refers to three-dimensional continua. This method is, in fact, used more often for one- and two-dimensional continua (frames, plates, shells) in terms of generalized stresses and strains used in (2) above.

Moreover, the second author has also rederived all COC type optimality criteria directly for discretized systems in terms of the usual matrix-notation [see, for example, (3)-(5) above] for such structures. These optimality criteria will be discussed

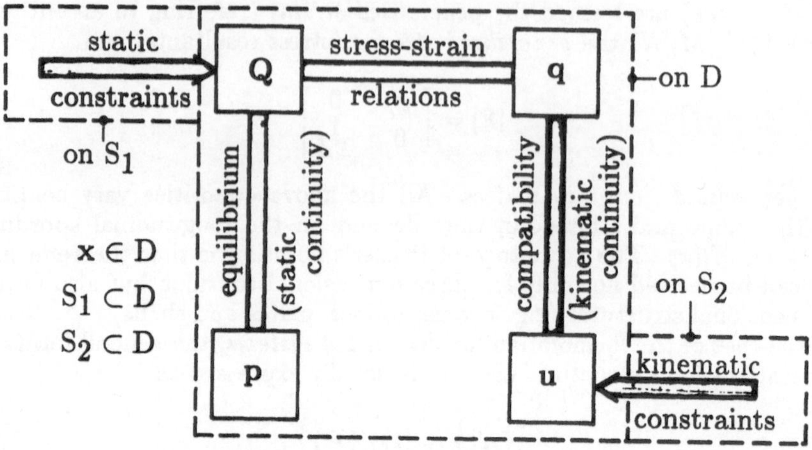

Fig. 1 Fundamental relations of structural mechanics.

in the second lecture (Sections 15-17) and termed DCOC (discretized continuum-based optimality criteria), where the term "continuum-based" refers merely to the historical origin of the approach. The real difference between DOC and DCOC was explained earlier, and lies in the role of variations in terms of real and virtual stress resultants in the formulation, which results in an "adjoint" or "modified virtual" system for the DCOC method.

A comprehensive list of optimality criteria for various design requirements and constraints is given in a recent book [2]. Continuum-based optimality criteria were also used in a number of publications by Olhoff, Taylor and Bendsøe.

2. Optimality Criteria for Freely Varying Cross-Sectional Dimensions of Linearly Elastic Structures with Deflection and Stress Constraints

Using Prager's terminology (e.g. [3]), the *basic variables of structural mechanics* are generalized stresses $\mathbf{Q}(\mathbf{x}) = [Q_1(\mathbf{x}), \ldots, Q_n(\mathbf{x})]$, generalized strains $\mathbf{q}(\mathbf{x}) = [q_1(\mathbf{x}), \ldots, q_n(\mathbf{x})]$, loads $\mathbf{p}(\mathbf{x}) = [p_1(\mathbf{x}), \ldots, p_m(\mathbf{x})]$ and displacements $\mathbf{u}(\mathbf{x}) = [u_1(\mathbf{x}), \ldots, u_m(\mathbf{x})]$, where $\mathbf{x}$ are the spatial coordinates. A generalized stress may represent a local stress or stress resultant (e.g. a bending moment or shear force) and a generalized strain may refer to a local strain or to an entire cross-section (e.g. curvature or twist of a bar).

Before stating the optimality conditions for the above class of problems, we review briefly the *fundamental relations of structural mechanics*. Considering an *elastic* system, for example, these relations are shown in Fig. 1. On the structural domain D, we must satisfy the equilibrium (or static continuity) conditions ($\mathbf{p}$, $\mathbf{Q}$), the compatibility (or kinematic continuity) conditions ($\mathbf{u}$, $\mathbf{q}$) and the generalized strain-stress relations ($\mathbf{Q}$, $\mathbf{q}$). In addition, on some subset S_1 of the domain D static constraints (or static boundary conditions) and on some other subset $S_2 \subset D$

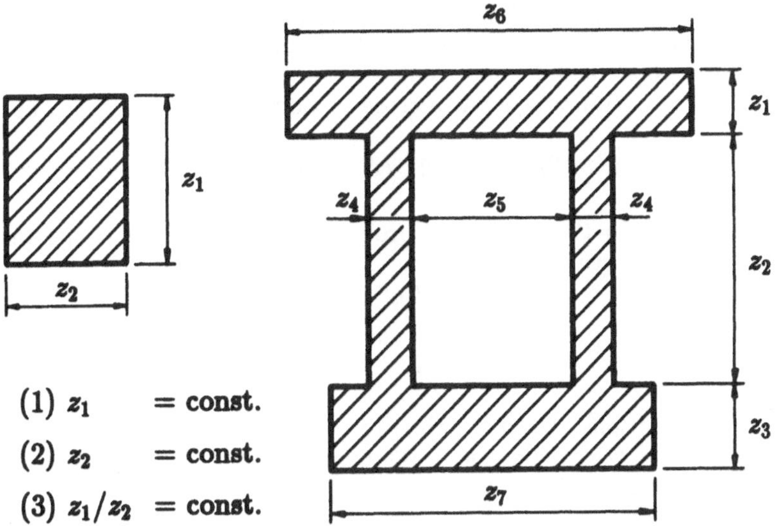

Fig. 2 Examples of cross-sectional parameters $\mathbf{z}$.

kinematic constraints (or kinematic boundary conditions) must be fulfilled.

A system of strains and displacements ($\mathbf{q}^K$, $\mathbf{u}^K$) are said to be *kinematically admissible* if they satisfy the kinematic continuity (compatability) and kinematic boundary conditions and a system of stresses and loads ($\mathbf{Q}^S$, $\mathbf{p}^S$) is called *statically admissible* if it satisfies the statical continuity (equilibrium) and statical boundary condition.

We consider structures for which the cross-sectional geometry is partially prescribed in such a way that the cross-section is fully defined by a finite number of variables

$$\mathbf{z}(\mathbf{x}) = [z_1(\mathbf{x}), \ldots, z_r(\mathbf{x})], \tag{6}$$

termed *cross-sectional parameters.* Examples of such parameters are given in Fig. 2.

The specific cost ψ (e.g. the weight, volume or material/construction costs per unit length or unit area) is a function of the cross-sectional parameters,

$$\psi(\mathbf{x}) = \psi[z_1(\mathbf{x}), \ldots, z_r(\mathbf{x})]. \tag{7}$$

The external loads $\mathbf{p}$ equilibrate the generalized stresses $\mathbf{Q}^S$, where the superscript S denotes statical admissibility. The structure is subject to a displacement constraint which can be expressed through the principle of virtual work in the form,

$$\int_D \mathbf{Q}^{S,K} \cdot [\mathbf{F}]\overline{\mathbf{Q}}_j^S \, \mathrm{d}x \leq \Delta_j \quad (j = 1, 2, \ldots, v), \tag{8}$$

where the superscript K denotes kinematic admissibility, $[\mathbf{F}]$ is the generalized flexibility matrix defined in (2) and $\overline{\mathbf{Q}}_j^S$ is the (statically admissible) virtual stress vector equilibrating some virtual load $\overline{\mathbf{p}}$. In the case of a displacement constraint at a single point, the virtual load is a unit ("dummy") load at that point in the direction of the prescribed displacement.

The local stress requirements at a cross-section can be represented as

$$S_\ell(\mathbf{z}, \mathbf{Q}) \leq 0 \quad (\ell = 1, 2, \ldots, t). \tag{9}$$

The "total cost" Φ for the considered problem is given by integrating the specific cost $\psi(\mathbf{x})$ for the structural domain D:

$$\Phi = \int_D \psi(\mathbf{x})\, \mathrm{d}\mathbf{x}. \tag{10}$$

Necessary conditions of cost minimality will be derived in Section 3 and are given in Fig. 3. The Lagrangians $\nu_j \geq 0$ (constant) and $\lambda_\ell(\mathbf{x})$ (function of $\mathbf{x}$) must also satisfy the conditions:

$$\begin{aligned} \nu_j \geq 0 \text{ and } \nu_j > 0 \quad &\text{only if} \quad \int_D \mathbf{Q}^{S,K} \cdot [\mathbf{F}]\overline{\mathbf{Q}}_j^S \,\mathrm{d}\mathbf{x} = \Delta_j\,, \\ \lambda_\ell(\mathbf{x}) \geq 0 \text{ and } \lambda_\ell(\mathbf{x}) > 0 \quad &\text{only if} \quad S_\ell[\mathbf{z}(\mathbf{x})\,,\, \mathbf{Q}^{S,K}(\mathbf{x})] = 0\,. \end{aligned} \tag{11}$$

It will be seen that Fig. 3 involves the usual fundamental relations for both the real structure and for a fictitious "adjoint" structure. For the latter, the equilibrium and compatability conditions are the same as for the real structure, but the generalized stress-strain relation is different, since the adjoint strains depend on both real and adjoint stresses. If only displacement constraints are active ($\lambda_\ell = 0$ for all ℓ), however, then the generalized stress-strain relations are the same for the real and adjoint systems. The static boundary conditions (constraints) are the same for both systems, but the kinematic boundary conditions can be different if the cost of reactions is non-zero or the real supports are non-rigid. The kinematic boundary conditions for the ajoint structure are given in the bottom right corner of Fig. 3, in which $\Omega(\mathbf{R})$ is the reaction cost function and $\mathbf{R} = (R_1, R_2, \ldots, R_n)$ the reactions. The kinematic boundary conditions for the real structure with elastic foundation is given at the bottom left corner where $[\mathbf{F}_R]$ is the generalized support flexibility matrix.

For relatively simple problems, all relations in Fig. 3 can be solved simultaneously in an analytical form. Many such solutions are summarized in a recent book [2]. However, for large, real systems an iterative procedure is necessary which consists of the following operations in each iteration:

(i) Analysis of the real structure.

(ii) Analysis of the adjoint structure. Since the adjoint strains also depend on the real stresses, but this influence is usually separable, the latter is included in the adjoint analysis in the form of pre-strains (e.g. temperature strains).

Both (i) and (ii) involve linear analysis and can be carried out conveniently by using existing FE software.

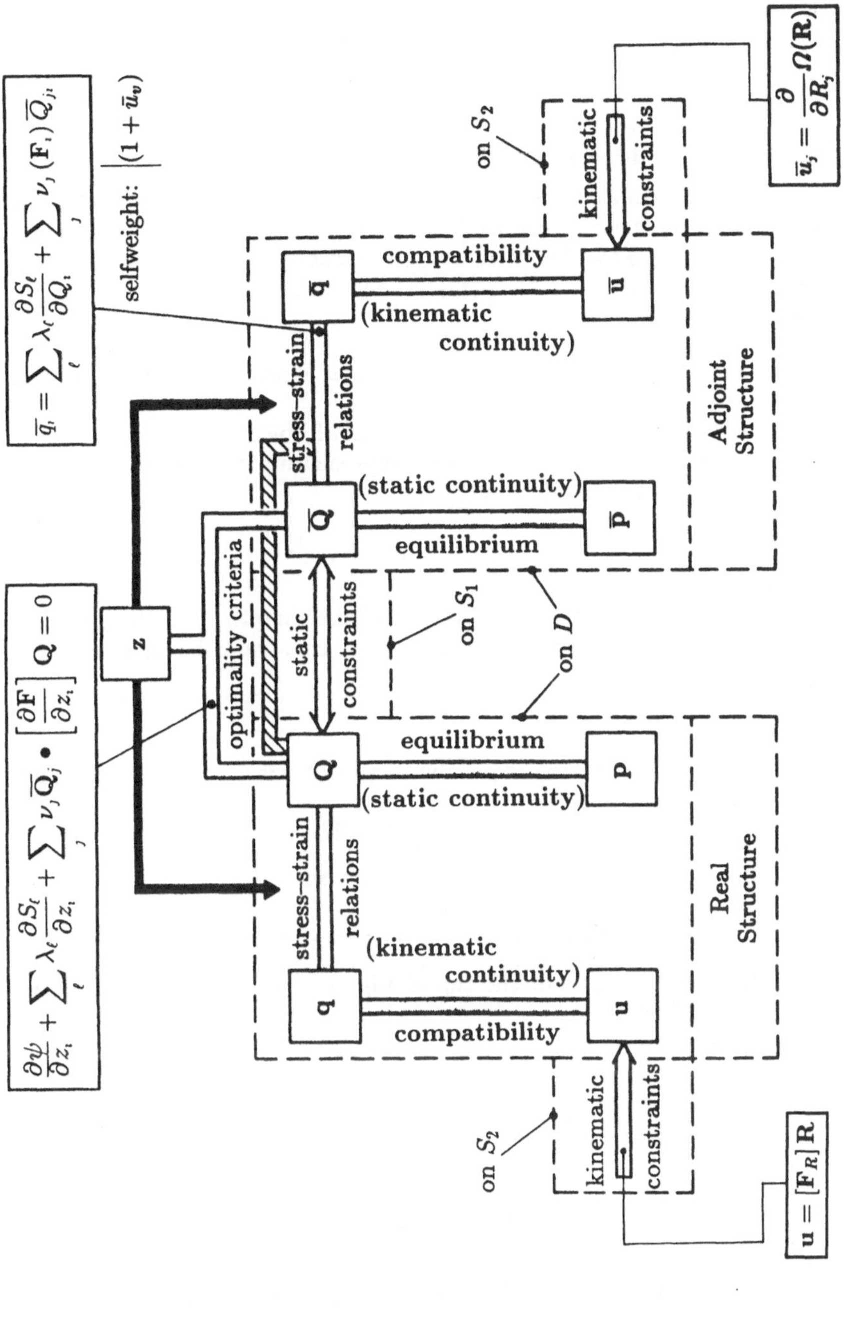

Fig. 3 Fundamental relations of optimal elastic design for stress and deflection constraints.

(iii) Updating of the Lagrangians (ν_i) for global constraints on the basis of the active global constraints.
(iv) Updating the cross-sectional dimensions ($\mathbf{z}$) on the basis of the optimality criteria.

2.1 Evaluation of the Lagrangians (λ_ℓ) for the Stress Constraints

The optimality criteria at the top of Fig. 3 furnish the same number of equations as the number (t) of design variables ($\mathbf{z}$). Additional equations are the active stress constraints, say $S_\ell(\mathbf{z}, \mathbf{Q}) = 0$ ($\ell = 1, 2, \ldots, w$), and by (11), w Lagrangians (λ_ℓ) are non-zero. This means that we have $(t + w)$ equations for the calculation of $(t + w)$ unknowns.

2.2 An Elementary Example: Bernoulli–Beam of Variable Width, Constraints on the Flexural Stresses and Deflection

We denote by $z(x)$ the variable width of a Bernoulli-beam with a rectangular cross-section. Then the specific cost function, representing the beam weight per unit length, and flexural stiffness $s(x)$ become

$$\psi = cz(x), \quad s(x) = rz(x), \tag{12}$$

with $c = h\gamma$ and $r = Eh^3/12$, where h is the beam depth, γ is the specific weight and E is Young's modulus. Moreover, the stress constraint becomes

$$S = (z - b|M|), \tag{13}$$

with $b = 6/(\sigma_p h^2)$, where σ_p is the permissible axial stress. The generalized stresses in this problem are

$$\mathbf{Q} = M, \quad \overline{\mathbf{Q}} = \overline{M}, \tag{14}$$

where M is the beam moment and the flexibility matrix together with its gradient reduces to

$$[\mathbf{F}] = 1/rz, \quad \mathrm{d}[\mathbf{F}]/\mathrm{d}z = -1/rz^2. \tag{15}$$

Then the optimality criterion at the top of Fig. 3 (with $z_1 \to z$, $r = 1$) implies

$$c - \frac{\nu M\overline{M}}{rz^2} + \lambda = 0, \tag{16}$$

and the adjoint strain-stress relation with $\overline{\mathbf{q}} = \overline{\kappa}$ becomes

$$\overline{\kappa} = -\lambda b \operatorname{sgn} M + \frac{\nu\overline{M}}{rz}. \tag{17}$$

The value of the Lagrangian λ can be determined from (16) as

$$\lambda = \frac{\nu M\overline{M}}{rz^2} - c. \tag{18}$$

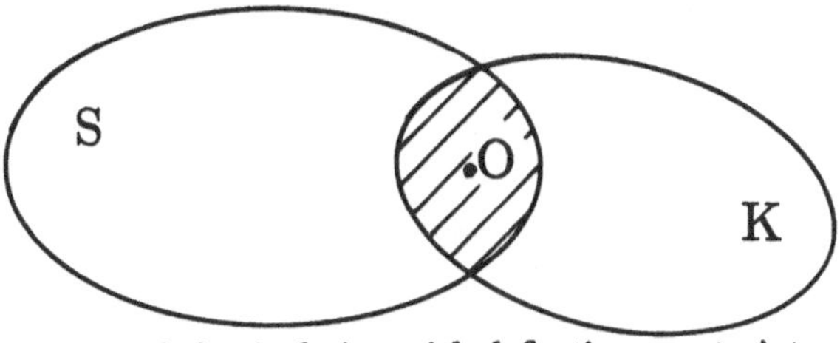

Fig. 4 Property of optimal elastic design with deflection constraints.

Then there are two possibilities:

(i) Stress constraint is active

This case implies

$$z = b|M| . \tag{19}$$

Substituting (18) and (19) into (17), we have

$$\overline{\kappa} = cb \operatorname{sgn} M . \tag{20}$$

(ii) Stress constraint is inactive

Then (11) implies $\lambda = 0$ and by (16) and (17) we have

$$z = \sqrt{\frac{\nu M \overline{M}}{rc}} , \tag{21}$$

$$\overline{\kappa} = \nu \overline{M} / rz . \tag{22}$$

The relation (21) was first derived for statically determinate beams by Barnett [4] in 1961.

3. Proof of the Optimality Criteria in Fig. 3

This general proof has the following two main features.

(i) Whilst the stresses and corresponding strains of the considered structure are required to be ***statically and kinematically*** admissible, the feasible set of solutions is imbedded in a larger set of solutions which are required to be *only statically* admissible. It is then shown that kinematic admissibility is ensured in the optimality conditions. This principle is indicated schematically in Fig. 4. If it is shown that the optimal solution (O) for all statically admissible solutions (S) is always contained by the set of kinematically admissible solutions (K), then the considered optimal solution is clearly optimal for the intersection ($S \cap K$) of the statically and kinematically admissible sets.

(ii) Equilibrium of the real and adjoint structures is expressed in terms of the virtual displacement theorem.

Note. Alternative proofs which use, instead of the virtual displacement theorem, the principle of stationary mutual energy (Shield and Prager [5], Huang [6]) or actual equilibrium equations for a special class of problems, are given in Sections 3.1 and 3.2. A further proof in the context of trusses is given in Section 12.1 (second lecture).

By the principle of virtual displacements, a necessary and sufficient condition for the equilibrium of the real loads and stresses (**p**, **Q**) is

$$\int_D \mathbf{p} \cdot \overline{\mathbf{u}}^K \, \mathrm{d}\mathbf{x} = \int_D \mathbf{Q} \cdot \overline{\mathbf{q}}^K \, \mathrm{d}\mathbf{x}, \tag{23}$$

where $\overline{\mathbf{u}}^K$ and $\overline{\mathbf{q}}^K$ represent any small kinematically admissible system of displacements and strains. Similarly, equilibrium of the virtual loads and stresses ($\overline{\mathbf{p}}_j$, $\overline{\mathbf{Q}}_j$) can be represented as

$$\int_D \overline{\mathbf{p}}_j \cdot \mathbf{u}_j^K \, \mathrm{d}\mathbf{x} = \int_D \overline{\mathbf{Q}}_j \cdot \mathbf{q}_j^K \, \mathrm{d}\mathbf{x} \quad (j = 1, 2, \ldots, v). \tag{24}$$

It is important to understand that at this stage ($\overline{\mathbf{u}}^K$, $\overline{\mathbf{q}}^K$) and ($\mathbf{u}_j^K$, $\mathbf{q}_j^K$) denote *any kinematically admissible small displacement systems* and *not* necessarily strains and displacements caused by the real or virtual loads.

Using the Lagrangians $\overline{\theta}$, θ_j ($j = 1, 2, \ldots, v$), ν_j ($j = 1, 2, \ldots, v$) and $\lambda_\ell(\mathbf{x})$ ($\ell = 1, 2, \ldots, t$) and the non-negative slack functions $\alpha_\ell(\mathbf{x})$ ($\ell = 1, 2, \ldots, t$) and slack variables β_j ($j = 1, 2, \ldots v$), the augmented functional for the considered problem becomes

$$\hat{\Phi} = \int_D \psi(\mathbf{x}) \, \mathrm{d}\mathbf{x} + \sum_j \nu_j \Big(\int_D \mathbf{Q} \cdot [\mathbf{F}] \overline{\mathbf{Q}}_j \, \mathrm{d}x - \Delta_j + \beta_j \Big) +$$

$$+ \int \sum_\ell \lambda_\ell(\mathbf{x}) [S_\ell(\mathbf{z}, \mathbf{Q}) - \alpha_\ell(\mathbf{x})] \, \mathrm{d}\mathbf{x} +$$

$$+ \overline{\theta} \Big[\int_D \mathbf{p} \cdot \overline{\mathbf{u}}^K \, \mathrm{d}\mathbf{x} - \int_D \mathbf{Q} \cdot \overline{\mathbf{q}}^K \, \mathrm{d}\mathbf{x} \Big] + \sum_j \theta_j \Big[\int_D \overline{\mathbf{p}}_j \cdot \mathbf{u}_j^K \, \mathrm{d}\mathbf{x} - \int_D \overline{\mathbf{Q}}_j \cdot \mathbf{q}_j^K \, \mathrm{d}\mathbf{x} \Big]. \tag{25}$$

Statical admissibility of **Q**, p, $\overline{\mathbf{Q}}_j$ and $\overline{\mathbf{p}}_j$ is no longer indicated in (25), because in the augmented problem these constraints are incorporated with the Lagrangians $\overline{\theta}$ and θ_j.

In general, a condition for the stationarity of a functional of the type

$$\hat{\Phi} = \int_D f[\mathbf{x}, y_1(\mathbf{x}), y_2(\mathbf{x}), \ldots, y_\omega(\mathbf{x})] \, \mathrm{d}\mathbf{x} \tag{26}$$

consists of the so-called Euler-Lagrange equations (e.g. [2], pp. 371-373)

$$f_{,y_i} = 0 \qquad (i = 1, 2, \ldots, \omega), \tag{27}$$

where a symbol after a comma denotes *partial* differentiation with respect to that symbol.

For variation of z_i, (25) and (27) imply

$$(\text{for } z_i > 0) \quad \frac{\partial \psi}{\partial z_i} + \sum_{\ell} \lambda_\ell \frac{\partial S_\ell}{\partial z_i} + \sum_j \nu_j \mathbf{Q} \cdot \frac{\partial [\mathbf{F}]}{\partial z_i} \overline{\mathbf{Q}}_j = 0 \qquad (i = 1, 2, \ldots, r), \quad (28)$$

which are the optimality conditions in Fig. 3.

Considering variations of Q_i, for linear systems we can adopt $\overline{\theta} = 1$ and then (25) and (27) furnish

$$\overline{q}_i = \sum_{\ell} \lambda_\ell \frac{\partial S_\ell}{\partial Q_i} + \sum_j \nu_j \mathbf{F}_i \overline{\mathbf{Q}}_j , \qquad (29)$$

where $\mathbf{F}_i$ is the i-th row of the flexibility matrix $[\mathbf{F}]$. The relation (29) is identical with the adjoint generalized strain-stress relation in Fig. 3.

Moreover, the Lagrangians ν_j and $\lambda_\ell(\mathbf{x})$ must satisfy the conditions (11) for the variation of *non-negative* slack functions $\alpha_\ell(\mathbf{x})$ and slack variables β_j.

Finally, for variations of $\overline{Q}_{ji}$, (25) and (27) imply

$$\theta_j q_{ji}^K = \nu_j \mathbf{F}_i \mathbf{Q} . \qquad (30)$$

This means that q_{ji}^K, which until now represented any small kinematically admissible displacement field, *through this optimality condition* has become equal to the (factored ν_j/θ_j) value of the elastic strains caused by the statically admissible real stress field $\mathbf{Q}$. It follows that the optimality condition in (30) requires the kinematical admissibility of the real elastic strains. This is the reason for not including kinematic admissibility in the original problem formulation.

Stationarity conditions for the augmented functional may not represent necessary conditions for the minimality of the orginal functional in some irregular cases, but this has little practical significance for well-posed problems of structural optimization.

3.1 Outline of a Proof Based on the Principle of Stationary Mutual Energy (Deflection Constraints)

The principle of stationary mutual energy states that the stationarity of the quantity

$$E_M = \int_D \mathbf{Q}^S \cdot \overline{\mathbf{q}}^S \, \mathrm{d}\mathbf{x} = \int_D \mathbf{Q}^S \cdot [\mathbf{F}] \overline{\mathbf{Q}}^S \, \mathrm{d}x , \qquad (31)$$

with respect to variations of $\mathbf{Q}^S$ and $\overline{\mathbf{Q}}^S$, which denote two unrelated *statically admissible* stress fields, implies kinematical admissibility of the corresponding elastic strains $\mathbf{q}$ and $\overline{\mathbf{q}}$.

The above principle was proved by Shield and Prager [5] and Huang [6].

The general validity of the above principle can be proved readily by considering (31) together with the equilibrium conditions in the form given in (23) and (24). Incorporating the latter with suitable Lagrangians into (31), the Euler-Lagrange equations for variations of $\mathbf{Q}$ and $\overline{\mathbf{Q}}$ establish the kinematic admissibility of the corresponding strains [cf. (30)]. It has also been established recently [7] that *the stationary point in the above theorem is a saddle point.*

Considering *the optimization of elastic structures with only displacement constraints*, the third integral (with λ_ℓ) in (25) can be deleted. Moreover, the last two expressions ensuring equilibrium can be omitted if $\mathbf{Q}$ and $\overline{\mathbf{Q}}$ are replaced by $\mathbf{Q}^S$ and $\overline{\mathbf{Q}}^S$ implying statical admissibility implicitly. Then the only remaining terms containing $\mathbf{Q}$ and $\overline{\mathbf{Q}}$ are represented by the second integral in (31). This means that for variations of $\mathbf{Q}$ and $\overline{\mathbf{Q}}$, the principle of stationary mutual energy gives directly the optimality condition that the corresponding strains $\mathbf{q}$ and $\overline{\mathbf{q}}$ must be kinematically admissible. For variations of $\mathbf{z}$, the usual variational equations furnish (28), but *without* the middle term λ_ℓ.

3.2 Derivation of Optimality Criteria for Bernoulli-Beams Using the Calculus of Variations

The considered problem can be stated as follows:

$$\min_{z(x)} \Phi = \int_D [cz + \nu M\overline{M}/rz + \lambda(z - k|M| + \alpha) + \overline{u}(M'' + p) + u(\overline{M}'' + \overline{p})]\, dx - \nu(\Delta - \beta)\,, \tag{32}$$

where ν is a Lagrangian multiplier (constant), $\overline{u}(x)$ and $u(x)$ are Lagrangian functions and $\alpha(x)$ is a slack function and β is a slack variable. The expressions in brackets after $\overline{u}$ and u, respectively, represent the equilibrium conditions for the real and virtual loads. Then for variation of z, M and $\overline{M}$, we obtain the following Euler-Lagrange equations (e.g. [2], pp. 371-373):

$$\delta z: \quad c - \nu MM/rz^2 + \lambda = 0\,, \qquad \delta M: \quad \nu\overline{M}/rz - \lambda k \operatorname{sgn} M = -\overline{u}''\,,$$
$$\delta\overline{M}: \quad \nu M/rz = -u''\,, \tag{33}$$

where the sign function sgn M has the usual meaning (sgn $M = 1$ for $M > 0$ and sgn $M = -1$ for $M < 0$). The relations (33) then imply the results in (16) and (17).

Considering the functional $\Phi = \int_D f(z, M, \overline{M})\, dx$ in (32), transversality conditions for variation of M and $\overline{M}$ give the same end conditions for u and $\overline{u}$ as in the case of an elastic beam, provided that the supports are rigid and costless. At a clamped end (B), for example, both the moment M and the shear force $V = -M'$ are variable and hence the following transversality conditions must be satisfied (e.g. [2], p. 378):

$$\delta M': \quad (f,_{M''})_B = 0\,, \qquad \delta M: \quad [f,_{M'} - (f,_{M''})']_B = 0\,, \tag{34}$$

where commas indicate partial derivatives with respect to the symbol in the subscript and primes denote differentiation with respect to x.

Combining (34) with (32), we have

$$\delta M' : \quad \overline{u}_B = 0 , \qquad \delta M : \quad \overline{u}'_B = 0 , \tag{35}$$

which are the usual end conditions for a clamped end. At a simple support, only the shear force $V = -M'$ is variable and hence only end condition (35) applies, as in the case of an elastic beam.

Similarly, transversality conditions for variations of $\overline{M}$ give the usual kinematic conditions for the elastic displacements of the real structure. This confirms the earlier general statement that kinematic admissibility is implied by an optimality condition for the class of problem considered.

4. Special Cases: Only Deflection Constraints or Only Stress Constraints Active (Elastic or Plastic Design)

4.1 Only Deflection Constraints are Active

As mentioned under Section 3.1, if only deflection constraints are active, then (28) reduces to

$$\frac{\partial \psi}{\partial z_i} + \sum_j \nu_j \mathbf{Q} \cdot \frac{\partial [\mathbf{F}]}{\partial z_i} \overline{\mathbf{Q}}_j = 0 , \tag{36}$$

and (29) is replaced by

$$\overline{q}_i = \sum_j \nu_j \mathbf{F}_i \overline{\mathbf{Q}}_j . \tag{37}$$

This implies that the adjoint strain $\overline{q}_i$ can be expressed as the usual elastic strains for the factored virtual loads $\sum_j \nu_j \overline{\mathbf{p}}_j$.

Finally, for a single displacement condition ($v = 1$, $\Delta_1 \to \Delta$, $\overline{\mathbf{Q}}_j \to \overline{Q}$), (36) and (37) are replaced by

$$\frac{\partial \psi}{\partial z_i} + \nu \mathbf{Q} \cdot \frac{\partial [\mathbf{F}]}{\partial z_i} \overline{\mathbf{Q}} = 0 , \quad \overline{q}_i = \nu \mathbf{F}_i \overline{Q} , \tag{38}$$

which means that the adjoint strains are the usual *elastic strains* associated with the virtual load $\overline{\mathbf{p}}$.

4.2 Only Stress Constraints are Active: the Difference between Optimal Plastic Design and Optimal Elastic Design

If only stress constraints are active, then by (11) we have $\nu_j = 0$ (for all j) and hence (28) and (29) reduce to

$$\frac{\partial \psi}{\partial z_i} + \sum_\ell \lambda_\ell \frac{\partial S_\ell(\mathbf{Q}, \mathbf{z})}{\partial z_i} = 0 , \tag{39}$$

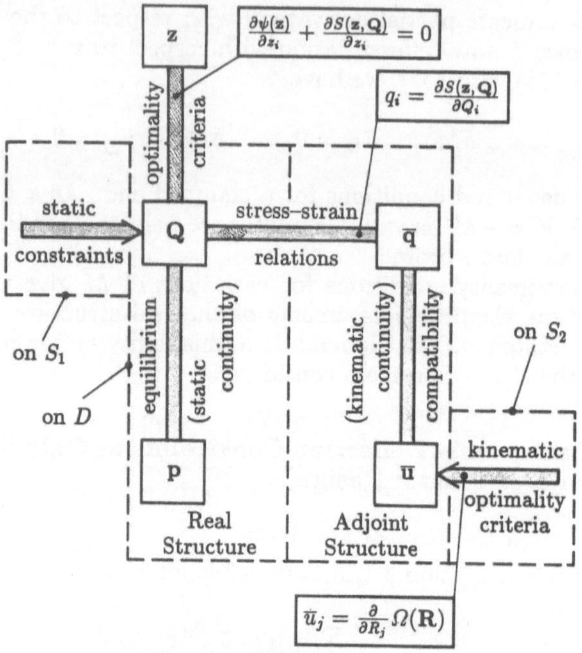

Fig. 5 Optimal plastic design: first COC formulation.

$$\overline{q}_i = \sum_{\ell} \lambda_\ell \frac{\partial S_\ell(\mathbf{Q}, \mathbf{z})}{\partial Q_i} . \tag{40}$$

Since the second sum of integrals (with $\nu_j = 0$) can be taken out of the augmented functional in (25) and hence the last sum of integrals (with θ_j, expressing the equilibrium of the now superfluous adjoint stresses $\overline{Q}_j$) also becomes unnecessary. It follows that (30) is not valid any more. This means that kinematic admissibility of the real structure is not assured by an optimality criterion. However, the second last term (with $\overline{\theta}$) in (25) remains intact and hence kinematic admissibility of $\overline{q}_i$ in (40) becomes an optimality condition.

4.2.1 Optimal plastic design. Since optimal plastic design requires only statical admissibility of the real stress field, (39) and (40) can be readily used for this case. The relations (39) and (40) were stated (for several alternative load conditions) in the first author's earlier book [8, p. 108, Eq. 3.85)] for plastic design with "generalized specific cost functions" in the form of $\psi(\mathbf{z})$. The above optimization procedure for a single stress constraint is given [with $t = 1$, $S_1(\mathbf{Q}, \mathbf{z}) \to S(\mathbf{Q}, \mathbf{z})$] in Fig. 5 in which we have only half a real system (with only static variables $\mathbf{Q}$ and $\mathbf{p}$) and half an adjoint system (with only kinematic variables $\mathbf{q}$ and $\mathbf{u}$) and a generalized strain-stress relation between the real stresses $\mathbf{Q}$ and the adjoint strains $\overline{\mathbf{q}}$.

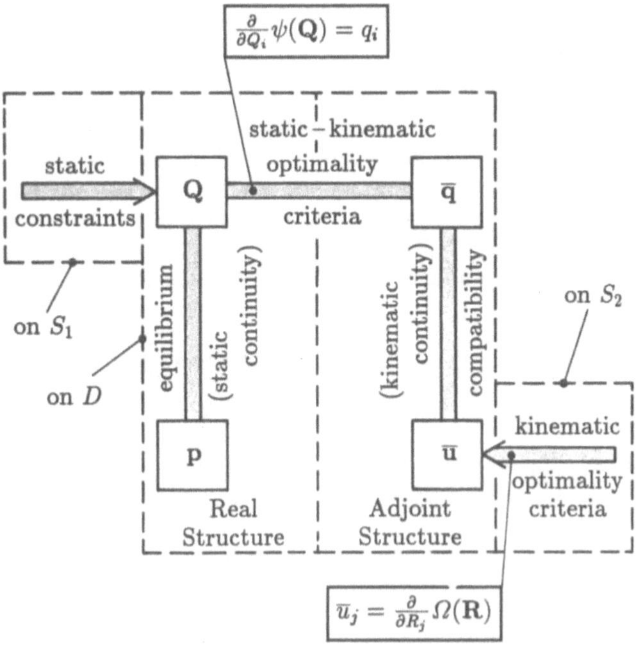

Fig. 6 Optimal plastic design: second COC formulation.

For a single stress constraint $S(\mathbf{Q}, \mathbf{z}) \geq 0$ and a single load condition, the problem of optimal plastic design can be reformulated by satisfying this constraint everywhere as an equality and then the terms $\psi(\mathbf{z})$ subject to $S(\mathbf{Q}, \mathbf{z})$ can be replaced by $\psi(\mathbf{Q})$. For this modified formulation the optimization procedure is shown in Fig. 6, which corresponds to the optimality criteria derived originally in 1967 by Prager and Shield [9] and extended later by the first author (e.g. [8]).

4.2.2 Optimal elastic design. As mentioned above, the optimal design given by (39) and (40) or Fig. 5 does not necessarily satisfy kinematic admissibility if all cross-sections are fully stressed. In some cases, as will be shown in the example under Section 5.1, the fully stressed design for a given stress constraint satisfies kinematic admissibility and then the optimal plastic design and optimal elastic design are identical. In other cases, as exemplified in Section 5.2, the fully stressed design (i.e. optimal plastic design) is kinematically inadmissible and then it is necessary to introduce at least one active displacement constraint which can be a compatibility condition at one of the redundancies (see Section 5.2).

5. Examples Illustrating the Differences between Optimal Plastic Design and Optimal Elastic Design — Reasons for Non-Fully-Stressed Designs

5.1 Example: Identical Optimal Plastic Design and Optimal Elastic Design (Fully Stressed Solution)

We consider a clamped rectangular beam (Fig. 7a) of given depth h and variable width z, for which the specific cost (weight per unit length) ψ and plastic moment capacity M_p are

$$\psi = \gamma bh , \quad M_p = \sigma_0 bh^2/4 , \tag{41}$$

where γ is the specific weight and σ_0 is the yield stress. It follows that for cross-sections satisfying the yield condition $M_p = |M|$ we have

$$\psi = a|M| , \quad a = 4\gamma/h\sigma_0 . \tag{42}$$

If we also impose a minimum cross-section condition ($M_p \geq M_0$), then we have the specific cost function

$$\psi = a|M| \qquad (\text{for } M \geq M_0) , \tag{43}$$
$$\psi = a|M_0| \qquad (\text{for } M \leq M_0) , \tag{44}$$

which is represented graphically in Fig. 7b. For the above cost function, the optimal generalized stress–adjoint strain relation in Fig. 6 gives

$$\overline{\kappa} = a \operatorname{sgn} M \qquad (\text{for } M > M_0) , \tag{45}$$
$$\overline{\kappa} = 0 \qquad (\text{for } M < M_0) , \tag{46}$$

(see Fig. 7c) where $\overline{\kappa} = -\mathrm{d}^2\overline{u}/\mathrm{d}x^2$ is the "curvature" (i.e. the negative second derivative) of the beam deflection. For the above problem, the optimal plastic design is based on the moment diagram in Fig. 7d. Using the optimal generalized stress-strain (moment-curvature) relations in (45) and (46), we have the adjoint curvature diagram in Fig. 7e. Since due to skew-symmetry those adjoint curvatures are clearly kinematically admissible, this design satisfies all requirements in Fig. 6 and hence it represents an ***optimal plastic design.***

We now consider the optimal elastic design of the beam in Fig. 7a with the stress conditions

$$S_1 = (-z + b|M|) \leq 0 , \quad S_2 = (-z + b|M_0|) \leq 0 , \tag{47}$$

where $b = 6/\sigma_p h^2$ and σ_p is the permissible stress (see Section 2.2). Moreover, we have again

$$\psi = cz , \qquad s = rz , \tag{48}$$

where c and r are given constants and s is the flexural stiffness (see Section 2.2). Then (11), (39) and (40) give the following results:

$$(\text{for } M > M_0) \quad \lambda_1 > 0 , \quad \lambda_2 = 0 , \quad c - \lambda_1 = 0 \Rightarrow \lambda_1 = c ,$$
$$\overline{\kappa} = b\lambda_1 \operatorname{sgn} M = cb \operatorname{sgn} M , \tag{49}$$
$$(\text{for } M < M_0) \quad \lambda_1 = 0 , \quad \lambda_2 = 1 , \quad c - \lambda_2 = 0 , \quad \lambda_2 = c , \quad \overline{\kappa} = \lambda_2 \cdot 0 = 0 \tag{50}$$

With the notation $a = cb$ and $\overline{q}_i = \overline{\kappa}$, (49) and (50) give the same generalized stress-adjoint strain relations as (45) and (46) or Figs. 7b and 7c. This means that

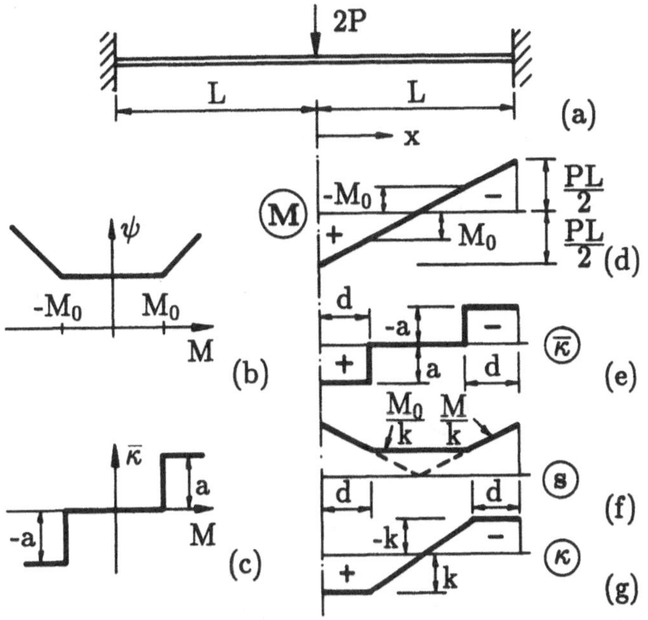

Fig. 7 Example with identical optimal elastic and optimal plastic design

the optimal moment diagram in Fig. 7d and the adjoint curvatures in Fig. 7e are also valid for optimal elastic design, if the above moments with the fully stressed cross-sections produce kinematically admissible generalized strains (i.e. curvatures) in the real structure. For $M > M_0$, we have $z = b|M|$ hence by (48) the flexural stiffness becomes $s = rb|M|$, or for $M < M_0$, $s = rb|M_0|$.

Introducing the notation $rb = 1/k$, we have the optimal stiffnesses in Fig. 7f and the corresponding curvatures of the real structure in Fig. 7g. Due to skew-symmetry, the latter are clearly kinematically admissible. Hence the optimal plastic design and optimal elastic design are identical in this example.

5.2 Example in Which the Optimal Plastic Design and Optimal Elastic Design are Different and the Latter is Not Fully Stressed

We consider again the clamped beam in Fig. 7a, but we assign a non-zero cost value of $\Omega = \alpha|M_A|$ to the supports where M_A is the clamping moment. The specific cost, stress constraint and flexural stiffness are the same as under Section 5.1.

For simplicity, it is assumed that the moment capacity M_p of the prescribed minimum cross-section is small ($M_0 \to 0$), but it still requires slope continuity of the real beam at points with $M_p = M_0$. As under 5.1, with the notation $rb = 1/k$ we have for fully stressed sections

$$s = |M|/k\,, \quad \kappa = M/s = k\,\mathrm{sgn}\,M\,, \tag{51}$$

and with the notation $a = cb$, we have again by (49)

$$\bar{\kappa} = a\,\mathrm{sgn}\,M\,. \tag{52}$$

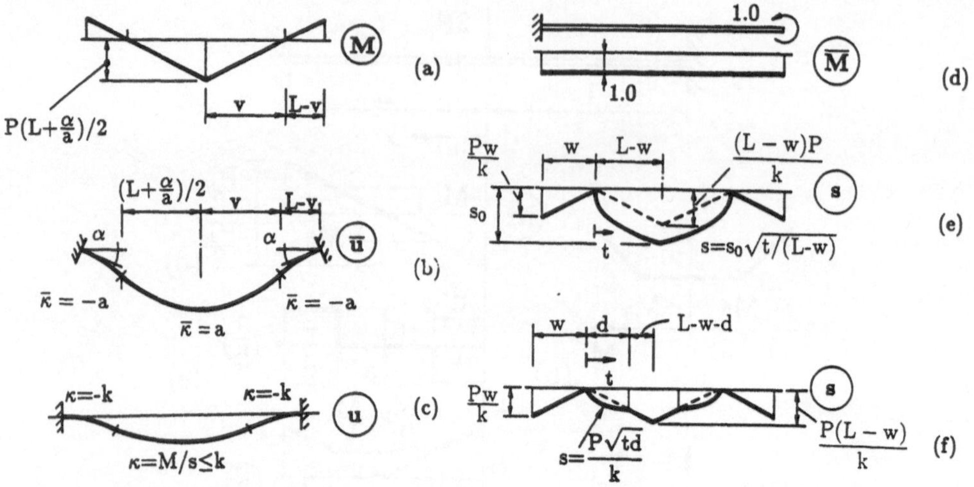

Fig. 8 Example with different optimal elastic and optimal plastic designs.

Due to the kinematic boundary condition for non-zero reaction costs at the bottom right corner of Fig. 3, we have a concentrated adjoint rotation of

$$\bar{\theta} = \frac{\partial \Omega(M_A)}{\partial M_A} = \frac{\partial \alpha |M_A|}{\partial M_A} = \operatorname{sgn} M_A \quad (\text{for } M_A \neq 0)\,, \tag{53}$$

at the beam ends.

It will be first shown that a fully stressed design satisfying the optimality criteria would lead to kinematic inadmissibility. Statically admissible moment fields are shown in Fig. 8a. The distance v can be calculated from kinematic admissibility of the adjoint displacements in Fig. 8b

$$\int_0^L \bar{\kappa}\, dx - \alpha = 0 = va - (L-v)a - \alpha \Rightarrow v = \frac{1}{2}\left(L + \frac{\alpha}{a}\right). \tag{54}$$

The above solution would make sense because, compared to the solution with zero reaction cost (Fig. 7), the negative end moments (now causing extra cost) in Fig. 8a have become smaller.

With the fully stressed elastic curvatures in (51), the moments in Fig. 8a would clearly cause kinematically inadmissible curvatures for the real beam since the region with $\kappa = -k$ is shorter than the region with $\kappa = k$. It follows that at least part of the central region with $M > 0$ must have curvatures smaller than k (Fig. 8c). In other words, the considered central region must be at least partially understressed.

Kinematical admissibility can be restored if we introduce an active displacement constraint [see (30) in the proof of COC]. This can always be done for statically indeterminate structures by removing a redundancy and replacing it with a zero displacement constraint.

In our example, the right-hand clamping is removed and the displacement constraint becomes $\int_D (M\overline{M}/s)\,\mathrm{d}x = 0$ where $\overline{M}$ is shown in Fig. 8d. Then we can use the optimality criterion in (21) with

$$M = Pt\,, \quad \overline{M} = 1\,, \tag{54a}$$

where the coordinate t is defined in Fig. 8e:

$$(\text{for } 0 < t < L-w) \quad z = \sqrt{\frac{\nu Pt}{rc}}\,, \quad s = rz = \sqrt{\frac{r\nu Pt}{c}} \quad s_0 = \sqrt{\frac{r\nu P(L-w)}{c}}\,, \tag{55}$$

$$\overline{\kappa} = -\mathrm{d}^2\overline{u}/\mathrm{d}x^2 = \nu\overline{M}/s = \nu/s\,. \tag{56}$$

Moreover, by (20) with $cb = a$, we have

$$(\text{for } -w < t < 0) \quad \overline{\kappa} = -a\,, \quad s = -M/k = -Pt/k\,. \tag{57}$$

The optimal values of ν and w can be calculated from kinematic admissibility of the real and adjoint fields (for details see [2, p. 258]), which give (in our notation)

$$\nu = \frac{4P(L-w)^3 c}{9k^2 r w^2}\,, \quad \frac{9}{4}\frac{ak}{c} - \frac{L^3}{w^2} + 3L + \frac{w}{4} = 0\,, \tag{58}$$

$$(\text{for } 0 < t < L-w) \quad s = s_0\sqrt{t/(L-w)}\,, \tag{59}$$

where

$$s_0 = \frac{2(L-w)^2 P}{(3wk)} \tag{60}$$

is defined in Fig. 8e.

The above solution topology is only valid if (Fig. 8e) $s_0 > (L-w)P/k$ or $ak/c \geq 1.4L$. If this is not the case, then the solution changes to the one shown in Fig. 8f.

Note. Sections 4 and 5 explain the reasons as to why optimal elastic designs with only stress constraints are in certain cases (allowance for reaction costs, not constant permissible stresses) not fully stressed. An early example with understressed optimal solutions (ten-bar truss) was given by Berke [10].

6. Problems with Nondifferentiable or Discontinuous Specific Cost Functions

The optimality criteria in Figs. 3, 5, and 6 are based on the assumption that the specific cost functions $\psi(\mathbf{z})$ or $\psi(\mathbf{Q})$ are differentiable. As can be seen from Fig. 7b, the above assumption is often not satisfied. Moreover, in the case of a discrete set of available cross-sections, the specific cost function may be subject to discontinuities. For these reasons, the original form of the optimality criteria discussed (e.g. [2])

uses generalized gradients instead of the partial derivatives in Figs. 3, 5, and 6. For example, in Fig. 7c, for $M = M_0$ the generalized gradient gives $0 < \overline{\kappa} < a$. These non-unique values of the adjoint strains will be particularly important in layout optimization by COC (see the fourth lecture).

7. Optimal Solutions for External Load Plus Selfweight

The power of the COC approach has been demonstrated by developing a number of extensions to special classes of problems. For example, in the case of allowance for selfweight one can use all the optimality criteria in Figs. 3, 5, and 6, but the adjoint strains must be multiplied by the term $(1+\overline{u})$, where $\overline{u}$ is the adjoint displacement in the vertical direction [2, p. 46 and pp. 72-73]. This simple modification enables us to handle this rather complex problem very efficiently. Examples were given in a recent paper [11].

8. Concluding Remarks: Initial Achievements of Iterative COC Methods

It will be shown in subsequent lectures that iterative COC methods have the following advantages:

- High efficiency for a large number of active stess constraints. Whilst the number of active global (deflection, system buckling, natural frequency, etc.) constraints is usually small, a very large number of stress constraints may be active. In one beam example, COC handled easily 75000 such constraints.
- Very high optimization capability for certain problems. In special examples (Bernoulli-beam with a single deflection constraint) up to one million variables have been optimized.
- The CPU requirement for COC can be considerably smaller than for some MP methods. In an example with 100 variables COC required 57 seconds (94 iterations) whilst sequential quadratic programming needed 1.5 million seconds (17.5 days, 77 thousand iterations).
- Simple extensions of COC for special conditions (allowance for selfweight or cost of reactions, various geometrical constraints) are available.
- COC can be readily extended to multiple deflection constraints and multiple load conditions.

References

1. Berke, L.; Khot, N.S. 1987: Structural optimization using optimality criteria. In: Mota Soares, C.A. (Ed.) *Computer Aided Optimal Design: Structural and Mechanical Systems*, pp. 271-312. Springer-Verlag, Berlin.

2. Rozvany, G.I.N. 1989: *Structural Design via Optimality Criteria (the Prager Approach to Structural Optimization)*. Kluwer, Dordrecht.

3. Prager, W. 1974: *Introduction to Structural Optimization*. (Course held in Int. Centre for Mech. Sci. Udine. CISM **212**). Springer-Verlag, Vienna.

4. Barnett, R.L. 1961: Minimum-weight design of beams for deflection. *J. Eng. Mech. ASCE*, **87**, EM1, 75-109.

5. Shield, R.T.; Prager, W. 1970: Optimal structural design for given deflection. *Zeit. ang. Math. Phys.* **21**, 513-523.

6. Huang, N.-C. 1971: On principle of stationary mutual complementary energy and its application to structural design. *Zeit. ang. Math. Phys.* **22**, 608-620.

7. Rozvany, G.I.N.; Zhou, M.; Kirsch, U. 1992: On the principle of stationary mutual energy (to be submitted).

8. Rozvany, G.I.N. 1976: *Optimal Design of Flexural Systems.* Pergamon Press, Oxford. Russian translation: Stroiizdat, Moscow, 1980.

9. Prager, W.; Shield, R.T. 1967: A general theory of optimal plastic design. *J. Appl. Mech.* **34**, 1, 184-186.

10. Berke, L.; Khot, N.S. 1974: Use of optimality criteria methods for large scale systems. *AGARD LS-70.*

11. Rozvany, G.I.N.; Zhou, M.; Gollub, W. 1990: Continuum-type optimality criteria methods for large finite element systems with a displacement constraint. Part II. *Struct. Optim.* **2**, 77-104.

12. Berke, L.; Khot, N.S. 1988: Performance characteristics of optimality criteria methods. In Rozvany, G.I.N.; Karihaloo, B.L. (Eds.) *Structural Optimization* (Proc. IUTAM Symposium, Melbourne, 1988), pp. 39-46. Kluwer, Dordrecht.

13. Rozvany, G.I.N.; Zhou, M.; Rotthaus, M.; Gollub, W. and Spengemann, F. 1989: Continuum-type optimality criteria methods for large discretized systems with a displacement constraint. *Struct. Optim.* **1**, 47-72.

14. Masur, E.F. 1975: Optimality in the presence of discreteness and discontinuity. In: Sawczuk, A.; Mróz, Z. (Eds.) *Optimization in Structural Design* (Proc. IUTAM Symp. held in Warsaw, Aug. 1973), pp. 441-453, Springer-Verlag, Berlin.

15. Rozvany, G.I.N.; Rotthaus, M.; Spengemann, F.; Gollub, W.; Zhou, M. 1990: The Masur paradox. *Mech. Struct. Mach.* **18**, 21-42.

16. Zhou, M.; Rozvany, G.I.N. 1991: *Iterative Continuum-Type Optimality Criteria Methods in Structural Optimization. Res. Rep.*, Univ. Essen.

17. Argyris, J.H.; Kelsey, S. 1960: *Energy Theorems and Structural Analysis.* Butterworth, London.

18. Gellatly, R.A.; Dupree, D.M.; Berke, L. 1974: Optim II: a magic compatible large scale automated minimum weight design program. Vol. 1. *AFFDL-TR-74-97.*

19. Berke, L. 1970: An efficient approach in the minimum weight design of deflection limited structures. *AFFDL-TM-70-FDTR.*

20. Khot, N.S.; Berke, L. 1984: Structural optimization using optimality criteria methods. In: Atrek, E.; Gallagher, R.H.; Ragsdell, K.M.; Zienkiewicz, O.C. (eds.): *New Directions in Optimum Structural Design*, pp. 47-74. Chichester, Wiley.

21. Venkayya, V.B. 1989: Optimality critera: a basis for multidisciplinary design optimization. *Comp. Mech.* **5**, 1-21.

22. Haftka, R.T.; Gürdal, Z.; Kamat, M.P. 1990: *Elements of Structural Optimization.* Dordrecht, Kluwer.

23. Gellatly, R.A.; Berke, L. 1971: Optimal structural design. *AFFDL-TR-70-165.*

24. McGuire, W.; Gallagher, R.H. 1979: *Matrix Structural Analysis.* John Wiley & Sons.

25. Gallagher, R.H. 1975: *Finite Element Analysis Fundamentals.* Prentice Hall Inc.

26. Morley, C.T. 1966: The minimum reinforcement of concrete slabs. *Int. J. Mech. Sci.* **8**, 305-319.

27. Prager, W.; Rozvany, G.I.N. 1977: Optimal layout of grillages. *J. Struct. Mech.* **5**, 1, 1-18.

28. Michell, A.G.M. 1904: The limits of economy of material in frame-structures. *Phil. Mag.* **8**, 47, 589-597.

29. Maxwell, J.C. 1890: On reciprocal figures, frames and diagrams of force. *Trans. Roy. Soc. Edinb.* **26**, 1. Also in: *Scientific Papers* **2** [Niven, W.D. (Ed.)] University Press, Cambridge, 174-177 (1872).

30. Kirsch, U. 1989: On the relationship between structural topologies and geometries. *Struct. Optim.* **2**, 39-45.

31. Prager, W.; Rozvany, G.I.N. 1977: Optimization of structural geometry. In: Bednarek, A.R.; Cesari, L. (Eds.) *Dynamical Systems*, pp. 265-293. Academic Press, New York.

32. Rozvany, G.I.N. 1981: Optimality criteria for grids, shells and arches. In: Haug E.J.; Cea, J. (Eds.) *Optimization of Distributed Parameter Structures* (Proc. NATO ASI held in Iowa City), pp. 112-151. Sijthoff and Noordhoff, Alphen aan der Rijn, The Netherlands.

33. Rozvany, G.I.N.; Ong, T.G. 1987: Minimum-weight plate design via Prager's layout theory (Prager memorial lecture). In: Mota Soares (Ed.) *Computer Aided Optimal Design: Structural and Mechanical Systems* (Proc. NATO ASI held in Troia, Portugal, 1986), pp. 165-179. Springer-Verlag, Berlin.

34. Rozvany, G.I.N.; Gollub, W.; Zhou, M. 1990: Layout optimization in structural design. In: Topping, B.H.V. (Ed.) *Proc. NATO ASI, Optimization and Decision Support Systems in Civil Engineering*, held 25 June - 7 July 1989, Edinburgh. Kluwer, Dordrecht.

35. Rozvany, G.I.N. 1984: Structural layout theory: the present state of knowledge. In: Atrek, E.; Gallagher, R.H.; Ragsdell, K.M.; Zienkiewicz, O.C. (Eds.) *New Directions in Optimum Structural Design*, pp. 167-195. Wiley & Sons, Chichester, England.

36. Hegemier, G.A.; Prager, W. 1969: On Michell trusses. *Int. J. Mech. Sci.* **11**, 209-215.

37. Rozvany, G.I.N.; Hill, R.H. 1978: A computer algorithm for deriving analytically and plotting optimal structural layout. In: Noor, A.K.; McComb, H.G. (Eds.) *Trends in Computerized Analysis and Synthesis* (Proc. NASA/ASCE Symp. held in Washington D.C., Oct. 1978), pp. 295-300. Wiley, New York. Also: *Comp. and Struct.* **10**, 1, 295-300.

38. Hill, R.H.; Rozvany, G.I.N. 1985: Prager's layout theory: a nonnumeric computer method for generating optimal structural configurations and weight-influence surfaces. *Comp. Meth. Appl. Mech. Engrg.* **49**, 1, 131-148.

39. Heymann, J. 1959: On the absolute minimum weight design of framed structures. *Quart. J. Mech. Appl. Math.* **12**, 3, 314-324.

40. Zhou, M. 1992: *A New Discretized Optimality Criteria Method in Structural Optimization.* Doctoral Thesis, Essen University.

41. Rozvany, G.I.N.: 1992: Optimal grillage layouts. In: Rozvany, G.I.N. (Ed.) *Proc. CISM Advanced School, Shape and Layout Optimization of Structural Systems* (held in Udine, July 1990). Springer-Verlag, Vienna.

42. Rozvany, G.I.N.; Gollub, W. 1990: Michell layouts for various combinations of line supports, Part I. *Int. J. Mech. Sci.* **32**, 12, 1021-1043.

43. Rozvany, G.I.N.; Gollub, W.; Zhou, M. 1992: Michell layouts for various combinations of line supports, Part II (to be submitted to *Int. J. Mech. Sci.*).

44. Rozvany, G.I.N.; Prager, W. 1979: A new class of structural optimization problems: optimal archgrids. *Comp. Meth. Appl. Mech. Engrg.* **19**, 1, 127-150.

45. Rozvany, G.I.N.; Nakamura, H.; Kuhnell, B.T. 1980: Optimal archgrids: allowance for selfweight. *Comp. Meth. Appl. Mech. Engrg.* **24**, 3, 287-304.

46. Rozvany, G.I.N.; Wang, C.-M.; Dow, M. 1982: Prager structures, archgrids and cable networks of optimal layout. *Comp. Meth. Appl. Mech. Eng.* **31**, 91-113.

47. Rozvany, G.I.N.; Wang, C.M. 1983: On plane Prager-structures (I). *Int. J. Mech. Sci.* **25**, 7, 519-527.

48. Wang, C.M.; Rozvany, G.I.N. 1983: On plane Prager-structures (II) — Non-parallel external loads and allowance for selfweight. *Int. J. Mech. Sci.* **25**, 7, 529-541.

49. Olhoff, N.; Rozvany, G.I.N. 1982: Optimal grillage layout for given natural frequency. *J. Engrg. Mech. ASCE* **108**, EM5, 971-975.

50. Kozlowski, W.; Mróz, Z. 1969: Optimal design of solid plates. *Int. J. Solids Struct.* **5**, 8, 781-794.

51. Rozvany, G.I.N.; Olhoff, N.; Cheng, K.-T.; Taylor, J.E. 1982: On the solid plate paradox in structural optimization. *DCAMM Report* **212** June 1981 and *J. Struct. Mech.* **10**, 1, 1-32.

52. Cheng, K.-T.; Olhoff, N. 1981: An investigation concerning optimal design of solid elastic plates. *Int. J. Solids Struct.* **17**, 305-323.

53. Rozvany, G.I.N.; Zhou, M. 1991: A new direction in cross-section and layout optimization. In: Hernandez, S.; Brebbia, C.A. (Eds.) *Optimization of Structural Systems and Industrial Applications*, pp. 39-50. Elsevier, London.

54. Wang, C.M.; Rozvany, G.I.N.; Olhoff, N. 1984: Optimal plastic design of axisymmetric solid plates with a maximum thickness constraint. *Comp. and Struct.* **18**, 4, 653-665.

55. Murat, F.; Tartar, L. 1985: Calcul des variations et homogénéisation. In: *Les méthodes de l'homogénéi-*
sation: théorie et applications en physique. Coll. del la Dir. des Etudes et recherches de Elec. de France, Eyrolles, Paris, pp. 319-370.

56. Lurie, K.A.; Cherkaev, A.V.; Fedorov, A.V. 1982: Regularization of optimal design problems for bars and plates. *J. Optimiz. Theory Appl.* **37**, 4, 499-522, 523-543.

57. Kohn, R.V.; Strang, G. 1986: Optimal design and relaxation of variational problems, I, II and III. *Comm. Pure Appl. Math.* **39**, 113-137, 139-182, 353-377.

58. Rozvany, G.I.N.; Ong, T.G.; Sandler, R.; Szeto, W.T.; Olhoff, N.; Bendsøe, M.P. 1987: Least-weight design of perforated elastic plates I. *Int. J. of Solids Struct.* **23**, 4, 521-536.

59. Khot, N.S. 1982: Optimality criterion methods in structural optimization. In: Morris, A.J. (Ed.) *Foundations of Structural Optimization: A Unified Approach*, pp. 99-236. Wiley, Chichester.

60. Hemp, W.S. 1973: *Optimum Structures.* Clarendon, Oxford.

61. Nagetegaal, J.C.; Prager, W. 1973: Optimal layout of a truss for alternative loads. *Int. J. Mech. Sci* **15**, 7, 583-592.

62. Spillers, W.R.; Lev. O. 1971: Design for two loading conditions. *Int. J. Solids Struct.* **7**, 1261-1267.

63. Rozvany, G.I.N. 1974: Analytical treatment of some extended problems in structural optimization, Part II. *J. Struct. Mech.* **3**, 4, 387-402.

64. Bendsøe, M.P.; Ben-Tal, A. 1990: Truss topology optimization, *Symp. Polish Acad. Sci. Optimal Design and Control of Structures*, Jablonna, Poland.

65. Olhoff, N.; Lurie, K.A.; Cherkaev, A.V.; Fedorov, A.V. 1981: Sliding regimes and anisotropy in optimal design of vibrating axisymmetric plates. *Int. J. Solids Struct.* **17**, 931-948.

66. Bendsøe, M.P. 1989: Optimal shape design as a material distribution problem. *Struct. Optim.* **1**, 4, 193-202.

67. Kikuchi, N.; Suzuki, K. 1992: Shape and topology optimization by the homogenization method. In: Rozvany, G.I.N. (Ed.), *Proc. CISM Advanced School, Shape and Layout Optimization of Structural Systems* (held in Udine, July 1990). Springer-Verlag, Vienna.

68. Olhoff, N.; Bendsøe, M.P.; Rasmussen, J. 1990: On CAD-integrated structural topology and design optimization. *Proc. 2nd World Congress on Computational Mechanics*, pp. 95-99. Int. Assoc. Comp. Mech., Stuttgart; also: *Comp. Meth. Appl. Mech. Eng.* **89**, 1-3, 259-280, 1991.

69. Rozvany, G.I.N.; Adidam, S.R. 1972: Rectangular grillages of least weight. *J. Eng. Mech. ASCE* **98**, 1337-1352.

70. Lowe, P.G.; Melchers, R.E. 1972/1973: On the theory of optimal constant thickness, fibre-reinforced plates I, II, III. *Int. J. Mech. Sci.* **14**, 5, 311-324; **15**, 2, 157-170; **15**, 9, 711-726.

ITERATIVE COC METHODS*

M. Zhou and G.I.N. Rozvany
FB 10, Essen University, D-4300 Essen 1, Germany

Abstract. In Part I of this lecture, the iterative COC algorithm is described in detail and then it is illustrated with beam and plate (plane stress) examples for a variety of design constraints. For trusses with stress constraints and a deflection constraint an independent derivation of the optimality criteria is presented and then the method is illustrated by an example involving a ten-bar truss. A detailed comparison with the DOC approach is given and some advantages of the COC approach are pointed out. Finally, extensions to multiple deflection constraints and multiple loading conditions are discussed.

In Part II, COC-type optimality conditions are derived directly for discretized (FE) structures in the usual matrix notation for such systems. The relative advantages and disadvantages of this approach, termed DCOC, are discussed and then it is illustrated by examples involving beams, frames and trusses.

Part I: Iterative Methods Based on Continuum-Formulation

9. The Simplest Case: Iterative COC Algorithm for a Single Deflection Constraint

9.1 An Elementary Analytical Example

We consider a uniformly loaded cantilever (Fig. 9a) of variable width z with a prescribed deflection $u_A = \Delta$ at its free end. The corresponding real moment diagram M, together with the adjoint load $\bar{p}$ and adjoint moment diagram $\overline{M}$, are shown in Figs. 9b, c and d. Using the optimality condition in (21) with $r = c = 1$, the optimal width distribution of the beam is given in Fig. 9e. The value of the Lagrangian can be calculated from the deflection conditions. Using a normalized problem with $\Delta = L = p = 1$, we have

$$1 = \int_0^1 \frac{M\overline{M}}{\sqrt{\nu M\overline{M}}}\,\mathrm{d}x = \int_0^1 \sqrt{\frac{M\overline{M}}{\nu}}\,\mathrm{d}x\,, \quad \sqrt{2\nu} = \int_0^1 x^{3/2}\,\mathrm{d}x = \frac{2}{5} \Rightarrow \nu = \frac{2}{25}\,. \tag{61}$$

9.2 An Iterative COC Algorithm for Large Discretized Systems

---

* In the five lectures given by Rozvany and Zhou, the section numbers, equation numbers, figure numbers and reference numbers are used continuously. All references are listed at the end of the first paper.

G. I. N. Rozvany (ed.), Optimization of Large Structural Systems, Vol. I, 27–58.

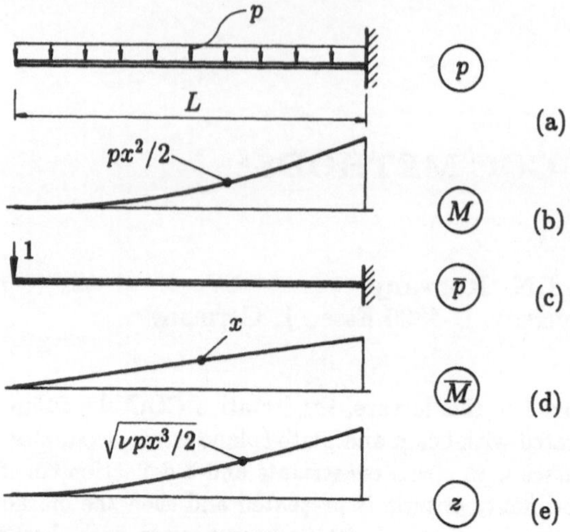

Fig. 9 Illustrative example: cantilever.

Figure 10 shows the flowchart of the iterative COC algorithm for a deflection constraint, in the context of Bernoulli beams of variable width. A similar procedure can be used for *any elastic structure* with a deflection constraint. In Fig. 10, the symbol i denotes the i–th element and z_i denotes its flexural stiffness. We stipulate a minimum stiffness constraint

$$z_i \geq z_a \qquad \text{(for all } i\text{)}. \tag{62}$$

Elements with $z_i > z_a$ shall be termed *active elements* and those with $z_i = z_a$ *passive elements*, as in earlier work by Berke and Khot [12]. The computational steps in Fig. 10 involve the following operations:

(A) The same constant initial stiffness value z_{in} can be adopted for all elements if a better estimate is not available.

(B, C) Any standard FE software can be used for analysing the real and adjoint systems. The adjoint beam is subject to the virtual load associated with the deflection constraint.

(D) The updated value of the Lagrangian ν can be calculated from the discretized work equation for the deflection Δ

$$\Delta = \sum_i \frac{M_i \overline{M}_i L_i}{r z_i}, \tag{63}$$

where M_i and $\overline{M}_i$ are the real and adjoint moment values at the middle of the beam element i, L_i is the length of the element i, and z_i is its width. The constant r was defined in Section 2.2. The RHS of (63) can be split into active (A) and passive

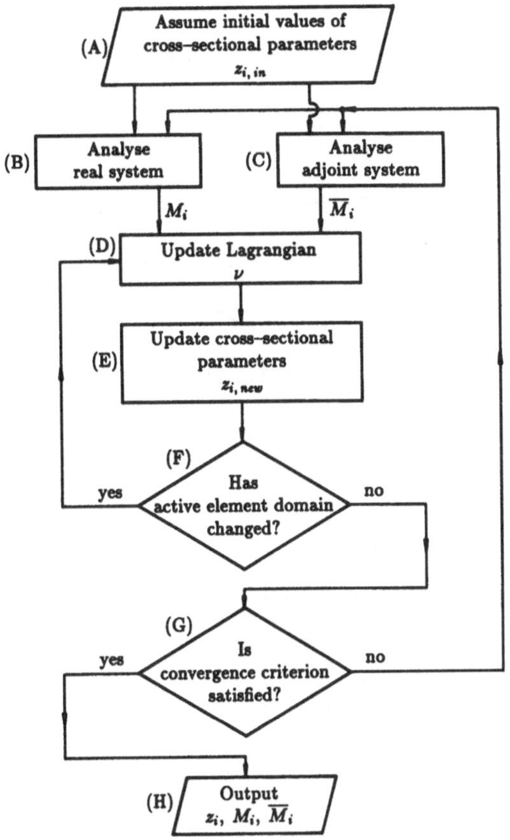

Fig. 10 Iterative COC method for a deflection and prescribed minimum stiffness constraint in the context of beam optimization.

(P) element domains, and then (63) and (21) imply

$$\sqrt{\nu} = \frac{\sqrt{\frac{r}{c}} \sum_{A} L_i \sqrt{M_i \overline{M}_i}}{\Delta - \sum_{P} \frac{M_i \overline{M}_i L_i}{r z_a}} . \tag{64}$$

The relation (64) represents an approximation because the moments in the numerator of the first fraction are from the *current cycle*, while the stiffness $\sqrt{\nu M_i \overline{M}_i}$ is the value in the *forthcoming cycle*. This error becomes negligibly small in later iterations because the change in the moment values becomes insignificant. The fact

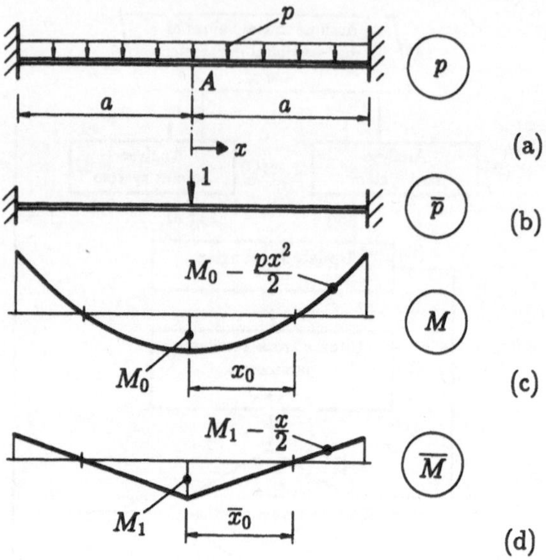

Fig. 11 Beam example – given deflection at midspan and no lower limit on the cross-sectional area.

that in (63) the expression $\sum_i \int_0^{L_i} [M(x)\overline{M}(x)/rz(x)]\mathrm{d}x$ is replaced by $\sum_i M_i\overline{M}_i L_i/rz_i$ is also an approximation, but the error involved is negligible when a large number of elements are used [13].

(E) The new beam stiffness is calculated from the relations based on an extended version of (21):

$$\left(\text{for } \frac{\nu M_i \overline{M}_i}{rc} \le z_a^2\right) \quad z_i = z_a \,, \qquad \left(\text{for } \frac{\nu M_i \overline{M}_i}{rc} > z_a^2\right) \quad z_i = \sqrt{\frac{\nu M_i \overline{M}_i}{rc}} \,. \tag{65}$$

(F) The reason for this step is as follows. If some elements have changed from the active to the passive set, or vice versa, then (64) would give an incorrect estimate of ν and hence steps (D) and (E) must be repeated.

(G) The tolerance test can be based on the following criteria:

$$\frac{\Phi_{\text{new}} - \Phi_{\text{old}}}{\Phi_{\text{new}}} \le \overline{E}\,, \qquad (\text{for all } i) \quad \frac{z_{i,\text{new}} - z_{i,\text{old}}}{z_{i,\text{new}}} \le E_1\,, \tag{66}$$

where $\overline{E}$ and E_1 are given tolerance values.

9.3 Illustrative Example: Beam with a Deflection Constraint, No Limits on the Cross-Sectional Parameters

9.3.1 Problem Description and Analytical Solution. We consider a clamped, uniformly loaded beam (Fig. 11a) of given depth h, but variable width z for which the deflection at midspan (A) has an upper limit

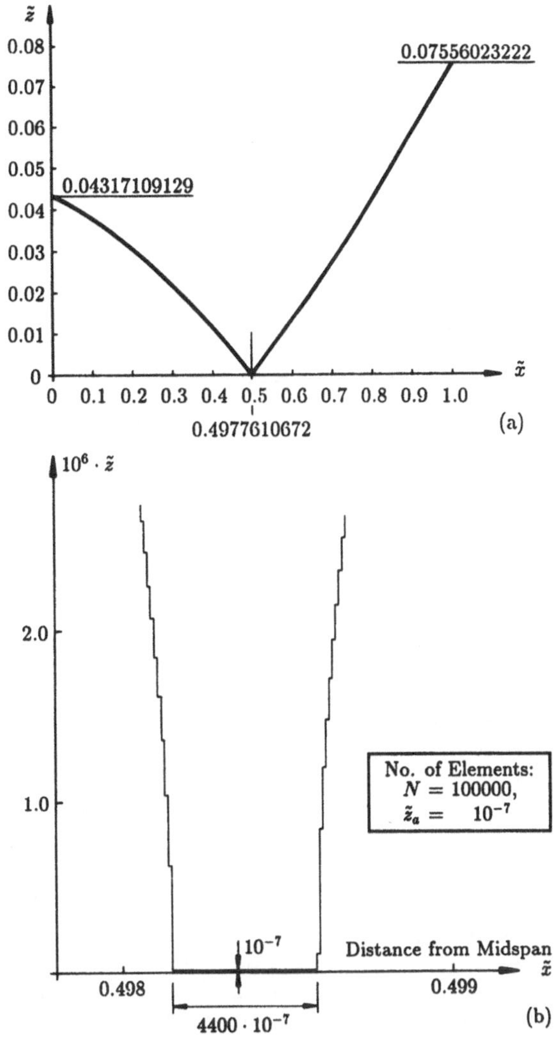

Fig. 12 (**a**) Optimal stiffness distribution in the beam example: analytical and COC solutions, and (**b**) enlarged detail of the COC solution in the vicinity of the singularity.

$$u_A \leq \Delta\,. \tag{67}$$

The virtual (adjoint) load $\overline{p}$, and the real and adjoint moment diagrams $(M, \overline{M})$ are shown, respectively, in Figs. 11b, c and d.

We introduce the nondimensional notation $\tilde{x} = x/a$, $\tilde{z} = z/a$, $\tilde{M} = M/pa^2$, $\tilde{\overline{M}} = \overline{M}/a$, $\tilde{u}_A = u_A/\Delta$, $\tilde{s} = 12s/(h^3aE) = \tilde{z}$, $\tilde{\Phi} = \Phi/(ha^2\gamma)$, where x is the longitudinal coordinate, E is Young's modulus, s is the flexural stiffness, Φ is the

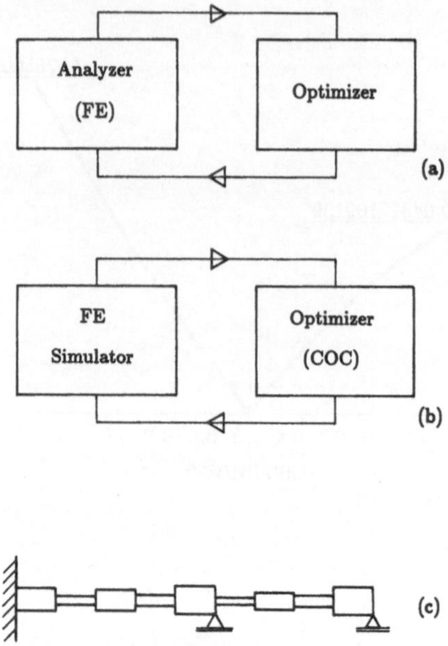

Fig. 13 The role of FE simulators in determining the optimization capability.

total beam weight, γ is the specific weight of the beam material and other symbols have been defined earlier. The nondimensional width $\tilde{z}$ and stiffness $\tilde{s}$ have been made identical since they are linearly interdependent in this problem. The real and adjoint bending moment diagrams (Figs. 11c and d) are represented by

$$\tilde{M} = \tilde{M}_0 - \tilde{x}^2/2, \quad \overline{\tilde{M}} = \tilde{M}_1 - \tilde{x}/2. \tag{68}$$

The considered problem is to be solved for the case when the prescribed minimum cross-sectional area approaches zero: $z_a \to 0$. This means that apart from regions of infinitesimal length with $z = z_a$, our relevant optimality condition is the nondimensional version of (21). In addition, the compatibility conditions for the real and adjoint systems and the nondimensional displacement condition imply:

$$\int_0^1 \frac{\tilde{M}}{\tilde{z}} \mathrm{d}\tilde{x} = 0, \qquad \int_0^1 \frac{\overline{\tilde{M}}}{\tilde{z}} \mathrm{d}\tilde{x} = 0, \qquad 2\int_0^1 \frac{\tilde{M}\overline{\tilde{M}}}{\tilde{z}} \mathrm{d}\tilde{x} = 1, \tag{69}$$

where $\tilde{M}$, $\overline{\tilde{M}}$ and $\tilde{z}$ are given by (68) and (21). Moreover, (21) gives a non-imaginary (real) value for $\tilde{z}$ only if in Figs. 11c and d

$$x_0 = \overline{x}_0 . \tag{70}$$

The relations under (69) and (70), after substitution of (68) and (21), represent four simultaneous equations and the unknown quantities involved are $\tilde{M}_0$, $\tilde{M}_1$ and

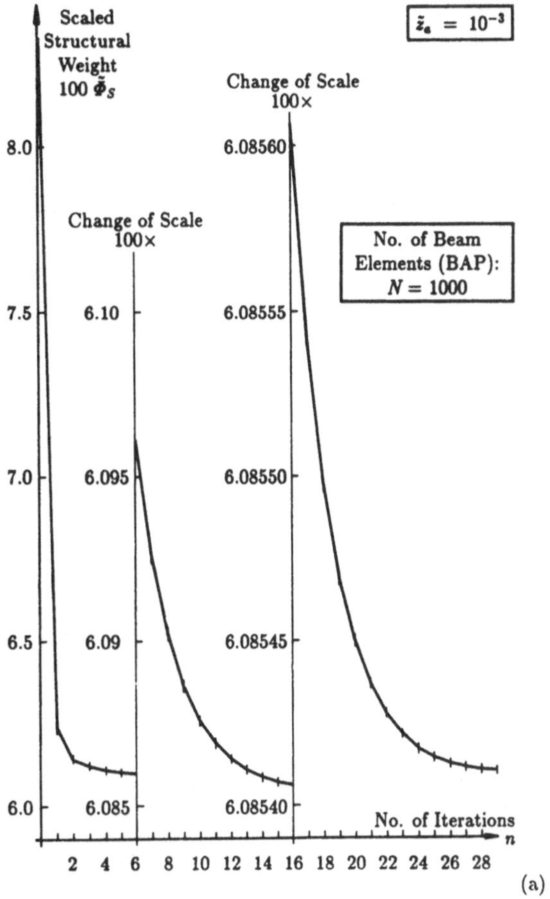

Fig. 14a Scaled total structural weight $\tilde{\Phi}$ as a function of the number of iterations (n), $N = 1000$.

ν. It can be shown easily that the above equations cannot be satisfied simultaneously. The paradoxical nature of this state of affairs was pointed out by Masur in a discussion of his paper [14] and is considered in detail elsewhere [15]. As is shown in the latter paper, the explanation of this paradox is that in the exact analytical solution for $z_a \to 0$, the beam stiffness (nondimensionally: $\tilde{s} = \tilde{z}$) takes on a second order infinitesimal value over a first order infinitesimal beam length with $z = z_a$, resulting in concentrated rotations in the real and adjoint deflection fields. It will be seen in Fig. 12b that discretized COC solutions with 100000 elements fully confirm the type of singularity predicted by the analytical solution.

The complete analytical solution [13, 15] yields the following optimal stiffness

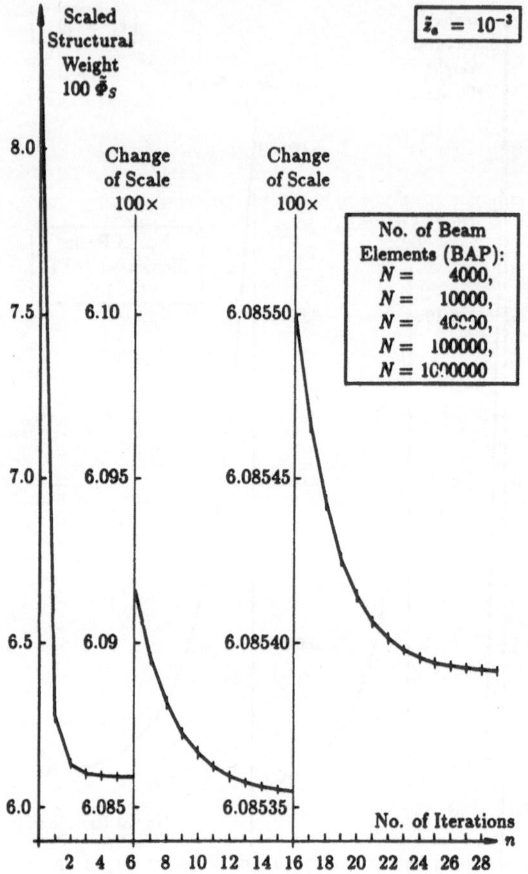

(b)

Fig. 14b Scaled total structural weight $\tilde{\Phi}$ as a function of the number of iterations (n), $N = 4000$ to 1000000.

distribution for the above problem

$$\tilde{z} = \Big[(\tilde{x}_0)^{5/2}(2^{9/2} - 7) - (\tilde{x}_0 + 1)^{3/2}(7\tilde{x}_0 - 3)\Big](\tilde{x}_0 - \tilde{x})\sqrt{\tilde{x}_0 + \tilde{x}}/15\,, \tag{71}$$

and the optimal $\tilde{x}_0$ value is furnished by the cubic equation

$$\tilde{x}_0^3\Big[2^9 - 7(2)^{11/2}\Big] - 63\tilde{x}_0^2 - 15\tilde{x}_0 - 1 = 0\,. \tag{72}$$

Moreover, the optimal total weight becomes

$$\Phi = 4\Big[\tilde{x}_0^{5/2}(2^{9/2} - 7) - (\tilde{x}_0 + 1)^{3/2}(7\tilde{x}_0 - 3)\Big]^2/225\,. \tag{73}$$

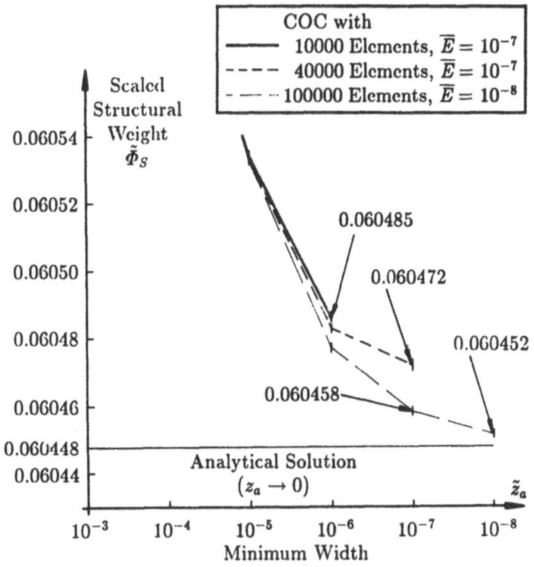

Fig. 15 A comparison of COC results using 10000, 40000 and 100000 elements with the analytical solution.

The relations (72) and (73) furnish the optimal values

$$\tilde{x}_0 = 0.49776107\,, \qquad \tilde{\Phi}_{\text{opt}} = 0.060448186\,. \tag{74}$$

The optimal variation of the stiffness $\tilde{z}(\tilde{x})$ is shown graphically in Fig. 12a.

9.3.2 Iterative COC Solutions. Using the COC procedure described in Section 9.2 (Fig. 10), discretized numerical solutions were obtained by using various numbers of elements ranging from fifty to one million. The structural analysis of the real and adjoint systems was carried out using the following programs:

ANSYS (number of elements $N \leq 150$), BAP (number of elements $N \geq 50$).

The analytically-based, special-purpose computer program BAP can analyse, for given boundary conditions, beams consisting of prismatic segments; it was found, as expected, to be much more efficient computationally than a general-purpose FE program. Both programs yielded identical results for $N = 50, 100$ and 150, and for this reason BAP and similar programs are termed *FE simulators*. The logic of using FE simulators in our investigation is as follows. The optimization procedure usually consists of some iterative use of an (FE) *analyser* and *optimizer* (Fig. 13a). However, we are testing here only the capability and efficiency of an optimizer and hence the analyser can be replaced by an FE simulator (Fig. 13b). The type of structure analysed exactly by BAP is shown in Fig. 13c. Using the COC approach, the following conclusions were reached:

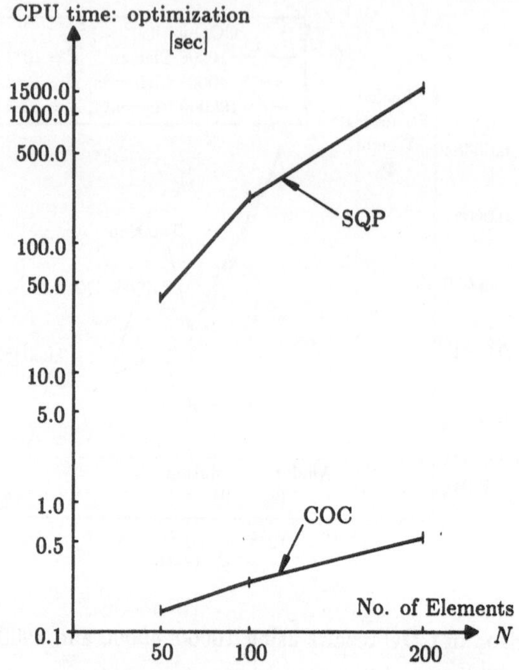

Fig. 16 A comparison of CPU times required for a structural optimization problem solved by the COC and SQP methods.

(i) The convergence in the considered example is found to be fully monotonic and almost uniform, and its rate largely independent of the number of elements (N) used (see Fig. 14).

(ii) In spite of an unusual ***singularity*** in the analytical optimal solution, the total weight in the best COC solution (100000 elements, $\tilde{z}_a = 10^{-8}$, $\tilde{\Phi} = 0.060452$) differs only by ***0.006 per cent*** from the analytical solution ($\tilde{\bar{\Phi}} = 0.060448$).

(iii) For problems with near-singularities in the discretized formulation, a very large number of elements are required for a reasonably high accuracy (see Fig. 15).

(iv) The type of singularity predicted by the analytical solution (stiffness of higher order infinitesimal over a length of first order infinitesimal) is indicated by the discretized COC solution in the vicinity of the singularity (see Fig. 12b).

(v) The considered problem was also investigated by using primal mathematical programming (MP) methods. Both Lawo (Karlsruhe, FRG) and Wang (Singapore) have found that with $\tilde{z}_a \rightarrow 0$, signs of instability appear when the number of elements (N) exceeds 50. The best result (with $N = 45$) yielded a total weight of $\tilde{\Phi} = 6.088672$, which represents an error of about *0.7 per cent* in comparison with the exact analytical solution (this error is over *hundred times* higher than that of the best COC solution).

(vi) The COC results were also verified by a *sequential quadratic programming*

(SQP) method which, however, was restricted to about 200 variables. For this reason, COC solutions with 50, 100 and 200 elements were obtained for $\tilde{z}_a = 0.01$ in this comparison. Whilst the agreement between the two sets of results was 8 to 9 significant digits, the SQP method required significantly more CPU time for optimization, as can be seen from Fig. 16. The relative time requirement difference increases with the number of variables, and for 200 variables *the SQP method requires approximately 3000 times more CPU time than the COC procedure.*

In order to assess the relative efficiency of the proposed method, a beam with 1000 elements and a prescribed minimum width of $\tilde{z}_a = 10^{-3}$ was optimized using the Berke-Khot recurrence relation [1, 12] with various values of the step size parameter r:

$$(z_i)_{n+1} = (z_i)_n \left(\nu \frac{M_i \overline{M}_i}{z_i^2} \right)_n^{1/r} . \tag{75}$$

For $r = 2$, (75) reduces to the relation (21) used by the authors.

The results of the above calculations are shown for convergence tolerance values of $\overline{E} = 10^{-7}$ and $\overline{E} = 10^{-6}$ in Fig. 17 which indicates the number of iterations (n) that fulfill the convergence criterion, as a function of the step size parameter (r).

10. COC Solutions for Other Classes of Problems

COC methods for the following other classes of structures with a deflection constraint have been considered by the authors [11]:

(a) prescribed lower and upper limits on the cross-sectional dimensions;
(b) segmentwise constant cross-section;
(c) allowance for the cost of supports;
(d) allowance for selfweight;
(e) non-linear and non-separable specific cost and stiffness functions;
(f) two-dimensional structural systems (plates).

In this lecture, only (e) and (f) above will be reviewed briefly.

10.1 Structures with Non-Linear and Non-Separable Specific Cost and Stiffness Functions

In all problems discussed in this lecture until now, it was possible to express the specific cost ψ as a sum of element costs ψ_i, which depended on only one cross-sectional parameter z_i associated with the element i. Moreover, each element specific stiffness s_i was a function of only one such parameter. In this section, we consider structures for which the element cost ψ_i and the specific element stiffnesses s_i are nonseparable functions of several cross-sectional parameters (dimensions) z_{ij} ($j = 1, \ldots, r$).

10.1.1 Optimality Criteria for Beams of Independently Variable Width z_1 and Depth z_2 with Bending in Both Horizontal and Vertical Directions. For the above class of

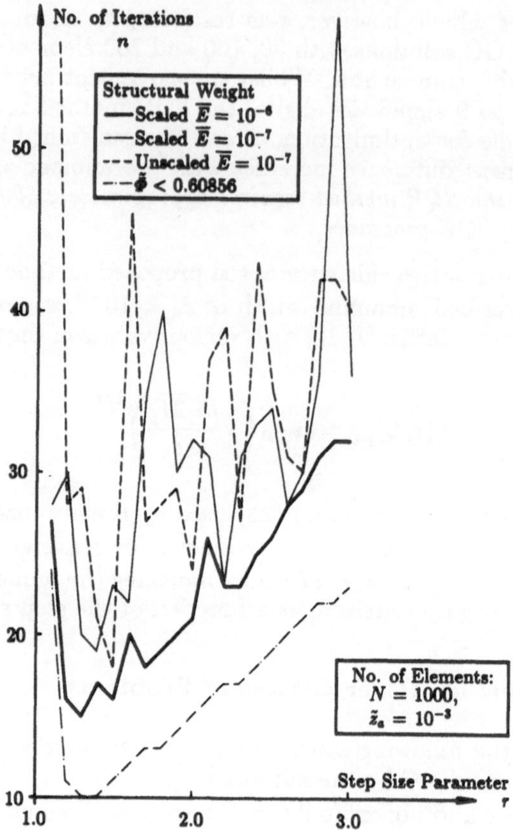

Fig. 17 Beam example: number (n) of iterations required to fulfill the convergence criterion as a function of the Berke-Khot step size parameter r.

problems we have

$$\mathbf{Q} = (M_1, M_2)\,,\ \overline{\mathbf{Q}} = (\overline{M}_1, \overline{M}_2)\,,\ [\mathbf{F}] = \begin{bmatrix} \dfrac{1}{r z_1^3 z_2} & 0 \\ 0 & \dfrac{1}{r z_1 z_2^3} \end{bmatrix}, \ \psi = c z_1 z_2\,,\ r = E/12, \tag{76}$$

where M_1 and M_2 are the real bending moments in the horizontal and vertical directions, $\overline{M}_1$ and $\overline{M}_2$ are the corresponding adjoint moments, $[\mathbf{F}]$ is the specific flexibility matrix, $c = \gamma$ is the specific weight and E is Young's modulus for the beam material. We consider a single deflection constraint which limits the sum of

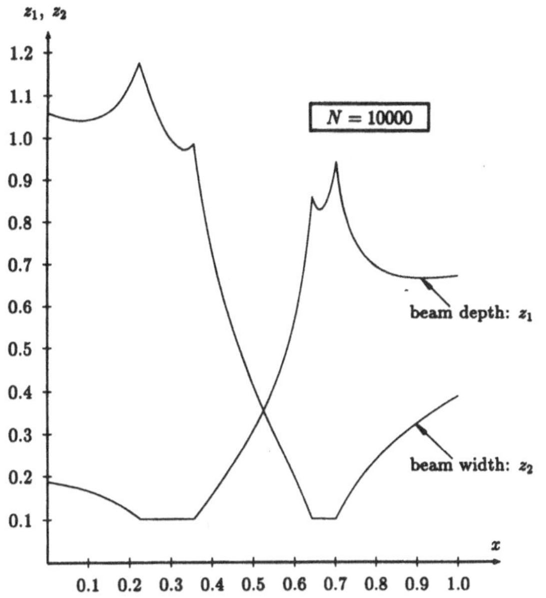

Fig. 18 A beam problem with a nonseparable specific cost function: optimal depth z_1 and width z_2 variations obtained by the COC procedure with 10000 elements.

the horizontal and vertical displacements

$$\Delta \geq \int_D \left(\frac{M_1\overline{M}_1}{r z_1^3 z_2} + \frac{M_2\overline{M}_2}{r z_1 z_2^3} \right) \mathrm{d}x \,, \tag{77}$$

and the cross-sectional dimensions are constrained from below

$$z_1 \geq z_{1a} \,, \qquad z_2 \geq z_{2a} \,. \tag{78}$$

For the above problem, the following optimality conditions have been derived from both the general formulae in Fig. 3 (with $\lambda_\ell = 0$ for all ℓ and $\nu_1 \to \nu$) and by a variational derivation [11]:

Case A: $z_1 > z_{1a}$, $z_2 > z_{2a}$.

$$z_1^6 = 4\nu \frac{(M_1\overline{M}_1)^2}{M_2\overline{M}_2} \,, \quad z_2^6 = 4\nu \frac{(M_2\overline{M}_2)^2}{M_1\overline{M}_1} \,. \tag{79}$$

The conditions for the validity of Case A and (79) are:

$$M_2\overline{M}_2 > 0 \,, \quad M_1\overline{M}_1 > 0 \,, \quad z_{1a}^6 < 4\nu \frac{(M_1\overline{M}_1)^2}{M_2\overline{M}_2} \,, \quad z_{2a}^6 < \frac{(M_2\overline{M}_2)^2}{M_1\overline{M}_1} \,. \tag{80}$$

Cases B and C: $z_i = z_{ia}$, $z_k > z_{ka}$ ($i = 1$, $k = 2$ and $i = 2$, $k = 1$, respectively).

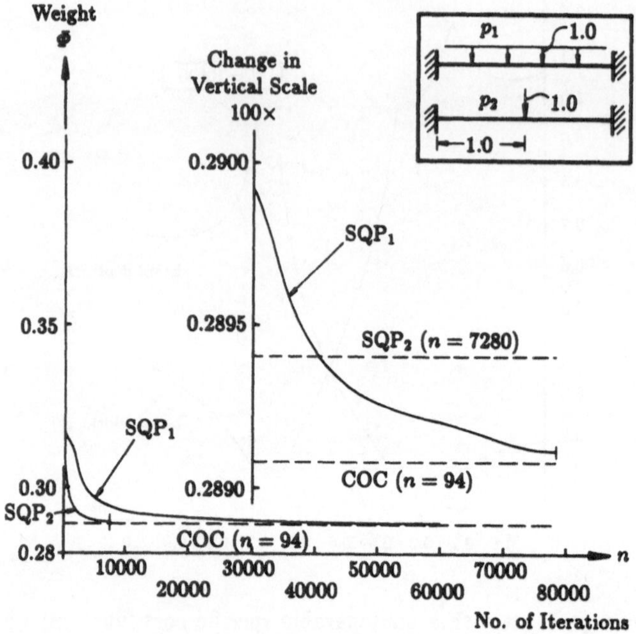

Fig. 19 A beam problem with a nonseparable specific cost function: a comparison of weights Φ obtained for 50 elements by unmodified and modified sequential quadratic methods (SQP_1 and SQP_2) after various iteration numbers and by the COC procedure after convergence.

$$z_k^2 = \frac{\nu M_i \overline{M}_i}{2z_{ia}^4} + \sqrt{\left(\frac{\nu M_i \overline{M}_i}{2z_{ia}^4}\right)^2 + \frac{3\nu M_k \overline{M}_k}{z_{ia}^2}}\,, \tag{81}$$

The conditions for the validity of Case B or C and (81) are:

$$z_{ia}^4 z_k^4 > \nu(3M_i\overline{M}_i z_k^2 + M_k\overline{M}_k z_{ia}^2)\,, \quad M_k\overline{M}_k > 0\,. \tag{82}$$

Case D: $z_1 = z_{1a}$, $z_2 = z_{2a}$. We have the following conditions for this case:

$$z_{1a}^4 z_{2a}^4 \geq \nu(3M_2\overline{M}_2 z_{1a}^2 + M_1\overline{M}_1 z_{2a}^2)\,, \quad z_{1a}^4 z_{2a}^4 \geq \nu(M_2\overline{M}_2 z_{1a}^2 + 3M_1\overline{M}_1 z_{2a}^2)\,. \tag{83}$$

The calculation of the Lagrangian from the deflection condition in (77) requires a Newton-Raphson procedure, which is described in a research report [16].

10.1.2 Solutions by the Iterative COC and SQP Methods. Considering a clamped beam having a span of 2.0 and a uniformly distributed load of unit intensity in

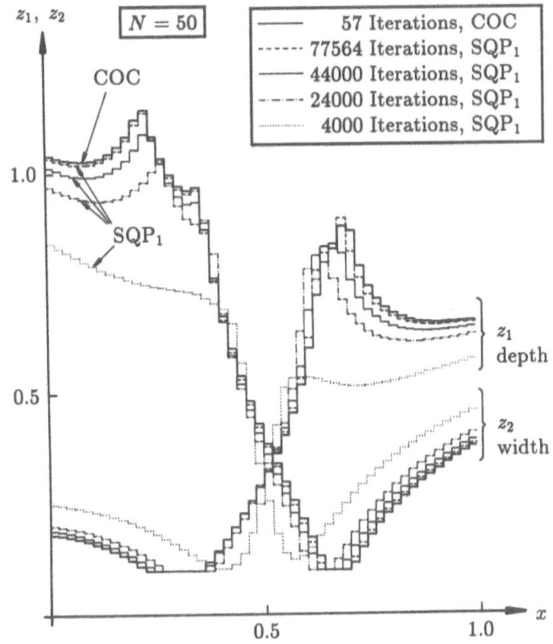

Fig. 20 A comparison of width and depth distributions obtained by the COC and unmodified SQP methods.

the vertical direction and a central unit point load in the horizontal direction (see insert in Fig. 19), a deflection value of $\Delta = 3$ was adopted in (77) and $c = r = 1$.

Using the COC method with only 50 elements and 100 variables (for a comparison with the SQP method), and a tolerance criterion of $(\Phi_{\text{new}} - \Phi_{\text{old}})/\Phi_{\text{new}} \leq 10^{-8}$, a weight value of $\Phi = 0.289081$ was obtained after 94 iterations, with a total CPU time (analyses plus optimization) of 57 sec. In repeating the COC procedure with 10000 elements and 20000 variables, the above convergence criterion was reached after 103 iterations with a CPU time of 12096 sec and a weight of $\Phi = 0.288782$. The corresponding optimal variations of the width and depth are shown in Fig. 18.

The same problem with 50 elements and 100 variables was also computed using the SQP method with reciprocal variables. The above convergence criterion was satisfied after 77564 iterations, requiring a CPU time of 1507451 sec (17.45 days). In Fig. 19, the iteration history of this calculation is compared with the result given by the COC method after convergence.

Since the convergence was extremely slow, a modified procedure was used in which the coupling effect was relaxed through multiplying the mixed second derivatives by various factors. After testing several such factors, a value of 0.2 was adopted which resulted in a much faster convergence (see SQP_2 in Fig. 19). Details of the modified SQP procedure used are described in a research report [16].

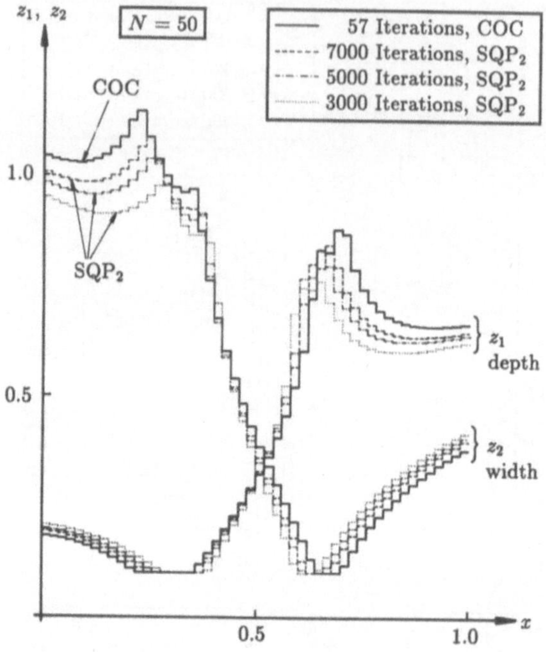

Fig. 21 A comparison of width and depth distributions obtained by the COC and modified SQP methods.

	Weight Φ	Δ%	No. of Iterations	CPU time sec
COC	0.289081	0	94	57
SQP_1	0.289114	0.011	77564	1507451 (17.45 days)
SQP_2	0.289408	0.113	7280	176746 (2.05 days)

Table 2. A comparison of weight values obtained by the SQP method after various numbers of iterations and by the COC method after convergence (= 94 iterations) and percentage differences.

The SQP_2 procedure reached the convergence criterion after 7280 iterations, requiring a CPU time of "only" 176746 sec (2.05 days). However, due to small instabilities in the convergence, satisfaction of the convergence criterion was somewhat accidental giving a fairly large error in comparison to the COC and SQP_1 results (see the enlarged right-hand part of Fig. 19). The results of the COC, as well as the unmodified and modified SQP methods, are compared in Table 2. The width and depth distributions given by the two SQP methods after various numbers of iterations are compared with those obtained by the COC method in Figs.

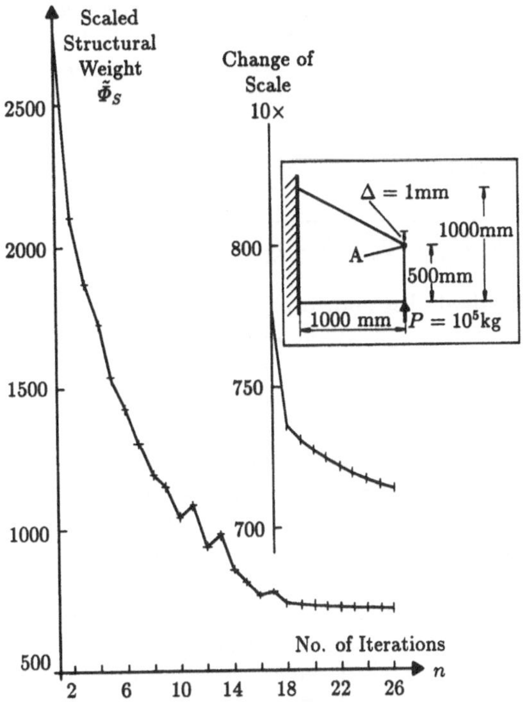

Fig. 22 Scaled structural weight $\tilde{\Phi}_s$ as a function of the iteration number for a plate of variable thickness subject to plane stress: optimization by the COC procedure with 1250 elements for a deflection constraint.

20 and 21. It can be concluded from the above diagrams and Table 2 that both the unmodified and modified SQP methods fully confirm the results by the COC algorithm. However, the more accurate SQP method (SQP_1) requires almost 30000 times as much time as the COC procedure.

10.2 Two-Dimensional Structural Systems

The proposed COC algorithm can be readily extended to two-dimensional systems (e.g. plates in plane stress or bending, and shells). In the case of a plate of variable thickness z in plane stress with a deflection constraint, we have

$$\mathbf{Q} = (N_x, N_y, N_{xy}), \quad \overline{\mathbf{Q}} = (\overline{N}_x, \overline{N}_y, \overline{N}_{xy}), \quad [\mathbf{F}] = \frac{1}{z}\begin{bmatrix} f_1 & -f_2 & 0 \\ -f_2 & f_1 & 0 \\ 0 & 0 & f_3 \end{bmatrix},$$

$$\psi = cz, \quad f_1 = \frac{1}{E}, \quad f_2 = \frac{\upsilon}{E}, \quad f_3 = \frac{1}{G}, \quad c = \gamma, \tag{84}$$

where N_x and N_y are the axial forces and N_{xy} is the shear force per unit width in the x and y directions, γ is the specific weight, E is Young's modulus, G is the modulus of rigidity and v is Poisson's ratio. Then the general formulae in Fig. 3 (with $\lambda_\ell = 0$ for all ℓ and $\nu_1 \to \nu$) imply the optimality criterion:

$$c = (\nu/z^2)[f_1(N_x\overline{N}_x + N_y\overline{N}_y) - f_2(N_x\overline{N}_y + \overline{N}_x N_y) + f_3 N_{xy}\overline{N}_{xy}],$$

$$z\sqrt{c/\nu} = \sqrt{[f_1 N_x\overline{N}_x + N_y\overline{N}_y) - f_2(N_x\overline{N}_y + \overline{N}_x N_y) + f_3 N_{xy}\overline{N}_{xy}]}. \qquad (85)$$

The problem investigated is shown in the insert of Fig. 22, with a point load at the bottom right corner and a prescribed deflection of 1 mm at the top right corner. For the minimum plate thickness a very low value ($z_a = 0.1$ mm) was prescribed. The material properties for mild steel have been adopted ($\gamma = 0.785 \times 10^{-4}$ N/mm^3, $E = 2.1 \times 10^5$ N/mm^2, $v = 0.3$).

Using the iterative COC method, the above problem was solved with discretizations having 32, 128, 512, 1250 and 3200 triangular (constant strain) elements. The results were confirmed independently by the SQP method. The thickness of two adjacent elements forming a quadrilateral was linked. The convergence was found not to be quite as smooth as for the beam problems discussed but, apart from small initial fluctuations, it was near-monotonic (see Fig. 22 which was plotted for the system with 1250 elements). For further details of this solution the reader is referred to Ref. [11] in which the optimal thickness variation is shown in coloured diagrams. The above solutions were also confirmed by the SQP method to an accuracy of 8 digits for up to 128 elements.

11. Structures with Stress Constraints and One Deflection Constraint

Up to this point, we discussed examples with only one displacement constraint. Problems with additional stress constraints on the basis of the optimality criteria in Fig. 3 will be treated in this Section.

11.1 Beams of Variable Width with a Deflection Constraint, a Flexural Stress and a Shear Stress Constraint (Taking the Shear Deformations into Consideration)

For this type of problems we have

$$\mathbf{Q} = (M, V), \quad \overline{\mathbf{Q}} = (\overline{M}, \overline{V}), \quad [\mathbf{F}] = \begin{bmatrix} f_1 & 0 \\ 0 & f_2 \end{bmatrix}, \quad S_1 = (k_1|M| - z),$$

$$S_2 = (k_2|V| - z), \quad \psi = cz, \quad f_1 = 1/(r_1 z), f_2 = 1/(r_2 z), \quad r_1 = Eh^3/12,$$
$$r_2 = 5hG/6, \qquad (86)$$

where M and V are the bending moment and the shear force on the real beam whilst $\overline{M}$ and $\overline{V}$ are the corresponding adjoint stress resultants, E is Young's modulus, h is the beam depth, $f_2 V$ is the generalized shear strain (average shear strain for an entire beam cross-section), G is the modulus of rigidity, $k_1 = 6/(\sigma_p h^2)$ where σ_p is the permissible flexural stress, $k_2 = 1.5/(\tau_p h)$ where τ_p is the permissible shear stress, z is the beam width, S_1 and S_2 are the two permissible stress conditions and $c = \gamma h$ where γ is the specific weight.

In the considered problem, the real and adjoint generalized strain vectors are

$$\mathbf{q} = (\kappa, \zeta), \quad \overline{\mathbf{q}} = (\overline{\kappa}, \overline{\zeta}), \tag{87}$$

where κ is the beam curvature and ζ is the generalized shear strain. Then the adjoint displacement $\overline{u}$, for example, is given by

$$\overline{u} = -\int_0^x \int_0^x \overline{\kappa}(x)\, \mathrm{d}x\, \mathrm{d}x - \int_0^x \overline{\zeta}(x)\, \mathrm{d}x\,. \tag{88}$$

On the basis of Fig. 3 and (86), the adjoint strains become (for $M \neq 0$, $V \neq 0$)

$$\overline{\kappa} = -\lambda_1 k_1 \operatorname{sgn} M + \nu\overline{M}/(r_1 z)\,, \quad \overline{\zeta} = -\lambda_2 k_2 \operatorname{sgn} V + \nu\overline{V}/(r_2 z)\,. \tag{89}$$

The relations in Fig. 3 and (86) imply the following optimality criterion:

$$c - \nu\left(\frac{M\overline{M}}{r_1 z^2} + \frac{V\overline{V}}{r_2 z^2}\right) + \lambda_1 + \lambda_2 = 0\,. \tag{90}$$

11.1.1 Optimality Criteria for Beams of Variable Width with a Deflection Constraint and Flexural Stress Constraints. Neglecting the shear terms in (89) and (90), i.e. with $k_2 \to 0$, $r_2 \to \infty$, we have

$$\lambda_1 = \frac{\nu M\overline{M}}{r_1 z^2} - c\,, \tag{91}$$

$$-\overline{u}'' = -\lambda_1 k_1 \operatorname{sgn} M + \frac{\nu\overline{M}}{r_1 z}\,. \tag{92}$$

Then there are two possibilities:

Case A: Stress constraint is active, $k_1|M| = z$.

$$\lambda_1 \geq 0\,, \quad (91)(92) \Rightarrow \overline{\kappa} = -\overline{u}'' = -\frac{\nu M\overline{M} \operatorname{sgn} M}{r_1 |M| z} + c k_1 \operatorname{sgn} M + \frac{\nu\overline{M}}{r_1 z} = c k_1 \operatorname{sgn} M\,. \tag{93}$$

Case B: Stress constraint is inactive $k_1|M| < z$.

$$\lambda_1 = 0\,, \quad \overline{\kappa} = -\overline{u}'' = \frac{\nu\overline{M}}{r_1 z}\,, z = \sqrt{\frac{\nu M\overline{M}}{c r_1}}\,, \tag{94}$$

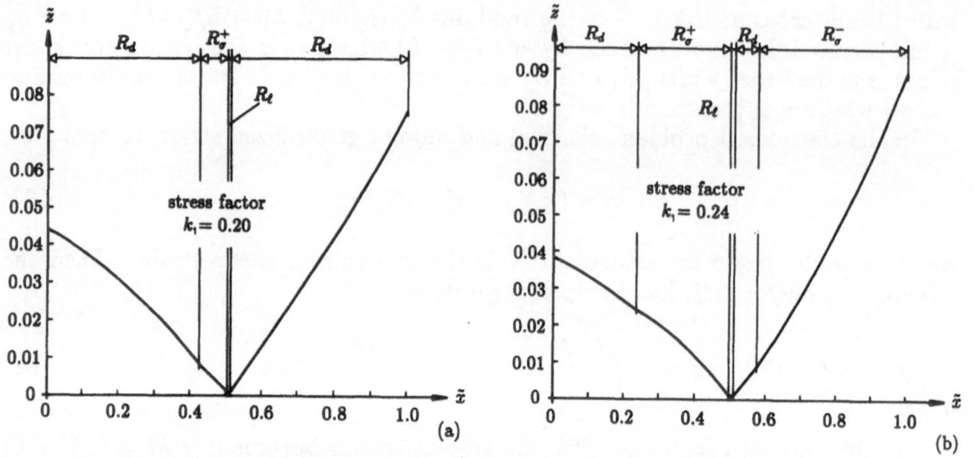

Fig. 23 Optimal variation of the width for a clamped, uniformly loaded beam with a deflection constraint and stress constraints; the value of the stress cost factor is (a) $k_1 = 0.20$, and (b) $k_1 = 0.24$.

as in (21).

Note. In the FE analysis of the adjoint system, a constant curvature given by (93) must be used for stress-controlled regions which can be done by specifying an initial curvature or curvature caused by ***nonuniform temperature distribution*** in the cross-section. Since the adjoint moment causes no curvature in stress-controlled regions, for the adjoint analysis a very large flexural stiffness value is adopted for such regions. If a minimum width z_a (with a flexural stiffness of $r_1 z_a$) is specified, then for regions controlled by such a constraint the adjoint curvature is $\nu\overline{M}/(r_1 z_a)$. A more general method for handling the effect of active stress constraints on the adjoint structure is given in terms of initial displacements in Section 15.2 [Eq. (168)].

11.1.2 Iterative COC Results for Beams of Variable Width with a Deflection Constraint and Stress Constraints. Using the optimality criteria given in Section 11.1.1 with $c = 1$, $r_1 = 1$ for a clamped beam with a uniformly distributed load and various values of the stress cost factor $k_1 \to k$, the solutions shown in Fig. 23 were obtained with 10000 elements. The regions R_d are controlled by the deflection constraint, the regions R_ℓ by the minimum width (= lower limit) constraint and the regions R_σ^+ and R_σ^- by the stress constraint with positive and negative moments, respectively.

In calculating the Lagrangian, the same equation [(64) in Section 3.6] can be used as for a deflection constraint without stress constraint, but the passive element set includes all elements in the regions R_ℓ, R_σ^+ and R_σ^-. The above results were verified by the semi-analytical method to a seven-digit accuracy [11, 15].

11.1.3 Iterative COC Results for Laminated Timber Beams with a Constraint on the Maximum Deflection and Limits on Both Flexural and Shear Stresses (in Accordance with the German Design Code DIN 1052. Optimality criteria similar to

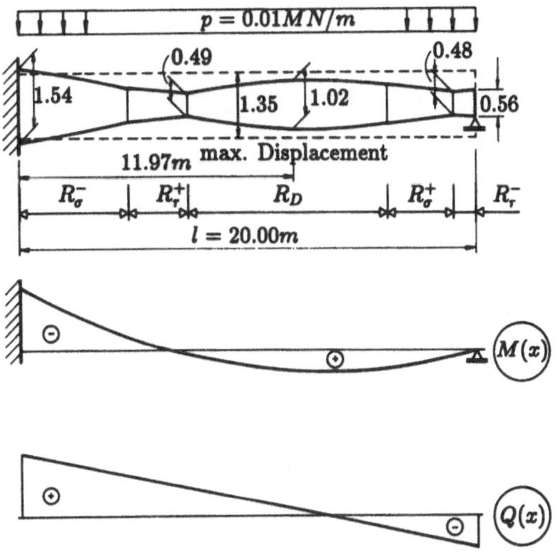

Fig. 24 Optimal depth distribution for a laminated timber beam designed in accordance with the German Design Code DIN 1052, having constraints on the maximum deflection, flexural stresses and shear stresses.

those derived in Section 11.1 were used for the optimization of long-span laminated timber beams of ***constant width and variable depth*** with the following deflection and stress constraints in accordance with DIN 1052:

$$\sigma_p = 11\text{N/mm}, \quad \tau_p = 1.2\text{N/mm}, \quad \Delta = L/300, \quad E = 11000\text{N/mm}^2, \qquad (95)$$

where σ_p is the permissible flexural stress, τ_p the permissible shear stress, Δ the permissible deflection, L the beam span, and E Young's modulus. The effect of the shear deformation on the above deflection was neglected. An example concerning a propped cantilever with a uniformly distributed load of $p = 0.01MN/\text{m}$ and a span of 20 m is given in Fig. 24, in which the given width of the beam was $b = 0.15$ m. The above problem was solved by the COC procedure using up to 100000 elements. As the location of the maximum deflection is not known, this was always based on the analysis of the real beam in the prior iterations. This procedure has resulted in a very satisfactory convergence, although the equivalent problem with an MP formulation would have 100000 deflection constraints (i.e. a prescribed deflection at each node), out of which only one is active.

The optimal variation of the depth for the above problem is shown in Fig. 24, in which the corresponding moment and shear force diagrams are also indicated. Regions controlled by the deflection, flexural stress and shear stress constraints, respectively, are denoted by R_d, R_σ and R_τ, and the signs in the superscripts refer to the sign of the corresponding bending moment or shear force. Figure 24 (broken lines) also shows the prismatic beam satisfying the same design constraints. The latter has a 58% higher volume than the optimal one.

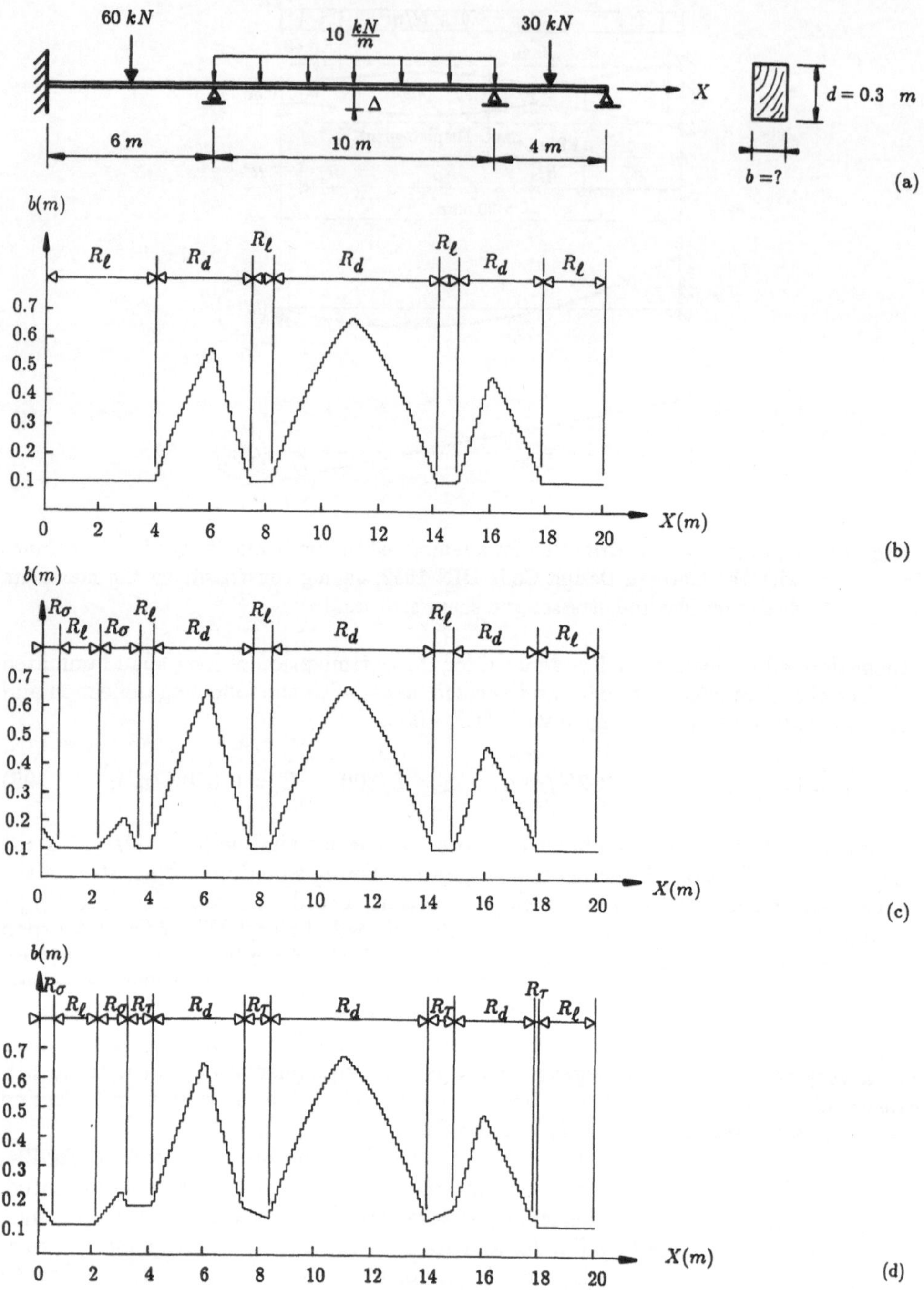

Fig. 25 Outputs from a software generating automatically optimal beam designs for deflection, flexural stress and shear stress constraints.

The same procedure was extended to beams with selfweight and external load. For computational details of this section, the reader is referred to a research report [16].

The first author also developed a program which couples COC with FE analysis for generating the optimal solution for beams of *variable width* with any support and loading conditions. The laminated timber beam shown in Fig. 25a was designed for the design constraints in (95) in accordance with the German design code DIN 1052. For the load shown in Fig. 25a and a given depth of $d = 0.3$ m, the width variations shown in Figs. 25b, c and d were obtained for deflection constraints only, for deflection and flexural stress constraints, and for deflection, as well as flexural and shear stress constraints, respectively, using 200 elements, 13 iterations and a convergence tolerance value of $\overline{E} = 10^{-6}$. The above program is based on the DCOC method to be discussed in Section 15.

12. Trusses with Stress Constraints and a Deflection Constraint — A Comparison of DOC and COC Methods

In this section, an independent derivation of optimality criteria for trusses is given, which may be easier to understand than the derivation for the general case. Moreover, important differences between DOC and COC are pointed out. Since trusses are inherently discretized structures, the derivation is based on the Kuhn-Tucker condition in a finite dimensional vector space.

12.1 Derivation of Optimality Criteria

Let the given external loads at joints of a linearly elastic truss be P_h $(h = 1, 2, \ldots, w)$, the forces in the members F_i $(i = 1, 2, \ldots, n)$, the corresponding cross-sectional areas A_i $(i = 1, 2, \ldots, n)$, the member lengths $L_i (i = 1, 2, \ldots, n)$ and the permissible stresses $\pm\sigma_{i0}$ $(i = 1, 2, \ldots, n)$. Moreover, for simplicity let the deflection u_A be constrained,

$$u_A \leq \Delta, \tag{96}$$

at only one joint (A) of the truss, where Δ is a given constant. All cross-sectional areas are to be constrained from below.

Expressing the deflection constraint in (96) by means of a virtual work equation, and denoting the prescribed minimum cross-sectional area, Young's modulus and the specific weight, respectively, for the member i by A_{i0}, E_i and γ_i, the truss weight (Φ) minimization problem under consideration can be stated as follows:

$$\min \Phi = \sum_i \gamma_i A_i L_i\,, \tag{97}$$

subject to

$$A_i \geq |F_i|/\sigma_{i0}\,, \tag{98}$$

$$\sum_i [F_i \overline{F}_i L_i/(A_i E_i)] \leq \Delta\,, \tag{99}$$

$$A_i \geq A_{i0}\,, \tag{100}$$

and to the equilibrium conditions for the loads P_h and member forces F_i as well as the virtual loads $\overline{P}_h$ and member forces $\overline{F}_i$. In the case of a constraint on a deflection at a single point, the virtual loads $\overline{P}_h$ reduce to a single point load at the joint whose deflection is constrained. If a linear combination of deflections at various joints and in various directions is constrained, then the virtual loads $\overline{P}_h$ represent the weighting factors for such a combination.

The above equilibrium conditions can be expressed by means of the theorem of virtual displacements (e.g. Argyris and Kelsey [17])

$$\sum_h \overline{u}_h P_h - \sum_i \overline{\varepsilon}_i L_i F_i = 0\,, \tag{101}$$

$$\sum_h u_h \overline{P}_h - \sum_j \varepsilon_i L_i \overline{F}_i = 0\,, \tag{102}$$

where $(\overline{u}_h,\, \overline{\varepsilon}_i)$ and $(u_h,\, \varepsilon_i)$ represent, at this stage, any kinematically admissible set of (small) virtual displacements which have no causal relationship with the force system in the same equations. The coupling between the strains $(\overline{\varepsilon}_i,\, \varepsilon_i)$ and forces $(\overline{F}_i,\, F_i)$ will be provided later by the Kuhn-Tucker conditions (105) and (106).

The Lagrangian function ($\overline{\Phi}$) in the augmented problem then becomes

$$\overline{\Phi} = \sum_i \Big\{\gamma_i A_i + \lambda_i[(|F_i|/\sigma_{i0}) - A_i + s_i] + \beta_i(A_{i0} - A_i + s_{i0}) + \nu F_i \overline{F}_i/(A_i E_i)\Big\} L_i - \nu\Delta +$$

$$+\nu s + \overline{\alpha}\Big(\sum_h \overline{u}_h P_h - \sum_i \overline{\varepsilon}_i L_i F_i\Big) + \alpha\Big(\sum_h u_h \overline{P}_h - \sum_i \varepsilon_i L_i \overline{F}_i\Big)\,, \tag{103}$$

where λ_i, ν, β_i, $\overline{\alpha}$ and α are Lagrangians, and s_i, s_0 and s are slack variables. It is not necessary to include kinematic admissibility in (103), because it will become an optimality condition for statically admissible forces.

Necessary conditions of a minimum in (103) become

$$\frac{\partial \overline{\Phi}}{\partial A_i} = 0 \Rightarrow \gamma_i - \lambda_i - \beta_i - \frac{\nu F_i \overline{F}_i}{A_i^2 E_i} = 0\,, \tag{104}$$

$$\frac{\partial \overline{\Phi}}{\partial F_i} = 0 \Rightarrow (\text{for } F_i \neq 0)\; \overline{\alpha}\, \overline{\varepsilon}_i = \frac{\lambda_i \operatorname{sgn} F_i}{\sigma_{i0}} + \frac{\nu \overline{F}_i}{A_i E_i}\,, \tag{105}$$

$$\frac{\partial \overline{\Phi}}{\partial \overline{F}_i} = 0 \Rightarrow \alpha \varepsilon_i = \frac{\nu F_i}{A_i E_i}\,. \tag{106}$$

The statement after (102) requires kinematical admissibility of $\overline{\varepsilon}_i$ and ε_i. The latter, by (106), is a factored (ν/α) value of the elastic member strains. Hence (for rigid supports) kinematical admissibility of the member strains follows from an optimality condition. This was the reason for not incorporating the compatibility of deformations into (103).

In other words, although we are aiming at the optimization of statically and kinematically admissible elastic trusses, we first imbedded our solution set into a larger set of only statically admissible trusses. It turned out, however, that for this enlarged solution set the optimal solution automatically fulfils kinematic admissibility.

For variations of the non-negative slack variables we have the minimality conditions

$$\lambda_i \geq 0\,, \quad \lambda_i \neq 0 \quad \text{only if} \quad |F_i|/\sigma_{i0} = A_i\,, \tag{107}$$

$$\beta_i \geq 0\,, \quad \beta_i \neq 0 \quad \text{only if} \quad A_i = A_{i0}\,. \tag{108}$$

Since linear scaling of the strains $\overline{\varepsilon}_i$ does not affect the solution, we adopt $\overline{\alpha} = 1$.

Then there exist various possibilities depending on which constraints are active. For reasons to be explained subsequently, $\overline{\varepsilon}_i$ is termed an "adjoint strain".

A. Active stress constraint

a. Cross-section in excess of prescribed minimum

$(A_i = |F_i|/\sigma_{i0},\ A_i > A_{i0})$

Then the second condition in the title and (108) imply $\beta_i = 0$ and by the first condition in the title together with (104) and (105) with $\overline{\alpha} = 1$ we have

$$\lambda_i = \gamma_i - \frac{\nu \overline{F}_i \operatorname{sgn} F_i \sigma_{i0}}{A_i E_i}\,, \tag{109}$$

$$\overline{\varepsilon}_i = \frac{\gamma_i \operatorname{sgn} F_i}{\sigma_{i0}} - \frac{\nu \overline{F}_i}{A_i E_i} + \frac{\nu \overline{F}_i}{A_i E_i} = \frac{\gamma_i \operatorname{sgn} F_i}{\sigma_{i0}}\,. \tag{110}$$

Note that in this case the adjoint strain $\overline{\varepsilon}_i$ *does not depend on the virtual force* $\overline{F}_i$ and it *involves no Lagrangian.*

b. Minimum prescribed cross-section

$(A_i = |F_i|/\sigma_{i0} = A_{i0})$

By (108) $\beta_i \geq 0$ for the above case and then by (104), (107) and (108) we have

$$0 \leq \lambda_i \leq \gamma_i - \frac{\nu \overline{F}_i \operatorname{sgn} F_i \sigma_{i0}}{A_{i0} E_i}\,. \tag{111}$$

Note that by (107) $\lambda_i \geq 0$ and hence the RHS of (111) must be positive.

This means that for this case *the adjoint strain* $\overline{\varepsilon}_i$ *is nonunique* and by (105) with $\overline{\alpha} = 1$ and $A_i = A_{i0}$ its limiting values are

$$\left(\text{for } \lambda_i = \gamma_i - \frac{\nu \overline{F}_i \operatorname{sgn} F_i \sigma_{i0}}{A_{i0} E_i}\right) \quad \overline{\varepsilon}_i = \frac{\gamma_i \operatorname{sgn} F_i}{\sigma_{i0}}\,, \tag{112}$$

and

$$(\text{for } \lambda_i = 0) \quad \overline{\varepsilon}_i = \frac{\nu \overline{F}_i}{A_{i0} E_i}\,. \tag{113}$$

However, $\overline{\varepsilon}_i$ can take on any convex combination of the limiting values in (114) and (115).

B. Inactive stress constraint

a. Cross-section in excess of prescribed minimum
$(A_i > |F_i|/\sigma_{i0}\ ,\ A_i > A_{i0})$
The first condition in the title and (107) imply $\lambda_i = 0$, by the second condition in the title and (108) $\beta_i = 0$ and then by (104) we have

$$A_i = \sqrt{\nu F_i \overline{F}_i/(E_i \gamma_i)} . \tag{114}$$

Moreover, (105) with $\overline{\alpha} = 1$ implies

$$\overline{\varepsilon}_i = \frac{\nu \overline{F}_i}{A_i E_i} . \tag{115}$$

b. Prescribed minimum cross-section
$(A_i > |F_i|/\sigma_{i0}\ ,\ A_i = A_{i0})$
The first condition in the title and (107) imply $\lambda_i = 0$, by the second condition in the title and (108) we have $\beta_i = 0$ and then (105) with $\overline{\alpha} = 1$ implies

$$\overline{\varepsilon}_i = \frac{\nu \overline{F}_i}{A_{i0} E_i} . \tag{116}$$

12.2 The "Adjoint Truss"

It can be seen from (115) and (116) that if the stress constraint is not active for a member then the values of $\overline{\varepsilon}_i/\nu$ are equal to the usual elastic strains corresponding to the member forces $\overline{F}_i$ equilibrating the virtual loads $\overline{P}_\ell$ ($\ell = 1, 2, \ldots, m$). For members with an active stress constraint, on the other hand, (110) implies that the value of $\overline{\varepsilon}_i$ does not depend on the virtual force $\overline{F}_i$, but only on the sign of the real member force F_i.

The values of $\overline{\varepsilon}_i$ for the above cases can, therefore, be calculated by considering a fictitious truss termed "adjoint truss". This truss is subject to the virtual loads $\overline{P}_\ell$ ($\ell = 1, 2, \ldots, m$), which can also be termed "adjoint" loads. For members with an inactive stress constraint the stiffness is the usual elastic stiffness $(A_i E_i)$ and the stiffness of members with an active stress constraint is infinity, in order to suppress the effect of virtual (or adjoint) forces $\overline{F}_j$ on the adjoint strains $\overline{\varepsilon}_j$. However, the latter members are subject to a pre-strain (or temperature strain) of $\gamma_i \operatorname{sgn} F_i/(\sigma_{i0}\nu)$.

Fully stressed members with a prescribed minimum cross-section need a special treatment owing to the non-uniqueness of the adjoint strain $\overline{\varepsilon}_i$. This problem will be discussed elsewhere.

12.3 A Comparison with the Traditional DOC Formulation

In the traditional discretized optimality criteria (DOC) method, the minimum weight truss design problem is formulated as

$$\min \Phi = \sum_{i=1}^{n} \gamma_i L_i A_i \,, \tag{117}$$

subject to

$$g_k = \sum_{i=1}^{n} \frac{F_i \overline{F}_{ki} L_i}{E_i A_i} - c_k \leq 0 \quad (k = 1, 2, \ldots, r)\,, \tag{118}$$

$$A_i \geq A_{i0} \quad (i = 1, 2, \ldots, n)\,, \tag{119}$$

where g_k are behavioural constraints. In the problem discussed in this note, a single displacement constraint and stress constraints for all the members are considered and, therefore, we have $r = 1 + n$. Moreover, $\overline{F}_k = \{\overline{F}_{k1}, \ldots, \overline{F}_{kn}\}^T$ is the virtual force vector corresponding to a virtual load $\overline{P}_{kj}$ $(j = 1, 2, \cdots, m)$ which consists of a unit dummy load for the displacement constraint and of two unit forces of opposite direction, applied at the two ends of a bar, for stress constraints. The value of c_k is simply the allowable displacement for the displacement constraint. For stress constraints, it takes the form

$$c_k = \frac{\sigma_{i0}}{E_i} L_i \,, \tag{120}$$

where i is the number of the member to which the constraint refers.

In (118), F_i is required to be statically and kinematically admissible and (by the principle of virtual work) $\overline{F}_{ki}$ only needs to be statically admissible. It has been shown [18], however, that if $\overline{F}_{ki}$ is also kinematically admissible, then we have

$$\sum_i \left(\frac{\partial e_i}{\partial F_i} \frac{\partial F_i}{\partial A_i} + \frac{\partial e_i}{\partial \overline{F}_{ki}} \frac{\partial \overline{F}_{ki}}{\partial A_i} \right) = 0\,, \tag{121}$$

where $e_i = \frac{F_i \overline{F}_{ki} L_i}{E_i A_i}$ in (118). This has the important consequence that in calculating the gradient of the expression in (118), we need to consider the partial derivative with respect to A_i only [see (122) later] and the effect of the dependence of F_i and $\overline{F}_{ki}$ on A_i cancels out.

Then the well-known optimality criteria can be derived from the Kuhn-Tucker condition (see e.g. [19])

$$\gamma_i L_i - \sum_{k=1}^{r} \lambda_k \frac{F_i \overline{F}_{ki} L_i}{E_i A_i^2} = 0 \quad (i = 1, 2, \ldots, n)\,, \tag{122}$$

$$\lambda_k \geq 0 \quad (k = 1, 2, \ldots, r)\,, \tag{123}$$

where λ_k are Lagrangians. The following recurrence formula can be obtained from (122):

$$A_i = \left(\sum_{k=1}^{r} \lambda_k \frac{F_i \overline{F}_{ki}}{\gamma_i E_i} \right)^{1/2} , \tag{124}$$

where $\lambda_k > 0$ for active constraints and $\lambda_k = 0$ for passive constraints. The recurrence formula (124) may be replaced with one containing an adjustable "step size parameter" to improve convergence. Many methods have been suggested in the literature to evaluate the Lagrangian multipliers λ_k (for a review see Khot and Berke [20], Berke and Khot [1], [12], Venkayya [21] and Haftka [20]).

The main difference between the proposed technique and the traditional DOC methods are as follows. If q represents the number of active constraints, then the traditional DOC methods require *the analysis of q virtual load systems and the evaluation of q Lagrangian multipliers.* Therefore, the capability of the above DOC methods is restricted by the number of active displacement *and* stress constraints. On the other hand, the recurrence relation for understressed members in the proposed technique is given by (114), in which *only one Lagrangian multiplier* appears and the effect of active stress constraints is represented by a *constant strain* [see (110)] *in the corresponding member of the adjoint truss*, which influences the adjoint force $\overline{F}_i$ in (114). It follows that the *capability of the proposed technique is not affected by the number of active stress constraints.*

12.4 Iterative Optimization Procedure

The iterative method outlined in Fig. 10 can also be used for trusses. For the updating of the cross-sectional areas (124) is employed and the Lagrangians are calculated from the expression

$$\sqrt{\nu} = \frac{\sum_A L_i \sqrt{\frac{F_i \overline{F}_i \gamma_i}{E_i}}}{\Delta - \sum_P \frac{F_i \overline{F}_i L_i}{A_i E_i}} . \tag{125}$$

For fully stressed elements, an overrelaxation relation (Gellatly and Berke [23]),

$$A_i^{\text{new}} = A_i^{\text{old}} \left(\frac{\sigma_i}{\sigma_{i0}} \right)^{\alpha} \quad (\alpha > 1) , \tag{126}$$

can be used.

12.5 A Numerical Example: A Ten-Bar Truss

We consider the truss in the insert of Fig. 26 with $A_{i0} = 0.1$ in^2, $\sigma_{i0} = \pm 25000$ psi, $\gamma = 0.1$ lb/in^3 and $E_i = 10^7$ psi for all members. The vertical deflection value at the joint A is limited to a maximum value of 5 in. Imperial (British) units are used because they are customary in the literature on this problem.

Using the iterative procedure described in Section 12.4 with $\alpha = 1.1$ in (30) and a convergence criterion $|\Phi_{new} - \Phi_{old}|/\Phi_{new} \leq 10^{-10}$, after 20 iterations a total weight value of $\Phi = 2139.104979981$ lb was obtained and the corresponding values of the cross-sectional areas, real and adjoint forces are given in Table 3.

	COC solution			dual programming	fully stressed design
member	real force F_i (lb)	adjoint force $\overline{F}_i$ (lb)	cross-sectional area A_i (in^2)	cross-sectional area A_i (in^2)	cross-sectional area A_i (in^2)
1	201709.24	1.9718038	12.1611740	12.1611740	12.1265762
2	-198290.76	-1.0281962	8.7070289	8.7070290	8.8274507
3	554.12	0.0132459	0.1000000	0.1000000	0.1000000
4	-99445.88	-0.9867542	6.0405799	6.0405799	6.0465853
5	139004.12	0.0398755	5.5601648	5.5601649	5.5643224
6	-143838.59	-1.3743381	8.5736403	8.5736402	8.4978822
7	140637.71	1.3954811	8.5426700	8.5426700	8.5511629
8	-783.64	-0.0187325	0.1000000	0.1000000	0.1000000
9	226.34	-0.0149504	0.1000000	0.1000000	0.1000000
10	554.12	0.0132459	0.1000000	0.1000000	0.1000000
Lagrangian ν	371.84462				
structural weight Φ (lb)	2139.104979981			2139.104979978	2139.197925707

Table 3. A comparison of solutions by COC, dual programming and fully stressed design.

Bars 3, 8, 9 and 10 are governed by the minimum cross-section constraint and bar 5 by the stress constraint. The convergence history is shown in Fig. 26. In both Fig. 26 and in the convergence criterion, the weight values were scaled linearly upwards, if the solution violated a constraint.

It can be seen from Fig. 26 that, apart from a small instability at iteration 3, the convergence looks near monotonic, in spite of repeated vertical scale magnifications. Note that the rate of convergence actually increases progressively, enabling an immense scale magnification (by 50000) towards the end of the iteration history (Fig. 26). Naturally, the twelve digit accuracy achieved is of interest only in test problems; in practical applications a few iterations would suffice.

The same problem was independently solved by a dual programming method which gave a total weight value of $\Phi = 2139.104979978$ lb, showing a twelve digits agreement with the COC solution.

A further comparison solution, in which the adjoint strain for fully stressed members was also calculated on the basis of (115) [$\overline{\varepsilon} = \nu \overline{F}_i/(A_i E_i)$] instead of (110),

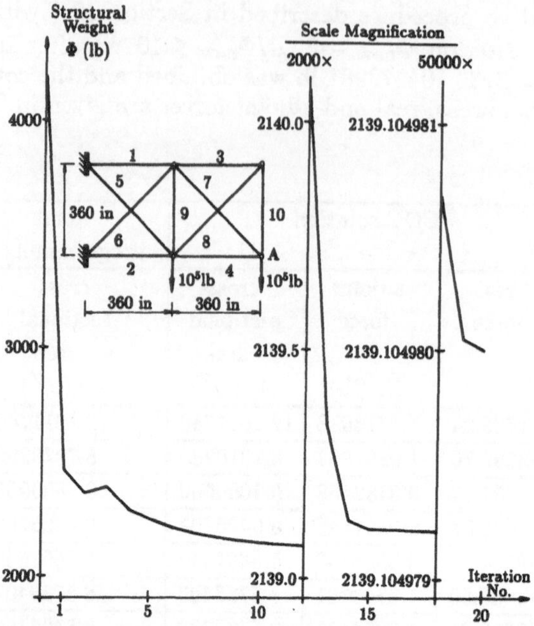

Fig. 26 Example: ten-bar truss.

resulted in a slightly higher weight than the optimal one ($\Phi = 2139.197925707$ lb), but the cross-sectional areas show significant differences (see Table 3, "fully stressed design"). This approach is an intuitive combination of the DOC and fully stressed design methods (Khot and Berke [20]).

All the above results were obtained with double precision (FORTRAN 77) on an HP 9000 work-station. This note also demonstrates the fact that in some *test examples* a three or four digit accuracy may not be sufficient; in the above example only the sixth digit differs in the weight values of the optimal and nonoptimal solutions.

12.6 The Treatment of Fully Stressed Members with a Prescribed Minimum Cross-Section

It was shown in (111)-(113) that for the above case, the adjoint strain $\bar{\varepsilon}_i$ is non-unique, which causes some initial computational difficulties in the iterative procedure. This problem can be solved by temporarily removing the considered member in the next iteration and replacing it with (i) two forces and (ii) a relative displacement constraint corresponding to the stress resultant and strain value in the fully stressed member of prescribed minimum cross-section. After each iteration using this modified problem, the adjoint strain is evaluated in the considered member. If this strain value passes through the limits in (112) and (113), then in the following cycle it is assumed, that the cross-section is fully stressed (with non-minimum cross-section) or has a minimum cross-section (not fully stressed), respectively.

13. Iterative COC Procedure for Several Displacement Constraints

In the optimality criteria and adjoint strain-stress relation in Fig. 3 or in (29) and (30), the cross-sectional parameters $\mathbf{z}$ and adjoint strains can be evaluated by considering elastic strains caused by the sum of the factored adjoint loads $\sum_j \nu_j \overline{\mathbf{p}}_j$. In the case of trusses, for example, (29) and (30) reduce to:
Active stress constraint

$$\overline{\varepsilon}_i = \frac{\gamma_i \operatorname{sgn} F_i}{\sigma_{i0}}, \quad A_i = |F_i|/\sigma_{i0}\,. \tag{127}$$

Inactive stress constraint

$$A_i = \sqrt{F_i(\sum_j \nu_j \overline{F}_{ji})/E_i\gamma_i}\,, \quad \overline{\varepsilon}_i = (\sum_i \nu_j \overline{F}_{ji})/A_i E_i\,, \tag{128}$$

where $\overline{F}_{ji}$ is the force in member i caused by the adjoint load condition j. This means that all quantities in (127) and (128) can be evaluated by a *single analysis of the adjoint system* with the combined load $\sum_j \nu_j \overline{p}_j = \sum_j \nu_j \overline{P}_{jh}$ $(h = 1, 2, w)$.

However, for *the calculation of the Lagrangians* it is convenient to analyse the adjoint structure for each adjoint load condition. For trusses, for example, the Lagrangians are calculated from the displacement conditions

$$\Delta_j = \sum_P \frac{F_i \overline{\overline{F}}_{ji} L_i}{A_i E_i} + \sum_A \frac{F_i \overline{\overline{F}}_{ji}}{E_i \sqrt{F_i(\sum_j \nu_j \overline{F}_{ji})/E_i\gamma_i}}\,, \tag{129}$$

where, by the principle of virtual work, $\overline{\overline{F}}_{ji}$ is required to be only statically admissible, whereas the term $(\sum_j \nu_j \overline{F}_i)$ represents the forces in the adjoint truss. These *combined* forces must satisfy kinematic admissibility. This means that both $\overline{F}_{ji}$ and $\sum_j \nu_j \overline{F}_i$ are non-unique.

It would be theoretically possible to start the iterative process by some initial set of ν_j values and then calculate the deflection values Δ_j as well as their numerical sensitivities by using only a single adjoint with the combined forces $(\sum_j \nu_j F_j)$. This sensitivity analysis, however, could become very uneconomical in the case of many active deflection constraints and could also cause numerical instabilities. For this reason, the following procedure has been adopted.

Due to their non-uniqueness, $\overline{\overline{F}}_{ji}$ and $\overline{F}_{ji}$ are made equal and are assigned a value which corresponds to the member forces caused by the separate factor adjoint load $\nu_j \overline{P}_{j\ell}$. For stress-controlled members, the *combined* adjoint strain is given by (127). This strain is distributed to the *separated* adjoint structures by using the factor $\nu_j/(\sum_j \nu_j)$. With these modifications, (129) provides a set of equations which yields the value of ν_j $(j = 1, 2, \ldots, v)$ by an iterative (e.g. Newton-Raphson) procedure. The above procedure was fully verified by numerical examples (see Section 17.2).

14. COC Procedure for Several Loading Conditions

In the case of alternative loading conditions $\mathbf{p}_k$ $(k = 1, 2, \ldots, \omega)$, the optimality conditions and adjoint stress-strain relations become [2, p. 64]:

$$\frac{\partial \psi}{\partial z_i} + \sum_{\ell} \sum_{k} \lambda_{\ell k}(\mathbf{x}) \frac{\partial}{\partial z_i} S_\ell(\mathbf{z}, \mathbf{Q}_k) + \sum_{j} \sum_{j} \nu_{jk} \mathbf{Q}_k \cdot \frac{\partial}{\partial z_i} [\mathbf{F}] \overline{\mathbf{Q}}_j \,, \tag{130}$$

$$\overline{q}_{ki} = \sum_{\ell} \lambda_{\ell k} \frac{\partial}{\partial z_i} S_\ell(\mathbf{z}, \mathbf{Q}_k) + \sum_{j} \nu_{jk} \mathbf{F}_i \overline{\mathbf{Q}}_j \quad (k = 1, 2, \ldots, \omega) \,. \tag{131}$$

The above relations do not cause any major difficulty in the iterative process, but by (131) a separate adjoint structure must be analysed for each load condition. A simple example involving two loading conditions is given in Section 26.2.

Part II: Iterative COC Methods Formulated Directly for Discretized Systems (DCOC Methods)

In the iterative COC methods discussed up to now, all optimality criteria and adjoint stress-strain relations are derived for a continuum in terms of generalized stresses (stress resultants), and the adjoint system is defined in terms of generalized strains. The relevant stress resultants for the real and adjoint system are then calculated by an FE program and subsequently substituted for each element into the analytical optimality criteria for updating the design parameters.

In this Part, the optimality criteria are derived directly for a discretized system consisting of finite elements, in the usual matrix notation for such systems. As pointed out in Section 1, the most important difference between COC and DOC is *not* the fact that one uses variational calculus and the other one differential calculus, but the way the problem is formulated explicitly in terms of the real and virtual stress resultants in COC. Although the present approach does not use a continuum formulation, it will be termed DCOC (*Discretized Continuum-Based Optimality Criteria*) method for historical reasons, since its derivation is a discretized equivalent of the usual COC formulation.

However, in applications of COC and DCOC there are significant differences with the following advantages and disadvantages.

The *disadvantage of COC* is that the analytical derivation of the optimality criteria and adjoint strain-stress relations is somewhat more demanding, although general formulae are available (e.g. Fig. 3). The optimality criteria are also more difficult to realize computationally.

The *advantage of COC* is that the solution provides an insight into fundamental properties of optimal structures. For example, in problems involving beams of constant depth, for regions with an active flexural stress constraint the curvature of the adjoint beam takes on a given *constant* value and for regions with an active shear stress constraint the adjoint shear strain is constant. Moreover, in layout problems the solution splits into certain optimal regions which makes the clear understanding of the optimal topology easy.

The great *advantage of DCOC* is that by substituting certain standard mechanical relations for particular classes of problems, the optimality criteria and the adjoint system are generated easily. Moreover, since the problem is already formulated in terms of nodal forces and displacements, it is not necessary to convert continuum-type quantities into discretized ones. For the above reasons, DCOC is suitable for *applications in conjunction with existing numerical algorithms and software*. In addition, most researchers and users in the field of structural optimization are familiar with the matrix notation of DCOC, whilst they are not used to Prager's notation of generalized stresses and strains used by COC.

The *disadvantage of DCOC* is that it partially obscures some of the basic features of optimal solutions.

Whereas DCOC has been formulated for elements of constant cross-section, iterative COC in its present form is only suitable for structures with continuously

G. I. N. Rozvany (ed.), Optimization of Large Structural Systems, Vol. I, 59–75.

varying cross-sections along the spatial coordinates. However, COC is available in an analytical form for "segmented" structures [2, pp. 47-48, 71-72] which should lead to the same results as DCOC.

15. Problem Formulation and Derivation of Optimality Criteria

15.1 Problem Statement for Stress Constraints and One Deflection Constraint

The class of problems considered in this section concern the optimum design of structural systems subject to a single displacement constraint at a specified degree of freedom D and stress constraints for all structural elements under one load case:

$$\min f(\mathbf{x}) \quad \text{subject to} \quad u_D - u_{Da} \leq 0\,,$$

$$g_j^e(\mathbf{x}) \leq 0 \quad [j = 1, \ldots, J(e)]\,,\ (e = 1, \ldots, NE)\,, \quad \mathbf{x} \in \mathbf{x}_F\,, \tag{132}$$

where $\mathbf{x}$ is the design variable vector of N components and $\mathbf{x}_F$ is its feasible domain. Generally, $\mathbf{x}$ can be expressed in NE partitions

$$\mathbf{x} = \{\mathbf{x}^1 \ldots \mathbf{x}^e \ldots \mathbf{x}^{NE}\}^T\,, \tag{133}$$

in which each partition $\mathbf{x}^e$ of $n(e)$ components corresponds to the cross-sectional dimensions of the e-th finite element

$$\mathbf{x}^e = \{x_1^e \ldots x_i^e \ldots x_{n(e)}^e\}^T\,. \tag{134}$$

NE is the total number of finite elements. The feasible domain $\mathbf{x}_F$ is usually defined with lower and upper bounds on the design variables

$$(x_i^e)^L \leq x_i^e \leq (x_i^e)^U\,, \quad [i = 1, \ldots, n(e)]\,,\ (e = 1, \ldots, NE)\,. \tag{135}$$

The symbols u_D and u_{Da} in (132) represent, respectively, the displacement at the specified degree of freedom D and the corresponding allowable value. Generally, u_D can also refer to a linear combination of several displacements. Moreover, g_j^e with $j = 1, \ldots J(e)$ represent $J(e)$ strength constraints of the e-th element.

15.2 Formulation in Terms of Forces

In order to derive the optimality criteria, the relevant behavioural constraints in the original problem (132) will be expressed explicitly in terms of the element nodal forces and the design variables. This can be achieved through the flexibility method of structural analysis (see, e.g. [24], [25]). The general form of the flexibility relationships for a structural element can be expressed as

$$\{\mathbf{u}_f^e\} = [\mathbf{f}^e]\{\mathbf{F}_f^e\}\,, \tag{136}$$

where $\{\mathbf{u}_f^e\}$ is the element relative displacement vector of n_f components, $[\mathbf{f}^e]$ the element flexibility matrix, $\{\mathbf{F}_f^e\}$ the element nodal force vector of length n_f.

The subscript f emphasizes that the nodal forces and associated displacements refer to the degrees of freedom that are free to displace on elements that are supported in a stable, statically determinate manner. Thus the number (n_f) of degrees of freedom for an element in the flexibility method is n_r fewer than that for the stiffness formulation, where n_r denotes the number of rigid body motions.

If initial strain or distributed loads are involved, the displacement-force relation becomes

$$\{\mathbf{u}_f^e\} = [\mathbf{f}^e]\{\mathbf{F}_f^e\} + \{\mathbf{u}_{fi}^e\}, \tag{137}$$

where $\{\mathbf{u}_{fi}^e\}$ are the initial displacements caused by the initial strain or distributed loads on the element with statically determinate supports.

Using the principle of virtual work, the displacement at the specified degree of freedom D can be expressed as

$$u_D = \sum_{e=1}^{NE} \{\overline{\mathbf{F}}_f^e\}^T \{\mathbf{u}_f^e\} = \sum_{e=1}^{NE} \left(\{\overline{\mathbf{F}}_f^e\}^T [\mathbf{f}^e]\{\mathbf{F}_f^e\} + \{\overline{\mathbf{F}}_f^e\}^T \{\mathbf{u}_{fi}^e\} \right), \tag{138}$$

in which $\{\overline{\mathbf{F}}_f^e\}$ are the virtual element nodal forces caused by the virtual unit dummy load $\{\overline{\mathbf{P}}^V\}$ applied at the specified degree of freedom D. If u_D refers to a linear combination of several displacements, $\{\overline{\mathbf{P}}^V\}$ represent the weighting factors for such a combination. It is important to note that *the real forces* $\{\mathbf{F}_f^e\}$ *must satisfy both equilibrium and compatibility conditions and the virtual forces* $\{\overline{\mathbf{F}}_f^e\}$ *need to satisfy only the equilibrium conditions.*

The principle of virtual displacements (e.g. Argyris and Kelsey [17]) states that for a deformable structure in equilibrium under the action of a system of applied loads $\{\mathbf{P}\}$, the external work W_{ext} due to *kinematically admisible* small virtual displacements $\{\delta\overline{\mathbf{u}}\}$ is equal to the virtual strain energy $\delta\overline{U}$ due to the same virtual displacements, i.e.

$$W_{\text{ext}} - \delta\overline{U} = 0. \tag{139}$$

The virtual external work and virtual strain energy can be expressed as

$$W_{\text{ext}} = \{\delta\overline{\mathbf{u}}\}^T \{\mathbf{P}\}, \tag{140}$$

$$\delta\overline{U} = \sum_{e=1}^{NE} \{\delta\overline{\mathbf{u}}_f^e\}^T [\mathbf{F}_f^e], \tag{141}$$

where $\{\delta\overline{\mathbf{u}}_f^e\}$ are the element relative displacements corresponding to $\{\delta\overline{\mathbf{u}}\}$. Equation (141) can also be expressed as

$$\delta\overline{U} = \{\delta\overline{\mathbf{u}}_f\}^T \{\mathbf{F}_f\}, \tag{142}$$

in which $\{\delta\overline{\mathbf{u}}_f\}$ represents all the element relative displacements

$$\{\delta\overline{\mathbf{u}}_f\} = \{\delta\mathbf{u}_f^1 \dots \delta\mathbf{u}_f^e \dots \mathbf{u}_f^{NE}\}^T, \tag{143}$$

and $\{\mathbf{F}_f\}$ represents all element nodal forces associated with the element flexibility relationships

$$\{\mathbf{F}_f\} = \{\mathbf{F}_f^1 \dots \mathbf{F}_f^e \dots \mathbf{F}_f^{NE}\}^T. \tag{144}$$

The relationship between global nodal displacements $\{\delta\overline{\mathbf{u}}\}$ and element relative displacements $\{\delta\overline{\mathbf{u}}_f\}$ can be expressed in the form

$$\{\delta\overline{\mathbf{u}}_f\} = [\mathbf{A}]\{\delta\overline{\mathbf{u}}\}, \tag{145}$$

where $[\mathbf{A}]$ is termed the kinematics matrix which is the transpose of the equilibrium (or statics) matrix $[\mathbf{B}]$:

$$\{\mathbf{P}\} = [\mathbf{B}]\{\mathbf{F}_f\}, \tag{146}$$

$$[\mathbf{A}] = [\mathbf{B}]^T. \tag{147}$$

The relation (146) is the standard form of the equilibrium condition.

On the basis of (140) and (141), (139) can be expressed as

$$\{\delta\overline{\mathbf{u}}\}^T\{\mathbf{P}\} - \sum_{e=1}^{NE}\{\delta\overline{\mathbf{u}}_f^e\}^T\{\mathbf{F}_f^e\} = 0, \tag{148}$$

which can also be represented through (142)-(147) in the following forms:

$$\{\delta\overline{\mathbf{u}}\}^T\{\mathbf{P}\} = \{\delta\overline{\mathbf{u}}_f\}^T\{\mathbf{F}_f\} = \{\delta\overline{\mathbf{u}}\}^T[\mathbf{A}]^T\{\mathbf{F}_f\} = \{\delta\overline{\mathbf{u}}\}^T[\mathbf{B}]\{\mathbf{F}_f\}. \tag{149}$$

Comparing (149) with (146) we can find that for a given vector $\{\delta\overline{\mathbf{u}}\}$, (149) is a weak form of the equilibrium equation (146).

Using the principle of virtual displacement mentioned above, the equilibrium conditions for the real and virtual systems in (138) can be expressed as follows:

$$\{\delta\overline{\mathbf{u}}^R\}^T\{\mathbf{P}\} - \sum_{e=1}^{NE}\{\delta\overline{\mathbf{u}}_f^{Re}\}^T\{\mathbf{F}_f^e\} = 0, \tag{150}$$

$$\{\delta\overline{\mathbf{u}}^V\}^T\{\overline{\mathbf{P}}^V\} - \sum_{e=1}^{NE}\{\delta\overline{\mathbf{u}}_f^{Ve}\}^T\{\overline{\mathbf{F}}_f^e\} = 0, \tag{151}$$

where $\{\delta\overline{\mathbf{u}}^R\}^T$ and $\{\delta\overline{\mathbf{u}}^V\}^T$ are *arbitrary kinematically admissible* virtual displacements, and $\{\delta\overline{\mathbf{u}}_f^{Re}\}^T$ and $\{\delta\overline{\mathbf{u}}_f^{Ve}\}^T$ are the corresponding element relative displacements.

The strength constraints g_j^e in (132) can be expressed as

$$g_j^e = \{\mathbf{S}_j^e\}^T\{\mathbf{F}_f^e\} - \sigma_{ja}^e\,, \tag{152}$$

where $\{\mathbf{S}_j^e\}^T$ is a vector converting the element nodal forces into a stress at a specified point j, and σ_{ja}^e is the allowable stress. In general, $\{\mathbf{S}_j^e\}$ is a function of the element cross-sectional dimensions $\mathbf{x}^e$. If the distributed load is applied on the element, then (152) is replaced by

$$g_j^e = \{\mathbf{S}_j^e\}^T\{\mathbf{F}_f^e\} + \{\mathbf{S}_{jd}^e\}^T\{\mathbf{F}_{jd}^e\} - \sigma_{ja}^e\,, \tag{153}$$

where $\{\mathbf{F}_{jd}^e\}$ are forces caused by the distributed load at the specified point j, $\{\mathbf{S}_{jd}^e\}$ is a vector converting $\{\mathbf{F}_{jd}^e\}$ into the stress. $\{\mathbf{F}_{jd}^e\}$ contains usually constants and $\{\mathbf{S}_{jd}^e\}$ are functions of the element cross-sectional dimensions $\mathbf{x}^e$.

By (138), (150), (151) and (153), the original optimization problem (132) can be expressed as follows:

$$\min f(\mathbf{x}) \quad \text{subject to} \quad \sum_{e=1}^{NE}\left(\{\overline{\mathbf{F}}_f^e\}^T[\mathbf{f}^e]\{\mathbf{F}_f^e\} + \{\overline{\mathbf{F}}_f^e\}^T\{\mathbf{u}_{fi}^e\}\right) - u_{Da} \le 0\,,$$

$$g_j^e = \{\mathbf{S}_j^e\}^T\{\mathbf{F}_f^e\} + \{\mathbf{S}_{jd}^e\}^T\{\mathbf{F}_{jd}^e\} - \sigma_{ja}^e \le 0 \quad [j = 1,\ldots,J(e)]\,,\ (e = 1,\ldots,NE)\,,$$

$$\{\delta\overline{\mathbf{u}}^R\}^T\{\mathbf{P}\} - \sum_{e=1}^{NE}\{\delta\overline{\mathbf{u}}_f^{Re}\}^T\{\mathbf{F}_f^e\} = 0\,, \quad \{\delta\overline{\mathbf{u}}^V\}^T\{\overline{\mathbf{P}}^V\} - \sum_{e=1}^{NE}\{\delta\overline{\mathbf{u}}_f^{Ve}\}^T\{\overline{\mathbf{F}}_f^e\} = 0\,,$$

$$\left.\begin{aligned} -x_i^e + (x_i^e)^L &\le 0 \\ x_i^e - (x_i^e)^U &\le 0 \end{aligned}\right\} \quad \begin{aligned} &[i = 1,\ldots,n(e)] \\ &(e = 1,\ldots,NE) \end{aligned}\,. \tag{154}$$

It is important to note that the compatibility condition of the real force system $\{\mathbf{F}_f\}$ is not included in (154), and therefore it can be regarded as a relaxed problem of (132). However, it will be shown that the stationary conditions for the problem expressed in (154) will include the compatibility conditions of $\{\mathbf{F}_f\}$. The basic variables involved in (154) are the element cross-sectional dimensions $\mathbf{x}$, the element nodal forces $\{\mathbf{F}_f\}$ of the real system and the element nodal forces $\{\overline{\mathbf{F}}_f\}$ of the virtual system. The initial displacements $\{\mathbf{u}_{fi}^e\}$ are usually functions of $\mathbf{x}$ and $\{\mathbf{F}_{fi}^e\}$ are constants. It is important to keep in mind that $\{\delta\overline{\mathbf{u}}^R\}$ and $\{\delta\overline{\mathbf{u}}^V\}$ are *arbitrary kinematically admissible virtual displacements.*

15.3 Optimality Criteria

The Lagrangian function of the optimization problem in (154) can be written as

$$\mathcal{L} = f(\mathbf{x}) + \nu\Big[\sum_{e=1}^{NE}\Big(\{\overline{\mathbf{F}}_f^e\}^T[\mathbf{f}^e]\{\mathbf{F}_f^e\} + \{\overline{\mathbf{F}}_f^e\}^T\{\mathbf{u}_{fi}^e\}\Big) - u_{Da}\Big] +$$

$$+\sum_{e=1}^{NE}\sum_{j=1}^{J(e)}\lambda_j^e\Big(\{\mathbf{S}_j^e\}^T\{\mathbf{F}_f^e\} + \{\mathbf{S}_{jd}^e\}^T\{\mathbf{F}_{jd}^e\} - \sigma_{ja}^e\Big) +$$

$$+\alpha^R\Big(\{\delta\overline{\mathbf{u}}^R\}^T\{\mathbf{P}\} - \sum_{e=1}^{NE}\{\delta\overline{\mathbf{u}}_f^{Re}\}^T\{\mathbf{F}_f^e\}\Big) + \alpha^V\Big(\{\delta\overline{\mathbf{u}}^V\}^T\{\overline{\mathbf{P}}^V\} - \sum_{e=1}^{NE}\{\delta\overline{\mathbf{u}}_f^{Ve}\}^T\{\overline{\mathbf{F}}_f^e\}\Big) +$$

$$+\sum_{e=1}^{NE}\sum_{i=1}^{n(e)}\Big[\beta_i^e(-x_i^e + (x_i^e)^L) + \gamma_i^e(x_i^e - (x_i^e)^U)\Big], \tag{155}$$

where ν, λ_j^e, α^R, α^V, β_i^e and γ_i^e are Lagrangian multipliers. The necessary conditions for a stationary point of the Lagrangian function are the well-known Kuhn-Tucker conditions which can be stated as

$$\frac{\partial \mathcal{L}}{\partial x_i^e} = 0 \quad [i = 1,\ldots,n(e)],\ (e = 1,\ldots,NE), \tag{156}$$

$$\frac{\partial \mathcal{L}}{\partial (F_f^e)_i} = 0 \quad (i = 1,\ldots,n_f),\ (e = 1,\ldots,NE), \tag{157}$$

$$\frac{\partial \mathcal{L}}{\partial (\overline{F}_f^e)_i} = 0 \quad (i = 1,\ldots,n_f),\ (e = 1,\ldots,NE). \tag{158}$$

All Lagrangian multipliers for the inequality constraints in (155) are nonnegative and if a constraint is not active the corresponding multiplier is zero. The symbols α^R and α^V denote non-zero multipliers for equality constraints.

By (155) and (156) we have

$$\frac{\partial f}{\partial x_i^e} + \nu\Big(\{\overline{\mathbf{F}}_f^e\}^T\Big[\frac{\partial \mathbf{f}^e}{\partial x_i^e}\Big]\{\mathbf{F}_f^e\} + \{\overline{\mathbf{F}}_f^e\}^T\Big\{\frac{\partial \mathbf{u}_{fi}^e}{\partial x_i^e}\Big\}\Big) + \sum_{j=1}^{J(e)}\lambda_j^e\Big(\Big\{\frac{\partial \mathbf{S}_j^e}{\partial x_i^e}\Big\}^T\{\mathbf{F}_f^e\} +$$

$$+\Big\{\frac{\partial \mathbf{S}_{jd}^e}{\partial x_i^e}\Big\}^T\{\mathbf{F}_{jd}^e\}\Big) - \beta_i^e + \gamma_i^e = 0 \quad [i = 1,\ldots,n(e)],\ (e = 1,\ldots,NE). \tag{159}$$

Moreover, (155), (157) and (158) imply

$$\nu[\mathbf{f}^e]\{\overline{\mathbf{F}}_f^e\} + \sum_{j=1}^{J(e)}\lambda_j^e\{\mathbf{S}_j^e\} - \alpha^R\{\delta\overline{\mathbf{u}}_f^{Re}\} = 0 \quad (e = 1,\ldots,NE), \tag{160}$$

$$\nu([\mathbf{f}^e]\{\mathbf{F}_f^e\} + \{\mathbf{u}_{fi}^e\}) - \alpha^V\{\delta\overline{\mathbf{u}}_f^{Ve}\} = 0 \quad (e = 1, \ldots, NE). \tag{161}$$

The relations (160) and (161) can be rewritten as

$$\frac{\alpha^R}{\nu}\{\delta\overline{\mathbf{u}}_f^{Re}\} = [\mathbf{f}^e]\{\overline{\mathbf{F}}_f^e\} + \frac{1}{\nu}\sum_{j=1}^{J(e)}\lambda_j^e\{\mathbf{S}_j^e\} \quad (e = 1, \ldots, NE), \tag{162}$$

$$\frac{\alpha^V}{\nu}\{\delta\overline{\mathbf{u}}_f^{Ve}\} = [\mathbf{f}^e]\{\mathbf{F}_f^e\} + \{\mathbf{u}_{fi}^e\} \quad (e = 1, \ldots, NE). \tag{163}$$

We shall assume that the displacement constraint in (154) is active and therefore $\nu > 0$. In (162) and (163) the terms $\{\delta\overline{\mathbf{u}}^R\}$ and $\{\delta\overline{\mathbf{u}}^V\}$ represent kinematically admissible virtual displacements which must satisfy compatibility and kinematic boundary conditions. For linear elastic structures, any constant scaling of $\{\delta\overline{\mathbf{u}}^R\}$ and $\{\delta\overline{\mathbf{u}}^V\}$ must also be kinematically admissible. Using the following notation:

$$\{\overline{\mathbf{u}}_f^e\} = \frac{\alpha^R}{\nu}\{\delta\overline{\mathbf{u}}_f^{Re}\} \quad (e = 1, \ldots, NE), \tag{164}$$

$$\{\mathbf{u}_f^e\} = \frac{\alpha^V}{\nu}\{\delta\overline{\mathbf{u}}_f^{Ve}\} \quad (e = 1, \ldots, NE), \tag{165}$$

(162) and (163) can be rewritten as

$$\{\overline{\mathbf{u}}_f^e\} = [\mathbf{f}^e]\{\overline{\mathbf{F}}_f^e\} + \frac{1}{\nu}\sum_{j=1}^{J(e)}\lambda_j^e\{\mathbf{S}_j^e\}, \tag{166}$$

$$\{\mathbf{u}_f^e\} = [\mathbf{f}^e]\{\mathbf{F}_f^e\} + \{\mathbf{u}_{fi}^e\}. \tag{167}$$

Since (167) is the standard form of the element flexibility relationships expressed in (137), the underlying compatibility conditions for $\{\mathbf{u}_f\}$ in (1667) give the compatibility conditions for the real force system $\{\mathbf{F}_f\}$, which were omitted in (154). Equation (166) together with the underlying compatibility conditions for $\{\overline{\mathbf{u}}_f\}$ defines a fictitious displacement field $\{\overline{\mathbf{u}}_f\}$ concerning the virtual forces $\{\overline{\mathbf{F}}_f\}$ equilibrating the virtual loads $\{\overline{\mathbf{P}}^V\}$. This fictitious system has been termed "adjoint" system in the continuum-based optimality criteria methods (see, e.g. Prager and Rozvany [27], Rozvany [2], [8]). It is important to note that the terminology of "adjoint" has also been used in the field of sensitivity analysis with a different meaning (see, e.g. Haftka [22]).

If all strength constraints are inactive, (166) reduces to

$$\{\overline{\mathbf{u}}_f^e\} = [\mathbf{f}^e]\{\overline{\mathbf{F}}_f^e\}, \tag{167a}$$

which refers to the standard element flexibility relationship of the virtual load system. Therefore, in this case the "adjoint" system is the real structure subject to the virtual loads $\{\overline{\mathbf{P}}^V\}$. If some of the strength constraints are active, then the following fictitious initial displacements for the corresponding elements are involved in the virtual load system

$$\{\overline{\mathbf{u}}^e_{fi}\} = \frac{1}{\nu}\sum_{j=1}^{J(e)} \lambda^e_j\{\mathbf{S}^e_j\}. \tag{168}$$

It follows that the so-called "adjoint" system is the *real structure subject to both the virtual loads* $\{\overline{\mathbf{P}}^V\}$ *and fictitious initial displacements* $\{\overline{\mathbf{u}}^e_{fi}\}$.

16. Solution Algorithm

16.1 General Scheme

In general, all equations governing the optimum design cannot be solved simultaneously. For this reason, (159), (166) and (167) must be solved iteratively. Each iteration consists of two steps:

(a) analysis of the real and adjoint systems defined by (166) and (167) together with the underlying equilibrium and compatibility conditions;

(b) updating the cross-sectional dimensions $\mathbf{x}$ on the basis of (159) and the forces $\{\mathbf{F}_f\}$ and $\{\overline{\mathbf{F}}_f\}$ obtained by step (a).

If the Lagrangian multiplier ν is given for the updating phase, we have NE uncoupled sets of equations for the cross-section parameters $\mathbf{x}^e$ and Lagrangian multipliers λ^e_j $[j = 1, \ldots, J(e)]$ β^e_i and γ^e_i $[i = 1, \ldots, n(e)]$. If all the local constraints are inactive for the e-th element, then the Lagrangian multipliers corresponding to the local constraints equal zero, and (159) provides $n(e)$ equations governing $n(e)$ unknowns $\mathbf{x}^e$. If $m(e)$ local constraints become active, then $m(e)$ additional equations become available in the e-th set of equations and $m(e)$ additional unknowns (the corresponding multipliers) appear. Therefore, we have $n(e) + m(e)$ equations and the same number of variables. This means that, in general, these equations can be solved satisfactorily. However, in some special situations where $m(e) > n(e)$, the equations provided by the active local constraints become linearly dependent and, therefore, the solutions of the e-th set of equations are not unique. This problem will appear in the application to trusses if both the element stress constraint and the lower side constraint are simultaneously active. In general, this problem can be solved by upgrading the considered local constraint into a global constraint and treated in the same way as a displacement constraint (for a further discussion, see Section 17.3).

The basic steps of the updating operation are

(1) calculate the Lagrangian multiplier ν associated with the global constraint on the basis of the equation provided by the active displacement constraint;

(2) update the design variables $\mathbf{x}^e$ $(e = 1, \ldots, NE)$ λ^e_j and $[j = 1, \ldots, J(e)]$ by NE uncoupled sets of equations.

The Lagrangian multipliers for stress constraints are useful for updating the adjoint initial displacements in (166). Generally, if several variables are used for each element, the above steps are coupled with each other and must be solved iteratively. For relatively simple problems the equations in each step can be solved explicitly, as will be shown later in the applications of the present method.

It is interesting to note that if we were to neglect the fictitious initial displacements given in (168), then the DCOC method would reduce to the semi-intuitive DOC-FSD method in which the displacement constraints are treated through rigorous optimality criteria but the fully stressed design concept is used for stress constraints (see, e.g. Khot [59]).

16.2 Analysis of the Real and Adjoint Systems

Since the optimization problem and optimality criteria were expressed in terms of forces, the resulting formulation of the real and adjoint systems expressed by (165) and (164) represent standard flexibility formulation. In the flexibility method, the element nodal forces $\{\mathbf{F}_f\}$ are the basic unknowns and therefore it is also termed force method. It is well known that in modern computerized numerical analysis, the force method is much less popular than the displacement method (also termed stiffness method) in which the nodal displacements are the basic unknowns. Therefore, it is preferable to solve the real and adjoint systems by the displacement method.

The stiffness equations governing the response of a linear structural system subject to a static loading condition is of the form

$$[\mathbf{K}]\{\mathbf{u}\} = \{\mathbf{P}\}, \tag{169}$$

where

$$\{\mathbf{P}\} = \{\mathbf{P}_N\} + \{\mathbf{P}_E\}, \tag{170}$$

where $[\mathbf{K}]$ is the system stiffness matrix, and $\{\mathbf{u}\}$ and $\{\mathbf{P}\}$ are the vectors of unknown displacements and known applied nodal loads. The load vector $\{\mathbf{P}\}$ includes two parts $\{\mathbf{P}_N\}$ and $\{\mathbf{P}_E\}$ which represent, respectively, the loads applied directly on nodes and the equivalent nodal loads caused by distributed loads applied within the elements or initial displacements or thermal strains. The system stiffness matrix $[\mathbf{K}]$ and load vector $\{\mathbf{P}\}$ can be generated from element level stiffness matrices $[\mathbf{K}^e]$ and equivalent load vector $\{\mathbf{P}^e_E\}$ using an assembly technique known as the direct stiffness method:

$$[\mathbf{K}] = \sum_{e=1}^{NE} [\mathbf{T}^e]^T [\mathbf{K}^e][\mathbf{T}^e], \tag{171}$$

$$\{\mathbf{P}\} = \{\mathbf{P}_N\} + \sum_{e=1}^{NE} [\mathbf{T}^e]^T \{\mathbf{P}^e_E\}, \tag{172}$$

where $[\mathbf{T}^e]$ is the orthogonal transformation matrix of the e-th element. The equivalent nodal loads $\{\mathbf{P}^e_E\}$ are the reversed fixed-end forces $[\mathbf{F}^e_F]$ caused by distributed loads, initial displacements and thermal strains within the e-th element, i.e.

$$\{\mathbf{P}^e_E\} = -\{\mathbf{F}^e_F\}. \tag{173}$$

The element stiffness relationships are

$$\{\mathbf{F}^e\} = [\mathbf{K}^e]\{\mathbf{u}^e\} + \{\mathbf{F}_F^e\}, \tag{174}$$

where $\{\mathbf{F}^e\}$ is the vector of element nodal forces and $\{\mathbf{u}^e\}$ the element nodal displacement vector. Both $\{\mathbf{F}^e\}$ and $\{\mathbf{u}^e\}$ have components of n_s representing the total number (n_s) of degrees of freedom of the element. Thus the number of degrees of freedom in the stiffness formulation is by n_r greater than that for the flexibility formulation, where n_r denotes the number of rigid motions, i.e. $n_s = n_f + n_r$.

The forces at a point j inside an element include two parts which are the forces caused by the nodal forces $\{\mathbf{F}^e\}$ and the forces caused by the distributed loads applied within the element.

The calculation of fixed-end forces for the real structural system is a standard operation. Therefore, we do not need to calculate $\{\mathbf{u}_{fi}^e\}$ in (167) and then transfer them to $\{\mathbf{F}_F^e\}$. However, the fixed-end forces for the adjoint system have to be generated from the fictitious displacements $\{\overline{\mathbf{u}}_{fi}^e\}$ given in (168). We designate $\{\overline{\mathbf{u}}_i^e\}$ of n_s components as the extended adjoint initial displacement vector of $\{\overline{\mathbf{u}}_{fi}^e\}$, in which the displacements of n_r supported degrees of freedom are set to zero. Then, the fixed-end forces corresponding to the adjoint initial displacements can be expressed as

$$\{\overline{\mathbf{F}}_F^e\} = [\mathbf{K}^e]\{\overline{\mathbf{u}}_i^e\}, \tag{175}$$

and the equivalent nodal loads are

$$\{\overline{\mathbf{P}}_E^e\} = -\{\overline{\mathbf{F}}_F^e\}. \tag{176}$$

Then, by (172), the load vector for the adjoint system can be expressed as

$$\{\overline{\mathbf{P}}\} = \{\overline{\mathbf{P}}^V\} + \sum_{e=1}^{NE}[\mathbf{T}^e]^T\{\overline{\mathbf{P}}_E^e\}, \tag{177}$$

where $\{\overline{\mathbf{P}}^V\}$ is the virtual load vector corresponding to the displacement constraint.

The solution of the stiffness equation (169) is based on the $[\mathbf{L}][\mathbf{D}][\mathbf{L}]^T$ decomposition and variable bandwidth technique of the stiffness matrix (see, e.g. McGuire and Gallagher [24]). Therefore, the analysis of the adjoint system once more concerns only the forward and backward substitution of the adjoint load vector $\{\overline{\mathbf{P}}\}$.

16.3 Illustrative Examples

16.3.1 Clamped Beams of Variable Width Subject to Deflection, Flexural and Shear Stress Constraints. This example has been described in Section 11.1.2. The flexural and shear stress constraints are represented as $x^e \le k_1|M_{\max}^e|$ and $x^e \le k_2|F_{y\max}^e|$ where $M_{\max}^e$ and $F_{y\max}^e$ are, respectively, the maximum moment and shear force of the e-th element, and $k_1 = 0.23$ and $k_2 = 0.03$ are given constants. A 100 element model of the beam was used. The optimum distribution of the variable width

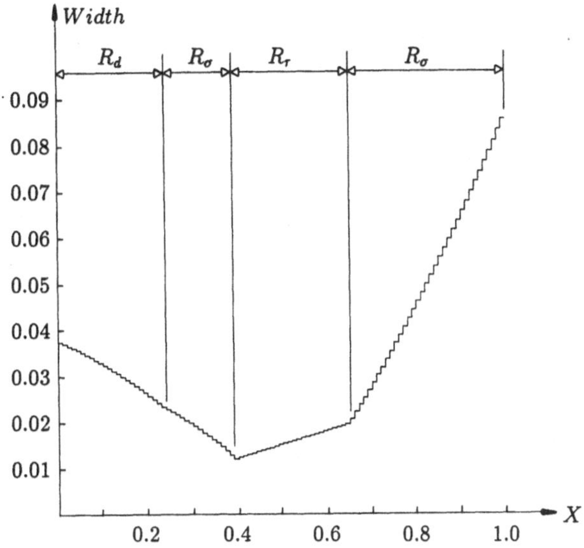

Fig. 27 Example: Beam with flexural and shear stress constraints and a deflection constraint.

obtained by the DCOC method is shown in Fig. 27. The convergence criterion $\frac{|f_{\text{new}}-f_{\text{old}}|}{f_{\text{new}}} \leq 10^{-8}$ was used and 29 analyses were required. The optimum weight obtained by the DCOC method is compared with those obtained by the DOC-FSD and dual methods in Table 4, in which the iteration numbers and CPU times required by the above three methods are also compared. It can be seen that the CPU time for the optimization phase of the DCOC method is only a fraction of that needed by the dual method. The optimum weight obtained by the DOC-FSD method is only slightly higher than that of the DCOC method in this example.

		DCOC	DOC–FSD	Dual
Optimum weight		0.064202389	0.064203018	0.064202871
Number of analyses		29	28	16
CPU times	optimization	7.84	3.63	2041.33
	analysis	116.95	112.70	305.69

Table 4. A comparison of results obtained by various methods in the beam example.

16.3.2 Portal Frame of Variable Width Subject to Deflection and Flexural Stress Constraints. The structural dimensions, deflection constraint and loading are shown in Fig. 28a and the optimum distribution width in Fig. 28b. In the above problem we have constant depth $d = 0.3$ m, $\gamma = 1.0$ (volume instead of weight), $E = 1.0 \times 10^7$ kN/m^3, $\sigma = 1.1 \times 10^4$ kN/m^2 and $\Delta = L/300 = 0.03333$ m, which are values taken from the relevant German Design Code for laminated timber beams (DIN 1052). The number of elements used was 2000 and the convergence criterion,

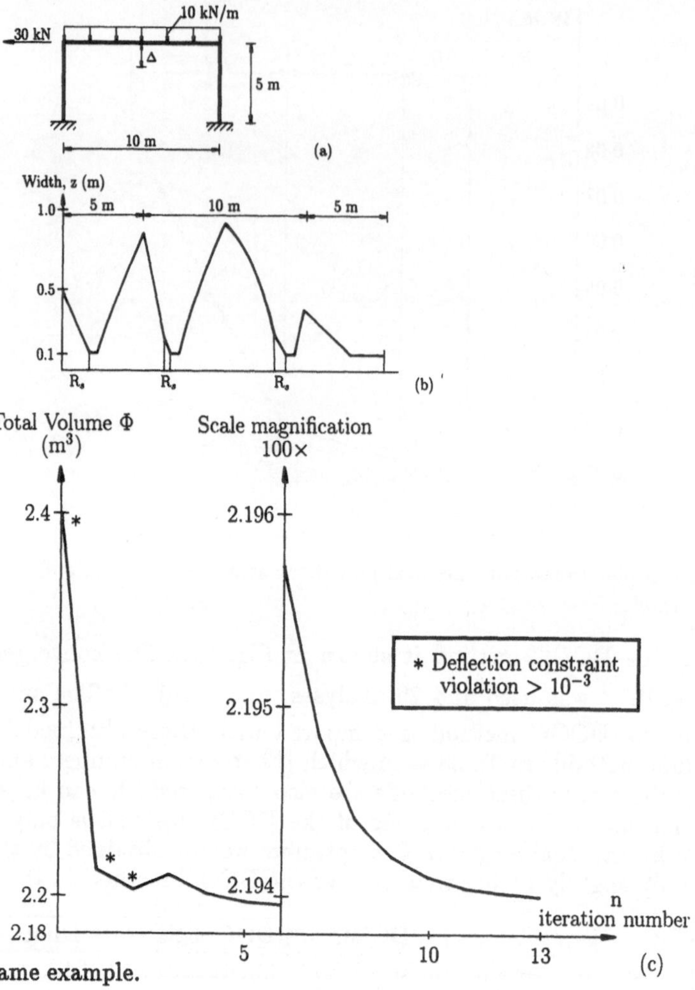

Fig. 28 Frame example.

$\frac{|f_{\text{new}}-f_{\text{old}}|}{f_{\text{new}}} \leq 10^{-5}$, was reached after 13 iterations, which yields an optimum value of 2.914 m^3. The iteration history is shown in Fig. 28c. The COC formulation of this problem was discussed at a recent meeting [53] and the DCOC formulation is given in the first author's thesis [40].

17. Generalization to Multiple Displacement Constraints and Multiple Load Conditions

17.1 Optimality Criteria and Solution Algorithm

For structural systems subject to multiple displacement constraints together with stress constraints under multiple load conditions, the optimality criteria (159) can be generalized as

$$\frac{\partial f}{\partial x_i^e}+\sum_{k=1}^{ND}\nu_k\left(\{\overline{\mathbf{F}}_{fk}^e\}^T\left[\frac{\partial \mathbf{f}^e}{\partial x_i^e}\right]\{\mathbf{F}_{f\ell_k}^e\}+\{\overline{\mathbf{F}}_{fk}^e\}\left\{\frac{\partial \mathbf{u}_{f\ell_k i}^e}{\partial x_i^e}\right\}\right)+$$

$$+\sum_{\ell=1}^{NL}\sum_{j=1}^{J(e)}\lambda_{j\ell}^e\left(\left\{\frac{\partial \mathbf{S}_j^e}{\partial x_i^e}\right\}\{\mathbf{F}_{j\ell}^e\}+\left\{\frac{\mathbf{S}_{jd}^e}{\partial x_i^e}\right\}\{\mathbf{F}_{j\ell d}^e\}\right)-\beta_i^e+\gamma_i^e=0$$

$$[i=1,\ldots,n(e)],\quad (e=1,\ldots,NE), \tag{178}$$

where ND is the number of displacement constraints, ℓ_k is the load case relevant to the k-th displacement constraint and NL is the total number of load conditions. Equations (160) and (161) can then be replaced by

$$\alpha_\ell^R\{\delta\overline{\mathbf{u}}_{f\ell}^{Re}\}=\sum_{k\in\ell}\nu_k[\mathbf{f}^e]\{\overline{\mathbf{F}}_{fk}\}+\sum_{j=1}^{J(e)}\lambda_{j\ell}^e\{\mathbf{S}_j^e\}\quad (e=1,\ldots,NE),\ (\ell=1,\ldots,NL), \tag{179a}$$

$$\alpha_k^V\{\delta\overline{\mathbf{u}}_{fk}^{Ve}\}=\nu_k\left([\mathbf{f}^e]\{\mathbf{F}_{f\ell_k}^e\}+\{\mathbf{u}_{f\ell_k i}^e\}\right)\quad (e=1,\ldots,NE),\ (k=1,\ldots,ND), \tag{179b}$$

in which $k\in\ell$ denotes the displacement constraints concerned in the ℓ-th load condition. Under the assumption of having at least one displacement constraint active in each load condition, neglecting constant scaling factors (179a) and (179b) imply

$$\{\overline{\mathbf{u}}_{f\ell}^e\}=\sum_{k\in\ell}\nu_k[\mathbf{f}^e]\{\overline{\mathbf{F}}_{fk}^e\}+\sum_{j=1}^{J(e)}\lambda_{j\ell}^e\{\mathbf{S}_j^e\}\quad (e=1,\ldots,NE),\ (\ell=1,\ldots,NL), \tag{180a}$$

$$\{\mathbf{u}_{f\ell}^e\}=[\mathbf{f}^e]\{\mathbf{F}_{f\ell}^e\}+\{\mathbf{u}_{f\ell i}^e\}\quad (e=1,\ldots,NE),\ (\ell=1,\ldots,NL), \tag{180b}$$

where $\{\overline{\mathbf{u}}_{f\ell}\}$ and $\{\mathbf{u}_{f\ell}\}$ are required to be kinematically admissible. Therefore, (180b) gives the standard flexibility relationships of the real structural system, and (180a) defines an adjoint displacement field for each load case which is a linear combination of the virtual displacements corresponding to the active displacement together with fictitious initial displacements caused by active stress constraints concerned in that load case. It is important to note that *whilst the virtual forces corresponding to the virtual load of a displacement constraint are required to be statically admissible, the corresponding virtual displacments are not required to be kinematically admissible.* However, a possibility to satisfy (180a) is to choose kinematically admissible virtual displacements together with the fictitious initial displacements, e.g.

$$\{\overline{\mathbf{u}}_{fk}^e\}=\nu_k[\mathbf{F}^e]\{\overline{\mathbf{F}}_{fk}^e\}+a_k\sum_{j=1}^{J(e)}\lambda_{j\ell_k}^e\{\mathbf{S}_j^e\}\quad (k=1,\ldots,ND), \tag{181}$$

with

$$\sum_{k\in\ell} a_k = 1 \quad (\ell = 1,\ldots,NL), \tag{181a}$$

in which $\{\overline{\mathbf{u}}_{fk}\}$ $(k = 1,\ldots,ND)$ are kinematically admissible displacements and a_k $(k = 1,\ldots,ND)$ are arbitrary factors satisfying (181a). It can be seen that $\sum_{k\in\ell}\{\overline{\mathbf{u}}^e_{fk}\} = \{\overline{\mathbf{u}}^e_{f\ell}\}$ and therefore $\{\overline{\mathbf{u}}_{f\ell}\}$ are also kinematically admissible. Equation (181) shows that ***arbitrary combinations of the fictitious initial displacements with the displacements caused by virtual forces can be used.***

The general solution algorithm described in Section 16.1 can also be applied to the multi-displacement and multi-load problems considered in this section. The evaluation of Lagrangian multipliers ν_k $(k = 1,\ldots,ND)$ becomes the most complex and time consuming task of the updating phase, which can be solved using the equations provided by the active displacement constraints and the relations between ν_k $(k = 1,\ldots,ND)$ and design variations $\mathbf{x}$ provided by the optimality criteria (178). Some techniques (e.g. Newton method) developed in DOC method for generating Lagrangian multipliers can be directly used for DCOC method. It can easily be seen that, as noted in Section 16.2, the DCOC method would reduce to the DOC-FSD method if we were to neglect the fictitious initial displacements of the adjoint system in (181). For details of the derivation, the reader is referred to Zhou's doctoral thesis [40].

17.2 Illustrative Example

The ten-bar truss considered in Section 12.5 is used here again as a demonstrating example for problems with multi-displacement constraints. The displacement constraints are $|u_A| \le 1.0$ (in) and $|v_A| \le 5.0$ (in), where u_A and v_A are, respectively, horizontal and vertical displacement of node A. The other conditions are the same as described in Section 12.5. It has been shown by a dual MP method that both displacement constraints and the stress constraint at 5-th bar are active at the optimum design. Therefore, it is an ideal example to examine the ***surprising feature of the adjoint systems*** outlined in 17.1.

For the considered problem, (180a) can be expressed as

$$\{\overline{\mathbf{u}}^e_f\} = \frac{\nu_1 \overline{F}^e_1}{E^e A^e} + \frac{\nu_2 \overline{F}^e_2}{E^e A^e} + \frac{\lambda^e \operatorname{sgn} F^e}{A^e} \quad (e = 1,\ldots,NE), \tag{182}$$

where $\{\overline{\mathbf{u}}^e_f\}$ is the elongation of a truss element, A^e is the element cross-sectional area, and the subscripts 1 and 2 refer to respectively the horizontal and vertical displacement constraints. For satisfying the kinematical admissibility of $\{\overline{\mathbf{u}}_f\}$, (178) can be arbitrarily split into two ***kinematically admissible*** systems

$$\{\overline{\mathbf{u}}^e_1\} = \frac{\nu_1 \overline{F}^e_1}{E^e A^e} + \alpha\frac{\lambda^e \operatorname{sgn} F^e}{A^e} \quad (e = 1,\ldots,NE), \tag{183a}$$

$$\{\overline{\mathbf{u}}^e_2\} = \frac{\nu_2 \overline{F}^e_2}{E^e A^e} + (1-\alpha)\frac{\lambda^e \operatorname{sgn} F^e}{A^e} \quad (e = 1,\ldots,NE), \tag{183b}$$

(a)	DCOC solution (Case A)				dual
	real force	adjoint force		cross-sectional area	cross-sectional area
member	F^e	$\overline{F}_1^e$	$\overline{F}_2^e$	A^e	A^e
1	200267.79	0.0246632	1.9403247	10.8278891	10.8278891
2	-199732.21	-0.9753368	-1.0596753	12.2950243	12.2950243
3	785.01	-0.0022185	0.0184593	0.1000000	0.1000000
4	-99215.00	-1.0022186	-0.9815407	8.6028430	8.6028430
5	141042.65	-0.0348790	0.0849365	5.6417060	5.6417060
6	-141800.06	-0.0348790	-1.3298199	7.6192547	7.6192547
7	140311.19	0.0031375	1.3881082	7.6052513	7.6052513
8	1110.17	0.0031375	-0.0261054	0.1000000	0.1000000
9	1052.79	0.0224446	-0.0412160	0.1000000	0.1000000
10	785.01	-0.0022185	0.0184593	0.1000000	0.1000000
Lagrangian multipliers		4544.5736	2959.4206		
structural weight (lb)	2220.352475375				2220.352475375

(b)	DCOC solution (Case B)				dual
	real force	adjoint force		cross-sectional area	cross-sectional area
member	F^e	$\overline{F}_1^e$	$\overline{F}_2^e$	A^e	A^e
1	200267.79	0.0065717	1.9680828	10.8278891	10.8278891
2	-199732.21	-0.9934128	-1.0319172	12.2950243	12.2950243
3	785.01	0.0008570	0.0137363	0.1000000	0.1000000
4	-99215.00	-0.9991430	-0.9862637	8.6028430	8.6028430
5	141042.65	-0.0093157	0.0451378	5.6417060	5.6417060
6	-141800.06	0.0093157	-1.3690758	7.6192547	7.6192547
7	140311.19	-0.0012120	1.3947875	7.6052513	7.6052513
8	1110.17	-0.0012120	-0.0194261	0.1000000	0.1000000
9	1052.79	0.0074442	-0.0181809	0.1000000	0.1000000
10	785.01	0.0008570	0.0137363	0.1000000	0.1000000
Lagrangian multipiers		4544.5736	2959.4206		
structural weight (lb)	2220.352475375				2220.352475375

Table 5. Truss example: comparisons of (a)-(c) DCOC results with various α-values for the adjoint field and (d) DOC-FSD results with dual programming results.

(c)	DCOC solution (Case C)				dual
	real force	adjoint force		cross-sectional area	cross-sectional area
member	F^e	$\overline{F}_1^e$	$\overline{F}_2^e$	A^e	A^e
1	200267.79	0.0175344	1.9512719	10.8278891	10.8278891
2	-199732.21	-0.9824656	-1.0487281	12.2950243	12.2950243
3	785.01	-0.0010056	0.0165967	0.1000000	0.1000000
4	-99215.00	-1.0010056	-0.9834033	8.6028430	8.6028430
5	141042.65	-0.0247974	0.0689119	5.6417060	5.6417060
6	-141800.06	-0.0247974	-1.3453016	7.6192547	7.6192547
7	140311.19	0.0014222	1.3907423	7.6052513	7.6052513
8	1110.17	0.0014222	-0.0234712	0.1000000	0.1000000
9	1052.79	0.0165288	-0.0321314	0.1000000	0.1000000
10	785.01	-0.0010056	0.0165967	0.1000000	0.1000000
Lagrangian multipiers		4544.5736	2959.4206		
structural weight (lb)	2220.352475375				2220.352475375

(d)	DOC–FSD solution				dual
	real force	adjoint force		cross-sectional area	cross-sectional area
member	F^e	$\overline{F}_1^e$	$\overline{F}_2^e$	A^e	A^e
1	200238.43	0.0065520	1.9400859	10.8362276	10.8278891
2	-199761.57	-0.9934481	-1.0599141	12.3310330	12.2950243
3	787.22	0.0008708	0.0184725	0.1000000	0.1000000
4	-99212.78	-0.9991292	-0.9815275	8.5691720	8.6028430
5	141084.17	-0.0092659	0.0847314	5.6433667	5.6417060
6	-141758.55	-0.0092659	-1.3294822	7.5675629	7.6192547
7	140308.06	-0.0012315	1.3880896	7.6484854	7.6052513
8	-1113.30	-0.0012315	-0.0261240	0.1000000	0.1000000
9	1025.65	0.0074228	-0.0414417	0.1000000	0.1000000
10	7872.23	0.0008708	0.0184725	0.1000000	0.1000000
Lagrangian multipiers		4453.1606	3007.6110		
structural weight (lb)	2220.390773364				2220.352475375

Table 5. Truss example: comparisons of (a)-(c) DCOC results with various α-values for the adjoint field and (d) DOC-FSD results with dual programming results.

in which α is an arbitrary constant. The following cases are tested for the ten-bar truss:

$$\text{Case A}: \quad \alpha = 1\,; \qquad \text{Case B}: \quad \alpha = 0\,; \text{ and}$$

$$\text{Case C}: \quad \alpha = \frac{\nu_1}{\nu_1 + \nu_2}\,.$$

The numerical results are given in Table 5a-c, which shows that all the three cases give exactly the same end design as the dual method. For a comparison, the results of DOC-FSD method are indicated in Table 5d. For details of the treatment, the reader is referred to Zhou's doctoral thesis [40].

17.3 Some Difficulties and Their Treatment

The optimality criteria and iterative algorithm given in Section 17.1 are based on the assumption that at least one displacement constraint is active in each load condition. It is apparent that this assumption is not always satisfied for any design problem. However, in order to restore the validity of the above assumption, one of the stress constraints concerned in ℓ-th load case can be upgraded into a displacement constraint if there is no displacement constraint active for the ℓ-th load case. It is well known that a stress constraint can be expressed in general as follows

$$g^e_{k\ell} = \overline{\mathbf{Q}}^T_k \mathbf{u}_\ell - \sigma^a_k\,, \tag{184}$$

where $\overline{\mathbf{Q}}_k$ is a vector converting the displacement vector into the concerning stress and σ^a_k is the allowable stress. If we regard $\overline{\mathbf{Q}}_k$ as a virtual load vector, the stress constraint $g^e_{k\ell}$ can be considered as a special kind of displacement constraint.

As already discussed in Section 16.2, certain difficulties occur if the number of local constraints of an element is higher than the number of design variables of this element. For example, this situation arises in truss design if the stress constraint and lower side constraint are simultaneously active for an element. The above problem can also be solved by upgrading the considered stress constraint into a displacement constraint using the described technique.

17.4 Concluding Remarks

The methods and results presented in this two-part lecture substantiate the claims made about the advantages of COC and DCOC at the end of the first lecture (Section 8).

LAYOUT OPTIMIZATION BY COC METHODS: ANALYTICAL SOLUTIONS

G.I.N. Rozvany, M. Zhou and W. Gollub
FB 10, Essen University, D-4300 Essen 1, Germany

Abstract. After discussing some general aspects of layout and generalized shape optimization, the theory of optimal structural layouts is outlined and then it is illustrated with several elementary examples. The difference between classical and advanced layout theories is explained and applications to least-weight grillages, trusses, shell grids, solid plates and perforated plates are briefly discussed.

18. Aims, Methods and Significance of Layout Optmization and Generalized Shape Optimization

General aspects of the COC algorithm and its applications to cross-section optimization were discussed in the first three lectures. Simultaneous optimization of the topology and geometry by means of COC methods, as well as generalized shape optimization are discussed in the next two lectures.

As mentioned in the first lecture, *layout optimization* consists of three simultaneous operations, namely

(a) topological optimization (spatial sequence or configuration of members and joints),
(b) geometrical optimization (location of joints and shape of member axes), and
(c) optimization of the cross-sectional dimensions.

One of the difficulties in *shape optimization* is that the optimal shape may represent a *multiply connected* set with internal boundaries whose topology is not known and is difficult to determine, because new internal boundaries cannot be easily generated. Moreover, in many optimization problems the theoretical optimal shape contains an infinite number of internal boundaries. The determination of the topology then becomes a *layout optimization* problem.

Generalized shape optimization is aimed at determining simultaneously the boundary topology and boundary shape. The optimal solution in these problems contains three types of regions, namely (a) solid (black), (b) empty (white), and (c) perforated (grey) regions.

Unconstrained cross-section (thickness) optimization of plates and shells often results in a, theoretically, infinite number of rib-like formations whose *layout* must also be optimized. This means that both cross-section and shape optimization may require, in effect, layout optimization.

Layout optimization is the most complex task in structural optimization because

- one has a choice of an infinite number of possible topologies, and

G. I. N. Rozvany (ed.), Optimization of Large Structural Systems, Vol. I, 77–102.

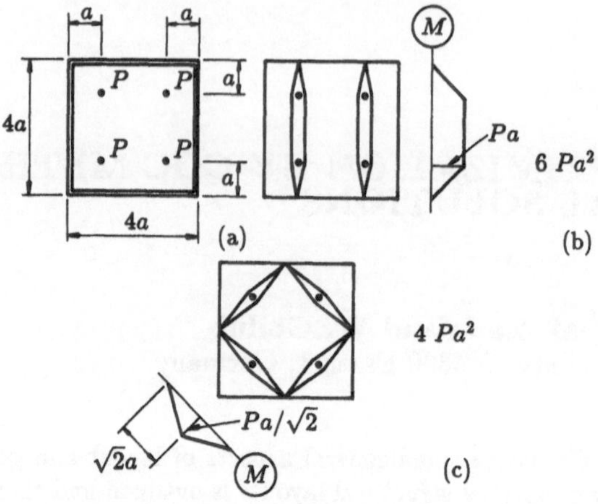

Fig. 29 Comparison of optimal and nonoptimal beam layouts.

- for each point of the structural domain, there exist an infinite number of member directions. The cross-sections of non-vanishing (optimal) members must be optimized simultaneously with the topology and geometry.

Layout optimization is important because it enables much greater material savings than pure cross-section optimization.

To illustrate this point on a simple example, we consider the problem in Fig. 29a in which four point loads P are required to be transmitted by beams of constant depth to simple supports along the boundary of a square domain. In the first solution shown (Fig. 29b), the total "moment area" (area of the moment diagrams), which is a measure of structural weight, is $6Pa^2$. The optimal beam layout in Fig. 29c gives a total moment area of $4Pa^2$. The weight difference of 50% is much greater than the usual savings achieved by cross-section optimization.

An even bigger difference between the weight of optimal and nonoptimal layouts is found for the boundary and loading condition in Fig. 30a, in which the transverse point load P is to be transmitted to two simply supported edges (double lines) of a rectangular domain. The other two edges are unsupported. Figure 30b shows the optimal beam layout and the corresponding moment diagrams with a total moment area of $8Pa^2$. The nonoptimal layout shown in Fig. 30c has a moment area of $17Pa^2$ which exceeds the optimal one by 112.5%. It has been shown [2, p. 17] that for long span grillages, for which the selfweight is a significant load condition, this difference between optimal and nonoptimal structural weights is often over 1000%.

The most systematic *analytical* method for layout optimization is the so-called *layout theory*, developed by Prager and Rozvany in the late seventies and extended by the research teams of the latter in the eighties. This theory is a generalization of an approach used around the turn of the century by Michell [28] (see Part 1, Section 2.2), who made use of some ideas by Maxwell [29]. Layout theory is based on two underlying concepts, namely,

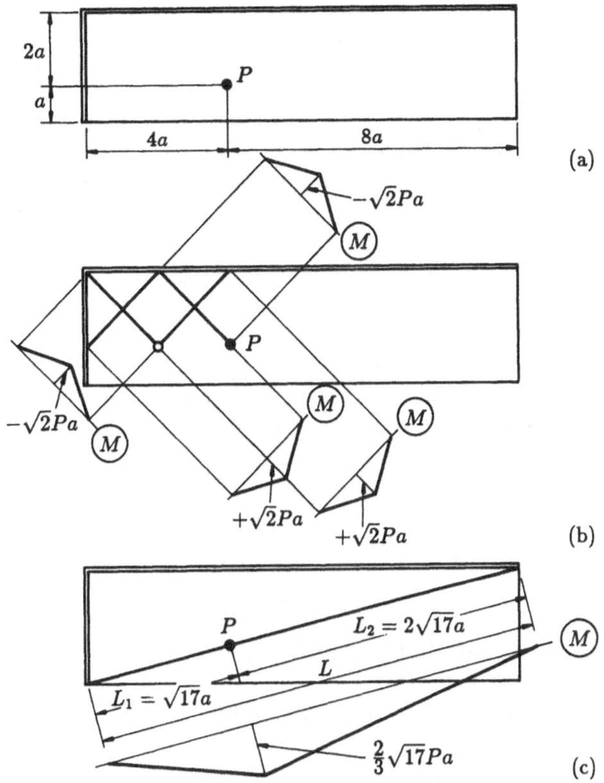

Fig. 30 Another comparison of optimal and nonoptimal beam layouts.

(a) the *structural universe*, which is the union of all feasible or potential members, and

(b) *continuum-based* or *static-kinematic optimality criteria*, which are mathematical conditions for the optimality of a structure, reinterpreted in terms of a fictitious *adjoint structure* (see Sections 2-16).

The adjoint strain-stress relations give a strain requirement, usually in the form of an inequality, also along *vanishing members* (having a zero cross-sectional area and zero generalized stress). This means that in *convex* problems, for which the optimality criteria represent sufficient conditions, their fulfilment for the entire structural universe also represents a *sufficient condition* of optimality *for the structural layout.*

It was shown by Prager and the first author (e.g. [3], [8], [31]) that the optimal layout often consists of an infinite number of members having an infinitesimal spacing. Prager termed the corresponding structures truss-like continua, grillage-like continua or shell-grid-like continua.

As was explained lucidly by Kirsch [30], layout optimization problems are solved by the *numerical school* in two stages. First, *topological optimization* is carried out by assuming a "highly connected" *ground structure* (in this lecture: "structural universe") and then removing non-optimal members. The second stage consists of *geometrical optimization*, in which the topology is assumed to be fixed (unless some joints coalesce during the solution process) and the design variables are the coordinates of the joints and the cross-sectional areas. The above design variables are usually optimized by mathematical programming methods.

Numerical methods of topological optimization in the past, even with simplifying assumptions and approximations, were restricted to a small number of potential members, because of the limited optimization capability of the mathematical programming methods used. Due to the introduction of iterative COC methods, to be described in the next lecture, it has become possible to investigate structural universes with *many thousand potential members*.

Earlier formulations of the layout theory for structural design were introduced by Prager and Rozvany [3, 8, 31], and more up-to-date reviews of this field were offered in principal lectures at NATO ASI's [32-34], in a book chapter [35] and, in particular, in a recent book [2] by Rozvany.

19. Simple Analytical Examples Illustrating Layout Optimization by the COC Method

The general ideas of layout theory will be explained through examples involving the following two *types of structures*:

- beams having rectangular cross-sections of given depth and variable width, and
- trusses or pin-jointed frames.

The two types of *design conditions* to be considered here will be

- optimal plastic design for a given ultimate load, and
- optimal elastic design for a given compliance (the sum of products of external forces and the corresponding deflections).

19.1 Optimality Criteria for Optimal Plastic Design

In our examples, we restrict our treatment to a specific cost function of the form (Fig. 34h in the context of beams)

$$\psi = k|Q|\,, \quad \Phi = \int_D k|Q|\,\mathrm{d}x\,, \tag{185}$$

where k is a given constant, ψ the member weight per unit length, Q is the relevant generalized stress, Φ is the total structural weight and D is the structural domain. For *trusses* we have $Q = N$ (where N is the axial member force) and $k = \gamma/\sigma_0$ (where γ is the specific weight of the truss material and $\pm\sigma_0$ is the yield stress in tension and compression).

For *beams of variable width z*, but given depth h, we have $Q = M$ (where M is the bending moment) $\psi = \gamma h z$ and $k = 4\gamma/h^2\sigma_0$.

Before evaluating the optimality conditions for the above class of problems, we clarify once more that the generalized gradient $\mathcal{G}$ for a cost function of one variable $\psi(Q)$ has the meaning ([2, 8] and Section 6 herein)

$$\mathcal{G}(\psi) = \mathrm{d}\psi/\mathrm{d}Q\,, \tag{186a}$$

if $\psi(Q)$ is differentiable at the considered Q-value, and it becomes

$$\mathcal{G}(\psi) = \nu(\mathrm{d}\psi/\mathrm{d}Q)^{-} + (1-\nu)(\mathrm{d}\psi/\mathrm{d}Q)^{+} \quad (0 \le \nu \le 1)\,, \tag{186b}$$

if $\psi(Q)$ has a cusp at the considered Q-value, where $(\mathrm{d}\psi/\mathrm{d}Q)^{-}$ and $(\mathrm{d}\psi/\mathrm{d}Q)^{+}$ represent the slope to the left and right of the cusp. Then it follows from the optimality condition in Fig. 6 (first lecture) (with $\mathcal{G}$ instead of $\frac{\partial}{\partial Q_i}$) that the adjoint strain for the specific cost function in (186) becomes

$$\overline{q} = k\,\mathrm{sgn}\,Q \quad (\text{for } Q \neq 0)\,, \tag{187a}$$

$$|\overline{q}| \le k \quad (\text{for } Q = 0)\,, \tag{187b}$$

in which $\overline{q} = \overline{\varepsilon}$ (where $\overline{\varepsilon}$ is the adjoint axial strain) for *trusses* and $\overline{q} = \overline{\kappa}$ (where $\overline{\kappa} = -\overline{u}''$ is the adjoint curvature) for *beams.*

These optimality conditions are represented graphically, in the context of beams, in Fig. 34h and i.

19.2 Optimality Criteria for Elastic Design - Given Compliance

In the literature, "compliance" means the product of the external loads and the corresponding displacements

$$C = \int_D \mathbf{p}\cdot\mathbf{u}\,\mathrm{d}\mathbf{x}\,. \tag{188}$$

In the case of a compliance constraint, the virtual load becomes identical with the real load $\overline{\mathbf{p}} = \mathbf{p}$ and hence in Fig. 3 (first lecture) the real and adjoint systems are also the same, with $\overline{\mathbf{Q}} = \mathbf{Q}$ and $\overline{\mathbf{q}} = \mathbf{q}$.

We consider the class of problems with single-component generalized stress and strain

$$\mathbf{Q} = Q\,, \quad \mathbf{q} = q\,, \quad \psi = cz\,, \quad [\mathbf{F}] = 1/rz\,, \tag{189}$$

where c and r are given constants and z is a cross-sectional variable. By Fig. 3 or (28), the optimality criterion in (21) is replaced by

$$z = \sqrt{\nu/rc}\,|Q|\,. \tag{190}$$

If a minimum value for z is prescribed ($z \ge z_a$), then we have [cf. (65) in the first lecture]:

$$(\text{for } \sqrt{\nu/rc}\,|Q| > z_a) \qquad z = \sqrt{\nu/rc}\,|Q|\,, \tag{191a}$$

$$(\text{for } \sqrt{\nu/rc}\,|Q| \le z_a) \qquad z = z_a\,. \tag{191b}$$

Since, by Fig. 3 or (29), in this class of problems $\overline{q} = \overline{Q}/rz = Q/rz$, (189) and (191) imply

$$(\text{for } \sqrt{\nu/rc}|Q| > z_a,\ z > z_a) \qquad \overline{q} = Q/rz = \sqrt{c/r\nu}\,\operatorname{sgn}|Q|\,, \tag{192a}$$
$$(\text{for } \sqrt{\nu/rc}|Q| \leq z_a,\ z = z_a) \qquad |\overline{q}| = |Q|/(rz_a) \leq \sqrt{c/r\nu}\,. \tag{192b}$$

After a limiting process with $z_a \to 0$ and replacing $\sqrt{c/r\nu}$ with k, we can see that (192a) and (192b) reduce to (187a) and (187b). This means that, within a constant factor, *the optimality criteria for optimal plastic design and optimal elastic design for compliance are identical* for both trusses and beams of given depth. This confirms earlier observations regarding Michell frames by Hegemier and Prager [36]. These authors have also shown that the same solutions (within a constant factor) are valid for given natural frequency and for given stiffness in stationary creep.

For the case of $z_a \to 0$, the value of the multiplier ν can be determined from the work equation and (190) as

$$C = \int_D pu\,\mathrm{d}x = \int_D \frac{Q^2}{rz}\,\mathrm{d}x = \int_D \frac{Q^2}{r\sqrt{\nu/rc}|Q|}\,\mathrm{d}x = \int_D \frac{|Q|}{\sqrt{\nu r/c}}\,\mathrm{d}x\,, \tag{193}$$

furnishing

$$\sqrt{\nu} = \int_D |Q|\,\mathrm{d}x/(C\sqrt{r/c})\,. \tag{194}$$

Moreover, for the total cost (weight) Φ we have by (189), (190) and (194) for $z_a \to 0$

$$\Phi = \int_D cz\,\mathrm{d}x = \int_D c\sqrt{\nu/rc}|Q|\,\mathrm{d}x = \frac{c}{r}\left(\int_D |Q|\,\mathrm{d}x\right)^2/C\,. \tag{195}$$

19.3 Two-Span Beam with a Central Point Load over One Span

This is a very simple layout problem in which the structural universe consists of two beam spans (Fig. 31). Over one of the spans the beam cross-section may take on a zero area. Since the optimal deflection diagrams can be subjected to a linear transformation, the constant k in (187a) and (187b) or $\sqrt{\nu/cr}$ in (192a) and (192b) with $z_a = 0$ will be replaced by unity and then for $Q = M$ and $\overline{q} = \overline{\kappa}$ the normalized adjoint beam curvatures for both plastic and elastic design are

$$(\text{for } |M| > 0)\,, \qquad \overline{\kappa} = \operatorname{sgn} M\,, \tag{196}$$
$$(\text{for } |M| = 0)\,, \qquad |\overline{\kappa}| \leq 1\,. \tag{197}$$

Moreover, the optimal values of the width in elastic compliance design are given by (191a) as

$$z = \sqrt{\nu/cr}|M|\,. \tag{198}$$

It will be shown that the solution depends on the ratio of the loaded and unloaded spans. In Fig. 31a, this ratio is 1:2. Assuming that the beam takes on a zero cross-sectional area over the longer span, we have the moment diagram in Fig. 31c and the adjoint deflection diagram in Fig. 31b satisfies the curvature conditions in (196) and (197). Note that for the vanishing beam region over the longer span we still have an adjoint deflection* diagram which plays an important role in determining the optimal solution. By (198) the diagram in Fig. 31c also represents the variation of $z/\sqrt{\nu/rc}$. Prescribing a compliance value of $C = 1$ (corresponding to a unit deflection at the loaded point), the moment area A in Fig. 31c and (195) with $C = 1$, $Q \to M$ furnish the beam weight

$$\Phi = (1/8)^2 = 1/64 = 0.015625\,. \tag{199}$$

If we now change the ratio of the loaded and unloaded spans to 2:1 (Fig. 31d) then the type of moment diagrams in Fig. 31f would require by (196) the adjoint deflection diagram in Fig. 31e. In the latter, the slope at the support C is $du/dx = 0.5$ and then the end condition $u_B = 0$ requires $2(0.5)+(2-x_0)(2+x_0)/2-x_0^2/2 = 0$, giving $x_0 = \sqrt{3}$. The moment diagram in Fig. 31f, whose absolute value also represents $z/\sqrt{\nu/rc}$, can then be determined from simple statical considerations. By (195), the normalized beam weight becomes

$$(A_1 + A_2)^2 = (2\sqrt{3} - 3)^2 = 0.215390309\,. \tag{200}$$

The above solutions will be verified by numerical (iterative COC) calculations in Section 24.1.

It is still interesting to know, at which span ratio the elementary topology in Figs. 31a-c changes to the topology in Figs. 31d-f.

Clearly, the topology represented by Figs. 31a-c is valid up to a span ratio of 1:1, at which the curvature in the unloaded span also reaches $\kappa = 1$. Beyond this span ratio, the condition in (197) does not allow a zero cross-sectional area in the unloaded span.

Note that all moment diagrams in this section had to be statically admissible and the corresponding adjoint curvatures kinematically admissible in order to fulfill the optimality conditions in Figs. 3 and 6.

19.4 Least-Weight Pin-Jointed Frame for a Point Load Parallel to a Supporting Line

In the case of pin-jointed frames, the generalized stress is the axial force $\mathbf{Q} = N$ and the generalized strain is the axial member strain $\mathbf{q} = \varepsilon$. Then the optimality conditions (187a) and (186b) for optimal plastic design in (192a) and (192b) with $z_a = 0$ for optimal elastic design with given compliance furnish after normalization $(k = 1)$

$$(\text{for } |N| > 0) \qquad \varepsilon = \operatorname{sgn} N\,, \tag{201}$$
$$(\text{for } |N| = 0) \qquad |\varepsilon| \le 1\,. \tag{202}$$

---

* In Fig. 31, the normalized values of the real and adjoint deflections are the same for non-vanishing cross-sections and hence κ refers to both real and adjoint curvatures.

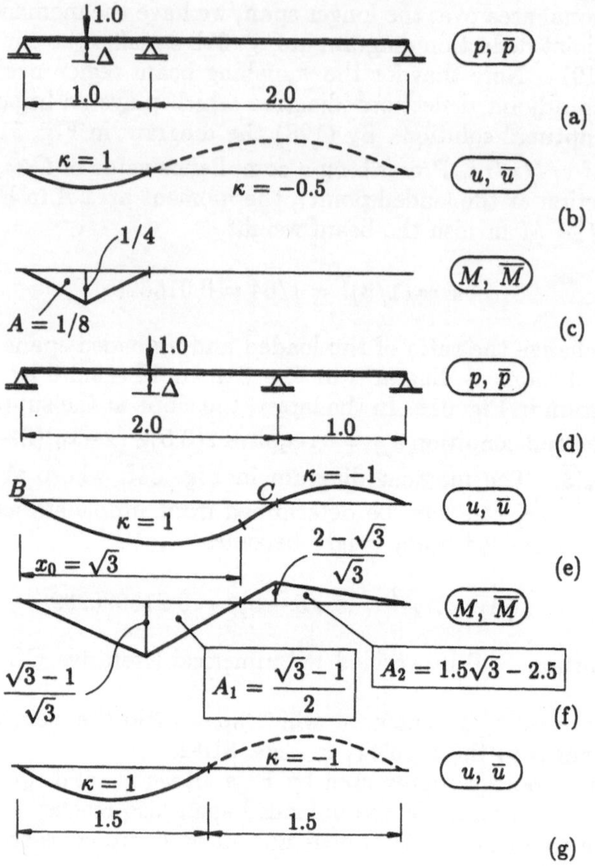

Fig. 31 A simple layout problem: two-span beam.

Figure 32a shows the loading and a structural universe for the considered problem and Fig. 32b the optimal solution which can be proved as follows. One must find a displacement field satisfying (201) and (202), as well as the kinematic boundary conditions, that is, zero displacements in all directions along the vertical support. This means that along the two bars in Fig. 32b the strains must be $\varepsilon = 1$ and $\varepsilon = -1$, respectively, and along all other members in Fig. 32a the absolute value of ε must not exceed unity. Denoting the displacements in the x and y directions in Fig. 32b by u and v, a displacement field satisfying the above conditions is

$$u(x, y) \equiv 0\,, \qquad v(x, y) = -2x\,, \tag{203}$$

which clearly satisfies the kinematic boundary conditions $u \equiv v \equiv 0$ along the support with $x = 0$. Moreover, the strains in the x and y directions are

$$\varepsilon_x = \frac{\partial u}{\partial x} = 0\,, \qquad \varepsilon_y = \frac{\partial v}{\partial y} = 0\,, \quad \gamma_{xy} = \frac{\partial u}{\partial y} + \frac{\partial v}{\partial x} = -2\,. \tag{204}$$

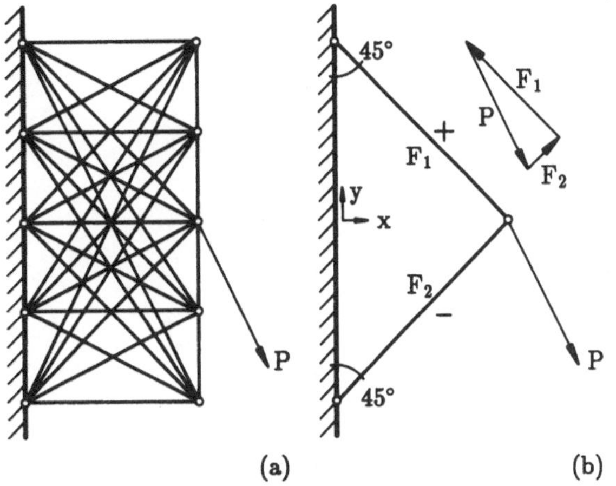

Fig. 32 Optimal layout of a simple pin-jointed frame.

Then in the principal directions at 45° to the vertical, the principal strain values become

$$\varepsilon_{1,2} = \frac{\varepsilon_x + \varepsilon_y}{2} \pm \sqrt{\left(\frac{\varepsilon_x - \varepsilon_y}{2}\right)^2 + \frac{\gamma_{xy}^2}{4}}\ , \ \varepsilon_1 = 1\ , \ \varepsilon_2 = -1\ ,$$

$$\alpha = \frac{1}{2}\text{arc tan}\,\frac{\gamma_{xy}}{\varepsilon_x - \varepsilon_y} = -\frac{1}{2}\text{arc tan}\,\infty = -45°. \qquad (205)$$

The above displacement field, therefore, satisfies (201) along the nonvanishing members in Fig. 32a, in which the forces are statically admissible. Moreover, since the principal strains represent the directionally highest absolute values of the strains, (202) is also fulfilled for any vanishing member in Fig. 32a. In fact, *the same solution would be still valid, if the structural universe consisted of all possible members contained in the half-plane* to the right of the vertical support in Fig. 32a.

19.5 Optimal Transmission of a Vertical Point Load to Supports Formed by a Horizontal and a Vertical Line

It can be shown that in the above problem, the optimal topology depends on the ratio of the distances of the vertical point load from the vertical and horizontal supporting lines. The analytical proof is based on the adjoint field given in Fig. 33b, in which the displacements satisfy all kinematic boundary conditions, as well as continuity conditions along the boundary of the two regions indicated. Arrow-like symbols indicate the direction and sign of principal strains having a unit value, along which, by optimality condition (201), a non-zero force is admitted. This means that for loads above the region boundary with a slope of 2:1, the optimal topology consists of two bars of 45° to the vertical and for loads below that boundary

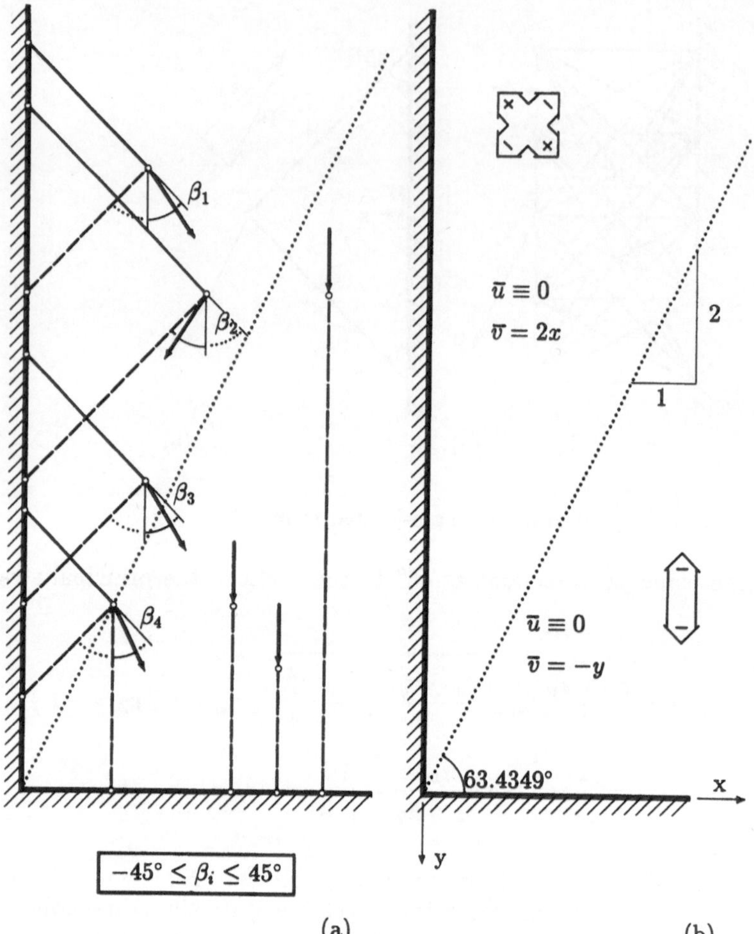

Fig. 33 Optimal layout (a) and adjoint field (b) for a least-weight truss in a corner region with a point load.

the optimal topology consists of a single vertical bar. Examples of the optimal load transmission are given in Fig. 33a.

19.6 A Simple Beam System

An extremely simple structural universe, consisting of two simply supported beams, is shown in Fig. 34a. The beam system is subject to a single point load P at the intersection of the two beams. In this problem, we shall use an inverse procedure, first assuming a statically admissible stress field and then showing that it satisfies all optimality criteria. The rather obvious optimal solution is given in Figs. 34b and

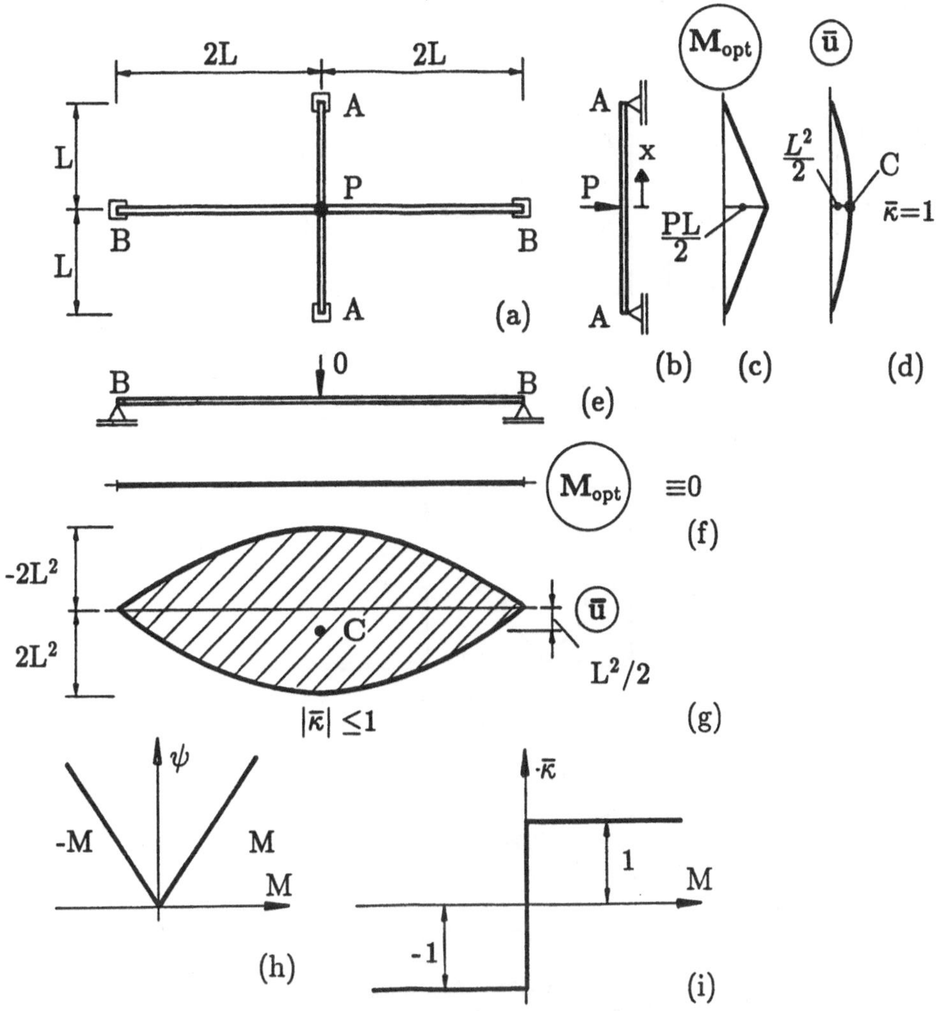

Fig. 34 Another elementary layout problem: grillage consisting of two beams.

e, in which the entire load P is carried by the short beam AA and the beam BB (with a zero cross-sectional area) is unloaded. The corresponding moment diagrams are shown in Figs. 34c and f.

The specific cost function and the optimality criteria [(196) and (197)] are represented graphically in Figs. 34h and i.

Since the deflection must be zero at the simple supports of the beams, the adjoint displacement field for the short beam AA becomes

$$\overline{u} = L^2/2 - x^2/2, \tag{206}$$

clearly satisfying the boundary conditions $\overline{u}(L) = \overline{u}(-L) = 0$ and the curvature

$$\bar{u}_4 = -\tfrac{1}{2}x^2 + \tfrac{1}{2}(y + \sqrt{2}a)^2$$

$$\bar{u}_2 = \tfrac{1}{2}(x - \sqrt{2}a)^2 - \tfrac{1}{2}y^2$$

$$\bar{u}_1 = \tfrac{1}{2}(a^2 - x^2 - y^2)$$

$$\bar{u}_5 = -\tfrac{1}{2}x^2 + \tfrac{1}{2}(y - \sqrt{2}a)^2$$

$$\bar{u}_3 = \tfrac{1}{2}(x + \sqrt{2}a)^2 - \tfrac{1}{2}y^2$$

(a)

(b)

(c)

(d)

(e)

(f)

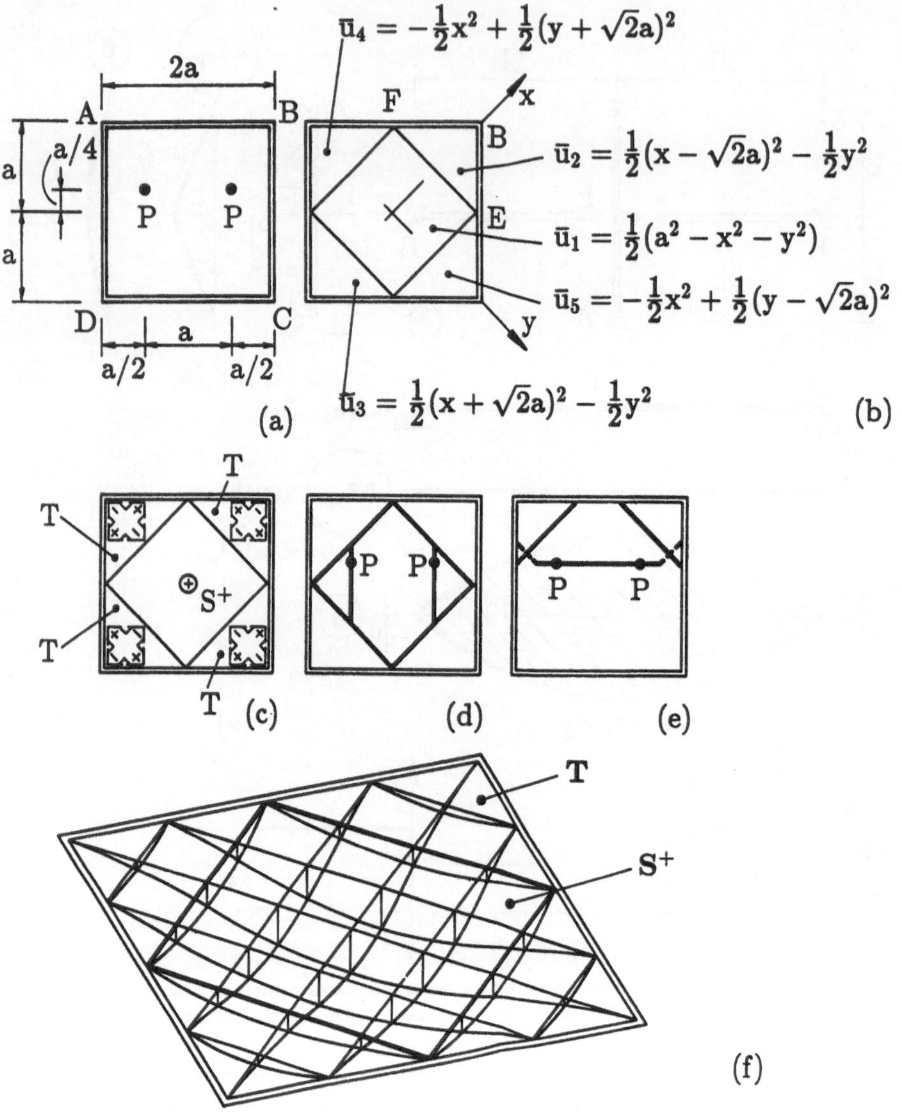

Fig. 35 A more advanced layout problem: square, simply supported grillage (after Morley [26]).

condition (196) $-\bar{u}'' = k$ (see Fig. 34i). For the long beam BB, the bending moment is throughout zero and hence (197) gives a non-unique curvature requirement which furnishes the non-unique adjoint deflection field indicated by the shaded area in Fig. 34g. As the latter does include a central deflection of $L^2/2$ (Point C in Fig. 34g), kinematic admissibility is also established. Moreover, since the specific cost function is convex (Fig. 34h), necessary and sufficient conditions for optimality have been fulfilled and thus the solution in Figs. 34b-f is optimal.

19.7 Square, Simply Supported Grillage

Figure 35a shows a square, simply supported area $ABCD$ with two point loads (P) which are to be transmitted to the supports by a system of intersecting beams. In this problem, the structural universe consists of an infinite number of potential members in any arbitrary direction, covering the entire area $ABCD$. The normalized specific cost function is again $\psi = |M|$ (Fig. 34h) and the corresponding optimality criteria are given in Fig. 34i. As the above optimality conditions imply that the directional maximum absolute value of the curvature is k, it follows that such maximal curvatures, and by (196) all optimal beams must have the same orientation as the ***principal*** directions of the adjoint displacement field. This means that in order to admit potential beams over the entire area, the latter must be covered by smoothly jointed regions on which at least in one principal direction the curvature of the adjoint displacement field has an absolute value of k. Figure 35b shows such a displacement field which consists of five distinct regions (u_1 to u_5). It can be checked easily that the kinematic boundary condition $u = 0$ is satisfied along the edges and the displacement field $u(x, y)$ is continuous and slope-continuous along all region boundaries. In the corner regions (in Section 21.1 termed type T) one principal curvature has a value of k and the other one $-k$ (Fig. 35c). In the central region (in Section 21.2 termed type S^+), the curvature takes on a value of k in all directions. It follows then from the optimality conditions (196) and (197) that in the central (S^+) region optimal beams may run in any arbitrary direction, so long as they are subject to positive bending. In the corner (T) regions, optimal beams may only run in two directions; in the direction of the diagonal passing through the considered corner the beams must be in negative bending and in the direction normal to that diagonal the beams must be in positive bending. The above conditions admit an infinite number of optimal beam layouts of equal structural weight. One of these, with only positive moments, is shown in Fig. 35d and another one, with beams in both positive and negative bending, in Fig. 35e. An oblique view of the adjoint field is given in Fig. 35f. The above optimal displacement field, which is also valid for any other non-negative (downward) loading, was first derived by Morley [26] in the context of optimized reinforced concrete plates.

20. Classical and Advanced Layout Theories

The introductory examples in Section 19 were based on the so-called "classical" layout theory, a generalization of Michell's [28] theorem, which has been used for the optimization of "low-density" structural systems whose structural material occupies only a small proportion of the feasible space. This theory has two fundamental features: (a) at any point of the structural domain potential members may run in any number of directions (Fig. 36a), but (b) the effect of the member intersections on both the cost and strength (or stiffness) is neglected. It follows that the specific cost function ψ is the sum of several terms each of which depends on a stiffness (or stress resultant) value s_i,

$$\psi = \psi_1(s_1) + \psi_2(s_2) + \ldots + \psi_n(s_n). \tag{207}$$

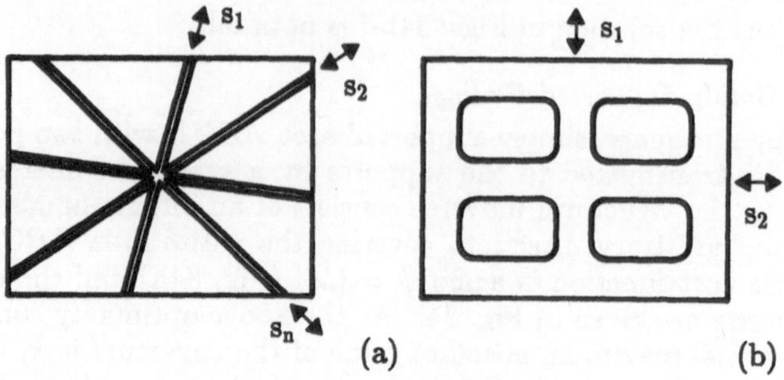

Fig. 36 Classical and advanced layout theories.

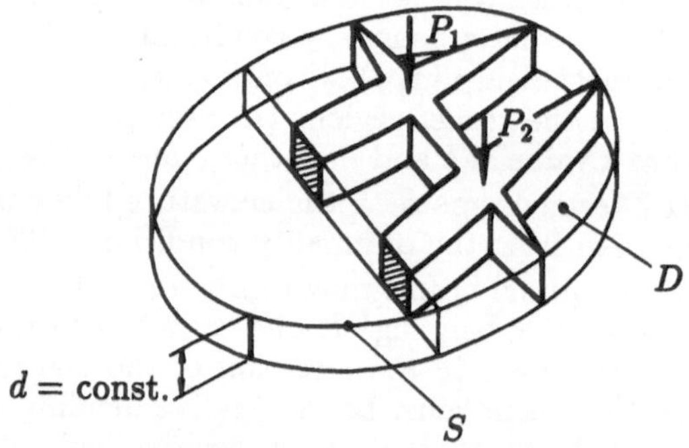

Fig. 37 The grillage layout problem.

"Advanced" layout theory is used for "high density" structures in which material occupies a high proportion of the feasible space or the structure consists of several materials whose interfaces are to be optimized. In this case, the microstructure of a perforated or composite structure is first optimized locally by minimizing, for given stiffnesses or stress resultants in the principal directions, the specific cost ψ (e.g. material volume per unit area or volume of the structural domain) for perforated structures and some factored combination of the material volumes per unit area or volume of the structural domain for composite structures. This means that the specific cost function, e.g. $\psi(s_1, s_2)$ in Fig. 36b, is in general a non-separable function of the principal stiffnesses or stress resultants.

Advanced layout theory results in substantial extra savings for "high density" structures, but the optimal solutions given by this theory tend to those of classical layout theory if the material volume/feasible volume ratio approaches zero.

Applications of the classical and advanced layout theory are given in Sections 21 and 22, respectively.

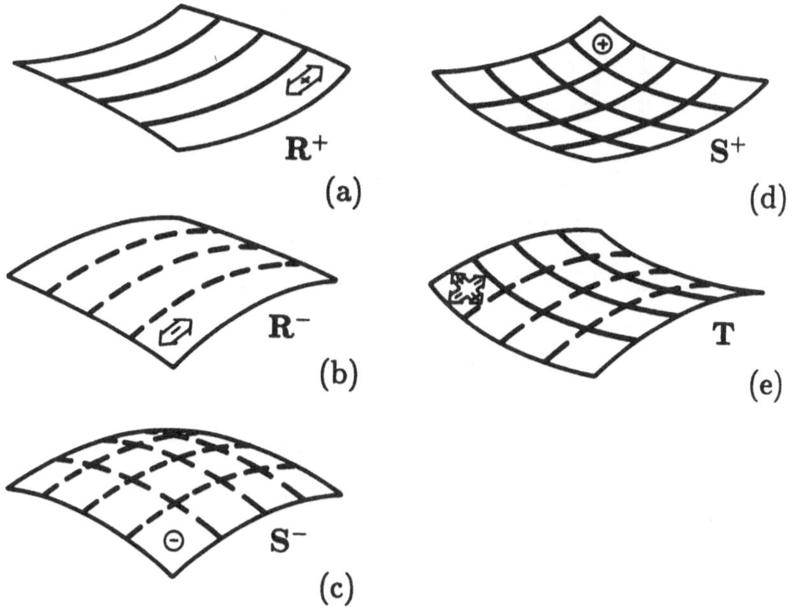

Fig. 38 Optimal regions for least-weight grillages.

21. Applications of Classical Layout Theory

21.1 Least-Weight Grillages (Beam Systems)

In order to check the validity and accuracy of optimal layouts derived by numerical methods, however, it is necessary to obtain *exact, analytical solutions* for relatively complex layout problems. One of the most successful applications of the exact layout theory was the optimization of grillage layouts, as can be seen from the following remark by Prager: "Although the literature on Michell trusses is quite extensive, the mathematically similar theory of grillages of least-weight was only developed during the last decade. Despite its late start, this theory advanced farther than that of optimal trusses. In fact, grillages of least-weight constitute the first class of plane structural systems for which the problem of optimal layout can be solved for almost all loadings and boundary conditions" (Prager and Rozvany, [27]). The problem of grillage optimization can be described as follows (Fig. 37). A structural domain D, bound by two horizontal planes and some vertical surfaces, is subject to a system of vertical loads which are to be transmitted to given supports by beams of rectangular cross-section having a given depth and variable width. The beams are to be contained in the structural domain and are to take on a minimum weight (or volume). The beam system is to be designed plastically or elastically for a given compliance (Sections 19.1 and 19.2).

As can be seen from the quotation above, Prager regarded the grillage optimization problem as particularly important because of the following unique features:

(a) Grillages constitute the first class of truly two-dimensional structural opti-

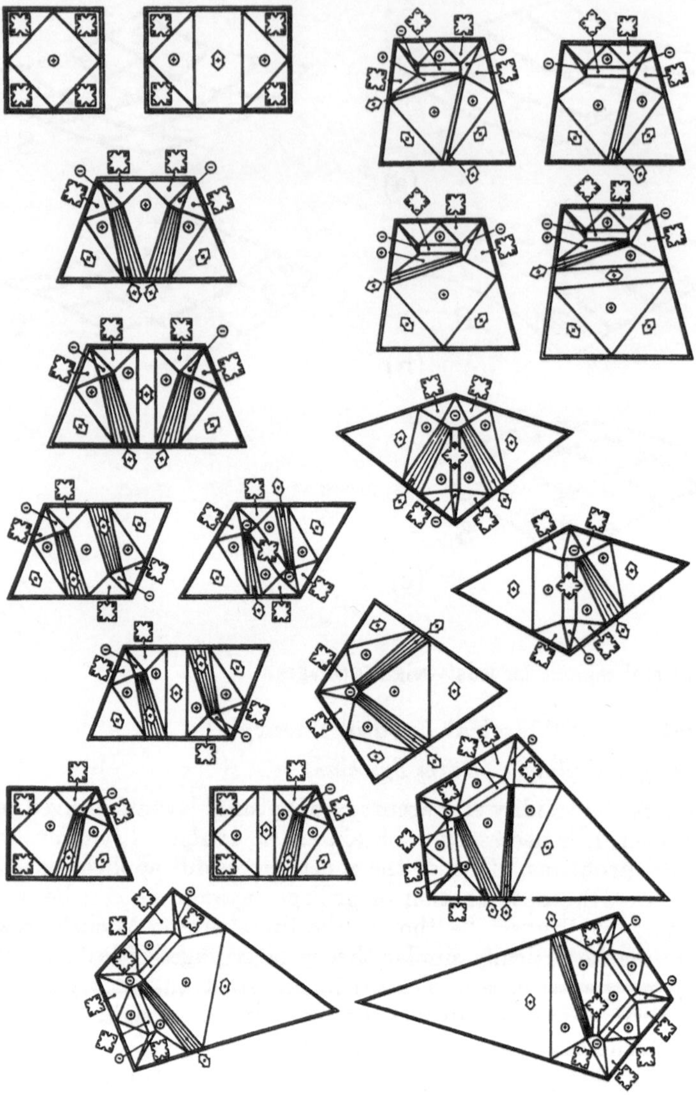

Fig. 39 Optimal grillage topologies for quadrilateral boundaries I.

mization problems for which closed form analytical solutions are available for most boundary and loading conditions.

(b) Optimal grillages are more practical than Michell structures (least-weight trusses), because the latter are subject to instability which is ignored in the formulation.

(c) The optimal rib layout of least-weight plates has been found to be similar to that of minimum weight grillages (see Section 22).

(d) A computer algorithm is available for generating analytically and plotting

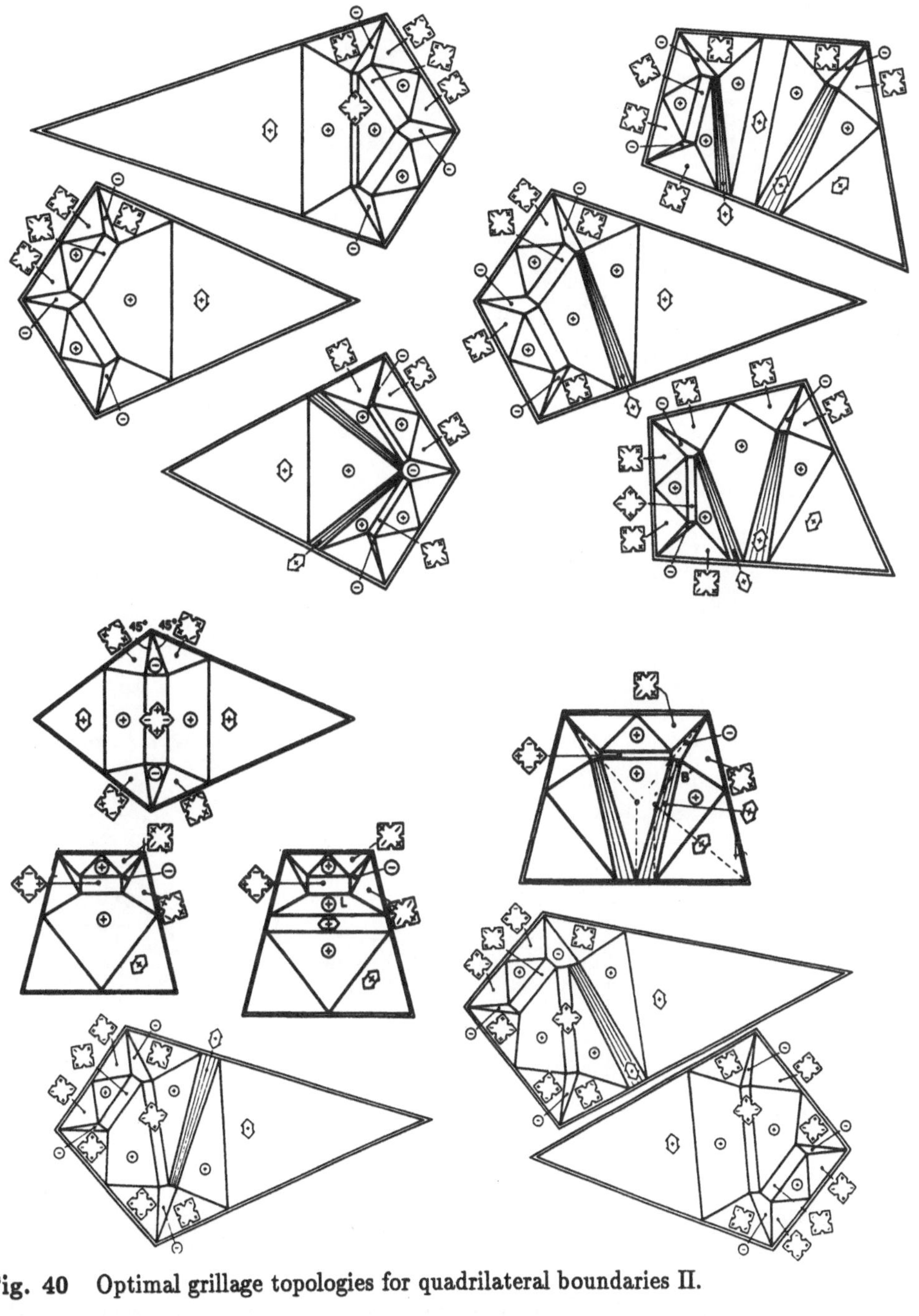

Fig. 40 Optimal grillage topologies for quadrilateral boundaries II.

optimal beam layouts for a wide range of boundary conditions (Rozvany and Hill [37]; Hill and Rozvany [38]).

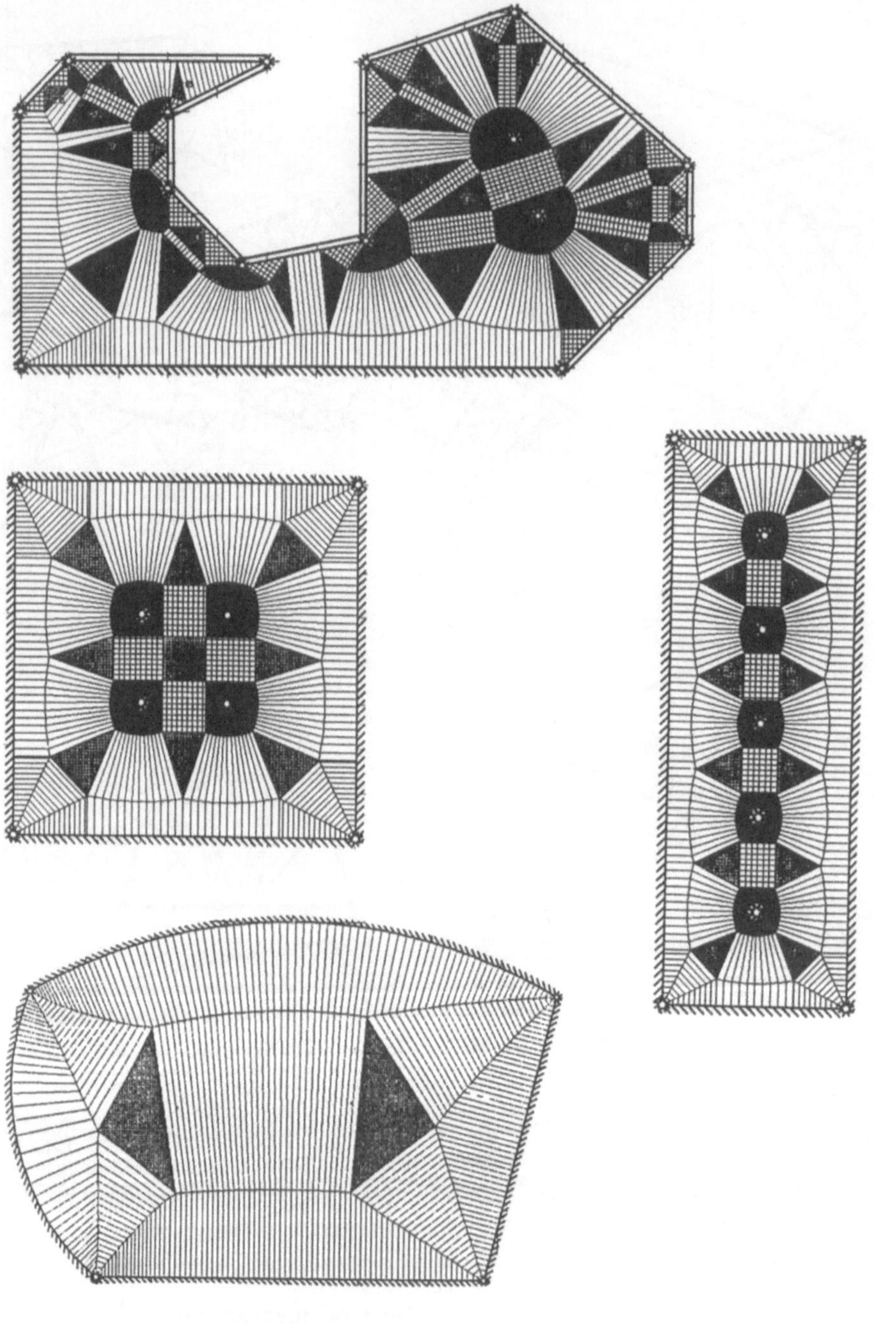

Fig. 41 Optimal grillage layouts derived analytically and plotted by a computer (after Gerdes and Gollub).

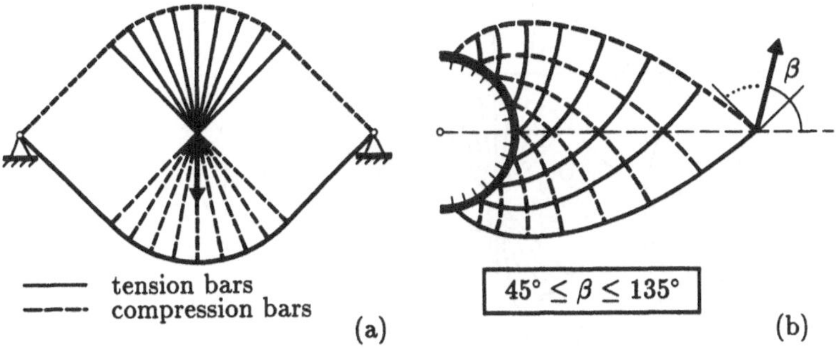

Fig. 42 Some earlier Michell layouts.

(e) It has been shown that the same grillage layout is optimal for plastic design and elastic design with a stress or compliance or natural frequency constraint (Rozvany [8]; Olhoff and Rozvany [49]).

(f) The optimal grillage layout is independent of the (non-negative) load distribution if no internal simple supports are present.

(g) The adjoint displacement field can be readily generated and it provides an *influence surface* for any (non-negative) loading (the total structural weight equals the integral of the product of loads and deflections).

(h) A number of additional refinements have been added to the optimal grillage theory.

The specific cost functions ψ for grillages of given depth is $\psi = k|M|$, where k is a given constant and M is the beam bending moment, and the optimality conditions are basically those of Heyman [39], given also in (196) and (197) herein. The latter imply (see Section 19.7) that all optimal beams must have the same orientation as the principal directions with a curvature $-\overline{u}'' = 1$ or $-\overline{u}'' = -1$ of an adjoint displacement field $\overline{u}(x, y)$.

This means that for fully loaded domains the optimal solution must consist of the following types of regions:

$$\left.\begin{array}{lllll} R^+: & M_1 > 0, & M_2 = 0, & \overline{\kappa}_1 = 1, & |\overline{\kappa}_2| \le 1, \\ R^-: & M_1 < 0, & M_2 = 0, & \overline{\kappa}_1 = -1, & |\overline{\kappa}_2| \le 1, \\ S^+: & M_1 > 0, & M_2 > 0, & \overline{\kappa}_1 = \overline{\kappa}_2 = 1, & \\ S^-: & M_1 < 0, & M_2 < 0, & \overline{\kappa}_1 = \overline{\kappa}_2 = -1, & \\ T\ : & M_1 > 0, & M_2 < 0, & \overline{\kappa}_1 = -\overline{\kappa}_2 = 1, & \end{array}\right\} \quad (|\overline{\kappa}_1| \ge |\overline{\kappa}_2|),\ (|M_1| \ge |M_2|)\ . \tag{208}$$

The above optimal regions are shown in Fig. 38 in which continuous and broken lines, respectively, indicate optimal beams with positive and negative bending moments. Arrows indicate the directions of principal curvatures. In the case of circles enclosing a sign all directions are equally optimal (the curvature is the same in all directions).

The above optimality conditions mean that we have replaced a complicated variational problem with a relatively simple geometrical task which requires that

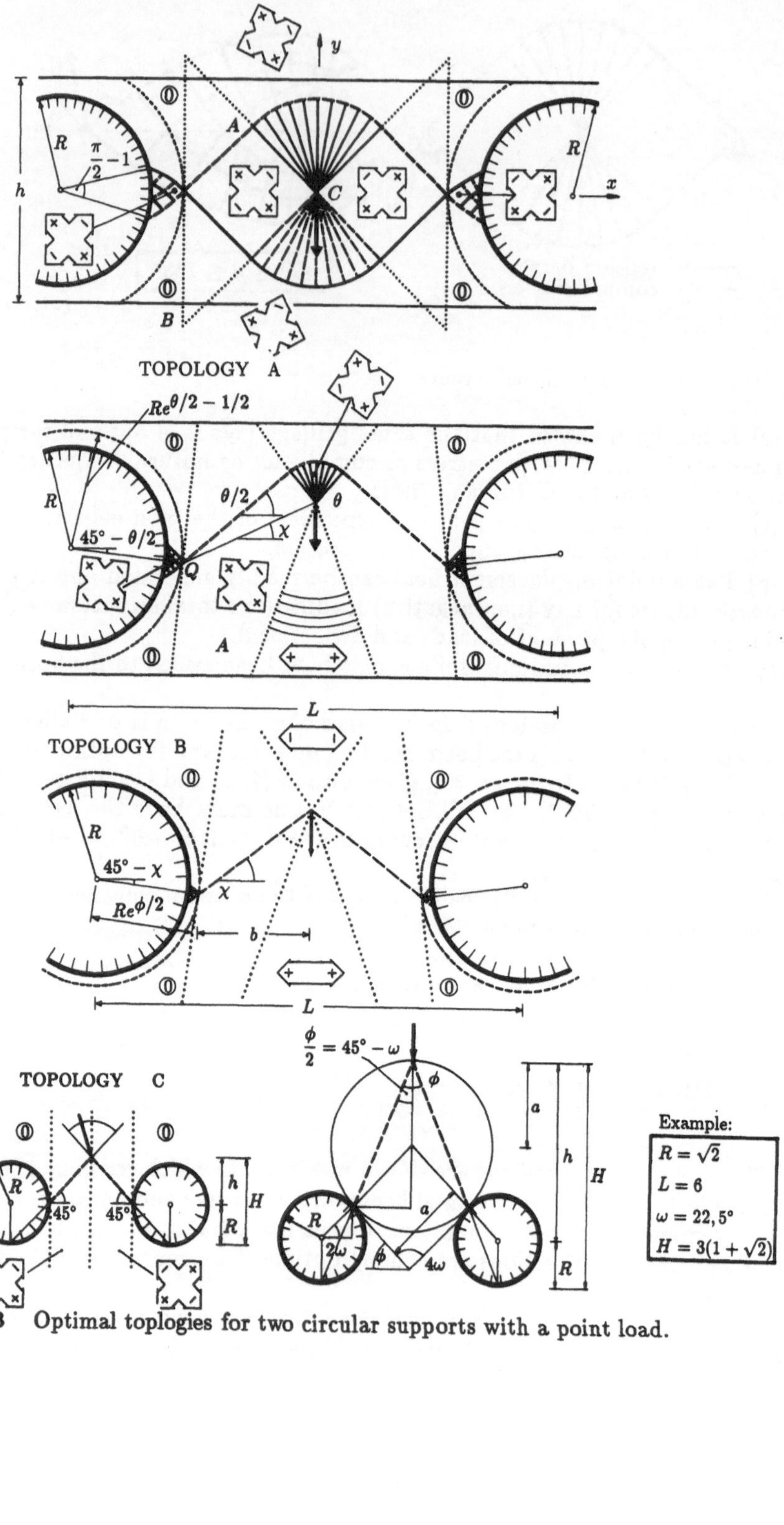

Fig. 43 Optimal toplogies for two circular supports with a point load.

	WITHOUT SELFWEIGHT	WITH SELFWEIGHT
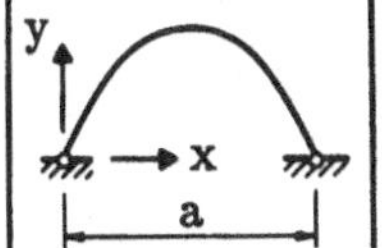	$\frac{\int_0^a (dy/dx)^2 dx}{a} = 1$	$\frac{\int_0^a (dy/dx)^2 e^{2ky} dx}{\int_0^a e^{2ky} dx} = 1$
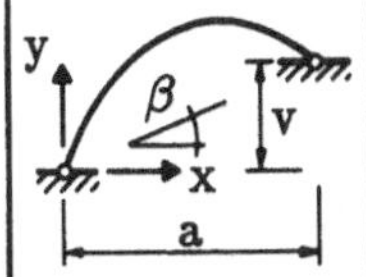	$\frac{\int_0^a (dy/dx)^2 dx}{a}$ $= 1 + 2\,tan^2\,\beta$	$\int_0^a k e^{2ky}\left[1 - (dy/dx)^2\right]\left[1 - C\int_0^x e^{-2ky} dx\right] dx - Cv = 0$ where $C = (e^{2kv} - 1)/(e^{2kv}\int_0^a e^{-2ky} dx)$

Optimal Geometry (a)

	WITHOUT SELFWEIGHT	WITH SELFWEIGHT
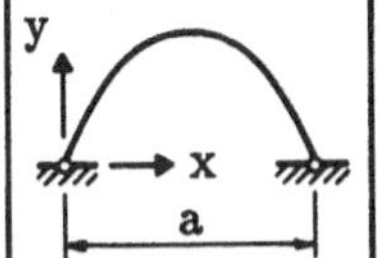	$u_y = 2ky$	$u_y = e^{2ky} - 1$
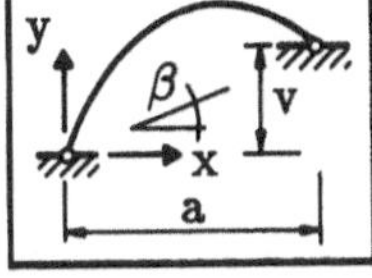	$u_y = 2k(y - x\,tan\,\beta)$	$u_y = e^{2ky}\left[1 - \frac{(e^{2kv} - 1)\int_0^x e^{-2ky} dx}{e^{2kv}\int_0^a e^{-2ky} dx}\right] - 1$

Influence Lines (b)

Fig. 44 Optimality conditions for Prager-structures in the plane.

(i) the structural domain is covered with the types of optimal regions given in Fig. 38, (ii) the above field is continuous and slope-continuous at region boundaries and satisfies the kinematic boundary conditions, and (iii) beams in the optimal directions given by such an adjoint field are capable of transmitting the loads to the supports (i.e. a statically admissible moment field is associated with the adjoint field).

The theory of least-weight grillages has been discussed elswhere in considerable detail [2, 3, 8, 27, 31-35, 41]. It is, therefore, perhaps sufficient here to give one example in Figs. 39 and 40, in which all 32 optimal grillage topologies for quadrilateral simply supported boundaries are given.

A quite recent development is a new computer program for deriving analytically and plotting optimal grillage layouts. This program was developed in Essen by D. Gerdes under the supervision of W. Gollub and represents an improved and

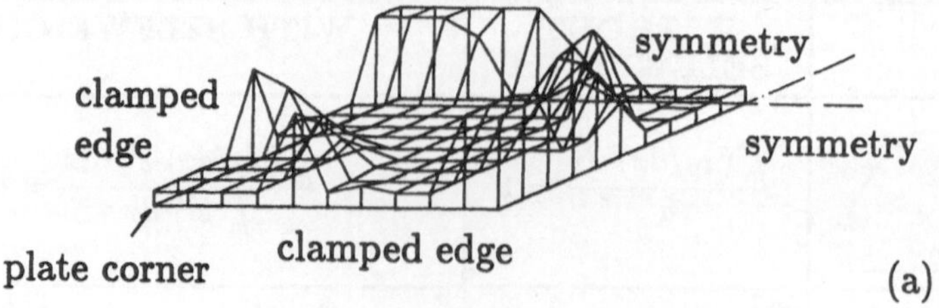

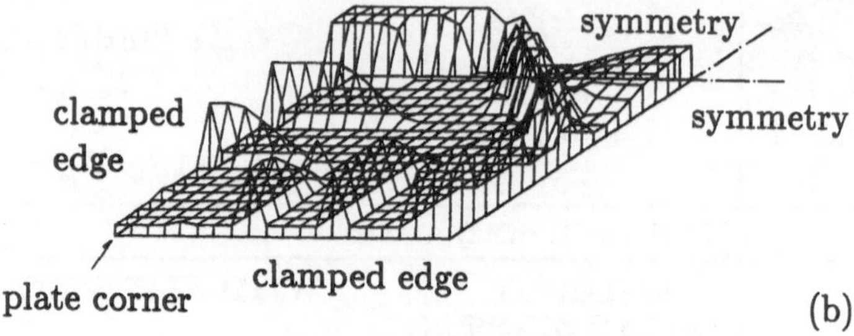

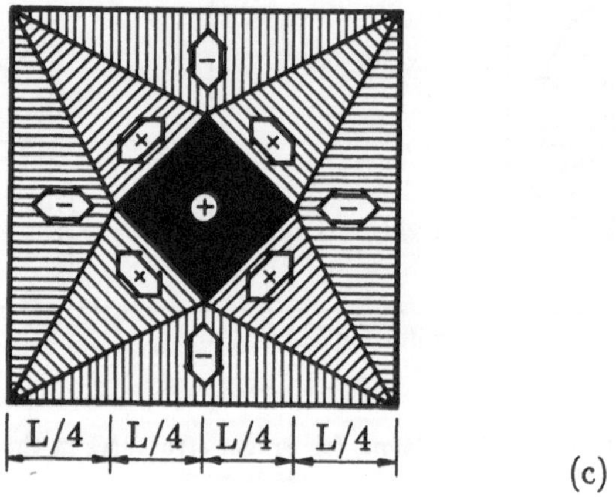

Fig. 45 (a and b) Numerical solutions by Cheng and Olhoff [52] showing rib-like formations and (c) optimal grillage of similar layout (Rozvany and Adidam [69] after Lowe and Melchers [70]).

extended version of an earlier program developed by Hill and Rozvany [37, 38]. The simple and clamped supports may consist of any straight or circular segment or internal clamped points. Figure 41 shows four outputs which are represented *analytically* in the computer, but the graphic representation has certain small inaccuracies due to the limited resolving power of the screen.

Note. Since topology optimization is a mathematically highly complex problem, it is rather remarkable that *exact analytical solutions* like the ones given in Figs.

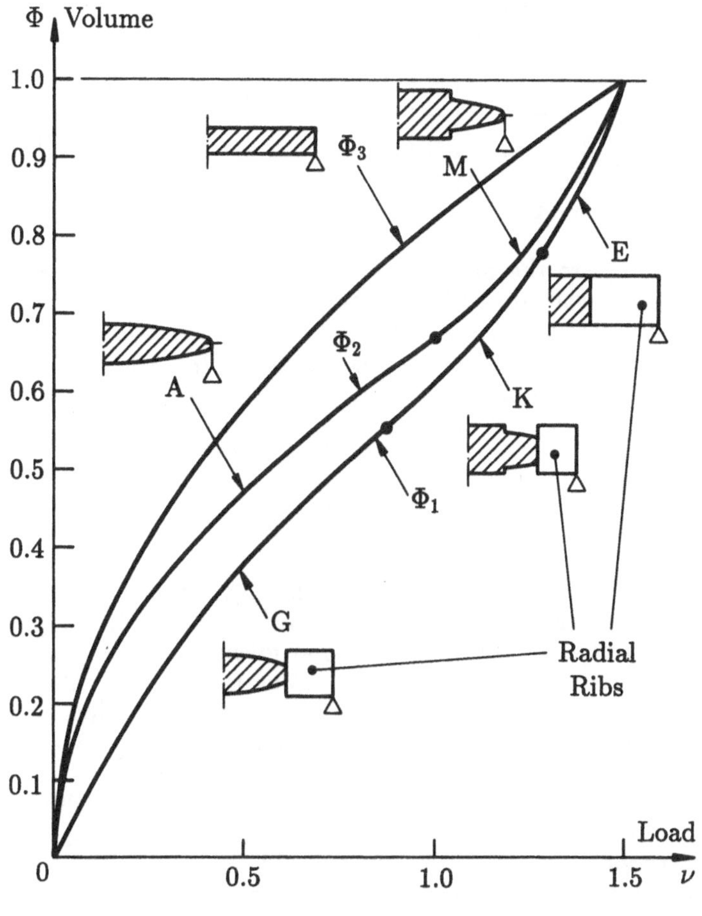

Fig. 46 Plastic plate design with a prescribed maximum thickness: a comparison of the weight of the absolute optimal (ribbed) solution Φ_1 with that of smooth Φ_2 and a constant thickness Φ_3 optimal solutions.

39-41 have been available since the mid-seventies.

21.2 Least-Weight Trusses (Michell Frameworks)

As mentioned earlier (Section 18), the theory of least-weight trusses was put forward in a most remarkable contribution around the turn of the century by an ingenious Australian inventor, A.G.M. Michell [28]. His general methodology has been influencing research into optimal structural layouts ever since.

The general optimality criteria (187a and b) imply that least-weight trusses *in the plane* must consist of the following types of optimal regions at points with non-vanishing members:

$$\left.\begin{array}{llll} R^+: & N_1>0, & N_2=0, & \bar{\epsilon}_1=k,\ |\bar{\epsilon}_2|\leq k, \\ R^-: & N_1<0, & N_2=0, & \bar{\epsilon}_1=-k,\ |\bar{\epsilon}_2|\leq k, \\ S^+: & N_1>0, & N_2>0, & \bar{\epsilon}_1=\bar{\epsilon}_2=k, \\ S^-: & N_1<0, & N_2<0, & \bar{\epsilon}_1=\bar{\epsilon}_2=-k, \\ T\ : & N_1>0, & N_2<0, & \bar{\epsilon}_1=-\bar{\epsilon}_2=k, \end{array}\right\} \quad (|\bar{\epsilon}_1|\geq|\bar{\epsilon}_2|),\ (|N_1|\geq|N_2|), \tag{208a}$$

where N_1 and N_2 are the principal forces, $\bar{\epsilon}_1$ and $\bar{\epsilon}_2$ are the principal strains, $k = 1/\sigma_p$ is a given constant and σ_p is the permissible stress.

If we have different permissible stresses for compression (σ_{pc}) and tension (σ_{pt}), then the corresponding cost factors become $k_c = 1/\sigma_{pc}$ and $k_t = 1/\sigma_{pt}$ and the stress values in (208a) become

$$\bar{\epsilon} = k_t \quad (\text{for } N > 0)\ , \tag{209}$$

$$\bar{\epsilon} = -k_c \quad (\text{for } N < 0)\ , \tag{210}$$

$$-k_c \leq \bar{\epsilon} \leq k_t \quad (\text{for } N = 0)\ . \tag{211}$$

Until recently, relatively few solutions for least-weight trusses were known. One possible reason for this was the fact that these solutions had highly restrictive statical boundary conditions with supports at given points or along a small circle (see Fig. 42). More recently, a systematic exploration of plane Michell-layouts for line supports has been undertaken. Whilst here only one example is reproduced in Fig. 43, a detailed description of this work is given elsewhere [42, 43].

21.3 Archgrids and Cable Nets of Optimal Layout (Prager Structures)

A Prager structure can be defined as a surface structure consisting of intersecting arches or cables for which the shape of the middle surface and the member layout are to be optimized. Moreover, the (usually vertical) loads are movable along their line of action. Alternatively, a Prager structure can be regarded as a special class of Michell frames for which (a) either the compressive or the tensile permissible stress tends to zero and (b) the position of (usually vertical) loads is unspecified and to be optimized. This special class of Michell structures has been shown to reduce always to a surface structure in 3D space (or a line structure in plane). On the basis of (209)-(211) the following optimality conditions apply to Prager structures:

$$\bar{\epsilon} = k\ (\text{for } N > 0), \quad -\infty < \bar{\epsilon} \leq k \quad (\text{for } N = 0)\ ,$$

$$N \geq 0, \quad \bar{\epsilon}_{\text{vertical}} \equiv 0\ . \tag{212}$$

The last condition is due to the fact that loads are "movable" vertically, which is equivalent to having weightless (i.e. costless) members in that direction. Closed form analytical solutions are now available for any vertical axisymmetric load in three-dimensional space and for any vertical load system in a plane, and also for additional selfweight [in which case k is replaced by $k(1+\bar{u})$, see Section 7]. Moreover, the above solutions are extended to "quasi-axisymmetric" loads (concentrated loads distributed in the circumferential direction at equal angular intervals and axisymmetric support conditions). Figure 44a shows the general form of the optimal solution for plane Prager structures with two supports at the same level and at different levels (at an angle β), both with or without selfweight. It can be seen that for two supports at the same level, the optimal form of the Prager structure is given by a funicular such that *the mean square slope is unity.* The above optimality criteria are also valid for several supports and for axisymmetric and quasi-symmetric systems.

Moreover, Fig. 44b gives optimal cost influence lines u_y for various types of plane Prager structures such that the total cost for any vertical load system is given by $\int_D p u_y \, dx$.

It can be seen from Fig. 44b that for the simplest case (two supports at equal elevation, no selfweight) the optimal cost (structural weight) is simply the *sum* (or integral) *of the products of loads $p(x)$ and their elevation $y(x)$* multiplied by a constant ($2k$).

For further information on Prager structures, the reader is referred to papers by Rozvany and Prager [44], Rozvany, Nakamura and Kuhnell [45], Rozvany, Wang and Dow [46], Rozvany and Wang [47], as well as Wang and Rozvany [48].

22. Applications of the Advanced Layout Theory

22.1 Optimal Plastic Design of Solid Plates

It was established already in the late sixties (e.g. Kozłowski and Mróz [50]) and confirmed more rigorously later (Rozvany, Olhoff, Cheng and Taylor [51]) that the weight of solid plates can be reduced to an arbitrarily small value by employing a system of sufficiently high and thin ribs. Naturally, the above solution is of purely theoretical interest because it ignores the lateral instability of such ribs. A finite value for the structural weight can be ensured, however, by introducing an *upper constraint* on the plate thickness. As can be seen from Fig. 44, Cheng and Olhoff [52] have discovered through numerical solutions that stiffener-like formations occur in the optimal solution for such systems. It was pointed out by Prager shortly before his death that the layout of the above stiffeners is similar to that of the corresponding grillage (Fig. 44c). Complete analytical solutions were obtained for plastically designed solid Tresca-plates with an upper limit on the plate thickness by Rozvany, Olhoff, Cheng and Taylor [51] and extended to other boundary conditions by Wang, Rozvany and Olhoff [54]. For simply supported circular plates, for example, the optimal solutions are compared with some non-optimal ones in Fig. 45 for various levels of the nondimensional load ν.

22.2 Optimal Elastic Design of Perforated Plates with a Compliance Constraint

It was shown in recent papers by several leading mathematicians (e.g. [55-57]) that optimal solutions in plane systems often contain regions with two sets of intersecting ribs (strips of material) at right angles: one such set has a first order infinitesimal spacing [of $O(\delta)$] with $\delta \to 0$] and the other set of second order infinitesimal spacing [of $O(\delta^2)$]. The implications of these results for plate optimization were investigated by Rozvany, Ong, Olhoff, Bendsøe, Szeto and Sandler [58], who derived several closed form analytical solutions for *perforated* plates.

It was found by the above authors that transversely loaded axially symmetric plates only the following *two types of regions* may occur in the optimal solution:

(i) unperforated solid regions;
(ii) regions consisting of radial ribs only;
(iii) empty regions.

The above research was also extended to composite plates consisting of two materials (for a review see [2], Chapter 8).

Iterative COC methods based on the layout theory outlined herein will be discussed in the next lecture.

LAYOUT AND GENERALIZED SHAPE OPTIMIZATION BY ITERATIVE COC METHODS

G.I.N. Rozvany and M. Zhou
FB 10, Essen University, D-4300 Essen 1, Germany

Abstract. On the basis of the optimal layout theory discussed in the preceding lecture, an iterative COC procedure is outlined and then applied to problems involving optimal plastic design and optimal elastic design with a given compliance. It is demonstrated that (i) the proposed method can handle up to several thousand potential members and (ii) the results show an excellent agreement with analytical solutions. An additional example with two alternate loading conditions is presented and it is found that for that case, the optimal plastic design differs from the optimal elastic design. Finally, the problem of generalized shape optimization is reviewed and an improved homogenization method is put forward.

23. General Formulation

In optimizing the layout of all elastic systems with a deflection constraint, first a structural universe (see Section 18) is defined and then basically the same procedure is adopted as the one described in Section 9.2. The main difference is that, for the prescribed minimum value z_{ia} of the parameter z_i, the smallest possible value (e.g. 10^{-12} times the average value of that parameter) is used which can still be handled by a program with double precision and does not cause ill-conditioning.

24. Elementary Examples of Layout Optimization by the Iterative COC Method

24.1 Two-Span Beams with a Central Unit Point Load over One Span

The problem solved analytically in Section 19.3 was also computed by the iterative COC procedure using a prescribed minimum width value of $z_{ia} = 10^{-8}$ with $N = 300$ and $N = 1800$ elements. For the span ratio in Fig. 31a, the COC method yielded total cost values of $\Phi_{300} = 0.015625018$ (22 iterations) and $\Phi_{1800} = 0.015625013$ (21 iterations), both showing a 7-digit agreement with the analytical result in (199). For the span ratio in Fig. 31d, the COC procedure yielded $\Phi_{300} = 0.21540290$ (22 iterations) and $\Phi_{1800} = 0.21539237$ (36 iterations), which represents four and five digits agreements, respectively, with the analytical result in (200). The above iteration numbers were required for a convergence tolerance value of $\overline{E} = 10^{-6}$ [see (66)], but a Φ-value with an error of less than 2% was obtained already after 3 iterations. The convergence was near-uniform, both the Φ-values and their differences decreasing monotonically.

G. I. N. Rozvany (ed.), Optimization of Large Structural Systems, Vol. I, 103–120.

24.2 Simple Pin-Jointed Frames

Figures 47a, b and c show some simple truss layout problems which were solved for given compliance by the iterative COC method. Members indicated in thick lines represent optimal bars in analytical solutions (see, for example, Fig. 33). Figure 47d shows the convergence history for the truss in Fig. 47b. The "weight" $\tilde{\Phi}$ here is nondimensional, $\tilde{\Phi} = \Phi E\Delta/(PL^2\gamma)$, where Φ is the total weight of the structure, E is Young's modulus, Δ is the prescribed deflection, P is the point load, L is the dimension shown in Fig. 47 and γ is the specific weight of the material used. The "scaled weight" $\tilde{\Phi}_s$ corresponds to a truss in which all cross-sections are linearly scaled to satisfy the compliance (deflection) condition exactly.

For these problems, the COC method was combined with either (a) a finite element (FE) program developed by the second author, or (b) ANSYS. Both FE programs yielded identical results.

Using the nondimensionalization given above for $\tilde{\Phi}$, the exact optimal weight for the 56-bar truss (Fig. 47b) is $\tilde{\Phi} = 16$ and the iterative COC procedure yielded after 126 iterations a weight of $\tilde{\Phi}_S = 16.000000000048$, which represents *an agreement of twelve significant digits*. In the COC solution, all non-optimal members took on the prescribed minimum cross-section ($\tilde{z}_a = 10^{-12}$), except members a and b (Fig. 47b) which had a cross-sectional area of $\tilde{z}_i = 2.88 \cdot 10^{-12}$ and $\tilde{z}_i = 1.74 \cdot 10^{-12}$. This indicates that all non-optimal members vanish when $\tilde{z}_a \to 0$, as in the analytical solution. The cross-sections of the optimal members agreed with the analytical solution for the first 12 significant digits.

In the 40-bar problem (Fig. 47c), *all* non-optimal members took on the value of $\tilde{z}_a = 10^{-12}$ in the iterative COC solution and the cross-sectional area for the optimal members agreed again with the analytical solution.

The problem in Fig. 47b (56-bar truss) was also run on the COC program with a structural universe having 114 members (Fig. 48a). The optimal members (thick lines) again took on a cross-sectional area of 2.8284271 ($\approx 2\sqrt{2}$) and all other members took on the prescribed minimum value of 10^{-12}, except the ones marked with a broken line, which were all under 10^{-11}. The above run yielded an optimal total cost value of 16.00000000016 after 231 iterations. The larger error compared to the analytical solution is due to the larger number and the greater average length of the members with the minimum cross-section.

The same problem was also investigated with a structural universe having 804 members (Fig. 48b). The optimal members are again shown in thick line (Fig. 48c) and the members having an area slightly greater than the prescribed minimum value ($10^{-12} < z_i < 10^{-11}$) in broken line. The unusually stringent convergence criterion of $(\Phi_{\text{new}} - \Phi_{\text{old}})/\Phi_{\text{new}} < 10^{-14}$ was satisfied after 627 iterations and gave a total cost value of 16.00000000081.

At the time of these investigations, the analytical solution was not available. It was obtained somewhat later [42] and is given in Fig. 33 herein.

25. More Advanced Examples of Layout Optimization by the Iterative COC Method

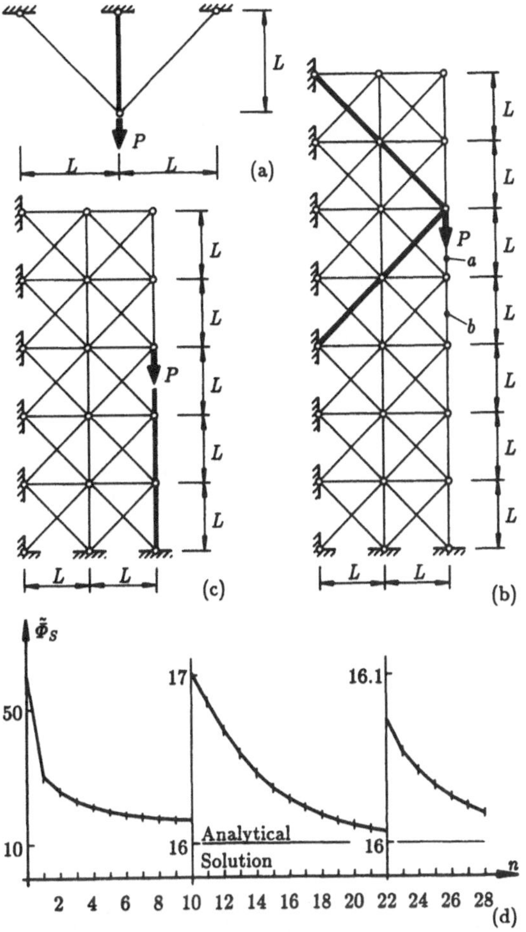

Fig. 47 (a)-(c) Various simple layout optimization problems solved by iterative COC methods; and (d) non-dimensional scaled structural weight $\tilde{\Phi}_S$ vs. number of iterations n for the problem under (b).

25.1 Truss-Like Continuum Containing a Hencky Net

It was pointed out by Prager (e.g. [3]) that many least-weight truss layouts consist of an infinite number of members having an infinitesimal spacing and hence they should be termed "truss-like continua". One such optimal layout is shown in Fig. 49a in which the truss is restricted to the rectangle $ABCD$ and a point load $P = 1$ is to be transmitted to the shorter side at the opposite end. The triangle ACE contains no members on its interior and the regions AEF and CEG consist of straight radial members. The region $EFGH$ is a Hencky net with curved members.

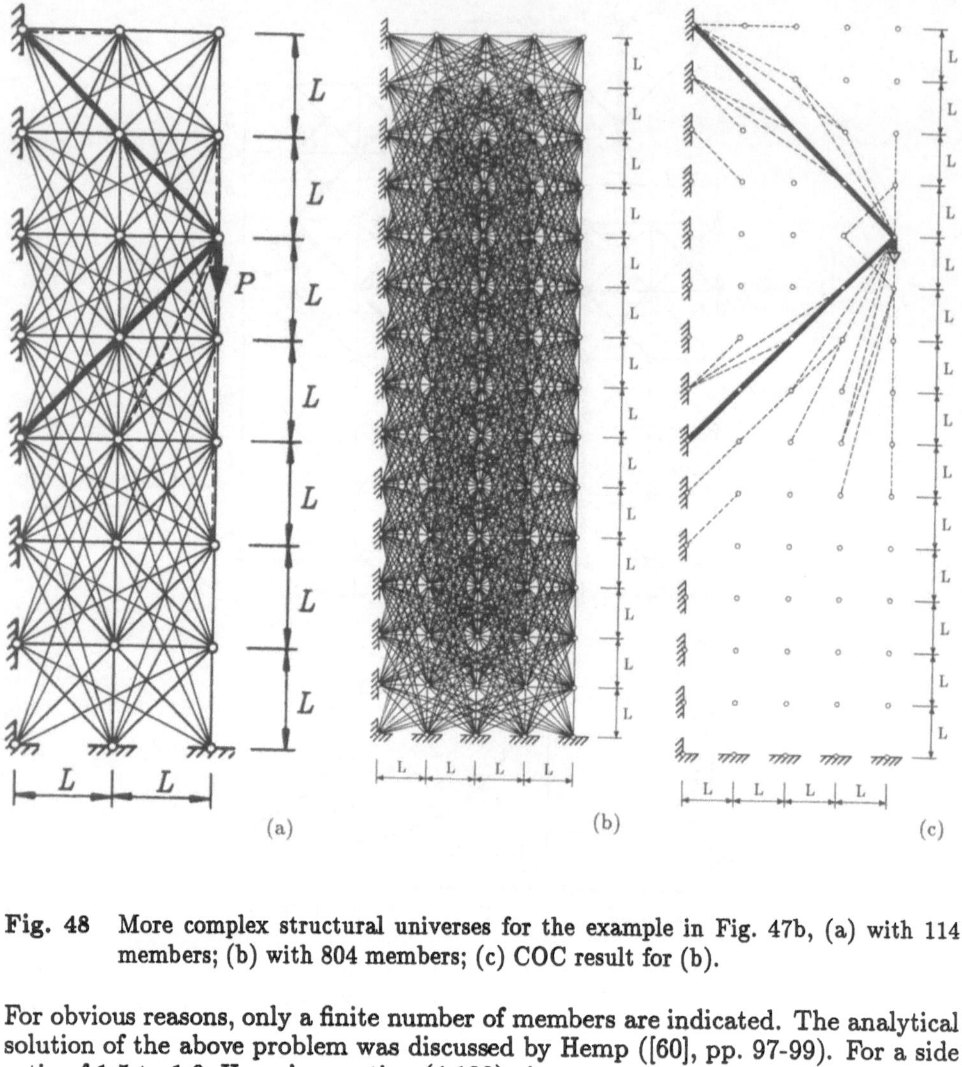

Fig. 48 More complex structural universes for the example in Fig. 47b, (a) with 114 members; (b) with 804 members; (c) COC result for (b).

For obvious reasons, only a finite number of members are indicated. The analytical solution of the above problem was discussed by Hemp ([60], pp. 97-99). For a side ratio of 1.5 to 1.0, Hemp's equation (4.120) gives

$$1.5 = \frac{1}{2}\int_0^{2\mu} [I_0(t) + I_1(t)]\,\mathrm{d}t = \frac{1}{2}\left[I_0(2\mu) - 1 + 2\sum_{n=0}^{\infty}(-1)^n I_{2n+1}(2\mu)\right], \quad (213)$$

yielding the angle $\mu = 82.690133°$. In (213), I_0, I_1, etc. denote modified Bessel functions. Then the total truss volume can be calculated from Hemp's equations (4.123) for a unit load and $k = 1$ (in Hemp's notation $\sqrt{2}FR/\sigma = 1$) as

$$\Phi = (1 + 2\mu)I_0(2\mu) + 2\mu I_1(2\mu)\,. \quad (214)$$

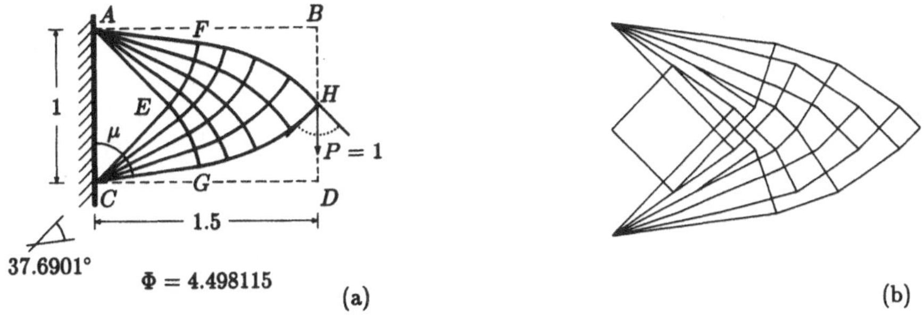

Fig. 49 Optimal layout consisting partly of a Hencky net: (a) analytical, and (b) COC solution.

For the above μ-value, (214) gives the optimal truss volume

$$\Phi_{\text{opt, plastic}} = 4.4498115\,. \tag{215}$$

The above value represents an optimal ***plastic design*** [(185) with $k = 1$]. For ***elastic design with a compliance constraint*** ($C = 1$), we have by (195) with $c = r = C = 1$

$$\Phi_{\text{opt, compliance}} = 4.498115^2 = 20.233042\,. \tag{216}$$

Using the iterative COC algorithm and structural universes with 5055 and 12992 members, respectively, truss volumes of $\Phi = 20.540807$ and $\Phi = 20.419699$ were obtained for elastic design with a compliance constraint. For plastic design, by (185) with $k = 1$ and (195) with $C = 1$, the square root of the above values must be taken, giving $\Phi = 4.532197$ and $\Phi = 4.518816$, which represent 0.76% and 0.46% errors compared to the analytical solution. In the above procedure, a minimum prescribed cross-sectional area of 10^{-12} was used. The system with 5055 members required over 3500 iterations with a convergence tolerance value of $\overline{E} = 10^{-8}$ [see (66)], but a Φ-value of 20.77 was reached already after 100 iterations (1% error compared to the value after satisfaction of the convergence criterion). The layout of members having a cross-sectional area over $z = 0.1$ in elastic compliance design is shown in Fig. 49b, which exhibits clear similarities with the analytical solution in Fig. 49a. The above limiting cross-sectional area [by (185) with $k = 1$, and by (190) and (194) with $c = r = C = 1$] corresponds $z = 0.1/\Phi_{\text{opt, plastic}} = 0.1/4.498115 \approx 0.0222$ in plastic design.

25.2 Single Point Load Parallel to a Supporting Line, Triangular Structural Domain

In the above problem the structural members are restricted to the triangle ABC in Fig. 50a, with a support along AB and free edges along AC and BC. The fairly obvious analytical solution consists of members running along the free edges of the domain (Fig. 50a). Before a rather complicated proof for the above solution was found ([43], see the correct adjoint field in Fig. 50b), the same problem was investigated by the iterative COC procedure, using a structural universe with 720

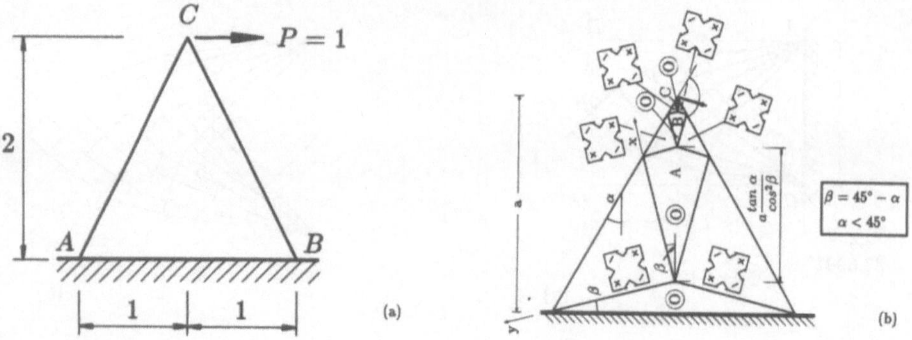

Fig. 50 (a) Loading and analytical solution, (b) adjoint field for a layout restricted to a triangular domain.

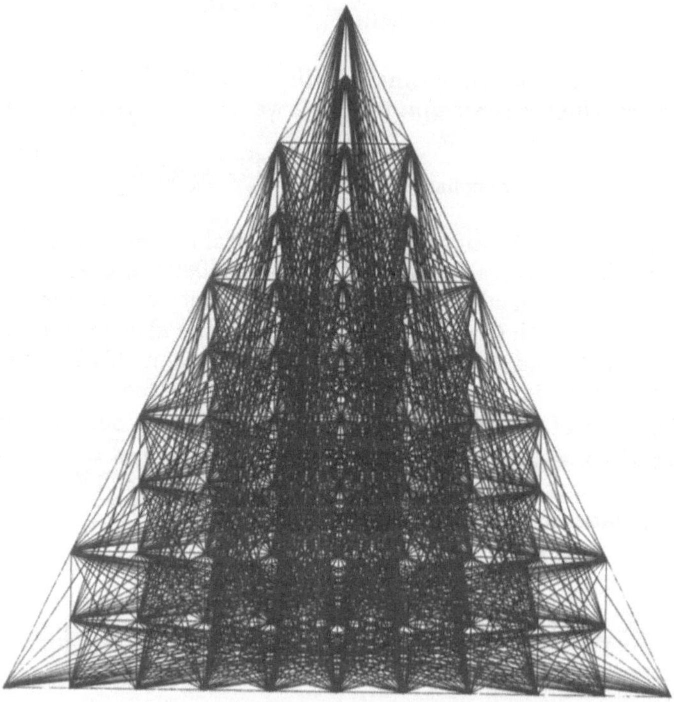

Fig. 51 Structural universe with 720 members for the problem in Fig. 50.

members (Fig. 51). The analytical solution for the above problem yields $\Phi = 25$ and the iterative COC solution with a prescribed minimum cross-sectional area of $z_a = 10^{-12}$ and a convergence tolerance value of $\overline{E} = 10^{-14}$ gave after 268 iterations $\Phi = 25.000000001168$.

26. Layout Problems with Alternate Loading Conditions

26.1 Plastic Design

If the structure is subjected to several alternative loading conditions ($\mathbf{p}_1$, $\mathbf{p}_2$, ..., $\mathbf{p}_j$, ..., $\mathbf{p}_n$) equilibrating the statically admissible stress fields $\mathbf{Q}_j$ $(1, 2, \ldots, n)$, our optimization problem in ***plastic design*** becomes

$$\min_{\mathbf{Q}_j^S} \Phi = \int_D \overline{\psi}\, d\mathbf{x} ,$$

subject to

$$(\text{for all } \mathbf{x} \in D) \quad \overline{\psi}(\mathbf{x}) = \max_j \psi \left[\mathbf{Q}_j^S(\mathbf{x}) \right] , \tag{217}$$

where ψ is the specific cost requirement for a given stress vector $\mathbf{Q}_j^S$ and $\overline{\psi}$ is the value of the specific cost (e.g. cross-sectional area) to be adopted in the optimal design.

For the above problem, the optimality criteria are ([2], (1.24) on p. 47):

$$(\text{for all } \mathbf{x} \in D) \quad \mathbf{q}_j^K = \lambda_j(\mathbf{x}) \mathcal{G} \left\{ \psi \left[\mathbf{Q}_j^S(\mathbf{x}) \right] \right\} ,$$

$$\lambda_j \geq 0 , \quad \lambda_j > 0 \quad \text{only if} \quad \overline{\psi} = \psi(\mathbf{Q}_j^S) , \quad \sum_j \lambda_j = 1 , \tag{218}$$

where $\mathcal{G}$ is the generalized gradient mentioned in Section 19.1.

Considering now the class of problems with a specific cost function $\psi = k|Q_j|$ where Q_j is a single-component generalized stress, and *two* loading conditions, the optimality criteria become

$$(\text{for } k|Q_1| = \overline{\psi}, \quad k|Q_2| < \overline{\psi}, \quad |Q_1| > 0) \qquad q_1 = k \operatorname{sgn} Q_1 , \quad q_2 = 0 , \tag{219}$$

$$(\text{for } k|Q_2| = \overline{\psi}, \quad k|Q_1| < \overline{\psi}, \quad |Q_2| > 0) \qquad q_2 = k \operatorname{sgn} Q_2 , \quad q_1 = 0 , \tag{220}$$

$$(\text{for } k|Q_1| = k|Q_2| = \overline{\psi} > 0) , \qquad q_1 = \lambda k \operatorname{sgn} Q_1 , \quad q_2 = (1 - \lambda) k \operatorname{sgn} Q_2 , \quad 1 \geq \lambda \geq 0 , \tag{221}$$

$$(\text{for } Q_1 = Q_2 = \overline{\psi} = 0) \qquad |q_1| + |q_2| \leq k . \tag{222}$$

It is relatively difficult to find, by a direct method, solutions satisfying the above optimality conditions. However, it was shown independently by Nagtegaal and Prager [61], Spillers and Lev [62], and Hemp [60] that one can employ a ***superposition principle*** consisting of the following steps. First, we construct the ***component load systems***

$$\mathbf{p}_1^* = (\mathbf{p}_1 + \mathbf{p}_2)/2 , \quad \mathbf{p}_2^* = (\mathbf{p}_1 - \mathbf{p}_2)/2 . \tag{223}$$

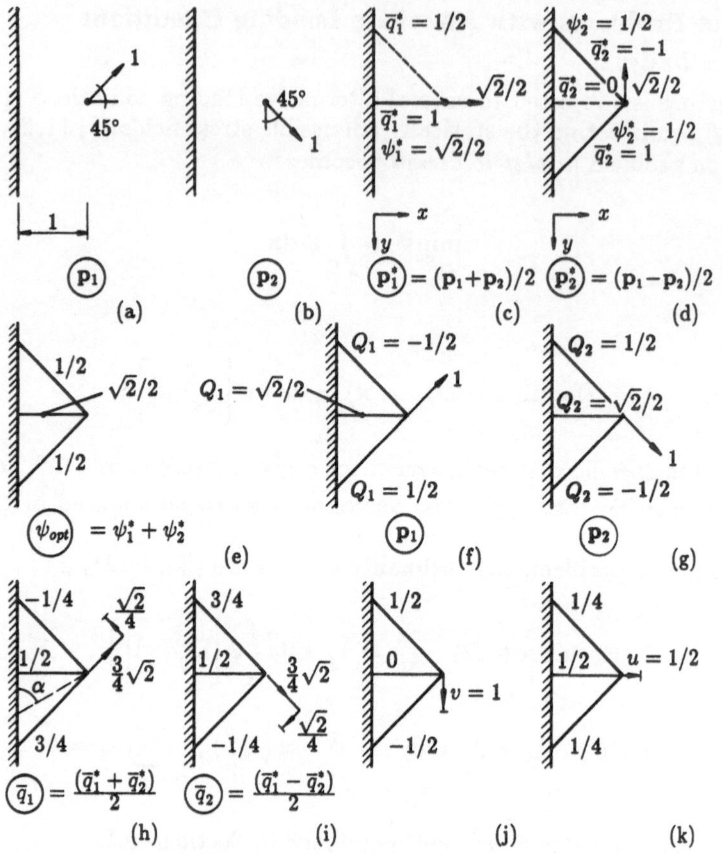

Fig. 52 Optimal layout for plastic design: two alternate loads.

Then we optimize the structure for $\mathbf{p}_1^*$ and $\mathbf{p}_2^*$ separately and add the corresponding specific costs ψ_1^* and ψ_2^*,

$$\psi = \psi_1^* + \psi_2^* . \tag{224}$$

The above specific cost (ψ) values represent the optimal solution for the two alternate loading conditions ($\mathbf{p}_1$ and $\mathbf{p}_2$). Denoting the adjoint strains of the optimal solutions for the component loads by $\overline{q}_1^*$ and $\overline{q}_2^*$, the adjoint strains for the original two loading conditions, satisfying (219)-(222), are given by

$$\overline{q}_1 = (\overline{q}_1^* + \overline{q}_2^*)/2, \quad \overline{q}_2 = (\overline{q}_1^* - \overline{q}_2^*)/2 . \tag{225}$$

The above superposition principle will be illustrated with a simple example. Determine the optimal truss layout for the two load conditions in Figs. 52a and b. The component loads and the corresponding optimal solutions are given in Figs. 52c and d, in which the adjoint displacements, respectively, are (with $k = 1$)

$$\overline{u}_1^* = x , \quad \overline{v}_1^* \equiv 0; \quad \overline{u}_2^* = 0 , \quad \overline{v}_2^* = 2x , \tag{226}$$

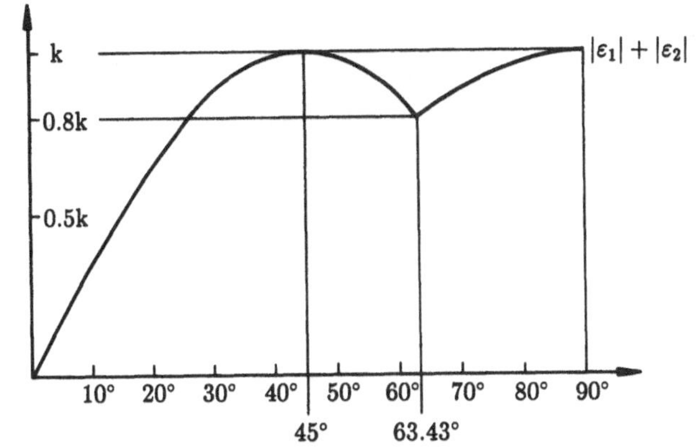

Fig. 53 Check on optimality condition (222).

where $\overline{u}$ and $\overline{v}$ denote displacements in the x and y directions. It can be readily checked that the displacements in (226) give the strains in Figs. 52c and d and that the latter satisfy the optimality conditions in (201) and (202) (Section 19.4). The superimposed optimal solution for the two alternate loads is shown in Fig. 52e and the member forces for the two load conditions in Figs. 52f and g. Note that only *statical* admissibility is required in plastic design. It can be seen from the above figures that all members are fully stressed ($k|Q_j| = \overline{\psi}$ with $k = 1$) for *both* loading conditions and hence optimality criterion (221) applies. The adjoint strains for the alternate loads are given in Figs. 52h and i, calculated on the basis of (225). Kinematic admissibility of the strains in Fig. 52i, for example, can be checked by superimposing the simple displacement fields in Figs. 52j and k. The strain fields in Figs. 52h and i satisfy the optimality criterion (221) with $\lambda = 1/4$, $\lambda = 1/2$ and $\lambda = 3/4$ for the top, middle and bottom members, respectively, and hence optimality of the considered solution is confirmed.

For layout optimality, it is still necessary to show that the strains for nonoptimal directions also satisfy conditions (222). For an arbitrary angle α, the strains become (Fig. 52h)

$$\varepsilon_1 = \frac{\sqrt{2}}{4} \sin\alpha[3\cos(135° - \alpha) - \cos(45° - \alpha)]\,,$$

$$\varepsilon_2 = \frac{\sqrt{2}}{4} \sin\alpha[\cos(135° - \alpha) + 3\cos(45° - \alpha)]\,. \tag{227}$$

By condition (222),

$$|\varepsilon_1| + |\varepsilon_2| \leq k\,, \tag{228}$$

for any value of α in an optimal solution. Indeed, relations (227) and (228) with $k = 1$ imply

$$\frac{\sin\alpha}{4}[|3(\sin\alpha - \cos\alpha) - (\sin\alpha + \cos\alpha)| + |3(\sin\alpha + \cos\alpha) - (\sin\alpha - \cos\alpha)|] \leq 1\,, \tag{229}$$

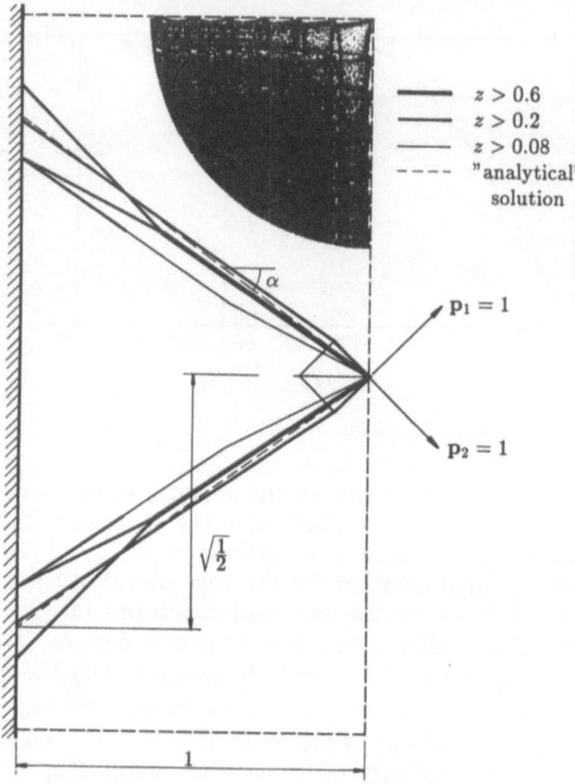

Fig. 54 Iterative COC solution for a compliance constraint with two alternate loading conditions and part of the structural universe.

which can readily be shown to be correct. The actual variation of the left-hand side of relation (229) is given in Fig. 53. It can be seen that $(|\varepsilon_1| + |\varepsilon_2|)$ only takes on the value k in the optimal directions and condition (228) is also satisfied for all other directions. The above problem was investigated in a paper in 1974 [63].

The superposition principle discussed here has been extended to an arbitrary number (2^n) of loading conditions by Rozvany and Hill [37].

26.2 Elastic Design for Compliance

For several loading conditions $(k = 1, 2, \ldots, \omega)$ and several displacement conditions $(j = 1, 2, \ldots v)$ without stress constraints, the optimality conditions change to:

$$\overline{\mathbf{q}}_k = \sum_j \nu_{jk} [\mathbf{F}] \overline{\mathbf{Q}}_j \quad (k = 1, 2, \ldots, \omega), \tag{230}$$

$$\mathcal{G}_{,\mathbf{z}}[\psi(\mathbf{z})] + \sum_j \sum_k \nu_{jk} \overline{\mathbf{Q}}_j^{S,K} \cdot \{\mathcal{G}[\mathbf{F}]\} \mathbf{Q}_k^{S,K} = 0\,, \tag{231}$$

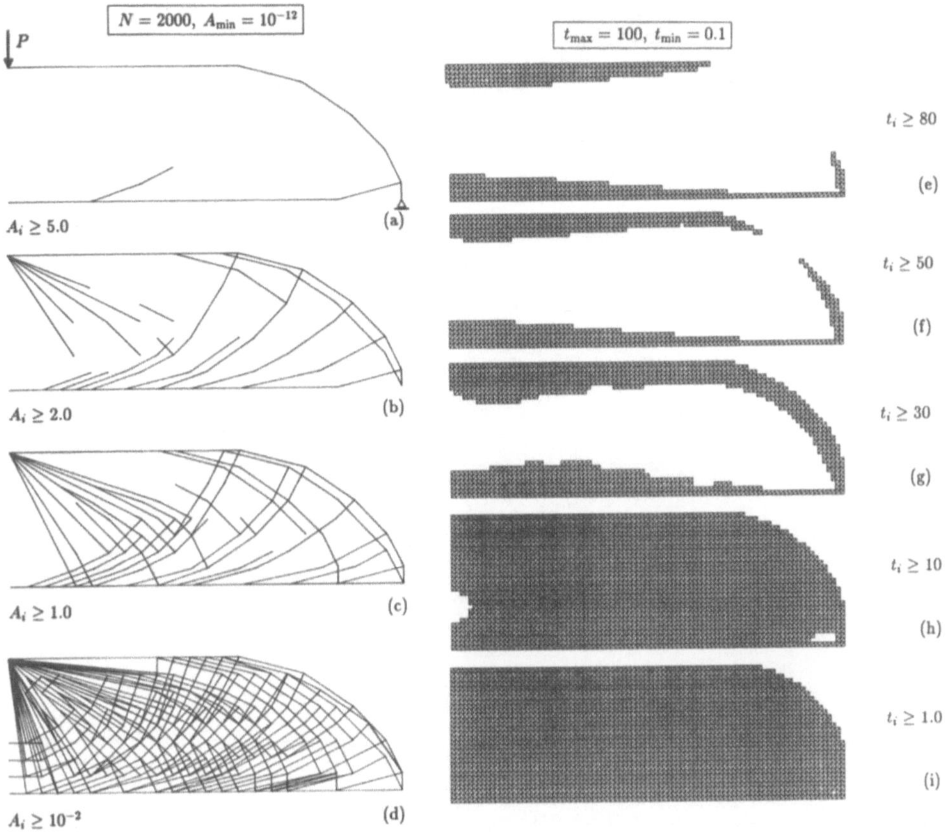

Fig. 55 A comparison of a truss layout and a plate of variable thickness obtained by the COC method.

where $\mathcal{G}$ is the generalized gradient mentioned in Section 19.1.

Considering the class of simple problems with $\psi = cz$, $[\mathbf{F}] = 1/rz$, and two compliance constraints [with scalar stresses $\overline{Q}_j \to Q_k$ in (230)] $\int_D (Q_k^2/rz)\,dx = C_k$ $(k = 1, 2)$, (230) and (231) reduce to

$$\overline{q}_k = \nu_{kk} Q_k / rz \quad (k = 1, 2)\,, \tag{232}$$

$$c - \sum_{k=1}^{2} \nu_{kk} Q_k^2 / rz = 0 \quad \Rightarrow \quad z = \sqrt{(\nu_{11} Q_1^2 + \nu_{22} Q_2^2)/rc}\,. \tag{233}$$

The above optimality criterion is difficult to handle in analytical derivations, but is highly suitable for an iterative COC procedure.

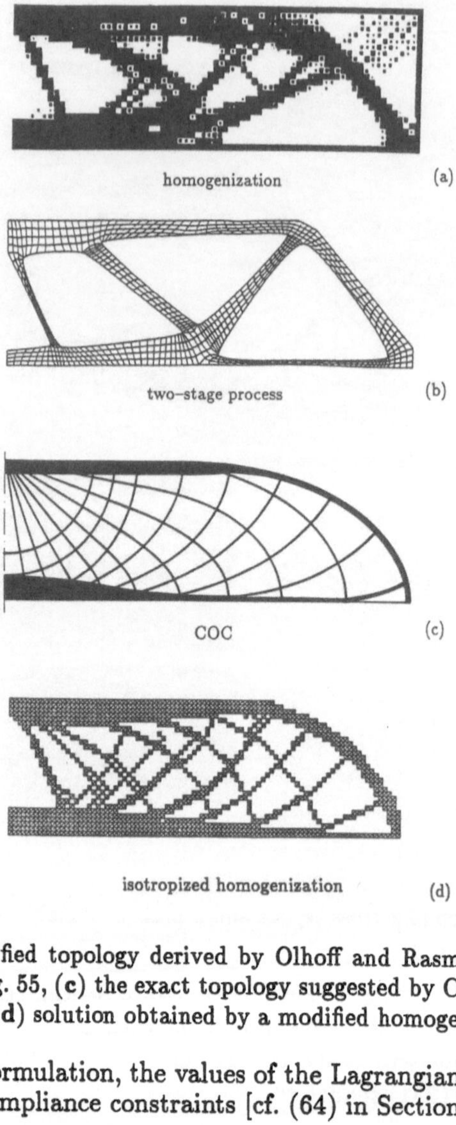

Fig. 56 (**a, b**) Simplified topology derived by Olhoff and Rasmussen for the problem treated in Fig. 55, (**c**) the exact topology suggested by COC layout solutions in Fig. 55, and (**d**) solution obtained by a modified homogenization method.

For a discretized formulation, the values of the Lagrangians ν_{11} and ν_{22} can be calculated from the compliance constraints [cf. (64) in Section 9.2]:

$$C_k = \sum_A \frac{Q_{ik}^2 \delta}{r\sqrt{(\nu_{11}Q_{i1}^2 + \nu_{22}Q_{i2}^2)/rc}} + \sum_P \frac{Q_{ik}^2 \delta}{r z_a} \qquad (k = 1, 2), \qquad (234)$$

where δ is the element length. Equations (234) can, in general, only be solved iteratively.

In the case of a symmetric boundary and two anti-symmetric load systems (as in Fig. 52), we have $\nu_{11} = \nu_{22} = \nu$. Moreover, we can carry out the analysis for one load condition only and denote by Q and $\hat{Q}$ the stresses in the corresponding elements of the two symmetric half-structures. Then (233) reduces to

$$z = \sqrt{\nu(Q^2 + \hat{Q}^2)/rc}\,. \tag{235}$$

If the entire structure has $2n$ elements and they are numbered in the same symmetric sequence for the two halves, then the discretized equivalent of (235) becomes

$$z_i = z_{i+n} = \sqrt{\nu(Q_i^2 + Q_{i+n}^2)/rc} \quad (i = 1, 2, \ldots, n)\,. \tag{236}$$

Moreover, for the considered symmetric problems with $C_1 = C_2$, (234) reduces to

$$\sqrt{\nu} = \sum_A \frac{Q_i^2 \delta}{\sqrt{r(Q_i^2 + Q_{i\pm n}^2)/c}} \Big/ \left(C - \sum_P \frac{Q_i^2 \delta}{r z_a}\right), \tag{237}$$

where in the subscript "$\pm$", we have "$+$" for $1 \le i \le n$ and "$-$" for $(n+1) \le i \le 2n$.

The problem in Figs. 52a and b was solved for an elastic compliance constraint using the above iterative COC procedure with 7170 potential bars. After 1500 iterations a total weight value of $\Phi = 3.49295726$ was obtained. The plot of bars having a cross-sectional area over $z = 0.08$ is shown in Fig. 54 (continuous lines), in which three different line thicknesses show various ranges of member sizes (z). The "structural universe" consisted of 11×21 grid-points and the connecting members were restricted to slopes of 0, 1:1, 1:2,..., 1:10, 2:3, 2:5, ..., 2:9 and their reciprocals (see part of the structural universe in the top right corner of Fig. 54).

The exact analytical solution for the above problem is, as yet, not known to the authors. However, if we assume a symmetric two-bar system (broken lines in Fig. 54), then the optimal solution within this topology can be determined easily. It can be shown from statical considerations that for $p_1 = 1$ the member forces are

$$N_1 = \frac{1}{2\sqrt{2}}\left(\frac{1}{\cos\alpha} + \frac{1}{\sin\alpha}\right), \quad N_2 = \frac{1}{2\sqrt{2}}\left(\frac{1}{\cos\alpha} - \frac{1}{\sin\alpha}\right), \tag{238}$$

and due to symmetry of the solution we have

$$C = 1 = \frac{N_1^2}{A}L + \frac{N_2^2}{A}L = \frac{1}{A\cos\alpha\sin^2(2\alpha)}, \tag{239}$$

where A is the cross-sectional area and L is the member length. The relation (239) implies:

$$A = \frac{1}{\cos\alpha\sin^2(2\alpha)}, \quad \Phi = 2AL = \frac{2}{\cos^2\alpha\sin^2(2\alpha)}\,. \tag{240}$$

Then the usual stationarity condition ($\mathrm{d}\Phi/\mathrm{d}\alpha = 0$) implies

$$2\cos\alpha\sin\alpha\sin^2(2\alpha) - 4\cos^2\alpha\sin(2\alpha)\cos(2\alpha) = 0\,, \qquad \tan^2\alpha = 1/2\,,$$

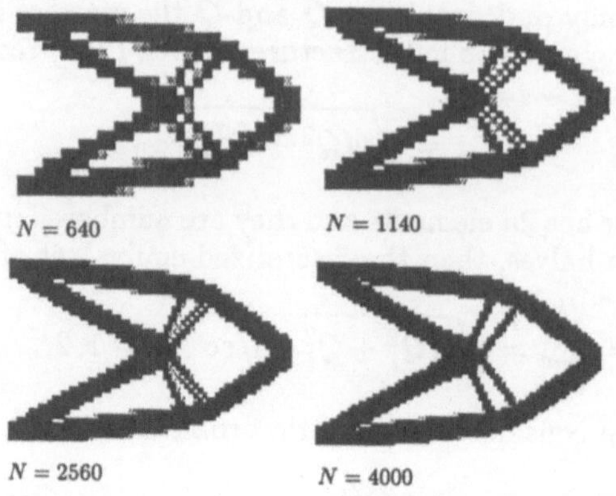

Fig. 57 Simplified topology obtained by Kikuchi [67] for the problem treated in Fig. 49.

$$\Phi = \frac{27}{8} = 3.375\,, \quad A = \frac{27}{16\sqrt{1.5}} = 1.37783798\,, \quad \alpha = 35.26438968^\circ\,. \qquad (241)$$

The optimal two-bar system, having a slope of $\tan\alpha = 1/\sqrt{2}$ is shown in broken line in Fig. 54. Since this system has 3.495% lower weight than the COC solution with 7170 potential members, it could be the absolute optimal layout. It can be seen from Fig. 54 that the COC procedure, within the limited range of admissible member slopes, is trying to achieve this solution.

Quite recently, another iterative COC calculation with 12202 members (an 11×21 grid as before, but with all possible connecting members within each half-grid) gave a weight of 3.375668, which is only 0.0198% above the assumed analytical solution. This numerical solution consisted of two heavy bars in the vicinity of the broken lines in Fig. 54, which confirms the assumed analytical solution.

Note. It can be seen from Sections 26.1 and 26.2 that unlike for a single load condition, *the solutions for plastic design and elastic compliance design differ significantly if several alternate loads are considered.*

The authors were informed recently that in an unpublished symposium paper Bendsøe and Ben-Tal [64], who used a different method, also presented optimal truss layouts involving several thousand potential members.

27. Generalized Shape Optimization

Before discussing the implications of layout optimization in shape optimization, we discuss one more application of the iterative COC method, concerning a simply supported truss with a central point load. Figures 55a-d show one half of the optimal truss for various ranges of cross-sectional areas. For a comparison, Figs. 55e-i show COC solutions for a plate of variable thickness (t) with the same support and load conditions.

It was mentioned in Section 18 that in generalized shape optimization one type

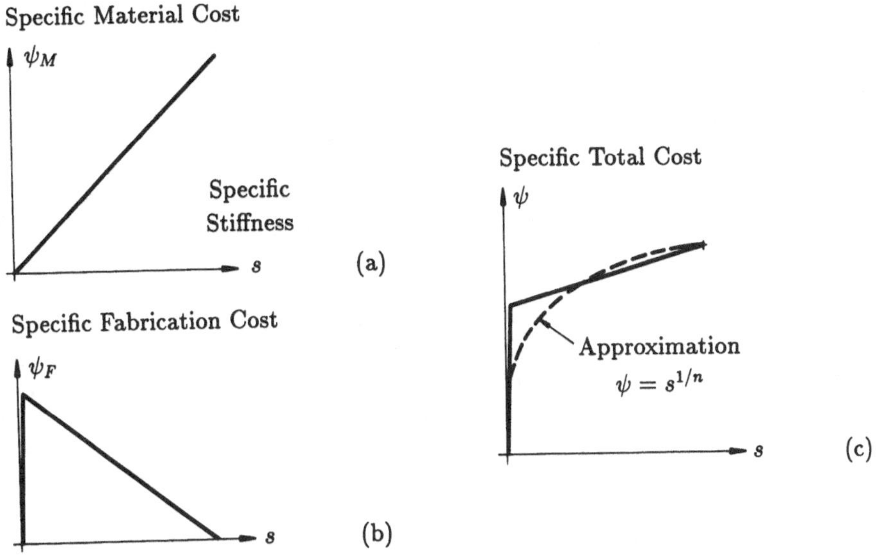

Fig. 58 Suppression of "grey" regions in isotropized solutions.

of region, termed ***perforated (grey) region***, contains a fine system of cavities or, theoretically, an infinite number of internal boundaries. A similar result was obtained in plate optimization where optimal solid and perforated plates were found ([58, 65], see also Section 22.2 herein) to contain systems of ribs of infinitesimal spacing and hence the thickness function has an infinite number of discontinuities over a finite width. Although the material in these problems is isotropic, the "homogenization" method consists of replacing ribbed or fibrous elements with homogeneous but anisotropic elements whose stiffness or strength is direction- but not location-dependent within the element (e.g. [65]).

From a historical point of view, the basic idea of homogenization was introduced already in the mid-seventies by Prager and the first author (e.g. [3, 31]), although they used the terms "grillage-like continua" and "truss-like continua". In these structures, a theoretically infinite number of bars or beams occur over a unit width but the above authors replaced this system with a continuum, whose specific cost, stiffness and strength depended on the "lumped" width of the bars or beams over a unit width. This concept was clearly equivalent to homogenization, but in an engineering rather than a mathematical context. Applications of homogenization in generalized shape optimization were discussed in pioneering contributions by Bendsøe (e.g. [66]) which represent one of the most important recent developments in structural optimization.

The main aim of the investigations by Bendsøe [66] and Kikuchi [67] is to come up with a practical topology, in which the perforated ("grey") areas disappear and the optimal structure consists of solid ("black") and empty ("white") regions only. This procedure can be seen from Figs. 56a and b, in which Olhoff, Bendsøe and Rasmussen [68] first determined the approximate optimal topology (Fig. 56a) for the support condition and loading considered in Fig. 55, and then carried out a separate shape optimization (Fig. 56b) for the more detailed design conditions. This "homogenization" method, indeed, gives negligibly small perforated (grey) areas, as can be observed in Fig. 56a, and also in Fig. 57 which was obtained by Kikuchi [67] for the load and support conditions in Fig. 49. The latter seems to give the same topology, irrespective of the number of elements (N) employed.

The following circumstantial evidence seems to indicate that the *exact* optimal topologies differ from those obtained by the homogenization method (e.g. Figs. 56b, 57):

- In the analytical solutions obtained for perforated plates [58], a high proportion of the plate area is covered by perforated (grey) areas.
- It was also shown previously [58] that the solution for very low volume fractions tends to that for grid-type structures (Michell frames or least-weight grillages). This was also observed by Prager who commented on some optimal solid plate designs by Cheng and Olhoff [52], see also Fig. 45 herein. It was also noted [51] that, as the volume fraction increased progressively in analytical solutions, solid (black) regions developed in areas where the ribs in the perforated regions had the greatest density at lower volume fraction levels. Making use of these observations, the grid-type solutions in Figs. 55a-d imply the topology in Fig. 56c (graphics by Dr. W. Gollub), in which the width of the solid regions is based on the cross-sectional areas of the "concentrated" bars along the top and bottom chords of the truss. The spacing of the members in the perforated region is theoretically infinitesimal, but a finite number of members would have to be used in any practical solution. In the neighbourhood of the top right corner, we have an empty (white) region. It can also be observed that the "homogenized" solution in Fig. 56a tries to achieve the solution in Fig. 56c, except that in areas of low rib density (e.g. right bottom region inside the chord) it comes up with empty regions. As Bendsøe pointed out at a recent meeting (Karlsruhe, 1990), this can be attributed to the fact that here non-optimal microstructures were used for the perforated (grey) regions. Whilst this does not change the cost of solid (black) and empty (white) regions, it does increase artificially the cost of perforated (grey) regions and hence it tends to suppress the latter.
- The solutions for plates of varying thickness represent an "isotropized" version of the exact solution. This can be observed by comparing Figs. 55e-i with Fig. 56c. The plate thickness in the former is roughly proportional to the average material density over the latter, with ribs occurring in Figs. 55e-i along the solid regions of Fig. 56c. This is a further confirmation of the improved topology in Fig. 56c. Moreover, a modified, isotropized homogenization method (Section 27.1) largely confirmed the solution in Fig. 56c, as can be seen from Fig. 56d.

Remark. The contention that existing homogenization methods give a *simplified* topology compared to the exact solution *by no means represents a criticism of these extremely important techniques.* Such "condensed" topologies are in fact very practical because, naturally, it is impossible to use an infinitesimal bar spacing in real structures. However, the exact optimal topology can also be of practical significance, because the client could be told by the designer that further weight savings can always be achieved by increasing the number of "holes" in the design and then former could decide as to how far he can go within realistic manufacturing capabilities.

27.1 An Alternative Method for Suppressing Perforated (Grey) Regions

Since the use of ***non-optimal*** microstructures homogenized into an ***anisotropic*** continuum introduces some ***unknown*** penalty for "grey" regions into shape optimization, the perforated (grey) regions could also be suppressed by using an ***isotropic*** microstructure but with a suitable penalty function for such regions. This can also be justified on practical grounds, as can be seen from the argument that follows.

As a first approximation, we could assume that the specific material cost (i.e. weight, ψ) is roughly proportional to the specific stiffness (s) of perforated regions (Fig. 58a, which is also valid for plates of variable thickness). On the other hand, the extra manufacturing cost of cavities would increase with the size of the cavities if we consider a casting process requiring some sort of formwork for the cavities (Fig. 58b). Note that for empty (white) macroscopic regions with $s = 0$ the manufacturing cost also becomes zero. The specific total cost and its suitable approximation is shown in Fig. 58c. The use of the above type of cost function promotes the suppression of perforated (grey) areas in isotropized designs which require *only one design parameter* (s) per element. The introduction of orthogonal cavities in the usual homogenization process [66, 67] requires 3 design variables per element for two-dimensional systems and 6 variables for three-dimensional ones. The solution in Fig. 56d was obtained with an n-value of 1.86 in Fig. 58c, and represents a topology closer to the "exact" optimal design than the design in Fig. 56b. As expected, simpler topologies can be obtained by adopting a higher n-value (i.e. by increasing the penalty for grey regions).

28. Current Research

The ongoing research by the authors and their research associates includes the following projects:

(a) Development of software for practical applications of the COC/DCOC methods.

(b) Extension of the above developments to
  - (i) simultaneous stress, deflection, system buckling and dynamic constraints;
  - (ii) multiple deflection constraints and multiple load conditions;
  - (iii) the treatment of complex cross-sectional topologies (e.g. I-sections with up to four variable dimensions).

(c) Topology and generalized shape optimization of two and three-dimensional systems for combined constraints [see (b)(i) above].

(d) Theoretical work showing the exact equivalence of DCOC methods and COC methods for "segmented" structures

(e) Extensions of DCOC to temperature stresses, support settlement, support costs and such control-parameters as variable prestrains or variable loading.

A relatively simple current sub-project concerns truss optimization for stress, deflection and member-buckling (Euler-buckling) constraints. The relation between member force F_i and cross-sectional area for this problem is shown in Fig. 59, in

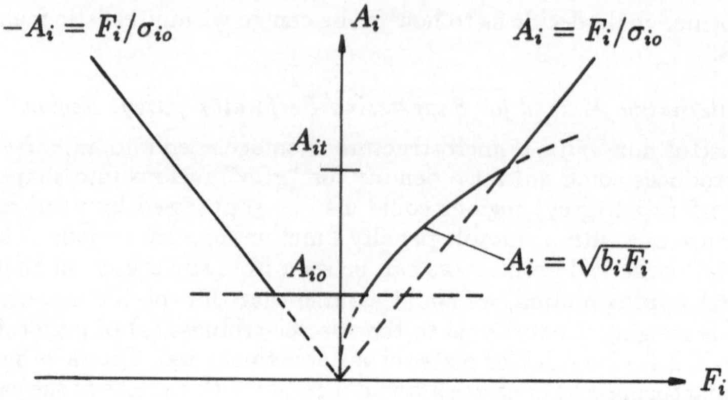

Fig. 59 Force/cross-sectional area relation for trusses with member buckling.

which A_{i0} is the prescribed minimum cross-sectional area, b_i is a given constant, and positive forces represent compression.

29. Conclusions

- The iterative COC method enables us, probably for the first time in the history of structural optimization
  - (a) to optimize *simultaneously* the topology and geometry of grid-type structures (trusses, grillages, shell-grids, etc.),
  - (b) for a compliance constraint (with future extensions to any combination of the usual design conditions such as stress, displacement, natural frequency, stability, etc. constraints),
  - (c) using a fully automatic method capable of handling many thousand potential members.
- With a suitable penalty formulation, COC methods can also be used for selectively suppressing perforated (grey) regions in generalized shape optimization.

Acknowledgements (Lectures by Rozvany and Zhou)

The authors are indebted to A. Fischer and S. Rozvany for the preparation of the manuscript in TEX and to the latter for editing of the text under extreme time pressure, to E. Becker and S. Liebermann for the preparation of the diagrams and to the Deutsche Forschungsgemeinschaft for financial support. The results described in these lectures represent joint work, except for Part II of the second lecture (discretized formulation of COC) which is due to M. Zhou.

DESIGN CONSIDERATIONS IN THE OPTIMIZATION OF STRUCTURAL TOPOLOGIES

U. KIRSCH*
and
G.I.N. ROZVANY
FB 10, Essen University, D-4300 Essen 1, Germany

Abstract. Some design considerations related to the optimization of structural topologies are discussed. The basic difficulties involved in the solution process are first reviewed and the ground structure approach is discussed. Formulating the problem as a nonlinear programming problem, the computational effort involved in the solution process might become prohibitive and solutions with a singular optimal topology might be encountered. Neglecting the implicit compatibility equations, an explicit approximate problem is obtained. However, the approximate displacement constraints might be meaningless in cases where members are eliminated from the structure. If the latter constraints are neglected, the resulting problem can be stated as linear programming under certain assumptions. The difficulties involved in topological optimization suggest that a two-stage design procedure might be useful. Several such procedures, where an approximate solution is evaluated at the first stage and modified at the second stage, are presented. It is shown that the first-stage solution can be viewed as the final optimum if a set of control forces or displacements is applied. The latter can readily be determined at the second stage by solving a linear programming problem.

1. Introduction

1.1 Problem Statement

Two classes of structures are often considered in layout optimization [1]:

a. *Grid-like structures* idealized as two-dimensional continua and treated by analytical methods (truss-like continua, grillage-like continua, etc.). This class of structures has been discussed in detail by Rozvany in other lectures at this meeting as well as in publications (e.g. [2]) and hence they will not be considered in this lecture.

b. *Discrete-skeletal structures* (trusses, grillages, etc.) usually idealized as one-dimensional continua and solved by numerical methods. The design considerations presented here are related to this class of structures.

It should be added however that the continuum approach under (a) above may also result in a small number of members in an optimal layout. In that case, the

* (Visiting Professor, on leave from Technion-Israel Institute of Technology)

G. I. N. Rozvany (ed.), Optimization of Large Structural Systems, Vol. I, 121–138.

above two approaches yield identical results.

The form of a discrete structure can usually be described by three types of design variables:

a. The *topological variables*, defining the pattern of connection of members or the number and spatial sequence of elements, joints, and supports in a structural system.

b. The *geometrical* or *configurational variables*, describing the structural geometry (nodal coordinates).

c. The *sizing variables*, representing the cross-sectional dimensions or member sizes.

Both the topological and the geometrical variables define the *layout* of the structure. While sizing variables may be optimized under either fixed or variable layout, the optimization of the layout must always be accompanied by sizing optimization. This necessity is due to changes in the analysis equations resulting from the change in structural layout.

Most of the work done on optimum structural design is related to optimization of cross-sections. Much less effort has been devoted to the optimization of the geometry and topology [1, 3]. It is recognized, however, that optimization of the structural layout (geometry and topology) can greatly improve a design. This is because, potential savings affected by layout optimization are generally more significant than those resulting from fixed-layout optimization.

Owing to the complexity of simultaneous optimization of the geometry, topology and cross-sections, two classes of problems are often considered in the optimization of the structural layout [4]:

a. *Geometrical optimization*, where coordinates of the joints and cross-sectional areas are treated as design variables and optimized simultaneously. In general, the design variables are assumed to be continuous and numerical search algorithms are used to find the optimum. The topology is usually assumed to be fixed, unless some of the joints coalesce during the solution process. The geometrical optimization problem is not considered in this lecture.

b. *Topological optimization*, where members are removed from a highly connected structure, called the *ground structure* (or structural universe [2]), to derive an optimum topology with the corresponding cross-sections. This means that in a topological optimization problem both the topological and the sizing variables are optimized simultaneously. The fact that topological optimization did not enjoy the same degree of progress as fixed-layout optimization may be attributed to some basic difficulties involved in the solution process [1]. The present article deals with some design considerations associated with topological optimization.

Most studies on optimal topologies deal with truss structures. This may be attributed to the fact that the truss by its very nature is most suitable for optimization of the topology. It possesses usually many nodes and elements that can be deleted or retained without affecting the functional requirements. In addition, the truss is a relatively simple, yet nontrivial, structure. It is therefore an ideal system for the investigation of some properties and characteristics associated with optimal topologies.

Whilst members of *trusses* have in general a constant cross-section along their entire length, *grillages* (beam grids) are not subject to this restriction and are therefore also highly suitable for demonstrating certain general aspects of layout problems, particularly for a continuum-type formulation [2]. However, the present lecture will be restricted to examples involving trusses.

While the displacement (stiffness) formulation is the prevalent structural analysis tool in current computational practice, the force (flexibility) formulation is adopted in many topological optimization studies due to several reasons:

a. The effect of some approximations can be studied directly.

b. The analysis model is convenient for investigating properties of optimal topologies.

c. A linear programming formulation can be obtained under certain assumptions.

The difficulties involved in topological optimization problems, make these problems perhaps the most challenging in structural optimization. One difficulty is that the structural model is itself allowed to vary during the design. Discretized structures are generally characterized by the fact that the finite element model of the structure is not modified during the optimization process. In topological design, however, both the finite element model and the set of design variables change during the design process when members are added or deleted. These phenomena greatly complicate the design and analysis interactions.

Another difficulty is that the number of possible element-joint connectivities grows dramatically as the number of possible joint locations is increased. This number might be very large particularly in structures of practical size. One solution to this problem is to specify a reduced set that does not include all possible element-joint connectivities. However, a fundamental disadvantage of this approach is that the optimal solution may not be included in the specified set.

An additional difficulty that might be encountered during elimination of members is that the problem can have singular global optima that cannot be reached by assuming a continuous set of variables. This suggests that it may be necessary to represent some design variables as integer variables and to declare the existence or absence of a structural element. While methods for mixed integer programming have been developed, these methods are still computationally costly for practical engineering applications.

The above-mentioned considerations have led to various approximations and simplifications in the problem formulation. These include:

— approximate analysis models (e.g. rigid plastic);

— consideration of only certain constraints (e.g. stress constraints);

— simplified objective function (e.g. weight);

— consideration of simple structural systems (e.g. trusses);

— simplified sizing variables (e.g. cross-sectional areas);

— consideration of a limited number of loadings.

1.2 The Ground Structure

Most topological optimization studies are based on the assumption of *an initial ground structure* that contains many joints and members connecting them. The analytical school has used for the same concept the term "structural universe" (e.g. [2]). Member areas are allowed to reach zero and hence can be deleted automatically from the ground structure. This permits elimination of uneconomical members during the optimization process [5–7].

Assume a grid of points that may be connected by many potential truss members to form a ground structure with a finite number of members. The objective function is assumed to be a linear function of the cross-sectional areas **X** and the constraints are related to stresses, displacements and bounds on **X**. Considering the force (flexibility) method of analysis, the topological optimization problem can be stated as the following nonlinear programming (NLP) problem: Find **X** such that

$$Z = \boldsymbol{\ell}^T\mathbf{X} \longrightarrow \min$$

$$\begin{aligned} \boldsymbol{\sigma}^L\mathbf{X} \le \mathbf{A}_p + \mathbf{A}_N\mathbf{N} \le \boldsymbol{\sigma}^U\mathbf{X} \qquad &\text{(stresses)} \\ \mathbf{D}^L \le \mathbf{D}_p + \mathbf{D}_N\mathbf{N} \le \mathbf{D}^U \qquad &\text{(displacements)} \\ \mathbf{X}^L \le \mathbf{X} \le \mathbf{X}^U \qquad &\text{(side constraints)} \end{aligned} \tag{1}$$

Here **N** is the vector of redundant forces; $\mathbf{A}_p$, $\mathbf{D}_p$, $\mathbf{A}_N$ and $\mathbf{D}_N$, are forces and displacements due to external loads and unit value of redundants, respectively, in the the primary structure; $\boldsymbol{\ell}$ is a vector of constants; $\boldsymbol{\sigma}$ and **D** are stresses and displacements, respectively; and L and U are superscripts denoting lower and upper bounds, respectively. The redundant forces are computed for any assumed design **X** by the implicit compatibility equations

$$\mathbf{F}\mathbf{N} = \boldsymbol{\delta} \tag{2}$$

where **F** is the flexibiliy matrix and $\boldsymbol{\delta}$ is the vector of displacements corresponding to redundants in the primary structure. In this formulation the elements of $\boldsymbol{\ell}$, $\boldsymbol{\sigma}^L$, $\boldsymbol{\sigma}^U$, $\mathbf{D}^L$, $\mathbf{D}^U$, $\mathbf{X}^L$, $\mathbf{X}^U$, $\mathbf{A}_p$ and $\mathbf{A}_N$ are constant. The elements of $\mathbf{D}_p$, $\mathbf{D}_N$, **F** and $\boldsymbol{\delta}$ are functions of **X**.

Neglecting temporarily the displacement constraints and the compatibility conditions (2), the problem can be stated in the following linear programming (LP) form in terms of *both* **X** *and* **N** *as independent variables:*

$$Z = \boldsymbol{\ell}^T\mathbf{X} \longrightarrow \min$$

$$\begin{aligned} \boldsymbol{\sigma}^L\mathbf{X} \le \mathbf{A}_p + \mathbf{A}_N\mathbf{N} \le \boldsymbol{\sigma}^U\mathbf{X} \\ \mathbf{X}^L \le \mathbf{X} \le \mathbf{X}^{\mathrm{U}} \end{aligned} \tag{3}$$

Since both $\mathbf{A}_p$ and $\mathbf{A}_N$ satisfy equilibrium conditions, any selection of $\mathbf{N}$ will give member forces satisfying these conditions. Assuming zero lower bounds on cross-sections

$$\mathbf{X}^L = \mathbf{O} \tag{4}$$

the LP method has the ability to make unnecessary members vanish from the structure to obtain the minimum weight design. The optimal structure in such cases will satisfy equilibrium and stress constraints, but it might not satisfy compatibility conditions or may represent unstable configuration under a general loading. In such cases the LP solution is not the final optimum and some modifications might be required. However this solution may be viewed as a lower bound on the optimum.

One advantage of the LP formulation is that the global optimum is reached in a finite number of steps. Thus, large structures with many members and joints can efficiently be solved. In addition, problems of singular optimal topologies are eliminated and some properties of the optimal solution can be studied directly. This subject will be discussed later in this article.

2. Types of Optimal Topologies

Assuming an arbitrary initial ground structure, the resulting optimal topology might represent one of the following classes of structure:

— statically determinate structure (SDS),

— unstable structure (USS),

— statically indeterminate structure (SIS).

Elimination of elements from the ground structure will change the numbers of variables and active stress constraints at the optimum. If buckling constraints are considered, the areas of compression elements might not converge to zero. In a complete nonlinear programming formulation, as the area of a compression element decreases, the buckling stress also decreases, until it becomes critical; then the area of the element will increase.

2.1 Statically Determinate Structures

Solving the LP for zero lower bounds on cross-sections, it is possible that the resulting optimal design will represent an SDS. That is, for a certain selection of redundant forces one can obtain at the optimum $\mathbf{N} = \mathbf{0}$. In this case the element forces $\mathbf{A}$ are constant, compatibility conditions can always be satisfied and the LP solution is the final optimum. (The LP solution must be modified only in cases where displacement constraints are not satisfied). It has been shown [5,16] that for structures subjected to a single loading condition with $\mathbf{X}^L = \mathbf{0}$, the optimal solution indeed will represent an SDS. This result, while not guaranteed, might be obtained also for structures subjected to multiple loading conditions.

2.2 Unstable Structures

The number of members eliminated during the LP solution might be larger than the degree of statical indeterminacy and the optimal topology might represent an

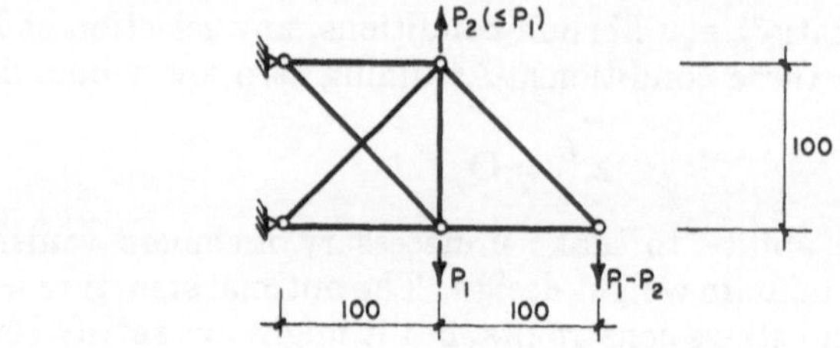

Fig. 1 Seven-bar truss, ground structure.

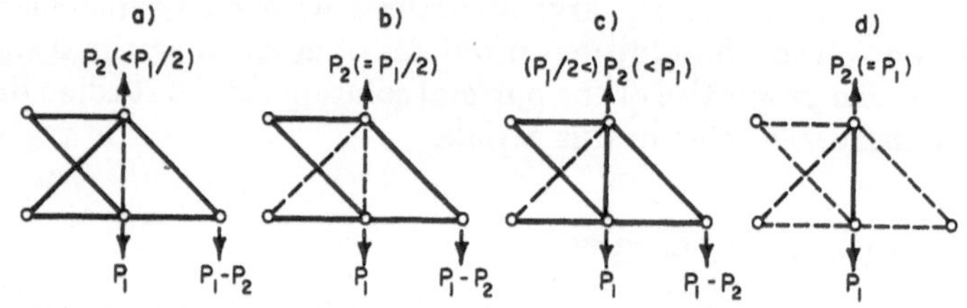

Fig. 2 Optimal topologies, seven-bar truss.

USS. The optimal structure in this case satisfies equilibrium and stress constraints, but there are unstable members. This situation may occur for certain geometries or loading conditions, where the forces in some elements change from tension to compression or vice versa. These elements are not required to maintain equilibrium for that particular geometry or loading condition, and will be eliminated during the LP solution. A possible solution to this problem is to add members to the optimal configuration. They are not needed to satisfy equilibrium, but they are required to satisfy the necessary relationship which exists between the joints and the members in a stable structure.

To illustrate situations where USS are obtained, consider the seven-bar ground structure shown in Fig. 1. The truss is subjected to three loads, P_1, P_2 and $P_1 - P_2$, acting simultaneously. Optimal topologies for various P_1/P_2 ratios are shown in Fig. 2. The main observations that can be made are as follows [8]:

— For $P_2 < P_1/2$ and $P_1/2 < P_2 < P_1$, the optimal topologies are SDS (Figs. 2(a) and 2(c)).

— For $P_2 = P_1/2$ (Fig. 2(b)) the optimal topology is a USS that can be viewed as a particular case of the former two topologies. The diagonal element (Fig. 2(a)) or the vertical element (Fig. 2(c)) become zero for this particular load ratio.

— For $P_2 = P_1$ (Fig. 2(d)) the optimal topology is reduced to a single unstable vertical element.

These results illustrate how unstable topologies might be obtained for particular loading cases.

2.3 Statically Indeterminate Structures

In cases where the optimal LP solution $\overset{*}{\mathbf{X}}_{\mathrm{LP}}$, $\overset{*}{\mathbf{N}}_{\mathrm{LP}}$, represents an SIS, compatibility conditions might not be satisfied. The vector Δ_{LP}, defined by

$$\Delta_{\mathrm{LP}} \equiv \overset{*}{\mathbf{F}}_{\mathrm{LP}} \overset{*}{\mathbf{N}}_{\mathrm{LP}} - \overset{*}{\delta}_{\mathrm{LP}} \tag{5}$$

indicates the discrepancy in satisfying the compatibility conditions by the optimal LP solutions. The asterisks and subscripts LP in (5) denote optimal values of the LP problem. For certain geometries or loading conditions it is possible that

$$\Delta_{\mathrm{LP}} = \mathbf{0} \tag{6}$$

In this case compatibility conditions are satisfied for the optimal SIS and the LP solution is the final optimum.

If the condition (6) is not satisfied for the optimal SIS, it is still possible to maintain compatibility by applying a set of prestressing forces [9]. This can always be done for structures subjected to a single loading condition. In the case of multiple loading conditions, a single set of prestressing forces does not ensure that compatibility conditions will be satisfied at the optimum for all loading conditions.

Analysis of the optimal LP structure will give the corresponding elastic force distribution $\overset{*}{\mathbf{N}}_{\mathrm{EL}}$. A certain deviation from elastic force distribution is often allowed on account of the inelastic behavior. The allowable deviation can be defined by the set of linear constraints

$$\boldsymbol{\alpha}^{L} \overset{*}{\mathbf{N}}_{\mathrm{EL}} \leq \boldsymbol{\alpha} \overset{*}{\mathbf{N}}_{\mathrm{LP}} \leq \boldsymbol{\alpha}^{U} \overset{*}{\mathbf{N}}_{\mathrm{EL}} \tag{7}$$

where $\boldsymbol{\alpha}, \boldsymbol{\alpha}^{L}$ and $\boldsymbol{\alpha}^{U}$ are matrices of given parameters. In cases where the constraints [7] are not satisfied for any loading condition, modifications of the LP optimum are required. Several procedures have been proposed for this purpose [1,7,10]. Alternatively, it is possible to maintain the LP optimum by applying control displacements, as will be shown in Section 4.2.

2.4 Multiple Optimal Topologies

Solving the LP problem, it is possible that for certain geometries identical optimal objective function values are obtained for an infinite number of force distributions in the structure. The various optimal force distributions usually correspond to several different optimal topologies [4]. The LP problem has in this case several basic optimal solutions, each corresponds to a different topology. Additional optimal topologies are obtained by linear combinations of the basic optimal solutions. One important property of such geometries is that new optimal topologies are introduced from existing basic optimal topologies by combination rather than the common ground structure approach of elimination.

To illustrate multiple optimal topologies, assume the fifteen-bar truss shown in Fig. 3 as a ground structure. The truss is subjected to three loads P acting

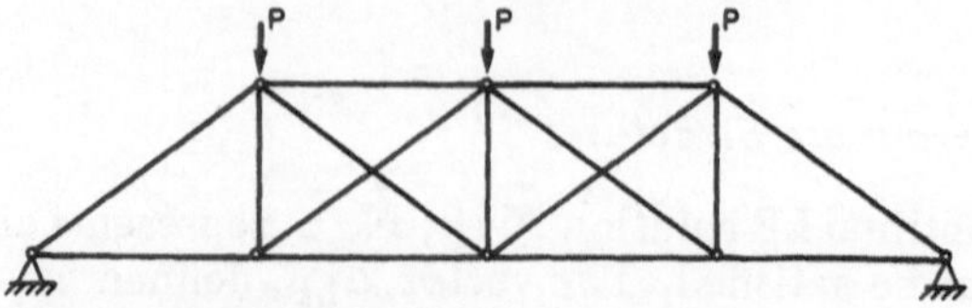

Fig. 3 Fifteen-bar truss, ground structure.

simultaneously and only stress constraints are considered. Five SDS basic optimal topologies are obtained for any assumed depth (Fig. 4(a)), four of which being USS and two nonsymmetric structures. These can be used to introduce an additional fourteen SIS combined optimal topologies (Fig. 4(b)), five of which are also USS. It has been found [4] that compatibility conditions are satisfied for the SIS combined optimal topologies.

Multiple optimal topologies play an important role in the optimal layout of trusses and grillages, where in certain regions (termed S-regions) any bar or beam direction gives the same optimal total weight for the structure (e.g. [2], pp. 329, 334)

3 Singular Optimal Topologies

One reason for excluding compatibility conditions from the optimization of the ground structure layout is that the computational effort in the solution process is considerably reduced. An additional difficulty in solving the NLP problem is that the optimal topology might correspond to a singular point in the design space. This occurs since a change in the structural topology (elimination of one or more elements) will result in a corresponding modification of the compatibility equations and elimination of some constraints previously included in the problem statement. That is, a change in the structural topology might change the design space. If the optimal solution is a singular point in the design space, it might be difficult or even impossible to arrive at the true optimum by numerical search algorithms. The singularity of the optimal topology in truss structures was first shown by Sved and Ginos [11]. Singular optima of grillages have been studied later [12,13].

To illustrate the effect of neglecting the compatability conditions on singular optimal topologies, consider two cases of stress and displacement constraints:

a. The accurate constraints in the complete NLP formulation, obtained by substituting **N** from Eq. (2) into Eq. (1)

$$\boldsymbol{\sigma}^L \mathbf{X} \leq \mathbf{A}_p + \mathbf{A}_N \mathbf{F}^{-1} \boldsymbol{\delta} \leq \boldsymbol{\sigma}^U \mathbf{X} \tag{8a}$$

$$\mathbf{D}^L \leq \mathbf{D}_p + \mathbf{D}_N \mathbf{F}^{-1} \boldsymbol{\delta} \leq \mathbf{D}^U \tag{8b}$$

b. The approximate constraints in the LP formulation, where the compatability conditions (2) are neglected and both **X** and **N** are treated as independent

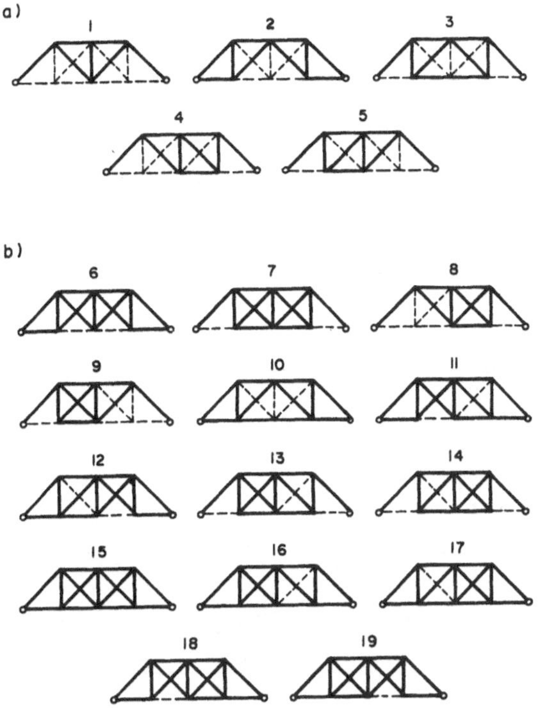

Fig. 4 Optimal topologies, fifteen-bar truss example.

variables

$$\sigma^L \mathbf{X} \leq \mathbf{A}_p + \mathbf{A}_N \mathbf{N} \leq \sigma^U \mathbf{X} \tag{9a}$$

$$\mathbf{D}^L \leq \mathbf{D}_p + \mathbf{D}_N \mathbf{N} \leq \mathbf{D}^U \tag{9b}$$

Assuming the accurate stress constraints (8a), elimination of a member might change the design space. Considering the approximate stress constraints (9a) on the other hand, problems of discontinuity due to deletion of members are eliminated. That is, in the case of linear stress constraints the design space is not changed if any cross-section is equal to zero.

Assuming the accurate displacement constraints (8b) and decreasing the cross-section of a specific member, its relative stiffness and internal force are reduced. Consequently, its contribution to the displacement expression might approach zero. That is, deletion of a member will not necessarily result in discontinuity or singular solutions similar to those obtained in the case of stress constraints. Considering the approximate displacement constraints (9b), the result might be meaningless in cases where members are eliminated from the structure [14].

The singularity phenomenon is illustrated for the three bar-truss shown in Fig. 5. The design variables are the cross-sectional areas of member 1 ($\mathbf{X}_1$) and members

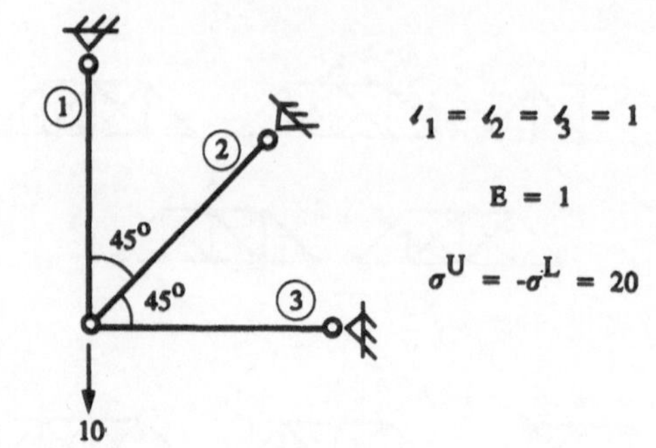

Fig. 5 Three-bar truss.

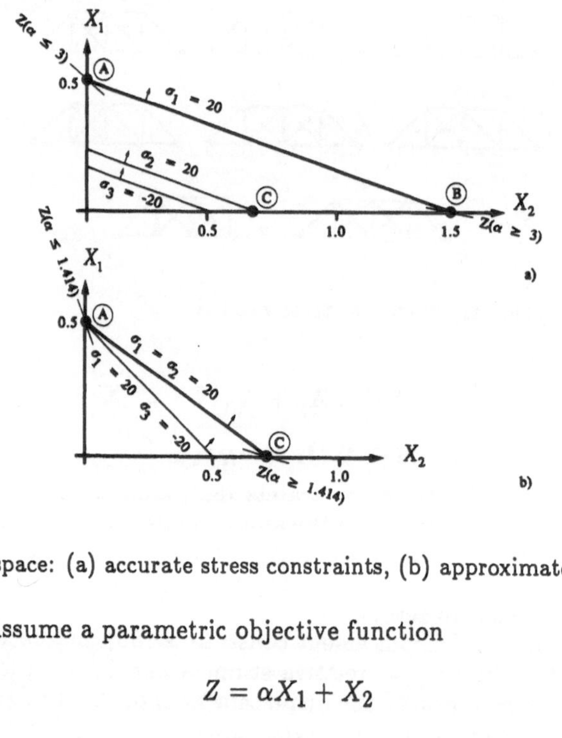

Fig. 6 Design space: (a) accurate stress constraints, (b) approximate stress constraints.

2 and 3 (X_2). Assume a parametric objective function

$$Z = \alpha X_1 + X_2 \tag{10}$$

where α is a non-negative parameter. Design spaces for the accurate and approximate stress constraints are shown in Fig. 6. Assuming first the accurate constraints and a numerical search procedure, the solution process will converge to the following points (Fig. 6(a))

$$\text{Point A}: X_1 = 0.5, \;\; X_2 = 0 \quad (\alpha \leq 3)$$

$$\text{Point B}: X_1 = 0,\ X_2 = 1.5 \quad (\alpha \geq 3) \tag{11}$$

However, at point B member 1 is eliminated and the true optimum for $\alpha \leq 3$ is at point C. The true optimum points are given by

$$\text{Point A}: X_1 = 0.5,\ X_2 = 0 \quad (\alpha \leq 1.414)$$

$$\text{Point C}: X_1 = 0,\ X_2 = 0.707 \quad (\alpha \geq 1.414) \tag{12}$$

That is, for $1.414 < \alpha \leq 3$, the solution process will converge to point A while the true optimum is at point C. For $\alpha \geq 3$ the solution process will converge to point B while the true optimum is again at point C. It should be noted that point A represents an unstable structure.

For the approximate constraints, a two-dimensional feasible region can be introduced for any assumed redundant force. Eliminating the latter variable the resulting design space is shown in Fig. 6(b) and the solution process converges to the true optimum points (12). Since both points A and C represent statically determinate structures, compatibility conditions can always be satisfied and the approximate solution is the final optimum.

4 Approximations and Two-Stage Procedures

4.1 Two-Stage Design Procedures

The difficulties involved in the solution of topological optimization problems have motivated two-stage design procedures. A typical procedure is to evaluate an approximate solution at the first stage and modify it at the second stage to achieve the final optimum.

Assuming lower bounds on cross-sections $\mathbf{X}^L > \mathbf{0}$ at the first stage, no members can be eliminated from the ground structure and the accurate NLP formulation will provide the true optimum. It has been noted that the approximate LP formulation will usually give a better solution since compatibility conditions and displacement constraints are not considered. Assuming mixed formulation, that is, approximate stress constraints and accurate displacement constraints, the resulting first-stage solution will be in between the former two cases.

Table 1. Alternative first-stage formulations

	Formulation	Side Constraints	Stress Constraints	Displacement Constraints
a.	Accurate	$\mathbf{X}^L > 0$	Accurate	Accurate
b.	Mixed		Approximate	Accurate
c.	Approximate		Approximate	Approximate
d.	Accurate	$\mathbf{X}^L = 0$	Accurate	Accurate
e.	Mixed		Approximate	Accurate
f.	Approximate		Approximate	Neglected

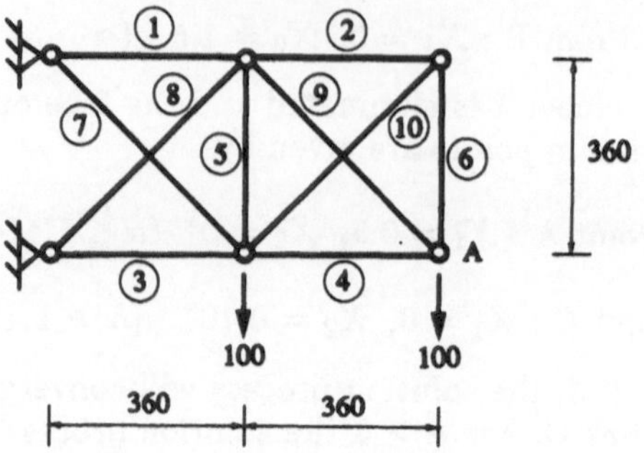

Fig. 7 Ten-bar truss, ground structure.

It has been shown that for $\mathbf{X}^L = \mathbf{0}$, the accurate constraints formulation might lead to incorrect results in cases of a singular optimum. To overcome this difficulty, it is possible to assume at the first stage mixed formulation with approximate stresses and accurate displacements. Alternatively, an approximate formulation, where displacement constraints are neglected, can be considered. As mentioned earlier, the approximate displacement constraints cannot be used for $\mathbf{X}^L = \mathbf{0}$.

The various possibilities for the first-stage formulations are given in Table 1. In summary, approximations assumed at the first stage may include:

— Approximations of stress constraints.

— Elimination of displacement constraints.

— Introduction of side constraints.

The criteria for selecting the type of approximation could reflect both the solution efficiency and the chance of arriving at the true optimum.

Several procedures might be assumed in modifying the first stage solution. For formulations with $\mathbf{X}^L > \mathbf{0}$, it is possible now to eliminate members with $\mathbf{X} = \mathbf{X}^L$ and modify the design accordingly. Accurate formulation of both stress and displacement constraints is needed at this stage. For formulations with $\mathbf{X}^L = \mathbf{0}$, modification of the first stage solution is required to account for displacement constraints and compatibility conditions. At this stage, the complete NLP problem (1) can be solved for the given topology obtained by the LP solution. Available NLP methods can be used for this purpose.

Consider the ten-bar truss shown in Fig. 7. The truss is subjected to two loads acting simultaneously. Bounds on stresses are $\sigma^U = -\sigma^L = 25$, the modulus of elasticity is 30,000 and the bound on the vertical displacement at joint A is $D^U = 2.0$. To illustrate the effect of various constraints on the optimum, the following cases have been solved [14]:

A) Only stress constraints.

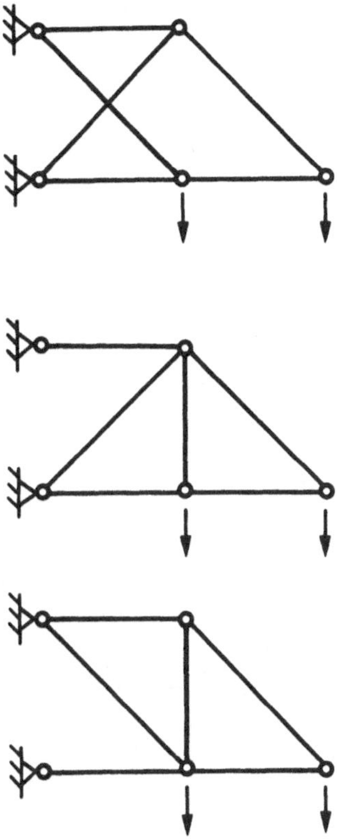

a)

b)

c)

Fig. 8 Various topologies, ten-bar truss example.

B) Stress and side constraints $\mathbf{X}^L = 0.1$.

C) Stress and displacement constraints.

D) Stress, displacement and side constraints $\mathbf{X}^L = 0.1$.

Results are given in Table 2 and the optimal topology for Cases A and C is shown in Fig. 8a. It can be noted that in this example:

— the displacement constraint significantly affects the optimum but does not change the optimal topology;

— the lower bound constraints do not appreciably affect the optimum.

To illustrate the effect of the topology on the optimal solutions, least-weight designs for various topologies (Fig. 8) are compared in Table 3. It will be noted that the optimal topology a (Fig. 8a) is much better than topologies b and c (Figs. 8b and 8c, respectively).

Table 2. Optimal cross-sections, effect of constraints, ten-bar truss

	Case			
Member	A	B	C	D
1	8.0	7.99	9.9	9.9
2	0	0.1	0	0.1
3	8.0	8.06	8.0	8.0
4	4.0	3.94	4.94	4.94
5	0	0.1	0	0.1
6	0	0.1	0	0.1
7	5.66	5.74	5.66	5.6
8	5.66	5.57	7.0	7.0
9	5.66	5.57	7.0	7.0
10	0	0.1	0	0.1
Z	15840	15932	18210	18343

Table 3. Optimal cross-sections, effect of topology, ten-bar truss

	Only Stress Constraints			Stress and Displacement		
Member	Fig. 8a	Fig. 8b	Fig. 8c	Fig. 8a	Fig. 8b	Fig. 8c
1	8.0	12.0	4.0	9.9	13.54	6.16
2	-	-	-	-	-	-
3	8.0	4.0	12.0	8.0	5.53	15.1
4	4.0	4.0	4.0	4.94	5.53	6.16
5	0	4.0	4.0	-	4.0	6.16
6	-	-	-	-	-	-
7	5.66	0	11.31	5.66	0	12.33
8	5.66	11.31	0	7.0	11.31	0
9	5.66	5.66	5.66	7.0	7.81	8.72
10	-	-	-	-	-	-
Z	15840	17276	17276	18210	20028	22803

4.2 Optimal Topologies of Controlled Structures

An alternative approach for the second-stage solution is to apply a set of control forces or displacements such that the accurate constraints will be satisfied at the LP optimum. The rationale of this approach is that the aproximate optimum is usually better than the accurate one, since some constraints (compatibility conditions and displacement constraints) are neglected. The required control forces or displacements can readily be determined by solving an LP problem [15]. The optimal solution will give the minimum magnitude and number of control forces needed to convert a nonfeasible first-stage solution into a feasible one.

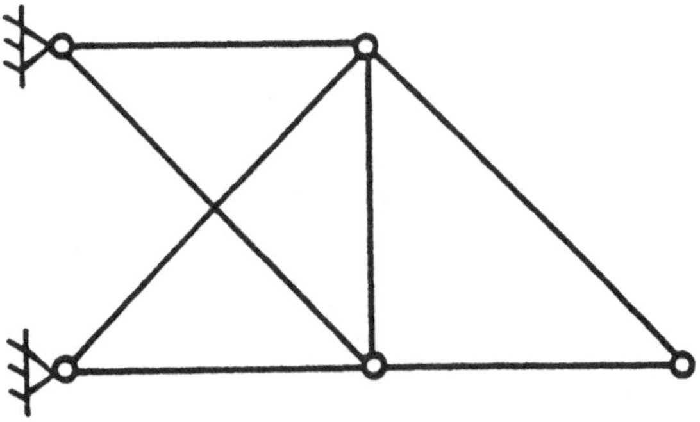

Fig. 9 Assumed topology, ten-bar truss example.

To illustrate this possibility consider again the ten-bar truss shown in Fig. 7. Two cases of constraints have been assumed:

A) Only stress constraints $\left(\sigma^U = -\sigma^L = 25\right)$.

B) Stress and displacement constraints $(D_A \leq 2.0)$.

Assume the topology of Fig. 9 with the given cross-sections $\overset{*}{\mathbf{X}}$ and stresses $\overset{*}{\sigma}$ as shown in Table 4. The resulting displacement at A is $D = 2.21$. That is, the stress constraint in member 8 and the displacement constraint at A are not satisfied for the given design. Solving the optimal control problem, it has been found that these constraints can be satisfied by a single control displacement. The final stresses for both cases are given in Table 4.

Table 4. Effect of control, ten-bar truss

Member	$\overset{*}{\mathbf{X}}$	$\overset{*}{\sigma}$	Final stresses	
			Case A	Case B
1	10.0	17.66	17.08	16.90
2	–	–	–	–
3	10.0	-22.23	-22.80	-22.98
4	4.0	-25.0	-25.0	-25.0
5	4.0	-5.86	-7.31	-7.75
6	–	–	–	–
7	8.0	21.82	22.84	23.15
8	4.0	-27.05	-25.0	-24.38
9	6.0	23.57	23.57	23.57
10	–	–	–	–

5. Conclusions

In this lecture, some design considerations related to the optimization of structural topologies were discussed. It was noted that some basic difficulties greatly complicate the design and analysis interactions and make the topological design problem perhaps the most challenging of the structural optimization tasks. To overcome the difficulties involved in the solution process, various simplifications and approximations can be assumed. These include approximate analysis models, consideration of only simple structural systems and a limited number of loadings, consideration of only certain constraints, simplified variables and the objective function. Despite all the difficulties, the effort is worthwile because substantial savings can be achieved in topological optimization compared with sizing optimization.

Numerical methods for topological optimization are usually based on the ground structure approach where uneconomical members are eliminated from an initial highly connected structure. Neglecting compatibility conditions and displacement constraints, the problem can be stated in an LP form, with the following advantages:

— The computational effort is considerably reduced.
— The global optimum is reached in a finite number of steps.
— Problems of singular optimal topologies are eliminated.
— Some properties of the optimal topologies can be studied directly.

In cases where the LP solution represents an SDS, compatibility conditions can always be satisfied and the solution is the final optimum, unless displacement constraints are not satisfied. It has been found that for particular geometries or loading conditions the optimal topology might represent a USS. A possible solution to this problem is to add members to the optimal structure. If the optimal LP topology is an SIS, compatibility conditions might not be satisfied and the solution should be modified accordingly. It has been shown that for certain geometries the LP solution might represent several optimal topologies. One important property of such geometries is that new optimal topologies are introduced from existing basic optimal topologies by combination rather than the common ground structure approach of elimination.

Another interesting finding is that singular optimal topologies might be obtained due to the nature of accurate stress constraints. This problem may be eliminated if compatibility conditions are neglected and approximate stress constraints are considered. Approximate displacement constraints, on the other hand, might lead to meaningless results if members are eliminated from the ground structure. Unlike stress constraints, accurate displacement constraints usually do not lead to discontinuities or singular solutions.

The difficulties involved in solution of the topological optimization problem suggest that a two-stage design procedure might be useful. Possible procedures, where an approximate solution is evaluated at the first stage and modified at the second stage, are presented. Several first-stage approximations are illustrated and two possibilities for the second-stage solution are considered. It is shown that the first-stage solution can be viewed as the final optimum if a set of control forces or displacements is applied. The latter can readily be determined at the second stage by solving an LP problem. The optimal solution will give the minimum magnitude and number of control forces needed to convert a nonfeasible first-stage solution into a feasible one.

Acknowledgement

The authors are indebted to the Deutsche Forschungsgemeinschaft for supporting this work during the first author's stay at Essen University as a Visiting Professor.

References

1. Kirsch, U. (1989) "Optimal topologies of structures", *Applied Mechanics Review* **42**, 223–239
2. Rozvany, G.I.N. (1988) *Structural design via optimality criteria.* Kluwer Academic Publishers, Dordrecht
3. Topping, B.H.V. (1983) "Shape optimization of skeletal structures: a review", *J. of Structural Engineering, ASCE* **109**, 1933–1951
4. Kirsch, U. (1990) "On the relationship between optimum structural geometries and topologies", *Structural Optimization* **2**, 39–45
5. Dorn, W.S., Gomory, R.E. and Greenburg, H.J. (1964) "Automatic design of optimal structures", *J. de Mecanique*, **3**, 25–52
6. Hemp, W.S. (1973) *Optimum Structures*, Clarendon Press, Oxford
7. Reinschmidt, K.F. and Russell, A.D. (1974) "Applications of linear programming in structural layout and optimization", *Computers and Structures* **4**, 855–869
8. Kirsch, U. (1989) "Optimal topologies of truss structures", *Computer Methods in Appl. Mech. and Engrg.* **72**, 15–28
9. Kirsch, U. (1989) "Effect of compatibility and prestressing on optimized structures", *J. of Struc. Engrg., ASCE* **115**, 724–737
10. Sheu, C.J. and Schmit, L.A. (1972) "Minimum weight design of elastic redundant trusses under multiple static loading conditions", *AIAA J.* **10**, 155–162
11. Sved, G. and Ginos, Z. (1968) "Structural optimization under multiple loading", *Int. J. Mech. Sci.* **10**, 803–805
12. Kirsch, U. and Taye, S. (1986) "On optimal topology of grillage structures", *Engineering with Computers* **1**, 229–243
13. Kirsch, U. (1987) "Optimal topologies of flexural systems", *Engrg. Optimization* **11**, 141–149
14. Kirsch, U. (1990) "On singular topologies in optimum structural design", *Structural Optimization* **2**, 133–142
15. Kirsch, U. and Topping, B.H.V. (1991) "Minimum weight design of structural topologies", *to be published, ASCE J. of Computing in Civil Engineering*

16. Sved, G. (1954) "The minimum weight of certain redundant structures", *Austral. J. Appl. Sci.* **5**, 1–18

TRUSS TOPOLOGY OPTIMIZATION BY A DISPLACEMENTS BASED OPTIMALITY CRITERION APPROACH

by

Martin P. Bendsøe
Mathematical Institute
Technical University of Denmark
Building 303, DK–2800 Lyngby
Denmark

Aharon Ben–Tal
Faculty of Industrial
Engineering and Management
Technion–Israel Institute of
Technology
Technion City, Haifa 3200,
Israel

ABSTRACT

In this paper we present a displacement based method for maximum stiffness truss topology design. The ground structure approach is used, and the problem is formulated in terms of displacements and bar areas. This large, non–convex optimization problem can be solved by identifying an equivalent, unconstrained and convex problem in the displacement which can be solved by a non–smooth, steepest algorithm. In this method we circumvent the explicit solving of the equilibrium equations and the assembly of the global stiffness matrix. A large number of examples have been studied, showing the attractive features of topology design as well as exposing interesting features of optimal topologies.

1. Introduction

The optimization of the geometry and topology of structural lay–out has great impact on the performance of structures, and recent years have seen a revived interest in this important area of structural optimization (Kirsch, 1989a). The last couple of years witnessed the development of the so–called homogenization method for generating optimal topologies of structural elements (e.g. Bendsøe and Kikuchi, 1988, Suzuki and Kikuchi, 1990, Diaz and Belding, 1990). For "thin" structures, that is, structures with a low fraction of available material compared to the spatial dimension of the structure, the homogenization method predicts grid and truss like structures. Thus the homogenization method supplements analytical methods for the study of fundamental properties of gridlike continua, as first treated by Michell, 1904, and also described in more recent monographs (e.g., Hemp, 1973, and Rozvany, 1989). Applications of numerical methods to truss problems and other discrete models are more recent, with contributions by, for example, Dorn, Gromory and Greenberg, 1964, Fleron, 1964 and Pedersen 1970; see also the review papers by Kirch, 1989a, and by Topping, 1983.

G. I. N. Rozvany (ed.), Optimization of Large Structural Systems, Vol. I, 139–155.

Truss topology design methods have been traditionally formulated in terms of member forces, ignoring kinematic compability in order to obtain a linear programming (LP) problem in member areas and forces. The resulting topology and force field is then often employed as a starting point for a more complicated design problem formulation (see, e.g., Ringertz, 1985, Vanderplaats, 1984), with heuristics, branch and bound techniques, etc. being used to link the two model problems. Alternatively, when displacement formulations are used, then nonzero lower bounds on the cross–sectional areas have been imposed in order to have a positive definite stiffness matrix, as exemplified in Rozvany and Zhou, 1990, where a traditional optimality criteria update scheme is used. In a little known paper by Taylor and Rossow, 1977, an algorithm for topology design based on a displacement formulation is presented; their method utilizes the optimality criteria to identify active members of the truss and is closely related to the work presented here.

The approach taken in this paper is known as the ground structure approach. This means that for an initially chosen layout of nodal points the optimum truss structure connecting the fixed boundary and the external loads is found as a subset of an initially chosen set of connections between the nodal points. As the positions of nodal points are not used as design variables, a high number of nodal points should be used in the ground structure in order to obtain efficient topologies. Figure 1 shows example ground structures with a common set of nodal points.

In parallel with developments in homogenization methods, we employ the compliance of the truss as the objective function, and the design problem is the simplest possible: The minimization of the compliance (maximizations of stiffness) for a given weight of the truss. The design variables are the cross–sectional areas of the individual truss members connecting the nodal points of the ground structures. As both weight and stiffness are linear in the design variables, *the layout of the optimal design is independent of the chosen weight constraint for the truss*, for the case of unconstrained design variables.

Allowing for zero cross–sectional areas of individual truss members implies that the stiffness matrix of the truss is not necessarily positive definite, and the optimization problem is non–differentiable. It is thus not possible to use gradient based methods of structural optimization. The standard way for circumventing this problem are either to shift to member forces as design variables, or to introduce a nonzero lower bound on the cross–sectional areas. A more general approach is to use both the displacement components and the cross–sectional areas as design variables, with the equations of equilibrium being applied as equality constraints, this being a so–called simultaneous analysis and design approach, cf. Haftka, Gürdal, Kamat, 1990 and Bendsøe, Ben–Tal, Haftka, 1991.

2. Problem formulation

In the ground structure approach a set of n chosen nodal points and m connections are given, and one seeks to find the optimal substructure of this structural universe. Let a_i, l_i denote the cross–sectional area and length of bar number i, respectively, and for simplicity we assume that all bars are made of the same linear elastic material, with Young's modulus E. The volume of the truss is

$$V = \sum_{i=1}^{m} a_i l_i \ ,$$

and in order to simply notation at a later stage, we introduce the bar volumes

$$t_i = a_i l_i \quad , \quad i = 1,...,m$$

as the fundamental design variables. Static compatibility is expressed as

$$B^T q = f$$

where q, f are the member force and nodal force vectors, respectively. The ground structure is chosen so that the compatibility matrix B has full rank excluding mechanisms and rigid body motions. With u denoting the vector of nodal displacements, the strain in bar number i is

$$\epsilon_i = \frac{1}{l_i} (Bu)_i$$

The stiffness matrix of the truss is written as

$$K = \sum_{i=1}^{m} t_i K_i$$

where $t_i K_i$ is the element stiffness matrix for bar number i. Equilibrium is thus written as

$$\sum_{i=1}^{m} t_i K_i \, u = f$$

in terms of displacements.

The problem of finding the minimum compliance truss for a given volume of material (the stiffest truss) is then formulated as

$$\begin{aligned} &\min_{u,t} \ f^T u \\ &\text{so: } \sum_{i=1}^{m} t_i K_i \, u = f \ ; \ \sum_{i=1}^{m} t_i = V \ ; \ t_i \geq 0 \ , \ i = 1,...,m \end{aligned} \qquad \text{(P1)}$$

where the design variables t_i and the displacements u appear as independent variables.

The zero lower bound on the variables t_i means that bars of the ground structure can be removed and the problem statement thus covers topology design. However, the zero lower bound also implies that the stiffness matrix is not necessarily positive definite and the state vector u cannot be removed by a standard adjoint method. Removing u from the formulation is actually not very important for the size of the problem, as typically, the number of bars m is much greater than the number of degrees of freedom. In the ultimate ground structure, we connect all nodes, having $m = n(n-1)/2$, while the degree of freedom only is of the order $2n$ or $3n$ (for planar and 3–D trusses); for this situation we have a fully–populated stiffness matrix lacking any sparsity and bandedness.

As with other truss topology formulations, problem (P1) can result in a optimal topology which is a mechanism; this mechanism is in equilibrium under the given loads, and infinitesimal bars can be added to obtain a stable structure (cf. Ringertz, 1985). Also, if the optimal topology has straight bars with inner nodal points, these nodal points should be ignored. The resulting truss maintains the stiffness and the equilibrium of the original optimal topology.

The problem formulation (P1) also covers the FEM discretized version of topology design for the restricted class of continuum structures that exhibit a linear relation between rigidity and design variable, as exemplified by the design of variable thickness sheets (ef. Rossow and Taylor, 1973) or design of sandwich plates.

3. A non–smooth, optimality criterion based method

In this section we shall present an alternative, equivalent formulation of problem P1, for which a computationally effective steepest descent algorithm can be devised. The new formulation is based on a suitable interpretation of the optimality conditions for problem P1. For complete treatment of this method we refer the reader to Ben–Tal and Bendsøe, 1991.

In order to obtain the necessary conditions for optimality for problem (P1) we introduce lagrange multipliers $\bar{y}$, Λ and μ_i, $i = 1,\dots,m$, for the equilibrium constraint, the volume constraint and the zero lower bound constraint respectively. The necessary conditions are thus found as the conditions of stationarity of the lagrangian:

$$L = f^T x - \bar{y}^T(\Sigma\, t_i K_i\, u - f) + \Lambda \left[\sum_{i=1}^{m} t_i - V\right] + \sum_{i=1}^{m} \mu_i(-t_i)\,.$$

By a standard development we obtain

$$\sum_{i=1}^{m} t_i K_i\, \bar{y} = f\ ;\ \ \bar{y}^T K_i\, u = \Lambda - \mu_i\,,\ \ \mu_i \ge 0\,,\ \ \mu_i t_i = 0\,,\ \ i = 1,\dots,m\,;\ \ \Lambda \ge 0\,.$$

Now let $\lambda(u)$ denote the maximal mutual energy $u^T K_i\, u$ of the individual bars, i.e.

$$\lambda(u) = \max \{u^T K_i u \mid i = 1,...,m\}$$

and let $J(u)$ denote the set of bars for which the mutual energy attains this maximum level:

$$J(u) = \{i \mid u^T A_i u = \lambda(u)\} ,$$

We also define nondimensional element volumes

$$\tilde{t}_i = t_i/V .$$

Then the necessary conditions are satisfied with

$$\bar{y} = u \; ; \; t_i = \tilde{t}_i V , i \in J \; ; \; t_i = 0 , i \notin J$$

$$\Lambda = \lambda(u) \; ; \; \mu_i = 0 , i \in J \; ; \; \mu_i = \lambda(u) - u K_i u > 0 , i \notin J ,$$

provided that

$$V \sum_{i \in J} \tilde{t}_i K_i u = f \; ; \; \sum_{i \in J} \tilde{t}_i = 1 , \tag{O1}$$

and the set J is the set of active bars. This implies that the optimal truss has bars with constant mutual energies and the set J is the set of active bars. The optimality conditions (O1) states that a convex combination of the gradients of the quadratic functions $\frac{V}{2} u^T K_i u$, $i \in J$, equals the load vector f, and this is precisely the necessary conditions for the non–smooth problem

$$\min_u \left[\max_{i=1,...,m} \left\{ \frac{V}{2} u^T K_i u - f^T u \right\} \right] \tag{P2}$$

which is an unconstrained, *convex* problem in the displacement variable u only.

It is possible to show existence of solutions to problem (P2) and to prove the equivalence between problem (P1) and (P2); the equivalence is understood in the sense that for a solution u to (P2) and the corresponding set J of active bars, there exists a corresponding set of bar volumes t satisfying the optimality condition

$$\sum_{i \in J} t_i K_i u = f \; , \; \sum_{i \in J} t_i = V$$

$$t_i = 0 , i \notin J ; t_i \geq 0 , i = 1,...,m$$

and the pair (u, t) is a global solution to problem (P1). Note also that the necessary conditions for problem (P1) are also sufficient, as shown in Taylor, 1969.

Problem (P2) is, up to a rescaling, equivalent to the convex problem

$$\min_u -f^T u$$

$$\text{so: } \frac{V}{2} u^T K_i u \leq 1 \,, \; i = 1,\ldots,m \qquad \text{(P3)}$$

The special form of the element stiffness matrix K_i for the truss implies that

$$u K_i u = \frac{E}{l_i^2} (b_i^T u)^2$$

where b_i^T is the ith row of the compatibility matrix B. Thus, (P3) can be written in LP – form as (Achtziger et.al., 1991):

$$\min_u -f^T u$$

$$\text{so: } -1 \leq \sqrt{\frac{VE}{2l_i^2}}\, b_i^T u \leq 1 \,, \; i = 1,\ldots,m \qquad \text{(LP u)}$$

which for a suitable stress constraint value σ is the dual of the traditional force formulations

$$\min_{q^+,q^-} \sum \frac{l_i}{\sigma} (q_i^+ + q_i^-)$$

$$\text{so: } B^T(q^+ - q^-) = f \qquad \text{(LP q1)}$$

$$q_i^+ \geq 0 \,, \; q_i^- \geq 0 \,, \; i = 1,\ldots,m$$

and

$$\min_{q,t} \sum t_i$$

$$\text{so: } B^T q = f \qquad \text{(LP q2)}$$

$$-\sigma t_i \leq l_i q_i \leq \sigma t_i \,, \; i = 1,\ldots,m$$

$$t_i \geq 0 \,, \; i = 1,\ldots,m$$

We thus see that the equivalence between problem (P1) and (P2) shows that the traditional force method results in minimum compliance designs. The equivalence between problems (LP q) and (LP u) can also be found in Dorn, Gromory and Greenberg, 1964 where it is also shown how the force formulations are convenient for studying an eventual statical determinancy of the solutions.

The linear programming formulations above hold only for the case of truss design with unconstrained design variables and a single load case. As a formulation in the lines of (P2) also holds for more general problems, we will here show how problem (P2) can be solved by a steepest descent method for non–smooth problems; the algorithm is simplest and most intuitive in the present setting. The algorithm generates both the solution u as well as the bar volumes t and it consists of the following steps.

0. Compute an initial guess of displacement field u, for example by solving $K u = f$ for a feasible set of bar volumes t.

1. For present u, compute $\lambda(u)$ and indices

$$J(u) = \{i \mid u^T K_i u \geq \lambda(u) - \epsilon\}$$

2. Compute descent direction d as

$$d = -\left[\sum_J t_i K_i u - f\right]$$

where t_i, $i \in J$ are found from

$$\min_t \| \sum_{i \in J} t_i K_i u - f \|^2 - \sum_{i \in J} t_i u^T K_i u$$

such that $\sum_{i \in J} t_i = V$, $t_i \geq 0$ $i \in J$.

3. If $\| d \| < \delta$, stop
Else go to 4.

4. Compute a step size α^* for update, $u := u + \alpha d$, by a line search with the function

$$F(u) = \max_i \left\{\frac{V}{2} u^T K_i u - f^T u\right\} .$$

5. Update, $u := u + \alpha^* d$, and go to 1.

Here, ϵ is a relaxation on the activity set J that is introduced to stabilize the algorithm, and δ determines the accuracy of the solution. Each iteration loop of the algorithm consists of first finding the set of almost active bars (Step 1). The descent direction is then found by first finding the bar volumes of these bars which minimizes the error in equilibrium for the given estimate of displacement; this is a quadratic programming problem. The error is measured in a least squares sense and the descent direction is given as the vectorial error with this best fit of bar volumes. For ϵ small enough, the set of almost active bars equals the set of actually active bars, so it is natural to work with a decreasing sequence of the relaxation parameter ϵ, as well as with a decreasing sequence of equilibrium errors δ.

The algorithm above is conceptually similar to the algorithm given in Taylor and Rossow, 1977, the difference being in the update scheme, which here is based on the formal identification of the equivalence between problems (P1) and (P2).

4. Multiple loads

It is very natural to introduce more complicated problem formulations for topology design than the one used hitherto in this paper. The optimality criteria based approach presented in the preceeding section employs the structure of problem (P1) to such an extent that generalizations to more complicated problem formulations are not evident. However, in certain cases it is possible to generalize this computationally attractive method, one case being multiple load optimal topology design formulated in terms of minimizing a weighted average of compliances, for a given volume. Such an approach is natural and has been shown to be effective for the homogenization method, cf. Diaz and Bendsøe, 1991.

For a set f_p , $p = 1,\ldots,M$, of M different load cases and weights W_p , $p = 1,\ldots,M$, we formulate the multiple load problem as

$$
\begin{aligned}
&\min_{u,t} \sum_{p=1}^{M} W_p f_p^T u_p \\
&\text{so: } \sum_{i=1}^{m} t_i K_i u_p = f_p \ , \ p = 1,\ldots,M \qquad \text{(P1m)} \\
&\sum_{i=1}^{m} t_i = V \ , \ t_i \geq 0 \ , \ i = 1,\ldots,m
\end{aligned}
$$

Introducing an extended displacement vector $u = (u_1,\ldots,u_M)$ for all the displacement vectors u_p , $p = 1,\ldots,M$, an extended force vector $\hat{f} = (W_1 f_1,\ldots,W_M f_M)$ of the weighted force vectors $W_p f_p$, $p = 1,\ldots,M$, and a set of extended element stiffness matrices as the block diagonal matrices

$$
\hat{K}_i = \mathrm{diag}(W_p K_i)_{p = 1,\ldots,M}
$$

problem (P1m) can be written as

$$
\begin{aligned}
&\min_{u,t} \hat{f}^T \hat{u} \qquad \text{(P}\hat{1}\text{m)} \\
&\text{so: } \sum_{i=1}^{m} t_i \hat{K}_i \hat{u} = \hat{f} \ ; \ \sum_{i=1}^{m} t_i = V \ ; \ t_i \geq 0
\end{aligned}
$$

which is precisely of the same form as problem P1. Thus, the algorithm for problem P2 can be readily invoked to solve the equivalent problem

$$\min_u \left[\max_i \left\{ \frac{V}{2} \sum_{p=1}^{M} W_p\, u_p^T \hat{K}_i\, u_p - \sum_{p=1}^{M} W_p\, f_p^T u_p \right\} \right] \qquad \text{(P2m)}$$

with p ranging form 1 to M and i from 1 to m . Notice that the matrices $\hat{K}_i$ can no longer be written as dyadic products and we do not have a LP formulation in this case.

The optimality criterion based method can thus be used also for multiple load problems. With respect to an increase in computational effort, the multiple loads affect the process of finding of the set of active bars as well as the line search, but these do not affect the size of the QP problem for finding the descent direction, except through a natural small increase in the number of active bars.

5. Bounded design variables

The topology design problem as formulated in problem (P1) allows for a total freedom in the assignment of material to the individual bars of the truss. Even for topology design, however, it is reasonable that an upper limit on bar volumes (or rather, bar areas) is prescribed. For topology design the lower limit on bar volumes should always be zero, but it is possible to design an optimality criterion based algorithm of the form presented in Section 4 with both upper *and* lower limits on bar volumes. With non–zero lower bounds on all bar volumes, standard optimality criterion based methods can be employed as the full stiffness matrix is positive definite (cf. Rozvany and Zhou, 1990). However, as the method presented here does not involve solving the equilibrium equations, we have a computational scheme which is usually faster, especially for large problems lacking sparsity and bandedness of the stiffness matrix (i.e. truss problems with most nodal connections in play).

The minimum compliance problem with upper and lower bounds on the bar volumes takes the form

$$\min_{u,t} f^T u$$

$$\text{so:} \sum_{i=1}^{m} t_i K_i u = f \qquad \text{(P1b)}$$

$$\sum_{i=1}^{m} t_i = V\,,\ 0 \le L_i \le t_i \le U_i < \infty\,,\ i = 1,\ldots,m$$

and the equivalent problem in displacements only takes the form

$$\min_{u,\lambda}\left[\lambda V - f^T u + \sum_{i=1}^{M} \max\left\{\left[\frac{1}{2}u^T K_i u - \lambda\right] U_i \ ,\right.\right.$$

$$\left.\left.\left[\frac{1}{2}u^T K_i u - \lambda\right] L_i\right\}\right] \tag{P2b}$$

where λ is an extra, auxiliary variable, being a threshold energy controlling which bars are at the lower and upper limit on bar volumes and which bars have an intermediate bar volume. Thus 2λ is the largest mutual energy for which the naive design with all bars of an energy greater or equal to 2λ set at the maximum volume and with all other bars set at the lower limit has a volume that exceeds the volume constraint value. The algorithm for solving (P2b) follows the same lines as the algorithm for solving (P2), and gives as a result the solution u, λ as well as the corresponding bar volume t, so that (u, t) solves (P1b).

The algorithm as well as convergence proofs etc. can be found in Ben–Tal and Bendsøe, 1991, and derivation of equivalent smooth, convex problems (cf. (P3)) and associated algorithms can be found in Achtziger et. al. 1991, and Kocvara, 1991.

6. The topology of reinforcement.

The problem of finding the optimal topology of the reinforcement of an already existing structure is a natural extension of the problem formulation presented above. Using the ground structure approach, we devide a given ground structure into the set S of bars of the given structure and the set R of possible reinforcing bars. Typically S and R will be chosen as disjoint, but in order to investigate the efficiency of the given topology R could be chosen so as to contain S as a subset.

The bars of the given structure have given bar volumes s_i, $i \in S$, and the optimal reinforcement t_i, $i \in R$, is the solution of the minimum compliance problem

$$\begin{aligned}&\min_{x,t} f^T u\\ &\text{so: } \sum_{i\in R} t_i K_i u + \sum_{i\in S} s_i K_i u = f \ , \ \sum_{i\in R} t_i = V \ , \ t_i \geq 0 \, , i \in R\end{aligned} \tag{R1}$$

The equivalent problem in displacements u only now takes the form

$$\min_{u}\left[\max_{i\in R}\left\{\frac{V}{2}u^T K_i u + \left[\sum_{i\in S}\frac{1}{2}s_i u^T K_i - f^T\right] u\right\}\right] \tag{R2}$$

and can be solved by the algorithm described in section 3. Note that the given structure S modifies the original topology design problem by given rise to a displacement dependent load $\sum_{i\in S} s_i K_i u$ in the equilibrium equation as well as in the algorithm for solving the problem. The reinforcement version of problem (P3) now reads

$$\min_{u,\mu} \frac{1}{2} \sum_{i \in S} s_i\, u^T K_i\, u - f^T u + \mu^2$$

$$\text{so: } \frac{V}{2} u^T K_i\, u \leq \mu^2 \quad , \; i \in R \qquad \text{(R3)}$$

and this problem is quadratic both in objective and constraints. For the single load case we can thus only obtain a QP problem and not a LP problem:

$$\min \frac{1}{2} u^T \left[\sum_{i \in S} s_i K_i \right] u - f^T u + \mu^2$$

$$\text{so: } -\mu \leq \sqrt{\frac{V E}{2 l_i^2}}\, b_i^T u \leq \mu \quad , \; i \in R \qquad \text{(QPr)}$$

Notice here that the matrix $\sum_{i \in S} s_i K_i$ is positive semidefinite, but usually not positive definite.

7. Results

In this section we present a number of optimal topologies obtained through the use of the optimality criteria based method. We concentrate the discussion on the single load problem with no constraints on the bar volumes, as defined by problem (P1).

Problem (P1) is made up of expressions which are element–wise linear in all variables, except geometric data. Thus, for a specific choice of ground structure geometry and load vector *direction*, the optimal topology needs only to be computed for one set of assigned values of Young's modulus E, volume V, load size, and one geometric scale; for any other values of these variables, the optimal values of the design variables t, the deformation u and the compliance $f^T u$ can be derived by a simple scaling. Thus, problem (P1) lends itself to the creation of a "catalogue of optimal topologies". This means that the optimal compliance may be given in terms of the nondimensional compliance ϕ

$$\phi = (f^T u)\, V E / (\| f \|^2 L^2)$$

where L is a typical length dimension (horizontal length of truss in the examples here).

Examples of optimal topologies are shown in Figures 2 through 7. The final topology and the performance of the optimal structure depend intimately on the choice of ground structure, and the most efficient structures are obtained with a large number of nodes and all possible connections included in the ground structure. Note that the method allows for the prediction of active bars as well as the *active* supports, and we do not assume external statically determinacy as in early papers on topology optimization of trusses (Dorn, Gromory and Greenberg (1964), Fleron (1964), Pedersen (1970)).

The optimality criteria based descent method work fastest if the optimal topology consists of only a few active bars. Such optimal topologies are possible, even if the ground structure consists of a very large number of potential bars, as exemplified by the optimal 2–bar 45° truss carrying a single load which is parallel to the line of possible support, cf. Fig. 2., where this 2–bar truss is a subset of the ground–structure. In many other cases, "nature's optimal topology" is a Michell–truss, and this is reflected in the optimal truss topologies that are generated, the topologies usually consisting of a rather high number of bars mimicking the curve linear lay—out of a Michell–truss and this number increases as the number of nodal points in the ground structure increases. This is illustrated in Fig. 3., and Figures 4 and 5 illustrate other interesting optimal topologies, for ground structures with all possible connections as well as a reduced set of connections, with multiple loads and with constraints on the bar areas.

The availability of efficient methods to solve large (sparse) LP problems makes it natural to solve the single load truss topology design problem using the formulations (LP u) or (LP q) . For problems with multiple loads and/or bounded bar areas, for the reinforcement problem as well as for the FEM case, we cannot obtain a linear programming formulation of the problem and we are forced to solve problems (P1), (P2) or (P3) directly. Problem (P1) generalizes most easily to more general design situations but is large scale and non–convex. Problems (P2) and (P3) are convex and have the size of the degrees of freedom of the ground structure; (P2) is non–differentiable and unconstrained and (P3) is differentiable, but at the cost of a high number of constraints. The algorithm presented in section 3 is a specialized algorithm for solving problem (P2) and has been implemented so as to take advantage of the sparsity of the matrices A_i . General purpose algorithms for min–max optimization or non–differentiable optimization can also be employed, but comparison is difficult for problem sizes where sparsity plays an important role; also most general purpose methods have enormous computer storage requirements. Likewise, problem (P3) can be solved by general purpose algorithms (SQP etc.), but again sparsity and the fact that the number of variables is much lower then the number of constraints should be utilized. Investigations into these questions are still being carried out, see Achtziger et. al. (1991).

8. Acknowledgement

The authors gratefully acknowledge the very fruitful collaboration with J. Zowe and W. Achtziger, University of Beyreuth, on the subject of this paper.

The work presented in this paper received support from the German–Israeli Foundation for Scientific Research and Development (A.B–T) and the Danish Technical Research Council (Programme of Research on Computer–Aided Design) (MPB).

9. References

W. Achtziger, M.P. BendsØe, A. Ben–Tal, J. Zowe (1991); "Equivalent Displacement Based Formulations for Maximum Strength Truss Topology Design". Preprint, Universität Bayreuth (in preparation).

M.P. BendsØe, N. Kikuchi (1988), "Generating Optimal Topologies in Structural Design using a Homogenization Method". Comp. Meth. Appl. Mech. Engrg., 71, pp. 197–224.

M.P. BendsØe, A. Ben–Tal, R.T. Haftka (1991), "New Displacement – Based Methods for Optimal Truss Topology Design". Proc. AIAA/ASME/ASCE/AHS/ASC 32nd. Structures, Structural Dynamics and Materials Conference, Baltimore, MD, USA, April 8–10, 1991.

A. Ben–Tal, M.P. BendsØe, (1991) "A New Method for Optimal Truss Topology Design". MAT–Report No. 1991–8, Mathematical Institute, The Technical University of Denmark, DK–2800 Lyngby. 35 pp.

A. Diaz, B. Belding (1990), "On Optimum Truss Lay–out by a Homogenization Method", ASME Transactions of Mechanical Design (to appear).

A. Diaz, M.P. BendsØe (1991), "Shape Optimization of structures for multiple loading conditions using a homogenization method", Structural Optimization (to appear).

W. Dorn, R. Gromory and M. Greenberg (1964), "Automatic Design of Optimal Structures", J. de Mechanique, 3, pp. 25–52.

P. Fleron (1964), "The minimum weight of trusses", Bygnings Statiske Meddelelser, 35, pp. 81–96.

R.T. Haftka, Z. Gürdal, M.P. Kamat (1990), Elements of Structural Optimization, 2nd edition, Kluwer Academic Publishers, Dordrecht, the Netherlands, p. 336.

W.S. Hemp (1973), Optimum Structures, Clarendon Press, Oxford, UK.

U. Kirsch (1989a), "Optimal Topologies of Structures". Applied Mechanics Reviews, vol. 42, pp. 223–239.

U. Kirsch (1989b), "Optimal Topologies of Truss Structures", Comp. Meth. Engrg., 72, pp. 15–28.

M. Kocvara (1991), "QP3 and CG0 – Programs for Solving Truss Optimization Problems", Internal Report, Math. Inst. Univ. of Bayreuth, FRG.

A.G.M. Michell (1904), "The Limits of Economy of Material in Frame Structures", Philosophical Magazine, Series 6, Vol. 8, pp. 589–597.

P. Pederson (1970), "On the Minimum Mass Layout of Trusses", AGARD Conf. Proc. No. 36, Symposium on Structural Optimization, AGARD–CP–36–70.

U. Ringertz (1985), "On Topology Optimization of Trusses", Engrg. Optimization, 9, pp. 21–36.

M.P. Rossow, J.E. Taylor (1973), "A finite element method for the optimal design of variable thichness sheets", AIAA J. 11, 1566–1569.

G.I.N. Rozvany (1989), Structural Design via Optimality Criteria, Kluwer, Dordrecht.

G.I.N., Rozvany, M. Zhou (1991), "Applications of the COC Algorithm in Layout Optimization". Proc. Int. Conf. Engrg. Optimization in Design Processes, Karlsruhe, 1990; Lecture Notes in Engineering vol 63, Springer Verlag, 59–70.

K. Suzuki and N. Kikuchi (1990), "A Homogenization Method for Shape and Topology Optimization," Computer Methods in Applied Mechanics and Engineering (to appear).

J.E. Taylor (1969), "Maximum Strength Elastic Structural Design", Proc. ASCE, 95, 653–663.

J.E. Taylor, M.P. Rossow (1977), "Optimal Truss Design Based on an Algorithm using Optimality Criteria", Int. J. Solids Struct., 13, 913–923.

B.M.V. Topping (1983), "Shape Optimization of Skeletal Structures: A Review", ACSE J. Struct. Engrg, 109, 1933–1951.

G.N. Vanderplaats (1984), "Numerical Methods for Shape Optimization: An Assessment of the State of the Art", in New Directions in Optimum Structural Design (E. Atrek et al., Eds.), Wiley, Chichester, UK.

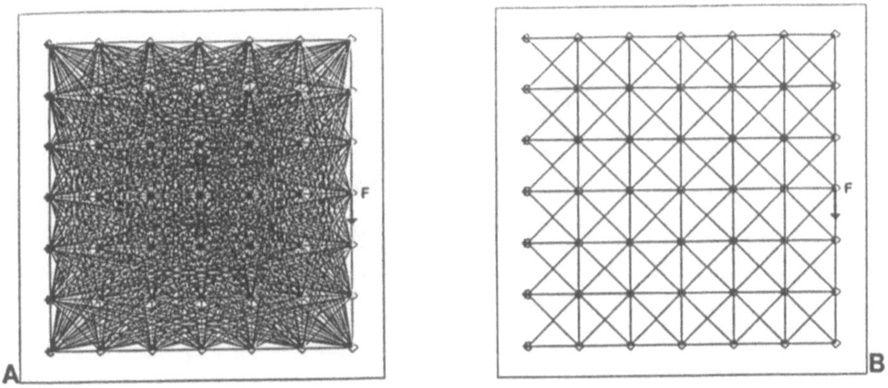

Figure 1. Two different ground structures for a 7 by 7 nodal lay-out in a square domain. A: Ground structure with all possible nodal connections. B: Ground structure with only neighbouring nodes connected.

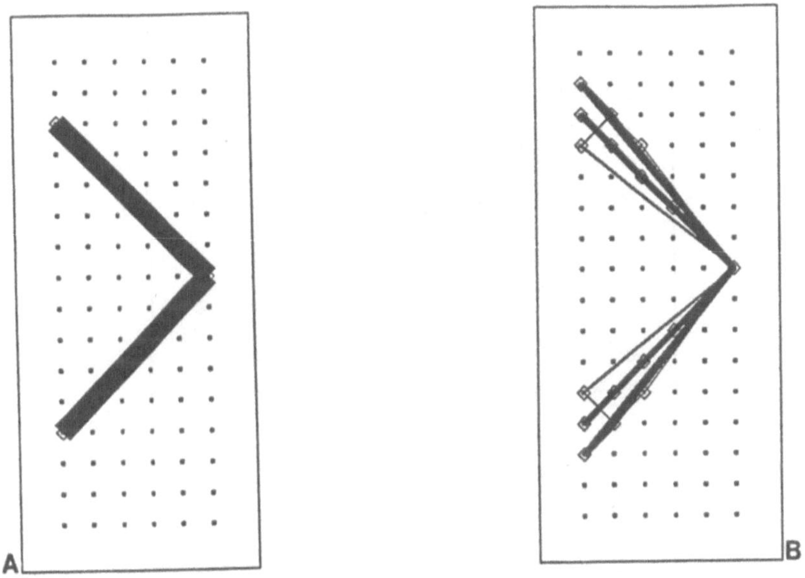

Figure 2. Optimal trusses for a ground structure with all connections between nodes in a 6 by 16 nodal lay-out in a 5 by 15 rectangular domain (2852 potential bars). Left hand nodes are potential supports, and truss is supposed to support a vertical load at the mid right hand node. A: The optimal truss with no constraints on design variables. The optimal nondimensional compliance is 4.0. B: The optimal truss, with upper bounds on the bar volumes ($u_i = 0.01 \cdot \ell_i \cdot V$). The optimal nondimensional compliance has increased to 4.1092 .

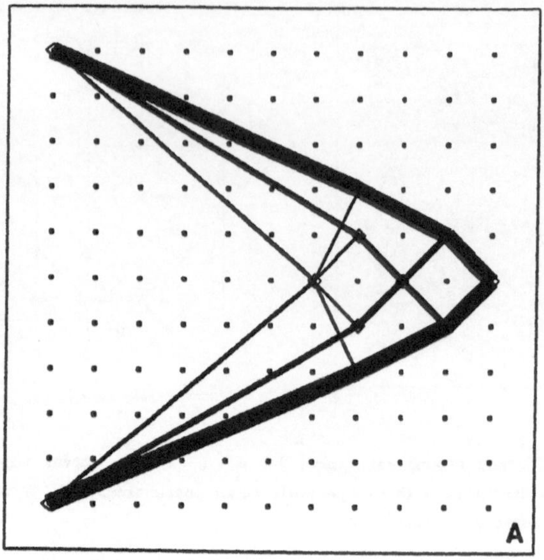

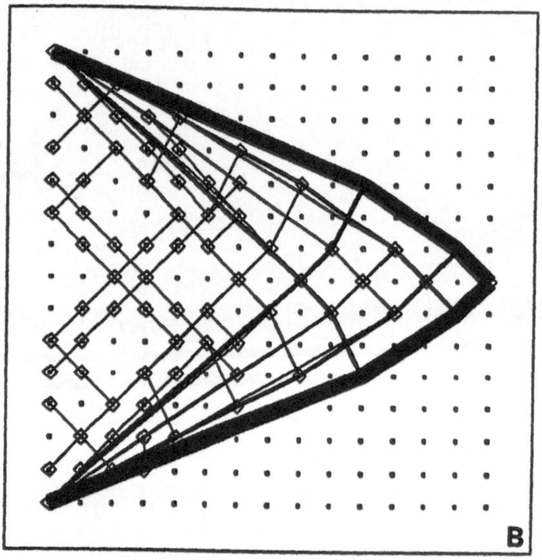

Figure 3. Optimal trusses for transmitting a vertical load to a vertical line of supports. Design area reduced to a square, with line of supports at left hand nodal points and force at mid right hand nodal point. A: The optimal truss for all possible nodal connections in a 11 by 11 nodal lay-out (4492 potential bars). Nondimensional compliance is 5.9646. B: The optimal truss for a ground structure with a 15 by 15 nodal lay-out (15,556 potential bars); optimal compliance is 5.9344 .

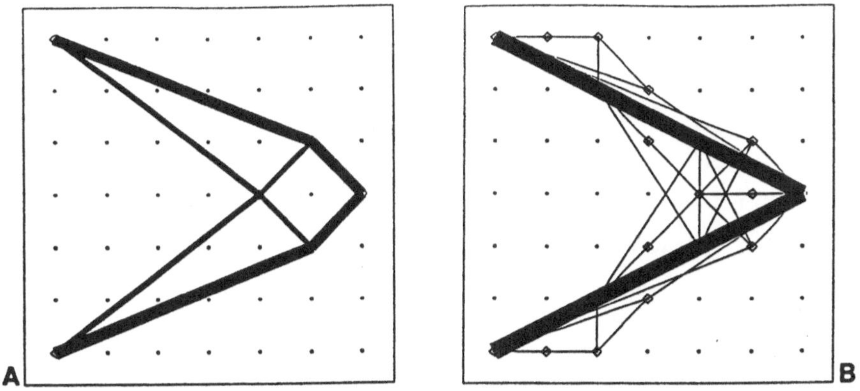

Figure 4. Optimal trusses for the ground structure of Figure 1A with all nodal connec-tions in a 7 by 7 nodal lay-out in a square domain, with possible support nodes at left hand side. A: The optimal truss with unconstrained design variables and a single vertical load at the mid right hand node. The nondimensional complicance is 6.0134 . (Compare with Figures 2 and 3.) B: The optimal multiload design of a truss with a vertical and a horizon-tal load at the mid right hand node. The loads are equal in size and the weights on the compliances are 2.0 and 1.0 , respectively. The average nondimensional compliance is 4.6943 and the compliances for each of the loads are 6.2541 and 1.5747 , respectively.

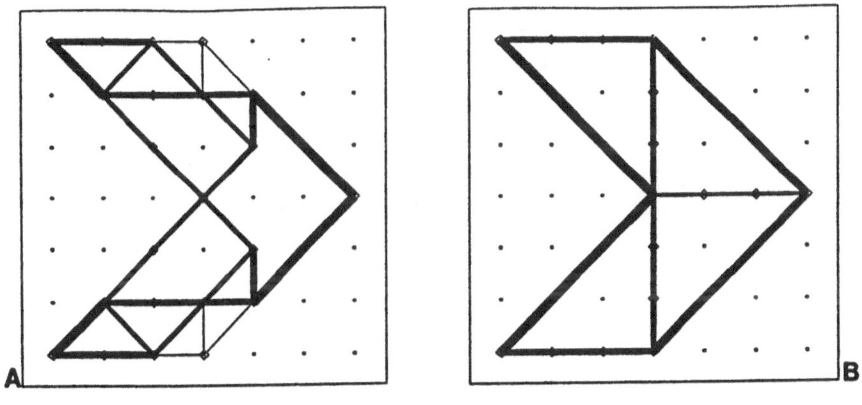

Figure 5. Optimal trusses for the ground structure of Figure 1B, with possible support nodes at left hand vertical line of nodes. A: The optimal truss for a single vertical load at the mid right hand node. The nondimensional compliance is 8.9865 . B: The optimal truss for a multiple load problem with three loads: A horizontal and a vertical load at the mid right hand node and a vertical load at the mid node, all of equal size and weighted 1.0, 2.0 and 1.0 , respectively. The average nondimensional compliance is 6.3737 and the individual compliances are 3.752 , 9.4577 , 2.8273 , respectively.

Layout Optimization using the Homogenization Method

Katsuyuki Suzuki
The University of Tokyo
Tokyo, JAPAN

and

Noboru Kikuchi
The University of Michigan
Ann Arbor, MI 48109-2215, U.S.A.

Abstract A generalized layout problem involving sizing, shape, and topology optimization is solved by using the homogenization method for three-dimensional linearly elastic shell structures in order to seek a possibility of establishment of an integrated design system of automotive car bodies, as an extension of the previous work by Bendsϕe and Kikuchi. A formulation of a three-dimensional homogenized shell, a solution algorithm, and several examples of computing the optimum layout are presented in this first part of the two articles.

1. Introduction

A modern theory of structural optimization based on mathematical programing and sensitivity analysis is first developed by Schmit [1] and Fox [2] in the 60s, although the concept of fully stressed design was widely applied in design practice without solid mathematical justification, but with engineers' intuition for a long time. Using variational methods such as Lagrange multipliers and calculus of variations, Prager and Taylor [3] made a justification of the fully stressed design for a class of structural optimization problems by deriving their optimality condition, whose direct use in constructing optimization algorithms leads the so-called optimality criteria method. Structural optimization in the 60s was restricted mostly to sizing problem of frame structures. Even some of topological layout optimization problems of a structure were solved as sizing problems by emphasizing the fact that unnecessary members of frames disappear in the optimization process as the area of the cross section of such members goes to zero. However, these restricted investigation of topology of the optimum structure could not provide significant advancement of the theory of the optimum layout of a structure which enhances the "classical" layout theory of Prager [4] for a very restricted class of structures as an extension of the concept of Michell trusses [5]. Even more recent work by Rozvany [6,7,8] in which Prager's idea is extensively developed, deals with layout problems of extremely complex, but still frame structures based on the analytically derived optimality criteria method, see also recent work by Bendsϕe, Ben-Tal, and Haftka [9] for complex plane truss layout problems.

In the present work, we shall extend our scope of layout of a structure involving size, shape, and topology, and shall solve layout problems for more general solid

G. I. N. Rozvany (ed.), Optimization of Large Structural Systems, Vol. I, 157–175.

structures without assuming frame with complex topology constructed by networking all possible combination of joints distributed in a given design space. In other words, we shall formulate the problem based on the assumption of *continuum* so that any topology of a structure can be generated without assuming any special structural elements and their combination. Furthermore, the shape and size of the optimum structure will also be determined as the result of the design problem without specifying any of parametric representation of the shape and size of a structure. In particular, the optimum layout of three-dimensional shell structures will be studied as an extension of the previous works of the homogenization method introduced in Bendsϕe and Kikuchi [10], and also in Suzuki and Kikuchi [11] for plane structures.

2. A Homogenization Method for Plate/Shell Structures

The basic concept of the homogenization method for the generalized layout problem for three-dimensional plate/shell structures in the present study, is the same to the one for plane elastic structures. Here we shall briefly describe a homogenization method for plate/shell structures. The following is the fundamental steps of the homogenization method of the layout optimization problem :

1) A base shell structure is assumed, the thickness of which is described as a function h_0. This is thickness can be zero, and then we shall consider "pure" layout design of a structure. If h_0 is not zero, a built-up shell structure is assumed, and its optimum reinforcement layout is obtained by solving the present optimization problem.

2) A design domain Ω is specified on the curved surface in which the middle surface Ω_0 of the initial shell structure is contained. This domain can be a subset of the middle surface of the initial shell, while it is also possible to be larger than the middle surface. It is, however, noted that the location of the middle surface must be on a *fixed* curved surface, and it is not a design variable.

3) Microscopic perforation (but it should be "sufficiently" smaller than the thickness of the shell structure) is assumed to find the optimum reinforcement of the initial shell by adding solid material whose volume is prescribed. Perforation is characterized by three distributed design variables $\{ a_1, a_2, \theta \}$, the first two are the sizes of the rectangular hole in the unit cell and the angle of rotation of the hole in the macroscopic shell structure. These functions may have different values at different points in the design domain. The "true" size of holes made in the shell structure is tiny and infinitely many, and they are described by εa_1 and εa_2 for a sufficiently small positive number $\varepsilon>0$, while they are still assumed to be much smaller than the thickness of a shell. Thus, it may consider there are infinitely many and "small" rectangular parallelopipes with holes, whose height is a half of a fixed value h_1, on the both sides of the initial shell in order to maintain symmetry of the cross section with respect to the middle surface. A schematic description of the design variables is given in Fig. 1.

4) Reinforcement is designed to be constructed by accumulation of appropriate scaling of the unit cell described in Fig. 2 with rotation θ about the normal line to the middle surface. Scaling is taken place only on the middle surface, while the height of the hollow rectangular parallelopipes keeps constant to be a half of h_1 in the both sides of the middle surface. The total volume of reinforcement is given by

$$V_R = \rho h_1 \int_{\Omega} (1 - a_1 a_2) d\Omega \tag{1}$$

where ρ is the mass density of the material for reinforcement.

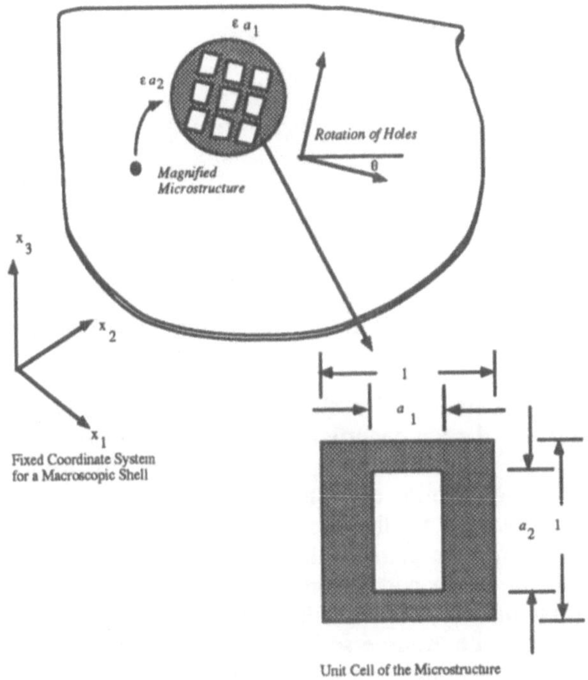

Figure 1 A Schematic Description of the Design Variables on a Three-Dimensional Shell Structure

5) The optimum design is defined by minimizing the mean compliance of a shell structure for a set of specified loading and support conditions under the volume constraint

$$V_R \leq V_{given} \, . \tag{2}$$

No other constraints on the stress, strain, and displacement of the shell are assumed in this formulation, while the standard formulation of the optimization is stated as minimizing the total weight of the shell under constraints of the stress, strain, and displacement. In this sense, this formulation may have significant limitation in design optimization in practice.

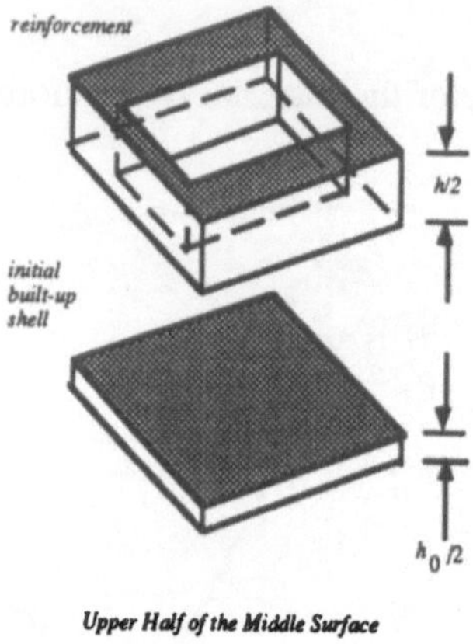

Figure 2 Upper Half Portion of the Unit Cell of the Initial Shell and its Reinforcement

6) To simplify formulation of a shell, let us assume that a curved shell is approximated by union of 4 node quadrilateral finite elements defined by four corner nodes placed in the three-dimensional space. Further, in order to neglect the curvature effect of a shell, finite element formulation assumes the flat 4 node quadrilateral element obtained by the projection of the original possible non-flat element onto the xy coordinate plane which is defined by minimizing the sum of squares of the distance of corner nodes from the plane. The coordinates x and y are then set up as the principal directions obtained by solving the associated eigenvalue problem. The thickness (i.e. transverse) direction is defined as the normal to the xy plane, and is identified with the z axis. The coordinates (x,y,z) can then define a local coordinate system in which a flat shell element is derived.

Suppose that the displacement field $\{u_x, u_y, u_z\}$ of an arbitrary point P in a shell element Ω_e is approximated by

$$
\begin{aligned}
u_x(x,y,z) &= u(x,y) + z\theta_y(x,y) \\
u_y(x,y,z) &= v(x,y) - z\theta_x(x,y) \\
u_z(x,y,z) &= w(x,y)
\end{aligned}
\tag{3}
$$

where $\{u, v, w\}$ is the displacement field of the point P' of the projection of P on the middle surface of the shell that coincides with the xy plane, θ_x and θ_y are the rotation of P about the x and y axes, respectively. This approximation of the displacement field yields the strains :

$$
\begin{aligned}
\{\varepsilon\}^T &= \{\varepsilon_x \quad \varepsilon_y \quad \varepsilon_y \quad \gamma_{yz} \quad \gamma_{zx} \quad \gamma_{xy}\} \\
&= \left\{\frac{\partial u_x}{\partial x} \quad \frac{\partial u_y}{\partial y} \quad \frac{\partial u_z}{\partial z} \quad \frac{\partial u_y}{\partial z}+\frac{\partial u_z}{\partial y} \quad \frac{\partial u_z}{\partial x}+\frac{\partial u_x}{\partial z} \quad \frac{\partial u_x}{\partial y}+\frac{\partial u_y}{\partial x}\right\} \\
&= \left\{\frac{\partial u}{\partial x} \quad \frac{\partial v}{\partial y} \quad 0 \quad 0 \quad 0 \quad \frac{\partial u}{\partial y}+\frac{\partial v}{\partial x}\right\} \\
&+z\left\{\frac{\partial \theta_y}{\partial x} \quad -\frac{\partial \theta_x}{\partial y} \quad 0 \quad 0 \quad 0 \quad \frac{\partial \theta_y}{\partial y}-\frac{\partial \theta_x}{\partial x}\right\} \\
&+\left\{0 \quad 0 \quad 0 \quad -\theta_x+\frac{\partial w}{\partial y} \quad \frac{\partial w}{\partial x}+\theta_y \quad 0\right\}
\end{aligned} \tag{4}
$$

Using the contracted notation, the internal virtual work in an arbitrary shell element Ω_e can be written by

$$
\delta U_e = \int_{\Omega_e}\int_{-h_1/2}^{h_1/2} \{\delta\varepsilon\}^T [D]\{\varepsilon\} dz d\Omega \tag{5}
$$

where $[D]$ is the elasticity matrix in the contracted notation obtained by assuming the plane stress condition in the xy plane. The matrix $[D]$ is defined by the elasticity tensor $\mathbb{E}^G$ obtained by rotation θ of the homogenized elasticity tensor $\mathbb{E}^H$ for plane stress problems. Applying the symmetry condition with respect to the middle surface, the internal virtual work in Ω_e can be written by

$$
\begin{aligned}
\delta U_e = &\int_{\Omega_e} \left(\{\delta\varepsilon_m\}^T [D_0]\{\varepsilon_m\} + \{\delta\kappa_B\}^T [D_2]\{\kappa_B\} + \{\delta\gamma\}^T [D_{ts}]\{\gamma\}\right) d\Omega \\
&+ \bar{\alpha}^2 \delta\theta_z \theta_z
\end{aligned} \tag{6}
$$

where

$$
[D_0] = \int_{-h_1/2}^{-h_0/2} [D_m(a_i,\theta)] dz + \int_{-h_0/2}^{h_0/2} [D_m(0,0)] dz + \int_{h_0/2}^{h_1/2} [D_m(a_i,\theta)] dz
$$

$$
[D_2] = \int_{-h_1/2}^{-h_0/2} [D_m(a_i,\theta)] z^2 dz + \int_{-h_0/2}^{h_0/2} [D_m(0,0)] z^2 dz + \int_{h_0/2}^{h_1/2} [D_m(a_i,\theta)] z^2 dz
$$

$$
[D_{ts}] = \int_{-h_1/2}^{-h_0/2} [D_s(a_i,\theta)] dz + \int_{-h_0/2}^{h_0/2} [D_s(0,0)] dz + \int_{h_0/2}^{h_1/2} [D_s(a_i,\theta)] dz
$$

$$[D_m(a_i,\theta)] = \begin{bmatrix} D_{11}(a_i,\theta) & D_{12}(a_i,\theta) & D_{16}(a_i,\theta) \\ & D_{22}(a_i,\theta) & D_{26}(a_i,\theta) \\ SYM & & D_{66}(a_i,\theta) \end{bmatrix} \qquad [D_s] = \beta \begin{bmatrix} D_{44} & 0 \\ 0 & D_{55} \end{bmatrix}$$

$$\{\varepsilon_m\} = \begin{Bmatrix} \partial u / \partial x \\ \partial v / \partial y \\ \partial u / \partial y + \partial v / \partial x \end{Bmatrix} \qquad \{\kappa_B\} = \begin{Bmatrix} -\partial\theta_y / \partial x \\ \partial\theta_x / \partial y \\ \partial\theta_x / \partial x - \partial\theta_y / \partial y \end{Bmatrix} \qquad \{\gamma\} = \begin{Bmatrix} -\theta_x + \partial w / \partial y \\ \theta_y + \partial w / \partial x \end{Bmatrix}$$

and

$$\overline{\alpha}^2 = \alpha^2 h_1 \int_{\Omega_e} \{(x - x_0)^2 D_{44} + (y - y_0)^2 D_{55}\} d\Omega_e$$

Here (x_0, y_0) may be identified with the centroid of Ω_e, α is a very small number to regularize with respect to spin of an element, and β is the so-called shear correction factor. Functions a_i are sizes and θ is angle of rotation of "microscopic" holes that only exists in $[-h_1/2, -h_0/2]$ and $[h_0/2, h_1/2]$. As shown in above the design variables $\{ a_1, a_2, \theta \}$ define the D matrices in the shell formulation, and thus, for a set of their fixed values we can regard the above a standard shell formulation that can be found in the literature of finite element analysis of shells, see, e.g., Noor, Belytschko, and Simo[9]. Since "torsional" rigidity is introduced artificially, this model involves 5 degrees of freedom per node. It is also noted that the homogenization process is applied at the level of the D matrix before integrating it in the thickness direction to compute appropriate rigidity for a plate/shell.

Approximating u, v, w, θ_x, and θ_y by bilinear polynomials in the parametric coordinates ξ and η, using the shape functions

$$N_\alpha(\xi,\eta) = \frac{1}{4}(1 + \xi_\alpha \xi)(1 + \eta_\alpha \eta) \qquad \alpha = 1,\ldots,4 \tag{7}$$

where $\{(\xi_\alpha, \eta_\beta)\}$ are the parametric coordinates of the four corner nodes of an element. Applying this approximation, we discretize the total strain energy, and then the internal virtual work of the shell is approximated. Indeed, in each finite element we obtain the element stiffness matrix in the local coordinate system (x,y,z) :

$$[K_e] = \int_{\Omega_e} [B_m]^T [D_0][B_m] d\Omega + \int_{\Omega_e} [B_B]^T [D_2][B_B] d\Omega + \int_{\Omega_e} [B_\gamma]^T [D_{ts}][B_\gamma] d\Omega + [k_\alpha]^T [k_\alpha] \tag{8}$$

Here

$$\{\varepsilon_m\} = \sum_{\alpha=1}^{4} \begin{bmatrix} \frac{\partial N_a}{\partial x} & 0 & 0 & 0 & 0 & 0 \\ 0 & \frac{\partial N_a}{\partial y} & 0 & 0 & 0 & 0 \\ \frac{\partial N_a}{\partial y} & \frac{\partial N_a}{\partial x} & 0 & 0 & 0 & 0 \end{bmatrix} \begin{Bmatrix} u_\alpha \\ v_\alpha \\ w_\alpha \\ \theta_{x\alpha} \\ \theta_{y\alpha} \\ \theta_{z\alpha} \end{Bmatrix} = [B_m]\{d_e\}$$

$$\{\kappa_B\} = \sum_{\alpha=1}^{4} \begin{bmatrix} 0 & 0 & 0 & 0 & -\frac{\partial N_a}{\partial x} & 0 \\ 0 & 0 & 0 & \frac{\partial N_a}{\partial y} & 0 & 0 \\ 0 & 0 & 0 & \frac{\partial N_a}{\partial x} & -\frac{\partial N_a}{\partial y} & 0 \end{bmatrix} \begin{Bmatrix} u_\alpha \\ v_\alpha \\ w_\alpha \\ \theta_{x\alpha} \\ \theta_{y\alpha} \\ \theta_{z\alpha} \end{Bmatrix} = [B_B]\{d_e\}$$

$$\{\gamma\} = \sum_{\alpha=1}^{4} \begin{bmatrix} 0 & 0 & \frac{\partial N_\alpha}{\partial y} & -N_\alpha & 0 & 0 \\ 0 & 0 & \frac{\partial N_\alpha}{\partial x} & 0 & N_\alpha & 0 \end{bmatrix} \begin{Bmatrix} u_\alpha \\ v_\alpha \\ w_\alpha \\ \theta_{x\alpha} \\ \theta_{y\alpha} \\ \theta_{z\alpha} \end{Bmatrix} = [B_\gamma]\{d_e\}$$

$$\theta_z = \sum_{\alpha=1}^{4} \{0 \quad 0 \quad 0 \quad 0 \quad 0 \quad \bar{\alpha}/4\} \begin{Bmatrix} u_\alpha \\ v_\alpha \\ w_\alpha \\ \theta_{x\alpha} \\ \theta_{y\alpha} \\ \theta_{z\alpha} \end{Bmatrix} = [k_\alpha]\{d_e\}$$

$$\delta U_e = \{\delta d_e\}^T [K_e]\{d_e\}$$

and $\{d_e\}$ is the vector of the degrees of freedom in an element. Applying appropriate numerical integration rules which yield equivalent effect of appropriately assumed stress or strain fields in an element, we can compute the element stiffness matrix of the shell. Details of such integration schemes can be found in Noor, Belytschko, and Simo[12].

7) If the shell is subject to external forces and moments on the lateral surfaces and its boundary, we can derive their approximation using the 4 node quadrilateral element. We shall represent such an approximation by

$$\delta P_e = \{\delta d_e\}^T \{f_e\} \tag{9}$$

where δP_e is the work done by the external forces and moment of an element Ω_e, and $\{ f_e \}$ is the set of equivalent nodal forces and moments with respect to the degrees of freedom in a finite element Ω_e.

Thus, the total potential energy of the shell structure is approximated by the discrete form

$$\Pi(\{d_e\}) = \sum_{e=1}^{E} \frac{1}{2}\{d_e\}^T[K_e]\{d_e\} - \{d_e\}^T\{f_e\} \tag{10}$$

where E is the number of finite elements covering the shell structure, i.e., design domain in the optimum reinforcement problem. It is clear that the element stiffness matrix $[K_e]$ depends on the design variables $\{ a_1, a_2, \theta \}$ which are distributed functions defined on the curved surface containing the middle surface of the shall.

8) Since the design problem is defined by minimizing the mean compliance under the volume constraint for the amount of reinforcement material, its finite element approximation is given as follows :

$$\underset{V_R=\rho h_1 \int_\Omega (1-a_1 a_2) d\Omega \le V_{given}}{Minimize} \quad \sum_{e=1}^{E} \{d_e\}^T \{f_e\} \tag{11}$$

Noting that

$$\Pi(\{d_e\}) = -\sum_{e=1}^{E} \frac{1}{2}\{d_e\}^T\{f_e\} = \underset{\{\bar{d}_e\}}{Minimize}\ \Pi(\{\bar{d}_e\})$$

we can define the optimization problem by

$$\underset{V_R=\rho h_1 \int_\Omega (1-a_1 a_2) d\Omega \le V_{given}}{Minimize} \quad \left(-2\,\underset{\{\bar{d}_e\}}{Minimize}\ \Pi(\{\bar{d}_e\})\right) \tag{12}$$

Introducing the Lagrange multiplier $\lambda \le 0$ to the volume constraint, and defining the Lagrangian

$$L = \Pi - \lambda\left(\rho h_1 \sum_{e=1}^{E} \int_{\Omega_e} (1 - a_1 a_2) d\Omega - V_{given}\right), \tag{13}$$

the first variation of this Lagragian with respect to the design variables, degrees of freedom in the finite element model, and the Lagrange multiplier, yields the optimality condition of the finite element approximation of the design problem :

$$\sum_{e=1}^{E} [K_e]\{d_e\} = \sum_{e=1}^{E} \{f_e\}$$

$$\frac{1}{2}\{d_e\}^T\left(\frac{\partial}{\partial a_1}[K_e]\right)\{d_e\} + \lambda\rho h_1 a_2 = 0$$

$$\frac{1}{2}\{d_e\}^T\left(\frac{\partial}{\partial a_2}[K_e]\right)\{d_e\} + \lambda\rho h_1 a_1 = 0 \qquad e = 1,\ldots,E$$

$$\frac{1}{2}\{d_e\}^T\left(\frac{\partial}{\partial \theta}[K_e]\right)\{d_e\} = 0$$

$$\lambda\left(\rho h_1 \sum_{e=1}^{E} \int_{\Omega_e} (1 - a_1 a_2) d\Omega - V_{given}\right) = 0 \tag{14}$$

Applying the optimality criteria method described in Bendsφe and Kikuchi [10] and also in Suzuki and Kikuchi [11], we can derive a computational scheme to determine the design variables { a_1, a_2, θ }. In the present work, we discretize the design variables by piecewise constant functions, i.e., within a finite element , a_1 , a_2 , and θ are assumed to be constant. Thus 3E discrete design variables are introduced in the discrete optimization problem.

9) It is noted that the height h_1 of the microscopic hollow rectangular parallelopipes is assumed to be constant in the optimization problem, while the sizes of hollowness, a_1 and a_2, are assumed to vary in design. The standard treatment of the optimum reinforcement of a plate/shell structure is defined by obtaining the thickness of added reinforcement to the initial thin plate/shell structure. In other words, the design variable is the thickness $h_1(x,y)$ of the reinforcement. In the present approach, we assume the constant height of the "ribs," but their width and the orientation will be determined so as to the optimum is achieved. The existence of microscopic ribs (*but whose size is smaller than the shell thickness*) may be

expected. Thus, the present optimization can be classified as a sizing optimization problem, despite that the choice of the sizes to be optimized is non-standard.

Another characteristics of the present approach is that the homogenization process is applied at the level of the D matrix of a solid, instead of applying the homogenization to the bending rigidity. In other words, we assume the size of microscale holes are much smaller than the plate/shell thickness, and then the Mindlin hypotheses that is a generalization of the Love-Kirchhoff assumption is applied to obtain the plate/shell formulation after the homogenization process is taken place in heterogeneous three-dimensional solids. If the size of microsctructure is larger than the plate/shell thickness, we must apply the homogenization process in the plate/shell formulation, see Bendsφe [13]. For this type of plates, the layout optimization is studied by Diaz and Soto [14] based on the Bendsφe and Kikuchi approach.

10) Thickness optimization of a plate/shell structure has been extensively studied in structural optimization by, e.g., Schmit et al. [15], Morrow and Schmit [16], Simitses [17], Banichuk [18], Haftka and Prasad [19], Cheng [20], Cheng and Olhoff [21,22], and others. Especially, majority of literature related to this subject can be found in the survey paper Haftka and Prasad [19] and a general theory of the thickness optimization of a plate can be found in Banichuk[18] for the case that the variation of the thickness is sufficiently small and smooth so that discrete stiffeners should not appear in the optimum. Delicate discussion on convergence and mechanical models of a plate/shell is discussed in Cheng [20], as well as importance of introduction of the homogenization method is demonstrated in Cheng and Olhoff [21,22] and also in Bendsφe [13]. A layout theory of plates are also discussed in Rozvany[23] with a different context from the present approach.

3. Optimum Layout of a Simply Supported Plate

For demonstration of the homogenization method to find the optimal layout of a plate/shell structure, we shall consider a simply supported plate subject to two loading conditions. The size of the plate is 60 cm × 60 cm as shown in Figs. 3 and 4. Using symmetry of the geometry and the loading condition considered here, one quarter of the domain need be modeled by 4 node finite elements to find the optimal layout of reinforcement of a thin plate of the initial thickness h_0=0.1cm. The height of "stiffeners" which are introduced for reinforcement is restricted to h_1=1cm. A quarter of the plate, i.e. the design domain is divided into 30x30 square and uniform finite elements in which 2,700 discrete design variables are involves while the total number of degrees of freedom of the finite element model is 4,805 because of 5 degrees of freedom per node. Young's modulus of the reinforcement and the initial plate material is assumed to be 200 GPa, while Poisson's ratio is 0.29. We shall apply two different loads, a point transverse force at the center and a uniformly distributed transverse load on the plate.

The optimal distribution of reinforcement of a thin plate is obtained as shown in Fig. 5 for the point load. If the volume of reinforcement (i.e. solid material added to the original thin plate) is very large, no reinforcement are assigned in the vicinity of the lines connected for the mid-points of two adjacent boundary edges of the plate. A square plate rotated 45 degree is formed as a part of stiffeners. If the thickness of the middle thin plate is assumed to be zero, i.e., $h_0=0$, the optimal layout of solid material to form a plate structure must yield line hinges along these four 45 degree inclined lines. It is noted that two hinges are generated for a beam when its thickness is optimized for the case of point load at the center of the clamped beam. Thus, line hinges in the optimal reinforcement may be expected. It should be noted that this does not mean discontinuity of the transverse displacement along these hinge lines.

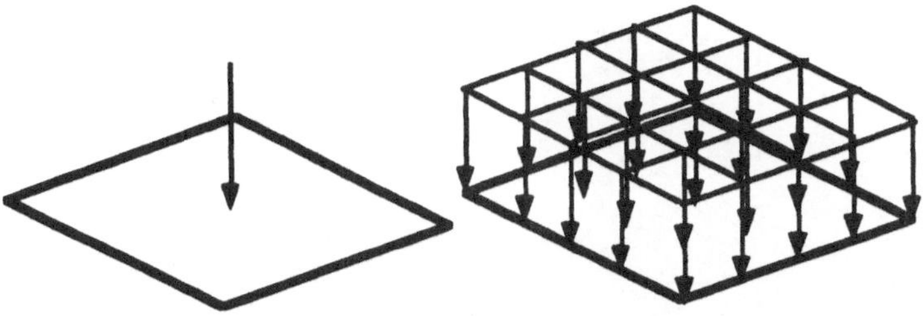

Figure 3 Simply Supported Square Plate Subject to Two Different Loading Conditions

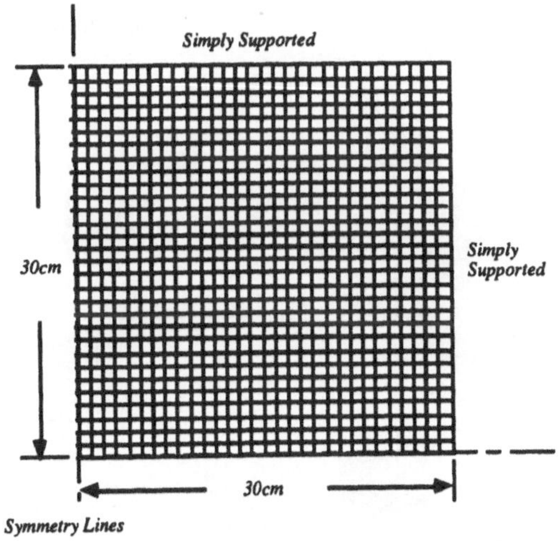

Figure 4 Finite Element Model of a Quarter Portion of the Plate

But only the slope may be discontinuous, and it is still admissible in the variational formulation of an elastic plate defined in the Sobolev space $H^2(\Omega)$. The slope, i.e., the normal derivative of the transverse deflection w along a finite number of curves in the plate can be discontinuous. In this sense, the optimal reinforcement or layout pushes the transverse deflection w to the limit of the admissible space, in other words, w is just in $H^2(\Omega)$ but not in $H^{2+\delta}(\Omega)$ for $\delta>0$. This, further, may imply that proof of convergence of finite element approximations requires very delicate argument when the representative mesh size goes to zero, since it expects extra regularity (smoothness) of the solution to obtain the rate of convergence of the finite element approximation. It should be possible to establish strong convergence of the finite element approximations to the optimal solution, but it may be difficult to establish an explicit rate of convergence.

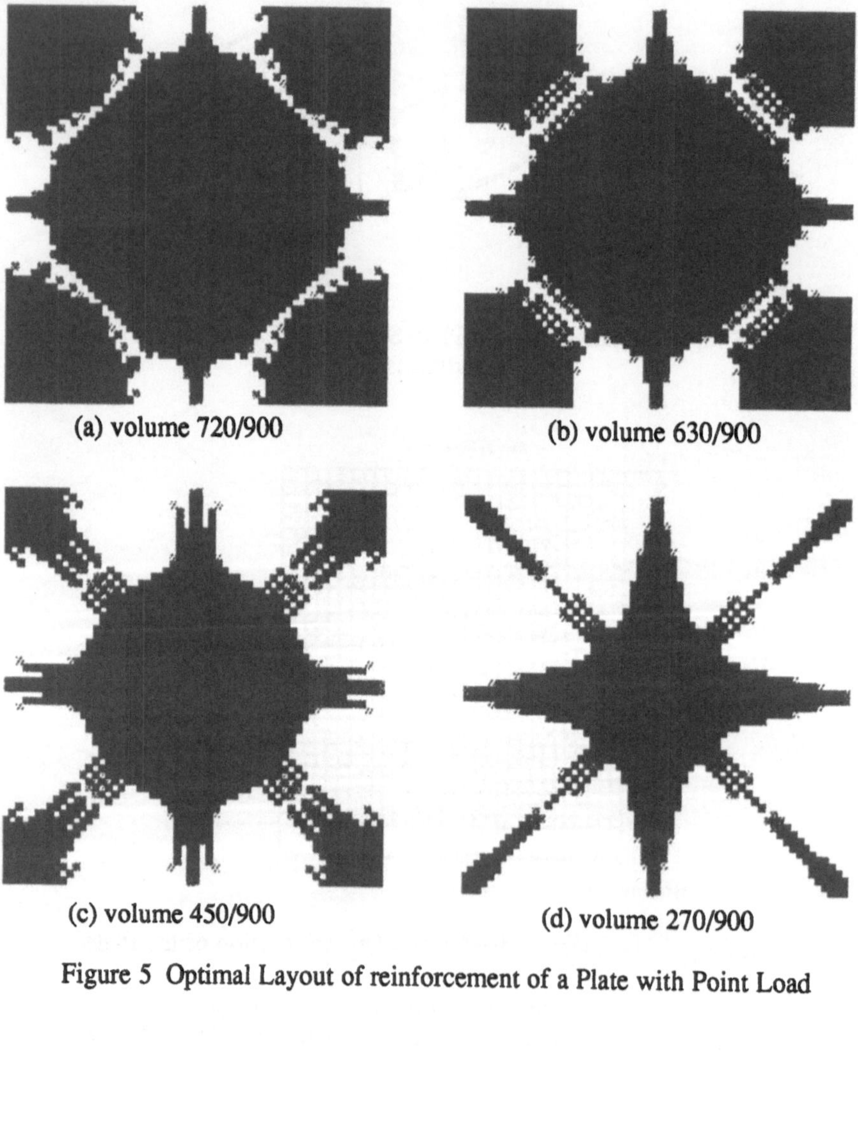

(a) volume 720/900 (b) volume 630/900

(c) volume 450/900 (d) volume 270/900

Figure 5 Optimal Layout of reinforcement of a Plate with Point Load

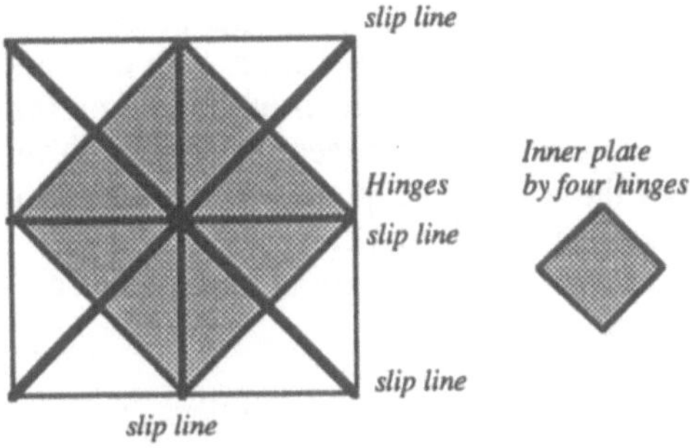

Figure 6 Slip Lines of Simply Supported Plates in Limit Analysis

To explain the result obtained in Fig.5 for the optimum layout of reinforcement, let us introduce a fact of limit line analysis of plasticity in which slip lines appear, see Figure 6. It can be easily understood that the reinforcement in layout optimization are formed on the slip lines of limit analysis obtained in Yang[24]. Here, because of formation of four hinge lines in the optimum layout, these behave so as to the simply supported boundary, and then two more slip lines are added by connecting diagonals of this "newly" formed simply supported plate. Most of reinforcement is formed along the slip lines, and this seems to be a natural consequence. When limit lines appears, displacement becomes infinite in linear theory, while our objective of reinforcement is to minimize the mean compliance which yields the minimum displacement for a specified load. To avoid plastic hinges in limit analysis, reinforcement in an elastic plate must be along these lines. A rather thick cross shape stiffener on the center (which corresponds to slip lines of the inner simply supported plate generated by appearance of four hinges in the reinforcement) and thin stiffeners along the diagonals of the original plate, are formed.

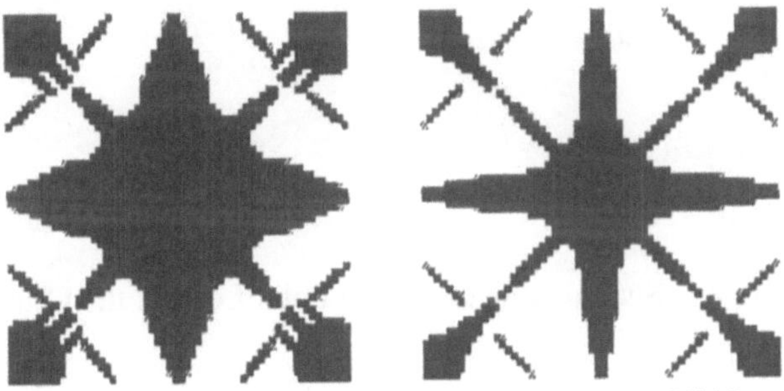

(a) volume 450/900 (b) volume 270/900

Figure 7 Optimal Layout of a Plate with Distributed Load

Note that the black portion in the figures indicates that no holes are generated over there in reinforcement i.e. stiffeners or plates whose height is fixed to be h_1-h_0.

Another interesting observation is that solid stiffeners or plates are generated in most of the portion. There is very little perforation in the optimal design, although it is allowed. In other words, fine microstructure by perforation may not be the optimum as far as the present result is concerned based on the formulation given in above. Figure 7 shows the result for a plate with the uniformly distributed load. It is noteworthy that in the distributed load case, additional stiffeners appear in the four corner area. The overall layout for this case is the same to the one for the point load.

4. Optimum Layout of a Simply Supported Shallow Shell

Let us consider a shallow shell whose projected plane to the xy plane is the same to the plate considered in the previous section. The shallow shell is defined by the curved surface

$$z(x,y) = z_{max} \sin \pi \frac{x}{x_{max}} \sin \pi \frac{y}{y_{max}} \qquad \textit{where} \qquad \frac{z_{max}}{x_{max}} = \frac{z_{max}}{y_{max}} = \frac{1}{12}.$$

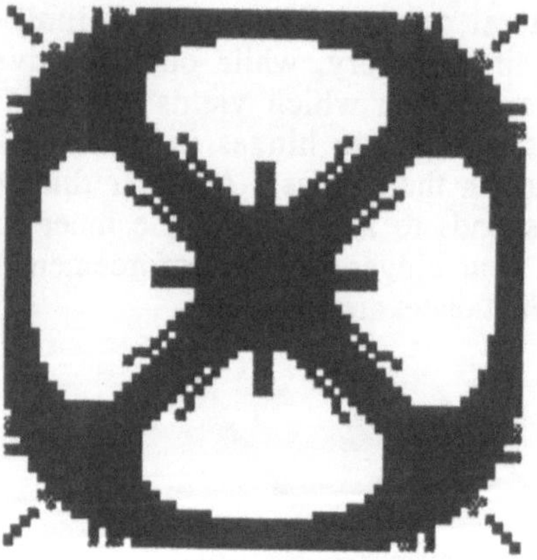

(a) volume 450/900

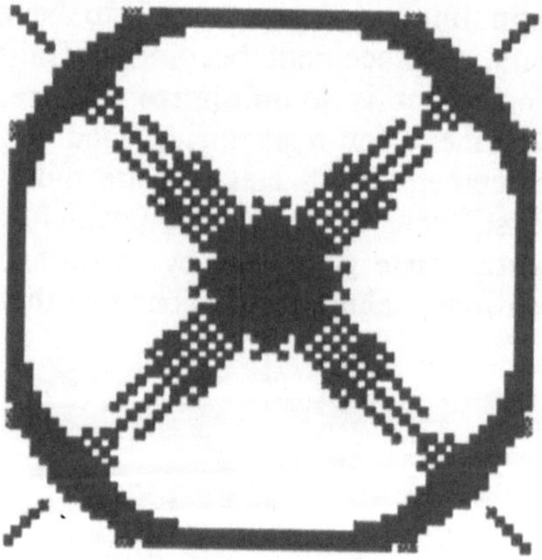

(b) volume 270/900

Figure 8 Shallow Shell with the Point Load at the Center

The results of a shell are very different from those of a plate as expected from the fact that the optimal layout of an arch is dramatically different from the one of a beam shown in Section 3. Figure 8 shows the optimal distribution of stiffener or reinforcement layout of a shell for the case of a point transverse load applied at the center top of the shallow shell surface. A ring shape reinforcement is formed that is never be generated in a plate. Inner cross shape center stiffener in a plate disappear in a shell. Rather thick stiffeners are assigned along the diagonals. This is again, very different from the case of the plate. Hinges appears on the diagonal stiffeners, but they are different from the plate, too. Another difference is that scattered reinforcement is observed inside a ring stiffener when the volume of solid material for reinforcement is reduced. This indicates possibility of the microstructure of perforation over there. For the plate, all of stiffeners are solid.

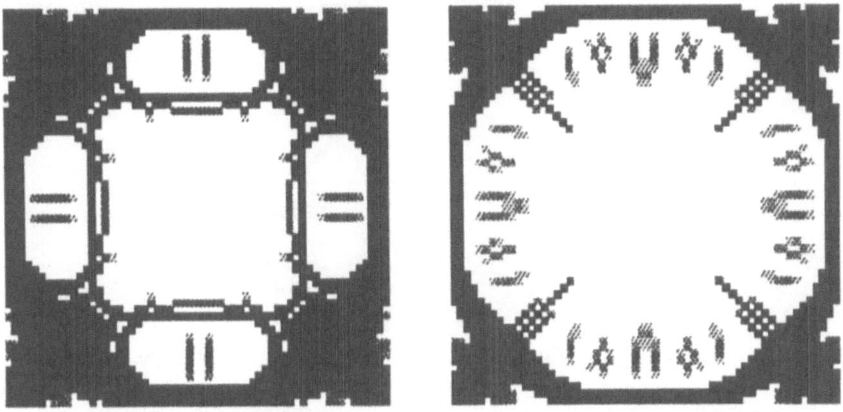

(a) volume 450/900 (b) volume 270/900

Figure 9 Shallow Shell with the Uniformly Distributed Load

Figure 9 shows the result of a shell for the distributed load. Basic pattern of distribution of stiffeners is similar to the one for the point load case, although they are more distributed in the four corners of a shell. Stiffeners in the center portion disappear in the distributed load case.

It is clear that very different layouts are obtained in a plate and a shell for the both loading cases. This difference may be explained by the fact that a shell is combination of a plate and a membrane. Quite large portion of applied forces is supported by membrane rather than plate in a shell structure. If pure flat plate is considered, all the loads must be carried as a plate. Thus, the mechanics nature of a shell is very different from that of a plate, and this difference implies different optimal layouts of reinforcement in design optimization.

5. Transition from a Plate to a Shallow Shell

As shown in above the layout for a shallow shell is different from the one for a plate. Thus, it may be interesting to investigate transition process from a shell to a plate by reducing the shell height z_{max} to zero. It is a natural question what happens in between. Does the basic pattern of the layout change suddenly at a certain point ? Figure 9 shows how the optimal distribution of reinforcement changes from Fig.7 (b) to Fig.9 (b). It is observed that sudden change does not occur, but changes are rather gradual. In other words, transition of the layout of a shell to that of a plate is "continuous." Since quantifying the change of topology is difficult, we shall display in Fig. 10 the change of the mean compliance in transition from a plate to a shallow shell. The uniformly distributed pressure is applied and the shape of the shell is the same to the one used in the previous section, while the shell height is varying from 0 to 4. Compliance change also shows very smooth transition from a plate to a shell, though there is a region in which compliance changes is very steep. It is natural to expect that in this region, the structure changes from "bending dominant" to "membrane dominant".

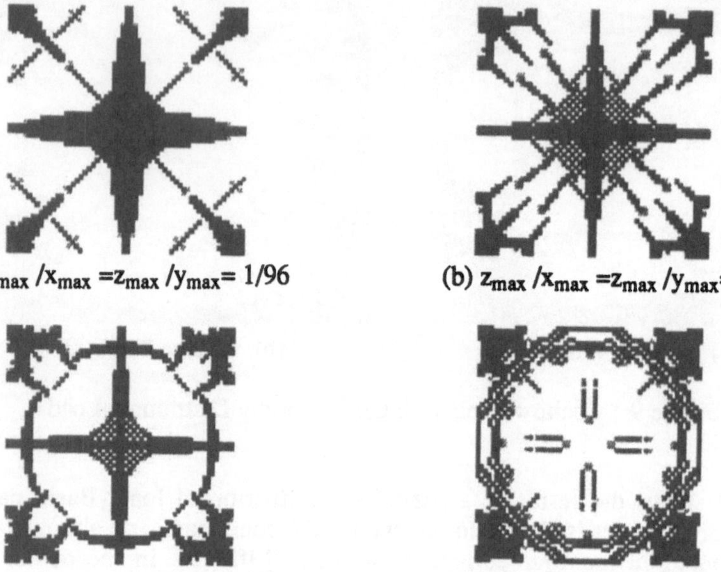

(a) $z_{max}/x_{max} = z_{max}/y_{max} = 1/96$ (b) $z_{max}/x_{max} = z_{max}/y_{max} = 2/96$

(c) $z_{max}/x_{max} = z_{max}/y_{max} = 3/96$ (d) $z_{max}/x_{max} = z_{max}/y_{max} = 4/96$

Figure 9 Optimal Reinforcement of Shells with Various Height
(Distributed Load, Volume 270/900)

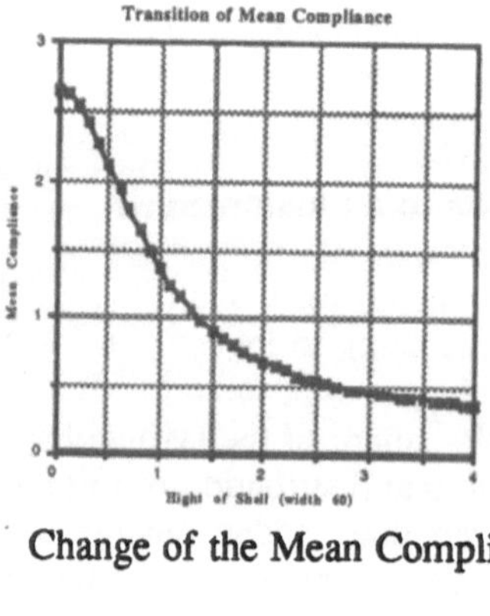

Figure 10 Change of the Mean Compliance

6. Conclusions

The homogenization method to find the optimum layout of reinforcement of three-dimensional shell structures is presented in this article together with several computational examples. As shown in above, without specifying the shape and topology of reinforcement of the shell initially built up, the optimum configuration can be obtained. The design variables are the size of rectangular parallelopipes in microscopic level as well as the rotation of these. Their height is held to be constant, while standard thickness optimization of a plate/shell utilizes the thickness as the design variable.

Acknowledgement The authors are partially supported by NSF DDM89-17430, NIH AR34399, NASA Lewis Research Center NAG3-1160, and FORD Foundation. Their generous support is sincerely appreciated by authors.

References

[1] Schmit, L.A., *Structural design by systematic synthesis*, Proceedings 2nd ASCE Conference on Electronic Computation (1960) pp.105-132, New York

[2] Fox, R. L., *Constraint surface normals for structural synthesis techniques*, AIAA J. 3-8 (1965) pp.1517-1518

[3] Prager, W and Taylor,J.E., *Problems of optimal structural design*, J. Appl. Mech. 35 (1968) pp.102-106

[4] Prager,W., *A note on discretized Michell structures*, Comput. Mechs. Appl. Mech. Engrg., 3-3 (1974) pp.349-355

[5] Michell, A.G.M., *The limits of economy of material in framed structures*, Phil. Mag. 6 (1904) pp.589-597

[6] Rozvany, G.I.N., *Structural Design via Optimality Criteria*, Kluwer, (1989), Dordrecht

[7] Rozvany, G.I.N., Zhou, M., Rotthaus, M., and Gollub, W., *Continuum optimality methods for large finite element systems with a displacement constraint, Part I*, Structural Optimization Vol.1-No.1, (1989) pp.47-71

[8] Rozvany, G.I.N., Zhou, M., and Gollub, W., *Continuum optimality methods for large finite element systems with a displacement constraint, Part II*, Structural Optimization Vol.2-No.2, (1990) pp.77-104

[9] Bendsφe, M.P., and Ben-Tal, A., and Haftka, R.T., *New displacement-based methods for optimal truss topology design*, in Proceedings of AIAA /ASME/ ASCE/ AHS/ ASC 32nd Structures, Structural Dynamics and Materials Conference, Baltimore, Maryland, April 8-10, 1991

[10] Bendsφe, M.P., and Kikuchi, N., *Generating optimal topologies in structural design using a homogenization method*, Comput. Meth. Appl. Mech. Engrg., 71 (1988) pp.197-224

[11] Suzuki K, Kikuchi N , *Homogenization Method for Shape and Topology Optimization*, Comp. Meth. Appl. Mech. Enginrg. to Appear in 1991

[12] Noor A.K., Belytschko, T., and Simo, J.C., *Analytical and Computational Models of Shells*, CED-Vol.3, American Society of Mechanical Engineers, (1989), New York

[13] Bendsφe, M.P., *Generalized Plate Models and Optimum Design*, in *Homogenization and Effective Moduli of Materials and Media*, ed. J.L. Ericksen et al., Springer-Verlag, (1986), New York, pp.1-26

[14] Diaz, A.R., and Soto, C.A., Optimum Shape and Layout of Plate Structures, Technical Report in Department of Mechanical Engineering, Michigan State University, (1991), East Lansing, Michiga, U.S.A.

[15] Schmit, L.A., Kicher, T.P., and Morrow, W.M., *Structural Synthesis capability for integrally stiffened waffle plates*, AIAA Journal, 1, (1963) pp.2820-2836

[16] Morrow, W.M., and Schmit, L.A., *Structural Synthesis of a Stiffened Cylinder*, NASA (1968) CR-1217

[17] Simitses, G.J., *Optimal versus the stiffened circular plate*, AIAA Journal, 11-10, (1973) pp.1409-1412

[18] Banichuk, N.V., *Problems and Methods of Optimal Structural Design*, Plenum Press, (1983), New York,

[19] Haftka, R.T., and Prasad, B., *Optimum structural design with plate bending elements - A Survey*, AIAA Journal, 19-4, (1981) pp.517-522

[20] Cheng, K.T., *On non-smoothness in optimal design of solid, elastic plates*, International Journal of Solids and Structures, 17, (1981) pp.795-810

[21] Cheng, K.T., and Olhoff N., *An investigation concerning optimal design of solid elastic plates*, International Journal of Solids and Structures, 17, (1981) pp.305-323

[22] Cheng, K.T., and Olhoff, N., *Regularized formulation for optimal design of axisymmetric plates*, International Journal of Solids and Structures, 18, (1982) pp.153-169

[23] Rozvany, G.I.N., *Structural Layout Theory - The Present State of Knowledge*, in *New Directions in Optimum Structural Design*, ed. E. Atrek et al, John Wiley & Sons, 1984, Chichester, pp.167-195

[24] Yang W.H., *Minimization Approach to Limit Solutions of Plate*, Comp Meth Appl Mech Enginrg 28 (1981) pp.265-274

Applications to Car Bodies :Generalized Layout Design of Three-Dimensional Shells

Junichi Fukushima
Toyota Technical Center, U.S.A., INC
Southfield, MI, U.S.A.

Katsuyuki Suzuki
The University of Tokyo
Tokyo, 113, JAPAN

and

Noboru Kikuchi
The University of Michigan
Ann Arbor, MI 48109-2215, U.S.A.

Abstract We shall describe applications of the homogenization method, formulated in Part 1, to design layout of car bodies represented by three-dimensional shell structures based on a multi-loading optimization.

1. Introduction

We shall apply the homogenization method described in Part 1 to find the optimum layout of reinforcement of three-dimensional shell-like car bodies. Since multiple loads are considered in design practice of car bodies, we have to modify the original formulation for a single objective.

The problem can be interpreted as a multi-objective (multi-purpose, multi-criteria optimization) problem in which there are several objective functions to be minimized or maximized. There are many approaches to find a Pareto optimum of the multi-objective problem, see a survey paper by Stadler [1] for literature in structural optimization, and Stadler [2] for a mathematical theory. The most typical approach is to use some combination of the objective functions to form a scalar single functional (i.e. scalarization of the vector optimum problem according to Stadler [1,2]) such as a modified global criteria approach in Chattopadhyay *et al* [3]. The most obvious scalarization consists of a linear combination of the objectives : $F(\boldsymbol{d}) = c_1 f_1(\boldsymbol{d}) + c_2 f_2(\boldsymbol{d}) + ... + c_N f_N(\boldsymbol{d})$, where $f_i(\boldsymbol{d})$ are individual objectives, $\boldsymbol{d}$ is the vector of design variables, and c_i are positive real numbers. Then the single functional F is minimized with respect to $\boldsymbol{d}$ and c_i. Another approach, the ß-method introduced by Taylor, reformulates the problem by introducing an additional design variable, the upper bound β of all the objective functions in order to construct a scalar function to be minimized, i.e., the multi-objective problem $min_{\boldsymbol{d}} \{ f_1(\boldsymbol{d}), ..., f_N(\boldsymbol{d}) \}$ is modified by $min_{\boldsymbol{d}}\, \beta$ subject to $f_i(\boldsymbol{d}) \le \beta$, i=1, ..., N, that is also equivalent to $min_{\boldsymbol{d}}\, max_{i=1,...,N}\, f_i(\boldsymbol{d})$. This upper bound formulation assures differentiability of the scalar functional. There are also other effective approaches such as the method based on a regret (

G. I. N. Rozvany (ed.), Optimization of Large Structural Systems, Vol. I, 177–191.

compromise) function by setting a goal to each objective to find utopia (or ideal) point, the goal programing approach, or more recently, the K-S function approach by Sobieskii *et al* [4] to solve the multi-objective optimization problem.

Here, we shall introduce a single objective function by using the maximum of the density of the mean compliance of multiple loadings to reformulate the original multiple loading optimization problem, although its mathematical justification has not been studied yet. In other words, we shall minimize the artificially constructed "compliance" by integrating the maximum of the density of the compliance of all the loadings. This formulation is a variation of the upper bound approach, and would find only a sub-optimum solution, while this may provide the safest design from the designer's point of view despite of lack of mathematical justification. That is, the solution of the present formulation for the multi-objective problem for multiple loadings need not be even a Pareto optimum. It is also noted that a more standard approach based on a linear combination of objectives is studied by Diaz and Bendsφe [5] in which the homogenization method introduced by Bendsφe and Kikuchi is applied to find the optimum layout of plane structures.

2. Formulation for the Multiple Load Optimization

We shall now describe a formulation of the multiple loading problem. Let a function F be defined by the maximum compliance among all the values of the compliance by the N set loadings $\{ (f_k, t_k) \}$, $k=1,...,N$,

$$F(u) = \underset{\text{load case } k}{Maximize} \left\{ \int_\Omega \rho f_k \bullet u_k d\Omega + \int_{\Gamma_T} t_k \bullet u_k d\Gamma \right\} \tag{1}$$

and we shall consider the optimization problem

$$\underset{\text{design variable } d}{Minimize} \quad F(u) \tag{2}$$

subject to equilibrium equations for multiple loads, and the weight constraint

$$\int_\Omega \rho(x) d\Omega \le \overline{W} \tag{3}$$

where ρ is the mass density of the structure.

The difficulty of this formulation is caused by non-differentiability of the function $F(\mathbf{u})$. This non-smoothness requires delicate handling for optimization, since most techniques assume computation of the sensitivity of the objective function. A method frequently used to overcome non-differentiability of the objective function is to introduce a dummy variable β that is an upper bound of all of the objective functions. In this case the problem is defined by

$$\underset{\text{design variables } \boldsymbol{d}}{\textit{Minimize}} \quad \beta \tag{4}$$

subject to

$$\int_{\Omega} \rho f_k \bullet u_k d\Omega + \int_{\Gamma_T} t_k \bullet u_k d\Gamma \leq \beta \text{ for all load cases } k \tag{5}$$

Equilibrium equations for multiple loads and the weight constraint

$$\int_{\Omega} \rho(x) d\Omega \leq \overline{W} \tag{6}$$

Since the scalar objective function β is differentiable with respect to design variable $\boldsymbol{d}$ = { ($\boldsymbol{a}$, $\boldsymbol{\theta}$) } in the homogenization method for the layout optimization, the Kuhn-Tucker conditions can be derived using the Lagrangian.

$$L(\boldsymbol{a},\boldsymbol{\theta},\boldsymbol{u},\boldsymbol{v},\boldsymbol{q},\Lambda_v) = \beta - \sum_{k=1}^{N} q_k \left(\beta - \int_{\Omega} \rho f_k \bullet u_k \, d\Omega - \int_{\Gamma_T} t_k \bullet u_k \, d\Gamma \right)$$
$$+ \sum_{k} \Pi(\boldsymbol{a},\boldsymbol{\theta},\boldsymbol{u}_k,\boldsymbol{v}_k) + \Lambda_v \left(\int_{\Omega} \rho \, d\Omega - \overline{W} \right) \tag{7}$$

where,

$$\Pi(\boldsymbol{a},\boldsymbol{\theta},\boldsymbol{u}_k,\boldsymbol{v}_k) = \int_{\Omega} \varepsilon(v_k) \bullet \mathbf{E}^G(\boldsymbol{a},\boldsymbol{\theta}) \varepsilon(u_k) \, d\Omega - \int_{\Omega} \rho f_k \bullet v_k d\Omega - \int_{\Gamma_T} t_k \bullet v_k d\Gamma \tag{8}$$

Here, $\mathbf{E}^G$ is the homogenized and rotated elasticity tensor that is a function of the design variable consisting of the vector $\mathbf{a} = \{a_1, \ldots, a_n\}$ and $\boldsymbol{\theta} = \{\theta_1, \ldots, \theta_n\}$ are the size and angle of rotation of the micro-scale holes. The Kuhn-Tucker conditions may be obtained as

$$\frac{\partial L}{\partial a_i} \delta a_i = 0 \,, \quad \frac{\partial L}{\partial \theta_i} \delta \theta_i = 0$$

$$\text{for all } \delta a_i = a_i^* - a_i \,, \ \delta\theta_i = \theta_i^* - \theta_i \,,$$

$$a_i^*, a_i, \theta_i^*, \theta_i \in H^1(\Omega) \,, \quad 0 \leq a_i^*, a_i \leq 1$$

$$\frac{\partial L}{\partial \mathbf{u}_k}\delta \mathbf{u}_k = 0, \quad \frac{\partial L}{\partial \mathbf{v}_k}\delta \mathbf{v}_k = 0 \quad \text{for each load case } k$$

$$\frac{\partial L}{\partial \beta} = 0$$

$$q_k\left(\beta - \int_\Omega \rho f_k \bullet u_k d\Omega - \int_{\Gamma_T} t_k \bullet u_k d\Gamma\right) = 0, \qquad k = 1,\ldots,N$$

$$q_k \geq 0, \beta - \int_\Omega \rho f_k \bullet u_k d\Omega - \int_{\Gamma_T} t_k \bullet u_k d\Gamma \geq 0 \quad k = 1,\ldots,N$$

$$\Lambda_V(\textstyle\int_\Omega \rho d\Omega - \overline{W}) = 0, \quad \Lambda_V \geq 0, \quad \int_\Omega \rho d\Omega - \overline{W} \leq 0 \tag{9}$$

Although this formulation is mathematically well-defined, it is difficult to develop an algorithm of the optimality criteria method that yields a resizing rule for $\boldsymbol{a}$ and $\boldsymbol{\theta}$, since there are too many constraints and Lagrange multipliers involved in this formulation, and hence too many equations must be satisfied in the optimality criteria method. Since there will be considerably many discrete design variables in the layout design problem, this situation is not desirable, especially to apply the optimality criteria method as a solution method.

In order to overcome this difficulty, another approach is introduced in this work by defining an artificial scalar objective function that is also smooth. To this end, note that v_k satisfies equilibrium for each load case :

$$\int_\Omega \varepsilon(v_k) \bullet \mathbf{E}^G(a,\theta)\varepsilon(u_k)\, d\Omega = \int_\Omega \rho f_k \bullet v_k d\Omega + \int_{\Gamma_T} t_k \bullet v_k d\Gamma \tag{10}$$

Hence our problem (1) - (3) can be written as

$$\underset{\text{design variables } d}{Minimize} \quad \underset{\text{load case } k}{Maximize} \quad \int_\Omega \varepsilon(u_k) \bullet \mathbf{E}^G(a,\theta)\varepsilon(u_k) d\Omega \tag{11}$$

subject to equilibrium equations of multiple loads and the weight constraint

$$\int_\Omega \rho(x) d\Omega \leq \overline{W} \tag{12}$$

Now, we shall introduce the new function

$$\overline{F}(u) = \int_\Omega \left\{ \underset{\text{load case } k}{Maximize} \ \ \varepsilon(u_k) \bullet \mathbf{E}^G(a,\theta)\varepsilon(u_k) \right\} d\Omega \tag{13}$$

that is defined by the maximum strain energy density of all the loads of the multiple load problem, and then we shall define a new approximation of the original multiple loading optimization problem {(11),(12)}:

$$\underset{\text{design variable } d}{Minimize} \quad \bar{F}(u)$$

subject to equilibrium equations of multiple loads, and

$$\int_\Omega \rho(x) d\Omega \le \overline{W} \tag{14}$$

It should be noted that this problem is not equivalent to the original one, but we have the following inequality relation:

$$F(u) \le \bar{F}(u)$$

i.e.,

$$\underset{\text{load case } k}{Maximize} \int_\Omega \varepsilon(u_k) \bullet \mathbf{E}^G(a,\theta)\varepsilon(u_k) d\Omega \le \int_\Omega \left\{ \underset{\text{load case } k}{Maximize} \ \varepsilon(u_k) \bullet \mathbf{E}^G(a,\theta)\varepsilon(u_k) \right\} d\Omega \tag{15}$$

Hence

$$\underset{\text{design variable } d}{Minimize} \ \underset{\text{load case } k}{Maximize} \int_\Omega \varepsilon(u_k) \bullet \mathbf{E}^G(a,\theta)\varepsilon(u_k) d\Omega$$

$$\le \underset{\text{design variable } d}{Minimize} \int_\Omega \left\{ \underset{\text{load case } k}{Maximize} \ \varepsilon(u_k) \bullet \mathbf{E}^G(a,\theta)\varepsilon(u_k) \right\} d\Omega \tag{16}$$

Since the original multi-loading problem leads a complex form of the optimality condition for the optimality criteria method, we have introduced a upper bound of the maximum of the mean compliance for all the loadings considered in the problem so that it yields the same form of the optimality condition to the single load case, and then without modifying the original program we can obtain a "sub-optimum" design for the multiple loading problem just by defining the new "strain energy" density. In other words, for numerical solution technique, the sensitivity of the load case which gives the maximum value of the strain energy density for each point in design domain is used in the update rule of sizes of holes, and the principal stress direction of the same load case is used in determining the rotational angles of microstructure.

It should be noted that a different formulation using a linear combination is introduced for the multi-loading layout optimization problem by Diaz and Bendosφe [5], while the similar homogenization method described here is applied. Details of such an "exact" treatment of the multi-loading problem can be found over there.

3. Three-Bar Frame Structure Comparison of the Two Objective Functions

In order to examine the formulation given in the previous section, we shall consider a three-bar frame problem solved by different methods in past. Three bars are placed as shown in Figure 1. Two bars are placed with 45 degree to the ceiling symmetrically, while a bar is vertical to the ceiling. Three loadings shown in Table 1 are applied. For this calculation circular cross section beam elements with bending and shear stiffness are used. This problem is first solved by Sheu and Schmit [4] as the three bar truss problem.

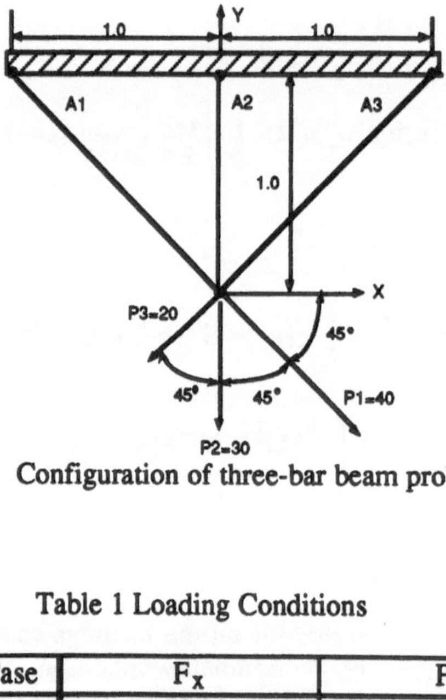

Figure 1 Configuration of three-bar beam problem

Table 1 Loading Conditions

Load Case	F_x	F_y
1	-14.14	-14.14
2	0	-30
3	28.28	-28.28

The problem with original formulation by (1),(2) and (3) is solved using mathematical programming method in DOT developed by Vanderplaats [7]. Design variables are the cross sectional areas of these three beam elements. A upper limit of the volume is set to be 10.0. Table 2 shows the initial and final values of the design variables, the objective functions, and the volume of the structure.

Table 2 Mathematical Programming Solution with the Original Problem Formulation

	A1	A2	A3	Objective Function	Volume
Initial	1.0	1.0	1.0	1320.4	3.828
Final	6.22	1.23	0.0	320.2	10.0

The third bar is eliminated at the optimum design.

We shall now solve this three-bar problem by using the homogenization method described in the previous section. Figure 2 shows the design domain, the loading conditions and the boundary condition. All loads, material constants and geometrical data are the same as the ones in above. The upper limit of the total volume is specified to be 20% of the initial one. Figure 3 shows the optimal material distribution in the design domain obtained by the shape and topology optimization method introduced here. Scale bar indicates material density in each element.

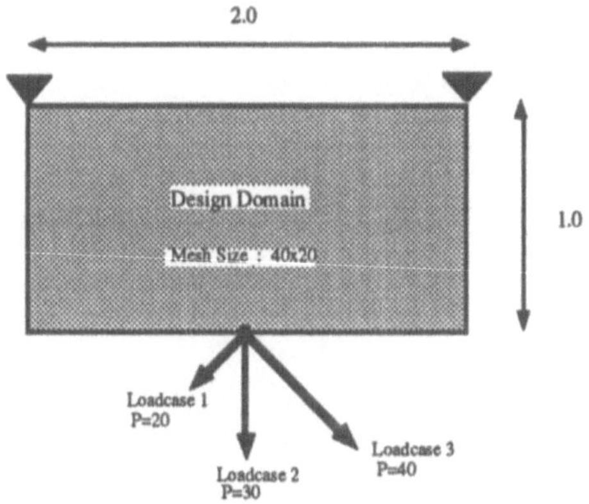

Figure 2 Design Domain with three different loads

The configuration obtained by the shape and topology optimization method is very similar to the mathematical programing solution using the beam model. The ratio of the size of two remained bards is 3/1 by the homogenization method, while it is about 5/1 by the multi-objective optimization problem using DOT. One of the reason of this difference may be stiffer response of the 4 node finite element in which there is no drilling degree of freedom in this plate element. Figure 4 shows the material rotational angle distribution in the final configuration. Orientation angles in each load case are continuous and these angles are the same as the directions of the bars. As the figure shows, the load case number which is chosen in each element in the formulation of the objective function is almost the same as the mathematical programming solution.

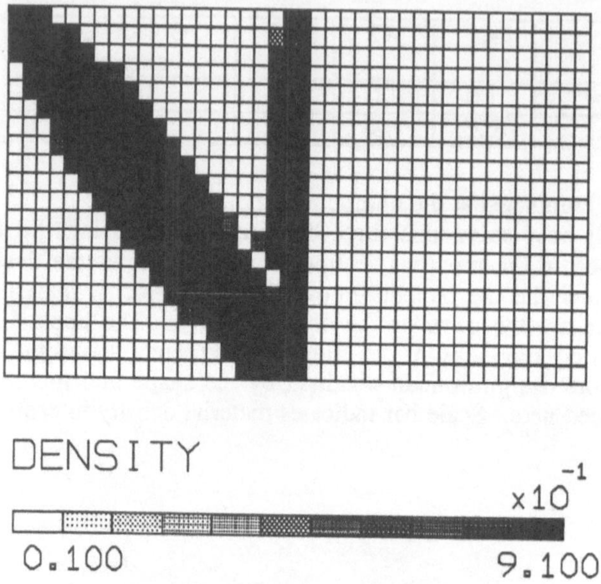

Figure 3 Configuration of Optimal Material Distribution in The Domain

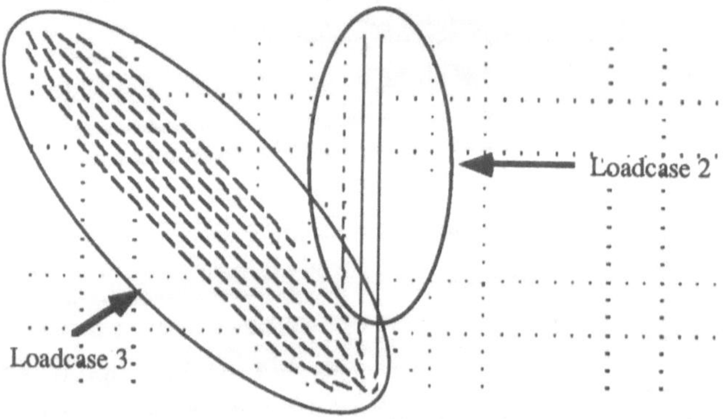

Figure 4 Material Orientation Angle Distribution in Final Configuration

4. Application to a Car Body

Figure 5 shows the design of the optimum layout of the reinforcement of a car body floor modeled by a plane problem whose design domain is placed in the 2D space. Axial direction forces and plane bending moments are applied along the front end of the domain, and only the axial direction force is applied to the tail. We also consider some possibility of loadings by car crash. Magnitudes of forces per unit length applied to the front end are the same in all load cases, and the half of forces to the front end is applied to the tail.

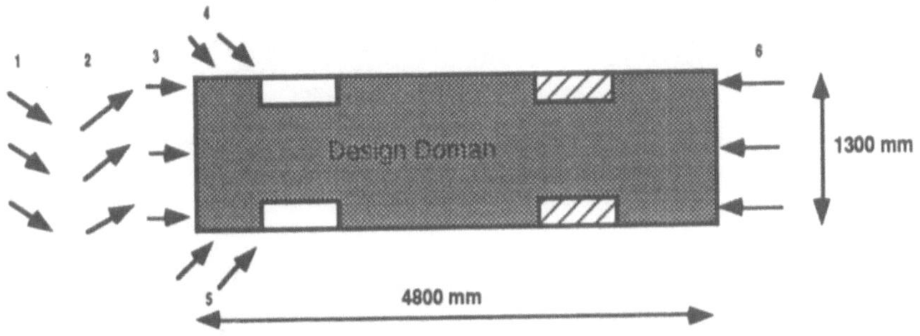

Figure 5 Design Domain & Applied Loads to Car Underbody

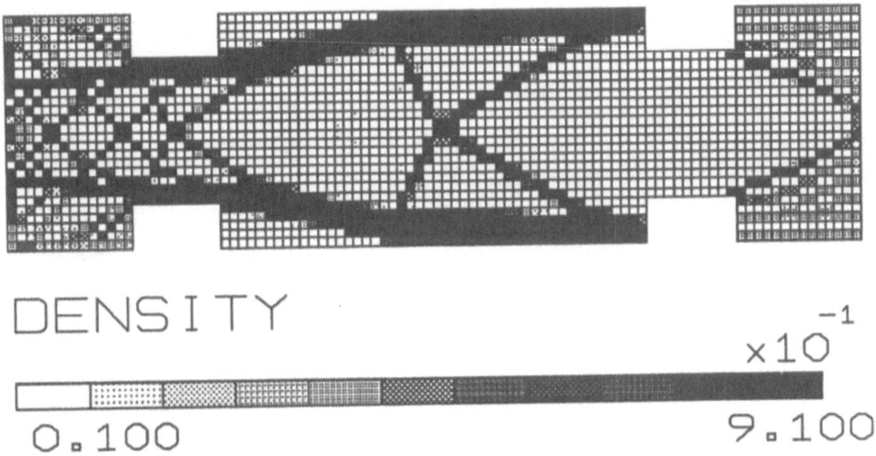

Figure 6 Configuration of Optimal Frame Structure

Solution obtain by the homogenization method is shown in Figure 6. Side frame layout is basically the same as the one which is used in usual passenger cars. But stiffener layout is significantly different from the standard one. This structure has larger stiffness in bending than that of usual type of reinforced structure which is usually a parallel cross frame.

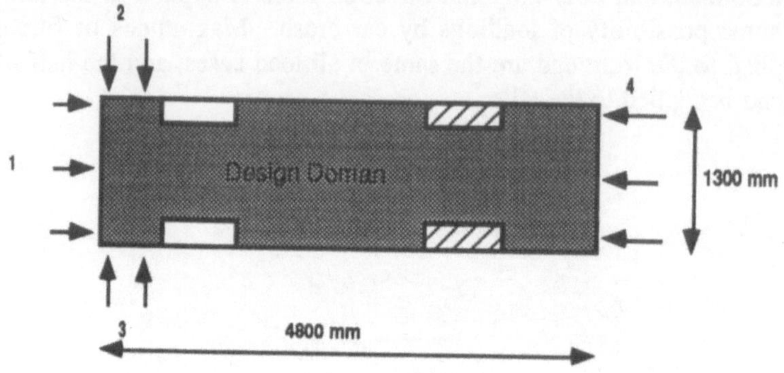

	Loadcase 1	: 0.1 mm
Prescribed Displacements	Loadcase 2,3	: 0.15 mm
	Loadcase 4	: 0.01 mm

Figure 7 Design Domain & Prescribed Displacements

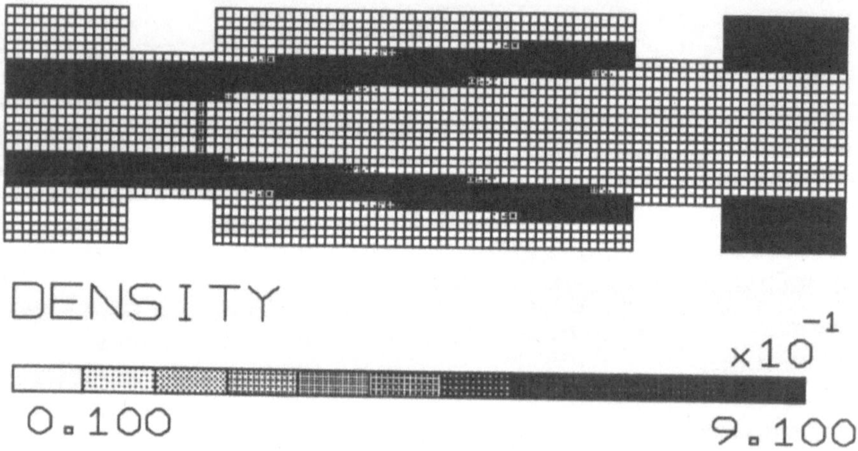

Figure 8 Configuration of Optimal Frame Structure

We have also solved the same problem by specifying loading differently from the previous study. We apply the prescribed displacements instead of tractions on the front and tail end of the body as shown in Figure 7. Design engineers sometimes should consider this kind of prescribed displacements instead of forces to examine the cases that different material or different weight structures contact/crash each other. We obtain the solution shown in Figure 8. It shows that simple parallel frame structure is formed for this enforced displacement loading conditions.

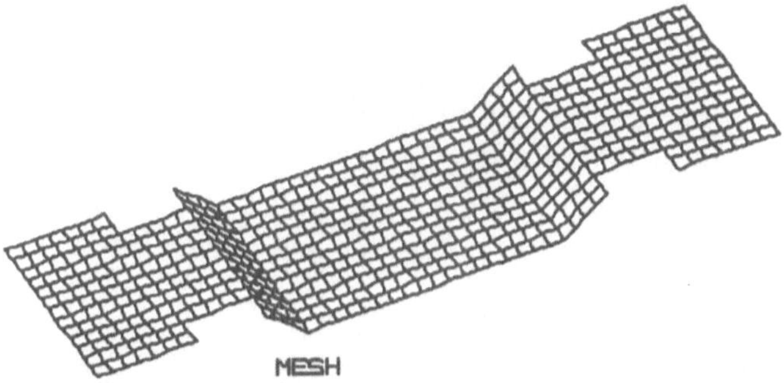

Figure 9 Configuration of a Folded Under Floor of a Car

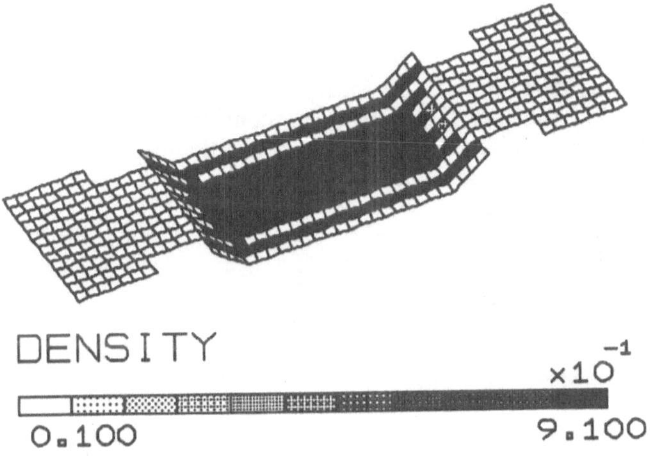

Figure 10 Optimum Reinforcement for the Traction Loading

If the under floor of a car is assumed to be a folded plate as shown in Fig. 9, we shall obtain entirely different "optimal" reinforcement from the case of the flat under floor, since the center lower plate is subject to fairly large transverse bending. For the flat under floor, no transverse bending is generated, thus the optimum reinforcement requires rather thick two main bars for large axial forces applied and orthogonal net of thin bars for the bending in the plane. For the folded case, axial forces generate large transeverse bending to the lower plate, and then main reinforcement should be distributed in this plate. Figures 10 and 11 shows the optimum reinforcement of the three-dimensional folded plate under the traction and prescribed displacement loading cases, respectively.

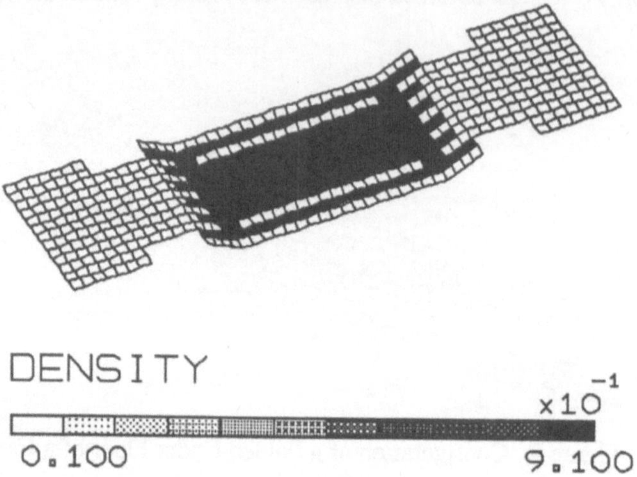

Figure 11 Optimum Reinforcement of a Folded Plate for the Displacement Loading

Reinforcement considered here is just for rigidity of a structure. If dynamical effects, especially vibration and impact problems are concerned, it is expected that entirely different reinforcement might be obtained as the optimum. Here, however, we only consider rigidity aspect of a structure.

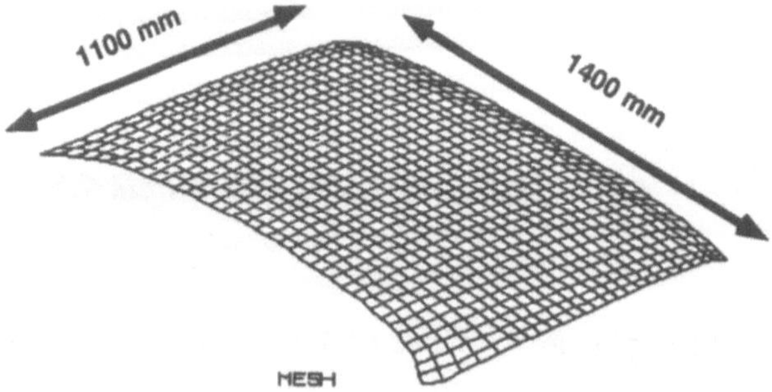

Figure 11 Design Domain for Automobile Engine Hood

Figure 12 shows a finite element model of an automobile engine hood as the design domain. Uniformly distributed and partly distributed loads are applied on the shell as shown in Figure 13.

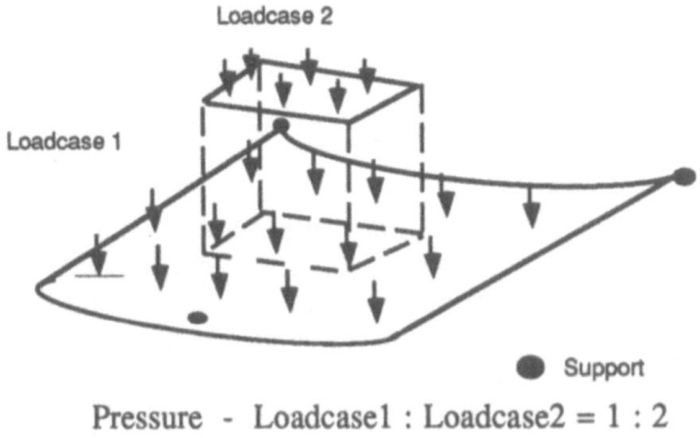

Figure 13 Loading and Support Condition for the Domain

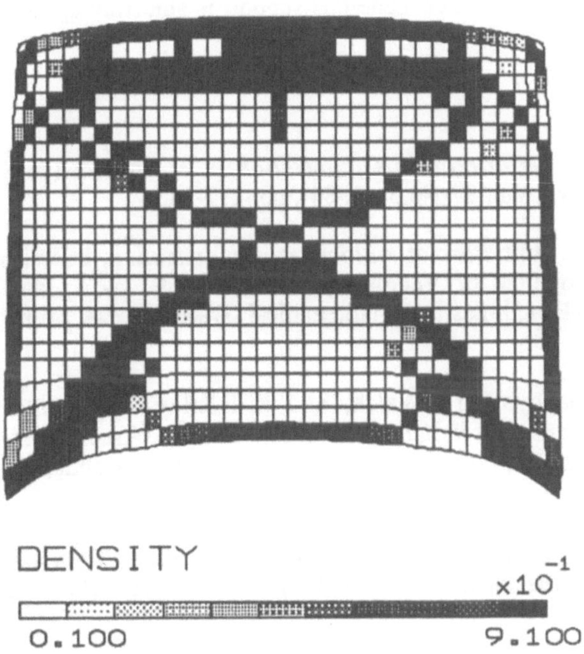

Figure 14 Configuration of The Material Distribution in The Domain

Figure 14 shows the optimum layout solution by the topology optimization. Ribs are distributed on the shell, and the x-shape structure is formed that is basically the same as the actual engine hood structure we can find in existing cars. This result is a little bit unsymmetrical because of the difference of material coordinate system defined in each finite element due to the node numbering order. The optimum layout should be considered to be symmetric.

5. Conclusion

A shape and topology optimization method for multiple loads is given for the homogenization approach to find the optimum layout of reinforcement of three-dimensional shell structures which are common in car bodies, and very few representable example problems are presented in this work. It is noted that there are many industrial applications of the present method to design appropriate layout of rib reinforcement of car bodies, but they are not described here because of space limitation.

Acknowledgement The last two authors (K.S. and N.K.) are partially supported by NSF DDM89-17430, NIH AR34399, NASA Lewis Research Center NAG3-1160, and FORD Foundation. Their generous support is sincerely appreciated by authors.

References

[1] Stadler, W., *Multicriteria Optimization in Mechanics (A Survey)*, Applied Mechanics Reviews, Vol. 37-No.3, (1984), pp. 277-286

[2] Stadler, W., *Initiators of multicriteria optimization*, in Recent Advances and Historical Development of Vector Optimization, eds. by J. Jahn and W. Krabs, Lecture Notes in Economics and Mathematical Systems #294, Spring-Verlag, (1987), pp. 3-25, Berlin

[3] A.Chattopadhyay, J.L.Walsh, and M.F.Riley, *Integrated Aerodynamic / Dynamic Optimization of Helicopter Blades*, Proc. AIAA/ASME/ASCE/AHS 30th Structures,Structural Dynamics and Material Conference, Mobile,Alabama, April 3-5, (1989)

[4] J.Sobieski, A.Dovi, and G Wrenn, *A New Algorithm for General Multiobjective Optimization*, NASA TM - 100536, March, (1988), Langley

[5] A. Diaz and M.P. Bendsϕe, *Shape Optimization of Multipurpose Structures by a Homogenization Method*, Technical Report , Computational Design Laboratory, Michigan State University, (1990), East Lansing, Michigan, U.S.A.

[6] L.A.Schmit and C.Y.Sheu , *Minimum Weight Design of Elastic Redundant Trusses under Multiple Static Loading Conditions*, AIAA Journal, Vol. 10, No. 2, (1972), pp. 155-162

[7] G.N.Vanderplaats, *DOT User's Manual*, Version 2.04, VMA Engineering, Goleta, CA

OPTIMIZATION BY DECOMPOSITION IN STRUCTURAL AND MULTIDISCIPLINARY APPLICATIONS

Jaroslaw Sobieszczanski-Sobieski
NASA Langley Research Center
Hampton, VA 23665 U.S.A.

ABSTRACT. An algorithm for a general, multilevel structural optimization by substructuring is derived, based on the linear decomposition concept that is rooted in the Bellman's Optimality Criterion enhanced with the optimum sensitivity derivatives used as a means to account for coupling among the subproblems, each of which is limited to optimization of a substructure. The algorithm applies also to those multidisciplinary problems whose subproblems form a hierarchy similar to that of substructures. In systems where the subproblems communicate with each other at the same level, the decomposition becomes non-hierarchic and the system may be optimized as a whole based on the derivatives of the system behavior with respect to the design variables computed by a method that bypasses finite differencing on the system analysis. When a multidisciplinary system includes a structure as its part, a hybrid, hierarchic/non-hierarchic decomposition applies. Numerical examples and references to computational experience accumulated to date illustrate the discussion.

Nomenclature

A	= cross-sectional area
B	= in Appendix A only, vector of parameters input and held constant in optimization; it is denoted by Y in Sec. 2.
C	= cumulative constraint, Eq. (2.22) in Algorithm A, redefined locally for the purposes of Algorithms B and C
c	= capacity, a limitation on the ability to meet a particular demand d (e.g., allowable stress)
d	= demand, a physical quantity the structure is required to have, to support, or to be subjected to in order to perform its function (e.g., stress)
F	= objective function
$f(\,)$	= functional relation
g^{ij}	= vector of constraint functions, g_w ($w = 1 \rightarrow W^{ij}$)
h^{ij}	= vector of partitions h_K^{ij}, h_M^{ij}, h_P^{ij}, Eq. (2.25b)
$h_K^{ij}, h_M^{ij}, h_P^{ij}$	= vectors of the equality constraints defined by Eqs. (2.13–2.15); the vector elements are: $h_{KS_1}^{ij}$ at 3, $h_{MS_2}^{ij}$ at 4, and $h_{PS_3}^{ij}$ at 5, where $S_1 = 1 \rightarrow S_1^{ij}$, $S_2 = 1 \rightarrow S_2^{ij}$, and $S_3 = 1 \rightarrow S_3^{ij}$
I	= cross-sectional moment of inertia
J	= matrix of first derivatives (the Jacobian matrix)
K^{ij}	= stiffness matrix of $SSijk$

G. I. N. Rozvany (ed.), Optimization of Large Structural Systems, Vol. I, 193–233.

K^{bij}	= boundary stiffness matrix for $SSijk$
J^{mk}	= set of subscripts j that identify all those $SSijk$ at the level $i = m+1$ which are the daughters of $SSmkl$
L^{ij}	= lower bound on X^{ij}, including move limits
M^{ij}	= mass of $SSijk$ (a scalar)
P^{ij}	= vector of the external loads applied to interior and/or boundary of $SSijk$
P^{bij}	= vector of P^{ij} transferred to the boundary of $SSijk$
Q^{ij}	= vector of the forces Q_r^{ij}, $r = 1 \rightarrow R^{ij}$, acting on the boundary of $SSijk$
S_K	= summation of stiffnesses contributed by substructures $SSijk$ assembled in a parent substructure $SSmkl$
S_P	= summation of the boundary loads contributed by substructures $SSijk$, assembled in a parent substructure $SSmkl$
R^{ij}	= length of Q^{ij}
R_i	= component of the objective function in Dynamic Programing contributed by the i-th box (stage)
S^{ij}	= length of a partition in h^{ij}
S_i	= vector of output from box i transmitted as input to box i-1 in Dynamic Programing
$SSijk$	= substructure (including the extremes of the assembled structure and a single structural element)
$SSmkl$	= substructure parent of $SSijk$, $m = i-1$, see Fig. 2
$SSSnlp$	= substructure parent of $SSmkl$, $n = m-1$, see Fig. 2
STOC	= subject to constraints
SUMT	= sequential unconstrained minimization technique
TOL^{ij}	= tolerance of the equality constraint satisfaction in $SSijk$
U^{ij}	= upper bound on X_t^{ik}
X^{ij}	= vector of design variables X_t, in $SSijk$, $t = 1 \rightarrow T^{ij}$
Y^{ij}	= vector of the entries in K^{bij}, M^{ij}, and the entries in P^{bij} that are held constant as parameters in optimization of $SSijk$; vector Y^{ij} contains V^{ij} elements Y_v^{ij}
$Y_\alpha, Y_\beta, Y_\gamma$	= vectors of the behavior variables for the black boxes designated α, β, and γ, in Fig. 3.1b; this notation is used only in Sec. 3
Z^{ij}	= vector of cross-sectional dimensions Z_b^{ij}, $b = 1 \rightarrow B^{ij}$, used as design variables in $SSijk$ that corresponds to a single structural element, also a submatrix defined by Eq. (A5b), and a generic variable in Appendix A
π	= vector defined by Eq. (26)
ρ	= user-controlled constant in the KS function, see Eq. (22)
Δ	= increment of a variable (see definition of subscript 0)

Indices, Subscripts, Superscripts

$(\bar{\ })$	= optimal quantity
b	= association with the SS boundary
e	= identification of an extrapolated value
0	= original (references) value from which an increment is measured

Introduction

Optimization by decomposition converts a large optimization problem that may be intractable because of the large number of design variables and constraints combined with computationally expensive analysis into a set of much smaller and separate but coordinated subproblems. This approach meshes well with the recent trend in computer technology toward computing distributed over a network of computers whose characteristics may be matched to individual subproblems for more efficiency and convenience. Moreover, the decomposition approach is natural in an engineering organization, because engineers work in teams, concentrating on parts of a project to develop a broad work front and, thus, to shorten the development time.

A number of procedures for implementing of the foregoing approach have been proposed for structural applications (e.g., [1–3]). A multilevel optimization with decomposition has also been formulated generally for use in engineering system design [4] concerned with the "tradeoffs" among various physical subsystems that different engineering disciplines may govern. The unique feature of the algorithm proposed in [4] is the use of the optimum sensitivity derivatives introduced in [5] and [6] as a means to approximate the coupling among the subsystems. Even though the algorithm proposed in [4] was developed by a heuristic approach, Sec. 1 of this paper shows it may be viewed as an extension and generalization of the well-established algorithm of Dynamic Programing due to Bellman [7]. In particular, this method alleviates to a large extent the algorithm limitation known as "the curse of dimensionality".

When the system optimization formulation established in [4] is applied to structural optimization, its analysis part coincides with a general, multilevel substructuring (see, e.g., [8–10]). In the simplest case, the system acquires the meaning of a complete structure and each subsystem corresponds to a single structural component that a finite element may represent. This is a two-level, structural optimization whose algorithm was illustrated by an example of a framework reported in [11]. It served as a verification of the general-purpose algorithm laid out in [4].

Because the general algorithm presented in [4] allows a theoretically unlimited number of hierarchical subsystem levels in the decomposition, its continuing development requires verification by applications with more than two levels. One such particular application is structural optimization with a three-level decomposition. In this decomposition, the highest level corresponds to the assembled structure, the level below corresponds to the substructures, and the third, bottom level represents the structural components that make up the substructures.

Extension of the scheme beyond the three levels would require more levels of nested substructures sandwiched between the top and bottom levels. Thus, such an extension would not add any qualitatively new subsystems to the scheme. Therefore, one may regard the three-level optimization as the simplest case of the most general multilevel optimization.

A general, multilevel algorithm for structural optimization is presented in Sec. 2. It is illustrated by a numerical example of a three-level optimization of a framework structure. A brief account of the operational experience with the algorithm published to date is also given, including applications showing that the algorithm may be generalized to those multidisciplinary systems that can be decomposed into a hierarchy similar to that of substructures in an assembled structure.

Finally, a more general problem of an engineering system whose constituent parts interact in such a way that the above hierarchic decomposition does not apply is addressed in Sec. 3. The proposed solution is based on the local explicit approximation of the problem based on the derivatives of the system behavior with respect to design variables and repeated optimization of the approximate problem within move limits. The numerical example illustrates the approach. It is also pointed out that a hybrid decomposition that is partially hierarchic and partially non-hierarchic may occur, particularly so if one of the subsystems is a structure.

Section 1: Dynamic Programing as Optimization by Decomposition

Dynamic Programing, originally due to Bellman [7], is perhaps the best-known mathematically established method for optimization by decomposition. Briefly summarized, the method applies to a system that may be represented by a train of the "black boxes" as in Fig. 1.1 (called the "stages" in [7]). The black boxes are mathematical models of the physical parts or conceptual aspects (engineering disciplines) of a large problem, and they are unified by a flow of information from the n-th box through the intermediate boxes ending with the 1-st box. An i-th box receives two input vectors: a vector S_{i+1} from its predecessor, and a vector of design variables X_i. It outputs a vector S_i that becomes an input to the successor box i-1. The i-th box also outputs a quantity R_i interpreted as a component of the objective function of the entire system. Each $\{S_i\}$ must be reducible to a function of a single variable, s_i, $S_i = f(s_i)$, and, because $S_i = f(S_{i+1}, X_i)$ and $R_i = f(S_{i+1}, X_i)$, it follows that $S_i = f(s_{i+1}, X_i)$, $R_i = f(s_{i+1}, X_i)$, and $s_i = f(s_{i+1}, X_i)$.

The problem of finding a set of vectors X_i, $i = 1 \ldots n$, that minimizes the sum of $R_i, i = 1 \ldots n$, is solved by starting with the 1-st box at the end of the train. The variable s_2 governing the input S_2 is assumed to vary within an interval of interest. Several values of s_2 distributed over that interval are set and an optimization problem, constrained or unconstrained, is solved to find $\{X_1\}$ so as to minimize R_1 at each value. That solution yields $\{X_1\}_{\text{opt}} = f(s_2)$ and $R_{1\text{min}} = f(s_2)$, either in a discrete (a look-up table) form or in a continuous form interpolated between the s_1 values assumed above, dependent on the nature of the problem.

Moving up to the 2-nd box, one seeks for each of the several values of s_3 in an interval of interest an optimal $\{X_2\}$ (denoted $\{X_2\}_{\text{opt}}$) that minimizes the sum $R_2 + R_{1\text{min}}$. One must consider that $s_2 = f(s_3, X_2)$ and that for each value of s_2, there are $\{X_1\}_{\text{opt}}$ and $R_{1\text{min}}$ already known from the optimizations that have been executed for box 1. This operation generates the values of $\{X_2\}_{\text{opt}} = f(s_3)$ and $(R_2 + R_{1\text{min}})_{\text{min}}$.

The procedure continues recursively from box i to box $i+1$, carrying forward $\{X_j\}_{\text{opt}}$, and $(R_j + (R_{j-1} + (R_{j-2} \cdots + (R_2 + R_{1\text{min}})_{\text{min}})_{\text{min}})_{\text{min}} \cdots)_{\text{min}}$, $j = 1 \ldots i$, through the initial box in the train, $i = n$, whereby the minimum sum of all R_i's and a complete set of $\{X_i\}_{\text{opt}}$'s gets established. The procedure rests on the fundamental principle formulated by Bellman which asserts that the set of $\{X_j\}$, $j = 1 \ldots n$, is optimal when its subset for $j = 1 \ldots i$ taken for any i minimizes the sum of R_j, $j = 1 \ldots i$, for S_{i+1} input given from the remainder of the train.

The computational cost of the procedure heavily depends on the aforementioned assumptions of $S_i = f(s_i)$, where s_i is a single variable. Indeed, if s_i were a vector of m elements, $R_{i\text{min}}$ would grow from a line plot into a hypersurface in m dimensions. For the lowest-order nonlinear representation of that hypersurface, one needs three values of each s_i spaced in its interval of interest. Hence, the number of discrete points at which optimizations would have to be performed for just one box would grow proportional to 3^m, and $n3^m$ for the entire train of n boxes, thus quickly destroying advantages of the procedure as a computational cost saver. Bellman called this the "curse of dimensionality" and regarded it as a barrier limiting applicability of the method.

OVERCOMING THE CURSE OF DIMENSIONALITY

Optimum sensitivity analysis formulated in [5] provides a means for generalization of the above procedure to include s_i defined as a vector of m elements. The optimum sensitivity analysis algorithm yields derivatives of the optimal $\{X\}$ and R with respect to the parameters of the optimization problem (unconstrained or constrained). Taking box 1 as an example, $\{S_2\}$ may now be defined as $\{S_2\} = f(\{s_2\})$, where $\{s_2\}$ is a vector of elements s_{2k}, $k = 1 \ldots m$. For $\{s_2\}$ given, one may find $\{X_1\}_{\text{opt}}$ and minimum of R_1 and their derivatives with respect to each s_{2k}, regarded as an optimization parameter. Using the notation $D(X_1, s_{2k})$ and $D(R_1, s_{2k})$ for these derivatives, the linear part of the Taylor series enables one to express $\{X_1\}_{\text{opt}}$ and $R_{1\min}$ as continuous, albeit approximate, functions of $\{s_2\}$:

$$\{X_1\}_{\text{opt}} = f(\{s_2\}) = (\{X_1\}_{\text{opt}})_o + [D(X_1, s_{2k})]\Delta\{s_2\} \tag{1.1}$$

$$R_{1\min} = f(\{s_2\}) = (R_{1\min})_o + \{D(R_1, s_{2k})\}'\Delta\{s_2\} \tag{1.2}$$

The Bellman's Dynamic Programing procedure may now be executed using the above approximations in place of $\{X_1\}_{\text{opt}} = f(s_2)$ and $R_{1\min} = f(s_2)$; otherwise, the procedure remains unchanged. The new component in the modified procedure is the optimum sensitivity analysis to be executed after each optimization involving boxes 1, $(2+1)$, $(3+2+1), \ldots n$, recursively. Because the linear relationships, Eqs. (1.1) and (1.2), introduce errors whose control requires move limits on design variables in each optimization, the entire procedure has to be repeated p times until satisfactory convergence is attained. In this case, the number p depends on the nonlinearities of the problem at hand. Consequently, because there is only one optimization in each box in one pass, the number of optimizations required to converge the procedure is pn. This is in contrast to $n3^m$, which is necessary for the original procedure. The curse of dimensionality with respect to m is removed. The ratio $pn/n3^m = p/3^m$ tends to be very small for large m and renders the modified procedure usable where the original one would be prohibitively expensive.

However, unlike the original procedure, the modified procedure relies on the continuity of the approximation function in Eqs. (1.1) and (1.2); hence, it cannot accommodate discrete design variables.

HIERARCHIC DECOMPOSITION

Further generalization of the modified procedure is possible if the boxes in the train may be partitioned internally as shown in Fig. 1.2. This figure shows the boxes split internally into smaller ones. In this scheme, the train of boxes that was horizontal in Fig. 1.1 is depicted vertically to form a pyramid whose levels correspond to the boxes in Fig. 1.1. A typical level is populated by several boxes that formed a single box in Fig. 1.1. The pyramidal arrangement emphasizes the hierarchic dependence of the boxes in level i ("daughters") on the information transmitted from a box located at the level above $j > i$ ("parent"), with the underlying assumption that the boxes at the same level ("sisters") do not exchange information with each other directly. Similar to the system shown in Fig. 1.1, the behavior information from each box flows in Fig. 1.2 from the parent to the daughters, or from the top level n down to level 1. The optimization information from each box flows in the opposite direction. This information includes the optimum sensitivity derivatives that enable optimization in each parent box to be performed taking into account the effect of its $\{X\}$ on the optimization results in all boxes descendent from that parent.

Thus, the above decomposition scheme first developed heuristically in [4] is shown to be a generalization of the Bellman's Dynamic Programing. The scheme became known as hierarchic, linear decomposition.

Section 2: Hierarchic, Linear Decomposition

In substructuring analysis, the substructures form a multilevel hierarchy shown in Fig. 2.1. Each box labeled "Substructure" is a mathematical model that accepts as input the boundary conditions in the form of either displacements or forces computed in the parent substructure, and outputs the similar boundary data for each substructure at the next lower level. However, it does not produce any data for the other substructures at the same level. This satisfies the condition for a generalized Dynamic Programing introduced in Sec. 1 and discussed in conjunction with Fig. 1.2, and suggests that a procedure for optimization by substructuring may be based on the idea formulated in that section. When comparing Fig. 1.2 and 2.1, it is important to note that in Fig. 1.2 the bottom level is designated as No. 1 consistent with the formalism of the Dynamic Programing; but in Fig. 1, level No. 1 is the top level—the numbering convention more often used in substructuring.

ONE-LEVEL OPTIMIZATION

An optimization formulation without decomposition serves as a reference from which the multilevel optimization algorithm is derived. The optimization is defined in terms of: the design variables Z_b (which are the cross-sectional dimensions of the structural components), the objective function $F(Z)$ that can be any computable function of these variables (structural mass is the frequent choice), and the constraints $g_w(Z)$ imposed on the behavior variables to account for the potential failure modes. Writing the constraint functions as

$$g = d/c - 1 \leq 0 \tag{2.1}$$

the optimization problem in a standard formulation is

$$\min_{Z} F(Z); \quad \text{STOC } g_w(Z) \leq 0 \tag{2.2}$$

and requires a search of the design hyperspace considering all of the design variables and constraints concurrently. In contrast, an algorithm presented in the next section breaks the problem into a number of search and analysis operations, each concerned with a smaller number of design variables and constraints.

MULTILEVEL OPTIMIZATION

Preliminary Definitions. The diagram in Fig. 2.1 shows a structure decomposed into several levels of substructures. The term "substructure" will refer to any entity in this decomposition scheme, including the extremes of the full, assembled structure the box on the top of the pyramid represents and single structural components representing the ultimate geometrical details appropriate to the problem at hand. The substructure levels are numbered from 1 on the top to i_{max} at the bottom. The hierarchical nature of the scheme instigates the use of the term "parent" to the structure at level i that, in turn, is decomposed into a number of "daughter" substructures at level $i+1$. A daughter may have only one parent and that parent must be at the level immediately above. Thus, it will be convenient to label each substructure $SSijk$, where i denotes the level, j defines the position at the level i counting from the left, and k identifies the parent's position at the level $i-1$. The substructure occupying the last position in a particular parent-daughter succession represents the ultimate level of detail at which the decomposition stops. There is no requirement that all such substructures must be at the same bottom level i_{max}. In discussions involving more than one substructure, the triplets nlp, mkl, and ijk are used to distinguish among the substructures forming the hierarchy shown in Fig. 2.2.

Substructuring analysis (e.g., [8–10]) establishes the following functional relations

$$Q^{ij} = f(K^{bij}, P^{bij}) \tag{2.3}$$

$$K^{bij} = f(K^{ij}) \tag{2.4}$$

$$K^{ij} = f(K^{bi+1,j}) = S_K(K^{bi+1,j}) \tag{2.5}$$

$$M^{ij} = f(M^{i+1,j}) = \sum_j M^{i+1,j} \tag{2.6}$$

$$P^{bij} = f(P^{ij}) \tag{2.7}$$

$$P^{ij} = S_P(P^{bi+1,j}) \tag{2.8}$$

These relations are computable in a manner prescribed by the particular substructuring algorithm chosen. For example, Eqs. (2.4) and (2.7) take the form of matrix equations given in [9], Chap. 9, Sec. 1, as Eqs. (9.13) and (9.14), respectively. The above equations are recursive because they equate input into substructures at level i to output from the next higher level in the same way the parameters S are defined in Dynamic Programing in Sec. 1.

For $SSijk$ at the ultimate level of detail, the distinctions between K^{bij}, P^{bij}, and K^{ij}, P^{ij} vanish; and K^{ij}, M^{ij} derive directly from Z^{ij}. Consequently

$$K^{bij} = K^{ij} \tag{2.9}$$

$$P^{bij} = P^{ij} \tag{2.10}$$

$$K^{ij} = f(Z^{ij}) \tag{2.11}$$

$$M^{ij} = f(Z^{ij}) \tag{2.12}$$

The local constraints that arise in $SSijk$ at the ultimate level of detail involve calculation of the stresses, strains, local buckling, etc., from Q^{ij} and Z^{ij}. In addition, constraints may be imposed on the internal forces, critical forces, and displacements of $SSijk$ to account fully for all the constraints that would have been included in the one-level optimization problem represented by Eq. (2.2).

Although the foregoing definition of substructuring analysis is based on the finite-element stiffness method, the use of a finite-element analysis is not mandatory for the multilevel optimization algorithm presented here. As far as that algorithm is concerned, the analysis is a "black box" where only the inputs and outputs are important.

Multilevel Optimization Algorithm. With the substructuring scheme and analysis established in the foregoing, this subsection describes the optimization algorithm itself. The essentials of the computer implementation are also given.

Basic Concept. The basic idea for the proposed multilevel optimization by substructuring stems from the elementary observation, based on Eqs. (2.3–2.8), that the effect of a daughter $SSijk$ on its parent $SSi-1,\ kl$ is felt only through K^{bij}, M^{ij}, and P^{bij}, which depend on $K^{b,i+1,j}$, $M^{i+1,j}$, and $P^{b,i+1,j}$, respectively. Consequently, the stiffness properties, the mass distribution, and the manner in which the interior loads are transferred to the boundary may be controlled in $SSijk$ without disturbing the results of the $SSi-1,\ kl$ analysis by manipulating the entries of $K^{b,i+1,j}$, $M^{i+1,j}$, and $P^{b,i+1,j}$ as generalized design variables in a manner somewhat restricted so that the entries of K^{bij}, M^{ij}, and P^{bij} are held constant. If the entries are held constant, then the boundary forces Q^{ij} acting on every $SSijk$ in $SSi-1,\ kl$ remain constant and the effect of manipulating the generalized design variables in a particular $SSijk$ is limited to that of $SSijk$ itself and its daughters. As will be explained later, the purpose of the above manipulation of the matrix entries is not to minimize the substructure mass M^{ij}, which, as stated above, remains constant. Instead, the purpose is to improve satisfaction of the constraints in the $SSijk$ and its daughters, while performing the task of the total mass optimization at the assembled structure level.

Invariance of the entries K^{bij}, M^{ij}, and P^{bij} can be enforced by rewriting Eqs. (2.4–2.8) as equality constraints

$$h_K^{ij} = K^{bij} - f(K^{bi+1,j}) = 0 \tag{2.13}$$

$$h_M^{ij} = M^{ij} - f(M^{i+1,j}) = 0 \tag{2.14}$$

$$h_P^{ij} = P^{bij} - f(P^{bi+1}) = 0 \tag{2.15}$$

Equations (2.13–2.15) establish the entries of K^{bij}, M^{ij}, and P^{bij} as parameters in optimization of $SSijk$. Simple replacement of indices renders these equations valid for $SSi-1,\ kl$ and redefines the optimization parameters of the daughter $SSijk$ as generalized design variables in the optimization of its parent $SSi-1,\ kl$, so that

$$\{X^{i-1,k}\} = \{Y^{ij}\} \tag{2.16}$$

$$\{Y^{ij}\} = \{K^{bij}|M^{ij}|P^{bij}\} \tag{2.17}$$

where parameters in Y^{ij} are analogous to the ones denoted as S in Sec. 1. These equations define a recursive relation of the variables and parameters that extends from the top of the substructuring scheme to the bottom. Of course, the total number T^{ij} of the design variables must exceed the total number V^{ij} of the parameters [V^{ij} equals the number of individual equations in the vector Eqs. (2.13–2.15)],

$$T^{ij} > V^{ij} \tag{2.18}$$

for a design freedom to exist, allowing for the symmetry of the stiffness matrices. Otherwise, if

$$T^{ij} \leq V^{ij} \tag{2.19}$$

then the equality constraints of Eqs. (2.13–2.15) either define the $SSijk$ design variables uniquely or overdetermine them.

The basic concept outlined above requires satisfaction of the equality constraints, Eqs. (2.13), (2.14), and (2.15). Satisfying these equality constraints can be accomplished directly by including the above constraints in optimization and forming an algorithm that was first introduced in [12] and will now be described in detail as Algorithm A. Two alternatives presented in [13] which satisfy the above constraints indirectly will be described subsequently as modifications to Algorithm A and referred to as Algorithms B and C.

Optimization at the Most Detailed Level—Algorithm A. Introduction of the optimization algorithm begins at the level of the most detailed substructures. Consequently, Eqs. (2.9–2.12) apply and the design variables are the cross-sectional dimensions so that

$$X^{ij} = Z^{ij} \tag{2.20}$$

and the parameters are

$$\{Y^{ij}\} = \{K^{ij}|M^{ij}|P^{ij}\} \tag{2.21}$$

It is assumed that a complete, top-down substructuring analysis for an initialized structure has been carried out so that for an $SSijk$ one has computed its Q^{ij}, while its M^{ij}, Z^{ij}, K^{ij}, and P^{ij} are given.

Optimization for improvement of the constraint satisfaction is achieved by minimizing a single measure representing all of the constraints—called the cumulative constraint, a concept similar to the use of a penalty term in the SUMT. A differentiable cumulative constraint function can be obtained (as it was in [11]) by means of the Kresselmeier-Steinhauser function (KS) defined in [14] as

$$\mathrm{KS}\left(g_w^{ij}\right) = (1/\rho)\ell n\left[\sum_w \exp\left(\rho g_w^{ij}\right)\right] \tag{2.22a}$$

or in an alternative form

$$\mathrm{KS}\left(g_w^{ij}\right) = g_{\max} + (1/\rho)\ell n\left[\sum_w \exp\left(\rho\left(g_w - g_{\max}\right)\right)\right] \tag{2.22b}$$

that keeps the exponential terms to conveniently small values.

The function has the property of approximating the maximum constraint so that

$$\max\left(g_w^{ij}\right) < \mathrm{KS}\left(g_w^{ij}\right) < \max\left(g_w^{ij}\right) + 1/\rho\ \ell n(W^{ij}) \tag{2.23}$$

with the factor ρ, which the user controls. Thus, the KS function serves as a convenient single measure of the degree of constraint violation (or satisfaction).

Analysis of $SSijk$ yields the local constraints as

$$g^{ij} = f(X^{ij}, Y^{ij}, Q^{ij}) \tag{2.24a}$$

and their cumulative constraint C^{ij} is expressed by means of the KS function

$$C^{ij} = \mathrm{KS}(g^{ij}) \tag{2.24b}$$

Based on the above definitions, the optimization problem is formulated

$$\min_{X^{ij}} C^{ij}\ (X^{ij}, Y^{ij}, Q^{ij})\mathrm{STOC} \tag{2.25a}$$

$$h_K^{ij} = 0, \quad h_M^{ij} = 0, \quad h_P^{ij} = 0 \tag{2.25b}$$

$$L^{ij} \leq X^{ij} \leq U^{ij} \tag{2.25c}$$

Solution of this optimization problem (by any technique available) yields a constrained optimum described by a vector π^{ij} composed of the minimum value of the cumulative constraint $\bar{C}^{(ij)}$ and the optimal vector of the design variables $\bar{X}^{(ij)}$

$$\pi^{ij} = \{\bar{C}^{ij} | \bar{X}^{ij}\} \tag{2.26}$$

This solution is sensitive to the parameters Y^{ij} and to Q^{kij} so that derivatives $\mathrm{d}\pi^{ij}/\mathrm{d}Y_z^{ij}$ and $\mathrm{d}\pi^{ij}/\mathrm{d}Q_r^{kij}$ exist and may be expressed by a chain differentiation to account for Eqs. (2.3) and (2.21) that tie Q^{ij} to Y^{ij},

$$\frac{\mathrm{d}\bar{C}^{ij}}{\mathrm{d}Y_z^{ij}} = \frac{\partial \bar{C}^{ij}}{\partial Y_z^{ij}} + \sum_r \left(\frac{\partial \bar{C}^{ij}}{\partial Q_r^{ij}}\right)\left(\frac{\partial Q_r^{ij}}{\partial Y_z^{ij}}\right) \tag{2.27}$$

$$\frac{\mathrm{d}\bar{X}^{ij}}{\mathrm{d}Y_z^{ij}} = \frac{\partial \bar{X}^{ij}}{\partial Y_z^{ij}} + \sum_r \left(\frac{\partial \bar{X}^{ij}}{\partial Q_r^{ij}}\right)\left(\frac{\partial Q_r^{ij}}{\partial Y_z^{ij}}\right) \tag{2.28}$$

In Eqs. (2.27) and (2.28), the partials of $\bar{C}^{ij}$ with respect to Y_z^{ij} and with respect to Q_r^{ij} are obtained from the algorithm described in [5] and the partial Q_r^{ij} with respect to the Y_z^{ij} by conventional structural sensitivity analysis. The algorithm which in [5] is called the analysis of optimum for sensitivity with respect to parameters (or the optimum sensitivity analysis), and its variant given in [6] are reproduced in an abridged form in Appendix A.

Optimization of a Parent Substructure—Algorithm A. As shown in Fig. 2.2, the parent substructure $SSmkl$, $m = i - 1$, receives from its daughters $SSijk$ the minimized values of their cumulative constraints C^{ij}, optimal values of their design variables $\bar{X}^{ij}$, and the optimum sensitivity derivatives of these quantities with respect to parameters Q^{ij} and Y^{ij}, according to Eqs. (2.26–2.28).

Preparing for the formulation of the optimization problem for the parent substructure, we consider the recursive relation between the design variables and parameters according to Eqs. (2.16) and (2.17) and recognize that Eqs. (2.9–2.12) do not apply. When optimizing the parent substructure, we want to improve the satisfaction of the assembled substructure constraints, such as its elastic deformations and stability that depend on the substructure stiffness, mass, and boundary forces, as

$$g^{i-1,k} = f(X^{i-1,k}, Y^{i-1,k}, Q^{i-1,k}) \tag{2.29}$$

At the same time, we want to improve that constraint satisfaction in all the substructure daughters. These can be approximated (as in [11]) by linear extrapolation of their cumulative constraints using the derivatives from Eq. (2.27) and taking into account Eq. (2.16)

$$\bar{C}_e^{ij} = \bar{C}_0^{ij} + \sum_t \left(\frac{\mathrm{d}\bar{C}^{ij}}{\mathrm{d}X_t^{mk}} \right) \Delta X^{mk} \tag{2.30}$$

This extrapolation plays a key role in the algorithm because it approximates the daughter-parent coupling without incurring the expense of reoptimizing the daughters [repeating Eq. (2.25)] for every change of the parent design variables.

Including the $\bar{C}_e^{ij}$ values together with $g^{i-1,k}$ in a cumulative constraint formed by the KS function we have

$$C^{mk} = 1/\rho \ell n \left[\sum_w \exp\left(\rho g_w^{mk}\right) + \sum_j \exp\left(\rho \bar{C}_e^{ij}\right) \right]; \quad j \in J^{mk} \tag{2.31}$$

The optimization problem to be solved for the parent $SSmkl$, $m = i-1$, is

$$\min_{X^{mk}} C^{mk}(X^{mk}, Y^{mk}, Q^{mk}) \text{STOC} \tag{2.32a}$$

$$h^{mk} = \left\{ h_K^{mk} | h_M^{mk} | h_P^{mk} \right\} = 0 \tag{2.32b}$$

$$L^{mk} \leq X^{mk} \leq U^{mk} \tag{2.32c}$$

$$L^{ij} \leq X_e^{ij} \leq U^{ij} \tag{2.32d}$$

where

$$X_e^{ij} = X_0^{ij} + \sum_t \left(\frac{\mathrm{d}\bar{X}^{ij}}{\mathrm{d}X_t^{mk}} \right) \Delta X_t^{mk} \tag{2.33}$$

The increment X^{mk} is defined as

$$\Delta X^{mk} = X^{mk} - X_0^m \tag{2.34}$$

The constraints of Eq. (2.32b) are analogous to Eqs. (2.25b) written in a compact format. The constraints of Eq. (2.32c) incorporate the side constraints to prevent the design variables from attaining physically impossible values (e.g., negative diagonal entries in a stiffness matrix) and include the move limits to control the extrapolation errors introduced by Eq. (2.30). The constraints of Eq. (2.32d) are introduced to keep the design variables in the daughters from exceeding their side constraints and will be discussed in more detail later.

Solution of the problem Eqs. (2.32) generates the result vector and its derivatives that are analogous to those of Eqs. (2.26–2.28) with the indices ij replaced by $m = i-1$ and k.

Optimization of the Next Parent Structure—Algorithm A. Moving up to $SSnlp$ that is shown in Fig. 2.2 as the parent of $SSmkl$, everything that was stated in the preceding subsection regarding the optimization of $SSmkl$ applies to $SSnlp$ directly, provided that the indexes m, k, and l are replaced according to the pattern illustrated in Fig. 2.2, so that $n = m-1$, and l in nlp is the same as l in mkl. For consistency, Eq. (2.32d) should be replicated to encompass fully each line of succession emanating downward from $SSnlp$. Beyond these changes, no new conceptual elements are introduced and no additional definitions or discussion are needed at the junctions between the levels until one arrives at the top level. Hence, any number of intermediate levels of substructuring can be inserted, if physically justified, into a line of succession extending downward from the assembled structure on the top; i.e., the algorithm is recursive.

Optimization of the Assembled Structure—Algorithm A. The assembled structure is designated $SS110$. Its optimization problem is similar to the one described for a parent substructure $SSmkl$ with the following differences:

1) No parameters are defined solely for the decomposition purposes; therefore, there is no need for the equality constraints to enforce constancy of the mass and the boundary stiffnesses.

2) The objective function is the mass of the assembled structure.

3) There is no need for a single cumulative constraint (unless one needs it to reduce the number of constraints to be processed at that level).

4) The boundary forces are the external loads on the assembled structure.

Accounting for these differences, the optimization problem for the top level is

$$\min_{X^{11}} M^{11}(X^{11})\text{STOC} \tag{2.35a}$$

$$g^{11} \leq 0 \tag{2.35b}$$

$$\bar{C}_e^{2j} \leq \text{TOL}^{2j}; \quad j \in J^{11} \tag{2.35c}$$

$$L^{11} \leq X^{11} \leq U^{11} \tag{2.35d}$$

$$L^{2j} \leq \bar{X}_e^{2j} \leq U^{2j} \tag{2.35e}$$

where $\bar{C}_e^{2j}$ is extrapolated by an equation similar to Eq. (2.30), and where Eq. (2.35e) is analogous to Eq. (2.32d). The limits L^{2j}, U^{2j} applied in conjunction with extrapolations of the type expressed by Eq. (2.33) are extended recursively to encompass all of the levels below as mentioned in the subsection on $SSnlp$. Similar to Eq. (2.32c), the side-constraints in Eq. (2.35d) include move limits to guard against excessive linearization errors. Unlike in the daughters $SSijk$, the optimization of $SS110$ does not have to be analyzed for optimum sensitivity. Information transmitted to the top level optimization problem is indicated in Fig. 2.2.

Iterative Procedure—Algorithm A. When the $SS110$ optimization is completed, the entire structure has acquired a new distribution of stiffness and mass within the move limits. Hence, the analysis must be repeated and followed by a new round and substructure optimizations in an iterative manner until convergence. Accordingly, the procedure follows these steps:

1) Initialize all cross-sectional dimensions.

2) Perform a substructuring analysis, including for each substructure at each level, the transformation of the stiffness matrix into the boundary stiffness matrix and the transformation of the forces applied to the interior degrees of freedom to the forces

coinciding with the boundary degrees of freedom. Calculations of the behavior derivatives needed for the ensuing optimizations and for optimum sensitivity analyses are included in the substructuring analysis.

3) Perform the operations of optimization and optimum sensitivity analysis as defined by Eqs. (2.25–2.34).

4) Optimize the assembled structure as defined by Eqs. (2.35).

5) Repeat from step 2 until all constraints g^{ij} are satisfied at all levels, and M^{11} has entered a phase of diminishing returns.

This procedure is illustrated in Fig. 2.3 by a flowchart in the Chapin format [15]. It is important to initialize by setting the cross-sectional dimensions of the most detailed substructures first, computing the corresponding parameters of the parent substructures next, and so on to the top level. This bottom-up approach assures that all the equality constraints of the type of Eqs. (2.13), (2.14), and (2.15), which arise at the parent/daughter junctions, are satisfied in the first pass through the procedure.

Modification Transforming Algorithm A to Algorithm B. In general, the equality constraints are harder to satisfy than inequality constraints, and their use in Algorithm A, Eqs. (2.25b), (2.32b), etc., has been identified in [16] as a cause of numerical difficulties that occasionally have been experienced with the algorithm, hence a motivation to improve the algorithm by elimination of these equality constraints even though its overall application record reported so far has been quite satisfactory (e.g., [11], [12], [17], [18], and [19]). The equality constraints may be removed from the algorithm by means of either of the two recently introduced techniques described in Appendix B. Both of them achieve satisfaction of the equality constraints not in every iteration as in Algorithm A, but when the entire procedure converges, and accomplish this indirectly by minimizing an objective function that measures the equality constraint violation. This approach is rooted in a similar technique that was shown in [3] to be effective in matching the stiffness properties of a finite element computed as a function of the element cross-sectional dimensions to the values prescribed for these properties by the design variables at the assembled structure level.

Implementation of the first technique from Appendix B requires only very few modifications to Algorithm A, as follows.

1) The objective function C^{ij} in Eq. (2.25a) changes from the one defined by Eq. (2.24b) to

$$C^{ij} = \mathrm{KS}\left(+h_K^{ij}, -h_K^{ij}, +h_M^{ij}, -h_M^{ij}, +h_P^{ij}, -h_P^{ij}\right) \tag{2.36}$$

whose minimum (see Appendix B) is zero when the equality constraints, Eq. (2.25b), are satisfied.

2) The equality constraints, Eq. (2.25b), are eliminated and replaced with inequality constraints

$$g^{ij} \leq 0 \tag{2.37}$$

3) Eq. (2.31) is not used and the objective function C^{mk} in Eq. (2.32a) changes from the one defined by Eq. (2.31) to

$$C^{mk} = \mathrm{KS}\left(h_K^{mk}, -h_K^{mk}, +h_M^{mk}, -h_M^{mk}, +h_P^{mk}, -h_P^{mk}\right) \tag{2.38}$$

4) The equality constraints in Eq. (2.32b) are eliminated and replaced by the following two inequality constraints

$$g_w^{mk} \leq 0 \tag{2.39a}$$

$$\bar{C}_e^{ij} \leq \mathrm{TOL}^{ij} \tag{2.39b}$$

where $\bar{C}_e^{ij}$ is extrapolated by Eq. (2.30).

5) Eq. (2.35c) is replaced by

$$\bar{C}_e^{2j} \leq \mathrm{TOL}^{2j} \tag{2.40}$$

where an equation similar to Eq. (2.30) is used to extrapolate $\bar{C}_e^{2j}$.

The above modifications convert Algorithm A to Algorithm B whose key feature is that in every iteration it satisfies the substructure local constraints and minimizes violation of the equality constraints that make the substructure conform to the stiffness and mass parameters input from the parent. If it is not possible to bring that violation to zero and to maintain at zero from the very beginning of the procedure, then the violation is gradually reduced to an acceptable level by decreasing TOL^{ij} from one iteration to the next while satisfying the local constraints in Eq. (2.37). In contrast, Algorithm A satisfies the equality constraints in every iteration and gradually reduces the local cumulative constraint violation.

<u>Modification Transforming Algorithm A to Algorithm C</u>. Algorithm C works on the same principle as Algorithm B above and achieves the same effect by somewhat different means provided by the second technique from Appendix B. Specifically, the following modifications to Algorithm A are required.

1) The objective function C^{ij} in Eq. (2.25a) changes from the one defined by Eq. (2.24b) to a variable, still designated C^{ij}, that together with the design variables X^{ij} form a set of independent variables in a following optimization problem that replaces the one formulated in Eq. (2.25a)

$$\text{find } C^{ij} \text{ and } X^{ij} \text{ such that } C^{ij} \text{ is at minimum STOC} \tag{2.41a}$$

2) The equality constraints, Eq. (2.25b), are eliminated and replaced with the following inequality constraints

$$-C^{ij} \leq \left\{ h_K^{ij} | h_M^{ij} | h_P^{ij} \right\} \leq C^{ij} \tag{2.41b}$$

$$C^{ij} > 0 \tag{2.41c}$$

$$g^{ij} \leq 0 \tag{2.41d}$$

3) Eq. (2.31) is deleted and the objective function C^{mk} in Eq. (2.32a) changes from the one defined by Eq. (2.31) to a variable, still denoted C^{mk}, that together with the design variables X^{ij} form a set of independent variables in the following optimization problem that replaces the one formulated in Eq. (2.32a)

$$\text{find } C^{ij} \text{ and } X^{ij} \text{ such that } C^{ij} \text{ is at minimum STOC} \tag{2.42a}$$

4) The equality constraints in Eq. (2.32b) are eliminated and replaced by the following inequality constraints

$$-C^{mk} \leq h^{mk} \leq C^{mk} \tag{2.42b}$$

$$C^{mk} > 0 \tag{2.42c}$$

$$\bar{C}_e^{ij} \leq \mathrm{TOL}^{ij} \tag{2.42d}$$

$$g^{mk} \leq 0 \tag{2.42e}$$

where $\bar{C}_e^{ij}$ is extrapolated by Eq. (2.30). The above modifications convert Algorithm A to Algorithm C.

Accounting in the Parent for the Daughter Side-Constraints. In Algorithm A one has to anticipate the following potential problem: the parent level optimization in $SSmkl$ may set its design variables X^{mk} at a combination of values for which there may be no feasible solution to the equality constraints in Eq. (2.25b) within the side-constraints on X^{ij} in a daughter $SSijk$ in the next iteration, when parameters $Y^{ij} = X^{mk}$ (Eq. (2.16)) are entered into the $SSijk$ optimization. This may prematurely terminate Algorithm A which requires satisfaction of the above equality constraints in each iteration. To alleviate that potential problem, the inequality constraints such as Eqs. (2.32d) and (2.35e) are included in each parent optimization. These constraints are applied to the daughter design variables approximated by X_e^{ij} using extrapolation such as Eq. (2.33), in which the derivatives of $\bar{X}^{ij}$ with respect to X_t^{mk} are obtainable from the optimum sensitivity analysis algorithm described in [5] (see Appendix A). However, that algorithm uses second derivatives of behavior that may be expensive to calculate. A variant of this algorithm introduced in [6] (see Appendix A) is computationally much cheaper because it avoids the second derivatives of behavior at the price of yielding the derivatives of the objective function but not of the design variables.

One may take advantage of the above less expensive algorithm and alleviate the daughter side-constraint problem at the same time by using a constraint relaxation technique described in [17] in the daughter $SSijk$ instead of the inequality constraint such as Eq. (2.32d) in the parent $SSmkl$.

In contrast to the above, in Algorithms B and C any difficulty to satisfy the equality constraints of the type represented by Eqs. (2.25b) and (2.32b) in Algorithm A translates simply in a larger value of the objective C^{ij} and may slow the convergence process but will not terminate the procedure for the lack of the feasible design. Hence, Algorithms B and C are well-suited for the use of the cheaper optimum sensitivity analysis algorithm from [6]. The constraint relaxation technique from [17] may still be useful in Algorithms B and C whenever the local constraints g^{ij} are the cause of difficulty in locating a feasible design.

Salient Features of the Algorithms. In perspective, the above three variants of the multi-level algorithm differ from a conventional one in a number of salient features outlined in this subsection.

A multitude of smaller problems, which may be processed concurrently, replace a single large problem. Although the subproblems are isolated, their coupling is preserved because the influence of the changes in the parent on the daughters is represented by linear extrapolation based on the optimum sensitivity and behavior sensitivity derivatives. With the exception of the most detailed level, the stiffness and mass distributions are controlled directly by generalized design variables. Mass is the objective at the top level, while at all the levels below the objective is either the improvement in the local constraint satisfaction in variant A or the improvement in the conformity of the daughter parameters to the parent design variables in variants B and C.

Selection of the generalized design variables is judgmental. In the extreme, one may choose to control all entries of the boundary stiffness matrix, boundary forces vector, and mass of each daughter; although, intuitively, this would seem impractical. Experience will probably show that a limited control, e.g., over the diagonal entries of the stiffness matrix only, will suffice in most cases.

The overall procedure building blocks, i.e., the operations of substructure analysis, constraint calculations, optimization, and behavior and optimum sensitivity analyses are "black boxes" whose algorithmic contents may be replaced freely, provided that the input/output definitions remain unchanged. For example, different types of structural analysis may be used at each level and even for each substructure, as it will be shown in the numerical example.

NUMERICAL EXAMPLE

Test Case. The subject algorithm was tested by optimizing, with and without decomposition, a framework structure similar to the one used in [11], [20], [21]. As shown in Figs. 2.4 and 2.5, the framework assembled at level 1 decomposes into three box beams, each beam being a substructure at level 2. Finally, each beam decomposes into three walls (the fourth wall is symmetric), each wall being the most detailed substructure at level 3. The external loads were applied at one corner of the framework as shown in Fig. 2.4. There were no interior loads on the substructures.

The objective was to minimize the structural material volume subject to constraints on the displacements of the loaded point, on in- and out-of-plane elastic stability of each beam treated as a column, and on the stresses and local buckling of the wall panels treated as stringer-reinforced plates. There were also minimum gage constraints and the physical realizability constraints on the cross-sectional dimensions.

The objective functions, design variables, parameters, and constraints are defined for the multilevel optimization in Table 2.1. A comprehensive description of all the physical and computational details of the test problem is given in [22].

Tools for Analysis and Design Space Search. A finite-element analysis was used to calculate the framework's displacements and beam end forces. Stresses in the beams loaded with the end forces were computed by ordinary engineering beam theory. The beams were treated as columns for stability analysis and the local buckling of the walls was based on closed form "designer handbook" formulas provided in [23] and [24], implemented as described in [25].

At each level, the design space search was conducted by the same general-purpose, nonlinear mathematical programming code based on the usable-feasible directions technique and documented in [26].

Three-Level Optimization. The framework was first optimized without decomposition to establish the reference results. Then, the multilevel optimization algorithm was applied to the structure decomposed as shown in Figs. 2.4 and 2.5. In the decomposition the stiffened panels are daughters clustered in triplets under a parent box beam. The beams, in turn, are daughters of the assembled structure.

As shown in Table 2.1, the top level optimization manipulates the beam extensional and bending stiffnesses through the cross-sectional areas and bending moments of inertia.

In this particular case, the cross-sectional area plays a dual role as it also controls the beam volume that contributes directly to the objective function.

At the middle level, the stiffnesses expressed by the area and moment of inertia become fixed parameters and the variables are the wall membrane stiffnesses controlled by the geometrical dimension variables. These variables and, consequently, the membrane stiffnesses become fixed parameters at the bottom level at which the ultimate detail dimensions are engaged as variables. The equality constraints arise between the parameters and variables. Owing to relative simplicity of the expressions involved, (see the Appendix of [22]), these constraints were solved explicitly.

Examination of Table 2.1 along with the previous description of the analysis tools illustrates the point that dissimilar analysis may be used as needed at different places in a decomposition scheme.

The sensitivity analysis of behavior has been carried out by a single-step forward finite-difference technique. The optimum sensitivity analysis was based on the algorithm given in [5].

Results and Remarks on the Method Performance. Figure 2.6 shows a sample of results obtained with and without decomposition. The starting points for both methods are the same. The normalized plots illustrate the objective function, a selected individual constraint, and a cumulative constraint containing the above individual constraint as they varied over the iterations. An iteration is defined in the optimization without decomposition as a usable-feasible directions iteration. In the three-level optimization, it is defined as one execution of the series of steps listed in the procedure definition in the previous section.

The results verified that the multilevel algorithm was capable of finding a feasible design having an objective function close to and, in some cases lower than, the reference optimization without decomposition. Similarly, as in [11], differences up to 72.1% were observed among the detailed design variables obtained by the two methods. However, these differences were no larger than those observed by comparing the designs obtained without decomposition starting from different initial design points. Therefore, these differences can be attributed to the problem nonconvexity. The jagged appearance of the graphs in Fig. 2.6 is a characteristic of the usable-feasible directions search algorithm, amplified in the multilevel optimization by the extrapolation errors. However, these errors never become excessive and the daughter substructure reactions to the changes in the parent design were effectively predicted by the optimum sensitivity derivatives. It was observed in at least one case that these predictions enabled the optimization at the middle level to remove the constraint violations at a bottom-level substructure without any change to the sizing of the bottom-level substructure. A detailed comparison of the results from both methods is given in [22].

Regarding computational efficiency, the main intrinsic advantage of the multilevel algorithm is in its capability to process the subproblems concurrently. Demonstration of this advantage would require a large application, distributed computing, and division of work among many people. Consequently, computational efficiency was not one of the goals in execution of the relatively small numerical example on a conventional serial computer. However, the example showed that the amount of computational labor per one iteration was less in the multilevel algorithm than in the single-level, conventional one and that both algorithms required about the same number of iterations for convergence. The example also showed that the multilevel algorithm was data-handling intensive; the operations of data moving and bookkeeping dominated the programming effort.

Other Applications. Several literature citations appeared reporting on the operational experience with the algorithm described in the foregoing. A sample selected from these reports is summarized in this section, including applications that extended beyond structures into multidisciplinary problems.

The algorithm in its variant A but reduced to two levels was applied to a portal framework example as reported in [11]. The overall geometry and loading of the framework was identical to that used in the above example and depicted in Fig. 2.4. However, instead of a box construction, a simple I-cross-section was used in each of the three beams of the framework. The three beams formed the bottom-level tier of substructures and their cross-sectional areas and moments of inertia were used as design variables at the next higher level representing the assembled framework. Because the loads were acting at the framework corners, there was no need for distributing the interior loads to the nodal points, hence Eq. (2.15) and those stemming from it fell out of the entire algorithm. The application resulted in the minimum mass of the framework falling within two percent of the benchmark value from the all-in-one, single-level optimization. Regarding computational cost, the numbers of iterations in the two-level and single-level optimizations were about equal. However, in one iteration of the single-level optimization one had to calculate gradients in the assembled structure analysis for all the detailed, cross-sectional variables in the structure, while the two-level optimization required to do so for only the much smaller number of the system level design variables. Hence, with the number of iterations being about equal in both methods, the two-level method resulted in computational cost reductions. An extensive investigation of the two-level optimization was also reported in [17] using a truss with thin-walled tubular rods as a test case.

Algorithm A also was adapted to multidisciplinary problems, consistent with [4] in which it was pointed out that such problems may be amenable to a hierarchic decomposition that converts the large system optimization problem into a number of smaller subproblems. The subproblems form a hierarchy similar to the one depicted in Fig. 2.1 if each subproblem receives parameters from its parent, returns optimized data to that parent, and exchanges no information directly with the siblings. The subproblems may then be the mathematical models of physical phenomena involved in the system design and/or of its physical parts. In the hierarchic scheme they simply perform as black box input-to-output converters comprising both the analytical and optimization functions. In the transport aircraft multidisciplinary optimization described in [18], the objective was the fuel consumption, design variables included airfoil-to-thickness ratio as a configuration variable in addition to the wing-box cross-sectional dimensions, and the constraints were drawn from structures, aerodynamics, and aircraft performance. A three-level, hierarchic decomposition illustrated in Fig. 2.7 entailed analyses of performance and aerodynamics, assembled wing-box structural analysis, and detailed analyses of the individual wing panels for local constraints, including buckling. The numbers of the design variables and constraints were well over 1000, and a satisfactory convergence after only 5 to 6 iterations was reported. Results corresponded well to the actual reference aircraft for feasible, as well as infeasible, initial design points.

Application to an actively controlled structure comprising of the subproblems of a structural optimization and an active-control synthesis was reported in [19]. Formal adaptation of Algorithm A to generic systems was presented in [27] and validated using a generic system simulation as described in [28]. In this generic adaptation, the physical meaning of the design variables, parameters, subproblems, and the analytical and optimization contents of the subproblems are application-dependent, but the algorithm operating on these entities is universal and the same as the one described for substructuring in the foregoing.

Finally, Algorithm B based on [30], tested ([29]) on the same framework optimization problem that was reported in [11] and discussed above, exhibited satisfactory convergence and produced the objective function results within 5% of the benchmark [11]. No verification of Algorithm C was reported so far, but effectiveness of its key ingredient—the technique for indirect satisfaction of the equality constraints—was established in [31].

Section 3: Non-Hierarchic Decomposition

In some systems, a structural mathematical model is only one of many constituent black boxes that exchange input/output data in a pattern whose example is illustrated in Fig. 3.1a. This example portrays an elastic, actively controlled wing as a system comprising mathematical models of structures, aerodynamics, and controls. The data exchange pattern precludes a hierarchic arrangement of the black boxes similar to the one depicted in Fig. 2.1 for substructures because no two black boxes can be found that would not send data to each other. Acknowledging that a generic system of this type may entail a rather large number of mathematical models representing its separate parts and physical phenomena involved in its behavior, it will be, nevertheless, convenient in the introductory discussion to limit the number of the black boxes in the system to only three, because it is a number small enough to foster comprehension and yet large enough to allow a generalization pattern to develop.

Accordingly, consider a generic system shown in Fig. 3.1b whose analysis yields the vector of the behavior variables $Y = \{Y\alpha|Y\beta|Y\gamma\}$ for a trial setting of the design variables X. If the derivatives $\mathrm{d}Y/\mathrm{d}X$, termed the system design derivatives (SDD), were available, then one could enter them into a gradient-guided search algorithm coupled to an extrapolation

$$Y_{\text{new}} = Y_{\text{old}} + \frac{\mathrm{d}Y}{\mathrm{d}X}\Delta X \tag{3.1}$$

and to perform one stage of the system optimization within move limits. As was shown in [32], the SDD may be calculated without the costly, "perturb-and-reanalyze", finite differencing on the system analysis by a technique based on the Implicit Function Theorem.

SYSTEM DESIGN DERIVATIVES

Ascribing a vector function representation to each black box, the set of equations representing the system of the black boxes α, β, and γ exchanging data as illustrated in Fig. 3.1b is

$$\begin{aligned} Y_\alpha &= Y_\alpha(X, Y_\beta, Y_\gamma) \\ Y_\beta &= Y_\beta(X, Y_\alpha, Y_\gamma) \\ Y_\gamma &= Y_\gamma(X, Y_\alpha, Y_\beta) \end{aligned} \tag{3.2}$$

The Y and X variables in the above equations are vectors entered in the black boxes selectively; e.g., some, but not necessarily all, elements of the vectors X and Y_α enter the black box β as inputs. Regarding $Y_\beta(X, Y_\alpha, Y_\gamma)$ as an example of a black box, the arguments, X, Y_α, Y_γ, are the inputs and Y_β is an output. The functions in Eq. (3.2) are coupled by their outputs appearing as inputs; hence, they form a set of simultaneous equations that can be solved for Y for given X. The act of obtaining such a solution is referred to as the system analysis (SA). In the presence of nonlinearities, SA is usually iterative.

For each function in Eq. (3.2), one can calculate derivatives of output with respect to any particular input variable, assuming that other variables are fixed. From the entire system perspective, these derivatives are partial derivatives because they measure only the local input-on-output effect, as opposed to SDD that are total derivatives because they include the effect of the couplings. To prepare for further discussion, the partial derivatives corresponding to the Y-inputs are collected in the Jacobian matrices designated by a pair of subscripts identifying the origins of the output and input, respectively. For example,

$$J_{\gamma\alpha} = [\partial Y_\gamma / \partial Y_\alpha] \tag{3.3}$$

is a matrix whose j-th column is made of the partial derivatives $\partial Y_{\gamma i}/\partial Y_{\alpha j}$. Assuming the length of Y_γ as N_γ and the length of Y_α as N_α, the dimensions of matrix $J_{\gamma\alpha}$ are $N_\gamma \times N_\alpha$. It will be mnemonic to refer to the partial derivatives in the Jacobian matrices as the cross-derivatives.

The remaining partial derivatives corresponding to the X-inputs are collected in vectors, one vector per each of the NX elements of the vector of design variables X; e.g,

$$\{\partial Y_\alpha / \partial X_k\} \quad k = 1, \ldots, NX \tag{3.4}$$

is a vector of the length N_α.

Calculation of the above partial derivatives may be accomplished by any means available for a particular black box at hand, and may range from finite differencing to quasi-analytical methods [33] and [34].

It was shown in [32] that differentiation of the functions in Eq. (3.2) as composite functions and application of the implicit function theorem leads to a set of simultaneous, linear, algebraic equations, referred to as the Global Sensitivity Equations (GSE), in which the above partial derivatives appear as coefficients and the SDD are the unknowns. For the system of Eq. (3.2), the GSE are

$$\begin{bmatrix} I & -J_{\alpha\beta} & -J_{\alpha\gamma} \\ -J_{\beta\alpha} & I & -J_{\beta\gamma} \\ -J_{\gamma\alpha} & -J_{\gamma\beta} & I \end{bmatrix} \begin{Bmatrix} \mathrm{d}Y_\alpha/\mathrm{d}X_k \\ \mathrm{d}Y_\beta/\mathrm{d}X_k \\ \mathrm{d}Y_\gamma/\mathrm{d}X_k \end{Bmatrix} = \begin{Bmatrix} \partial Y_\alpha/\partial X_k \\ \partial Y_\beta/\partial X_k \\ \partial Y_\gamma/\partial X_k \end{Bmatrix} \tag{3.5}$$

These equations may be formed only after the SA was performed for a particular X, a particular point in the design space because the computation of the partial derivatives requires that all the X and Y values be known. For a given X, the matrix of coefficients depends only on the system couplings and is not affected by the choice of X for the right-hand side. Hence, that matrix may be factored once and reused in a backsubstitution operation to compute as many sets of SDD's as many different X_k variables are represented in the set of multiple right-hand side vectors.

As recommended in [32], numerical solution of Eq. (3.5) and interpretation of the SDD values will be facilitated by normalization of the coefficients in the matrix and in the right-hand sides by the values of Y_o and X_o of the Y and X variables for which the partial derivatives were calculated. The normalized coefficients take on the following form, illustrated by a few examples from i-th row in the β partition in Eq. (3.5)

$$-\frac{\partial Y_{\beta i}}{\partial Y_{\alpha j}} q_{\beta\alpha ij}; \quad -\frac{\partial Y_{\beta i}}{\partial Y_{\gamma j}} q_{\beta\gamma ij}; \quad \frac{dY_{\beta i}}{dX_k} q_{\beta Xik} \tag{3.6}$$

where the normalization coefficients q are

$$q_{\beta\alpha ij} = \frac{Y_{\alpha jo}}{Y_{\beta io}}; \quad q_{\beta\gamma ij} = \frac{Y_{\gamma jo}}{Y_{\beta io}}; \quad \bar{q}_{\beta Xik} = \frac{X_{ko}}{Y_{\beta io}}$$

Solving the normalized Eq. (3.5) yields normalized values of the SDD's from which the unnormalized values may always be recovered given the above definitions.

Forming the GSE and their solution for a set of SDD's will be referred to as the System Sensitivity Analysis (SSA).

UTILITY OF THE SYSTEM DESIGN DERIVATIVES

The SDD carry the trend information that under a conventional approach would be sought by resorting to statistical data or to the parametric studies. The former have the merit of capturing a vast precedent knowledge but may turn out to be ineffective if the vehicle at hand is advanced far beyond the existing experience. The latter provide an insight into the entire interval of interest but only for a few variables at a time, and that insight tends to be quickly lost if there are many design variables; in which case, the computational cost of the parametric studies also may become an impediment.

In contrast, the SDD information is strictly local but it reflects the influences of all the design variables on all aspects of the system behavior. Therefore, the SSA should not be regarded as a replacement of the above two approaches but as their logical complement whose results are useful in at least two ways.

Ranking Design Variables for Effectiveness. A full set of SDD for a system with NY variables in Y and NX variables in X is a matrix $NY \times NX$. The j-th column of the matrix describes the degree of influence of variable X_j on the behavior variables Y. Conversely, the i-th row shows the strength of influence of all the design variables X on the i-th behavior variable Y_i. For normalized SDD's, comparison of these strengths of influence becomes meaningful and may be used to rank the design variables by the degree of their influence on the particular behavior variable. This ranking may be used as a basis for judgmentally changing the design variable values and for deciding which design variables to use in a formal optimization.

An example of such ranking is illustrated for the wing of a general aviation aircraft shown in Fig. 3.2. The design variables are thicknesses t of the panels in the upper cover of the wing box, and the behavior variable is the aircraft range R. The chain of influences leading from a panel thickness to the range calculated by means of the Breguet formula is depicted in Fig. 3.3a. In the Breguet formula, W_e denotes the zero-fuel weight and W_p stands for the fuel weight. Increasing t in one of the panels increases the weight W_e and, in general, reduces the drag of a flexible wing by stiffening its structure. Consequently, the range is influenced in conflicting ways that would make prediction by judgment difficult. However, the corresponding SSA yields the SDD's for the upper row of the wing cover panels illustrated by the heights of the vertical bars over the upper wing cover panels in

Fig. 3.3b. The bars show that among all the wing cover panels, increasing t in the extreme outboard panel would increase range the most.

Gradient-Guided Formal Optimization. Most of the formal optimization methods applicable in large engineering problems use the first derivative information to guide the search for a better design. Because the SDD values provide such information for all the Y and X variables of interest, the SSA may be incorporated, together with SA, in a system optimization procedure (SOP) based on the well-known piecewise approximate analysis approach (e.g., [35]). The SOP flowchart is depicted in Fig. 3.4. An important benefit of the SOP organization is the opportunity for parallel processing seen in the flowchart operation immediately following the SA. In that operation, one computes concurrently the partial derivatives of input with respect to output for all the system black boxes, to form the Jacobian matrices (Eq. (3.3)) and the right-hand-side vectors (Eq. (3.4)) needed to form the GSE (Eq. (3.5)) whose solution yields the SDD's. In a conventional approach, these SDD's would be computed by finite differencing on SA. The SDD values are subsequently used in Approximate Analysis (extrapolation formulas) that supplies the optimizer (a design space search algorithm) with information on the system behavior for every change of the design variables that optimizer generates, and does it at a cost negligible in comparison with the cost of SA.

A generic hypersonic aircraft was used as a test for the above optimization. The geometrical design variables for the case are shown in Fig. 3.5. Additional design variables were the deflections of the control surfaces and the cross-sectional structural dimensions of the forebody. The propulsive efficiency measured by the I_{sp} index, defined as thrust minus drag divided by the fuel mass flow rate, was chosen as the objective function to be maximized. The aircraft take-off gross weight (TOGW) for a given mission is very sensitive to that index, thus maximization of the index effectively minimizes TOGW. The problem requires consideration of a system composed of aerodynamics, propulsion, performance analysis, and structures. The optimization included constraints on the aircraft as a whole and on behavior in the above disciplines. Results are shown in Table 3.1 in terms of the initial and final values of the design variables (cross-sectional dimensions omitted) and of the objective function, all normalized by the initial values. Considering that the initial values resulted from an extensive design effort using a conventional approach, the nearly 13% improvement in the propulsive efficiency was regarded as very significant indeed.

MERITS AND DEMERITS

Before discussion of the ramifications of the above sensitivity-based optimization in a system design process, it may be useful to examine briefly the merits and demerits of the proposed approach relative to the conventional technique of generating SSD by finite differencing on the entire SA.

Accuracy and Concurrent Computing. The SSA based on Eq. (3.5) has two unique advantages. First, the accuracy of SDD is intrinsically superior to that obtainable from finite differencing whose precision depends on the step length in a manner that is difficult to predict. As pointed out in [36] it is particularly true in the case of an iterative SA whose result often depends on an arbitrary, "practical" convergence criterion. Second, there is an opportunity for concurrent computing in the generation of the partial derivatives that exploits the technology of parallel processing offered by multiprocessor computers and computer networks. Concurrent computing also enables the engineering workload to be distributed among the specialty groups in an engineering organization to compress the project execution time.

Computational Cost. Experience indicates that in large engineering applications, most of the optimization computational cost is generated by the finite difference operations. Therefore, relative reduction of the cost of these operations translates into nearly the same relative reduction of the cost of the entire optimization procedure.

The computational cost of the SSA based on Eq. (3.5), designated C_1, may be reduced, in most cases very decisively, below that of finite differencing on the entire SA, denoted by C_2, but to achieve that reduction the analyst should be aware of the principal factors involved. To define these factors, let the computational cost of one SA be denoted by CSA while CBA_i will stand for the computational cost of one analysis of the i-th black box in the system composed of NB black boxes. The i-th black box receives an input of NX_i design variables X, and NY_i variables Y from the other black boxes in the system. Assuming for both alternatives the simplest, one-step finite difference algorithm that requires one reference analysis and one perturbed analysis for each input variable, the costs C_1 and C_2 may be estimated as

$$C_1 = \sum_{i=1}^{i=NB} (1 + NX_i + NY_i)CBA_i \qquad (3.7)$$

$$C_2 = (1 + NX)CSA$$

Even though one may expect $CBA_i < CSA$, a sufficiently large NY_i may generate $C_1 > C_2$ and render SSA based on Eq. (3.5) unattractive compared with finite differencing on the entire SA. This points to NY_i, termed the interaction bandwidth, as the critical factor whose magnitude should be reduced as much as possible. Reducing the interaction bandwidth requires judgment as illustrated by an example of an elastic, high-aspect ratio wing treated as a system whose aeroelastic behavior is modeled by interaction of aerodynamics and structures, represented by a CFD analysis and Finite Element analysis codes. If one lets the full output from each of these black boxes be transmitted to the other, there might be hundreds of pressure coefficients entering the structural analysis and thousands of deformations sent to the aerodynamic analysis. With the NY_i values in the hundreds and thousands, respectively, it would be quite likely that $C_1 > C_2$. However, one may condense the information flowing between the two black boxes by taking advantage of the high-aspect ratio wing slenderness. For a slender wing, it is reasonable to represent the entire aerodynamic load by, say, a set of 5 concentrated forces at each of 10 separate chords, and to reduce the elastic deformation data to, say, elastic twist angles at 7 separate chords. This condensation reduces the NY_i values to 50 for structures and 7 for aerodynamics. In the finite element code, that implies 50 additional loading cases, all of which can be computed very efficiently by the multiple loading case option—a standard feature in finite element codes. The CFD code would have to be executed only 7 additional times. Thus, the advantage of the interaction bandwidth condensation is evident. In general, a condensation such as the one described above for a particular example may be accomplished by the reduced-basis methods, among which the Ritz functions approach is, perhaps, the best known one.

Potential Singularity. One should be aware when using SSA based on Eq. (3.5) that, in some cases, the matrix of coefficients in these equations may be singular. In geometrical terms, a solution in SA may be interpreted geometrically as a vertex of hyperplanes on which the residuals of the governing equations for the black boxes involved are zero. As shown in [32], Eq. (3.5) are well-conditioned if these hyperplanes intersect at large angles, ideally when they are mutually orthogonal. For two functions of two variables the zero-residual hyperplanes reduce to the zero-residual contours, and an example of a nearly-orthogonal solution intersection is shown in Fig. 3.6a. In some cases, the intersection angles are very acute; in the limit they may be zero, in which case a solution exists by virtue of tangency of two curved contours as illustrated in Fig. 3.6b. It is shown in [32] that Eq. (3.5) imply local linearization of these contours in the vicinity of the intersection point so that the solution point is interpreted as an intersection of the tangents. Consequently, in the situation depicted in Fig. 3.6b, the tangents coincide and the matrix of Eq. (3.5) becomes singular. In such a case, Eq. (3.5) should be replaced by an alternative formulation of the system sensitivity equations in [32] based on residuals.

There were no cases of singularity reported so far in any applications, probably because the system solutions of the type illustrated in Fig. 3.6b characterize an ill-posed system analysis usually avoided in practice.

Discrete Variables. Neither the reference technique nor the SSA based on Eq. (3.5) can accommodate truly discrete design variables. Truly discrete design variables are defined for the purposes of this discussion as those with respect to which SA is not differentiable. These are distinct from quasi-discrete variables with respect to which SA is differentiable but which may only be physically realizable in a set of discrete values. An example of the former is an engine location on the aircraft, either under the wing or at the aft end of the fuselage. An example of the latter is sheet metal thickness available in a set of commercial gages.

In the case of truly discrete design variables, different combinations of such variables define different design concepts (alternatives) and each concept may be optimized in its own design space of the remaining continuous variables, to bring it up to its true potential. Then, one may choose from among the optimal alternatives. Occasionally, a continuous transformation might be possible between two concepts that seem to be discretely different. For example, a baseline aircraft with a canard, a wing, and a conventional tail may be reshaped into any configuration featuring all, or only some of these three lifting surfaces. This is so because a sensitivity-guided SOP may eliminate a particular feature, if a design variable is reserved for that feature and if the feature is present in the initial design (however, a feature initially absent cannot, in general, be created).

Non-utilization of Disciplinary Optimization. Organization of the SOP discussed above may be described as "decomposition for sensitivity analysis followed by optimization of the entire, undecomposed system". It may be regarded as a shortcoming that the procedure leaves no clear place for the use of the vast expertise of optimization available in the individual black boxes representing engineering disciplines. Examples of such local, disciplinary optimization techniques are the optimality criteria for minimum weight in structures, and shaping for minimum drag for a constant lift in aerodynamics. It appears that combining these local, disciplinary optimization techniques with the overall system optimization should benefit the latter. Indeed, one way in which these techniques may be used without changing anything in the SOP organization described above is in the

SOP initialization. Obviously, starting SOP from a baseline system composed of the black boxes already preoptimized for minimum weight, minimum drag, maximum propulsive efficiency, etc. should accelerate the SOP convergence and improve the end result. Such local optimizations could be accomplished separately for each black box, assuming X and guessing at the Y inputs.

Beyond that, the issue of incorporating the local, disciplinary optimization in SOP remains to be a challenge for further development. Some solutions were proposed in [37] and [38] but their effectiveness is yet to be proven in practice.

HYBRID DECOMPOSITIONS

If one of the black boxes in a non-hierarchic, coupled system represents a structure, the hierarchic decomposition depicted in Fig. 2.1 may be embedded in that box. Conversely, a hierarchic system may comprise one or more black boxes, each containing a non-hierarchic decomposition of the type illustrated in Fig. 3.1. An example of the latter is an aircraft wing that appears as a substructure of the entire airframe but may be regarded as a non-hierarchic system within itself composed of the wing mathematical model for structural analysis, and its aerodynamic and control models. Such hybrid decompositions may occur simultaneously in various combinations suggesting that development of the corresponding optimization procedures based on those discussed in Secs. 2 and 3 may be a fruitful direction for future research.

Concluding Remarks

An algorithm for solving the structural optimization problem as a set of smaller subproblems corresponding to levels of nested substructures has been presented as a generalization of the Bellman's optimality principle that underlies the method of Dynamic Programming. The algorithm was presented in three variants that differed in the means by which the substructure stiffness and mass properties were matched to those set in the parent substructure. In all variants, the matching was assisted by the use of the behavior and optimum sensitivity derivatives. The algorithm is inherently compatible with distributed computing because the subproblems can be processed concurrently. It also can be generalized to those multidisciplinary systems whose subsystems may be arranged into a hierarchy of dependencies similar to those exhibited by substructures.

Numerical tests performed on a conventional, serial computer for a framework decomposed into a three-level hierarchy of 13 subproblems demonstrated satisfactory convergence to the close vicinity of the reference solutions obtained by an optimization without decomposition. A significant part of the effort to program the algorithm went into the data handling, indicating that a systematic data management system would be required in extending the applications to more levels. Because the addition of more levels beyond three would introduce no new qualitative elements into the algorithm, the three-level test case has been sufficient as a proof of the concept. The algorithm has been also extensively verified by applications reported in literature, including demonstrations of the algorithm adaptability to multidisciplinary problems.

Optimization of those multidisciplinary systems that are not amenable to a hierarchic decomposition may be performed using a procedure based on the derivatives of such non-hierarchic systems behavior with respect to design variables. The procedure embeds a technique for calculation of these derivatives bypassing finite differencing of the system analysis and is illustrated by an application example.

The optimization procedures reviewed in the paper form a toolkit for use in design of large structures and engineering systems that incorporate structure as one of subsystems.

Acknowledgment

Contribution of the NASP configuration optimization example (Fig. 3.5 and Table 3.1) by Dr. F. Abdi and Mr. J. Tulinius of Rockwell International—North American is gratefully acknowledged.

References

[1]Kirsch, U., Reiss, M., and Shamir, U., "Optimum Design by Partitioning into Substructures," *Journal of Structural Division ASCE*, Jan. 1972, p. 249.

[2]Sobieszczanski, J., "Sizing of Complex Structure by the Integration of Several Different Optimal Design Algorithms," *Structural Optimization*, AGARD-LS-70, Sept. 1974.

[3]Schmit, L. A. and Ramanathan, R. K., "Multilevel Approach to Minimum Weight Design Including Buckling Constraints," *AIAA Journal*, Vol. 16, Feb. 1973, pp. 97–104.

[4]Sobieszczanski-Sobieski, J., "A Linear Decomposition Method for Large Optimization Problems—Blueprint for Development," NASA TM 83248, Feb. 1982.

[5]Sobieszczanski-Sobieski, J., Barthelemy, J.-F., and Riley, K. M., "Sensitivity of Optimum Solutions to Problem Parameters," *AIAA Journal*, Vol. 20, Sept. 1982, p. 1291.

[6]Barthelemy, J.-F. M. and Sobieszczanski-Sobieski, J., "Optimum Sensitivity Derivatives of Objective Function in Nonlinear Programming," *AIAA Journal*, Vol. 21, June 1983, pp. 913–915.

[7]Bellman, R., Adaptive Control Processes: A Guided Tour; Princeton University Press, 1961.

[8]Noor, A. K., Kamel, H. A., and Fulton, R. E., "Substructuring Techniques—Status and Projections," *Computers and Structures*, Vol. 8, May 1978, pp. 621–632.

[9]Przemieniecki, J. S., *Theory of Matrix Structural Analysis*, McGraw-Hill, New York, 1968, Chap. 9.

[10]Aaraldsen, P. O., "The Application of the Superelement Method in Analysis and Design of Ship Structures and Machinery Components," Paper presented at National Symposium on Computerized Structures Analysis and Design, George Washington University, Washington, DC, March 1972.

[11]Sobieszczanski-Sobieski, J., James, B., and Dovi, A., "Structural Optimization by Multi-Level Decomposition," *AIAA Journal*, Vol. 23, Nov. 1983, pp. 1775–1782.

[12]Sobieszczanski-Sobieski, J., James, B. B., and Riley, M. F., Structural Sizing by Generalized, Multilevel Optimization, AIAA J. Vol. 25, No. 1, January 1987, p.139.

[13]Sobieszczanski-Sobieski, J., Two Alternative Ways for Solving the Coordination Problem in Multilevel Optimization, NASA TM 104036, August 1991.

[14]Kreisselmeier, G. and Steinhauser, R., "Application of Vector Performance Optimization to a Robust Control Loop Design for a Fighter Aircraft," *International Journal of Control*, Vol. 37, No. 2, 1983, pp. 251–284.

[15]Chapin, N., "New Format for Flowcharts," *Software—Practice and Experience*, Vol. 4, 1974, pp. 341–357.

[16]Thareja, R., and Haftka, R. T., Numerical Difficulties Associated with Using Equality Constraints to Achieve Multi-level Decomposition in Structural Optimization, AIAA Paper No. 86-0854, AIAA/ASME/ASCE/AHS 27th Structures, Structural Dynamics, and Materials Conference, San Antonio, Texas, May 1986.

[17]Barthelemy, J-F. M., and Riley, M. F., Improved Multilevel Optimization Approach for Design of Complex Engineering Systems; AIAA J., Vol. 26, No. 3, March 1988, pp. 353–360.

[18]Wrenn, G. A., and Dovi, A. R., Multilevel Decomposition Approach to the Preliminary Sizing of a Transport Aircraft Wing; AIAA Journal of Aircraft, Vol. 25, No. 7, July 1988, pp. 632–638.

[19]Zeiler, T. A., and Gilbert, M. G., Integrated Control/Structure Optimization by Multilevel Decomposition, NASA TM 102619, March 1990.

[20]Lust, R. V. and Schmit, L. A., "Alternative Approximation Concepts for Space Frame Synthesis," AIAA Paper 85-0696-CP, April 1985.

[21]Haftka, R. T., "An Improved Computational Approach for Multilevel Optimum Design," *Journal of Structural Mechanics*, Vol. 12, No. 2, 1984, pp. 245–261.

[22]Sobieszczanski-Sobieski, J., James, B. B., and Riley, M. F., "Structural Optimization by Generalized Multilevel Optimization," AIAA Paper 85-0698, April 1985 (also NASA TM 87605, Oct. 1985).

[23]Angermayer, K., "Structural Aluminum Design," Reynolds Metals Co., Richmond, VA, 1965.

[24]Timoshenko, S. P. and Gere, J. M., *Theory of Elastic Stability*, McGraw-Hill, New York, 1961.

[25]Sobieszczanski-Sobieski, J., "An Integrated Computer Procedure for Sizing Composite Air Frame Structures," NASA TP-1300, Feb. 1979.

[26]Vanderplaats, G. N., "CONMIN—A Fortran Program for Constrained Function Minimization: User's Manual," NASA TM X-62282, Aug. 1973.

[27]Sobieszczanski-Sobieski, J., and Barthelemy, J-F. M., Improving Engineering System Design by Formal Decomposition, Sensitivity Analysis, and Optimization. Proceedings of International Conference on Engineering Design, ICED 1985, Hamburg, Germany, Aug. 1985, Vol. 1, pp. 314–321. Published by Heurista, Zurich.

[28]Padula, S. L., and Sobieszczanski-Sobieski, J., A Computer Simulator for Development of Engineering System Design Methodologies; NASA TM 89109, Feb. 1987. Presented at International Conference on Engineering Design, ICED 1987, Boston, Massachussetts, U.S.A., Aug. 1987.

[29]Kumar, V., General Electric Research Center, Schenectady, N.Y.; Private Communication.

[30]Sobieszczanski-Sobieski, J., A Technique for Locating Function Roots and for Satisfying Equality Constraints in Optimization, NASA TM 104037, August 1991.

[31]Vanderplaats, G., Newsletter of VMA-Engineering, May 1990.

[32]Sobieszczanski-Sobieski, J., On the Sensitivity of Complex, Internally Coupled Systems; AIAA/ASME/ASCE/AHS 29th Structures, Structural Dynamics and Materials Conference, Williamsburg, VA, April 1988; AIAA Paper No. CP-88-2378, and AIAA J., Vol. 28, No. 1, Jan. 1990, also published as NASA TM 100537, January 1988.

[33]Proceedings of the Symposium on Sensitivity Analysis in Engineering, NASA Langley Research Center, Hampton, VA, Sept. 1986; Adelman, H. M.; and Haftka, R. T.—editors. NASA CP-2457, 1987.

[34]Adelman, H. A. and Haftka, R. T., Sensitivity Analysis of Discrete Structural Systems, AIAA J., Vol. 24, No. 5, May 1986, pp. 823–832.

[35]Sobieszczanski-Sobieski, J., From a Black-Box to a Programing System, Ch.11 in Foundations for Structural Optimization—A Unified Approach; Morris, A. J., ed.J.; Wiley & Sons, 1982.

[36]Haftka, R. T., "Sensitivity Calculations for Iteratively Solved Problems," *International Journal for Numerical Methods in Engineering*, Vol. 21, 1985, pp. 1535–1546.

[37]Sobieszczanski-Sobieski, J., Optimization by Decomposition: A Step from Hierarchic to Non-Hierarchic Systems; Second NASA/Air Force Symposium on Recent Advances in Multidisciplinary Analysis and Optimization; Hampton, Virginia, September 28–30 1988; Proceedings published as NASA CP - No. 3031; editor: Barthelemy, J. F.

[38]Bloebaum, C. L., Non-Hierachic System Decomposition in Structural Optimization Formal and Heuristic System Decomposition Methods in Multidisciplinary Synthesis; Ph.D. Dissertation, School of Engineering, Department of Aerospace Engineering, University of Florida, Gainesville, FL, 1991.

Appendix A: Sensitivity of Optimum to Parameters

Consider the following standard optimization problem

$$\min_{X} F(X, B) \quad \text{STOC} \tag{A1a}$$

$$g_j(X, B) \leq 0; \quad j = 1 \rightarrow m \tag{A1b}$$

where X is a vector of design variables, and B is a vector of constant parameters.

Solution of Problem A1 yields the optimum values of X and F, denoted by $\bar{X}$ and $\bar{F}$, which depend on B

$$\bar{X} = f(B); \quad \bar{F} = f(B) \tag{A2}$$

Sensitivity of $\bar{X}$ and $\bar{F}$ to B is measured by the derivatives termed optimum sensitivity derivatives. Capability to calculate such derivatives is useful not only as an ingredient in the algorithm described in the body of the paper but also in its own right as a means to assess the impact of the parameter change on the optimum as illustrated by the following two examples.

Example 1.: A structure was optimized for minimum weight for an allowable stress given as a parameter. The derivatives of the minimum weight and optimal cross-sectional variables with respect to the allowable stress may be used to predict by extrapolation the changes to the minimum weight and to the optimal variables caused by a change in the allowable stress, without repeating the entire optimization.

Example 2.: An aircraft was optimized for minimum take-off weight for a given payload and range. Without repeating the optimization, one may use the optimum sensitivity derivatives to estimate the optimal configuration changes corresponding to the changes in the required range and payload, taken separately or simultaneously.

The optimum sensitivity derivatives could be obtained by finite differencing of the solution to Problem A1 for the elements B_k of B perturbed one at a time. The algorithm introduced in [5] computes the above derivatives without finite differencing and is summarized in this Appendix.

In the following it is assumed that the set of active constraints has been identified when solving Problem A1 and that the membership of that set remains the same for small numerical changes of B_k. For brevity, a dot and prime notation is defined to replace the standard differential notation:

$$(\,)^{\cdot} = \frac{\partial(\,)}{\partial X_i}; \quad (\,)' = \frac{\partial(\,)}{\partial B_k} \tag{A3}$$

The algorithm is derived from the familiar Lagrange multiplier equations that are satisfied at the constrained minimum obtained as a solution to Problem A1.

$$\dot{F}^i + \dot{g}^i \lambda = 0, \quad i = 1 \rightarrow n \tag{A4a}$$

$$g_j = 0, \quad j = 1 \rightarrow \bar{m} \tag{A4b}$$

where Eq. (A4b) comprises the subset of $\bar{m} < m$ active constraints.

The above equations are differentiated with respect to parameter B_k using a chain-rule reflecting the composite function relationships implied by Eqs. (A1) and (A2). This leads to a set of simultaneous, linear, algebraic equations

$$\begin{bmatrix} \underset{n\times n}{[\ddot{F}+Z]} & \underset{n\times\bar{m}}{[\dot{g}]} \\ \underset{\bar{m}\times n}{[\dot{g}]^T} & \underset{\bar{m}\times\bar{m}}{[0]} \end{bmatrix} \begin{Bmatrix} \bar{X}' \\ \bar{\lambda}' \end{Bmatrix} + \begin{Bmatrix} \underset{n\times 1}{\{\dot{F}'\}} \; \underset{n\times\bar{m}\;\bar{m}\times 1}{+[\dot{g}']\{\lambda\}} \\ \underset{\bar{m}\times 1}{\{g'\}} \end{Bmatrix} = 0 \tag{A5a}$$

$$(n+\bar{m})\times(n+\bar{m}) \quad (n+\bar{m})\times 1 \qquad (n+\bar{m})\times 1$$

where the dimensions of the vectors and matrices are inscribed for clarity and where Z stands for a square matrix, $n \times n$, whose i, q element is

$$Z^{iq} = \sum_{j=1}^{m} \frac{\partial^2 g_j}{\partial X_i \partial X_q} \lambda_j \tag{A5b}$$

In Eq. (5a), the unknown $\bar{X}'$ and $\bar{\lambda}'$ are the total derivatives of the optimum with respect to B_k that are being sought. All the other derivatives are partial derivatives that are precomputed and entered into these equations as given coefficients. The equations have many right-hand-side vectors, one per each B_k. The Lagrange multipliers entering the equations may be obtained from

$$\lambda = -\left[\dot{g}^T \dot{g}\right]^{-1} \dot{g}^T \dot{F} \tag{A6}$$

containing only the active constraints, if they are not produced by the optimization algorithm used to solve Problem A1.

Solution of Eq. (5a) yields the values of $\bar{X}'$ and $\bar{\lambda}'$ for each B_k represented on the right-hand side. The derivative of the objective function is then calculated as

$$\mathrm{d}\bar{F}/\mathrm{d}p = \bar{F}' + \sum_i \dot{\bar{F}}^i \bar{X}_i' \tag{A7}$$

If the side-constraints are present among the active constraints in the solution of Problem A1, they ought to be included in the constraint functions whose derivatives enter as coefficients in Eqs. (5a) and (5b). That requires to write these side-constraints in the following standard form

$$g_j = 1 - X_i/X_{il} \leq 0$$

or

$$g_j = X_i/X_{iu} - 1 \leq 0 \tag{A8}$$

where X_{iu} and X_{il} are upper and lower side-constraints on X_i.

The second partial derivatives in Eqs. (5a) and (5b) may be expensive computationally. According to [6], they may be avoided by replacing Eqs. (5a) and (5b) with

$$\frac{\mathrm{d}\bar{F}}{\mathrm{d}p} = \bar{F}' + \bar{\lambda}^T \bar{g}_a' \tag{A9}$$

to obtain the optimum sensitivity derivative of the objective function. However, to do so one has to give up the derivatives of $\bar{X}'$.

When using the optimum sensitivity derivatives in extrapolation, one should remember that membership of the active constraint set may change when departing from the constrained minimum. Any such change may introduce large errors into the extrapolation. Hence, one must validate the extrapolation results by analysis to ascertain whether the constraints active at the end of extrapolation are still the same as those that were active at the original constrained minimum.

Appendix B: Two Techniques for Satisfying Equality Constraints

Technique 1

The algorithm developed in [30] is based on the Kreisselmeier-Steinhauser function (KS function) that was defined for a set of constraints in Eqs. (22) and (23). Redefined for a set of generic functions, the KS function is a differentiable envelope function for a set of functions of the form $Y = F_k(X)$, $k = 1 \ldots NK$. It is assumed that each of the Y functions is continuous in X but not necessarily continuous in the derivatives of Y with respect to X. The KS function is expressed in two alternative but completely equivalent formats

$$\text{KS}(F_k) = (1/\rho)\ell n\left[\sum \exp(\rho F_k)\right]; \quad k = 1 \ldots NK \tag{B1}$$

$$\text{KS}(F_k) = F_{\max} + (1/\rho)\ell n\left[\sum \exp(\rho(F_k - F_{\max}))\right]; \quad k = 1 \ldots NK \tag{B2}$$

The format in Eq. (B2) is recommended if the standard format in Eq. (B1) generates too large values of the exponential function.

The KS function has a property that

$$F_{\max} \le \text{KS}(F_k) \le F_{\max} + \ell n(NK)/\rho \tag{B3}$$

where ρ is a user-controlled parameter.

Example of the KS function is plotted in Fig. A1 for one independent variable and a set of three functions. As implied by Eq. (B3), the user may draw KS closer to $F_{\max}$ by increasing the value of parameter ρ and vice versa. In other words, the ρ parameter is a means for controlling how close the KS follows the piecewise envelope of the set of functions F_k. The KS function is a differentiable equivalent of the non-differentiable "selective" function MAX(F_k) available in many high-level programming languages.

The KS function may be used as a tool for root finding because it has the following property defined using Z as a generic, single independent variable:

> "If a function $F_k(Z)$ has a root $F_k(Z) = 0$ for $Z = Zr$, the KS function of F_k and $-F_k$ is at minimum at $Z = Zr$ for any value of nonzero and nonnegative ρ."

The above property is illustrated in Fig. A2 and was proven as a theorem in [30].

Extension to a set of functions $F_k(Z)$, $k = 1 \ldots NK$, in a hyperspace of NN dimensions (Z is then a vector of length NN) is straightforward owing to the property represented by Eq. (B3). The previous theorem extended to the above general case is:

> "The KS function of a set of functions made up of the original set of $F_k(Z)$ and their negatives, $-F_k(Z)$, (mirror images),
>
> $$\text{KS}(Z) = \text{KS}(F_1(Z), -F_1(Z), F_2(Z), -F_2(Z), \ldots$$
>
> $$F_k(Z), -F_k(Z), \ldots F_{NK}(Z), -F_{NK}(Z)) \tag{B4}$$
>
> has a local minimum at each point in the Z hyperspace where all the functions $F_k(Z)$ attain zero value simultaneously."

It follows that a set of equality constraints $h(X)$ may be satisfied by solving the following optimization problem

$$\text{find } X \text{ such that } \mathrm{KS}(h) = (h_1, -h_1, h_2, -h_2, \ldots, h_j, -h_j, \ldots h_s, -h_s), \quad j = 1 \to s, \text{ is at minimum STOC} \tag{B5a}$$

$$g(X) \leq 0 \tag{B5b}$$

$$L \leq X \leq U \tag{B5c}$$

where Eqs. (B5b) and (B5c) are added to provide an option of including the inequality and side constraints.

Applying the above technique discussed in the body of this paper, there are parameters $B = Y$ embedded in h and g. For a particular setting of these parameters it may be impossible to reduce KS to its minimum (Fig. B2), which means that the violation of h is minimized but not eliminated. For that reason, the tolerance TOL is introduced whose gradual reduction in the overall iterative procedure forces adjustments to B and eventually allows the KS function to reach its minimum where the h constraint violations are decreased to an acceptable level.

Technique 2

An algorithm introduced in [31] for fitting an empirical function to a set of experimental data points is adapted as a means for satisfying equality constraints $h(X)$. The algorithm requires augmentation of the vector of design variables X by an additional independent variable, C, that also doubles for the objective function in the following optimization problem

$$\text{find } C \text{ and } X \text{ such that } C \text{ is at minimum STOC} \tag{B6a}$$

$$g(X) \leq 0 \tag{B6b}$$

$$-C \leq h_j \leq +C; \quad j = 1 \to s; \tag{B6c}$$

$$L \leq X \leq U; \tag{B6d}$$

$$C > 0 \tag{B6e}$$

Solving the above problem produces the values of $\bar{C}$ and $\bar{X}$ that reduce the maximum absolute violation of h to a minimum achievable within the inequality and side constraints in Eqs. (B6b) and (B6d). This is accomplished by "squeezing" the violation of h_j in Eq. (B6c) simultaneously from the positive and negative sides, owing to non-negativity of C enforced in Eq. (B6e). When $\bar{C}$ is decreased to zero, the h constraints are satisfied. In the application discussed in the body of this paper, the value of $\bar{C}$ may have to be reduced to zero gradually using TOL, in a way analogous to the iterative reduction of KS described at the end of the definition of Technique 1, above.

Table 2.1 Quantities defined for the multilevel test case optimization

	Top level
Objective	Framework material volume
Design variables	A and I of the beams
Constraints	Displacements of the loaded corner and $\bar{C}_e$ for the beams
	Middle level
Objective	Cumulative constraint C representing the column buckling and $\bar{C}_e$ for the walls
Parameters	Beam cross-sectional area and moment of inertia
Design variables	Wall membrane stiffness contributing to the beam axial and bending stiffnesses controlled through the dimensions shown in Fig. 3 (Sec. A-A)
Constraints	Equality-beam cross-sectional area and moment of inertia
	Bottom level
Objective	Cumulative constraint C representing a set of stress and local buckling constraints of the wall
Parameters	Membrane stiffness of the wall
Design variables	Cross-sectional dimensions shown in Fig. 3 (Detail B)
Constraints	Inequality-minimum gages, geometrical proportions, and geometrical realizability Equality-membrane stiffnesses for tension-compression and bending of the wall in its own plane

Table 3.1 Hypersonic aircraft optimization results

Optimization parameter	Baseline value	Optimization results
Design variable		
1. Forebody length	1.000	1.0209
2. Cone angle	1.000	0.9693
3. Upper surface height	1.000	1.0029
4. Geometric transition length	1.000	1.0760
5. Elevon deflection	1.000	0.8620
6. Bodyflap deflection	1.000	1.0320
Objective		
Effective trimmed I_{sp}	1.000	1.1259

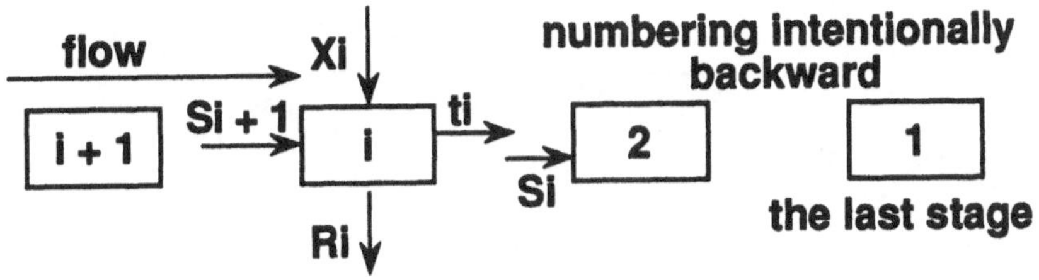

Fig. 1.1 Train of black boxes.

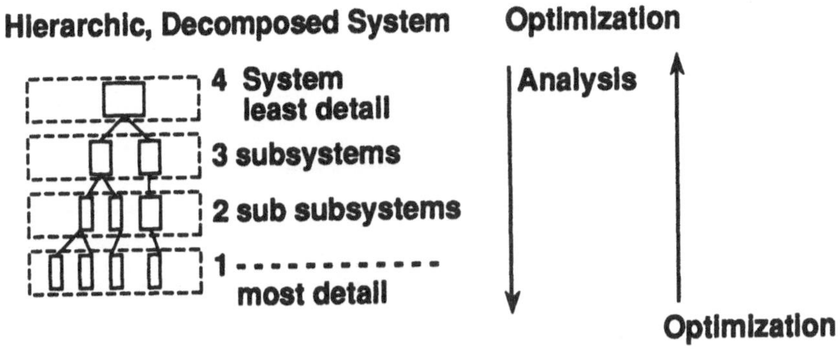

Fig. 1.2 Hierarchic decomposition.

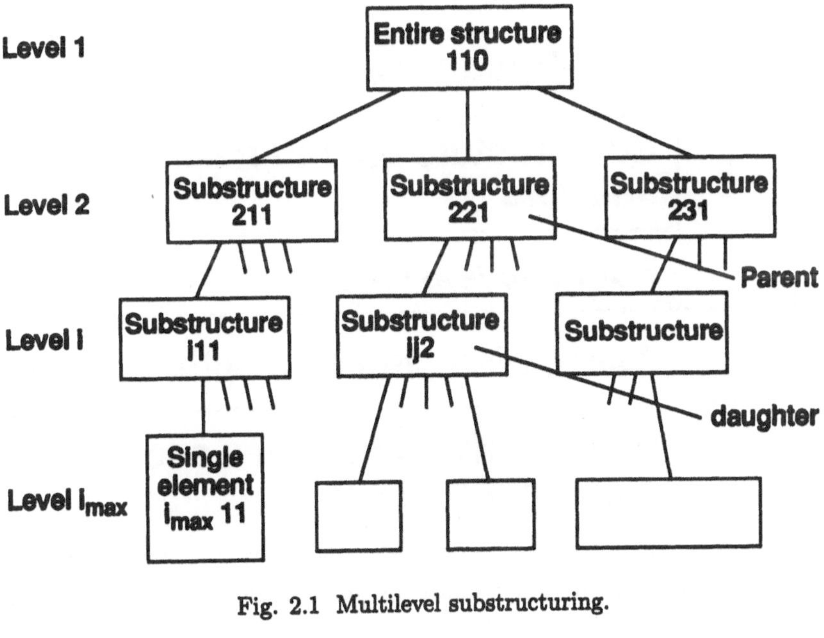

Fig. 2.1 Multilevel substructuring.

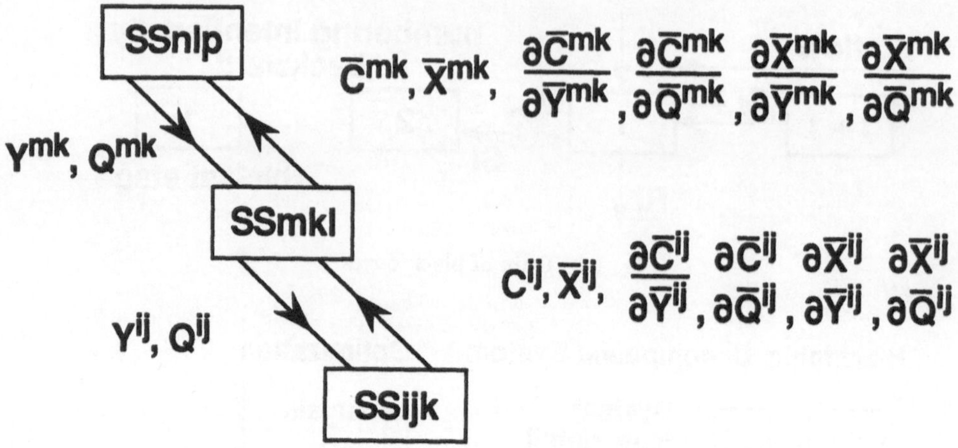

Fig. 2.2 Flow of information.

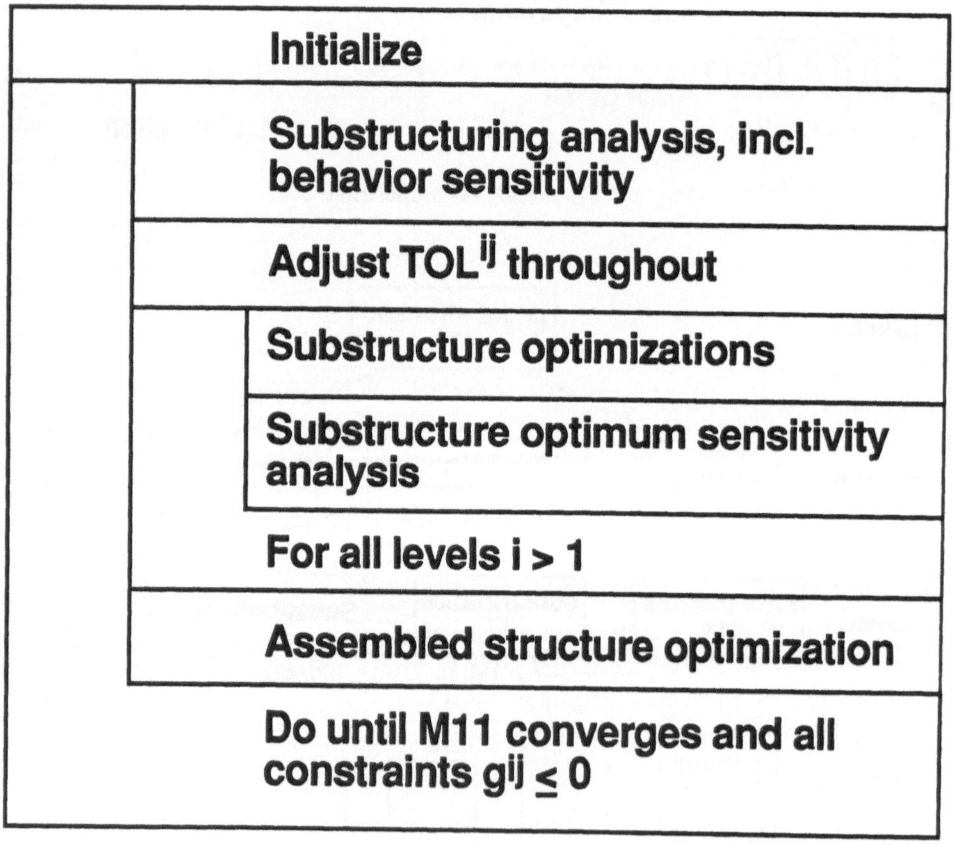

Fig. 2.3 Multilevel optimization procedure flowchart.

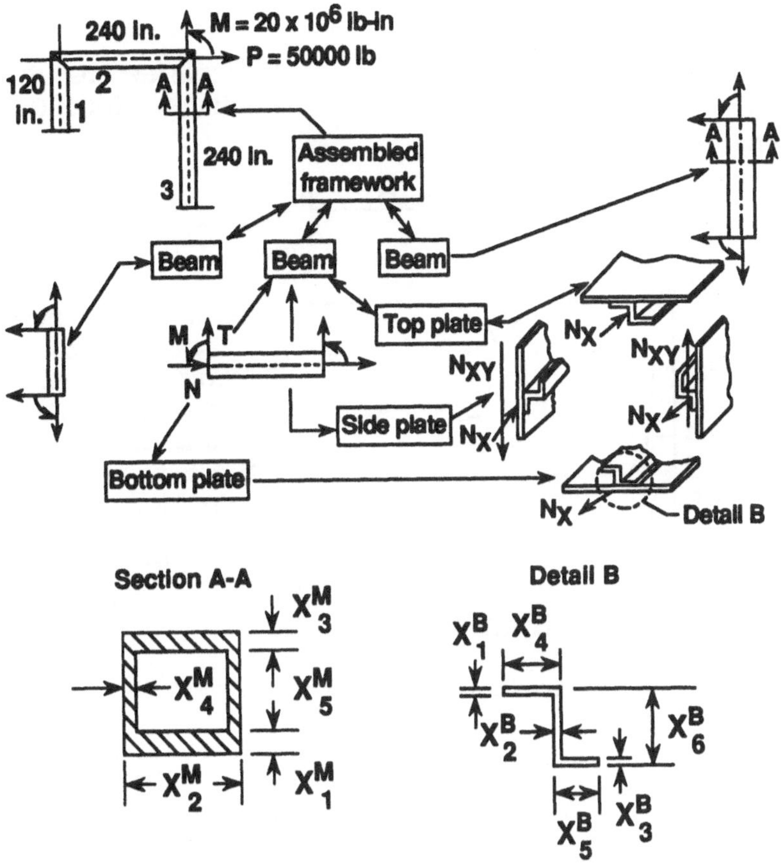

Fig. 2.4 Portal framework.

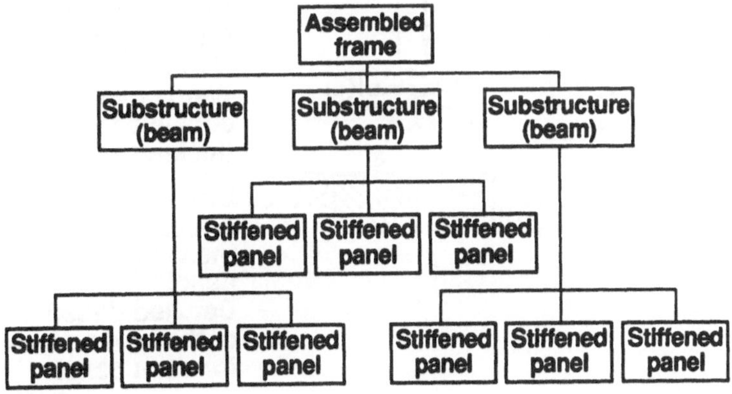

Fig. 2.5 Hierarchical decomposition of the framework structure shown in Fig. 2.4.

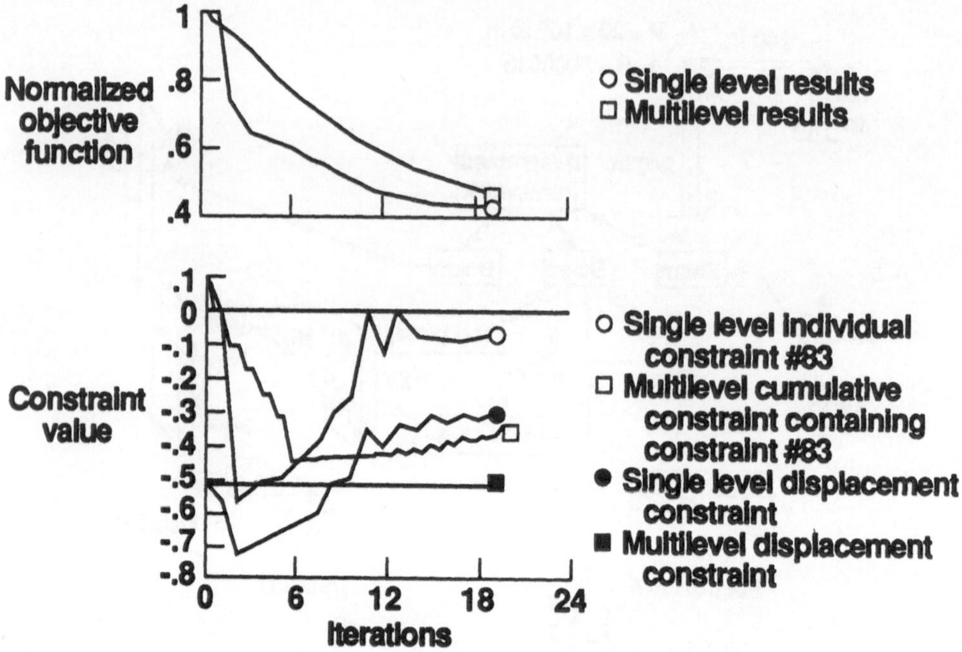

Fig. 2.6 Representative results.

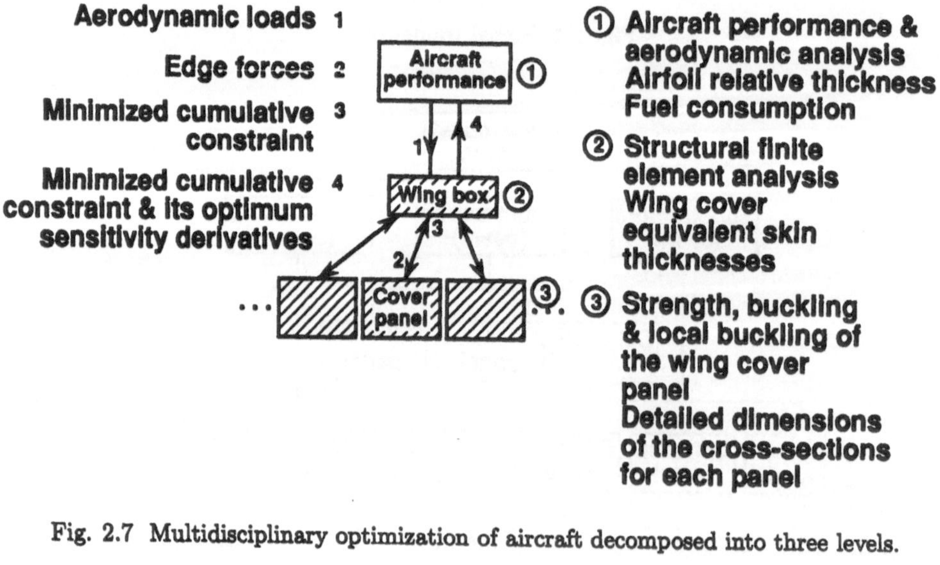

Fig. 2.7 Multidisciplinary optimization of aircraft decomposed into three levels.

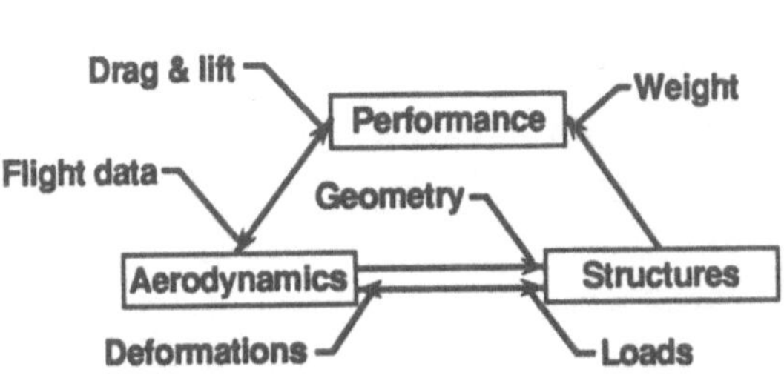

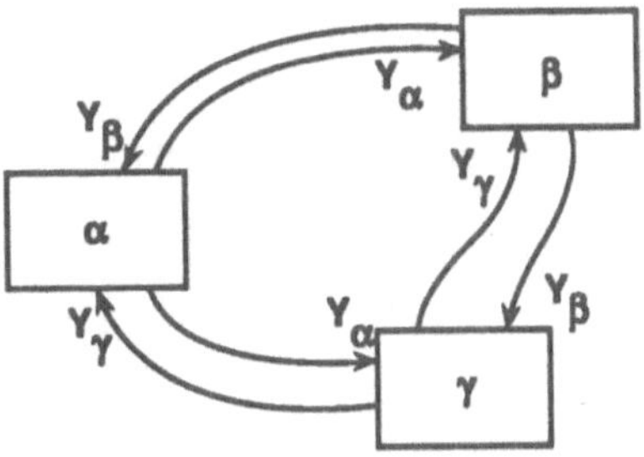

Fig. 3.1 a) Wing as a three-component system.

b) Abstract representation of a three-component system.

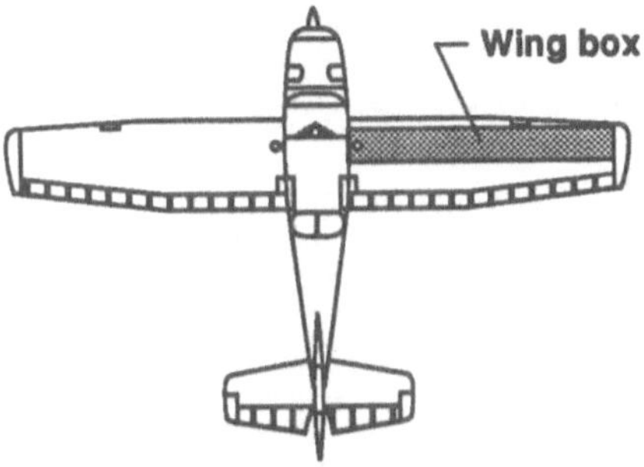

Fig. 3.2 Wingbox in aircraft wing.

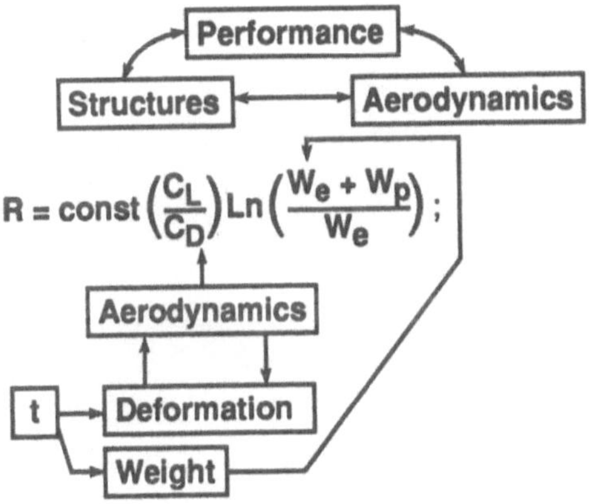

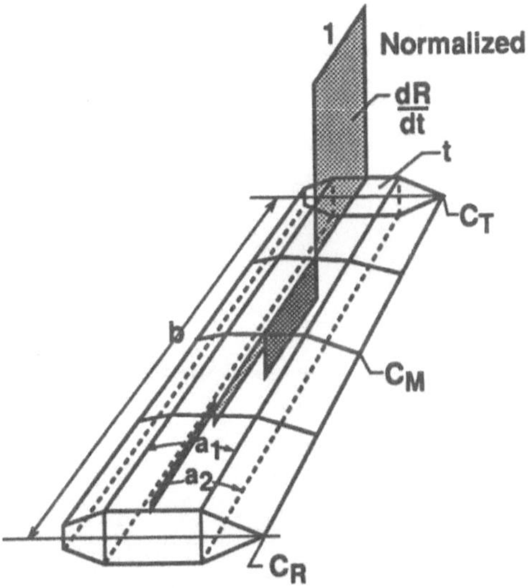

Fig. 3.3 a) System of mathematical models, the Breguet formula, and the channels of influence for the wing cover thickness;

b) Vertical bars illustrate magnitude of derivatives of range with respect to thickness.

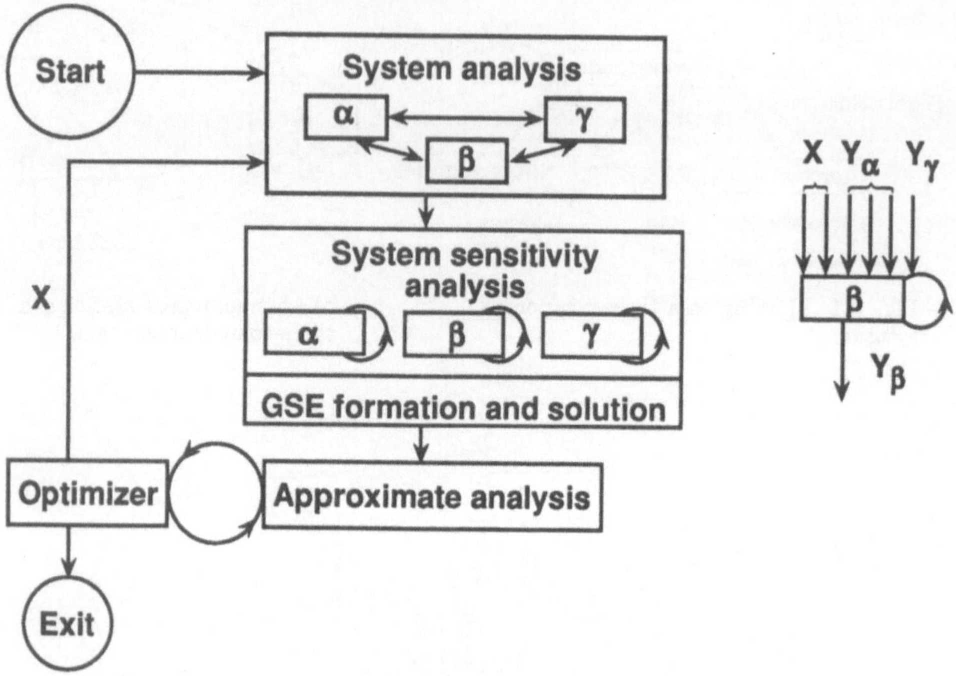

Fig. 3.4 Flowchart of the System Optimization procedure (SOP).

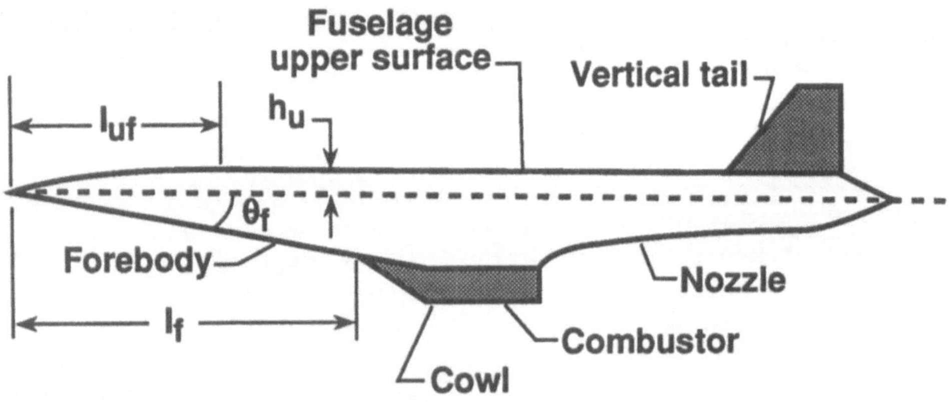

Fig. 3.5 Hypersonic aircraft; some of the configuration design variables.

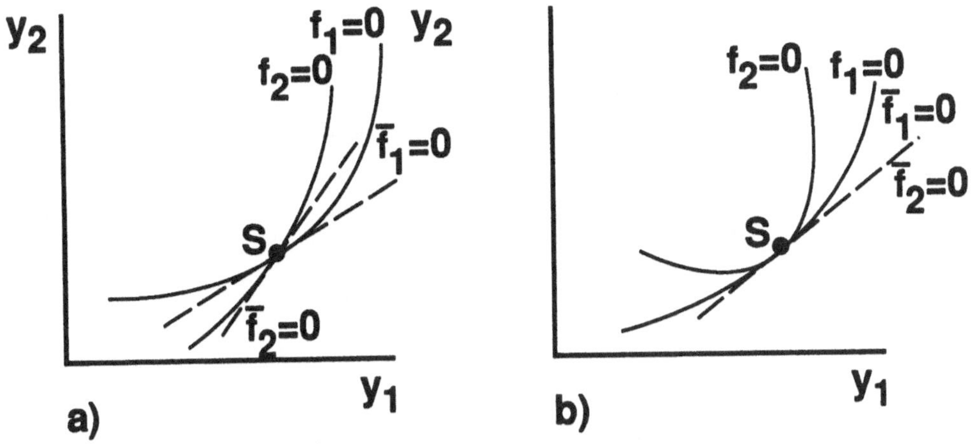

Fig. 3.6 System solution: a) Intersection point; b) Tangency point.

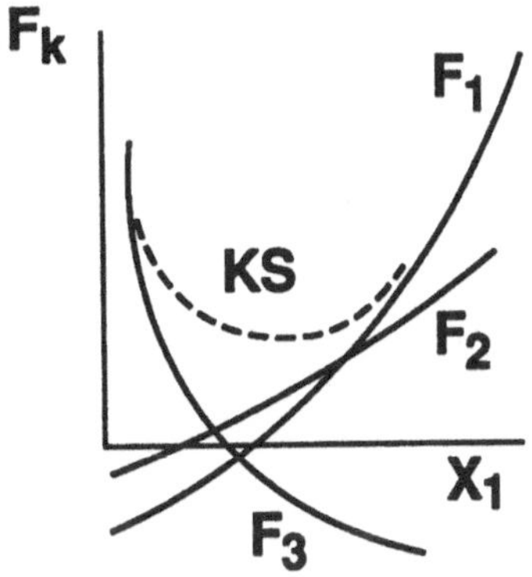

Fig. A1 KS-function envelope of a set of functions.

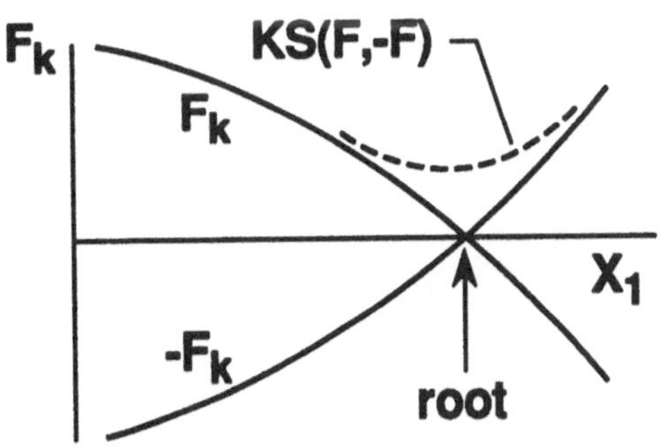

Fig. A2 The minimum of $KS(F, -F)$ coincides with $F = 0$.

RECENT ADVANCES IN APPROXIMATION CONCEPTS FOR OPTIMUM STRUCTURAL DESIGN

J.-F.M. BARTHELEMY[1]
NASA/Langley Research Center
Hampton, VA 23665-5225
U.S.A.

R.T. HAFTKA[2]
Virginia Polytechnic Institute and State University
Blacksburg, VA 24061
U.S.A.

ABSTRACT. This paper reviews the basic approximation concepts used in structural optimization. It also discusses some of the most recent developments in that area since the introduction of approximation concepts in the mid-seventies. The paper distinguishes between local, medium-range and global approximations; it covers functions approximations and problem approximations. It shows that, although the lack of comparative data established on reference test cases prevents an accurate assessment, there have been significant improvements. The largest number of developments have been in the areas of local function approximations and use of intermediate variable and response quantities. It appears also that some new methodologies emerge which could greatly benefit from the introduction of new computer architectures.

1. Introduction

In the mid-seventies, Schmit and his coworkers showed that applications of nonlinear programming methods to large structural design problems could prove cost effective, provided that suitable approximation concepts were introduced (Schmit and Farshi (1974), Schmit and Miura (1976)). They combined the now familiar techniques of intermediate variable definition, explicit approximation, reduced basis and design variable linking as well as constraint deletion and regionalization. This paper reviews the basic ideas underlying approximation concepts as well as some recent results. The emphasis is on methods that are generic in that they are applicable to any engineering discipline and are largely independent of the details of the analysis methodology used. As a consequence, the paper will not cover the closely connected field of solution of perturbed analysis equations also known as approximate reanalysis techniques. There are many excellent reviews of this field including Arora (1976), Kirsch (1984) on the static equilibrium equation, and Murthy and Haftka (1988) on the eigenvalue problem.

We identify three general categories of approximations. In a *global approximation*, the approximation concept is valid for the whole design space or, at least, large regions of it. In *local approximation* the approximation is only valid in the vicinity of a point in the design space. Finally, some approaches attempt to give global qualities to local approximations, and we

[1] Senior Aerospace Engineer, Interdisciplinary Research Office, Structural Dynamics Division

[2] Christopher Kraft Professor of Aerospace and Ocean Engineering

G. I. N. Rozvany (ed.), Optimization of Large Structural Systems, Vol. I, 235–256.

will refer to those as to *mid-range approximations*. We will also distinguish between *function approximation*, in which an alternate and explicit expression is sought for the objective function and/or constraints of the problem, and *problem approximation* where the focus is on replacing the original statement of the problem by one which is approximately equivalent but which is easier to solve. Finally, approximation concepts can be combined to make for a very efficient problem formulation.

2. Local Approximations

Local approximations are valid in the vicinity of the point at which they are generated. Typically, they are used to generate an approximate problem formulation which is solved for an optimum solution point. A new approximate problem is then generated at that point, and the process continues until convergence. Local function approximations are variations on the Taylor series expansion; local problem approximations try to reduce the size of the active constraint set.

2.1. LOCAL FUNCTION APPROXIMATIONS

Local function approximations are probably the most popular approximations used in optimization. One of the first robust optimization algorithms is the Simplex algorithm developed by Dantzig in 1947 for the solution of linear optimization problems (Linear Programs). It was natural that people attempted to use this highly successful algorithm for nonlinear programs by linearizing the constraints and objective function about a trial design. This led to the Sequential Linear Programming (SLP) method and the wide use of the linear Taylor series approximations. Applications can be found in Zienkiewicz and Campbell (1973) who optimize the shape of dams and in Pedersen (1981) who finds the optimum design of space trusses, for example. Another reason for the popularity of local approximations is that, as discussed later, some global approximations become very expensive computationally when the dimensionality of the design space exceeds about 10.

2.1.1. Approximations Based on Zeroth- and First-Order Function Information. The most commonly used local approximations to objective functions and constraints are based on the use of the function and its first derivatives at a single design point, say $\mathbf{X}_0$. The simplest is the *linear approximation* based on the Taylor series. Given a function $g(\mathbf{X})$, the linear approximation $g_L(\mathbf{X})$ is

$$g_L(\mathbf{X}) = g(\mathbf{X}_0) + \sum_{i=1}^{n}(x_i - x_{0i})\left(\frac{\partial g}{\partial x_i}\right)_{\mathbf{X}_0}. \tag{1}$$

For some applications the linear approximation is inaccurate even for design points $\mathbf{X}$ that are close to $\mathbf{X}_0$. Accuracy can be increased by retaining additional terms in the Taylor series expansion. This, however, requires the costly calculation of higher-order derivatives. Instead, many researchers tried to obtain other approximations that use only first derivatives but which can be more accurate than the linear approximation.

One approximation of this type is the *reciprocal approximation* which is a linear approximation in y_i, the reciprocal of x_i,

$$y_i = \frac{1}{x_i}. \tag{2}$$

Its frequent use reflects the fact that many of the early structural optimization studies were performed on structures consisting of truss or plane stress elements. The design variables in these studies were usually the cross-sectional areas of the truss elements and the thicknesses of the plane-stress elements. For statically determinate structures, stress and displacement constraints are linear functions of the reciprocals of these design variables. Even for statically indeterminate structures, using the reciprocals of the design variables still proved to be a useful device in making the constraints more linear (see, for example, Storaasli and Sobieszczanski, 1974, and Noor and Lowder, 1975a). The reciprocal approximation can be written in terms of the original design variables as

$$g_R(\mathbf{X}) = g(\mathbf{X}_0) + \sum_{i=1}^{n}(x_i - x_{0i})\frac{x_{0i}}{x_i}\left(\frac{\partial g}{\partial x_i}\right)_{\mathbf{X}_0} . \tag{3}$$

One of the attractive features of the reciprocal approximation, even for statically indeterminate structures, is that it preserves the property of scaling. That is, when the stiffness matrix is a homogeneous function of order h in the components of $\mathbf{X}$ the displacements are a homogenous function of order $-h$ in the components of $\mathbf{X}$. For truss and membrane elements $h = 1$, so that the displacements are a homogeneous function of the reciprocal of the design variables. If all the design variables are scaled by a factor, the displacement vector is scaled by the reciprocal of that factor. The reciprocal approximation preserves this scaling property, and therefore it is exact for scaling the design. Fuchs (1980) has investigated the importance of the homogeneity property, and Fuchs and Haj Ali (1990) have proposed a family of constraints that generalize the reciprocal approximation to any order of homogeneity. The approximate constraints are obtained by taking the envelope of the direct linear Taylor polynomials along points on a given scaling line. By using intermediate variables of the type $(1/x_i)^m$ it was shown that improved results are obtained for values $m = 1/2$ and $m = 1/4$.

One problem with the reciprocal approximation is that it becomes unbounded when one of the variables approaches zero. This is acceptable when the design variables are bounded away from zero, as is the case in many structural problems. However, it can result in large errors when one of the design variables becomes very small. To correct this deficiency Haftka and Shore (1979) proposed a modified reciprocal approximation given as

$$g_{mR}(\mathbf{X}) = g(\mathbf{X}_0) + \sum_{i=1}^{n}(x_i - x_{0i})\frac{(x_{0i} + x_{mi})}{(x_i + x_{mi})}\left(\frac{\partial g}{\partial x_i}\right)_{\mathbf{X}_0} , \tag{4}$$

where the values of x_{mi}'s are typically small compared to representative values of the corresponding x_i's. It is possible, however to take large values for some x_{mi}'s, and this results in an approximation which is closer to the linear approximation than the reciprocal approximation in these variables.

Another approximation, called the *conservative approximation* (Starnes and Haftka, 1979), is a hybrid form of the linear and reciprocal approximations which is more conservative than either. It is particularly suitable for interior and extended interior penalty function methods which do not tolerate constraint violations well. To obtain the conservative approximation, we start by subtracting the reciprocal approximation from the linear approximation

$$g_L(\mathbf{X}) - g_R(\mathbf{X}) = \sum_{i=1}^{n}\frac{(x_i - x_{0i})^2}{x_i}\left(\frac{\partial g}{\partial x_i}\right)_{\mathbf{X}_0} . \tag{5}$$

The sign of each term in the sum is determined by the sign of the ratio $(\partial g/\partial x_i)/x_{0i}$ which is also the sign of the product $x_{0i}(\partial g/\partial x_i)$. Design variables for which this product is negative contribute to make the reciprocal approximation more positive than the linear approximation, and vice versa. Since the constraint is usually expressed as $g(\mathbf{X}) \leq 0$, a more positive approximation is more conservative. The conservative approximation, g_C, is created by selecting for each design variable the more positive contribution

$$g_C(\mathbf{X}) = g(\mathbf{X}_0) + \sum_{i=1}^{n} G_i(x_i - x_{0i})\left(\frac{\partial g}{\partial x_i}\right)_{\mathbf{X}_0}, \tag{6}$$

where

$$G_i = \begin{cases} 1 & \text{if } x_{0i}(\partial g/\partial x_i) \geq 0, \\ x_{0i}/x_i & \text{otherwise.} \end{cases} \tag{7}$$

Note that $G_i = 1$ corresponds to a linear approximation, and $G_i = x_{0i}/x_i$ corresponds to a reciprocal approximation in x_i.

The conservative approximation is not the only hybrid linear-reciprocal approximation possible. Sometimes physical considerations may dictate the use of linear approximation for some variables and the reciprocal for others (see, for example, Haftka and Shore, 1979, and Prasad 1984b). However, as can be easily checked, the conservative approximation has the advantage of being convex. If the objective function and all the constraints are approximated by the conservative approximation, the approximate optimization problem is convex. Convex problems are guaranteed to have only a single optimum, and they are amenable to treatment by dual methods. Fleury and Braibant (Braibant and Fleury, 1985, Fleury and Braibant, 1986) suggested taking advantage of this property and solving the approximate problem by dual methods. They also introduced the term convex linearization for the process of approximating the optimization by the conservative-convex approximation for the objective function and constraints. This approach has been used by many others (e.g., Ding and Esping, 1986, Ding, 1987).

There has been a systematic investigation of approximations based on using powers of the design variables (Prasad, 1983, 1984a,b, Woo, 1987). Many of these are conservative and/or convex approximations, but it is important to note that the one presented here and the others are not guaranteed to be conservative in an absolute sense (that is, we do not know that the approximation is more conservative than the exact constraint, $g_C(\mathbf{X}) \geq g(\mathbf{X})$). The conservative approximation presented here is only more conservative than either the linear or reciprocal approximations. Finally, it has been shown (e.g., Haftka, 1988) that the conservative-convex approximation tends to be less accurate than either the linear or the reciprocal approximation. Therefore, it should not be used unless its convexity or conservativeness are needed.

The reciprocal and conservative approximation destroy the linearity of the approximation, and the possibility of using sequential linear programming (SLP). Chan and Turlea (1978) used a nonlinear approximation that still permits the use of SLP. This is the *posynomial approximation* (Duffin, Peterson and Zener, 1967).

$$g_P(\mathbf{X}) = g(\mathbf{X}_0)\prod_{i=1}^{n}\left(\frac{x_i}{x_{0i}}\right)^{a_i}, \tag{8}$$

where a_i is the logarithmic derivative of g,

$$a_i = \frac{x_{0i}}{g(\mathbf{X}_0)}\left(\frac{\partial g}{\partial x_i}\right)_{\mathbf{X}_0}. \tag{9}$$

The logarithm of g_P is a linear function of the design variables, so if the objective function and constraints are approximated by this posynomial approximation, the optimization problem can be transformed into a linear problem by taking the logarithm of the constraint equations and objective function.

The posynomial approximation can be used without any transformation when the optimization method is geometric programming which actually requires such form. Applications of the posynomial approximations to structural optimization using geometric programming include Morris (1972, 1974), Templeman and Winterbotton (1974) and Hajela (1986).

2.1.2. Differential Equation Based Approximation. Pritchard and Adelman (1990) recently introduced a new method that begins with the equation for the sensitivity of the quantity being approximated. It takes it as a constant coefficient differential equation, integrates it and derives a high-quality nonlinear approximation. For example, in a dynamical system, the sensitivity of an eigenvalue ω^2 to change in a single design variable x is given by

$$\frac{d\omega^2}{dx} = \Phi^T \left[\frac{d\mathbf{K}}{dx} - \omega^2 \frac{d\mathbf{M}}{dx} \right] \Phi \tag{10}$$

provided that the eigenvectors are normalized with respect to the mass matrix ($\Phi^T M \Phi = 1$). Assuming that

$$a = \Phi^T \frac{d\mathbf{K}}{dx} \Phi \text{ and } b = \Phi^T \frac{d\mathbf{M}}{dx} \Phi \tag{11}$$

are constant, the following differential equation results

$$\frac{d\omega^2}{dx} = a - b\omega^2 \tag{12}$$

which, upon integration and specification of the boundary condition that for the original value x_0 of the variable, the eigenvalue is ω_0^2, yields the following approximation

$$\omega^2 = \left(\omega_0^2 - \frac{a}{b} \right) e^{-b(x-x_0)} + \frac{a}{b} \tag{13}$$

They extended this approach to several design variables as well as to approximation of eigenvectors, and displacements. They generally showed good approximation quality, by comparison with conventional linear Taylor series expansions.

2.1.3. Approximations Based on Higher-Order Function Information. Higher order approximations are also used occasionally. For example, the quadratic approximation, g_Q is obtained by including the quadratic terms in the Taylor series expansion

$$g_Q(\mathbf{X}) = g(\mathbf{X}_0) + \sum_{i=1}^{n} (x_i - x_{0i}) \left(\frac{\partial g}{\partial x_i} \right)_{\mathbf{X}_0} + \frac{1}{2} \sum_{i=1}^{n} \sum_{j=1}^{n} (x_i - x_{0i})(x_j - x_{0j}) \left(\frac{\partial^2 g}{\partial x_i \partial x_j} \right)_{\mathbf{X}_0} . \tag{14}$$

The reciprocal quadratic approximation g_{QR} is obtained by using the quadratic approximation in terms of the reciprocal design variables to obtain (Haftka *et al.*, 1990)

$$\begin{aligned} g_{QR}(\mathbf{X}_0) = g(\mathbf{X}_0) &+ \sum_{i=1}^{n} \left(\frac{x_{0i}}{x_i} \right) \left(2 - \frac{x_{0i}}{x_i} \right) (x_i - x_{0i}) \left(\frac{\partial g}{\partial x_i} \right)_{\mathbf{X}_0} \\ &+ \frac{1}{2} \sum_{i=1}^{n} \sum_{j=1}^{n} \left(\frac{x_{0i}}{x_i} \right) \left(\frac{x_{0j}}{x_j} \right) (x_i - x_{0i})(x_j - x_{0j}) \left(\frac{\partial^2 g}{\partial x_i \partial x_j} \right)_{\mathbf{X}_0} \end{aligned} \tag{15}$$

Quadratic approximations have been used primarily for eigenvalue problems where the linear approximation tends to be particularly limited in applicability. For example, Rommel (1983) used a quadratic approximation for flutter speeds and damping factors, while Miura and Schmit (1978) used a quadratic approximation for vibration frequencies. However, it should be noted that for eigenvalue problems the quadratic approximation is not efficient, because for nearly the same cost it is possible to obtain a cubic approximation by using a linear approximation to the eigenvectors in the Rayleigh quotient (Murthy and Haftka, 1988). Often the high cost of obtaining these derivatives dictates a compromise based on using only the diagonal second derivatives (e.g., Fleury, 1988, 1989a,b, Renwei and Peng, 1987). For stress constraints, the use of only diagonal second derivatives can be justified by invoking St. Venant principle and assuming that changes in the property of an element would affect only the stress in that element (Renwei and Peng, 1987). The use of diagonal second derivatives is also sometimes motivated by the desire for a separable approximation (e.g., Smaoui *et al.*, 1988, Fleury and Smaoui, 1988, Fleury, 1989a,b). Haftka (1988) compared the performance of first- and second-order approximations in structural optimization. He found that the second-order approximations reduce the number of required optimization cycles by about 10-50 percent. This is marginal improvement when the cost of second derivatives is high. Jawed and Morris (1984, 1985) suggest the calculation of approximate second derivatives, which can make the use of quadratic approximations more attractive.

When second derivatives are available and it is still desirable to use SLP, it is possible to use the diagonal second derivatives to construct a better linear approximation to the constraint near the critical surface $g = 0$ (Mistree *et al.* 1981).

2.2. LOCAL PROBLEM APPROXIMATIONS

Technically, algorithms based on sequential approximations (linear programming, quadratic programming or convex linearization) can be viewed as problem approximation concepts as they replace the given nonlinear problem by a sequence of subproblems that are easier to solve. However, in this section, we will focus on techniques which have to do with reduction in the number of constraints or the number of design variables.

2.2.1. Constraint Number Reduction. In order to improve the optimization algorithm efficiency, one may reduce the number of constraints retained at each iteration to those which are active or nearly active. This limits the necessary constraint gradient calculations and greatly reduces the cost of optimization per iteration. In addition to *constraint deletion*, Schmit and Miura (1976) also have advocated the use of *regionalization*. This is a technique where for each region of the structure, and for each load case, only a few (two or three) of the most active stress and displacement constraints are retained. Assuming that the most active constraints do not change during one iteration, then only those gradients must be found for that region. A region can be identified as the area described by a single variable in a design variable linking scheme (see subsection on global problem approximation).

2.2.2. Design Variable Reduction. Hajela and Sobieski offered an interesting local design variable linking scheme (for more on design variable linking see the section on Global Problem Approximations) which they termed *controlled growth method* (Hajela and Sobieszczanski-Sobieski (1981), Hajela (1982)). This approach is applied at each optimization iteration. All the design variables are ranked according to their *combined measure of effectiveness (CME)*. Variables with low effectiveness are held constant during the current iteration, The remaining variable variations are linked to that of the variable with the highest CME, in effect replacing the original multi-variable optimization problem by a sequence of single variable subproblems. Hajela and Sobieszczanski-Sobieski (1981) showed reduction in analysis and gradient counts on conventional problems with up to 13 variables.

3. Global Approximations

Global approximations are valid for the whole design space or large areas of it. As such, they are used to modify the formulation of the problem from the outset and generate an alternate formulation that is more tractable. Global function approximations techniques include the generation of response surfaces; global problem approximations include the introduction of intermediate variable or response quantities as well methods to reduce the number of design variables in the problem.

3.1. GOBAL FUNCTION APPROXIMATIONS

3.1.1. Response Surface Approach. A natural approach to solving an optimization problem is to first build approximate analytical response surfaces giving the dependent variables as functions of the independent variables. Then an optimization algorithm can be used to optimize the approximate problem. Typically, these response surfaces are global in that they cover the whole design space, although, this is not necessary. Depending on the quality of the response surfaces, the resulting design can be used as a final solution or high-quality starting point for an ultimate optimization with direct coupling to a structural analysis. The main challenge in generating response surfaces is to do so without an excessive number of exact analyses.

Schoofs (1987) gave the following rationale for combining nonlinear optimization with response surface generation: i) both techniques can use the same variables, ii) both techniques aim at minimizing the number of expensive analyses, iii) once a global model is derived for a particular design problem, multiple optimization studies can be performed without additional analyses (as required for example when a multiobjective problem is solved or when a design problem formulation is being fine-tuned). In addition, gradients are not generally required for response surface generation although, if they are available, they can be used to enhance the process.

Construction of response surface (or model-building) relies heavily on the theory of experiments (see Box and Draper (1987)). It is an iterative process. A typical application begins with postulating a model for the relationship between dependent and independent variables. Although linear or quadratic polynomial approximations are by far the most common forms employed, other forms (e.g., polynomials in powers of trigonometric functions) have been used as well. The approximation contains a number of unknown parameters (such as polynomial coefficients) that must be adjusted for it to match the behavior of the system. To do so, analyses are performed at a number of carefully selected design points and a least-square solution is typically used to extract the parameter values from the analysis results. Then the approximate model is used to predict the response of the system at a number of selected test points and statistical measures are used to assess the goodness-of-fit, or the accuracy of the response surfaces. If the fit is not satisfactory, the process is restarted and further experiments are made or the postulated model is improved by removing and/or adding terms.

There is a limited number of examples of applications of response surface techniques in the structural optimization literature. Brown and Nachlas (1985) selected the orientation of the layers of composite fibers in the three sections of a missile exit cone. They used polynomials in trigonometric functions of the orientations to approximate the safety factors under selected load cases. With 4 design variables per section and up to 7 possible values for these variables, an exhaustive search of the design space would require 2401 analyses per section. Instead they selected 28 design points per section and generated a final design with a 37 lb weight reduction with respect to a metal baseline. White *et al.* (1985 and 1986) used the response surface approach in their study of passenger car crashworthiness. Most of the dependent variables optimized in their study were derived from deceleration data generated by simulation programs. They first approximated the deceleration time histories by polygonal profiles; they then constructed response surfaces to relate the parameters describing the polygonal profiles to their structural design variables. Using polynomials of up to third power, they required over 200 sets of simulation results for fitting 11 crash signature parameters with 7 design variables. After optimizing the problem they retained the model to conduct inexpensive univariate sensitivity studies. Schoofs (1987) described small mechanical engineering problems (design of pin joints, bearing joints, beam cross-sectional shape and heart leaflet valves). He also described at length the challenging design of the shape of a carillon bell to prescribed natural frequency ratios. The shape design variables were 7 radii, describing the bell cross-section in a vertical plane. Polynomial approximations of up to third power in the variables were used. After several attempts, he concluded that 1220 analyses must be performed to obtain an adequate fit. In the process, he reported the design of bells with frequency ratio distributions that had eluded professional bell-founders for centuries. Lawson *et al.* (1989) discussed the design of a moving head disk drive actuator arm. They reviewed two approaches to selecting the sample analysis results and optimized a 4-variable model using quadratic response surfaces derived from as little as 15 full analyses and requiring as little as 27% of the time required for conventional optimization.

The experience with response surfaces in structural optimization is limited to problems of relatively small size. This is because the number of analyses required to construct response surfaces increases dramatically with the number of design variables. The usage of high-quality approximations has reduced the number of analyses required for the solution of most structural optimization problem to 15-20 regardless of the number of design variables. Assuming finite-difference calculations this translates to about $15n$ to $20n$ analyses. In the theory of experiments, the basic set of design points considered is called a *full factorial design*. Each variable (factor) is assumed to take on a number of possible discrete values (levels) and each possible combination of variable values is considered. A full factorial design with only two values per variables requires 2^n analyses. It permits to generate response surfaces linear in each design variables (and this would probably be inadequate for most structural responses). On this basis alone, response surface methodology is not competitive for problems with more than 8 design variables. However, direct comparison is not easy. First, in a typical design exercise, many conventional optimizations must be performed to develop a satisfactory problem formulation, to try and isolate a global optimum or to solve a multiobjective problem. With each additional optimization, a completely new set of analyses and sensitivity analyses must be performed. Second, in the theory of experiments, numerous methods exist to construct *fractional factorial designs* where the number of analyses is reduced drastically from the full factorial design while maintaining a sufficient level of accuracy (see Box and Draper (1987), for example). Finally, a third point is that constructing response surfaces is an inherently parallel operation. While analysis (and, possibly, sensitivity analysis) results are needed sequentially in applications of conventional nonlinear programing methodology, they are needed all at one time when response surface methodology is used, thereby enabling better usage of multi-processor computers. It must be noted that the conventional analysis and sensitivity analysis process lends itself to parallel implementation if derivatives are found by finite difference. However, constructing a response surface offers even more parallelism since all the analyses are required at the outset. Further investigation of response surface methodology may show it to be a competitive alternative to using conventional optimization method for some problems.

3.1.2. Other Global Function Approximations. Hajela and Berke (1990) have proposed using neural networks in optimization to provide fast approximate structural analysis. A neural network is a computer that attempts to mimic neurobiological processes. It is a massively parallel network of interconnected computing elements that processes input data and generates output. A neural network is trained by presenting it pairs of input and output data and then iteratively adjusting weights in the connections between computing elements so that its output matches the known output data. Once trained, the network can be used to replace complex and time-consuming analysis procedures. In that sense, neural networks can be thought of as an alternate approach to global function approximation. Hajela and Berke described how neural networks are able to abstract key information and patterns present in their input sets. Also, they showed that networks are fault tolerant in that they are relatively insensitive to degeneration of a few computing elements or to corruption of a few data sets. Although very different conceptually, neural networks and response surfaces provide the same type of information and present a lot of the same advantages and disadvantages. On the one hand both methodologies i) require for input a number of analyses of the system considered, ii) can accommodate input from different sources (including analytical and experimental), iii) are adaptive in that they can be improved as more information becomes available, and, finally, iv) provide a rapid analysis capability that is global and can be reused many times at little or no additional cost. On the other hand, both require a significant amount of up-front computations. Hajela and Berke demonstrated the applicability of neural networks in optimization by considering simple truss and wing structures of 16 design variables or less. With as little as 100 training sets (complete analyses), and a significant amount of iterative training time, they showed optimization results that were comparable to those generated by conventional methods.

For problems in which some constraints are quite expensive to calculate, bounds can be developed for the constraints which are significantly easier to calculate than the constraints and which help provide bounds on the optimum solution. These bounds then replace the original constraints in the optimization problem. One such example was given by Mills-Curran and Schmit (1983) in an application of optimization under dynamic behavior constraints. In that application, they developed time-dependent upper bounds for dynamic displacement and stress constraints which are valid for lightly damped systems away from resonant forcing conditions.

3.2. GLOBAL PROBLEM APPROXIMATIONS

One of the most direct approaches to approximating a problem formulation is the use of simplified analysis models. On the one hand, the simplification can be to obtain the numerical solution with, for example, a coarser finite element mesh discretization. Haftka and Starnes (1976) looked at the effect of varying the number of degrees-of-freedom and design variables in wing models. They showed CPU times increasing linearly with the number of design variables for a given number of degrees of freedom and almost quadratically with the number of degrees of freedom for a given number of design variables. Salama *et al.* (1984) also examined the benefit of varying the number of degrees-of-freedom and design variables in problems of optimization of simple beams and built-up truss structures. On the other hand, the analysis can be performed with a simpler model, for example a plate model of a wing instead of a complex built-up finite element model. McCullers and Lynch (1972) used such a model to develop the program TSO which has remained a classic preliminary design tool for flexible wings. Sobieszczanski-Sobieski and Loendorf (1972) developed a simple lumped property model of a fuselage for use in a two-level sizing procedure; Ricketts and Sobieszczanski-Sobieski (1977) developed a complete lumped model for aircraft sizing under flutter constraints. The focus of this section, however, remains on methods that are more generic in nature.

3.2.1. Intermediate Variables and Response Quantities. In their quest for improving the quality of function approximations, researchers occasionally resort to using intermediate response quantities and intermediate variables. When a particular response $\mathbf{R}(\mathbf{X})$ can be written in terms of an *intermediate response* (or response vector) $\mathbf{R}_\mathbf{I}$ and of an *intermediate variable* (or variable vector) $\mathbf{X}_\mathbf{I}$, then, for example,

$$\mathbf{R}(\mathbf{X}) = \mathbf{R}(\mathbf{R}_\mathbf{I}(\mathbf{X}_\mathbf{I}(\mathbf{X}))) \tag{16}$$

If the relationships $\mathbf{R}_\mathbf{I} = \mathbf{R}_\mathbf{I}(\mathbf{X}_\mathbf{I})$ and $\mathbf{X}_\mathbf{I} = \mathbf{X}_\mathbf{I}(\mathbf{X})$ are known analytically and if a very accurate approximation $\widetilde{\mathbf{R}}_\mathbf{I}(\mathbf{X}_\mathbf{I})$ exists, then, a very accurate approximation $\widetilde{\mathbf{R}}(\mathbf{X})$ is given by

$$\mathbf{R}(\mathbf{X}) \cong \widetilde{\mathbf{R}}(\mathbf{X}) = \mathbf{R}\left(\widetilde{\mathbf{R}}_\mathbf{I}(\mathbf{X}_\mathbf{I}(\mathbf{X}))\right) \tag{17}$$

In general, all three nested relationships may be approximated. In this case,

$$\mathbf{R}(\mathbf{X}) \cong \widetilde{\mathbf{R}}(\mathbf{X}) = \widetilde{\mathbf{R}}\left(\widetilde{\mathbf{R}}_\mathbf{I}\left(\widetilde{\mathbf{X}}_\mathbf{I}(\mathbf{X})\right)\right) \tag{18}$$

The examples presented in this section are all local approximations. However, the idea of resorting to intermediate design variables or intermediate response quantities to improve approximation quality is applicable to all types of approximations, whether local, mid-range or global.

Schmit and Farshi (1974) were the first to propose the use of reciprocal variable as intermediate variables (see previous section on local approximations). For trusses, the intermediate variable of choice for displacement constraints is the reciprocal of the cross-sectional area (see Bennett (1981), for example). This reciprocal approximation is exact for statically determinate trusses and, in general, very good for all trusses. Similarly, the reciprocal of the thickness is an appropriate intermediate variable for approximations of the response of built up structures made of membrane elements.

For built-up structures made of beams of simple cross-section, Fleury and Sander (1983) recommended i) the reciprocal of the wall thickness for structures made of thin walled beam, tubes or sandwich beams with constant cross-section but variable thickness, ii) the reciprocal of the square of the dimensions for beams with uniformly varying cross-sections and iii) the reciprocal of the cube of the height for built-up structures made of beams of varying height. For plates in bending, they suggested i) the reciprocal of the face sheet thickness for sandwich plates of constant core depth and ii) the reciprocal of the cube of the plate thickness for solid plates. When both extensional and bending behaviors are present, they derived approximations based on the reciprocal of the thickness, its second and third power for displacement, stress, frequency and buckling constraints. In this latter case, however, the approximations were not simple linearizations any more but were derived on the basis of energy considerations.

In works on frameworks made of beams of complex cross-sections, researchers recognized that while a beam cross-section is fully described by all its *cross-section dimensions (CSD)*, (wall thickness, beam height and width, for example), the behavior of the assembled framework (nodal displacements, member forces) is best described in terms of *reciprocal section properties (RSP)* (reciprocal of area, moment of inertia). Therefore, they recommended to include the latter as intermediate design variables. Mills-Curran *et al.* (1983) restated the conventional frame design problem in terms of the RSP taken as intermediate variables, eliminating the CSD during optimization and recovering them through linear approximations. Lust and Schmit (1986) compared results where the problem was stated in terms of CSD only to results where intermediate variables including the RSP were included. Their results were slightly better when RSP were used as intermediate variables. Yoshida and Vanderplaats (1988) linearized framework responses in terms of *reciprocal cross-section dimensions (RCSD)*, showing slightly faster convergence than with CSD. Salajegheh and Vanderplaats (1986, Vanderplaats and Salajegheh 1988) recommended linearizing framework eigenvalues (squared frequencies) in terms of the *direct section properties (DSP)*. Zhou and Xhia (1990) proposed a two-level approximation to solve framework problems. At the first level, the relevant behavior variables (e.g. stresses, displacements) are linearly approximated in terms of intermediate variables (called generalized variables) which include the RSPs as well as entries in the stress matrices. At the second level, the resulting approximate problem is replaced by a sequence of quadratic programming problems in terms of the CSDs which they solve using a dual approach. They showed good analysis count improvement over traditional implementations.

The first notable introduction of intermediate response quantity was due to Salajegheh and Vanderplaats (1986/1987, Vanderplaats and Salajegheh 1989) who recommended the use of linear approximation of forces in terms of areas for trusses and in terms of section properties for frameworks. Stress constraints were then derived exactly from the approximate forces. Kodiyalam and Vanderplaats (1989) extended this idea to the problem of shape optimization of a three-dimensional continuum and reported improved convergence of the optimization process. However, the repeated exact calculation of element stresses is quite expensive and offsets somewhat the improvement in approximation quality. Vanderplaats and Kodiyalam (1990) resolved the issue of stress recovery cost by using a two-level approximation. For each structural analysis, a linear force approximation is constructed. A sequence of linear problems is then solved. At the beginning of each problem a new linear stress approximation is derived from the linear force approximation and the sequence of inner linear problems is continued until convergence. At that point, a new structural analysis is conducted and a new linear force approximation is constructed and the process is continued until convergence. Hansen and Vanderplaats (1990) applied the force approximation in configuration optimization of trusses and demonstrated a markedly reduced analysis count. Moore and Vanderplaats (1990) showed that the linear force approximation yields stress approximates that include higher order terms than the conventional linear stress approximation. They then developed a simplified force approximation including only a subset of these higher-order terms and demonstrated promising results in the design of plates and shells. Thomas and Vanderplaats (1991) improved this formulation for plate structures by using as intermediate response quantities the 6 elements forces as opposed to the 24 nodal forces, reducing the computational difficulty, in the process. Vanderplaats and Han (1990) applied the force approximation to the shape optimal design of arches, demonstrating a much lower iteration count than when not using approximations.

In his optimization with crashworthiness and static constraints, Lust (1990) needed to calculate the nonlinear crash response of a car. He used a spring-mass model with nonlinear component force-displacement characteristics. He selected those force-displacement characteristics as intermediate response quantities and then used a linear scaling relation for them in terms of the design variables. He demonstrated that the resulting approximation permits very efficient calculations, while the same problem cannot be solved with conventional Taylor series expansions because of the low quality of the resulting approximations.

Canfield (1990) used a Rayleigh quotient approximation to improve approximations of frequencies in a cantilever beam and several truss examples. Each frequency is replaced by its Rayleigh quotient and the corresponding modal strain and kinetic energies serve as intermediate response quantities. They are in turn linearized with respect to the design variables, assuming that the normal modes remain constant. He demonstrated stable and smooth convergence in less iterations than other existing approaches. In Thomas *et al* (1991) the method is generalized to complex eigenvalues. A more general discussion of approximating eigenvalues of modified matrices was given by Murthy and Haftka (1988) who covered approximations based on eigenvalue derivatives, generalized Rayleigh quotient and the trace theorem.

For steady-state dynamic response, Manning *et al* (1986) introduced the idea of approximating steady-state response amplitudes in terms of the individual components of the complex dynamic displacements. Miura and Chargin (1991) introduced a product form of approximation of displacements where the intermediate responses included the system eigenvalues approximated by the Rayleigh quotient approximations. For steady-state dynamic response calculated by modal superposition, Mills-Curran and Schmit (1983) took eigenvalues and eigenvectors as intermediate response quantities and approximated those linearly. Thomas *et al.* (1990) used modal participation factors as intermediate responses in their approximation of dynamic displacements near resonance, approximating the eigenvalues with the Rayleigh quotient approximation and assuming that the eigenvectors remain constant.

For transient dynamic response, Sepulveda *et al* (1991) propose approximations of peak displacements in terms of constant modes, linear approximations to peak times and pseudo-modal energies and Rayleigh quotient approximations to the modal frequencies.

Thomas *et al* (1991) used both DSP's and RSP's as intermediate design variables for control augmented structural synthesis of trusses and frameworks. Their intermediate response quantities included the individual components of actuator forces and dynamic displacements, and the components of the complex pseudo-modal strain and kinetic energies for constraints on complex frequencies (see above). They showed that using this combination of approximations for a control force minimization problem reduces significantly the number of iterations and improves the final solution.

3.2.2. Design Variable Linking and Reduced Basis Method. Schmit and Miura (1976) proposed a two-step approach to reducing the number of design variables in an optimization problem by combining design variable linking and reduced basis vectors. *Design variable linking* is an approach introduced initially when optimizing structures modelled by finite element models. In this approach, many finite elements may be controlled by one or several design variables. The choice of what elements are linked is based on considerations of symmetry, manufacturability or even some preexisting optimization results which show that some areas of the structures appear to converge to similar designs. In effect, design variable linking amounts to writing

$$\mathbf{X} = [\mathbf{L}]\mathbf{Y} \tag{19}$$

where X is the original unlinked variable vector, Y is the new linked variable vector (of size smaller than X) and [L] the linking matrix. It should be noted that, for a minimization problem, the design obtained after variable linking is an upper bound to that obtained with the full set of design variables.

A further reduction in the number of design variables may be achieved by the introduction of *reduced basis vectors*. The linked variable vector may now be written

$$\mathbf{Y} = [\mathbf{B}]\mathbf{Z} \tag{20}$$

where [B] is a matrix whose columns are the basis vectors. **Z** is now the final variable vector, much smaller in size than **X**. Picket *et al.* (1973) introduced the reduced basis concept. They recommended to include among the basis vectors i) quasi-fully stressed designs generated for each load case of the initial problem using a few cycles of fully stress design, ii) quasi-fully displaced designs generated similarly with a few cycles of fully displaced design, iii) the vector of minimum gauge for the variables. For design problems for trusses with up to 200 members, they showed computer time reductions by factors of at least 7-8 with little weight penalty. Rajamaran and Schmit (1981) suggested a different type of basis vectors. They recommended to create one auxiliary design problem for each type of behavior (stresses, displacements, buckling) constrained in the initial problem. Then, the basis vectors are generated with a few iterations in each of the auxiliary subproblems. When designing a truss with 132 members, they showed a 33% time reduction with a 26% increase in optimum objective value. A reduced basis representation is an excellent approach to modelling shapes for optimization, although it is highly dependent on a proper choice of basis vectors. The initial implementation of this idea is due to Vanderplaats (1979) who used it to model an airfoil shape by superposition of shapes of basis airfoils. The same ideas has since been used in structural optimization; Barthelemy *et al.* (1991), for example, compare different trigonometric functions as basis for the representation for the shape of a hole in a plate.

3.2.3. Envelope Function. The use of envelope functions is one approach to reducing the number of functions in an optimization problem. It essentially reduces the number of constraints handled by the optimizer and makes it more efficient by reducing the total number of computations and the storage requirements. It also makes it easier to get global understanding of the problem. It can be combined with conventional constraint deletion and regionalization techniques and can reduce the number of gradients to calculate if an adjoint method is used for sensitivity analysis. One such envelope function was introduced by Kreisselmeier and Steinhauser (1979). It replaces the constraints $g_j(\mathbf{X}), j = 1, n_g$ by the function

$$K(\mathbf{X}) = \frac{1}{\rho} ln\left(\sum_{j=1}^{n_g} exp(-\rho g_j(\mathbf{X}))\right) \tag{21}$$

where ρ is a user-specified parameter which controls how close the envelope is to the original constraints. It can be shown that this function closely tracks the maximum constraint and:

$$g_{max}(\mathbf{X}) \leq K(\mathbf{X}) \leq g_{max}(\mathbf{X}) + \frac{ln(n_g)}{\rho} \tag{22}$$

Sobieski first proposed the use of this envelope function and used it in multilevel optimization applications to combine all level subproblem constraints (Sobieszczanski-Sobieski *et al.* (1983)). Hajela (1982) first applied it in a single level optimization problem. In the shape optimization of a three-dimensional solid, Barthelemy *et al.* (1988a) used this formulation to reduce the total number of constraints handled by the optimizer from 421 to 14. Additional details about this envelope function can be found in Barthelemy *et al.* (1988b).

4. Mid-Range Approximations

Mid-range approximations are an attempt to endow local function approximations with a wider range of applicability. Two general devices are used for this purpose. The first is the use of information at several points, and the second is the combination of a local approximation and a global approximation. There are no mid-range problem approximations.

4.1.1. Multipoint Approximations. Because the optimization process requires the calculation of constraints and their derivatives at more than one point, it makes sense to try and use the information and construct approximations based on that data, that would have a wider range of accuracy than approximations based on information at a single point. Early work in this area was limited to using the values of the constraint functions generated in a line search to construct a polynomial approximation along that line (e.g., Kirsch and Toledano, 1983). More recent work investigates the use of data generated during several optimization iterations for the purpose of generating approximations in an entire region of the design space. Haftka *et al.* (1989) examined approximations based on two and three points. One approach that they employed was based on the modified reciprocal approximation, Eq. (4), where the information on the derivatives at a second point was used to estimate the best values of the x_{mi}'s. However, the results indicated that while the approximation was good when it represented interpolation (for example, at points inside the triangle formed by three data points in a three-point approximation), the improvement in accuracy was marginal when it represented extrapolation.

A two-point approximation that shows more promise was proposed by Fadel *et al.* (1990). The approximation is a linear approximation in the variables

$$y_i = x_i^{p_i}, \tag{23}$$

as suggested by Prasad (1983). However, while Prasad suggested the choice of arbitrary exponentials, here the exponentials p_i are selected to match the data at a second point. The linear approximation in terms of y_i may be written in terms of the original variables x_i as

$$g_{tp}(\mathbf{X}) = g(\mathbf{X}_0) + \sum_{i=1}^{n} \left[\left(\frac{x_i}{x_{0i}} \right)^{p_i} - 1 \right] \left(\frac{x_{0i}}{p_i} \right) \left(\frac{\partial g}{\partial x_i} \right)_{\mathbf{X}_0}. \tag{24}$$

If we have the value of the derivatives at another design point, $\mathbf{X}_1$, we can now impose the condition that

$$\left(\frac{\partial g_{tp}}{\partial x_i} \right)_{\mathbf{X}_1} = \left(\frac{\partial g}{\partial x_i} \right)_{\mathbf{X}_1} = \left(\frac{x_i}{x_{0i}} \right)^{p_i - 1} \left(\frac{\partial g}{\partial x_i} \right)_{\mathbf{X}_0}. \tag{25}$$

From this equation p_i can be extracted

$$p_i = 1 + \frac{\ln \left\{ \left(\frac{\partial g}{\partial x_i} \right)_{\mathbf{X}_1} / \left(\frac{\partial g}{\partial x_i} \right)_{\mathbf{X}_0} \right\}}{\ln \left(x_{1i} / x_{0i} \right)} \tag{26}$$

To avoid large exponentials, the value of p_i is limited to +1 or -1. In addition, in the limit where $p_i = 0$, it can be shown (using l'Hospital rule or a Taylor series expansion about small values of p_i) that

$$\lim_{p_i \mapsto 0} \frac{\left[\left(\frac{x_i}{x_{oi}} \right)^{p_i} - 1 \right]}{p_i} = \ln \left(\frac{x_i}{x_{oi}} \right) \tag{27}$$

Belegundu *et al.* (1990) developed a two-point posynomial approximation (Eq. 8). They used a least-square approach to find the approximation parameters, matching not only the value of the function and its gradient at the current point but also the value of the function at a second point. Except for one set of frequency approximations, they showed significantly improved approximations for small conventional problems when comparing their results with linear, quadratic, reciprocal and one-point posynomial approximations.

Another multipoint approximation has been proposed by Rasmussen (1990). The so called cumulative approximation assumes that values of the constraint function g are available at m points $\mathbf{X}_1, \ldots, \mathbf{X}_m$, and seeks to improve the linear approximation at $\mathbf{X}_0$ based on this data. The influence of $\mathbf{X}_p$ on the approximation is determined by an exponentially decaying influence function, ϕ_p, given as

$$\phi_p(\mathbf{X}) = e^{-||\mathbf{X}-\mathbf{X}_p||^2/s_p}, \tag{28}$$

where s_p is a positive number that defines the range of the influence, and the Euclidean norm is used. It is suggested that a good choice for s_p is

$$s_p = \alpha||\mathbf{X}_0 - \mathbf{X}_p||^2, \tag{29}$$

where α is a constant. The cumulative approximation g_c is given as

$$g_c(\mathbf{X}) = \frac{g_L(\mathbf{X})\prod_{p=1}^{m}\left[1-\phi_p(\mathbf{X})\right] + \left[1-\phi_0(\mathbf{X})\right]\sum_{p=1}^{m}\phi_p(\mathbf{X})g(\mathbf{X}_p)}{\prod_{p=1}^{m}\left[1-\phi_p(\mathbf{X})\right] + \left[1-\phi_0(\mathbf{X})\right]\sum_{p=1}^{m}\phi_p(\mathbf{X})}. \tag{30}$$

where $g_L(x)$ is the linear approximation (Eq. 1) based on data at $\mathbf{X}_p$. The exponential decay is an attractive feature of the approximation because it limits the influence of far away points. Rasmussen (1991) suggested that previous gradients may even be included by replacing $g(\mathbf{X}_p)$ in the numerator of Eq. 30 by the linear approximation

$$g_{Lp}(\mathbf{X}) = g(\mathbf{X}_p) + \Sigma_{i=1}^{n}\left(\frac{\partial g}{\partial x_i}\right)_{\mathbf{X}_p}(\mathbf{X}_p - \mathbf{X}) \tag{31}$$

although no result has been published that validates this proposal.

Finally, it is worth noting that in many cases in optimization, it may be unnecessary to combine the data from several point into a single approximation. Instead a constraint $g \leq 0$ is replaced not only by its most recent linearization, but by several of its previous linearizations (see, for example, Mistree *et al.* (1981)).

4.1.2. Scaling or Local-Global Approximations. Because of computational constraints, optimization is often performed on the basis of a model of the structure that is simpler than the one which is used for analysis of the same structure. Such a simpler model is based on a simpler theory (e.g., beam theory versus shell analysis), or a coarser discretization of the numerical model associated with the same theory. This simpler model can be viewed as a global approximation, as discussed earlier. Here we consider the complex model to provide the exact value of the function $g(\mathbf{X})$, while the simple model is assumed to provide a global approximation $g_G(\mathbf{X})$. A local flavor can be injected into the approximation by calculating a scale factor associated with the simple and complex model at a point $\mathbf{X}_0$. That is the scale factor s_{c0} is given as

$$s_{c0} = g(\mathbf{X}_0)/g_G(\mathbf{X}_0)\,. \tag{32}$$

Then the scaled global approximation, g_{s0}, is given as

$$g_{s0}(\mathbf{X}) = s_{c0}\, g_G(\mathbf{X})\,. \tag{33}$$

The scaling is likely to improve the quality of the global approximation near $\mathbf{X}_0$, but it may increase the error far from $\mathbf{X}_0$ if the scale factor varies significantly with $\mathbf{X}$. Haftka (1991) suggested using a linear approximation to the scale factor. That is,

$$s_{c1} = s_{c0} + \sum_{i=1}^{n} (x_i - x_{i0}) \left(\frac{\partial s_c}{\partial x_i}\right)_{\mathbf{X}_0} . \tag{34}$$

and then the linear-scale-factor approximation is given as

$$g_{s1} = s_{c1}\, g_G(\mathbf{X})\,. \tag{35}$$

where, using Eq. 32,

$$\left(\frac{\partial s_c}{\partial x_i}\right)_{\mathbf{X}_0} = s_{c0} \left(\frac{1}{g}\frac{\partial g}{\partial x_i} - \frac{1}{g_G}\frac{\partial g_G}{\partial x_i}\right)_{\mathbf{X}_0} \tag{36}$$

Chang *et al.*. (1991) have applied this approach to a plate theory approximation of a built-up wing structure where the refined analysis is based on a finite element model. Their results indicate that the linear scale factor is better than either the constant scale factor or the linear approximation based on the refined model.

4.1.3. Other Mid-Range Approximations. In a departure from the use of response surfaces as global approximations, Free *et al.* (1987) used the theory of experiments to develop an optimization algorithm which constructs local approximations in the vicinity of the current design point, solves the approximate design problem and then updates the current approximation, contracting or expanding its range as appropriate. Using this algorithm on standard nonlinear programming problems and a ten-bar truss design they showed performance comparable to that of a very efficient SQP/GRG algorithm for noise-free problems. When noise was present in the problem formulation, they showed significantly improved performance. Toropov (1989) used a similar approach, developing response surfaces on progressively smaller subsets of the design space. He also proposed a two-level approach to constructing response surfaces. At the global structural level, a response surface is built that gives the behavior of the structure as a function of the response of the individual structural members. At the structural member level, the member responses can be related to the design variables either analytically or through response surfaces.

5. Concluding Remarks

This paper reviews the main approximation concepts in applications of nonlinear programming to structural optimization. It shows that approximation concepts can be classified according to their range into local, medium-range, and global approximations. Approximations can be designed to approximate the functional relationship between dependent variables (objective function and constraints) and independent variables. Also, they can aim at approximating the problem formulation by reducing the number of problem variables or constraints, effecting changes of variable or response definition to improve functional relationships or simplifying structural model.

There is not enough data in the open literature to establish a comprehensive comparison of the effectiveness of various approximation concepts. Often, a particular approximation is compared for accuracy only in extrapolations about a given design point. The question of overall performance, as indicated by the number of reanalyses required in reference test problems or that of overall cost, as indicated by comparison of CPU requirements are not always addressed. However, available results show that computational cost can generally be reduced by the use of approximation concepts, particularly if several of them are combined. Recently, structural optimization programs have begun to offer as options combinations of the approximation concepts described herein.

The more recent contributions found for this review focus on two specific areas: i) careful selection of intermediate variable and response quantity and ii) improvement of the traditional Taylor series expansion by inclusion of some higher-order terms or of information at different points in the design space.

However, newer methods may emerge and older approaches may be revisited in light of development of new computer hardware. For example, the construction of response surfaces may become cost effective even for large problems with the advent of massively parallel computers. On the other hand, neural networks may also provide a unique opportunity to build inexpensive approximations to expensive analytical models.

6. Acknowledgements

The authors are grateful to the following researchers who accepted to review this paper and offered very constructive comments: H. Adelman, A. Belegundu, H. Eschenauer, M. Fuchs, H. Miura, J. Rasmussen, H. Thomas, J. Sobieski, and G. Vanderplaats.

The work of the second author was supported in part by NASA grant NAG-1-224.

7. References

1. Arora, J.S., (1976) "Survey of Structural Reanalysis Techniques," *J. Structural Division, ASCE*, Vol 102, No. ST4, pp. 783-802
2. Barthelemy, B., Haftka, R.T., Madapur, U., and Sankaranarayanan, S., (1991), "Integrated Analysis and Design using 3D Finite Elements," *AIAA J.*, in press.
3. Barthelemy, J.-F.M., Chang, K.J., and Rogers, J.L.Jr., (1988a) "Shuttle Solid Rocket Booster Bolted Field Joint Shape Optimization," *J. Spacecraft and Rockets*, Vol. 25, No. 2, Mar-Apr 1988, pp. 117-124.
4. Barthelemy, J.-F.M., Riley, M.F., (1988), "Improved Multilevel Optimization Approach to the Design of Complex Engineering Systems," *AIAA J.*, Vol. 26, No. 3, pp. 353-360.

5. Belegundu, A.D., Rajan, S.D., and Rajgopal, J. (1990) "Exponential Approximations in Optimal Design," work-in-progress paper at AIAA/ASME/ASCE/AHS/ASC 31st Structures, Structural Dynamics and Materials Conference, Apr. 2-4, Long Beach, CA, in NASA CP 3064, pp. 137-150.
6. Bennett, J.A. (1981), "Application of Linear Constraint Approximation to Frame Structures," *Proc. of International Symposium on Optimum Structural Design*, Tucson, AZ, Oct. 19-22, pp. 7.9-7.15.
7. Box, G.E.P., and Draper, N.R., (1987), *Empirical Model-Building and Response Surfaces*, Wiley, New York.
8. Braibant, V., and Fleury, C., (1985), "An Approximation Concepts Approach to Shape Optimal Design," *Computer Methods in Applied Mechanics and Engineering*, Vol. 53, pp. 119-148.
9. Brown, R.T., and Nachlas, J.A., (1985), "Structural Optimization of Laminated Conical Shells," *AIAA J.*, Vol, 23, No., 5, pp. 781-787.
10. Canfield, R.A., (1990), "High-Quality Approximation of Eigenvalues in Structural Optimization," *AIAA J.*, Vol. 28, No. 6, pp. 1116-1122.
11. Chan, A.S.L., and Turlea, E., (1978), "An Approximate Method for Structural Optimization," *Computers and Structures*, Vol. 8, pp. 357-363.
12. Chang, K.-J., Haftka, R.T., Giles, G.L., and Kao, P.-J., "Sensitivity Based Scaling for Correlating Structal Response from Different Analytical Models," AIAA Paper 91-0925, *Proc. of AIAA/ASME/ASCE/AHS/ASC 32nd Structures, Structural Dynamics and Materials Conference*, Baltimore, MD, April 8-10.
13. Ding, Y., (1987), "Optimum Design of Sandwich Constructions," *Computers and Structures*, Vol. 25, No. 1, pp. 51-68.
14. Ding, Y., and Esping, B.J.D., (1986), "Optimum Design of Frames with Beams of Different Cross-Sectional Shapes," *Proc. of AIAA/ASME/ASCE/AHS 27th Structures, Structural Dynamics and Materials Conference*, San Antonio, TX, May 19-21, Part I, pp. 262-275.
15. Duffin, R.J., Peterson, E.L., and Zener, C.M.,(1967), *Geometric Programming*, John Wiley.
16. Fadel, G.M., Riley, M.F., and Barthelemy, J.-F.M., (1990), "Two Point Exponential Approximation Method for Structural Optimization," *Structural Optimization*, Vol. 2, pp. 117-124.
17. Fleury, C., (1988), "A Convex Linearization Method using Second Order Information," *Proc. of Fourth SAS-World Conference*, Oct. 17-19, Vol. 2, pp. 374-383.
18. Fleury, C., (1989a), "Efficient Approximation Concepts using Second Order Information," *International Journal for Numerical Methods in Engineering*, Vol. 28, pp. 2041-2058.
19. Fleury, C., (1989b), "First and Second Order Convex Approximation Strategies in Structural Optimization," *Structural Optimization*, Vol. 1, pp. 3-10.
20. Fleury, C., and Braibant, V., (1986), "Structural Optimization: a New Dual Method Using Mixed Variables," *Int. J. Num. Meth. Eng.*, Vol. 23, pp. 409-428.
21. Fleury, C., and Sander, G., (1983), "Dual Methods for Optimizing Finite Element Flexural Systems," *Computer Methods in Applied Mechanics and Engineering*, Vol. 37, pp. 249-275.
22. Fleury, C., and Smaoui, H., (1988), "Convex Approximation Strategies in Structural Optimization," *Discretization Methods and Structural Optimization Procedures and Applications*, Eschenauer, H.A., and Thierauf, G., Eds., Springer-Verlag, Berlin, pp. 118-126.
23. Free, J.W., Parkinson, A.R., Bryce, and G.R., Balling, R.J., (1987), "Approximation of Computationally Expensive and Noisy Functions for Constrained Nonlinear Optimization," *Journal of Mechanisms, Transmissions, and Automation in Design*, Vol. 109, pp. 528-532.

24. Fuchs, M.B., (1980), " Linearized Homogeneous Constraints in Structural Design," *Int. J. Mech. Sci.*, Vol. 22, pp. 33-40.
25. Fuchs, M.B., and Haj Ali, R.M., (1990), "A Family of Homogeneous Analysis Models for the Design of Scalable Structures," *Structural Optimization*, Vol. 2, pp. 143-152.
26. Haftka, R.T., (1988), "First- and Second-Order Constraint Approximations in Structural Optimization," *Comp. Mech.*, Vol. 3, pp. 89-104.
27. Haftka, R.T., (1991) "Combining Local and Global Approximations," *AIAA Journal* Vol. 29, no. 9, pp. 1523-1525.
28. Haftka, R.T., Gurdal, Z., and Kamat, M.P. (1990) *Elements of Structural Optimization*, 2nd Ed., Kluwer Academic Publishers Group, the Netherlands.
29. Haftka, R.T., Nachlas, J.A., Watson, L.T., Rizzo, T., and Desai, R., (1989), "Two-Point Constraint Approximation in Structural Optimization," *Comp. Meth. Appl. Mech. Eng.*, Vol 60, pp. 289-301.
30. Haftka, R.T., and Shore, C.P., (1979), "Approximation Method for Combined Thermal/Structural Design", NASA TP-1428.
31. Haftka, R.T., and Starnes, J. (1976), "Applications of a Quadratic Extended Interior Penalty Function for Structural Optimization," *AIAA J.*, Vol. 14, No. 6, pp. 718-724.
32. Hajela, P., (1982), "Further Developments in the Controlled Growth Approach for Optimal Structural Synthesis," Paper 82-DET-62, *Proc. of ASME 1982 Design Automation Conference*, Arlington, VA.
33. Hajela, P., (1986), "Geometric Programming Strategies in Large-Scale Structural Synthesis," *AIAA Journal*, Vol. 24, No. 7, pp. 1173-1178.
34. Hajela, P., and Berke, L. (1990), "Neurobiological Computational Models in Structural Analysis and Design," *Proc. of 31st AIAA/ASME/ASCE/AHS/ASC Structures, Structural Dynamics and Materials Conference*, Apr. 2-4, Long Beach, CA, Part I, pp. 345-355.
35. Hajela, P., and Sobieszczanski-Sobieski, J., (1981) "The Controlled Growth Method - A Tool for Structural Optimization," *Proc. of AIAA/ASME/ASCE/AHS 22nd Structures, Structural Dynamics and Materials Conference*, Apr. 6-8, Atlanta, GA, Part I, pp. 206-215.
36. Hansen, S.K., and Vanderplaats, G.N., (1990), "Approximation Method for Configuration Optimization of Trusses," *AIAA J.*, Vol. 28, No. 1, pp. 161-168.
37. Jawed, A.H., and Morris, A.J., (1984), "Approximate Higher-Order Sensitivities in Structural Design," *Engineering Optimization*, Vol. 7, No. 2, pp. 121-142.
38. Jawed, A.H., and Morris, A.J., (1985), "Higher-Order Updates for Dynamic Responses in Structural Optimization," *Computer Methods in Applied Mechanics and Engineering*, Vol 49, pp. 175-201.
39. Kirsch, U., (1984), "Approximate Behaviour Models for Optimum Structural Design," *New Directions in Optimal Structural Design*, Atrek, E. et al., Eds., John Wiley and Sons, New York, pp. 365-384.
40. Kirsch, U., and Toledano, G. (1983), "Approximate Reanalysis for Modifications of Structural Geometry," *Computers and Structures*, Vol 16, No. 1-4, pp. 269-277.
41. Kodiyalam, S., and Vanderplaats, G.N., (1989), "Shape Optimization of 3D Continuum Structures Via Force Approximation Technique," *AIAA J.*, Vol. 27, No. 1, pp. 161-168.
42. Kreisselmeier, G., and Steinhauser, R., (1979), "Systematic Control Design by Optimizing a Vector Performance Index," *Proc. of IFAC Symposium on Computer Aided Design of Control Systems*, Zurich, Switzerland, pp. 113-117.

43. Lawson, J.S., Batchelor, C., Parkinson, A.R., and Talbert, J., (1989) "Consideration of Variance and Bias in the Choice of a Saturated Second-Order Design for use in Engineering Optimization," Report EDML 89-7, Engineering Design Methods Laboratory, Brigham Young University.
44. Lust, R.V., (1990), "Structural Optimization with Crashworthiness Constraints," to appear in *Proc. of III Air Force/NASA Symposium on Recent Advances in Multidisciplinary Analysis and Optimization*, San Francisco, CA, Sep. 24-26.
45. Lust, R.V., and Schmit, L.A., (1986), "Alternative Approximation Concepts for Space Frame Synthesis," *AIAA J.*, Vol. 24, No. 10, pp. 1676-1684.
46. Manning, R.A., Lust, R.V., and Schmit, L.A., (1986), "Behavior Sensitivities for Control-Augmented Structures," in *Proc. of NASA-Va. Tech. Symposium on Sensitivity Analysis in Engineering*, Hampton, VA, 25-26 Sep. 1986, H.M. Adelman and R.T. Haftka Compilers, NASA CP 2457, pp. 33-57.
47. McCullers, L.A., and Lynch, R.W., (1972) "Composite Wing Design for Aeroelastic Requirements," in *Proc. of Conference on Fibrous Composites in Flight Vehicle Design*, AFFDL TR-72-130, pp. 951-972.
48. Mills-Curran, W.C., and Schmit, L.A.Jr., (1983), "Structural Optimization with Dynamic Behavior Constraints," *Proc. of AIAA/ASME/ASCE/AHS 24th Structures, Structural Dynamics and Materials Conference*, Lake Tahoe, NV, May 2-4, Part 1, pp. 369-382.
49. Mills-Curran, W.C., Lust, R.V., and Schmit, L.A., (1983), "Approximations Method for Space Frame Synthesis," *AIAA J.*, Vol. 21, No. 11, pp. 1571-1580.
50. Miura, H., and Chargin, K.L. (1991), "New Approximation of Frequency Response for Structural Synthesis and Parameter Identification," *Proc. of Ninth International Modal Analysis Confernce and Exhibit*, Florence, Italy, 15-18 Apr. 1991.
51. Miura, J., and Schmit, L.A., (1978) "Second Order Approximation of Natural Frequency Constraints in Structural Synthesis," *International Journal of Numerical Methods in Engineering*, Vol. 13, No. 2, pp. 337-351.
52. Mistree, F., Hughes, O.F., Phuoc, H.B., (1981), "An Optimization Method for the Design of Large, Highly Constrained Complex Systems," *Eng. Opt.*, Vol. 5, pp. 179-197.
53. Moore, G.J., and Vanderplaats, G.N., (1990), "Improved Approximations for Static Stress Constraints in Shape Optimal Design of Shell Structures," *Proc. of AIAA/ASME/ASCE/AHS/ASC 31st Structures, Structural Dynamics and Materials Conference*, Apr. 2-4, Long Beach, CA., Part I, pp. 161-170.
54. Morris, A.J., (1972), "Structural Optimization by Geometric Programming," *Int. J. Solids and Structures*, Vol. 8, pp. 847-874.
55. Morris, A.J., (1974), "The Optimization of Statically Indeterminated Structures by Means of Approximate Geometric Programming," *Second Symposium on Structural Optimization*, AGARD-CP-123, pp. 6.1-6.15.
56. Murthy, D.V., and Haftka, R.T. (1988) "Approximations to Eigenvalues of Modified General Matrices," *Computers and Structures*, Vol. 29, No. 5, pp. 903–917.
57. Noor, A.K., and Lowder, H.E., (1975a), "Structural Reanalysis via a Mixed Method," *Computers and Structures*, Vol. 5, pp. 9-12.
58. Pedersen, P., (1981), "The Integrated Approach of FEM-SLP for Solving Problems of Optimal Design," *Optimization of Distributed Parameters Structures*, Vol. 1, pp. 757-780 Haug, E.J., and Cea, J., Eds., Stijthoff and Noordhoff, Amsterdam.
59. Pickett, R.M.Jr., Rubinstein, M.F., and Nelson, R.B. (1973), "Automated Structural Synthesis using a Reduced Number of Design Coordinates," *AIAA J.*, Vol. 11, No. 4, pp. 489-494.

60. Prasad, B., (1983), "Explicit Constraint Approximation Forms in Structural Optimization, Part I: Analyses and Projections," *Comp. Meth. Appl. Mech. Eng.*, Vol. 40, pp. 1-26.
61. Prasad, B., (1984a), "Explicit Constraint Approximation Forms In Structural Optimization. Part 2: Numerical Experiences," *Computer Methods in Applied Mechanics and Engineering*, Vol. 46, pp. 15-38.
62. Prasad, B., (1984b), "Novel Concepts for Constraint Treatments and Approximations in Efficient Structural Synthesis," *AIAA J.*, Vol. 22, No. 7, pp. 957-966.
63. Pritchard, J.I., and Adelman, H.M., (1990), "Differential Equation Based Method for Accurate Approximations in Optimization," *Proc. of AIAA/ASME/ASCE/AHS/ASC 31st Structures, Structural Dynamics and Materials Conference*, Apr. 2-4, Long Beach, CA, to appear AIAA J. see also NASA TM-102639.
64. Rajamaran, A., and Schmit, L.A.Jr., (1981), "Basis Reduction Concepts in Large Scale Structural Synthesis," *Engineering Optimization*, Vol. 5, pp. 91-104.
65. Rasmussen, J., (1990), "Accumulated Approximations - A New Method for Structural Optimization by Iterative Improvements," To appear in: *Proc. of IIIrd Air Force/NASA Symposium on Recent Advances in Multidisciplinary Analysis and Optimization*, Sep. 24-26, San Francisco, CA.
66. Rasmussen, J. (1991), private communication.
67. Renwei, X., and Peng, L., (1987), "Structural Optimization based on Second-Order Approximations of Functions and Dual Theory," *Computer Methods in Applied Mechanics and Engineering*, Vol. 65, pp. 101-114.
68. Ricketts, R.H., and Sobieszczanski-Sobieski, J., (1977) "Simplified and Refined Structural Modeling for Economical Flutter Analysis and Design," AIAA Paper 77-421, Presented at AIAA/ASME/SAE 18th Structures, Structural Dynamics and Materials Conference, San Diego, CA.
69. Rommel, B.A., (1983) "The Developement of FAST-FLOW (A Program for Flutter Optimization to Satisfy Multiple Flutter Requirements)," in AGARD Conference Proceedings 354, *Aeroelastic Considerations in the Preliminary Design of Aircraft*, pp. 8.1-8.17.
70. Salajegheh, E., and Vanderplaats, G.N., (1986/1987), "An Efficient Approximation Method for Structural Synthesis with Reference to Space Structures," *Space Struct. J.*, Vol 2, pp. 165-175.
71. Salama, M., Ramanathan, R.K., Schmit, L.A.Jr., and Sarma, I.S. "Influence of Analysis and Design Models on Minimum Weight Design," in *Proc. of NASA Symposium on Recent Experiences in Multidisciplinary Analysis and Optimization*, Apr. 24-26, 1984, Hampton, VA, NASA CP 2327, Part 1, pp. 329-342.
72. Schmit, L.A.Jr., and Farshi, B., (1974), "Some Approximation Concepts for Structural Synthesis,", *AIAA J.*, Vol. 12, No. 5, pp. 692-699.
73. Schmit, L.A. Jr., and Miura, H., (1976), "Approximation Concepts for efficient Structural Synthesis," NASA CR-2552.
74. Schoofs, A.J.G. (1987), "Experimental Design and Structural Optimization," PhD Dissertation, Technical University of Eindhoven, 1987.
75. Sepulveda, A.E., Thomas, H.L., and Schmit, L.A.Jr., (1991) "Improved Transient Response Approximations for Control Augmented Structural Optimization," *Proc. of PACAM II*, Valparaiso, Chile, Jan. 2-4, 1991, pp. 611-614.
76. Smaoui, H., Fleury, C., and Schmit, L.A.Jr., (1988), "Advances in Dual Algorithms and Convex Approximations Methods," *Proc. of AIAA/ASME/ASCE/AHS 29th Structures, Structural Dynamics and Materials Conference*, Williamsburg, VA, Apr. 18-20, Part 3, pp. 1339-1347.

77. Sobiesczcanski, J., and Loendorf, D., (1972) "A Mixed Optimization Method for Automated Design of Fuselage Structures," *J. Aircraft*, Vol. 9, No. 12, pp. 805-811.
78. Sobieszczanski-Sobieski, J., James, B.B., and Dovi, A.R., (1985), "Structural Optimization by Multilevel Decomposition," *AIAA J.*, Vol. 23, No. 11, pp. 1775-1782.
79. Starnes, J.H. Jr, and Haftka, R.T., (1979) "Preliminary Design of Composite Wings for Buckling, Stress, and Displacement Constraints," *J. Aircraft*, Vol. 16, pp. 564-570.
80. Storaasli, O.O., and Sobieszczanski-Sobieski, J. (1974), "On the Accuracy of the Taylor Approximation for Structure Resizing," *AIAA J.*, Vol. 12, pp. 231-233.
81. Templeman, A.B., Winterbottom, S.K., (1974), "Structural Design Applications of Geometric Programming," *Second Structural Optimization Symposium*, AGARD-CP-123, pp. 5.1-5.16.
82. Thomas, H.L., Sepulveda, A.E., and Schmit, L.A. Jr., (1990), "Improved Approximations for Dynamic Displacements using Intermediate Response Quantities," To appear in: *Proc. of IIIrd Air Force/NASA Symposium on Recent Advances in Multidisciplinary Analysis and Optimization*, Sep. 24-26, San Francisco, CA.
83. Thomas, H.L, Sepulveda, A.E., and Schmit, L.A. Jr., (1991), "Improved Approximations for Control Augmented Structural Synthesis," *AIAA J.*, To appear.
84. Thomas, H.L., and Vanderplaats, G.N., (1991), "An Improved Approximation for Stresses Constraints in Plate Structures," *Proc. of Opti91*, Boston, MA, 25-27 June 1991.
85. Toropov, V.V., (1989), "Simulation Approach to Structural Optimization," *Structural Optimization*, Vol. 1, pp. 37-46.
86. Vanderplaats, G.N., (1979), "Efficient Algorithm for Numerical Airfoil Optimization," *J. Aircraft*, Vol. 16, No. 2, pp. 842-847.
87. Vanderplaats G.N., and Han, S.H., (1990), "Arch Shape Optimization using Force Approximation Methods," *Structural Optimization*, Vol. 2, pp. 193-201.
88. Vanderplaats, G.N., and Kodiyalam, S., (1990), "Two-Level Approximation Method for Stress Constraints in Structural Optimization," *AIAA J.*, Vol. 28, No. 5, pp. 948-951.
89. Vanderplaats, G.N., and Salajegheh, E., (1988), "An Efficient Approximation Technique for Frequency Constraints in Frame Optimization," *International Journal for Numerical Methods in Engineering*, Vol. 26, pp. 1057-1069.
90. Vanderplaats, G.N., and Salajegheh, E. (1989), "A New Approximation Method for Stress Constraints in Structural Synthesis," *AIAA J.*, Vol. 27, No. 3., pp. 352-358.
91. White, K.P.Jr., Hollowell, W.T., Gabler, H.C.III, Pilkey, W.D., (1985), "Simulation Optimization of the Crashworthiness of a Passenger Vehicle in Frontal Collision using Response Surface Methodology," *SAE Transactions*, Sec. 3, pp. 3.798-3.811.
92. White, K.P.Jr., Gabler, H.C.III, and Pilkey, W.D., (1986), "Approximating Dynamic Response in Small Arrays using Polynomial Parameterizations and Response Surface Methodology," *The Shock and Vibration Bulletin*, Vol. 55, Part 3, pp. 167-173.
93. Woo, T.H. (1987), "Space Frame Optimization Subject to Frequency Constraints," *AIAA J.*, Vol. 25, No. 10, pp. 1396-1404.
94. Yoshida, N., and Vanderplaats, G.N., (1988), "Structural Optimization using Beam Elements," *AIAA J.*, Vol. 26, No. 4, pp. 454-462.
95. Zhou, M., and Xhia, R.W. (1990), "Two-Level Approximation Concept in Structural Synthesis," *Int. J. Num. Meth. Engr.*, Vol 29, pp. 1681-1699.
96. Zienkiewicz, O.C., and Campbell, J.S., (1973), "Shape Optimization and Sequential Linear Programming," in *Optimum Structural Design, Gallagher*, R.H. and Zienkiewicz, O.C., Eds., Wiley, New York.

THE STATE OF THE ART OF APPROXIMATION CONCEPTS IN STRUCTURAL OPTIMIZATION

H. L. THOMAS and G. N. VANDERPLAATS
VMA Engineering
5960 Mandarin Ave., Suite F
Goleta, CA 93117

ABSTRACT. Recent methods in the approximation concepts approach to structural optimization are presented. The approximate optimization problem that is constructed using these methods is closer to the actual problem than one constructed using older methods. This increased accuracy results in faster design convergence. The two methods presented in this paper are approximations with respect to intermediate design variables and approximations of intermediate response quantities. These methods are applied to truss, frame, and plate structures subject to static displacement and stress constraints as well as frequency constraints. Examples that show the increased design convergence rate are presented.

1. Introduction

The use of math programming to solve structural optimization problems was first presented in [1] in 1960. In this work a numerical optimization program was used directly with a finite element code to solve the structural optimization problem. Although this approach was viable, the large number of finite element analyses required made it impractical.

In the mid 1960's structural optimization problems where solved by linearizing the objective function and constraints with respect to the design variables (see for example [2]) or their reciprocals (see [3]) and solving the linear program with the Simplex Method. Since the actual optimization problems are nonlinear this method required careful selection of move limits and exhibited slow design convergence.

In the mid 1970's approximation concepts were first applied to structural optimization in [14]. In this work the static displacements and stresses in truss structures were approximated using a linear Taylor Series expansion in the design variables:

$$u \approx \tilde{u} = u_0 + \sum_{i=1}^{NDV} \frac{\partial u}{\partial Y_i}(Y_i - Y_{0i}) \tag{1}$$

where NDV is the number of design variables Y_i. The first use of approximations concepts in conjunction with a general purpose finite element capability was presented in [5]. In this work it

G. I. N. Rozvany (ed.), *Optimization of Large Structural Systems, Vol. I*, 257–270.

was pointed out that more accurate displacement approximations would be generated if they were made with respect to the inverse of the design variables:

$$\tilde{u} = u_0 + \sum_{i=1}^{NDV} \frac{\partial u}{\partial(1/Y_i)}\left(\frac{1}{Y_i} - \frac{1}{Y_{0i}}\right) \tag{2}$$

This is because the displacements in statically determinate truss and frame structures are linear functions of the reciprocals of the truss areas and bending moments of inertias. This can be called the first use of approximations with respect to intermediate design variables (the reciprocals of the design variables). Also in [5] the concept of the intermediate response quantities was first presented in the approximation of radial displacement of a grid point. The displacement of the grid point in the three coordinate directions was approximated using Eq. 2 and the the displacement in any radial direction was calculated using:

$$\tilde{u}_r = \sqrt{\tilde{u}_x^2 + \tilde{u}_y^2 + \tilde{u}_z^2} \tag{3}$$

Note in Eq. 3 the nonlinearities associated with the square and square root operations are captured. While the use of inverse design variables was popular, few other major advances in approximation concepts were made until the mid 1980's. It was at this time that the development of applications of intermidate design variables and response quantities to the approximation concepts approach to structural optimization blossomed.

2. Intermediate Design Variables

In the approximation concepts approach a linear Taylor series expansion of the response is made with respect to some variable. It is important to note that this variable does not have to be the one that the designer is using. However, the variable used in the approximation should be able to be calculated explicitly from the variables that the designer is using. For example the bending moment of inertia of a beam can be explicitly calculated from the beam's cross sectional dimensions.

2.1 TRUSS STRUCTURES

The variable used in the approximation, called the intermediate design variable, should be chosen as the one that the response is most nearly a linear function of. As stated above, the inverse of the truss element areas is chosen as the intermediate design variable in approximations of displacements in truss structures. This is because the displacement is exactly a linear function of this intermediate design variable in statically determinate structures. An intuitive argument can be made that the inverse of the areas should be used in indeterminate structures. The error in this approximation is a function of the indeterminatcy of the structure.

2.2 FRAME STRUCTURES

The obvious choice for intermediate design variables in the design of frame structures subject to displacement constraints are the inverse of section properties (area, bending moment of inertia, and torsional stiffness). The actual design variables in frame structures are the member cross sectional dimensions (CSD's). An explicit relationship between the CSD's and the inverse of the section properties (SP's) must be available during the solution of the approximate optimization problem. This is facilitated using a "design element library" which contains these relationships. The use of a design element library and approximations with respect to the inverse of the section properties was first presented in [6]. In this work the intermediate variables were called the "sensitivity variables." As an example consider design of a frame structure composed of rectangular members of height H and width B. The approximation for the displacements in this structure would be (neglecting torsion):

$$\tilde{u} = u_0 + \sum_{k=1}^{NDE}\left[\frac{\partial u}{\partial(1/A_k)}\left(\frac{1}{B_kH_k} - \frac{1}{A_{0k}}\right) + \frac{\partial u}{\partial(1/I_k)}\left(\frac{12}{B_kH_k^3} - \frac{1}{I_{0k}}\right)\right] \tag{4}$$

where NDE is the number of design elements. Note that this approximation is a nonlinear, but explicit, function of B and H. In general this approximation can be written as

$$\tilde{R} = \tilde{R}(X(Y)) = R_0 + \sum_{j=1}^{NIDV}\frac{\partial R}{\partial X_j}(X_j(Y) - X_{0j}) \tag{5}$$

where NIDV is the number of intermediate design variables X_j.

In many structural optimization problems the objective function is the volume or mass. It is obvious that these objective function quantities should be approximated by a Taylor Series expansion in the direct SP's (area) rather than the CSD's. In this case the approximate value of the objective function will be exact.

It was shown in [7] that the use of SP's as intermediate design variables leads to improved approximations for structural eigenvalues and element forces. In the case of structural eigenvalues the intermediate variables decouple the stiffness and mass terms in the approximation. While a change in a CSD effects both the stiffness and mass of the structure, a change in the area only effects the mass and changes in the bending and torsional moments of inertia only effect the stiffness. It was reported in [8] that this type of approximation was even more accurate when there the was large amounts of nonstructural mass. In this case the eigenvalues are almost a linear function of the stiffness and therefore the bending and torsional moments of inertia.

2.3 PLATE STRUCTURES

In [9] intermediate design variables were used to improve approximations for displacements and element forces in plate structures. In this case the intermediate design variables are the membrane thickness (t_m) and the plate bending stiffness (D). Note that for a homogeneous plate the thickness is both a design variable and an intermediate design variable. As with frame structures, the resulting displacement approximations are exact for statically determinate structures.

3. Intermediate Response Quantities

3.1 ELEMENT FORCES: SIZING DESIGN

In [10] it was shown that improved approximations for stress constraints in truss and frame structures could be developed by using the element forces as intermediate response quantities. If the stresses are approximated directly the approximation is:

$$\tilde{\sigma} = \sigma_0 + \sum_{j=1}^{NIDV} \frac{\partial \sigma}{\partial X_j}(X_j - X_{0j}) \tag{6}$$

Using the forces as intermediate response quantities the new approximation for stresses in truss structures is (noting that the truss areas are the intermediate design variables):

$$\tilde{\sigma}_i = \frac{F_i}{A_i} = \frac{F_{0i} + \sum_{j=1}^{NDE} \frac{\partial F_i}{\partial A_j}(A_j - A_{0j})}{A_i} \tag{7}$$

where A_i, F_i, and σ_i are the area, force, and stress of element i respectively. Note that $A_j = X_j$ and NDE=NIDV for truss structures. Equation 7 produces a more accurate approximation than Eq. 6 because it captures the cross coupling between the effect of A_i and A_j on the stress. This cross coupling is present because the stress is a nonlinear function of the element force F_i, which is a function of the A_j, and the element area A_i. This can be seen mathematically by examining the second partial derivatives of Eqs. 6 and 7. The second partial derivative of Eq. 7 is:

$$\frac{\partial^2 \tilde{\sigma}_i}{\partial A_i \partial A_j} = \frac{1}{A_i^2} \frac{\partial F_i}{\partial A_j} A_{0j} \tag{8}$$

It is observed that the derivative is a nonlinear function of the area of the element. Note that the second partial of Eq. 6 is zero.

An intuitive argument can also be made for using the element force as the intermediate response quantity. Note that in a statically determinate structure the element forces are constant while the stresses are not. This implies that in a nondeterminate structure the stresses are more nonlinear functions of the truss areas than element forces. Note that for statically determinate structures the approximation presented in Eq. 7 is exact.

In frame structures the 12 element forces; two bending moments, two shear forces, axial force, and torque at each end, are used as the intermediate response quantities. The stresses in each element are then calculated from these approximate element forces. For example, consider the approximation for the normal stress in a planer element:

$$\tilde{\sigma} = \frac{\tilde{M}c}{I} + \frac{\tilde{P}}{A} \tag{9}$$

where the bending moment M and the axial force P are approximated by linear Taylor Series expansions in the SP's A and I. Note that the nonlinear relationship between the element forces and the element SP's is captured explicitly. Also note that since the element SP's A, I, and c are calculated from the CSD's when Eq. 8 is evaluated, the nonlinearity associated with these calculations is also captured explicitly.

3.2 ELEMENT FORCES: SHAPE DESIGN

It is shown in [11] that use of element forces as intermediate response quantities leads to improved approximations for stresses in the configuration optimization of truss structures. The reason for the improvement can be seen by examining the approximations for the stress in an element of a statically determinate structure. The direct stress approximation is:

$$\tilde{\sigma} = \sigma_0 + \sum_{j=1}^{NDE} \frac{\partial \sigma}{\partial A_j}(A_j - A_{0j}) + \sum_{k=1}^{NCDV} \frac{\partial \sigma}{\partial C_k}(C_k - C_{0k}) \tag{10}$$

where the C_k are the coordinate design variables and NCDV is the number of coordinate design variables.

The improved approximation is:

$$\tilde{\sigma} = \frac{\tilde{F}}{A} \tag{11}$$

where

$$\tilde{F} = F_0 + \sum_{j=1}^{NDE} \frac{\partial F}{\partial A_j}(A_j - A_{0j}) + \sum_{k=1}^{NCDV} \frac{\partial F}{\partial C_k}(C_k - C_{0k}) \tag{12}$$

In statically determinate structures the partial derivatives $\frac{\partial \sigma}{\partial A_j}$ and $\frac{\partial \sigma}{\partial C_k}$ are, in general, functions of both the A_j and C_k. However the partial derivative $\frac{\partial F}{\partial A_j}$ is zero and $\frac{\partial F}{\partial C_k}$ is only a function of C_k in these structures. Therefore the problem is uncoupled when the approximation presented in Eq. 11 is used for statically determinate problems. It can be argued intuitively that for nondeterminate structures, the approximation presented in Eq. 11 is still more decoupled than that of Eq. 10.

The use of frame member forces as intermediate response quantities for arch shape optimization was reported in [12]. As with truss structures the decoupling of effect of the approximation leads to increased accuracy. Once again the approximation is exact for statically determinate structures.

Internal forces were also used used as intermediate design variables in [13] and [14] in the design of arch dams. It was found the using the internal force at a point in the dam normalized by the radius of curvature, or its square, at that point produced high quality approximations. The stresses in the dam where calculated using these quantities and the formulae of the strength of materials.

3.3 ELEMENT LENGTH

In shape design problems the length of the elements can be used as intermediate response quantities. Instead of approximating an elements volume as a function of its end point coordinates it is better to directly calculate its length:

$$L = \sqrt{(x_1 - x_2)^2 + (y_1 - y_2)^2 + (z_1 - z_2)^2} \tag{13}$$

The element length can be used to exactly calculate the element volume and mass in this case. Note that the area of a frame element must be calculated from its CSD's.

Euler bucking constraints on truss members will also be more accurate if the element length is used. In order to prevent Euler buckling the quantity $\frac{FL^2}{A^2}$ must be greater than some critical value. The best approximation for this type of constraint is to use a linear Taylor Series expansion of the force with respect to the intermediate design variables and use the exact values of L and A, which are calculated for the end coordinates and CSD's explicitly.

3.4 PLATE ELEMENT FORCES

The stresses on the surface of a homogeneous isotropic plate are related to the element membrane and bending forces by the equation:

$$\{\sigma\} = \begin{Bmatrix} \sigma_x \\ \sigma_y \\ \tau_{xy} \end{Bmatrix} = \begin{Bmatrix} \frac{N_x}{t} - \frac{M_x z}{D} \\ \frac{N_y}{t} - \frac{M_y z}{D} \\ \frac{N_{xy}}{t} - \frac{M_{xy} z}{D} \end{Bmatrix} \tag{14}$$

where N are the membrane forces, M are the bending moments, t is the plate thickness, z is the distance from the centerline to the surface, and the plate bending stiffness is $D = \frac{t^3}{12}$. As with the truss and frame structures, the element forces should be approximated using a linear Taylor series expansion in the intermediate design variables (t and D), and the stresses calculated explicitly from these forces using Eq. 14. Once again, this approximation is exact for statically determinate structures. In many plate design problems the Von Mises failure criterion is used. In this case the approximate Von Mises stress is calculated from the approximate surface stresses using the explicit nonlinear relation:

$$\tilde{\sigma}_{VM} = \sqrt{\tilde{\sigma}_x^2 + \tilde{\sigma}_y^2 - \tilde{\sigma}_x\tilde{\sigma}_y + 3\tilde{\tau}_{xy}^2} \tag{15}$$

3.5 RAYLEIGH QUOTIENT APPROXIMATION

It was shown in [11] that the accuracy of eigenvalue constraints can be improved if the modal potential and kinetic energies are approximated by a Taylor Series rather than the eigenvalues themselves. The Rayleigh quotient approximation is defined as:

$$\tilde{\lambda} = \frac{\tilde{U}}{\tilde{T}} = \frac{\{\phi\}^T[\tilde{K}]\{\phi\}}{\{\phi\}^T[\tilde{M}]\{\phi\}} \tag{16}$$

where [K] and [M] are the system stiffness and mass matrices, $\{\phi\}$ is the eigenvector for λ, and U and T are the modal potential and kinetic energies. This approximation is more accurate than the direct approximation of the eigenvalue because the effects of changes in the mass and stiffness on the eigenvalue are decoupled. In this approximation it is assumed that the mode shapes (eigenvectors) are invariant. Note that with this assumption the modal energies are linear functions of the intermediate design variables (truss area, frame area and inertias, plate thickness and bending stiffness, etc.). Therefore if the modal energies are approximated as linear functions of the intermediate design variables and the mode shapes are invariant, the approximation is exact.

In most practical design problems the mode shapes are not invariant. This resulting effect is that the Rayleigh quotient approximation tends to overestimate the eigenvalue. This causes convergence problems when lower bound frequency constraints are imposed on the design. In order to overcome these convergence problems it is recommended that a hybrid approximation [12] be used for the modal energies. The result of the hybrid approximation is that the modal potential energy is approximated by a Taylor series expansion in the reciprocals of the intermediate design variables for lower bound constraints. For upper bound constraints the modal kinetic energy should be approximated using the reciprocals of the intermediate design variables.

4. Examples

The following is a collection of examples that demonstrate convergence characteristics of design problems solved using the approximation concepts described above. All of the results were generated with the GENESIS structural optimization program [17].

4.1 TEN BAR TRUSS

The first example is the mass minimization of the ten bar truss [10] shown in Figure 1 subject to stress constraints of ±25,000 psi in each member, except for member 9 which has limits of ±50,000. Youngs modulus is 10^7 psi and the mass density is 0.1 lbm/in^3. The initial truss areas are 20.0 in^2 and the minimum allowable area is 0.1 in^2. Two cases are considered here. In the case A there are only stress constraints. In the case B displacement constraints of ± 2.0 in were imposed at each node. The final designs for both cases as well as those from [5] are shown in Table 1. Note that the improved stress constraint approximation reduces the number of design cycles from 16 to 7 in case A.

4.2 EIGHTEEN BAR TRUSS

This example is the mass minimization of the eighteen bar truss shown in Figure 2 subject to stress constraints of ±20,000 psi and Euler buckling constraints. Youngs modulus is 10^7 psi and the mass density is 0.1 lbm/in^3. The truss member areas are linked so that there are four sizing design variables. The X and Y locations of the bottom four nodes are controlled by design variables for a total of 12 design variables. The final design is shown in Table 2 along with that achieved in [18]. Note that the improved approximations reduce the number of design cycles from 78 to 8.

4.3 PORTAL FRAME

This example is the mass minimization of the portal frame [19] shown in Figure 3 subject to stress and local buckling constraints. There are six design variables per element for a total of 18 design variables. The final design is shown in Table 3 along with that from [19]. Note that the improved approximations reduces the number of required design cycles from 11 to 5.

4.4 CANTILEVERED BEAM

This example is the mass minimization of the cantilever beam [20] shown in Figure 4 subject to a lower bound frequency constraint of 20.0 Hz. The cap areas and web thicknesses for each element are design variables for a total of six design variables. The improved frequency approximation results allows the optimum design to be achieved in just four design cycles. Seven design cycles were required in [20].

4.5 PLATE WITH A HOLE

This example is the mass minimization of a square plate with a central hole, shown in Figure 5, subject to Von Mises stress constraints of 10,000 psi. The hole is subjected to balanced biaxial tension. Only one quarter of the plate needs to be modeled due to symmetry. The seven design variables control the radial locations of the grid points in the mesh. The optimum design is achieved in just two design cycles in this example. In [21] it is reported that if a direct approximation of the stress is used, 11 design cycles are required. The final radius of the hole is 3.6 in.

4.6 CLAMPED PLATE

This example consists of a quarter plate model of a uniformly loaded clamped plate. The mass is minimized subject to stress and displacement constraints described in [22]. Maximum and minimum principle stress constraints of 12,000 and -12,000 psi, respectively, are used. Deflections of all points in the plate are constraint to be less than 0.1 in. The finite element model consists of 100 elements, but due to symmetry considerations only 55 design variables need to be used. Using the approximation concepts reported in this paper the optimum design is reached in 10 design cycles. This problem was solved using optimality criteria in [22] and required 39 finite element analysis.

5. Conclusions

Recent developments in the approximation concepts approach to structural synthesis have led to a reduction in the number of design cycles required to achieve optimum designs. By examining the underlying analysis equations it is possible to recognize many of the explicit nonlinearities associated with the analysis. These explicit an nonlinear relationships are then used to increase the accuracy of the approximations.

References

1. Schmit, L.A. (1960) 'Structural Design by Systematic Synthesis', Proc. 2nd Conference on Electronic Computation, ASCE, New York, 105-122.

2. Moses, F. (1964) 'Optimum Structural Design using Linear Programming', ASCE, Journal of the Structural Division, Vol. 90, No. ST6, pp. 89-104.

3. Reinschmidt, K. F., Cornell, A. C. and Brotchie, J. F. (1966) 'Iterative Design and Structural Optimization', ASCE, Journal of the Structural Division, Vol. 92, No. ST6, pp. 281-318.

4. Schmit, L. A., and Farshi, B. (1974) 'Some Approximation Concepts for Structural Synthesis', AIAA J., Vol. 12, No. 5, 692-699.

5. Schmit, L.A., and Miura, H. (1976) 'Approximation Concepts for Efficient Structural Synthesis', NASA CR-2552.

6. Yoshida, N. and Vanderplaats, G. N. (1988) 'Structural Optimization Using Beam Elements', AIAA J., Vol. 26, No. 4, 454-462.

7. Salajegheh, E. and Vanderplaats, G.N. (1986-87) 'An Efficient Approximation Method for Structural Synthesis with Reference to Space Structures', International Journal of Space Structures, Vol 2, No 3, 165-175.

8. Vanderplaats, G. N. and Salajegheh, E. (1988) 'An Efficient Approximation Technique for Frequency Constraints in Frame Optimization', Int. Journal for Numerical Methods, Vol. 26, 1057-1069.

9. Thomas, H. L. and Vanderplaats, G. N., (1991) 'An Improved Approximation Function for Stress Constraints in Plate Structures', Proc. Computer Aided Optimum Design of Structures, Boston MA, June 25-27.

10. Vanderplaats, G. N. and Salajegheh, E. (1989) 'A New Approximation Method for Stress Constraints in Structural Synthesis', AIAA J., Vol. 27, No. 3, 352-358.

11. Hansen, S.R. and Vanderplaats, G.N. (1990) 'An Approximation Method for Configuration Optimization of Trusses', AIAA Journal, Vol 28, No. 1, 161-172.

12. Vanderplaats, G. N. and Han, S. H., (1990) 'Arch Shape Optimization Using Force Approximation Methods', Structural Optimization, Vol. 2, No. 4, 193-201.

13. Bofang, Z. and Zhanmei, L. (1981) 'Optimization of Double-Curvature Arch Dams' (In Chinese), Chinese Journal of Hydraulic Engineering, No. 4, pp. 11-21.

14. Bofang, Z. (1987) 'Shape Optimization of Arch Dams', Water Power and Dam Construction, March 1987, pp. 43-51.

15. Canfield, R. A. (1990) 'High-Quality Approximations of Eigenvalues in Structural Optimization', AIAA J., Vol. 28, No. 6, 1116-1122.

16. Starnes, J. H. and Haftka, R. T. (1979) 'Preliminary Design of Composite Wings for Buckling Stress and Displacement Constraints', J. Aircraft, Vol. 16, 564-470.

17. GENESIS Users Manual (1991), VMA Engineering, Goleta, CA.

18. Felix, J. and Vanderplaats, G. N. (1987) 'Configuration Optimization of Trusses Subject to Strength, Displacement, and Frequency Constraints', J. of Mechanisms, Transmissions, and Automation in Design, Vol. 109, 233-241.

19. Lust, R. V. and Schmit, L. A. (1986) 'Alternative Approximation Concepts for Space Frame Synthesis', AIAA J., Vol. 24, No. 10, 1676-1684.

20. Woo, T. H. (1986) 'Space Frame Optimization Subject to Frequency Constraints', Proc. of the 27th AIAA/ASME/ASCE/AHS Structures, Structural Dynamics and Materials Conf., San Antonia, TX, May 19-21, 103-115.

21. Moore, G. J. and Vanderplaats, G. N. (1990) 'Improved Approximations for Shape Optimal Design of Shell Structures', Proc. of the 31st AIAA/ASME/ASCE/AHS/ACS Structures, Structural Dynamics and Materials Conf., Long Beach, CA, 161-170.

22. Grandhi, R. V., Venugopal, N. S., and Venkayya, V. B. (1991) 'Optimality Criteria Method for Minimum Weight Design of Plate Structures', Proc. of the 32nd AIAA/ASME/ASCE/AHS/ACS Structures, Structural Dynamics and Materials Conf., Baltimore, MD, 213-226.

Table 1 Final Designs for Ten Bar Truss

	Case A		Case B	
Member	GENESIS	Ref. 3	GENESIS	Ref. 3
1	7.92	7.90	30.64	30.67
2	0.10	0.10	0.10	0.10
3	8.12	8.10	24.04	23.76
4	3.91	3.90	14.71	14.59
5	0.10	0.10	0.10	0.10
6	0.10	0.10	0.10	0.10
7	5.81	5.80	8.70	8.58
8	5.53	5.52	20.90	21.07
9	3.68	3.68	20.89	20.96
10	0.14	0.14	0.10	0.10
Mass	1502	1498	5084	5077

Table 2 Final Designs for Eighteen Bar Truss

Design Variable	GENESIS	Ref. 14
A1	12.51	11.34
A2	17.76	19.28
A3	5.39	10.97
A4	3.75	5.30
X3	910.5	994.6
Y3	184.4	162.3
X5	641.9	747.4
Y5	146.0	102.9
X7	412.4	482.9
Y7	97.1	33.0
X9	201.4	221.7
Y9	30.1	17.1
Mass	4515	5713

Table 3 Final Designs for Portal Frame

Member	Design Variable	GENESIS	Ref. 15
1	B1	8.14	11.26
1	T1	0.35	0.41
1	H	84.32	78.21
1	TW	0.49	0.52
1	B2	10.00-	10.17
1	T2	0.32	0.46
2	B1	6.70	11.69
2	T1	0.79	0.42
2	H	100.00+	99.47+
2	TW	0.40	0.44
2	B2	11.93	10.94
2	T2	0.44	0.45
3	B1	5.02-	5.00-
3	T1	0.14	0.14
3	H	25.00-	25.00-
3	TW	0.10-	0.10-
3	B2	10.03-	10.00-
3	T2	0.27	0.28
Volume		80,348	84,058

\- Lower Bound

\+ Upper Bound

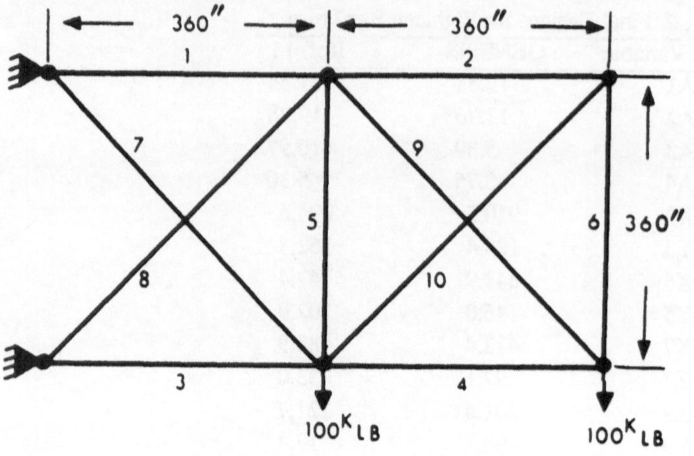

Figure 1 Ten Bar Truss

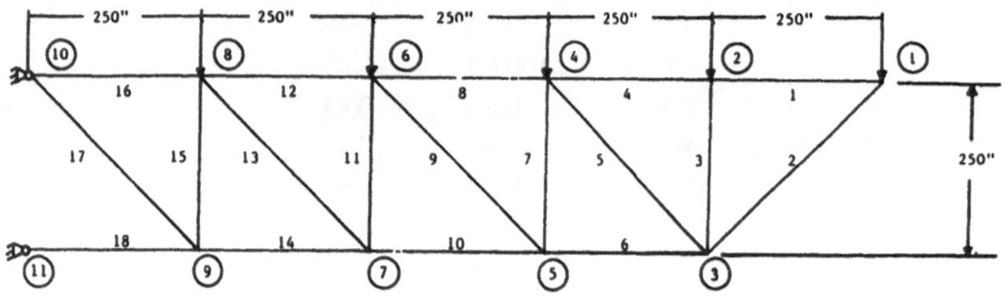

Figure 2 Eighteen Bar Truss

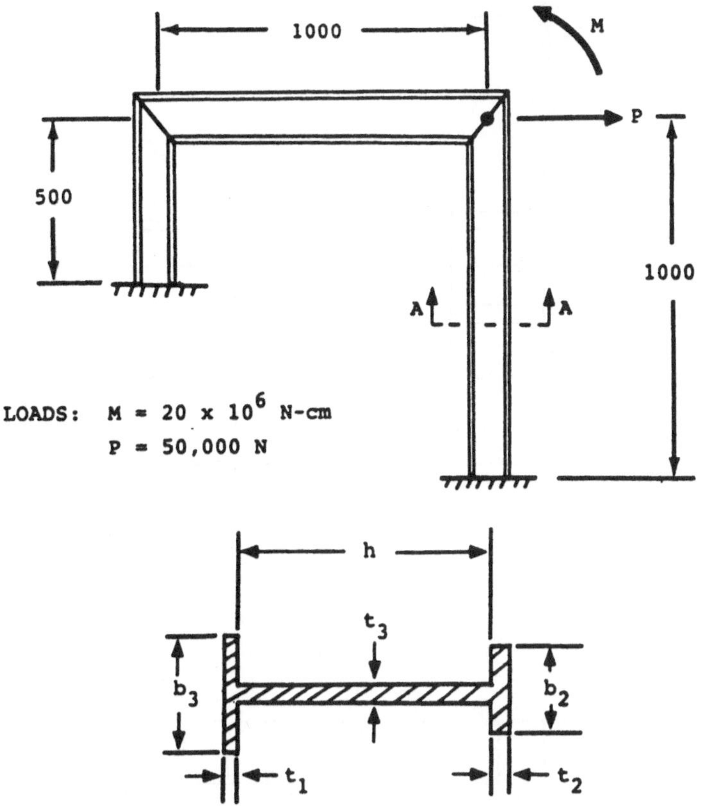

MATERIAL: Aluminum Alloy

$E = 7.06 \times 10^6$ N/cm^2

$\sigma_a^+ = 20{,}000$ N/cm^2

$\sigma_a^- = -12{,}000$ N/cm^2

Figure 3 Portal Frame

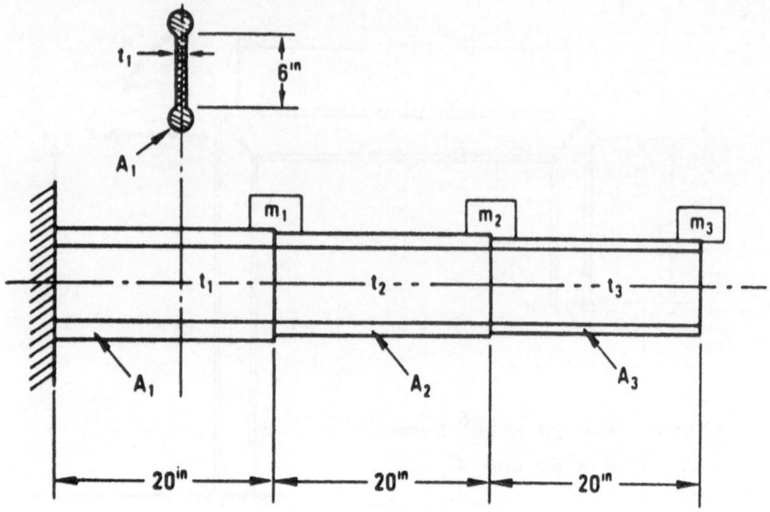

Figure 4 Cantilevered Beam

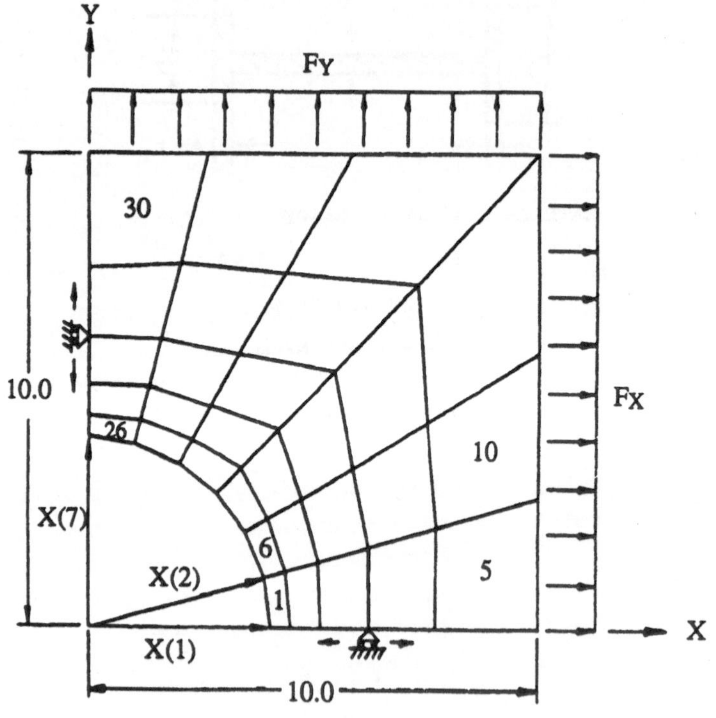

Figure 5. Plate with a Hole

APPROXIMATE MODELS FOR THE OPTIMIZATION OF LARGE STRUCTURAL SYSTEMS

URI KIRSCH
Department of Civil Engineering
Technion, Israel Institute of Technology
Haifa 32000, Israel

ABSTRACT. Approximate reanalysis methods, based on results of a single precise analysis, are discussed. The basic approximations of Taylor series, the Binomial series, and the reduced basis method are first presented. Combined approximations, intended to improve the quality of the results are then introduced. A general solution procedure, for improved approximate reanalysis of structures is proposed. It is shown that the quality of the approximations can greatly be improved by combining the computed terms of a series expansion, used as high quality basis vectors, and coefficients of a reduced basis expression. The latter coefficients can readily be determined by solving a reduced set of the analysis equations. The presented procedure is suitable for various types of design variables and can be used with a general finite element model. Numerical examples illustrate the effectiveness of the solution process. It is shown that high quality approximations can be obtained with a small computational effort for very large changes in the design variables

1. Introduction

1.1. PROBLEM STATEMENT

In most structural optimization problems the implicit behavior constraints must be evaluated for successive modifications in the design. For each trial design the analysis equations must be solved and the multiple repeated analyses usually involve extensive computational effort. Consequently, optimization of large scale structures might become prohibitive. This difficulty motivated several studies on explicit approximations of the structural behavior (i.e. displacements and stresses) in terms of the design variables [1-4].

In general, two conflicting factors should be considered in choosing an approximate behavior model for a specific optimal design problem:

(a) the accuracy of the calculations, or the quality of the approximation; and
(b) the computational effort involved, or the efficiency of the method.

Various approximate reanalysis models have been proposed in recent years. Two of the commonly used methods are: a) series expansion; and b) the reduced basis method.

First order Taylor series expansion is perhaps the most commonly used approximation in structural optimization, but other series expansions, such as the Binomial series, have also been used. Although series expansions can considerably reduce the amount of computations during reanalysis, the quality of the approximations might be insufficient.

G. I. N. Rozvany (ed.), Optimization of Large Structural Systems, Vol. I, 271–287.

Most of the approximate series expansions proposed in the past are valid only for relatively small changes in the design variables. For large changes in the design, the accuracy of the approximations is deteriorated and might become meaningless.

The reduced basis method is based on expressing the structural behavior as a linear combination of a reduced number of basis vectors. One problem in using this approach is that, in general, several precise analyses must be carried out to introduce the approximations. In addition, it is not always clear how to choose high quality basis vectors.

Various means have been proposed to improve the quality of the approximations. One of the early studies on structural optimization [5] showed that assuming the inverse (reciprocal) cross-sectional areas as design variables might considerably improve the results. Since then further studies confirmed this property and clarified some of the reasons for this phenomenon [1,6,7]. The inverse variables formulation can be viewed as a special case of the general approach of applying intervening variables [6,8,9]. The main problem in using intervening variables is that it might be difficult to select appropriate variables for cases of general optimization where geometrical or shape design variables are considered.

Another approach to improve the quality of the results is to scale the initial design such that the changes in the design variables are reduced [10,11]. It has been shown that the scaling operation is useful for various types of design variables and behavior functions.This approach has successfully been used for homogeneous functions [8]. Several criteria for selecting the scaling multiplier have been proposed. These include geometrical considerations [11] and mathematical criteria [10]. The concept of scaling has been extended recently to include not only the initial design but also a fictitious set of loads [12-14]. This approach has been found most effective for various reanalysis problems.

The approximate reanalysis methods discussed in this paper are based on results of a single precise analysis. The basic approximations of Taylor series, the Binomial series, and the reduced basis method are first presented. Combined approximations, intended to improve the quality of the results are then introduced. These include combined series expansion and scaling; combined first order approximations and two types of scaling; transformation of the two types of scaling into a reduced basis form; and a general approach based on combined series expansion and the reduced basis method.

A general solution procedure, for improved approximate reanalysis of structures is presented. It is shown that the quality of the approximations can greatly be improved by combining the computed terms of a series expansion, used as high quality basis vectors, with coefficients of a reduced basis expression. The latter coefficients can readily be determined by substituting the reduced basis expression into the analysis equations. An alternative criterion, based on minimizing the errors in satisfying the analysis equations is demonstrated. The two criteria are compared in terms of the quality of the approximations, and some computational considerations are discussed.

The presented procedure is suitable for various types of design variables and can be used with a general finite element model. Numerical examples illustrate the effectiveness of the solution process. It is shown that high quality approximations can be obtained with a small computational effort for very large changes in the design variables

1.2 MATHEMATICAL FORMULATION

The problem under consideration can be stated as follows: Given an initial design variables vector $\mathbf{X}^*$, the corresponding stiffness matrix $\mathbf{K}^*$, and the displacements $\mathbf{r}^*$, computed by the equilibrium equations

$$\mathbf{K}^* \mathbf{r}^* = \mathbf{R} \tag{1}$$

The elements of the load vector $\mathbf{R}$ are often assumed to be independent of the design variables and the stiffness matrix $\mathbf{K}^*$ is usually given from the initial analysis in the decomposed form

$$\mathbf{K}^* = \mathbf{U}^{*T}\mathbf{U}^* \tag{2}$$

where $\mathbf{U}^*$ is an upper triangular matrix. Assume a change $\Delta\mathbf{X}$ in the design variables so that the modified design is

$$\mathbf{X} = \mathbf{X}^* + \Delta\mathbf{X} \tag{3}$$

and the corresponding stiffness matrix is

$$\mathbf{K} = \mathbf{K}^* + \Delta\mathbf{K} \tag{4}$$

where $\Delta\mathbf{K}$ is the change in the stiffness matrix due to the change $\Delta\mathbf{X}$.

The object is to find efficient and high quality approximations of the modified displacements $\mathbf{r}$ due to various changes in the design variables $\Delta\mathbf{X}$, without solving the modified analysis equations

$$\mathbf{K}\,\mathbf{r} = (\mathbf{K}^* + \Delta\mathbf{K})\mathbf{r} = \mathbf{R} \tag{5}$$

The elements of the stiffness matrix are not restricted to certain forms and can be general functions of the design variables. That is, the design variables $\mathbf{X}$ may represent coordinates of joints, the structural shape, geometry, members cross-sections, etc.

Once the displacements are evaluated, the stresses can readily be determined by the stress-displacement relations. Thus, the presented approximations of $\mathbf{r}$ are intended only to replace the set of implicit equations (5).

2. Basic Approximations

2.1. SERIES EXPANSION

A common approach is to consider the first terms of a series expansion, to obtain the approximate displacements $\mathbf{r}_a$

$$\mathbf{r}_a = \mathbf{r}_1 + \mathbf{r}_2 + \mathbf{r}_3 + \ldots \tag{6}$$

Taylor series expansion is one of the most commonly used approximations in structural optimization. The first three terms, obtained by expanding $\mathbf{r}$ about $\mathbf{X}^*$, are given by

$$
\begin{aligned}
r_1 &= r^* \\
r_2 &= \nabla r_x^* \Delta X \\
r_{3j} &= 1/2\, \Delta X^T H_j^* \Delta X
\end{aligned} \tag{7}
$$

where the displacements $\mathbf{r}^*$, the matrix of first derivatives $\nabla \mathbf{r}_x^*$, and the matrices of second derivatives $\mathbf{H}_j^*$, are computed at $\mathbf{X}^*$. The scalar r_{3j} is the j-th component of the vector $\mathbf{r}_3$.

Since the calculation of high-order derivatives is usually not practicable, linear approximations are often used. These require evaluation of the first derivatives which can readily be calculated by sensitivity analysis methods . It should be noted, however, that the first-degree approximations may be insufficient in many cases and better models are needed. A possible approach is to consider only the diagonal elements of matrices $\mathbf{H}_j$, to obtain more accurate results. Neglecting the off-diagonal elements of matrices $\mathbf{H}_j$ will considerably reduce the computational effort for the second order approximations.

Alternative series approximations are obtained by rearranging Eq. (5) to read

$$\mathbf{K}^* \mathbf{r} = \mathbf{R} - \Delta\mathbf{K}\, \mathbf{r} \tag{8}$$

Writing this equation as the recurrence relation

$$\mathbf{K}^* \mathbf{r}^{(k+1)} = \mathbf{R} - \Delta\mathbf{K}\, \mathbf{r}^{(k)} \tag{9}$$

where $\mathbf{r}^{(k+1)}$ is the value of $\mathbf{r}$ after the k-th cycle, and assuming the initial value $\mathbf{r}^{(1)} = \mathbf{r}^*$, the following Binomial series expansion is obtained

$$\mathbf{r}_a = (\mathbf{I} - \mathbf{B} + \mathbf{B}^2 - \dots)\, \mathbf{r}^* \tag{10}$$

where $\mathbf{B}$ is defined by

$$\mathbf{B} \equiv \mathbf{K}^{*-1} \Delta\mathbf{K} \tag{11}$$

That is, the first three terms of the series are given by

$$
\begin{aligned}
\mathbf{r}_1 &= \mathbf{r}^* \\
\mathbf{r}_2 &= -\mathbf{B}\, \mathbf{r}^* \\
\mathbf{r}_3 &= \mathbf{B}^2\, \mathbf{r}^*
\end{aligned} \tag{12}
$$

Calculation of $\mathbf{r}_a$ by Eq. (10) involves only forward and backward substitutions if $\mathbf{K}^*$ is given in the decomposed form of Eq. (2). The calculation of $\mathbf{r}_2$, for example, is carried out from

$$\mathbf{K}^* \mathbf{r}_2 = -\Delta\mathbf{K}\, \mathbf{r}^* \tag{13}$$

We first solve for $\mathbf{t}$ by the forward substitution

$$\mathbf{U}^{*T}\mathbf{t} = -\Delta\mathbf{K}\ \mathbf{r}^{*} \tag{14}$$

$\mathbf{r}_2$ is then calculated by the backward substitution

$$\mathbf{U}^{*}\mathbf{r}_2 = \mathbf{t} \tag{15}$$

Similarly, $\mathbf{r}_3$ is calculated from

$$\mathbf{K}\mathbf{r}_3 = -\Delta\mathbf{K}\ \mathbf{r}_2 \tag{16}$$

Problems of slow convergence or divergence may be encountered in applying the series of Eq. (10) . The series converges if and only if

$$\lim_{k\to\infty} \mathbf{B}^{k} = \mathbf{0} \tag{17}$$

A sufficient criterion for the convergence of the series is that

$$\|\mathbf{B}\| \le 1 \tag{18}$$

where $\|\mathbf{B}\|$ is the norm of $\mathbf{B}$.

It has been shown that the terms of the Binomial series (Eq. (12)) are equivalent to those of Taylor series (Eq. (7)) for some homogeneous displacement functions.

Series expansions are based on information of a single design. As a result, the quality of the approximations might be sufficient only for limited changes. Several methods have been proposed to improve the series convergence. These include the Jacobi iteration, Block Gauss-Seidel iteration, dynamic acceleration methods and scaling of the initial design [4,11,15]. These means may considerably improve the results with a moderate computational effort. Since series expansions are based on analyzing the structure at a single point, they may be classified as ***local approximations***.

2.2. THE REDUCED BASIS METHOD

In this approach [16] it is assumed that the displacement vector $\mathbf{r}$ of a new design can be approximated by a linear combination of s linearly independent vectors $\mathbf{r}_1, \mathbf{r}_2,...,\mathbf{r}_s$ of previously analyzed designs (where s is assumed to be much smaller than the number of degrees of freedom m)

$$\mathbf{r}_a = y_1\mathbf{r}_1 + y_2\mathbf{r}_2 + ... + y_s\mathbf{r}_s \tag{19}$$

or in matrix form

$$\mathbf{r}_a = \mathbf{r}_B\ \mathbf{y} \tag{20}$$

where

$$\underset{m\times s}{\mathbf{r}_B} = \mathbf{r}_1,\ \mathbf{r}_2,\ \cdots,\mathbf{r}_s \qquad \mathbf{y} = \begin{Bmatrix} y_1 \\ y_2 \\ \vdots \\ y_s \end{Bmatrix} \tag{21}$$

y is a vector of coefficients to be determined. Substituting Eq. (20) into the modified analysis equations (5) and premultiplying by $\mathbf{r}_B^T$ yields

$$\underset{s\times m}{\mathbf{r}_B^T}\ \underset{m\times m}{\mathbf{K}}\ \underset{m\times s}{\mathbf{r}_B}\ \underset{s\times 1}{\mathbf{y}} = \underset{s\times m}{\mathbf{r}_B^T}\ \underset{m\times 1}{\mathbf{R}} \tag{22}$$

Introducing the notation

$$\mathbf{K}_R = \mathbf{r}_B^T\,\mathbf{K}\,\mathbf{r}_B \qquad\qquad \mathbf{R}_R = \mathbf{r}_B^T\,\mathbf{R} \tag{23}$$

and substituting into Eq. (22) we obtain

$$\underset{s\times s}{\mathbf{K}_R}\ \underset{s\times 1}{\mathbf{y}} = \underset{s\times 1}{\mathbf{R}_R} \tag{24}$$

For cases of $s << m$, the approximate displacement vector can be obtained by solving the smaller ($s\times s$) system in Eq. (24) for **y** instead of computing the exact solution by solving the large ($m\times m$) system in Eq. (5). $\mathbf{r}_a$ is then computed for the given **y** by Eq. (20).

A basic question in using the reduced basis method lies in the choice of an appropriate set of the linearly independent vectors $\mathbf{r}_1, \mathbf{r}_2, .., \mathbf{r}_s$ which span the design variables space. Displacement vectors of previously analyzed designs can be used, but it should be emphasized that an ad hoc or intuitive choice may not lead to satisfactory approximations. In addition, calculation of the basis vectors requires several precise analyses of the structure for the basis design points, which involve extensive computational effort. Since such an approach is based on analyzing the structure at a number of points, it may be classified as ***global approximations***.

3. Combined Approximations

3.1. COMBINED SERIES EXPANSION AND SCALING

It has been noted that the quality of the results obtained by series expansion might be insufficient. Poor approximations might be obtained for large changes in the design and the series might even diverge. Scaling of the initial design can greatly improve the quality of the approximations.

Scaling of the initial stiffness matrix $\mathbf{K}^*$ is defined by [11]

$$\mathbf{K} = \mu\mathbf{K}^* \tag{25}$$

where μ is a positive scalar multiplier. From Eqs. (1), (5) and (25) it is clear that the precise displacements after scaling can be calculated directly by

$$\mathbf{r} = \mu^{-1}\,\mathbf{r}^* \tag{26}$$

It should be noted that Eq. (25) does not require linear dependence of **K** on **X**. Furthermore, in many cases of general changes in **K**, the elements of $\mu\mathbf{K}^*$ do not correspond to an actual design. That is, the matrix **K** computed by Eq. (25) does not have the usual physical meaning. Scaling of the initial stiffness matrix (Eq. (25)) will improve the quality of the approximations, if the known modified displacements $\mu^{-1}\,\mathbf{r}^*$ (Eq. (26)) provide better initial data than the original displacements $\mathbf{r}^*$.

The modified stiffness matrix $\mathbf{K}$ (Eq. (4)) can be expressed in terms of μ by (see Fig.1)

$$\mathbf{K} = \mathbf{K}^* + \Delta\mathbf{K} = \mu\mathbf{K}^* + \Delta\mathbf{K}_\mu \tag{27}$$

That is, if an initial design $\mu\mathbf{K}^*$ is assumed instead of $\mathbf{K}^*$, the modified stiffness matrix $\mathbf{K}$ is expressed in terms of the corresponding changes in the matrix $\Delta\mathbf{K}_\mu$ instead of $\Delta\mathbf{K}$.

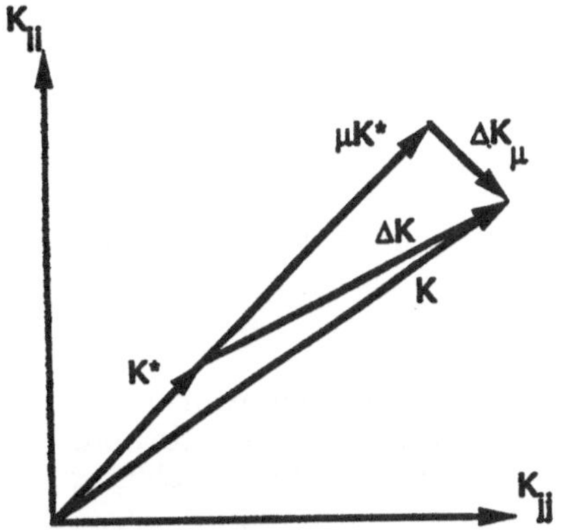

Figure 1. Scaling of the initial stiffness matrix

From Eq. (27), $\Delta\mathbf{K}_\mu$ is given by

$$\Delta\mathbf{K}_\mu = (1 - \mu)\,\mathbf{K}^* + \Delta\mathbf{K} \tag{28}$$

Consider the recurrence relation of Eq. (9), with $\mu\mathbf{K}^*$, $\Delta\mathbf{K}_\mu$ and $\mu^{-1}\,\mathbf{r}^*$ assumed as initial values instead of $\mathbf{K}^*$, $\Delta\mathbf{K}$ and $\mathbf{r}^*$, respectively. The resulting Binomial series (Eq. (10)) is

$$\mathbf{r}_a = \frac{1}{\mu}(\mathbf{I} - \mathbf{B}_\mu + \mathbf{B}_\mu^2 - \ldots)\mathbf{r}^* \tag{29}$$

where $\mathbf{B}_\mu$ is given by

$$\mathbf{B}_\mu = \frac{1-\mu}{\mu}\,\mathbf{I} + \frac{1}{\mu}\,\mathbf{B} \tag{30}$$

and $\mathbf{B}$ is defined by Eq. (11). For $\mu=1$ we find $\mathbf{B}_\mu=\mathbf{B}$ and Eq.(29) is reduced to Eq. (10).

Several criteria for selecting the value of μ have been proposed [10,11]. One possible criterion is to choose μ such that the distance between $\mathbf{X}^*$ and $\mathbf{X}$ is minimized. Another approach is to select μ such that the Euclidean norm of $\mathbf{B}_\mu$ is minimized. The drawback of using this criterion is that the elements of matrix $\mathbf{B}$ must be calculated. Since this operation involves much computational effort, we may use an alternative criterion which minimizes the Euclidean norm of the first term in the series (29), that is [10]

$$\|\mathbf{B}_\mu \mathbf{r}^*\| \rightarrow \min. \tag{31}$$

Differentiating with respect to μ and setting the result equal to zero, yields

$$\mu = \frac{a}{b} \tag{32}$$

where

$$a = \sum_{i=1}^{m} (r_{1i} - r_{2i})^2 \tag{33}$$

$$b = \sum_{i=1}^{m} (r_{1i}^2 - r_{1i} r_{2i}) \tag{34}$$

Here, r_{1i} are the given elements of $\mathbf{r}_1 = \mathbf{r}^*$, and r_{2i} are the elements of $\mathbf{r}_2$ that can readily be computed by Eq. (13).

3.2. FIRST ORDER APPROXIMATIONS

Due to efficiency considerations, first order series approximations are often used . If only two terms of the series (6) are considered, the following first order approximations of Taylor series (Eq. (7)) and the Binomial series (Eq. (12)) are obtained

$$\mathbf{r}_a = \mathbf{r}^* + \nabla \mathbf{r}_x^* \Delta \mathbf{X} \tag{35}$$

$$\mathbf{r}_a = (\mathbf{I} - \mathbf{B})\mathbf{r}^* \tag{36}$$

Equation (36) can also be introduced by substituting $\mathbf{r} = \mathbf{r}^* + \Delta\mathbf{r}$ in the right hand side of Eq. (8), giving

$$\mathbf{K}^*\mathbf{r} = \mathbf{R} - \Delta\mathbf{K}\,\mathbf{r}^* - \Delta\mathbf{K}\,\Delta\mathbf{r} \tag{37}$$

Neglecting the second order term $\Delta\mathbf{K}\,\Delta\mathbf{r}$, premultiplying by $\mathbf{K}^{*-1}$ and substituting Eqs. (1) and (11) gives Eq. (36).

It will be shown now that by combining two types of scaling:

a) scaling of the initial stiffness matrix $\mathbf{K}^*$; and
b) scaling of a fictitious set of loads $\mathbf{R}_a$;

the first order Binomial series approximations of Eq. (36) can be transformed into a reduced basis form . A similar procedure can be used for Taylor series approximations.

Based on Eqs. (25), (26) and (28), it is possible to assume $\mu\mathbf{K}^*$ and $\mu^{-1}\mathbf{r}^*$ as initial values instead of $\mathbf{K}^*$ and $\mathbf{r}^*$, respectively, and the corresponding changes $\Delta\mathbf{K}_\mu$ instead of $\Delta\mathbf{K}$. Substituting into Eq. (36) yields

$$\mathbf{r}_a = \mu^{-2}(2\mu - 1)\mathbf{r}^* - \mu^{-2}\mathbf{B}\mathbf{r}^* = \mu^{-2}(2\mu - 1)\mathbf{r}_1 + \mu^{-2}\mathbf{r}_2 \tag{38}$$

where $\mathbf{r}_1$ and $\mathbf{r}_2$ are defined by Eq. (12). It can be noted that for $\mu = 1$ Eq. (36) is obtained.

It has been shown that μ can be selected by the criterion of Eq. (32). The scaling of loads presented here is intended to introduce more effective approximations. Define a fictitious load vector $\mathbf{R_a}$ [15]

$$\mathbf{R_a} = (\mathbf{K}^* + \Delta\mathbf{K})\, \mathbf{r_a} = \mathbf{K}\, \mathbf{r_a} \tag{39}$$

Note that $\mathbf{r_a}$ are precise displacements for the stiffness matrix $\mathbf{K}=\mathbf{K}^*+\Delta\mathbf{K}$ and the fictitious load vector $\mathbf{R_a}$. The fictitious loads can be scaled by

$$\mathbf{R_s} = \Omega \mathbf{R_a} \tag{40}$$

where Ω is a scalar. The precise displacements $\mathbf{r_s}$ corresponding to the modified stiffness matrix $\mathbf{K}$ and the scaled fictitious loads $\mathbf{R_s}$ are therefore given by

$$\mathbf{r_s} = \Omega \mathbf{r_a} \tag{41}$$

Evaluating $\mathbf{r_a}$ for any given μ by Eq. (38), the resulting $\mathbf{R_a}$ can readily be calculated by Eq. (39). The latter fictitious loads can then be scaled by Eq. (40) such that the final displacements $\Omega\mathbf{r_a}$ (Eq. (41) are improved. Substituting Eq. (38) into Eq. (41) yields

$$\mathbf{r_s} = \Omega[\mu^{-2}(2\mu-1)\mathbf{r}_1 + \mu^{-2}\mathbf{r}_2] \tag{42}$$

That is, each evaluation of the displacements can be viewed as the following two steps:

a) Selecting μ-scaling of the initial stiffness matrix $\mathbf{K}^*$ and evaluation of the approximate displacements $\mathbf{r_a}$ for the given loads $\mathbf{R}$.
b) Selecting Ω-scaling of the fictitious loads $\mathbf{R_a}$ and the corresponding displacements $\mathbf{r_a}$ for the given modified design $\mathbf{K}$.

Assuming the transformation

$$\begin{aligned} y_1 &= \Omega[\mu^{-2}(2\mu-1)] \\ y_2 &= \Omega\mu^{-2} \end{aligned} \tag{43}$$

and substituting Eqs. (43) into Eq. (42) gives

$$\mathbf{r_a} = y_1\, \mathbf{r}_1 + y_2\, \mathbf{r}_2 = \mathbf{r_B}\, \mathbf{y} \tag{44}$$

where

$$\mathbf{r_B} = \{\mathbf{r}_1\, ,\, \mathbf{r}_2\} \tag{45}$$

$$\mathbf{y}^T = \{y_1\, ,\, y_2\} \tag{46}$$

That is, Eq. (42) which is based on combining the two types of scaling is equivalent to the two-terms expression of the reduced basis method (Eq. (44)). It is instructive to note that μ and Ω are determined uniquely for any assumed $\mathbf{y}$ by Eq. (43).

The combination of first order series expansion and the reduced basis method can be generalized to any number of terms in the series, as will be shown in the next section.

3.3. COMBINED SERIES EXPANSION AND REDUCED BASIS

The drawbacks of series expansion and the reduced basis method motivated combination of the two approaches to achieve an improved solution procedure [12-14]. In this procedure, the computed terms of a series expansion are used as high quality basis vectors in a reduced basis expression. The advantage is that the efficiency of local (series expansion) approximations and the improved quality of global (reduced basis) approximations are combined to obtain an effective solution procedure. The solution process involves the following steps:

a) The modified stiffness matrix **K** is introduced.
b) The vectors $\mathbf{r}_i$ of a series expansion (i.e. Eqs. (7) or Eqs. (12)) are calculated, and matrix $\mathbf{r}_B$ is introduced. To maintain efficiency of the calculations, only two or three basis vectors might be considered.
c) The elements of $\mathbf{K}_R$ and $\mathbf{R}_R$ (Eqs. (23)) are determined.
d) The coefficients **y** are calculated by solving the set of (2×2 or 3×3) equations (24).
e) The displacements are evaluated by Eq. (20).

To evaluate the quality of the results, we substitute the approximate displacements expression (Eq. (20)) into Eq. (39) to obtain the errors in the modified equations (5)

$$\Delta\mathbf{R}(\mathbf{y}) = \mathbf{R}_a - \mathbf{R} = \mathbf{K}\mathbf{r}_B\mathbf{y} - \mathbf{R} \tag{47}$$

It has been noted that if $\mathbf{r}_a$ is the vector of the precise displacements, then $\Delta\mathbf{R} = \mathbf{0}$. Thus, $\Delta\mathbf{R}$ can be used to evaluate the quality of the approximations. Define the common measure of smallness of $\Delta\mathbf{R}(\mathbf{y})$ by the quadratic form

$$q(\mathbf{y}) = \Delta\mathbf{R}^T\Delta\mathbf{R} \tag{48}$$

Substituting Eq. (47) into Eq. (48) , differentiating with respect to **y** and setting the result equal to zero, we obtain a set of linear equations in the form of Eq. (24)

$$\mathbf{a}\ \mathbf{y} = \mathbf{b} \tag{49}$$

where **a** and **b** are given by

$$\mathbf{a} = (\mathbf{K}\mathbf{r}_B)^T(\mathbf{K}\mathbf{r}_B) \qquad \mathbf{b} = (\mathbf{K}\mathbf{r}_B)^T\mathbf{R} \tag{50}$$

This alternative criterion for determining **y** can be used instead of Eq. (23) in step c. The two criteria have been compared by several numerical examples [12]. It has been found that although the method of Eq. (50) provides smaller q(y) values, better results might be obtained by the method of Eq. (23).

3.4. COMPUTATIONAL CONSIDERATIONS

The quality of the results and the efficiency of the calculations are two conflicting factors that should be considered in selecting an approximate reanalysis model. That is, better approximations are often achieved at the expense of more computational effort. In this section, some computational considerations associated with the presented approximate models are discussed. Consider first the two methods of calculating the basis vectors, Taylor series and the Binomial series.

Assuming the common first order Taylor series expansion, once the matrix $\nabla \mathbf{r}_x^*$ is available each redesign involves only calculation of the product $\nabla \mathbf{r}_x^* \Delta \mathbf{X}$. This is probably the most efficient reanalysis model. However, it has been noted that the quality of the results might be insufficient for large changes in the design variables. Second order Taylor series expansion is usually not practicable due to the large computational effort involved in calculation of the second order derivative matrices $\mathbf{H}_j$. An exception to this is the common case of homogeneous displacement functions, where Taylor series and the Binomial series become equivalent.

The advantage of using the Binomial series is that, unlike Taylor series, calculation of derivatives is not required. This makes the method more attractive in general applications where derivatives are not available. Calculation of each term of the Binomial series involves only forward and backward substitutions, if $\mathbf{K}^*$ is given in the decomposed form of Eq. (2). Thus, the second order terms can readily be calculated. In the case of first order approximations, calculation of only $\mathbf{Br}^*$ must be repeated for each trial design. This requires calculation of a single vector by forward and backward substitutions. Moreover, in the common case of homogeneous displacement functions, once the series coefficients are available, each redesign involves only calculation of simple products [11]. However, the quality of the results obtained even by second order approximations of both Taylor series and the Binomial series might be insufficient in cases of large changes in the design variables.

As to the selection of the number of basis vectors to be considered in the combined solution procedure, it has been noted that, in general, second order approximations (three basis vectors) provide better results than first order approximations (two basis vectors). However, in cases of poor selections of the first order vectors, the second order terms will not significantly improve the results.

To evaluate the computational effort involved in the combined solution procedure, compared with conventional basic methods, assume first order approximations. The basic series approximations require only calculation of the second basis vector (Eq. (7) or Eq. (12)). The combined procedure requires, in addition, introduction of the modified stiffness matrix $\mathbf{K}$, calculation of the products given by Eqs. (23) or Eqs. (50), determination of $\mathbf{y}$ by solving the set of 2×2 equations (24) or (49) and multiplying the two basis vectors by $\mathbf{y}$. Certainly, these operations increase the computational cost. However, the result is often considerably better approximations, particularly in cases of large changes in the design variables. That is, high quality approximations can be obtained in cases where the basic series approximations provide meaningless results. Consequently precise analyses, which involve more computational effort, are not required in cases where the basic approximations provide insufficient results.

Finally, the basic approximations may be viewed as a special case of the combined procedure where $\mathbf{y} = \mathbf{1.0}$ is selected. An additional advantage of the combined procedure is that, unlike the basic approximations, the errors involved in the approximations can be evaluated by $\Delta\mathbf{R}$ and q.

4. Examples

4.1. THIRTEEN-BAR TRUSS

Consider the thirteen-bar truss with the initial geometry and loading shown in Fig. 2a. The modulus of elasticity is 10000, the initial cross-sections are $\mathbf{X}^* = 1.0$, and the unknown displacements are r_h and r_v in joints B, C, D, F, G and H, respectively.
Assume first the modified cross-sectional areas $\mathbf{X}^T = \{10,10,10,10,8,8,8,8,6,6,6,6,6\}$. Results obtained for these large changes in cross-sections (up to 900%) by various approximate methods are shown in Table 1. It can be observed that:

- Results obtained by the basic first and second order series expansion (Eq. (7) or Eq. (12), which are equivalent in this case) are meaningless. That is, the series diverges due to the large changes in the design.
- Very good results have been obtained by first and second order approximations for both methods of Eq. (23) and Eq. (50). The high quality of the results can be explained by the scaling procedure. That is, the modified design is relatively close to the design line through the initial design $\mathbf{K}^*$. Therefore, the large changes in the design variables do not affect the quality of the results.

Consider a single geometric variable Y representing the span (Fig. 2b), the optimal design for this example [10], Y = 153, and the modified cross-sectional areas **X** as in the previous case. Results obtained for these large changes in both geometry (155% in Y) and cross-sections (up to 900%) by the proposed approximations are shown in Table 2. Improved results have been obtained for second order approximations (three terms) by both methods of Eqs. (23) and (50). Specifically, assuming three terms for the method of Eq. (23) the errors in most displacements are less than 5%.

TABLE 1. First and second order approximations, thirteen-bar truss, modified cross-sections.

Displace.	Eq. (7) or (12)		Eq. (50)		Eq. (23)		Precise
Number	2 Terms	3 Terms	2 Terms	3 Terms	2 Terms	3 Terms	Method
1	-3.90	35.7	0.051	0.048	0.048	0.048	0.048
2	-2.12	19.3	0.028	0.026	0.026	0.026	0.026
3	-11.85	103.8	0.178	0.167	0.168	0.167	0.167
4	-3.1	26.3	0.050	0.047	0.047	0.047	0.047
5	-20.7	177.2	0.338	0.319	0.320	0.319	0.319
6	-3.3	27.4	0.058	0.055	0.055	0.055	0.055
7	-4.0	36.5	0.054	0.051	0.050	0.051	0.051
8	2.1	-19.6	-.029	-.027	-.027	-.027	-.027
9	-11.7	102.5	0.173	0.162	0.162	0.162	0.162
10	3.1	-26.4	-.050	-.046	-.047	-.047	-.047
11	-21.0	178.4	0.348	0.329	0.329	0.329	0.329
12	3.3	-27.5	-.059	-.056	-.056	-.056	-.056

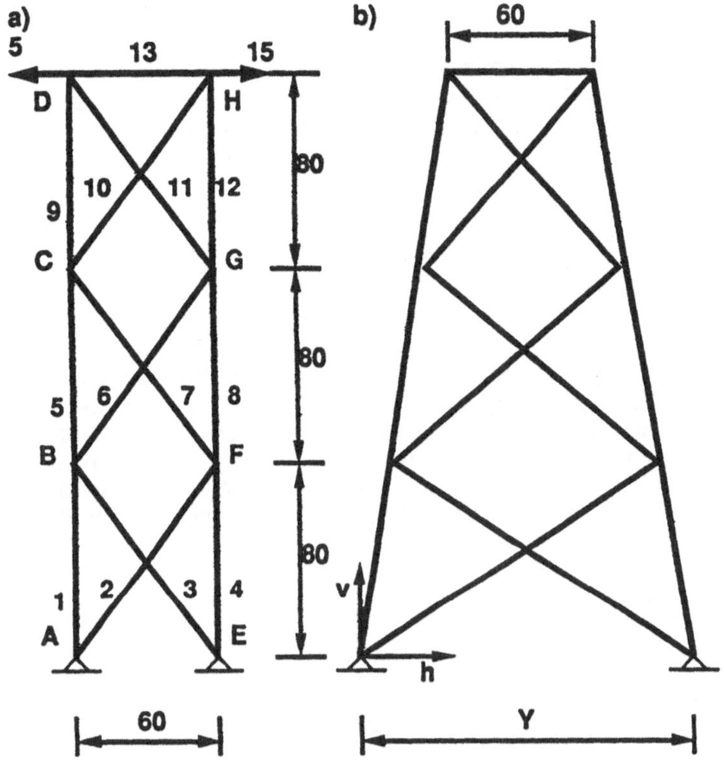

Figure 2. Thirteen-bar truss .

TABLE 2. First and second order approximations, thirteen-bar truss, modified geometry

Displa. Number	Precise Method	Eq. (50) 2 Terms	Eq. (50) 3 Terms	Error (%)	Eq. (23) 2 Terms	Eq. (23) 3 Terms	Error (%)
1	0.0098	0.0043	0.0105	7	0.0089	0.0112	14
2	0.0101	0.0059	0.0091	10	0.0084	0.0099	2
3	0.0400	0.0230	0.0364	9	0.0378	0.0396	1
4	0.0162	0.0105	0.0142	12	0.0146	0.0155	4
5	0.0811	0.0518	0.0726	10	0.0788	0.0795	2
6	0.0134	0.0095	0.0114	15	0.0040	0.0127	5
7	0.0114	0.0061	0.0108	5	0.0108	0.0117	3
8	-.0101	-.0063	-.0089	12	-.0088	-.0096	5
9	0.0360	0.0185	0.0339	6	0.0330	0.0367	2
10	-.0166	-.0104	-.0141	15	-.0145	-.0154	7
11	0.0902	0.0603	0.0817	9	0.0879	0.0895	1
12	-.0133	-.0102	-.0120	10	-.0147	-.0133	0

4.2. FIFTY-BAR TRUSS

Consider the fifty-bar truss with the geometry and loading shown in Fig. 3. The assumed modulus of elasticity is 1.0, the initial cross-sectional areas are given by $\mathbf{X}^*=1.0$, and the initial geometry is given by $X_H=X_V=1.0$. The following three cases of modified geometry have been considered:

Case A: $X_H = 1.0$ $X_V = 1.2$
Case B: $X_H = 2.0$ $X_V = 1.9$
Case C: $X_H = 2.0$ $X_V = 2.0$

Results obtained by first order approximations of Eq. (23) are summarized in Table 3. The two rows given for each joint are related to the horizontal displacement and the vertical displacement, respectively. Similar results have been obtained for the bottom chord joints. In case A, relatively small errors have been obtained for the left hand joints horizontal displacements, and right hand joints vertical displacements. In case B, much better results have been obtained for significantly larger change in the geometry. This can be explained by scaling of the geometry. That is, the modified geometry is relatively close to the design line through the initial geometry. Indeed, scaling of the initial geometry will give the precise solution by the proposed method (case C).

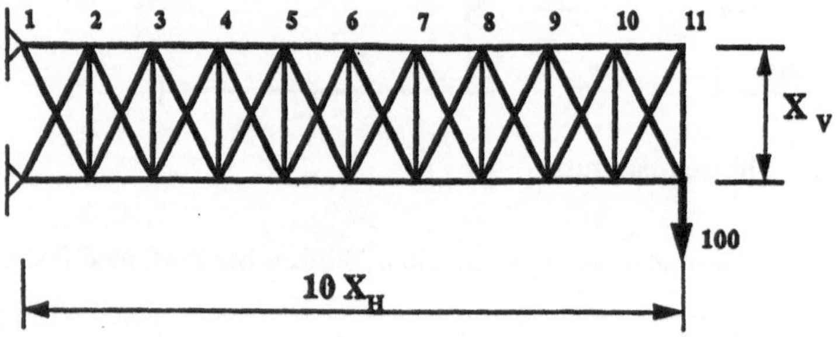

Figure 3. Fifty-bar truss

TABLE 3. Results, fifty-bar truss, modified geometries

Joint	Case A			Case B			Case C
	Precise	Appr.	Error (%)	Precise	Appr.	Error (%)	**
2	792	890	12	2000	2037	2	1900
	792	1064	34	2396	2471	3	2184
3	1500	1600	7	3790	3839	1	3600
	2834	3473	22	8781	8969	2	7966
4	2125	2161	2	5369	5414	1	5100
	5987	6926	16	18712	19013	2	16950
5	2667	2596	3	6737	6767	*	6400
	10112	11175	10	31746	32131	1	28734
6	3125	2927	6	7895	7903	*	7500
	15071	16020	6	47439	47864	1	42920
7	3500	3172	9	8842	8827	*	8400
	20724	21301	3	65348	65762	1	59104
8	3792	3348	12	9579	9542	*	9100
	26933	26889	*	85029	85382	*	76888
9	4000	3465	13	10105	10050	*	9600
	33558	32683	3	106040	106290	*	95874
10	4126	3533	14	10422	10356	*	9902
	40464	38613	5	127940	128050	*	115664
11	4162	3543	15	10517	10446	*	9992
	47477	44537	6	150240	150190	*	135800

* - Error less than 1%.
** - Approximate solution = precise solution.

5. Conclusions

Approximations of the structural behavior in terms of the design variables are essential in optimization of large scale structures, where the time consuming analysis must be repeated many times. A major problem is that the quality of the commonly used approximations might be insufficient, particularly in cases of large changes in the design.

A general solution procedure for effective approximations is presented in this article. The quality of displacement approximations is greatly improved by combining the computed terms of a series expansion, used as high quality basis vectors, and the coefficients of a reduced basis expression. It is shown that the latter coefficients can readily be determined such that a reduced set of the analysis equations are satisfied. An alternative criterion for selecting these coefficients, based on minimizing the errors in satisfying the analysis equations, is demonstrated. The two criteria provide similar results.

It is shown that first order approximations for the proposed procedure can be introduced by combining scaling of the initial stiffness matrix and scaling of a fictitious set of loads. Integrating these two types of scaling, the approximate displacements can be expressed in a reduced basis form as functions of two coefficients. The accuracy of the presented first order approximations is often sufficient and calculation of higher order terms is not necessary.

The presented approach is general and can be applied with different types of design variables (i.e. geometrical variables, cross-sectional variables etc.). It can be used with various finite element programs considering different versions, such as:
-Various selections of the basis vectors (i.e. Eq. (7) or Eq. (12)).
-Various criteria for selecting the parameters **y** (i.e. Eq. (23) or Eq. (50)).

The computational effort involved in the proposed procedure is larger, compared with conventional Taylor series or Binomial series approximations. However, the result is considerably better approximations, particularly in cases of large changes in the design variables. Consequently, precise analyses which involve more computational effort are not required in cases where conventional approximations provide insufficient results. It has been noted that conventional approximations may be viewed as a special case of the proposed procedure where **y** =**1.0** is selected. An additional advantage of the presented procedure is that, unlike conventional approximations, the errors involved in the approximations can be evaluated by $\Delta\mathbf{R}$ and q.

Several simple examples illustrate the effectiveness of the procedure. Specifically, the following observations have been made:
- Good results have been obtained for very large changes in the design with a relatively small computational effort.
- Similar results have been obtained by the two criteria of Eqs. (23) and (50). This means that the proposed basis vectors are most effective for various criteria of selecting **y**.

In summary, the presented solution procedure is a powerful tool to achieve efficient and high quality approximations. It also provides insight and better understanding of the behavior of structural models.

Acknowledgements

The author is indebted to the Ministry of Construction and Housing in Israel for supporting this work.

References

1. Schmit, L. A. and Farshi, B.(1974) 'Some approximation concepts for structural synthesis', AIAA J., 11, 489-494.
2. Schmit, L. A. (1984) 'Structural optimization, some key ideas and insights' in E. Atrek. et al (eds.), New directions in optimum structural design, John Wiley & Sons, New York .
3. Abu Kassim, A.M. and Topping, B.H.V. (1987) 'Static reanalysis of structures: A review', J. Struct. Eng. ASCE, 113, 1029-1045.
4. Kirsch, U. (1989) 'Approximate models for structural optimization', presented in NATO ASI on Optimization and decision support systems in civil engineering, Edinburgh, U.K. .
5. Reinschmidt, K. F. , Cornell, A. C. and Brotchie, J. F. (1966) 'Iterative design and structural optimization', J. Struct, Div. ASCE, 92, 281-318.
6. Haftka, R. T. and Kamat, M. P. (1985) Elements of structural optimization, Martinus Nijhoff, The Hague.
7. Fuchs, M. B. (1980) 'Linearized homogeneous constraints in structural design', Int. J. Mech. Scien. 22, 333-40.

8. Hjali, R.M. and Fuchs, M.B. (1989) 'Generalized approximations of homogeneous constraints in optimal structural design', in C.A. Brebbia and S. Hernandez (eds.), Computer aided optimum design of structures, Springer-Verlag,Berlin .
9. Prasad, B. (1983) 'Explicit constraint approximation forms in structural optimization', Comp. meth. appl. mech. engrg., 40, 1-26.
10. Kirsch, U. and Toledano, G. (1983) 'Approximate reanalysis for modifications of structural geometry', Computers and structures, 16, 269-279.
11. Kirsch, U. (1984) 'Approximate behavior models for optimum structural design', in E. Atrek, et al (eds.), New directions in optimum structural design, John Wiley &Sons, NY.
12. Kirsch, U. (1991) 'Reduced basis approximations of structural displacements for optimal design', to be published AIAA Journal,.
13. Kirsch, U. (1990) 'Improved Approximations of Displacements for Structural Optimization'; presented at the Third Air Force / NASA Symposium on Recent Advances in Multidisciplinary Analysis and Optimization San Francisco California.
14. Kirsch, U. and Fisenberger, M. (1991) 'Approximate interactive design of large structures', Proc. of the Int. Conf. on Computational structures technology, Edinburgh, U.K.
15. Kirsch, U. (1981) Optimum structural design, McGraw-Hill, New York,.
16. Fox, R. L. and Miura, H. (1971) 'An Approximate analysis technique for design calculations', AIAA J., 9, 177-179.

Sensitivity of Discrete Systems

Raphael T. Haftka
Virginia Polytechnic Institute and State University
Blacksburg, Virginia
and

Howard M. Adelman
NASA Langley Research Center
Hampton, Virginia

Abstract: Sensitivity techniques for structures discretized by finite element analysis are described. In particular, static response, eigenvalue problems and linear transient response are covered.

Introduction

Sensitivity analysis is emerging as a fruitful area of engineering research. The reason for this interest is the recognition of the variety of uses for sensitivity derivatives. In its early stages, sensitivity analysis found its predominant use in assessing the effect of varying parameters in mathematical models of control systems; see, for example, the texts of Tomovic,[1] Brayton and Spence,[2] Frank,[3] and Radanovic[4] for discussions of the early development of sensitivity theory. Interest in optimal control in the early 1960s (see, for example, Ref. 5) and automated structural optimization (see for example, Ref. 6) led to the use of gradient-based mathematical programming methods in which derivatives were used to find search directions toward optimum solutions. More recently, there has been strong interest in promoting systematic structural optimization as a useful tool for the practicing structural design engineer on large problems—a process still under way. Early attempts to use formal optimization for large structural systems resulted in excessively long and expensive computer runs. Examination of the optimization procedures indicated that the predominant contributor to the cost and time was the calculation of derivatives. As a consequence, emerging interest in sensitivity analysis has emphasized efficient computational procedures. In addition, researchers have developed and applied sensitivity analysis for approximate analysis, analytical model improvement, and assessment of design trends—so that structural sensitivity analysis has become more than a utility for optimization and is a versatile design tool in its own right. Most recently, researchers in disciplines such as physiology,[7] thermodynamics,[8] physical chemistry,[9] and aerodynamics,[10–12] have been using sensitivity methodology to assess the effects of parameter variations in their analytical models and to create designs insensitive to parameter variation.[13,14]

Derivatives of structural response can be calculated analytically at three stages. We can differentiate the continuum equations defining the response of the structure. We can differentiate the equations obtained when the continuum equations are discretized which is the topic of the present chapter. Finally, we can differentiate directly the computer program used to solve the structural response, such as a finite element computer program. This third approach is not discussed in this textbook, but the interested reader is referred to Refs 15 and 16. Analytical derivative calculations typically entail a substantial effort of analysis and software development. In many cases it is better to use derivatives obtained from a finite difference approximation. These notes therefore start with the discussion of the calculation of derivatives by finite differences.

G. I. N. Rozvany (ed.), Optimization of Large Structural Systems, Vol. I, 289–311.

Finite Difference Sensitivities

Truncation and Condition Errors

The simplest finite difference approximation is the first-order forward-difference approximation. Given a function $g(v)$ of a design variable v, the forward-difference approximation $\Delta g/\Delta v$ to the derivative dg/dv is given as

$$\frac{\Delta g}{\Delta v} = \frac{g(v+\Delta v) - g(v)}{\Delta v} \tag{1}$$

Another commonly used finite-difference approximation is the second-order central-difference approximation

$$\frac{\Delta g}{\Delta v} = \frac{g(v+\Delta v) - g(v-\Delta v)}{2\Delta v} \tag{2}$$

Higher-order finite-difference approximations are available but rarely used in structural optimization applications because of the associated high computational cost. If we need to find the derivatives of the structural response with respect to n design variables, the forward-difference approximation requires n additional analyses, the central-difference approximation $2n$ additional analyses, and higher order approximations are even more expensive.

The key to the selection of the approximation and the step size Δx is an estimate of the required accuracy. This topic is discussed in Ref. 17, and is summarized next.

Whenever finite-difference formulae are used to approximate derivatives, there are two sources of error: truncation and condition errors. The *truncation error* $e_T(\Delta v)$ is a result of the neglected terms in the Taylor series expansion of the perturbed function. For example, the Taylor series expansion of $g(v+\Delta v)$ can be written as

$$g(v+\Delta v) = g(v) + \Delta v \frac{\Delta g}{\Delta v}(v) + \frac{(\Delta v)^2}{2}\frac{d^2 g}{dv^2}(v+\zeta\Delta v), \qquad 0 \le \zeta \le 1 \tag{3}$$

From Eq. (3) it follows that the truncation error for the forward-difference approximation is

$$e_T(\Delta v) = \frac{\Delta v}{2}\frac{d^2 g}{dv^2}(v+\zeta\Delta v) \qquad 0 \le \zeta \le 1 \tag{4}$$

Similarly, by including one more term in the Taylor series expansion we get that the truncation error for the central difference approximation is

$$e_T(\Delta v) = \frac{\Delta v^2}{6}\frac{d^3 g}{dv^3}(v+\zeta\Delta v) \qquad 0 \le \zeta \le 1 \tag{5}$$

The *condition error* is the difference between the numerical evaluation of the function and its exact value. One contribution to the condition error is round-off error in calculating dg/dv from the original and perturbed values of g. This contribution is comparatively small for most computers unless Δv is extremely small. However if $g(x)$ is computed by a lengthy or ill-conditioned numerical process, the round-off contribution to the condition error can be substantial. Additionally, condition errors may result if $g(x)$ is calculated by an iterative process which is terminated early. If we have a bound ϵ_g on the absolute error in the computed function g, we can estimate the condition error. For example, for the forward-difference approximation the condition error $e_C(\Delta v)$ is (very!) conservatively estimated from Eq. (1) as

$$e_C(\Delta v) = \frac{2}{\Delta v}\epsilon_g \tag{6}$$

Equations (4) and (6) present us with the so called "step-size dilemma." If we select the step size to be small, so as to reduce the truncation error, we may have an excessive condition error. In some cases there may not be any step size which results in acceptable error!

A bound e on the total error, the sum of the truncation and condition errors, for the forward-difference approximation is obtained from Eqs. (4) and (6) as

$$e = \frac{\Delta v}{2}|s_b| + \frac{2}{\Delta v}\epsilon_g \tag{7}$$

where s_b is a bound on the second derivative in the interval $[v, v + \Delta v]$. When ϵ_g and s_b are available it is possible to calculate an optimum step-size that minimizes e as

$$\Delta v_{opt} = 2\sqrt{\frac{\epsilon_g}{|s_b|}} \tag{8}$$

Procedures for estimating s_b and ϵ_u are given in Refs. 17 and 18.

Iteratively solved Problems

Condition errors can become important when iterative methods are used for performing some of the calculations. Consider a simple example of a single displacement component u which is obtained by solving a nonlinear algebraic equation which depends on one design variable v

$$f(v, u) = 0 \tag{9}$$

The solution of Eq. (9) is obtained by an iterative process which starts with some initial guess of u and terminates when the iterant $\tilde{u}$ is estimated to be within some tolerance ϵ of the exact u (Note that ϵ is a bound on the condition error in u). To calculate the derivative du/dv assume that we use the forward-difference approximation. That is, we perturb v by Δv and solve Eq. (9) for u_Δ

$$f(v + \Delta v, u_\Delta) = 0 \tag{10}$$

The iterative solution of Eq.(10) yields an approximation $\tilde{u}_\Delta$, and then du/dv is approximated as

$$\frac{du}{dv} \approx \frac{\tilde{u}_\Delta - \tilde{u}}{\Delta v} \tag{11}$$

To start the iterative process for obtaining u_Δ, two initial guesses come to mind. The first is to start with the same initial guess that was used to solve for u. If the convergence of the iterative process is monotonic there is a good chance that when we use Eq. (11) the errors in $\tilde{u}$ and $\tilde{u}_\Delta$ will almost cancel out, and we will get a very small condition error. The other logical initial guess for u_Δ is $\tilde{u}$. This initial guess is known to be good because Δx is typically small, and so we may get fast convergence. Unfortunately, this time we cannot expect the condition errors to cancel. As we iterate on $\tilde{u}_\Delta$, the original error (the difference between u and $\tilde{u}$) will be reduced at the same time that the change due to Δx is taking effect (consider, for example, what happens if Δx is set to zero, or an extremely small number).

Reference 19 suggests a strategy which allows us to start the iteration for u_Δ from $\tilde{u}$ without worry of excessive condition errors. The approach is to pretend that $\tilde{u}$ is the exact rather than approximate solution by changing the problem that we want to solve. Indeed, $\tilde{u}$ is the exact solution of

$$f(v,u) - f(v,\tilde{u}) = 0 \tag{12}$$

which is only slightly different from our original problem (because $f(v,\tilde{u})$ is almost zero). We now find the derivative du/dv from Eq.(12), by obtaining u_Δ as the solution of

$$f(v+\Delta v, u_\Delta) - f(v,\tilde{u}) = 0 \tag{13}$$

Because $\tilde{u}$ is the exact solution of this equation for $\Delta v = 0$ the iterative process will only reflect the effect of Δv, and we will obtain a good approximation from Eq. (11).

Because of the high cost and the accuracy problems associated with finite-difference derivatives, there has been much effort put into developing analytical derivative approximations. The rest of these notes is devoted to such analytical expressions for sensitivity derivatives.

Sensitivity of Static Response

First Derivatives of Linear Response

This section of the notes focuses on the calculation of first derivatives of static linear structural response (displacements and stresses) computed from finite element models. The governing equation for displacement is

$$KU = F \tag{14}$$

where K is the symmetric stiffness matrix of order nn, U the vector of displacements, and F the vector of applied forces. Both K and F are, in general, functions of design variables v. A typical function of displacement (e.g., a constraint) will be represented as

$$g = g(U,v), \qquad U = U(v) \tag{15}$$

Analytical calculations of derivatives of displacements and their functions have been performed by three methods: the direct or design space method, the atjoint variable or state space method, and the virtual load method. The virtual load method is a special case of the direct method. Both the direct and adjoint methods begin with the differentiation of Eqs. (14) and (15).

$$K\frac{\mathrm{d}U}{\mathrm{d}v} = \frac{\partial F}{\partial v} - \frac{\partial K}{\partial v}U \equiv R_v \tag{16}$$

$$\frac{\mathrm{d}g}{\mathrm{d}v} = \frac{\partial g}{\partial v} + \left(\frac{\partial g}{\partial U}\right)^T \frac{\mathrm{d}U}{\mathrm{d}v} \tag{17}$$

The direct method is to solve Eq. (16) for $\mathrm{d}U/\mathrm{d}v$ and substitute $\mathrm{d}U/\mathrm{d}v$ into Eq. (17). Equation (16) needs to be solved once for each design variable v so that the direct method is costly when the number of design variables is large.

The adjoint variable method starts by defining a vector of adjoint variables that satisfies the equation

$$K\lambda = \frac{\partial g}{\partial U} \tag{18}$$

where $\partial g/\partial U$ is sometimes referred to as the dummy load vector. (If g is a particular displacement component, then $\partial g/\partial U$ corresponds to a force of unit magnitude in the direction of the component.) Then using Eqs. (16–18),

$$\frac{dg}{dv} = \frac{\partial g}{\partial v} + \lambda^T R_v \tag{19}$$

The adjoint variable method requires the solution of Eq. (18) once for each function g. Therefore, if the number of functions is smaller than the number of design variables, the adjoint variable method is more efficient and, conversely, if the number of design variables is smaller, the direct approach is more efficient. Both the direct and adjoint methods involve fewer computations than the finite difference approach, which requires repeated factorization of the stiffness matrix, whereas the direct and adjoint methods require a single factorization with several right-hand sides.

Calculation of $\partial K/\partial v$

An important computational task in the adjoint and direct methods is the calculation of $\partial K/\partial v$. If the structural model contains only elements whose stiffness matrix is proportional to v (such as rods where v is the cross-sectional area or membranes and shear panels where v is the thickness), $\partial K/\partial v$ is a constant matrix. But for elements having bending stiffness such as beams and plates, the stiffness matrix is a nonlinear function of the cross-sectional dimensions, and the stiffness matrix derivatives are not easily evaluated. The difficulties associated with shape design variables are even more severe. Analytical expresssions for derivatives of the stiffness matrix are cumbersome and more expensive to evaluate than the stiffness matrix itself. Furthermore, coding analytical derivatives of stiffness matrices with respect to all possible design variables is a formidable task, especially that in many cases users of structural analysis software that does not have sensitivity capabilities do not have access to the source code of the software. For these reasons, the preferred approach by most analysts is to compute $\partial K/\partial v$ by finite differences. This combination of analytical derivative experssions such as Eq. (19) coupled with finite-difference evaluation of the stiffness matrix is known as the *semi-analytical* method.

Unfortunately, the semi-analytical method is prone to large errors for some shape design variables. The problem was explained in Ref. 20 by noting that Eq. (16) treats the sensitivity of the displacement vector as the solution of a structural analysis problem with the load replaced by the pseudo-load vector R_v. This presupposes that the derivative of the displacement vector is a legitimate displacement vector itself, which is not always the case. A simple example when the derivative of the displacement vector is not a legitimate displacement is a nearly incompressible material (Poisson's ratio close to 0.5). The derivative of the displacement with respect to shape changes, treated as a displacement field, would typically represent large volume changes. Thus the pseudo-load vector, R_v, would have extremely large components to extract such large volume changes from a nearly incompressible material. In such a case, the small truncation errors associated with the finite-difference calculation of the pseudo load are greatly amplified with a resulting very poor accuracy of the semi-analytical sensitivities. A similar phenomenon can occur for shape changes in bending problems, such as those associated with beams, plates and shells. The sensitivity field is often dominated by shear deformations. Since it is difficult to force a slender beam or a thin shell to undergo large shearing deformations, we again require very large pseudo loads with disastrous effects of small errors in these loads.[21]

Calculation of Second Derivatives

Second derivatives of displacement and constraint functions are used for approximate analysis, and for the calculation of derivatives of optimal solutions. Such derivatives may be obtained by differentiating Eqs. (16) and (17), for example,

$$K\frac{d^2U}{dv^2} = \frac{\partial R_v}{\partial v} + \frac{\partial R_v}{\partial U}\frac{dU}{dv}$$

$$\frac{d^2g}{dv^2} = \frac{\partial^2 g}{\partial v} + 2\left(\frac{\partial^2 g}{\partial U \partial v}\right)^T \frac{dU}{dv} + \left(\frac{\partial g}{\partial U}\right)^T \frac{d^2U}{dv^2} \tag{20}$$

However, for m design variables there are $m(m+1)/2$ second derivatives, and Eqs. (20) need to be solved for that many right-hand sides. It is possible to proceed with a more efficient approach to use Eq. (18) to obtain

$$\frac{d^2g}{dv^2} = \frac{\partial^2 g}{\partial v^2} + \left(\frac{\partial^2 g}{\partial U \partial v}\right)\frac{dU}{dv} + \lambda^T\left(\frac{\partial R_v}{\partial v} + \frac{\partial R_v}{\partial U}\frac{dU}{dv}\right) \tag{21}$$

This approach requires the solution of Eq. (16) for all the first derivatives and Eq. (6) for all adjoint vectors.

Stress Derivatives

The stresses in an element may be obtained from the displacements using

$$\sigma = SU - GT \tag{22}$$

where σ is a vector of element stresses, T is an element temperature, and S and G are stress-displacement and stress-temperature matrices, respectively.

Derivatives of stresses may be obtained by differentiating Eq. (22).

$$\frac{d\sigma}{dv} = S\frac{dU}{dv} + \frac{\partial S}{\partial v}U - \frac{\partial G}{\partial v}T \tag{23}$$

For finite elements such as rods, membranes, and shear panels, S and G are independent of v, and stress derivatives are obtained by simply substituting dU/dv for U and $T = 0$ in Eq. (22). For bending-type elements, S and G may be functions of v and the complete expression must be used; see Camarda and Adelman.[22]

Derivatives of Nonlinear Response

In the case of nonlinear analysis, the equations of equilibrium may be written as

$$P(U, v) = \mu F(v) \tag{24}$$

where P is the internal force generated by the deformation of the structure, and μF is the external applied load. The load scaling factor μ is typically used in nonlinear analysis procedures for tracking the evolution of the solution as the load is increased. This is useful because the equations of equilibrium may have several solutions for the same applied loads. By increasing μ gradually we make sure that we obtain the solution that corresponds to the structure being loaded from zero.

Differentiating Eq. (24) with respect to the design variable v we obtain

$$J\frac{\mathrm{d}U}{\mathrm{d}v} = \mu\frac{dF}{dv} - \frac{\partial p}{\partial v} \tag{25}$$

where J is the Jacobian of P at U,

$$J_{kl} = \frac{\partial F_k}{\partial U_l} \tag{26}$$

often called the tangential stiffness matrix.

The direct method for obtaining dg/dv is to solve Eq. (25) for dU/dv and substitute into Eq. (17). The matrix J is often available from the solution of the equations of equilibrium when these are solved by using Newton's method. Newton's method is based on a linear approximation of the equations of equilibrium about a trial solution $\tilde{U}$

$$P(\tilde{U}, v) + J(\tilde{U}, v)(U - \tilde{U}) \approx \mu F(v) \tag{27}$$

Equation (27) solved for U, typically provides a better approximation to U than $\tilde{U}$. This new approximation replaces $\tilde{U}$ in Eq. (27) for the next iteration, either with an updated J (Newton's method) or with the old J (modified Newton's method). The iteration continues until convergence to a desired accuracy is achieved. If the last iterate $\tilde{U}$, for which J was calculated, is close enough to U, then that J can be used for calculating the derivative of U.

The adjoint approach is very similar to that used in the linear case. The adjoint vector λ is the solution of the equation

$$J^T\lambda = \frac{\partial g}{\partial U} \tag{28}$$

Then it is easy to check that we obtain

$$\frac{\mathrm{d}g}{\mathrm{d}v} = \frac{\partial g}{\partial v} + \lambda^T(\mu\frac{dF}{dv} - \frac{\partial P}{\partial v}) \tag{29}$$

Sensitivity of Limit Loads

For sensitivity of limit loads we need to consider a general path parameter p, which can be a load parameter, a design variable, or a combination of both—a parameter that controls both structural design and loading simultaneously. We denote differentiation with respect to p by a dot. Differentiating Eq. (24) with respect to p

$$J\dot{U} + \frac{\partial P}{\partial v}\dot{v} = \dot{\mu}F + \mu\frac{dF}{dv}\dot{v} \tag{30}$$

If the path parameter does not affect the design we denote differentiation by a prime, and Eq. (30) becomes

$$JU' = \mu' F \tag{31}$$

At a critical load μ^*, the tangential stiffness matrix becomes singular. At a limit point $\mu' = 0$, and U' is the buckling mode. At a bifuraction point $\mu' \neq 0$, and the load F must satisfy a consistency requirement

$$Y^T F = 0, \qquad \text{where} \quad Y^T J = 0. \tag{32}$$

That is, the loading is orthogonal to the left buckling mode (which is equal to the buckling mode if the tangential stiffness matrix is symmetric). To calculate the sensitivity of the limit load, $d\mu^*/dv$, we need a path parameter that changes the design and the load simultaneously so that the design remains at a limit load. We use asterisks to denote quantities evaluated at the limit load, and then Eq. (30) becomes

$$J^*\frac{\partial U^*}{\partial v} + \frac{\partial P^*}{\partial v} = \frac{d\mu^*}{dv}F + \mu^*\frac{\partial F^*}{\partial v} \tag{33}$$

Premultiplying Eq. (33) by the transpose of the left eigenvector Y and solving for the derivative of the limit load

$$\frac{d\mu^*}{dv} = \frac{Y^T\left[\frac{\partial P^*}{\partial v} - \mu^*\frac{dF^*}{dv}\right]}{Y^T F} \approx \frac{Y^T\frac{\Delta R}{\Delta v}}{Y^T F} \tag{34}$$

where R is the residual of the equations of equilibrium

$$R = P - \mu F \tag{35}$$

Equation (34) indicates that the calculation of the derivative of the limit loads can be computed by a semi-analytical approach, where the design variable is perturbed and the residual vector is recalculated. Once the derivative of the residual is estimated, the computation of the derivative of the limit load requires only two scalar products.

Sensitivity of Eigenvalues and Eigenvectors

Distinct Eigenvalues

The general problem is to compute derivatives of eigenvalues and eigenvectors with respect to design variables or system parameters. For reference purposes, the most general case considered is the following eigenvalue problem:

$$AX = \lambda BX \tag{36}$$

$$Y^T A = \lambda Y^T B \tag{37}$$

$$Y^T BX = 1 \tag{38}$$

where λ is an eigenvalue (generally complex). The generally nonsymmetric real nn matrices A and B are assumed to be explicit functions of a set of design variables v, and X and Y are right and left eigenvectors, respectively. The first result on eigenvalue derivatives was published by Jacobi,[23] who developed the result for the special case of symmetric A, and $B = I$

$$\frac{\partial \lambda}{\partial v} = Y^T \frac{\partial A}{\partial v} X \tag{39}$$

Wittrick[24] applied Jacobi's formula for the case of a symmetric matrix to the derivatives of buckling eigenvalues and presented results for the change in buckling loads of plates with respect to aspect ratio and thickness.

Fox and Kapoor[25] and Fox[26] considered the special case of symmetric A and B matrices. For eigenvalues their formula is

$$\frac{\partial \lambda}{\partial v} = X^T \left(\frac{\partial A}{\partial v} - \lambda \frac{\partial B}{\partial v} \right) X \tag{40}$$

in which it is assumed that the eigenvectors are normalized such that

$$X^T B X = 1 \tag{41}$$

For eigenvector derivatives, two methods are presented by Fox and Kapoor. The first is to differentiate Eq. (36), giving a set of simultaneous equations for the eigenvalue and eigenvector derivatives. Differentiating the eigenvalue problem of Eq. (36) gives

$$(A - \lambda B) \frac{\partial X}{\partial v} = - \left(\frac{\partial A}{\partial v} - \frac{\partial \lambda}{\partial v} B - \lambda \frac{\partial B}{\partial v} \right) X \tag{42}$$

The matrix $A - \lambda B$ is singular since λ is an eigenvalue. The set is solvable only after algebraic manipulation, which destroys the banded nature of the equations, a point that arises later in connection with another method. The second method for eigenvector derivatives, developed by Fox and Kapoor, is to expand the derivative as a series of eigenvectors. Thus, for the ith eigenvector

$$\frac{\partial X_i}{\partial v} = \sum_{k=1}^{n} a_{ik} X_k \tag{43}$$

The coefficients a_{ik} are obtained by substituting Eq. (43) into Eq. (42). In principle, it is necessary to use all n modes in the expansion of Eq. (43). However, as with the modal method generally, it should be possible to obtain reasonable results with fewer than n eigenvectors. A modification of the method of Fox and Kapoor which has exhibited faster convergence [27] is denoted the modified modal method[28]. This method represents the eigenvector derivative as

$$\frac{\partial X_i}{\partial v} = \left. \frac{\partial X_i}{\partial v} \right)_s + \sum_{k=1}^{n} \bar{a}_{ik} X_k \tag{44}$$

where $\left.\frac{\partial X}{\partial v}\right)_s$ is denoted a "psuedo static" solution which satisfies the equation

$$A\left.\frac{\partial X_i}{\partial v}\right)_s = \left(\frac{\partial \lambda}{\partial v} B - \frac{\partial A}{\partial v} + \lambda \frac{\partial B}{\partial v}\right) X_i \tag{45}$$

The coefficients $\bar{a}_{ik}$ are obtained by substituting eq. (44) into eq. (42).

Rogers[29] and Stewart[30] derived sensitivity formulas for eigenvalues and eigenvectors of the general problem [Eqs. (36) and (37)]. For eigenvalues the equation is

$$\frac{\partial \lambda}{\partial v} = Y^T \left(\frac{\partial A}{\partial v} - \lambda \frac{\partial B}{\partial v}\right) X \tag{46}$$

Rogers expressed the eigenvector derivatives as an expansion in terms of the eigenvectors

$$\frac{\partial X_i}{\partial v} = \sum_{k=1}^{n} a_{ik} X_k, \qquad \frac{\partial Y_i}{\partial v} = \sum_{k=1}^{n} b_{ik} Y_k \tag{47}$$

The coefficients a_{ik} and b_{ik} are computed by substituting Eqs. (47) into an expression obtained by differentiating the eigenvalue problem and combining it with appropriate orthogonality conditions.

An alternate method for calculation of eigenvector derivatives for the symmetric problem is due to Nelson.[31] The method of Nelson is to represent the eigenvector derivative as

$$\frac{\partial X}{\partial v} = V + cX \tag{48}$$

where V is the solution of a reduced version of Eq. (42) obtained by deleting the kth row and column from $A - \lambda B$ (where k is chosen to correspond to the maximum component of X) and setting the kth component of V equal to zero. The multiplier c is evaluated by substituting Eq. (48) into Eq. (42). This method has certain advantages over previous eigenvector derivative techniques: it requires only the eigenvalue and eigenvector for the mode being differentiated, and the equation for V retains the banded character of coefficient matrix unlike the algebraic methods (e.g., Fox and Kapoor).

Repeated Eigenvalues

The sensitivity of repeated eigenvalues has been a focus of recent interest, even though the eigenvalues are not differentiable and only directional derivatives can be found. For the real symmetric case, a generalization of Nelson's method which preserves the bandedness of the matrix was obtained by Ojalvo[32] and amended by Mills-Curran[33] and Daily[34]. These methods compute the derivatives of the m eigenvectors corresponding to eigenvalues of multiplicity m. As stated by Dailey (see also Lancaster[35]), when the eigenvalues are repeated and a design variable is perturbed, the eigenvectors "split" into as many as m distinct eigenvectors. We seek the derivatives of these distinct eigenvectors which "appear" with design variable perturbation. Using Dailey's notation, define the eigenvalue problem

$$KX = MX\Lambda, \tag{49}$$

where X contains the m eigenvectors cited previously, and

$$\Lambda = \lambda I, \tag{50}$$

where λ is the repeated eigenvalue and I is the identity matrix of order m. The normalization condition, Eq. (41) is now

$$X^T M X = I. \tag{51}$$

The eigenvectors which appear as a result of the splitting are contained in a matrix denoted Z which is related to X as follows

$$Z = X\Gamma \tag{52}$$

where Γ is a set of orthogonal vectors to be determined. To simplify the notation we consider a single design variable v, and denote derivatives with respect to that design variable by a prime. The technique for calculating Z' as contained in Daily is outlined next. The vector Γ and the derivative of the multiple eigenvalues Λ' are obtained as solutions of the following eigenvalue problem

$$D\Gamma = \Gamma\Lambda', \tag{53}$$

where

$$D = X^T(K' - \lambda M')X, \tag{54}$$

with a normalization condition

$$\Gamma^T\Gamma = I. \tag{55}$$

Next in a manner analogous to Nelson[31] let

$$Z' = V + ZC, \tag{56}$$

where V is the solution to

$$(K - \lambda M)V = (\lambda M' - K')Z + MZ\Lambda', \tag{57}$$

(numerically obtained by removing m rows and columns from $K - \lambda M$ using the strategy described in Reference 32) and C is a matrix which is obtained as the solution to the equation

$$C\Lambda' - \Lambda'C + \frac{1}{2}\Lambda'' = -Z^T(K' - \lambda M')V - Z^T(M'Z + MV)\Lambda' + \frac{1}{2}Z^T(K'' - \lambda M'')Z \equiv R. \tag{58}$$

Equation (58), which requires substantial algebraic manipulations for its derivation, determines the matrix C and the matrix of second derivatives of the eigenvalues Λ''. Fortunately Λ'' is diagonal and $C\Lambda' - \Lambda'C$ always has zero on the diagonal. Therefore, we can solve for the matrix C separate from Λ'', and the latter matrix only needs to be calculated if it is needed for some other purpose.

Using Eqs. (51), (52) and (55) we have

$$Z^T MZ = \Gamma^T X^T M X \Gamma = I. \tag{59}$$

Differentiate Eq. (59) and use Eq. (56) to obtain

$$C + C^T = -V^T MZ - Z^T MV - Z^T M'Z \equiv Q, \tag{60}$$

from which

$$c_{ii} = \frac{1}{2} q_{ii} \tag{61}$$

The non-diagonal elements of C are

$$c_{ij} = \frac{r_{ij}}{\lambda'_i - \lambda'_j} \qquad i \neq j \qquad \lambda'_i \neq \lambda'_j. \tag{62}$$

For the case where $\lambda'_i = \lambda'_j \quad i \neq j$, Eq. (62) may not be used. The situation here is that the eigenvalues are not "splitting" when the design variable is perturbed because the design variable is affecting both in exactly the same way. In such a case, Z' is not unique and any values of c_{ij} and c_{ji} satisfying Eq. (60) may be used. Dailey proposes the choice $c_{ij} = c_{ji} = \frac{1}{2}q_{ij}$ whenever $\lambda'_i = \lambda'_j$.

Before leaving the topic of derivatives associated with repeated eigenvalues, we note the limited utility of such derivatives. For example, the eigenproblem is differentiable in terms of a single parameter, but not as a function of several. This may be demonstrated by the example where the matrix

$$A = \begin{bmatrix} 2+y & x \\ x & 2 \end{bmatrix}. \tag{63}$$

The eigenvalues of A are

$$\lambda_{1,2} = 2 + y/2 \pm \sqrt{x^2 + y^2/4}. \tag{64}$$

At $x = y = 0$, the eigenvalues are repeated and $\partial\lambda/\partial x = \pm 1, \quad \partial\lambda/\partial y = 0, 1$. However, the eigenvalues are not differentiable as a function of both x and y, that is the relation

$$\mathrm{d}\lambda = \frac{\partial\lambda}{\partial x}\mathrm{d}x + \frac{\partial\lambda}{\partial y}\mathrm{d}y \tag{65}$$

does not hold. Therefore, the utility of the partial derivatives is questionable. The eigenvectors are also discontinuous at $(0,0)$. This can be checked by noting that at $(\epsilon, 0)$, the eigenvectors are $(1,0)$ and $(0,1)$ and at $(0,\epsilon)$ they are $(1,1)$ and $(1,-1)$ no matter how small ϵ is.

Sensitivity derivatives for nonlinear eigenvalue problems

In flutter and nonlinear vibration problems, we encounter eigenvalue problems where the dependence on the eigenvalue is not linear. For example, Bindolino and Mantegazza[36] consider aeroelastic response problem which produces a transcendental eigenvalue problem of the form

$$A(\lambda, v)X = 0. \tag{66}$$

Differentiating Eq. (66) we get

$$A\frac{\partial X}{\partial v} + \frac{\partial\lambda}{\partial v}\frac{\partial A}{\partial\lambda} = -\frac{\partial A}{\partial v}X. \tag{67}$$

Using the normalizing condition $X_m = 1$ we can solve Eq. (67) for $\partial X/\partial v$ and $\partial\lambda/\partial v$. Instead, it is also possible to use the adjoint method, employing the left eigenvector Y satisfying

$$Y^T A = 0, \qquad Y_m = 1 \tag{68}$$

we obtain

$$\frac{\partial\lambda}{\partial v} = -\frac{Y^T\dfrac{\partial A}{\partial v}X}{Y^T\dfrac{\partial A}{\partial\lambda}X}. \tag{69}$$

A common treatment of flutter problems is to have two real parameters representing the approach of the frequency and speed as an eigenpair instead of one complex eigenvalue. For example in Murthy[37], Eq. (66) is replaced by

$$A(M,\omega,v)X = 0, \tag{70}$$

where the Mach number, M, and the frequency, ω, are real parameters. Using this approach, differentiate Eq. (70) and premultiply by Y^T to get

$$f_M\,\frac{\partial M}{\partial v} + f_\omega\,\frac{\partial\omega}{\partial v} = -\,f_v, \tag{71}$$

where

$$f_M = Y^T\frac{\partial A}{\partial M}X, \qquad f_\omega = Y^T\frac{\partial A}{\partial\omega}X, \qquad f_v = Y^T\frac{\partial A}{\partial v}X. \tag{72}$$

Multiplying Eq. (71) by $\bar{f}_\omega$ (the complex conjugate of f_ω) we get

$$f_M\bar{f}_\omega\,\frac{\partial M}{\partial v} + \mid f_\omega\mid^2\,\frac{\partial\omega}{\partial v} = -\bar{f}_\omega f_v \tag{73}$$

The second term in Eq. (73) as well as $\frac{\partial M}{\partial v}$ are real, so by taking the imaginary part of Eq. (73) we get

$$\frac{\partial M}{\partial v} = -\,\frac{Im(\bar{f}_\omega f_v)}{Im(f_M\bar{f}_\omega)} = -\frac{Im\left[\left(Y^T\frac{\partial A}{\partial v}X\right)\left(\bar{Y}^T\frac{\partial\bar{A}}{\partial\omega}\bar{X}\right)\right]}{Im\left[\left(Y^T\frac{\partial A}{\partial M}X\right)\left(\bar{Y}^T\frac{\partial\bar{A}}{\partial\omega}\bar{X}\right)\right]}. \tag{74}$$

Next, multiplying Eq. (71) by $\bar{f}_M$ and following a similar procedure gives

$$\frac{\partial\omega}{\partial v} = -\frac{Im\left[\left(Y^T\frac{\partial A}{\partial v}X\right)\left(\bar{Y}^T\frac{\partial\bar{A}}{\partial M}\bar{X}\right)\right]}{Im\left[\left(Y^T\frac{\partial A}{\partial M}X\right)\left(\bar{Y}^T\frac{\partial\bar{A}}{\partial\omega}\bar{X}\right)\right]} \tag{75}$$

Rudisill and Bhatia[38] have a derivation of the flutter eigenpair that employs the reduced frequency and flutter speed as the eigenpair and provides also second derivatives.

It is possible to treat in a similar manner the case where the nonlinearity is in X instead of in λ. For example, Hou et al.[39] treated the nonlinear vibration problem

$$K(X)X - \lambda MX = 0 \tag{76}$$

Differentiating Eq. (76) with respect to v we obtain

$$(J - \lambda M)\frac{\partial X}{\partial v} - \frac{\partial \lambda}{\partial v}MX = -[\frac{\partial K}{\partial v} - \lambda\frac{\partial M}{\partial v}]X \tag{77}$$

where J is the tangent stiffness matrix whose components are given as

$$J_{ij} = K_{ij} + \sum_k \frac{\partial K_{ik}}{\partial X_j}X_k \tag{78}$$

Equation (77) can now be solved for eigenvector deriviatives using Nelson's method. For eigenvalue of derivatives use the left eigenvector satisfying

$$Y^T(J - \lambda M) = 0 \tag{79}$$

Premultiply Eq. (77) by Y^T to obtain

$$\frac{\partial \lambda}{\partial v} = \frac{Y^T(\frac{\partial K}{\partial v} - \lambda\frac{\partial M}{\partial v})X}{Y^TMX} \tag{80}$$

Sensitivity of Linear Transient Response

Methodology.

The equations of motion are usually written as

$$M\ddot{U} + C\dot{U} + KU = F(t) \tag{81}$$

Most often the problem is reduced in size by expressing U in terms of m basis functions Φ^i, $i = 1,\ldots m$ where m is usually much less than the number of degrees of freedom of the original system Eq.(81)

$$U = \Phi Q \tag{82}$$

where Φ is a matrix with Φ^i as columns. Then a reduced set of equations can be written as

$$\bar{M}\ddot{Q} + \bar{C}\dot{Q} + \bar{K}Q = \bar{F}(t) \tag{83}$$

where

$$\bar{M} = \Phi^TM\Phi, \quad \bar{C} = \Phi^TC\Phi, \quad \bar{K} = \Phi^TK\Phi, \quad \bar{F} = \Phi^TF \tag{84}$$

When the basis functions are the first m natural vibration modes of the structure scaled to unit modal masses, Φ satisfies the equation

$$K\Phi - M\Phi\Omega^2 = 0 \tag{85}$$

where Ω is a diagonal matrix with the *ith* natural frequency ω_i in the *ith* row. In that case $\bar{K} = \Omega^2$ and $\bar{M} = I$ are diagonal matrices. For special forms of damping, the damping matrix $\bar{C}$ is also diagonal so that the system Eq. (83) is uncoupled. After Q is calculated from Eq. (83) we can use Eq. (82) to calculate F. This method is known as the *mode-displacement method.*

When the load F has spatial discontinuities the convergence of the modal approximation, Eq. (82), can be very slow. The convergence can be dramatically accelerated by using the mode acceleration method. The mode acceleration method can be derived by rewriting Eq. (81) as

$$U = K^{-1}F - K^{-1}C\dot{U} - K^{-1}M\ddot{U} \tag{86}$$

The first term in Eq. (86) is called the quasi-static solution because it represents the response of the structure if the loads are applied very slowly. The second term and third terms are approximated in terms of the modal solution. It can be shown (e.g., Greene[40]) that K^{-1} can be approximated as

$$K^{-1} = \Phi\Omega^{-2}\Phi^T \tag{87}$$

Using this approximation for the second and third terms of Eq. (86) we get

$$U \approx K^{-1}F - \Phi\Omega^{-2}\bar{C}\dot{Q} - \Phi\Omega^{-2}\ddot{Q} \tag{88}$$

This approximation is exact when Φ contains the full set of vibration modes. Note that $\dot{Q}$ and $\ddot{Q}$ in Eq. (88) are obtained from the mode-displacement solution, Eq. (83). Therefore, there is no difference in velocities and accelerations between the mode-displacement and the mode acceleration.

In considering the calculation of sensitivities we treat first the mode-displacement method. The direct method of calculating the response sensitivity is obtained by differentiating Eq. (83) to obtain

$$\bar{M}\frac{d\ddot{Q}}{dv} + \bar{C}\frac{d\dot{Q}}{dv} + \bar{K}\frac{dQ}{dv} = R \tag{89}$$

where

$$R = \frac{d\bar{F}}{dv} - \frac{d\bar{M}}{dv}\ddot{Q} - \frac{d\bar{M}}{dv}\dot{Q} - \frac{d\bar{K}}{dv}Q \tag{90}$$

The first step in forming this equation is the calculation of the derivatives of $\bar{F}$, $\bar{M}$, $\bar{C}$, and $\bar{K}$ with respect to v. Differentiation of $\bar{K}$ yields

$$\frac{d\bar{K}}{dv} = \Phi^T\frac{dK}{dv}\Phi + \frac{d\Phi^T}{dv}K\Phi + \Phi^T K\frac{d\Phi}{dv} \tag{91}$$

with similar expressions for the derivatives of $\bar{M}$, $\bar{C}$, and $\bar{F}$. The calculation is simplified considerably by using a fixed set of basis functions Φ or neglecting the effect of the change in the modes. In many cases the error associated with neglecting the effect of changing modes is small. When this error is unacceptable we have to face the costly calculation of the derivatives of the modes needed for calculating the derivatives of the reduced matrices, such as Eq. (91). Fortunately it was found by Greene[40] that the cost of calculating the derivatives of the modes can be substantially reduced by using the modified modal method[27], keeping only the first term in this equation. This approximation to the derivatives of the modes may not always be accurate, but it appears to be sufficient for calculating the sensitivity of the dynamic response.

For the adjoint method we consider a function in the form

$$g(Q,v)=\int_0^{t_f} p(Q,v,t)dt \tag{92}$$

so that

$$\frac{dg}{dv}=\int_0^{t_f}\left(\frac{\partial p}{\partial v}+\frac{\partial p}{\partial Q}\frac{dQ}{dv}\right)dt \tag{93}$$

To avoid the calculation of dQ/dv we use an adjoint vector λ, and start by multiplying Eq. (89) by λ^T and integrating to obtain

$$\int_0^{t_f}\lambda^T\left(\bar{M}\frac{d\ddot{Q}}{dv}+\bar{C}\frac{d\dot{Q}}{dv}+\bar{K}\frac{dQ}{dv}\right)dt=\int_0^{t_f}\lambda^T R dt \tag{94}$$

Integrating by parts we get

$$\lambda^T\bar{M}\frac{d\dot{Q}}{dv}\Big|_0^{t_f}-\dot{\lambda}^T\bar{M}\frac{dQ}{dv}\Big|_0^{t_f}+\lambda^T\bar{C}\frac{dQ}{dv}\Big|_0^{t_f}+\int_0^{t_f}(\ddot{\lambda}^T\bar{M}-\dot{\lambda}^T\bar{C}+\lambda^T\bar{K})\frac{dQ}{dv}=\int_0^{t_f}\lambda^T R dt \tag{95}$$

Assuming that the initial conditions do not depend on the design variable v, Eq. (95) suggests the following definition for λ

$$\bar{M}\ddot{\lambda}-\bar{C}\dot{\lambda}+\bar{K}\lambda=\left(\frac{\partial p}{\partial Q}\right)^T, \qquad \lambda(t_f)=\dot{\lambda}(t_f)=0 \tag{97}$$

and then Eq. (93) becomes

$$\frac{dg}{dv}=\int_0^{t_f}\left(\frac{\partial p}{\partial v}-\lambda^T R\right)dt \tag{98}$$

For the mode-acceleration method we consider only the direct method. We start by differentiating Eq. (81) and rearranging it as

$$\frac{dU}{dv}=K^{-1}\left[\frac{dF}{dv}-\frac{dK}{dv}U-C\frac{d\dot{U}}{dv}-\frac{dC}{dv}\dot{U}-M\frac{d\ddot{U}}{dv}-\frac{dM}{dv}\ddot{U}\right] \tag{99}$$

Next we use Eq. (88) to approximate the second term, and the modal expansion, Eq. (82) to approximate the other terms to get

$$\frac{dU}{dv} \approx K^{-1}\left[\frac{dF}{dv} - \frac{dK}{dv}[K^{-1}F - \Phi\Omega^{-2}\bar{C}\dot{Q} - \Phi\Omega^{-2}\ddot{Q}] - C\Phi\frac{d\dot{Q}}{dv} - \frac{dC}{dv}\Phi\dot{Q} - M\Phi\frac{d\ddot{Q}}{dv} - \frac{dM}{dv}\Phi\ddot{Q}\right] \tag{100}$$

Finally we use the modal approximation to K^{-1}, Eq. (87) to obtain

$$\frac{dU}{dv} \approx K^{-1}\left[\frac{dF}{dv} - \frac{dK}{dv}K^{-1}F\right] + \Phi\Omega^{-2}\Phi^T\left[\frac{dK}{dv}\Phi\Omega^{-2}\bar{C}\dot{Q} - \frac{dC}{dv}\Phi\dot{Q} - C\Phi\frac{d\dot{Q}}{dv}\right] + K^{-1}\left[\frac{dK}{dv}\Phi\Omega^{-2} - \frac{dM}{dv}\Phi\right]\ddot{Q} - \Phi\Omega^{-2}\frac{d\ddot{Q}}{dv} \tag{101}$$

Note that the calculation involves the solution of Eqs. (83) and (89) for Q and dQ/dv, followed by Eq. (101) for retrieving dU/dv. Additional details can be found in Ref. 40.

Example of use of transient sensitivity

The effects of updating modes are demonstrated with a delta wing example taken from Ref. 40. The wing has a composite material skin modeled by membrane finite elements and titanium spars and ribs modeled by shear web elements. The skin is divided into 16 regions, and the shear webs are grouped into 12 groups as shown in Fig. 1.

The wing is loaded by uniform pressure applied as a step function, and the stresses at various points on the wing are calculated at critical times where these stresses reach their peak values.

Figure 2 shows the convergence of selected stresses in the skin and the shear webs for the mode displacement method. The superscripts on the stresses indicate the region while the subscript indicate the ply direction. It is seen that convergence is very rapid.

The convergence of selected derivatives of the stresses with respect to skin and shear web thicknesses is shown in Figure 3. The derivatives are calculated by the finite difference method, with the modes updated for the perturbed design. The convergence with increasing number of modes is again excellent. However, if the modes are not updated in the perturbed design the convergence of the derivatives is poor as is shown in Fig. 4.

One way of accounting for the change in modes is to calculate the derivatives of the modes. Figure 5 shows that the convergence is again good when the semi-analytical direct method is used with derivatives of the mode approximated by the modified modal method. The most important term in the modified modal method approximation to the derivcative of the mode is the first (static) term. Figure 6 shows that if the derivative of the mode is approximated by this term only, the convergence of the stress derivative is still good. This first term approximation reduces the cost of mode shape derivatives substantially. Finally, the effect of changing modes is also taken into account by the mode acceleration method given by Eq. (101). Figure 7 shows good convergence using this method.

References

[1] Tomovic, R., *Sensitivity Analysis of Dynamic Systems*, McGraw-Hill Book Co., New York, 1963.

[2] Brayton, R. K. and Spence R., *Sensitivity and Optimization*, Elsevier, New York, 1980.

[3] Frank, P. M., *Introduction to Sensitivity Theory*, Academic Press, Orlando, FL, 1978.

[4] Radanovic, L. (ed.), *Sensitivity Methods in Control Theory*, Pergamon Press, Oxford, England, 1966.

[5] Kelley, H. J., "Method of Gradients," *Optimization Techniques with Applications to Aerospace Systems*, edited by George Leitmann, Academic Press, Orlando, FL, 1962.

[6] Schmit, L. A. Jr., Structural Synthesis—Its Genesis and Development," *AIAA Journal*, Vol. 19, Oct. 1981, pp. 1249–1263.

[7] Leonard, J. I., "The Application of Sensitivity Analysis to Models of Large Scale Physiological Systems," NASA CR-160228, 1974.

[8] Irwin, C. I. and O'Brien, T. J., "Sensitivity Analysis of Thermodynamic Calculations," U.S. Dept. of Energy Rept. DOE/METC/82-53, 1982.

[9] Hwang, J. T., Dougherty, E. P., Rabitz, S., and Rabitz, H., "The Green's Function Method of Sensitivity Analysis in Chemical Kinetics," *Journal of Chemical Physics*, Vol. 69, Dec. 1978, pp. 5180–5191.

[10] Dwyer, H. A. and Peterson, T., "A Study of Turbulent Flow with Sensitivity Analysis," AIAA Paper 80-1397, 1980.

[11] Dwyer, H. A., Peterson, T., and Brewer, J., "Sensitivity Analysis Applied to Boundary Layer Flow," *Proceedings of the 5th International Conference on Numerical Methods in Fluid Dynamics*, Lecture Notes in Physics, Vol. 59 Springer-Verlag, pp. 179-184, NY, 1976.

[12] Bristow, D. R. and Hawk, J. D., "Subsonic Panel Method for Designing Wing Surfaces from Pressure Distributions," NASA CR-3713, 1983.

[13] Schy, A. A. and Giesy, D. P., "Multiobjective Insensitive Design of Airplane Control Systems with Uncertain Parameters," paper presented at AIAA Guidance and Control Conference, Albuquerque, NM, 1981.

[14] Schy, A. A. and Giesy, D. P., "Tradeoff Studies in Multiobjective Insensitive Design of Airplane Control Systems," AIAA No. 83-2273, Proceedings paper, pp. 719-728, AIAA Guidance and Control Conference, Gatlinburg, TN, Aug. 1983.

[15] Wexler, A.S., "Automatic Evalutation of Derivatives," Appl. Math. Comp., Vol. 24, pp. 19–46, 1987.

[16] Oblow, E.M., Pin, F.G., and Wright, R.Q., "Sensitivity Analysis Using Computer Calculus: A Nuclear Waste Isolation Application," Nucl. Sci. Eng., Vol. 94, p. 46, 1986.

[17] Gill, P. E., Murray, W., Saunders, M. A., and Wright, M. H., "Computing Forward Difference Intervals for Numerical Optimization," *SIAM Journal of Scientific and Statistical Computing*, Vol. 4, June 1983, pp. 301–321.

[18] Iott, J., Haftka, R. T.; and Adelman, H. M., "Selecting Step Sizes in Sensitivity Analysis by Finite Differences", NASA TM 86382, 1985.

[19] Haftka, R.T., "Sensitivity Calculations for Iteratively Solved Problems," International Journal for Numerical Methods in Engineering, Vol. 21, pp. 1535–1546, 1985.

[20] Barthelemy, B., and Haftka, R.T., "Accuracy Analysis of the Semi-Analytical Method for Shape Sensitivity Analysis," Mechanics of Structures and Machines, Vol 18, No. 3, pp. 407–432, 1990.

[21] Haftka, R.T., "Stiffness-Matrix Condition Number and Shape Sensitivity Errors,", AIAA Journal, Vol. 28, No. 7, pp. 1322-1324, 1990.

[22] Camarda, C.J.; Adelman, H. M., "Static and Dynamic Structural Sensitivity Derivative Calculations in the Finite-element Based Engineering Analysis Language (EAL) system. NASA TM-85743, 1984.

[23] Jacobi, C. G. J., "Uber ein leichtes Verfahren die in der Theorie der Saecularstoerungen vorkommenden Gleichungen numerisch aufzuloesen," *Zeitschrift für Reine und Angewandte Mathematik*, Vol. 30, 1846, pp. 51–95; also NASA TT.F-13,666, June 1971.

[24] Wittrick, W. H., "Rates of Change of Eigenvalues, with Reference to Buckling and Vibration Problems," *Journal of the Royal Aeronautical Society*, Vol. 66, Sept. 1962, pp. 590–591.

[25] Fox, R. L. and Kapoor, M. P., "Rate of Change of Eigenvalues and Eigenvectors," *AIAA Journal*, Vol. 6, Dec. 1968, pp. 2426–2429.

[26] Fox, R. L., *Optimization Methods for Engineering Design*, Addison-Wesley, New York, 1971, pp. 242–249.

[27] Sutter, T.R.; Camarda, C.J.; Walsh, J.L.; Adelman, H.M. "A Comparison of Several Methods for the Calculation of Vibration Modes Shape Derivatives. AIAA J. 26, 1506-1511.

[28] Wang, B., P. "Improved Approximate Methods for Computing Eigenvalue Derivatives Unstructural Dynamics". AIAA Journal, Vol. 29, No. 6, pp. 1018-1020, 1991.

[29] Rogers, L. C., "Derivatives of Eigenvalues and Eigenvectors," *AIAA Journal*, Vol. 8, May 1970, pp. 943–944.

[30] Stewart, G. W., "On the Sensitivity of the Eigenvalue Problem $Ax = \lambda Bx$," *SIAM Journal of Numerical Analysis*, Vol. 9, No. 4, 1972, pp. 669–686.

[31] Nelson, R. B., "Simplified Calculation of Eigenvector Derivatives," *AIAA Journal*, Vol. 14, Sept. 1976, pp. 1201–1205.

[32] Ojalvo, I. U., "Efficient Computation of Mode-Shape Derivatives for Large Dynamic Systems". *AIAA Journal* 25, 1386-1390, 1987.

[33] Mills-Curran, W. C., "Calculation of Derivatives for Structures with Reapeated Eigenvalues," *AIAA Journal* 26, 867-871, 1988.

[34] Dailey, A. L., "Eigenvector Derivatives with Repeated Eigenvalues," *AIAA Journal* 27, 486-491, 1989

[35] Lancaster, P., "On eigenvalues of matrices dependent on a parameter. Numerische Mathematik 6, 377-387, 1964.

[36] Bindolino, G., and Mantegazza, P., "Aeroelastic Derivatives as Sensitivity Analysis of Nonlinear Systems," *AIAA Journal*, 25, pp. 1145-1146, 1987.

[37] Murthy, D.V., "Solution and Sensitivity of a complex Transcendental Eigenproblem with Pairs of Real Eigenvalues", Proceedings of the 12th Biennral ASME Conference on Me-

chanical Vibration and Noise (DE-Vol. 18-4) Montreal Canada, September 17-20, 1989, pp. 229-234 (to be published in the international Journal for Numberical Methods in Engineering).

[38] Rudisill, C. S., and Bhatia, K. G., "Second Derivatives of the Flutter Velocity and the Optimization of Aircraft Structures", *AIAA Journal*, Vol. 10, No. 12, pp. 1569-1572, 1972.

[39] Hou, J. W., Mei, C., and Xue, Y. X., "On the Design Sensitivity Analysis of Beams under Free and Forced Nonlinear Vibrations", *AIAA Paper* 87-0936, 1987.

[40] Greene, W.H., Computational Aspects of Sensitivity Calculations in Linear Transient Analysis, Ph.D. dissertation, Virginia Polytechnic Institute and State University, August 1989. Also, NASA TM-4156, May, 1990.

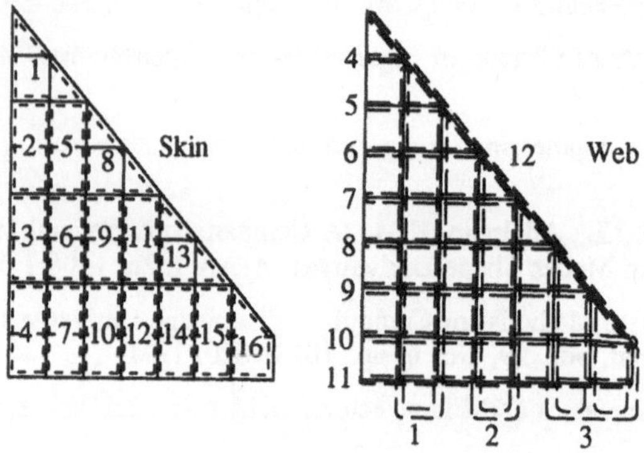

Figure 1: Delta wing design variables.

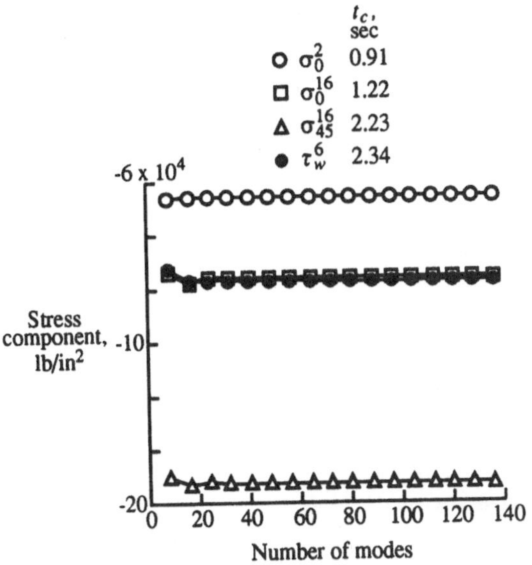

Figure 2: Modal convergence of selected stresses for delta wing calculated with modal displacement method.

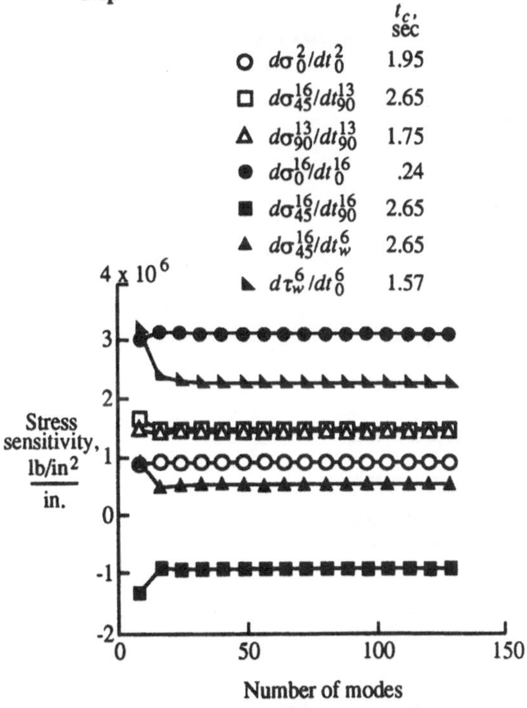

Figure 3: Modal convergence of selected stress sensitivities for delta wing calculated with forward difference method with updated modes.

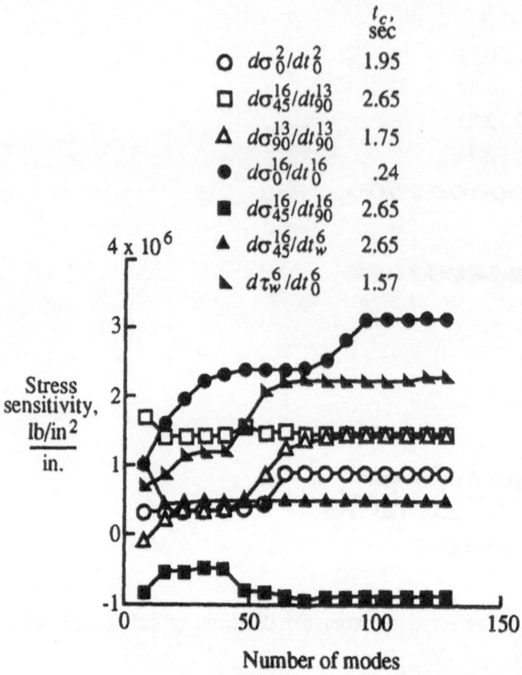

Figure 4: Modal convergence of selected stress sensitivities for delta wing calculated with forward difference method with fixed modes.

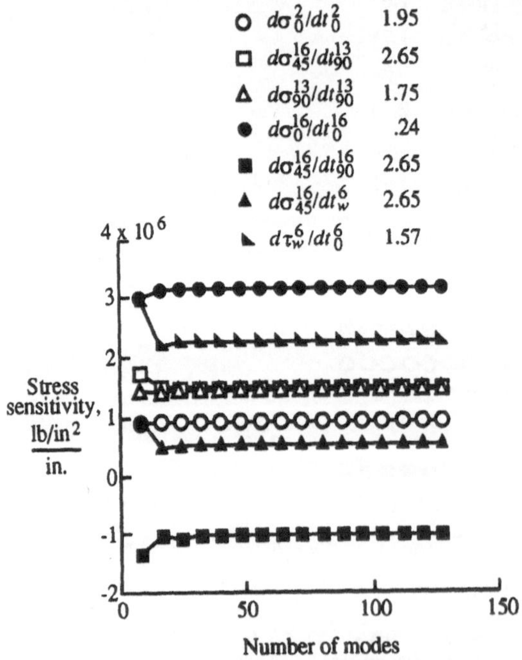

Figure 5: Modal convergence of selected stress sensitivities for delta wing calculated with semianalytical method with $d\Phi/dx$ approximated with modified modal method.

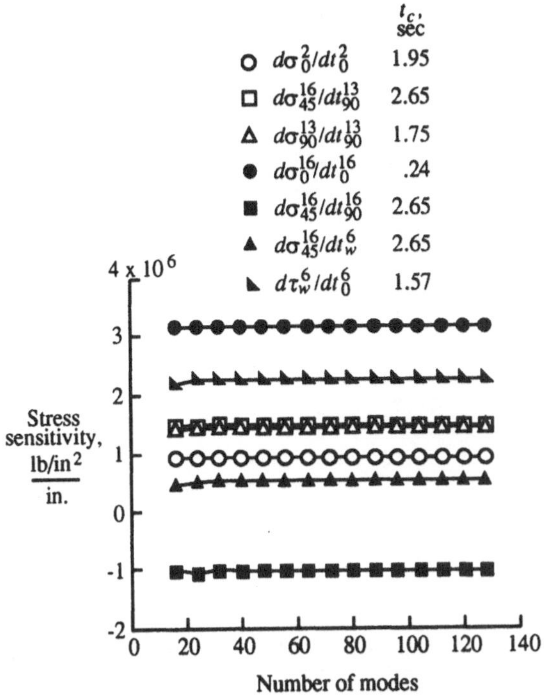

Figure 6: Modal convergence of selected stress sensitivities for delta wing calculated with semianalytical method with $d\Phi/dx$ approximated with only pseudostatic solution.

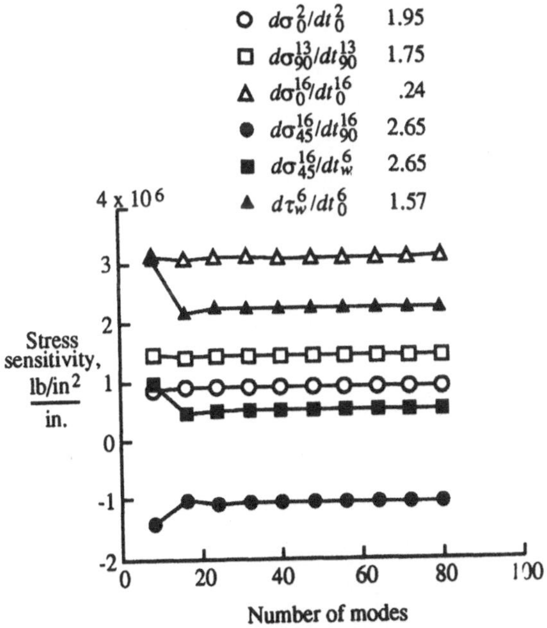

Figure 7: Modal convergence of selected stress sensitivities for delta wing calculated with semianalytical mode acceleration method.

DESIGN SENSITIVITY ANALYSIS OF DYNAMICS OF BUILT-UP STRUCTURES USING RITZ AND MODE ACCELERATION METHODS

Kyung K. Choi and Semyung Wang
Department of Mechanical Engineering
and
Center for Simulation and Design Optimization
of Mechanical Systems
The University of Iowa
Iowa City, Iowa 52242, U. S. A.

ABSTRACT. In this paper, a unified continuum-based sizing design sensitivity analysis (DSA) method is developed for transient dynamic responses and eigenvectors of large-scale built-up structures. Using the direct differentiation method, the variational equation of the built-up structure is differentiated with respect to design variables to obtain a variational design sensitivity equation in which the design sensitivity of the transient response is unknown. For large scale built-up structures, the same superposition method that is used to reduce the dimension of the finite element matrix equation can also be used to reduce the dimension of the design sensitivity equation. A continuum-based sizing DSA method for eigenvectors is developed by taking design derivatives of the variational equation of eigenvalue problems. Accuracy and efficiency is improved by adding load dependent Ritz vectors (LDRV) to the minimal eigenbasis. The continuum-based DSA methods can be implemented outside established finite element analysis (FEA) codes by post-processing analysis data, and it does not require information on the stiffness, damping, and mass matrices and their derivatives. Moreover the method is efficient since it does not require derivatives of basis vectors. Either the LDRV method or mode acceleration method (MAM) can be used for both analysis and design sensitivity analysis. Examples show that the same basis vectors that are used for analysis of the built-up structure are suitable for analysis of the design sensitivity equation. Also, numerical examples for eigenvector DSA indicate that adding two LDRVs to the minimal eigenbasis is recommended.

1. Introduction

Since transient dynamic analysis of large scale built-up structures requires a large amount of computation time, development of an efficient method is especially desirable for transient dynamic analysis and optimization. To achieve this goal, a substantial literature has emerged in the areas of structural dynamic analysis and DSA during the past two decades.

In structural dynamic analysis, significant improvements have been made for efficiently obtaining transient dynamic response, using using superposition methods such as the MAM [1,2], LDRV method [3-6], and Lanczos algorithm [7,8]. The MAM is a variation of the mode displacement method (MDM) in which a truncated set of

G. I. N. Rozvany (ed.), Optimization of Large Structural Systems, Vol. I, 313–328.

eigenvectors is used as a basis to compensate for the effect of neglected high frequency modes. The LDRV method, proposed by Wilson et. al. [3], uses a sequence of orthogonal Ritz vectors. The LDRVs, generated from the externally applied load, are orthonormalized, using the Gram-Schmidt orthogonalization procedure. Thus LDRVs can be generated at a fraction of the cost that is required to calculate eigenvectors. The LDRV method has the same effect as MAM since its initial vector is the static deflection of the structure. Kline [4] extended the use of LDRVs to construct a basis by combining LDRVs with eigenvectors. Leger, Wilson, and Clough [5] proposed a modified version of the LDRV method, called the LWYD algorithm, which eliminates loss of orthogonality, as in the Lanczos method.

For DSA of eigenvectors, Fox and Kapoor [9] presented two methods for computing derivatives of eigenvectors for self-adjoint systems: the first consists of solving N linear equations, and the second is an eigenvector expansion. Nelson presented [10] a numerically efficient method for obtaining derivatives of eigenvectors of general matrices by modifying Fox and Kapoor's first method. Nelson's method requires only one eigenvalue and eigenvector. Moreover, the bandwidth character and sparsity are not destroyed. Wang [11,12] proposed modified modal methods which are later called the explicit and implicit methods. In the explicit method, the first LDRV [3-6] is used in the solution but in the implicit method, it is added to the basis vectors. Wang [12] demonstrated numerically that the implicit method is superior to the explicit method.

Continuum methods of structural sizing and shape DSA for static responses and eigenvalues, have been developed in Ref. 13. A numerical method has been developed to evaluate continuum DSA results outside established FEA codes, using postprocessing data. In this paper, continuum-based sizing DSA methods for transient dynamic responses and eigenvectors of built-up structures presented in Refs. 14-16 are summarized. The continuum-based DSA methods utilize the LDRVs and MAM.

In the continuum-based DSA method, the variational equation of the built-up structure is differentiated with respect to design u to obtain a variational equation of design sensitivity z' of the transient response z. Just as the transient response z of the large scale built-up structure is approximated using the superposition method, design sensitivity z' is approximated using a linear combination of the same basis vectors. The reduced governing equation for z' is then solved using the direct integration method. Thus, the design sensitivity z' can be obtained without computing derivatives of basis vectors. Moreover use of LDRVs as basis vectors improves accuracy of analysis results [3-6]. If necessary, more LDRVs can be added to the basis to approximate the design sensitivity z', thereby improving accuracy of design sensitivity results without significantly increasing computation time. Also, a different time step can be used for analysis of the design sensitivity equation. The direct differentiation method is more attractive for DSA of transient dynamic response since it is not necessary to store the transient response of the built-up structure during forward integration and to use it during backward integration as in the adjoint variable method [13].

2. Variational Equations of Structural Dynamics and Eigenvalue Problems

The dynamic equation of a built-up structure can be written in the form [13]

$$m(x,u)\, z_{tt} + C_u\, z_t + A_u\, z = f(x,t,u), \quad x \in \Omega, \quad t \geq 0 \tag{1}$$

with initial conditions

$$z(x,0,u) = z^{0}(x,u), \quad z_t(x,0,u) = z_t^{0}(x,u), \qquad x \in \Omega \tag{2}$$

where Ω is the domain of the structure; t is time; u(x) is the time independent design variable; z(x,t,u) is the transient response; m(x,u) represents mass effect in the structure; and f(x,t,u) is the applied dynamic load. In Eq. 1, A_u is the spatial partial differential operator and C_u is the linear differential operator that corresponds to the damping effect in the structure [13]. Also z(x,t,u) must satisfy homogeneous boundary conditions specified for the built-up structure.

A variational form of Eq. 1 can be obtained by multiplying both sides by the transpose of an arbitrary virtual displacement $\bar{z}(x)$ and integrating over the physical domain as

$$d_u(z_{tt}, \bar{z}) + c_u(z_t, \bar{z}) + a_u(z, \bar{z}) = \ell_u(\bar{z}), \qquad t \geq 0 \tag{3}$$

which must hold for all kinematically admissible virtual displacements $\bar{z} \in Z$ that satisfy the homogeneous boundary conditions. In Eq. 3,

$$d_u(z_{tt}, \bar{z}) \equiv \iint_{\Omega} \bar{z}^T\, m(x,u)\, z_{tt}\, d\Omega \tag{4}$$

$$a_u(z, \bar{z}) \equiv \iint_{\Omega} \bar{z}^T A_u z\, d\Omega \tag{5}$$

$$c_u(z_t, \bar{z}) \equiv \iint_{\Omega} \bar{z}^T C_u z_t\, d\Omega \tag{6}$$

and

$$\ell_u(\bar{z}) \equiv \iint_{\Omega} \bar{z}^T f(x,t,u)\, d\Omega \tag{7}$$

are the kinetic, strain, and dissipative energy bilinear forms, and the load linear form, respectively.

The variational equation of the eigenvalue problems can be written as [13]

$$a_u(y^i, \bar{y}) = \zeta_i\, d_u(y^i, \bar{y}) \tag{8}$$

which must hold for all kinematically admissible virtual eigenfunction $\bar{y} \in Z$. In Eq. 8, ζ_i and y^i are the ith eigenvalue and the corresponding eigenfunction, respectively. Since Eq. 8 is homogeneous in y^i, orthonormalizing conditions

$$d_u(y^i, y^j) = \delta_{ij} \tag{9}$$

for all i,j=1,2,... are added to uniquely define eigenfunctions and δ_{ij} is the Kronecker delta.

3. Approximate Dynamic Analysis

Since analytical solutions of Eq. 3 cannot be obtained for complex built-up structures, the FEA method is usually used for an approximate solution. Using the FEA method, an approximate matrix equation of Eq. 3 is

$$\mathbf{M}\ddot{\mathbf{z}}(t) + \mathbf{C}\dot{\mathbf{z}}(t) + \mathbf{K}\,\mathbf{z}(t) = \mathbf{f}(t) \tag{10}$$

for $t \geq 0$, where **M**, **C**, and **K** are n×n mass, damping, and stiffness matrices, respectively, and **f**(t) is the external load vector. In this paper, upper-case bold letters denote matrices and lower-case bold letters are vectors. Initial conditions **z**(0) and $\dot{\mathbf{z}}(0)$ can be obtained by evaluating the initial conditions of Eq. 2 at nodal points. The n×1 vector **z**(t) in Eq. 10 is a discretization of the transient response z(x,t,u) of Eq. 3.

Likewise, an approximate matrix eigenvalue problem of Eq. 8 is

$$\mathbf{K}\mathbf{y}^i = \zeta_i \mathbf{M}\mathbf{y}^i \tag{11}$$

with **M**-orthonormality conditions, from Eq. 9,

$$\mathbf{y}^{j^T}\mathbf{M}\mathbf{y}^i = \delta_{ij} \tag{12}$$

where the n×1 vector $\mathbf{y}^i$ in Eq. 11 is a discretization of the eigenfunction y(x,u) in Eq. 8 and the same notation ζ_i is used for the eigenvalue in Eq. 12 as in Eq. 8 even though it is an approximation of the eigenvalue ζ_i in Eq. 8.

For many cases, the loading **f**(t) can be written in the form **f**(t)=**p**h(t) where **p** is the time independent vector describing spatial distribution of forces and h(t) is the time functions [3]. Let φ^i be the LDRVs [3-6] that are generated from the following recurrence formula:

$$\mathbf{K}\varphi^{i^*} = \mathbf{M}\varphi^{i-1},\ i=2,\dots,r-q \tag{13}$$

The first vector is obtained from

$$\mathbf{K}\varphi^{1^*} = \mathbf{p} \tag{14}$$

The vectors are then **M**-orthogonalized, including eigenvectors, by

$$\varphi^i = \varphi^{i^*} - \sum_{k=1}^{q} d_k \mathbf{y}^k - \sum_{j=1}^{i-1} c_j \varphi^j \tag{15}$$

where

$$c_j = \varphi^{j^T}\mathbf{M}\varphi^{i^*} \quad \text{and} \quad d_k = \mathbf{y}^{k^T}\mathbf{M}\varphi^{i^*} \tag{16}$$

After each vector is found, it is **M**-normalized. The final LDRVs then satisfy

$$\varphi^{j^T}\mathbf{M}\varphi^i = \delta_{ij} \quad \text{and} \quad \varphi^{j^T}\mathbf{M}\mathbf{y}^i = 0 \tag{17}$$

For practical analysis of large scale built-up structures, it is often necessary to reduce the dimension of Eq. 10. In the MDM, MAM, and LDRV methods, the transient response **z**(t) is approximated by a linear combination of eigenvectors and LDRVs as

$$\mathbf{z}(t) = \sum_{i=1}^{r} v_i(t)\,\phi^i = \Phi\,\mathbf{v}(t) \tag{18}$$

where r<n, $\Phi=[\phi^1,\phi^2,\dots,\phi^r] = [\mathbf{y}^1,\dots,\mathbf{y}^q,\varphi^1,\dots,\varphi^{r-q}]$, and $\mathbf{v}(t)\in R^r$. Thus, $\Phi^T\mathbf{M}\Phi=\mathbf{I}$. Substituting Eq. 18 into Eq. 10 and premultiplying the resulting matrix equation by Φ^T,

$$\ddot{\mathbf{v}}(t) + \Phi^T\mathbf{C}\Phi\,\dot{\mathbf{v}}(t) + \Phi^T\mathbf{K}\Phi\,\mathbf{v}(t) = \Phi^T\mathbf{f}(t) \tag{19}$$

for $t \geq 0$, with initial conditions

$$\mathbf{v}(0) = \Phi^T\mathbf{M}\mathbf{z}(0), \qquad \dot{\mathbf{v}}(0) = \Phi^T\mathbf{M}\dot{\mathbf{z}}(0) \tag{20}$$

which can be obtained from Eq. 18. Once the solution **v**(t) of Eq. 19 is obtained, the approximate transient response **z**(t) can be obtained from Eq. 18. If all of the ϕ^i are eigenvectors, then the MDM or MAM is applicable. On the other hand, if all ϕ^i are LDRVs, then the Wilson LDRV method [3] is applicable. For the Kline LDRV method [4,6], ϕ^i is a combination of eigenvectors y^i and LDRVs φ^i.

4. Design Sensitivity Analysis of Transient Responses and Eigenvectors

4.1 VARIATIONAL EQUATIONS OF DESIGN SENSITIVITIES

For design sensitivity analysis of the transient dynamic response of built-up structures, two methods are generally used: the direct differentiation and adjoint variable methods [13]. The latter method has been more commonly used because it can handle a large number of design variables; however, it suffers from a few drawbacks when used for the transient dynamic response of built-up structures. First, during forward integration of the state equations, information must be saved to be retrieved at an appropriate time step during the backward integration of the adjoint equations. It is undesirable to store all transient responses because of computer disk space. The second drawback of the method is the difficulty of positive error control during numerical integration of the state and adjoint equations. Since backward integration can be started after forward integration is completed, it is hard to estimate how the error in the forward integration will influence the solution during the backward integration.

With the development of parallel computing techniques, the direct differentiation method has become attractive since each set of sensitivity equations corresponding to a design variable is independent, and can be assigned to a processor in a parallel computer. The adjoint variable method, on the other hand, may not be suitable for parallel computing because of the sequential computation procedure and data dependency during the forward and backward integration.

In this section, a continuum-based DSA method for transient dynamic response that does not require design sensitivity of basis vectors is presented. In this method, approximation is not involved until a variational sensitivity equation for z' is obtained. This approach is computationally attractive since derivatives of the basis vectors are not required. The method is applicable for the MDM, MAM, and LDRV methods, since only the basis vectors and the corresponding responses are required in the design sensitivity formulation.

Take the the first variation of Eq. 3 with respect to design and rearrange to obtain

$$d_u(z'_{tt}, \bar{z}) + c_u(z'_t, \bar{z}) + a_u(z', \bar{z}) = -d'_{\delta u}(z_{tt}, \bar{z}) - c'_{\delta u}(z_t, \bar{z}) - a'_{\delta u}(z, \bar{z}) + \ell'_{\delta u}(\bar{z}), \quad t \geq 0 \qquad (21)$$

which must hold for all $\bar{z} \in Z$. In Eq. 20, the first variations of the energy bilinear and load linear forms with respect to explicit dependence on design u are

$$\left.\begin{aligned} a'_{\delta u}(z, \bar{z}) &\equiv \frac{d}{d\tau}[a_{u+\tau\delta u}(z, \bar{z})]\,\Big|_{\tau=0} \\ c'_{\delta u}(z_t, \bar{z}) &\equiv \frac{d}{d\tau}[c_{u+\tau\delta u}(z_t, \bar{z})]\,\Big|_{\tau=0} \\ d'_{\delta u}(z_{tt}, \bar{z}) &\equiv \frac{d}{d\tau}[d_{u+\tau\delta u}(z_{tt}, \bar{z})]\,\Big|_{\tau=0} \end{aligned}\right\} \qquad (22)$$

and

$$\ell'_{\delta u}(\tilde{z}) = \frac{d}{d\tau}[\ell_{u+\delta u}(\tilde{z})]\,|_{\tau=0} \tag{23}$$

where $\tilde{z}$ denotes the state z with dependence on τ (design variable) suppressed. Also,

$$z' = z'(x,t,u) = \frac{d}{d\tau}\, z(x,\ t,u+\tau\,\delta u)|_{\tau=0} \tag{24}$$

is the first variation of z with respect to design u in the direction δu of design change. From Eq. 2,

$$z'(x,0,u) = z^{0'}(x,u), \qquad z_t'(x,0,u) = z_t^{0'}(x,u) \tag{25}$$

where $z'(x,t,u)$ satisfies the same homogeneous boundary conditions as $z(x,t,u)$. Equation 21 is a variational equation in which the design sensitivity $z'(x,t,u)$ is the unknown. If the solution $z(x,t,u)$ of Eq. 3 is obtained, the right side of Eq. 21 can be computed to obtain the fictitious load.

For design sensitivity of eigenvectors, taking the first design variation of Eq. 8 yields

$$a'_{\delta u}(y^i,\bar{y}) + a_u(y^{i'},\bar{y}) = \zeta_i'\, d_u(y^i,\bar{y}) + \zeta_i\, d'_{\delta u}(y^i,\bar{y}) + \zeta_i\, d_u(y^{i'},\bar{y}) \tag{26}$$

which must hold for all $\bar{y} \in Z$. It is shown in Ref. 13 that the simple eigenvalue ζ_i and the corresponding eigenfunction y^i are differentiable. Letting $\bar{y}=y^i$ in Eq. 26 and using Eq. 9,

$$\zeta_i' = a'_{\delta u}(y^i,y^i) - \zeta_i\, d'_{\delta u}(y^i,y^i) + [a_u(y^{i'},y^i) - \zeta_i\, d_u(y^{i'},y^i)] \tag{27}$$

Since $y^{i'} \in Z$, the last two terms of Eq. 27 cancel each other and

$$\zeta_i' = a'_{\delta u}(y^i,y^i) - \zeta_i\, d'_{\delta u}(y^i,y^i) \tag{28}$$

From Eq. 26,

$$a_u(y^{i'},\bar{y}) - \zeta_i\, d_u(y^{i'},\bar{y}) = -\, a'_{\delta u}(y^i,\bar{y}) + \zeta_i\, d'_{\delta u}(y^i,\bar{y}) + \zeta_i'\, d_u(y^i,\bar{y}) \tag{29}$$

which must hold for all $\bar{y} \in Z$. Like in the transient response case of Eq. 21, Eq. 29 is a variational equation in which the design sensitivity $y^{i'}(x,u)$ is the unknown.

4.2 APPROXIMATE SOLUTIONS OF VARIATIONAL DESIGN SENSITIVITY EQUATIONS

Just as the continuum Eq. 3, since the analytical solution of Eq. 21 cannot be obtained for complex structures, the FEA method is used for an approximate solution. Using the FEA method, an approximate matrix equation of Eq. 21 is

$$\mathbf{M}\ddot{\mathbf{z}}'(t) + \mathbf{C}\dot{\mathbf{z}}'(t) + \mathbf{K}\,\mathbf{z}'(t) = \mathbf{f}_f(t) \tag{30}$$

with initial conditions $\mathbf{z}'(0)$ and $\dot{\mathbf{z}}'(0)$ that can be obtained by evaluating the initial conditions of Eq. 25 at nodal points.

For efficient computation of $\mathbf{z}'(t)$, the same superposition method used in Section 3 can be used; i.e., let $\mathbf{z}'(t)$ be approximated by

$$\mathbf{z}'(t) = \sum_{i=1}^{k} w_i(t)\, \phi^i = \Phi\, \mathbf{w}(t), \qquad r \le k < n \tag{31}$$

where r is the number of basis vectors used for the original analysis, and k is the number of basis vectors used for the sensitivity analysis. In Eq. 31, ϕ^i is either eigenvector y^i or the LDRV φ^i that satisfies Eqs. 12 and 17, $\Phi=[\phi^1,\phi^2,\ldots,\phi^k]$, $\mathbf{w}(t) \in R^k$, and $\Phi^T\mathbf{M}\Phi=\mathbf{I}$.

Substituting Eq. 31 into Eq. 30 and premultiplying the resulting matrix equation by Φ^T,

$$\ddot{w}(t) + \Phi^T C \Phi \, \dot{w}(t) + \Phi^T K \Phi \, w(t) = \Phi^T f_f(t) \tag{32}$$

for t≥0, with initial conditions

$$w(0) = \Phi^T M \, z'(0), \qquad \dot{w}(0) = \Phi^T M \, \dot{z}'(0) \tag{33}$$

Note that dimension k of the sensitivity Eq. 32 is not necessarily the same as the dimension r of the original Eq. 19. The coupled Eq. 32 can be solved by the direct integration method to obtain **w**(t). To evaluate $f_f(t)$ in Eq. 30, $d'_{\delta u}(z_{tt}, \bar{z})$, $c'_{\delta u}(z_t, \bar{z})$, $a'_{\delta u}(z, \bar{z})$, and $\ell'_{\delta u}(\bar{z})$ in Eq. 21 must be computed for each design parameter. Once **w**(t) is obtained, the approximate design sensitivity **z'**(t) can be obtained from Eq. 31.

If the MAM was used for the original analysis, Eq. 31 is modified to

$$z'(t) = \Phi \, w(t) + K^{-1} f_f(t) - \Phi \Lambda^{-1} \Phi^T f_f(t) \tag{34}$$

where the basis vectors are all eigenvectors and Λ is a diagonal matrix with eigenvalues at diagonal entries. If we have proportional damping, **w**(t) can be obtained from the uncoupled equation, instead of the coupled Eq. 32,

$$\ddot{w}_i + 2\zeta_i \omega_i \dot{w}_i + \omega_i^2 w_i = \phi^{i^T} f_f(t) \tag{35}$$

For Rayleigh damping, $2\zeta_i\omega_i$ in Eq. 35 is replaced with $\alpha+\beta\omega_i^2$, where α and β are damping coefficients that define the damping matrix, **C**=α**M**+β**K**.

For design sensitivity of eigenvectors, using the FEA method, an approximate matrix equation of Eq. 29 is

$$(K - \zeta_i M) y^{i'} = g_f \tag{36}$$

Let the design sensitivity of the ith eigenvector be approximated by

$$y^{i'} = \sum_{i=1}^{r} c_i \phi^i = \Phi \, c \tag{37}$$

where r<n, $\Phi=[\phi^1,\phi^2,\ldots,\phi^r] = [y^1,\ldots,y^q,\varphi^1,\ldots,\varphi^{r-q}]$, and $c \in R^r$. LDRVs are generated as the same in Eqs. 13~17 replacing **p** with g_f. Substituting Eq. 37 into Eq. 36 and premultiplying the transpose of Φ,

$$\Phi^T (K - \zeta_i M) \Phi \, c = \Phi^T g_f \tag{38}$$

Equation 38 can be written in a partitioned form as

$$\begin{bmatrix} A_{11} & A_{12} \\ A_{12}^T & A_{22} \end{bmatrix} \begin{bmatrix} c_1 \\ c_2 \end{bmatrix} = \begin{bmatrix} Y^T g_f \\ \Psi^T g_f \end{bmatrix} \tag{39}$$

where $Y = [y^1,\ldots,y^q]$, $\Psi = [\varphi^1,\ldots,\varphi^{r-q}]$, and

A_{11}= qxq diagonal matrix with diagonal terms ζ_j-ζ_i, j = 1 to q

A_{12}= qx(r-q) zero matrix

A_{22}= (r-q)x(r-q) nonsingular full matrix $\varphi^{k^T} K \varphi^j - \zeta_i \delta_{kj}$

From Eq. 39, since A_{11} is diagonal,

$$c_j = \frac{y^{j^T} g_f}{\zeta_j - \zeta_i}, \quad j \neq i, \; j = 1 \text{ to } q \tag{40}$$

and c_k (k=q+1 to r) can be obtained using a linear equation solver. By taking the first variation of the normalizing condition of Eq. 9 and using Eq. 37,

$$c_i = -\frac{1}{2} d'_{\delta u}(y^i, y^i) \tag{41}$$

Note that the third term on the right side of Eq. 29 is not necessary in computing $y^{j^T} g_f$ in Eq. 40 since $d_u(y^i, y^j)=0$ for $j \neq i$. That is, design sensitivity of eigenvalues is not required to compute design sensitivity of eigenvectors. It is shown in Ref.13 that eigenfunctions corresponding to repeated eigenvalues are not Frechet differentiable.

5. Design Components

To use the continuum-based DSA method, the first variations of the energy bilinear and load linear forms must be derived for each design component so that these can be used to compute f_f and g_f in Eqs. 30 and 36, respectively. In this section, these terms are presented for the truss/beam design component. For the plane elastic solid/plate design component, see Refs. 14 or 15.

The kinetic and strain energy bilinear forms of the truss/beam design component, shown in Fig. 1, are

$$d_u(z_{tt}, \bar{z}) = \int_0^L \rho \left(h \sum_{i=1}^{3} z_{i,tt} \, \bar{z}_i + J \, z_{4,tt} \, \bar{z}_4 \right) dx_1 \tag{42}$$

and

$$a_u(z, \bar{z}) = \int_0^L (Ehz_{1,1}\bar{z}_{1,1} + EI_3 z_{2,11}\bar{z}_{2,11} + EI_2 z_{3,11}\bar{z}_{3,11} + GJ z_{4,1}\bar{z}_{4,1})\, dx_1 \tag{43}$$

where z_1, z_2, and z_3 are the axial displacement and two orthogonal lateral displacements, respectively; z_4 is the angle of twist; and $z=[z_1, z_2, z_3, z_4]^T$. In Eqs. 42 and 43, ρ is the mass density; E is Young's modulus; G is the shear modulus; and h, I_2, I_3, and J are the cross-sectional area, two moments of inertia and the torsional moment of inertia, respectively. Also, the dissipative energy due to Rayleigh damping is

$$c_u(z_t, \bar{z}) = \alpha \, d_u(z_t, \bar{z}) + \beta \, a_u(z_t, \bar{z}) \tag{44}$$

where α and β are damping coefficients.

The load linear form of the external loads shown in Fig. 1 is

$$\ell_u(\bar{z}) = \int_0^L \left(\sum_{i=1}^{3} f_i \, \bar{z}_i + T_1 \bar{z}_4 + M_2 \bar{z}_{3,1} + M_3 \bar{z}_{2,1} \right) dx_1 \tag{45}$$

where f_1, f_2, and f_3 are the axial and two orthogonal lateral loads, respectively, T_1 is the torque, and M_2 and M_3 are two moments.

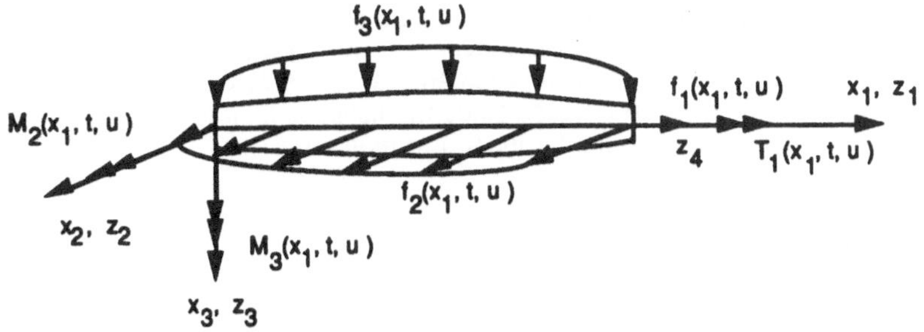

Figure 1 Truss/Beam Design Component

The first variations of the energy bilinear and load linear forms of Eqs. 42~45 are [13]

$$d'_{\delta u}(z_{tt}, \bar{z}) = \int_0^L \rho \left(h_{,u} \sum_{i=1}^{3} z_{i,tt} \bar{z}_i + J_{,u} z_{4,tt} \bar{z}_4 \right) \delta u \, dx_1 \tag{46}$$

$$a'_{\delta u}(z, \bar{z}) = \int_0^L (Eh_{,u} z_{1,1} \bar{z}_{1,1} + EI_{3,u} z_{2,11} \bar{z}_{2,11} + EI_{2,u} z_{3,11} \bar{z}_{3,11} + GJ_{,u} z_{4,1} \bar{z}_{4,1}) \delta u \, dx_1 \tag{47}$$

$$c'_{\delta u}(z_t, \bar{z}) = \alpha \, d'_{\delta u}(z_t, \bar{z}) + \beta \, a'_{\delta u}(z_t, \bar{z}) \tag{48}$$

and

$$\ell'_{\delta u}(\bar{z}) = \int_0^L \left(\sum_{i=1}^{3} f_{i,u} \bar{z}_i + T_{1,u} \bar{z}_4 + M_{2,u} \bar{z}_{3,1} + M_{3,u} \bar{z}_{2,1} \right) \delta u \, dx_1 \tag{49}$$

6. Numerical Examples

6.1 DESIGN SENSITIVITY ANALYSIS OF TRANSIENT RESPONSES

The design sensitivity of pointwise displacements of two undamped structures [17], a 25-member transmission tower with 18 D.O.F. and 108 member helicopter tail boom with 72 D.O.F., are discussed in Refs. 14 and 18, respectively. The helicopter tail boom with Rayleigh damping is studied here using the LDRV method and MAM. The damping coefficients are α=0.1318d+02 and β=0.2132d-04. This structure is subject to suddenly applied loads. For this structure, design variables are the cross-sectional areas and performance measures are the displacements at critical points [19]. ANSYS [20] is used to obtain stiffness and mass matrices and the eigenvalue problem is solved using EISPACK. A Fortran program is written to generate LDRVs, and the Wilson θ method is used for direct integration of the coupled matrix equation in the LDRV method. In the MAM, the Newmark β method is used for direct integration of the uncoupled equation. In this study, an Alliant FX/8 [21] is used to obtain the design sensitivity using parallel computation.

LDRV vectors generated using the original load of Eq. 10 may not be appropriate basis vectors for the fictitious load of the sensitivity Eq. 30. However, regenerating LDRV vectors at every time step would be very expensive. If the MAM of Eq. 34 is used, the basis vectors do not have to be updated since the fictitious load is taken care of by Eq. 34. The tail boom truss is used to compare these two methods.

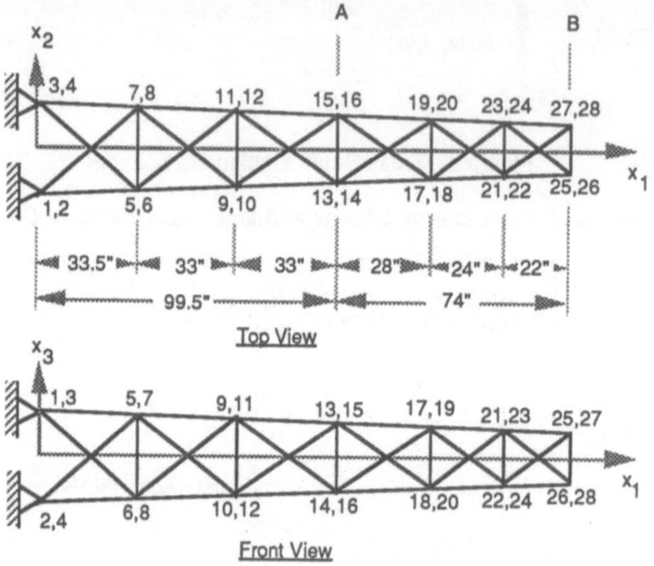

Figure 2 Helicopter Tail Boom Truss

The geometry and dimensions of a 108-member helicopter tail boom truss are shown in Fig. 2; and the load data, design data and initial design are given in Ref. 17. The first 5 natural frequencies, computed using EISPACK, are 34.20, 35.51, 69.93, 98.13, and 101.44 Hz. The time step of 0.0001 sec is used for analyses of both original and sensitivity equations. Sizing design sensitivity of displacements with respect to the first design variable, using the LDRV method and MAM at 0.016 second, which is the first peak time in Fig. 3, are given in Tables 1~4. In Tables 1~4, the y-direction displacement at the last four nodes are given since these are the maximum displacements. The first design variable is the cross-sectional areas of members 2 and 3.

The central finite difference is denoted by $\Delta\psi = (\psi(u+\delta u) - \psi(u-\delta u))/2$, and ψ' is the computed design sensitivity prediction. The ratio between $\Delta\psi$ and ψ' times 100 is used as a measure of accuracy of the design sensitivity computation. That is, 100% agreement means that the predicted change is exactly the same as the finite difference result. When $\Delta\psi$ is too small, this accuracy measure may fail to give correct information because $\Delta\psi$ may lose numerically significant digits. On the other hand, if $\Delta\psi$ is too large, the finite difference may contain nonlinear terms. Subscript E denotes the solution which was obtained using full basis vectors with analytical integration. In these tables, $\Delta\psi_E/\psi'$ and $\Delta\psi_E/\Delta\psi$ are used to check the accuracy of design sensitivity with the central finite difference of the solution with full basis. Also, ψ_E/ψ, $\dot{\psi}_E/\dot{\psi}$ and $\ddot{\psi}_E/\ddot{\psi}$ represent agreements between displacements, velocities and accelerations obtained with full and less than full bases, respectively.

Results of the LDRV method using 10 eigenvectors and 5 Ritz vectors are shown in Table 1. Table 2 contains results of the MAM using 15 eigenvectors. Results of both methods are excellent: the error in $\Delta\psi_E/\psi'$ is less than 3% for the LDRV method and is less than or equal to 1% for the MAM. Results of the LDRV method using 6 eigenvectors and 3 Ritz vectors are shown in Table 3. Table 4 contains results of the MAM using 7 eigenvectors. These are the minimum numbers of bases that produce less than 10% error in $\Delta\psi_E/\psi'$ for each method. For the LDRV method, results of these tables indicate that the basis vectors for the original load are good for the fictitious load of the sensitivity equation. However, results of the MAM are slightly better than those of the LDRV method because the displacement using the MAM is updated at every time step using the fictitious load. The time histories of displacement, velocity, acceleration and sensitivity prediction in the y-direction at node 27, where the maximum displacement occurs, are given in Figs. 3~6, respectively. In Figs. 3~6, responses of the LDRV method using 6 eigenvectors and 3 Ritz vectors are solid lines; responses of the MAM using 7 eigenvectors are dot-dashed lines; and responses of the full basis are dashed lines. In the beginning, responses of both methods using the minimum bases do not agree with responses of the full basis, but become very close after 0.04 second.

6.2 DESIGN SENSITIVITY ANALYSIS OF EIGENVECTORS

10-Member Cantilever Truss

The geometry, dimension, and design data of a 10-member cantilever truss are given in Fig. 7. The first 3 natural frequencies, computed using EISPACK, are 14.17, 39.73, and 41.45 Hz. Design sensitivity of the first eigenvector with respect to the seventh design variable, which is the cross-section area of the seventh element, is shown in Table 5, using a combination of eigenvectors and LDRVs.

It is shown in Table 5 that the accuracy of sensitivity is improved by adding 1 LDRV to 1 eigenvector. This basis is better than 4 eigenvectors. However, adding 2 LDRVs to 1 eigenvector is more accurate than adding 1 LDRV to 2 eigenvectors.

Helicopter Tail Boom Truss

For simplicity, the first 15 components of the design sensitivity of eigenvectors are shown in Tables 6 and 7. In Table 6, the design sensitivity of the first eigenvector with respect to the third design variable is given using various basis vectors. Adding 2 LDRVs to 4 eigenvectors gives excellent results. Adding three LDRVs, however, does not yield much improvement.

Design sensitivity of the fifth eigenvector with respect to the third design variable is given in Table 7. Adding 2 LDRVs to the eigenbasis gives accurate results. In this example, each design variable has to be perturbed to different amounts since perturbing all design variables to the same amount gives poor finite difference results.

Table 1 Sizing Design Sensitivity of Transient Displacements w.r.t. the First Design Variable for Tail Boom, Using 10 Eigenvectors and 5 Ritz Vectors at 0.016 second (α = 0.1318D+02, β = 0.2132D-04, 2% perturbation)

Node	DOF	$\psi(u)$	$\Delta\psi$	ψ'	$\Delta\psi/\psi'$ (%)	$\Delta\psi_E/\psi'$ (%)	$\Delta\psi_E/\Delta\psi$ (%)	ψ_E/ψ (%)	$\dot{\psi}_E/\dot{\psi}$ (%)	$\ddot{\psi}_E/\ddot{\psi}$ (%)
25	UY	0.7429D+0	-0.1998D-2	-0.2030D-2	98.4	102.6	104.2	100.0	200.2	88.7
26	"	0.5787D+0	-0.2477D-2	-0.2460D-2	100.7	100.4	100.0	100.1	103.5	121.5
27	"	0.7430D+0	-0.1997D-2	-0.2029D-2	98.5	102.5	104.1	100.0	201.3	89.1
28	"	0.5787D+0	-0.2477D-2	-0.2460D-2	100.7	100.5	99.8	100.1	103.6	121.1

Table 2 Sizing Design Sensitivity of Transient Displacements w.r.t. the First Design Variable for Tail Boom, Using 15 Eigenvectors at 0.016 second (α = 0.1318D+02, β = 0.2132D-04, 2% perturbation)

Node	DOF	$\psi(u)$	$\Delta\psi$	ψ'	$\Delta\psi/\psi'$ (%)	$\Delta\psi_E/\psi'$ (%)	$\Delta\psi_E/\Delta\psi$ (%)	ψ_E/ψ (%)	$\dot{\psi}_E/\dot{\psi}$ (%)	$\ddot{\psi}_E/\ddot{\psi}$ (%)
25	UY	0.7410D+0	-0.2086D-2	-0.2062D-2	101.2	101.0	99.8	99.7	123.1	61.3
26	"	0.5775D+0	-0.2466D-2	-0.2483D-2	99.3	99.4	100.1	99.9	91.8	97.4
27	"	0.7410D+0	-0.2085D-2	-0.2060D-2	101.2	101.0	99.8	99.7	124.3	61.4
28	"	0.5775D+0	-0.2466D-2	-0.2484D-2	99.3	99.5	100.2	99.9	91.1	97.2

Table 3 Sizing Design Sensitivity of Transient Displacements w.r.t. the First Design Variable for Tail Boom, Using 6 Eigenvectors and 3 Ritz Vectors at 0.016 second (α = 0.1318D+02, β = 0.2132D-04, 2% perturbation)

Node	DOF	$\psi(u)$	$\Delta\psi$	ψ'	$\Delta\psi/\psi'$ (%)	$\Delta\psi_E/\psi'$ (%)	$\Delta\psi_E/\Delta\psi$ (%)	ψ_E/ψ (%)	$\dot{\psi}_E/\dot{\psi}$ (%)	$\ddot{\psi}_E/\ddot{\psi}$ (%)
25	UY	0.7268D+0	-0.2108D-2	-0.2114D-2	99.8	98.5	98.8	97.8	76.4	-59.5
26	"	0.5855D+0	-0.2303D-2	-0.2320D-2	99.2	106.4	107.2	101.3	264.0	189.8
27	"	0.7269D+0	-0.2108D-2	-0.2113D-2	99.8	98.4	98.7	97.8	76.7	-59.1
28	"	0.5855D+0	-0.2303D-2	-0.2321D-2	99.2	106.5	107.3	101.3	261.4	189.2

Table 4 Sizing Design Sensitivity of Transient Displacements w.r.t. the First Design Variable for Tail Boom, Using 7 Eigenvectors at 0.016 second (α = 0.1318D+02, β = 0.2132D-04, 2% perturbation)

Node	DOF	$\psi(u)$	$\Delta\psi$	ψ'	$\Delta\psi/\psi'$ (%)	$\Delta\psi_E/\psi'$ (%)	$\Delta\psi_E/\Delta\psi$ (%)	ψ_E/ψ (%)	$\dot{\psi}_E/\dot{\psi}$ (%)	$\ddot{\psi}_E/\ddot{\psi}$ (%)
25	UY	0.7361D+0	-0.2073D-2	-0.2077D-2	99.8	100.2	100.4	99.1	132.6	26.2
26	"	0.5811D+0	-0.2339D-2	-0.2342D-2	99.9	105.4	105.6	100.5	246.7	133.9
27	"	0.7361D+0	-0.2072D-2	-0.2076D-2	99.8	100.2	100.4	99.1	134.6	26.3
28	"	0.5811D+0	-0.2340D-2	-0.2343D-2	99.9	105.5	105.6	100.5	243.6	133.8

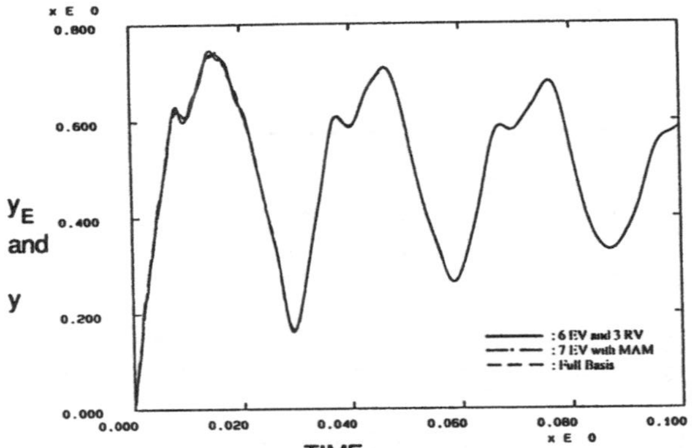

Figure 3 Comparison of y-Displacement at Node 27 Using the LDRV method and MAM

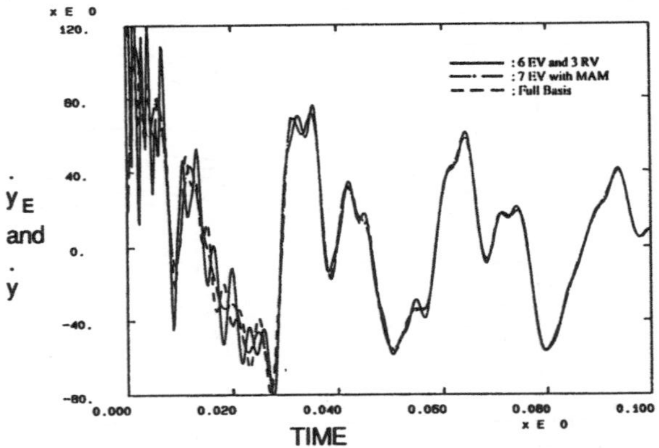

Figure 4 Comparison of y-velocity at Node 27 Using the LDRV method and MAM

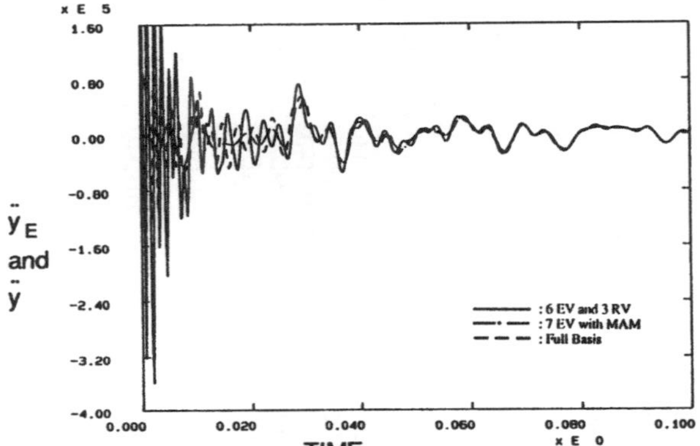

Figure 5 Comparison of y-Acceleration at Node 27 Using the LDRV method and MAM

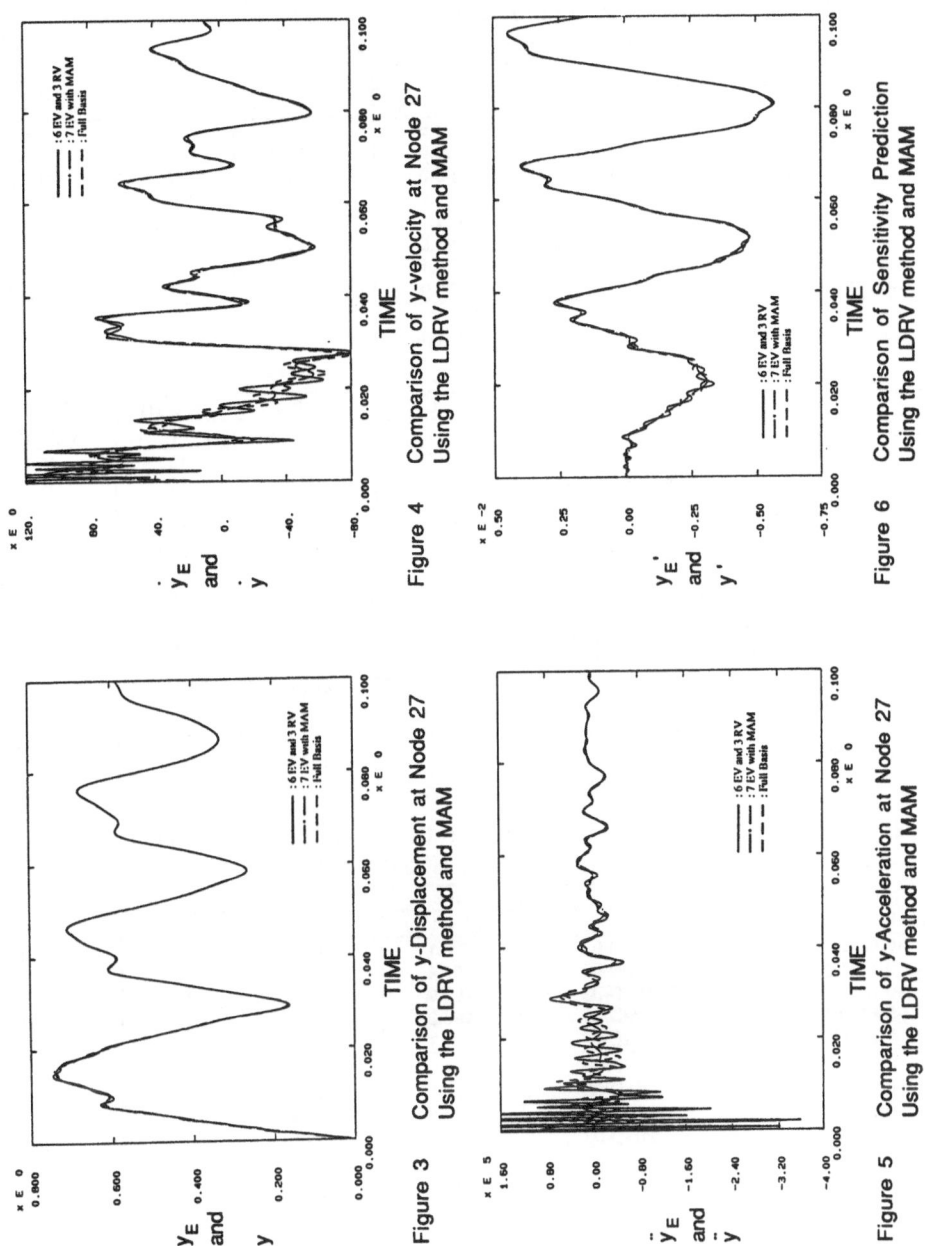

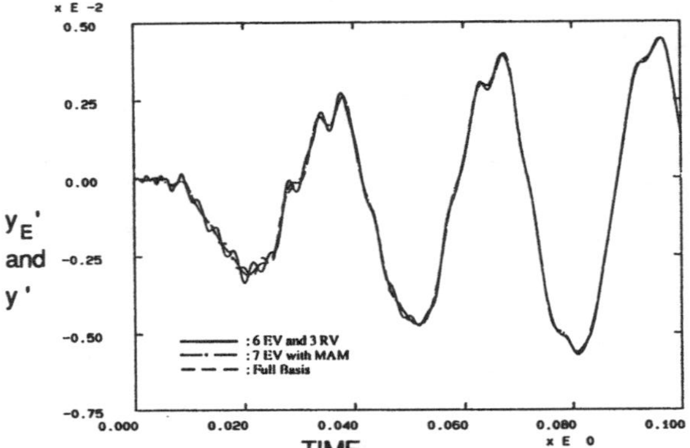

Figure 6 Comparison of Sensitivity Prediction Using the LDRV method and MAM

Table 5 Design Sensitivity of y_1 w.r.t. Seventh Design Variable where Design Sensitivity of λ_1 w.r.t. the Seventh Design Variable is 118.7 (1% Perturbation)

		4 E	1E & 1R	2E & 1R	1E & 2R
1st Eigenvector	Δy_i	$\Delta y_i/y_i'$(%)	$\Delta y_i/y_i'$(%)	$\Delta y_i/y_i'$(%)	$\Delta y_i/y_i'$(%)
-0.9543D-01	-0.1700D-04	84.6	63.5	110.0	97.4
-0.2152D+00	-0.2450D-03	130.8	96.1	96.0	100.3
-0.1212D+00	0.2900D-04	165.3	125.5	82.7	103.2
-0.4951D+00	0.1090D-03	97.4	98.7	96.8	101.5
0.1212D+00	-0.3400D-03	124.1	95.6	98.1	100.6
-0.4951D+00	0.1240D-03	94.6	97.9	96.8	101.1
0.9543D-01	-0.3066D-03	76.1	99.7	101.3	100.4
-0.2152D+00	-0.3520D-03	75.0	101.0	99.6	100.6

Table 6 Design Sensitivity of y_1 w.r.t. Third Design Variable where Design Sensitivity of λ_1 w.r.t. the Third Design Variable is 357.1 (10% Perturbation)

		20 E	4E & 1R	4E & 2R	4E & 3R
1st Eigenvector	Δy_i	$\Delta y_i/y_i'$(%)	$\Delta y_i/y_i'$(%)	$\Delta y_i/y_i'$(%)	$\Delta y_i/y_i'$(%)
0.8111D-01	0.1707D-02	154.8	96.4	100.4	100.3
-0.4063D+00	0.2586D-01	100.1	101.0	100.3	100.3
-0.2891D-01	-0.3726D-03	114.3	97.4	99.3	99.3
-0.8095D-01	-0.1714D-02	154.5	96.4	100.4	100.3
-0.4063D+00	0.2586D-01	100.1	101.0	100.3	100.3
0.2932D-01	0.4162D-03	114.3	98.8	100.5	100.6
-0.7942D-01	-0.1693D-02	155.1	96.4	100.3	100.3
-0.4062D+00	0.2567D-01	100.1	101.0	100.3	100.3
-0.2922D-01	-0.6350D-04	1501.6	92.1	99.5	99.3
0.7975D-01	0.1702D-02	154.7	96.4	100.4	100.3
-0.4063D+00	0.2566D-01	100.1	101.0	100.3	100.3
0.2974D-01	0.1069D-03	252.0	99.9	104.8	104.7
0.1390D+00	0.2150D-02	116.5	101.4	100.3	100.3
-0.1027D+01	0.1833D-01	100.1	107.5	100.3	100.3
-0.1947D-01	-0.2710D-03	81.6	110.7	98.0	98.1

Table 7 Design Sensitivity of y_5 w.r.t. the Third Design Variable where Design Sensitivity of λ_5 w.r.t. the Third Design Variable is -267.8 (10% Perturbation)

		30 E	5E & 1R	5E & 2R	5E & 3R
5th Eigenvector	Δy_i	$\Delta y_i/y_i'$(%)	$\Delta y_i/y_i'$(%)	$\Delta y_i/y_i'$(%)	$\Delta y_i/y_i'$(%)
-0.4732D-01	-0.2127D-03	202.1	99.1	102.2	101.5
0.1111D-01	-0.5340D-03	97.4	86.2	96.4	97.0
-0.1880D+01	-0.8186D-03	102.9	127.0	104.3	104.0
-0.4750D-01	-0.2399D-03	193.3	98.9	102.3	101.6
-0.1525D-01	0.1342D-02	101.1	98.0	101.1	101.1
-0.1886D+01	-0.8142D-03	104.6	123.2	105.3	105.6
0.4748D-01	0.2245D-03	201.1	97.9	101.3	100.6
0.1459D-01	-0.6622D-03	97.5	92.9	97.6	97.6
-0.1885D+01	-0.8286D-03	103.7	121.8	104.5	104.9
0.4735D-01	0.2153D-03	187.8	98.5	101.4	100.7
-0.8317D-02	0.1153D-02	101.7	95.4	101.1	101.5
-0.1880D+01	-0.8342D-03	103.1	126.7	104.5	104.3
-0.1842D+00	-0.1817D-04	-44.4	230.6	119.3	112.5
0.7958D-01	0.5744D-03	106.5	107.4	105.7	105.5
-0.2668D+01	0.1247D-02	100.7	95.3	100.0	100.4

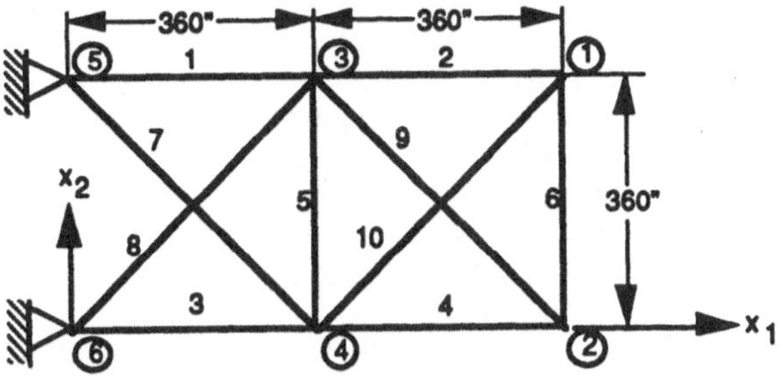

Design Data

Modulus of elasticity	=	10^4 ksi
Material density	=	0.10 lb/in^3
cross-section area	=	10 in^2

Figure 7. Ten-Member Cantilever Truss

7. Conclusion

A unified continuum-based sizing DSA method for the transient dynamic response and eigenvectors of built-up structures, utilizing the direct differentiation and superposition methods, has been proposed. This method is very efficient since derivatives of basis vectors which demand large computational effort are not required. Moreover, derivatives of stiffness and mass matrices are not required. Lastly, use of the LDRV method or the MAM makes this method inexpensive and improves the accuracy of analysis results and, thereby, the accuracy of sensitivity. A helicopter tail boom truss with Rayleigh damping using the LDRV method and MAM, is studied and very good design sensitivity results for transient responses are obtained. The minimum bases which produce less than 10% error in the sensitivity prediction are 6 eigenvectors and 3 Ritz vectors for the LDRV method, and 7 eigenvectors for the MAM. For this numerical example, the same number of basis vectors that are used for analysis of the original structure is enough for analysis of the design sensitivity equation.

Two structures, a 10-member cantilever truss and a helicopter tail boom truss, are studied and very good sensitivity results of eigenvectors are obtained by adding 2 LDRVs to the existing eigenvectors.

References

1. Cornwell, R.E., Craig, R.R., and Johnson, C. P., "On the Application of the Mode Acceleration Method to Structural Engineering Problems," Earthquake Engineering and Structural Dynamics, Vol. 11, 1983, pp. 679-688.
2. Leger, P. and Wilson, E.L., "Modal Summation Methods for Structural Dynamic Computations," Earthquake Engineering and Structural Dynamics, Vol. 16, 1988, pp. 23-27.
3. Wilson, E.L., Yuan, M.W., and Dickens, J.M., "Dynamic analysis by Direct Superposition of Ritz Vectors," Earthquake Engineering and Structural Dynamics, Vol. 10, 1982, pp. 813-821.

4. Kline, K.A., "Dynamic Analysis Using a Reduced Basis of Exact Modes and Ritz Vectors," AIAA Journal, Vol. 24, 1986, pp. 2022-2029.
5. Leger, P., Wilson, E.L., and Clough, R.W., "The Use of Load Dependent Vectors for Dynamic and Earthquake Analysis," Earthquake Engineering Research Center, Report UBC/EERC-86/04, 1986.
6. Drucker, D.S., Lou, M., Wang, S., and Kline, K.A., "A Comparison of Mode Acceleration and Ritz Vector Reduced Basis Procedures in Transient Analysis," Proceedings of 7th International Conference on Vehicle Structural Mechanics, Detroit, MI,April, 1988, SAE 880908.
7. Nour-Omid, B. and Clough, R.W., "Dynamic Analysis of Structures Using Lanczos Coordinate," Earthquake Engineering and Structural Dynamics, Vol. 12, 1984, pp. 499-505.
8. Kim, H.M. and Craig, R.R. Jr., "Structural Dynamics Analysis Using an Unsymmetric Block Lanczos Algorithm," International Journal for Numerical Methods in Engineering, Vol. 26, 1988, pp. 2305-2318.
9. Fox, R.L. and Kapoor, M.P., "Rate of Change of Eigenvalues and Eigenvectors," AIAA Journal, Vol. 18, pp. 1511-1514, 1968.
10. Nelson, R.B., "Simplified Calculation of Eigenvector Derivatives," AIAA Journal, Vol. 14, pp. 1201-1205, 1976.
11. Wang, B.P., "An Improved Approximate Method for Computing Eigenvector Derivatives," Presented at Work-in-progress session, AIAA/ASME/ASCE/AHS/ASC 26th Structures, Structural Dynamics and Materials Conference, 1985.
12. Wang, B.P., "Improved Approximate Methods for Computing Eigenvector Derivatives in Structural Dynamics," Submitted to AIAA Journal, March 1990.
13. Haug, E.J., Choi, K.K., and Komkov, V., Design Sensitivity Analysis of Structural Systems, Academic Press, 1986; also translated to the Russian by N. V. Banichuk and published in the USSR, 1988.
14. Choi, K.K. and Wang, S., "Continuum Design Sensitivity of Structural Dynamic Response Using Ritz Sequence," AIAA/ASME/ASCE/AHS/ASC 31st Structures, Structural Dynamics and Materials Conference, Paper No. 90-1137, 1990, pp. 385-393.
15. Wang, S. and Choi, K.K., "Continuum Design Sensitivity Analysis of Transient Responses Using Ritz and Mode Acceleration Methods," AIAA Journal, to appear 1991.
16. Wang, S. and Choi, K.K., "Continuum Design Sensitivity Analysis of Eigenvectors Using Ritz Vectors," AIAA/ASME/ASCE/AHS/ASC 32nd Structures, Structural Dynamics and Materials Conference, Paper No. 91-1092, 1991, pp. 406-412.
17. Haug, E.J. and Arora, J.S., Applied Optimal Design, John Wiley & Sons, New York, 1979.
18. Choi, K.K. and Wang, S., "Continuum Design Sensitivity of Structural Dynamic Reponse Using Ritz Sequence," Technical Report R-73, Center for Simulation and Design Optimization of Mechanical Systems, University of Iowa, May 1990.
19. Green, W.H. and Haftka, R.T., "Computational Aspects of Sensitivity Calculations in Transient Structural Analysis," Computers & Structures, Vol. 32, No. 2, 1989, pp. 433-443.
20. DeSalvo, G.J. and Swanson, J.A., ANSYS Engineering Analysis System, User's Manual, Vols. I and II, Swanson Analysis Systems, Inc., Houston, PA, 1987.
21 FX/Fortran Programmer Handbook, Alliant Computer Systems Corporation, July 1988.

DESIGN SENSITIVITY ANALYSIS OF DYNAMIC FREQUENCY RESPONSES OF ACOUSTO-ELASTIC BUILT-UP STRUCTURES

Kyung K. Choi, Inbo Shim, and Jaehwan Lee
Department of Mechanical Engineering
and
Center for Simulation and Design Optimization
of Mechanical Systems
The University of Iowa
Iowa City, Iowa 52242, U. S. A.

Harihar T. Kulkarni
Body Structures Department
Product and Manufacturing Engineering Staff
Ford Motor Co.
Dearborn, Michigan 48124, U. S. A.

ABSTRACT. A continuum design sensitivity analysis (DSA) method for dynamic frequency responses of acousto-elastic built-up structures is developed using the adjoint variable and direct differentiation methods. A variational approach with non-selfadjoint operators for complex variable is used to retain the continuum elasticity formulation throughout derivation of design sensitivity results. Sizing design variables such as thickness and cross-sectional area of structural components are considered for the design sensitivity analysis. A numerical implementation method for continuum DSA results is developed by postprocessing analysis results from established finite element analysis (FEA) codes to obtain the design sensitivity of displacement and pressure performance measures of the acousto-elastic built-up structures. The numerical method is tested using passenger vehicle problems. Accurate design sensitivity results are obtained for analysis results obtained from MSC/NASTRAN and ABAQUS FEA codes.

1. Introduction

Dynamic frequency response is used in various industries for noise, vibration, and harshness (NVH) analysis of mechanical and structural systems that are subject to harmonically varying external loads caused by reciprocating powertrains or other rotating machineries such as motors, fans, compressors, and forging hammers [1]. For example, airplane body and wing structures are subject to a harmonic load transmitted from the propulsion system. Also, ship vibrations resulting from propeller and engine excitations can cause noise problem, cracks, fatigue failure of tailshaft, and discomfort to crew. When a machine or any structure oscillates in some form of periodic or random motion, the motion generates alternating pressure waves that propagate from the moving surface at the velocity of sound. For instance, the interior sound pressure

G. I. N. Rozvany (ed.), Optimization of Large Structural Systems, Vol. I, 329–343.

in an automobile compartment can occur when the input forces transmitted from road and power train excite the vehicle compartment boundary panels. These motions with frequencies between 20 Hz and 20 KHz stimulate the human hearing mechanism [2].

There are several published works in the area of optimization of structural systems with dynamic frequency responses. Mroz used a variational principle to derive necessary and sufficient conditions for optimal design [3]. Lekszycki and Olhoff [4] derived a general set of necessary conditions for optimal design of one-dimensional, viscoelastic structures acted on by harmonic loads. A non-selfadjoint operator is used by means of variational analysis and the concept of complex stiffness modulus is adopted. Yoshimura performed [5] DSA of the frequency response of machine structures and presented a numerical example of design sensitivity using a simplified structural model of a lathe. Lekszycki and Mroz [6] extended their previous work [4] to find necessary conditions for optimal support reactions to minimize stress and displacement amplitudes. A variational approach with non-selfadjoint operator is used to consider a one-dimensional viscoelastic structure that is subject to harmonic loads. Choi and Lee developed a continuum DSA method of dynamic frequency responses of structural systems [7] using the adjoint and direct differentiation methods. A variational approach with a non-selfadjoint operator for complex variables is used to retain the continuum elasticity formulation throughout derivation of design sensitivity expressions. For numerical implementation, the COSMIC/NASTRAN FEA code is used for analysis of the original and adjoint structures. Since NASTRAN does not provide shape functions for some finite elements, external shape functions that have the same degrees of freedom are used to compute the adjoint loads and evaluate the continuum design sensitivity expressions. It is shown in Ref. 7 that use of external shape functions do not affect accuracy of the design sensitivity results.

The objective of this paper is to develop a continuum DSA method for dynamic frequency responses of acousto-elastic built-up structures. As in Ref. 7, the continuum DSA method can be implemented outside established FEA codes [8,9] using postprocessing data only, since it does not require derivatives of the stiffness, damping, and mass matrices.

2. Variational Formulation of Acousto-Elastic Built-up Structure

A typical acousto-elastic built-up structure is shown in Fig. 1. All members of the built-up structure are assumed to be plates (two-dimensional components) and/or beams (one-dimensional components) in three dimensional space. The built-up structure encloses a three-dimensional cavity that contains a medium (fluid) that transmits linear acoustic wave.

The coupled dynamic motion of the built-up structure and acoustic medium can be described using the following system of differential equations:

Structure

$$m(x,u)z_{tt}(x,u,t) + C_u z_t(x,u,t) + A_u z(x,u,t) = f(x,u,t) + p^s(x,u,t)n, \quad x \in \Omega^s, \; t \geq 0 \qquad (1)$$

with boundary and initial conditions

$$Gz = 0, \quad x \in \Gamma^s \qquad (2)$$

$$z(x,u,0) = z_t(x,u,0) = 0, \quad x \in \Omega^s \qquad (3)$$

Fluid

$$\frac{1}{\beta}p_{tt}(x,u,t) + Tp_t(x,u,t) - \frac{1}{\rho_0}\nabla^2 p(x,u,t) = 0, \qquad x \in \Omega^a, \;\; t \geq 0 \tag{4}$$

with boundary and initial conditions

$$\nabla p^T n = 0, \qquad x \in \Gamma^{ar} \tag{5}$$

$$p(x,u,0) = p_t(x,u,0) = 0, \quad x \in \Omega^a \tag{6}$$

Interface Conditions

$$p^s = p, \qquad x \in \Gamma^{as} \equiv \Omega^s \tag{7}$$

$$\nabla p^T n = -\rho_0 z_{tt} n, \qquad x \in \Gamma^{as} \equiv \Omega^s \tag{8}$$

where Ω^s is the domain of the structure; m(x,u) is the mass effect in the structure; C_u is the linear differential operator that corresponds to the damping effect in the structure; A_u is the fourth order partial differential operator; f(x,t,u) is the time dependent applied load ; p^s is the acoustic pressure at the interface between the structure and the acoustic medium; and n is the outward unit normal vector on the boundary of acoustic medium. The design variable u(x) is independent of time and dynamic response $z(x,u,t)=[z_1, z_2, z_3]^T$ is the displacement field of the structure. Boundary conditions of Eq. 2 are imposed on the structural boundary Γ^s using the trace operator G.

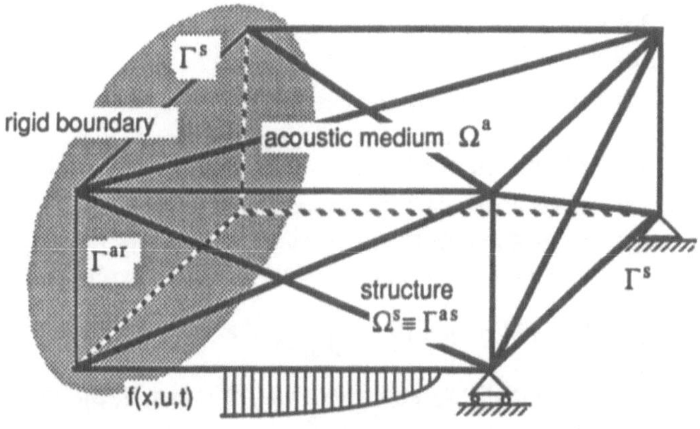

Figure 1 Acousto-Elastic Built-up Structure

In Eq. 4, Ω^a is the domain of the acoustic medium; β is the adiabatic bulk modulus; and ρ_0 is the equilibrium density. The dynamic response p(x,u,t) is the acoustic or excessive pressure and the linear operator T corresponds to dissipation of acoustic energy. The normal gradient of pressure vanishes on the rigid wall Γ^{ar} as shown in Eq. 5.

Structure-fluid interaction between two systems can be seen in Eq. 7 in the form of structural load p^s, and in Eq. 8 in the form of acoustic boundary condition with the relation between the structural acceleration z_{tt} and the gradient ∇p of the acoustic pressure at the interface. Note that, as can be seen in Fig. 1, the acoustic-structure interface Γ^{as} is the domain Ω^s of the structure.

When the harmonic force f(x,u,t) with a single frequency ω is applied to the built-up structure of the coupled linear system, the corresponding dynamic responses z(x,u,t) and p(x,u,t) are also harmonic functions with the same frequency ω. These can be represented using complex harmonic functions as

$$\left.\begin{aligned} f(x,u,t) &= \mathrm{Re}\{\, \mathbf{f}(x,u)\, e^{j\omega t} \} \\ z(x,u,t) &= \mathrm{Re}\{\, \mathbf{z}(x,u)\, e^{j\omega t} \} \\ p(x,u,t) &= \mathrm{Re}\{\, \mathbf{p}(x,u)\, e^{j\omega t} \} \end{aligned}\right\} \tag{9}$$

where **f**, **z**, and **p** are complex phasors that are independent of time. Then Eqs. 1~8 can be reduced to the following time independent system of equations:

Structure

$$D_u\mathbf{z} \equiv -\omega^2 m(x,u)\mathbf{z} + i\omega C_u\mathbf{z} + A_u\mathbf{z} = \mathbf{f}(x,u) + \mathbf{p}^s n, \qquad x \in \Omega^s \tag{10}$$

with boundary conditions

$$G\mathbf{z} = 0, \qquad x \in \Gamma^s \tag{11}$$

Fluid

$$B\mathbf{p} \equiv -\frac{\omega^2}{\beta}\mathbf{p} + i\omega T\mathbf{p} - \frac{1}{\rho_0}\nabla^2\mathbf{p} = 0, \qquad x \in \Omega^a \tag{12}$$

with boundary conditions

$$\nabla\mathbf{p}^T n = 0, \qquad x \in \Gamma^{ar} \tag{13}$$

Interface Conditions

$$\mathbf{p}^s = \mathbf{p}, \qquad x \in \Gamma^{as} \equiv \Omega^s \tag{14}$$

$$\nabla\mathbf{p}^T n = \omega^2 \rho_o \mathbf{z}^T n, \qquad x \in \Gamma^{as} \equiv \Omega^s \tag{15}$$

The differential operator D_u in Eq. 10 depends on design u, while the differential operator B in Eq. 12 does not because the acoustic medium is assumed to be fixed.

A variational form of Eqs. 10 and 12 can be obtained by multiplying both sides of Eqs. 10 and 12 by the transpose of complex conjugates $\bar{\mathbf{z}}^*$ and $\bar{\mathbf{p}}^*$ of kinematically admissible virtual states $\bar{\mathbf{z}} \in Z$ and $\bar{\mathbf{p}} \in P$, respectively, integrating by parts over each physical domain, adding them, and using boundary and interface conditions,

$$b_u(\mathbf{z}, \bar{\mathbf{z}}) - \iint_{\Gamma^{as}} \mathbf{p}\bar{\mathbf{z}}^{*T} n \, d\Gamma + d(\mathbf{p}, \bar{\mathbf{p}}) - \omega^2 \iint_{\Gamma^{as}} \bar{\mathbf{p}}^* \mathbf{z}^T n \, d\Gamma = \ell_u(\bar{\mathbf{z}}) \tag{16}$$

which must hold for all kinematically admissible virtual states $\{\bar{\mathbf{z}}, \bar{\mathbf{p}}\} \in Q$ where

$$Q = \{\, \mathbf{z} \in Z,\ \mathbf{p} \in P \mid \mathbf{p}^s = \mathbf{p} \text{ and } \nabla\mathbf{p}^T n = \omega^2 \rho_o \mathbf{z}^T n, \quad x \in \Gamma^{as} \} \tag{17}$$

and

$$\left.\begin{aligned} Z &= \{\mathbf{z} \in [H^2(\Omega^s)]^3 \mid G\mathbf{z} = 0, \quad x \in \Gamma^s\} \\ P &= \{\mathbf{p} \in H^1(\Omega^a) \mid \nabla\mathbf{p}^T n = 0, \quad x \in \Gamma^a\} \end{aligned}\right\} \tag{18}$$

and H^1 and H^2 are Sobolev spaces of orders one and two, respectively [10]. In Eq. 16, the sesquilinear forms b_u and d, and semilinear form ℓ_u are defined, using L_2-inner product on complex function spaces, as

$$b_u(\mathbf{z}, \bar{\mathbf{z}}) \equiv (D_u\mathbf{z}, \bar{\mathbf{z}}) = -\iint_{\Omega^s} \omega^2 m \bar{\mathbf{z}}^{*T} \mathbf{z} \, d\Omega + i\omega c_u(\mathbf{z}, \bar{\mathbf{z}}) + a_u(\mathbf{z}, \bar{\mathbf{z}}) \tag{19}$$

where

$$c_u(z, \bar{z}) \equiv \iint_{\Omega^s} \bar{z}^{*T} C_u z \, d\Omega \quad \text{and} \quad a_u(z, \bar{z}) \equiv \iint_{\Omega^s} \bar{z}^{*T} A_u z \, d\Omega \tag{20}$$

$$d(p, \bar{p}) \equiv (Bp, \bar{p}) = \iiint_{\Omega^a} \left[-\frac{\omega^2}{\beta} p \, \bar{p}^* + i\omega T p \, \bar{p}^* + \frac{1}{\rho_0} \nabla p^T \nabla \bar{p}^* \right] d\Omega \tag{21}$$

and

$$\ell_u(\bar{z}) \equiv \iint_{\Omega^s} f^T \bar{z}^* \, d\Omega \tag{22}$$

If the built-up structure does not have an acoustic medium, then the variational Eq. 16 can be simplified by dropping all terms corresponding to the acoustic medium, including interface conditions, and the result will be the same as in Ref. 7.

3. Design Sensitivity Analysis of Dynamic Frequency Response

3.1 DIRECT DIFFERENTIATION METHOD

To develop the direct differentiation method of DSA, take the first variation of Eq. 16 with respect to design u and rearrange to obtain

$$b_u(z', \bar{z}) - \iint_{\Gamma^{as}} p' \bar{z}^{*T} n \, d\Gamma + d(p', \bar{p}) - \omega^2 \iint_{\Gamma^{as}} \bar{p}^* z'^T n \, d\Gamma = \ell'_{\delta u}(\bar{z}) - b'_{\delta u}(z, \bar{z}) \tag{23}$$

which must hold for all kinematically admissible virtual states $\{\bar{z}, \bar{p}\} \in Q$. In Eq. 23,

$$z' \equiv \frac{d}{d\tau} z(x, u + \tau\delta u)\Big|_{\tau=0} \tag{24}$$

$$p' \equiv \frac{d}{d\tau} p(x, u + \tau\delta u)\Big|_{\tau=0} \tag{25}$$

are the first variations of z and p with respect to design u in the direction δu of design change. Also, the first variations of the sesquilinear form b_u and semilinear form ℓ_u with respect to explicit dependence on design u are

$$b'_{\delta u}(z, \bar{z}) \equiv \frac{d}{d\tau} b_{u+\tau\delta u}(\tilde{z}, \bar{z})\Big|_{\tau=0} \tag{26}$$

$$\ell'_{\delta u}(\bar{z}) \equiv \frac{d}{d\tau} \ell_{u+\tau\delta u}(\bar{z})\Big|_{\tau=0} \tag{27}$$

where $\tilde{z}$ denotes the state z with dependence on τ (design variable) suppressed. Equation 23 is a variational equation in which the design sensitivities z' and p' are unknowns. If the solution z of Eq. 16 is obtained, the right side of Eq. 23 can be computed, using the shape functions of the finite element to evaluate integrands at Gauss points and numerical integration, to obtain the fictitious load. This yields the direct differentiation method of DSA.

3.2 ADJOINT VARIABLE METHOD

Harmonic performance measures of the acousto-elastic built-up structure can be expressed in terms of complex phasors of the structural displacement and the acoustic pressure. For the adjoint variable method, first consider the pressure at a point $\hat{x}$ in the acoustic medium enclosed by the built-up structure under harmonic excitation

$$\psi_p = \iiint_{\Omega^a} \hat{\delta}(x - \hat{x}) p \, d\Omega \tag{28}$$

The pressure can be correlated to noise at the human ear in the passenger vehicle, aircraft, or ship. The first variation of the performance measure is

$$\psi_p' = \iiint_{\Omega^a} \hat{\delta}(x - \hat{x}) p' \, d\Omega \tag{29}$$

To use the adjoint variable method, define non-selfadjoint operators D_u^a and B^a corresponding to the operators D_u and B of Eqs. 10 and 12, respectively, as

$$(D_u z, \lambda) \equiv (z, D_u^a \lambda), \qquad \text{for all } z, \lambda \in Z \tag{30}$$

and

$$(Bp, \eta) \equiv (p, B^a \eta), \qquad \text{for all } p, \eta \in P \tag{31}$$

Then by the definition of $b_u(\cdot,\cdot)$ of Eq. 19 and $d(\cdot,\cdot)$ of Eq. 21,

$$b_u(\bar{\lambda}, \lambda) = (\bar{\lambda}, D_u^a \lambda) = -\iint_{\Omega^s} \omega^2 m \bar{\lambda}^T \lambda^* \, d\Omega + i\omega c_u(\bar{\lambda}, \lambda) + a_u(\bar{\lambda}, \lambda) \tag{32}$$

and

$$d(\bar{\eta}, \eta) = (\bar{\eta}, B^a \eta) = \iiint_{\Omega^a} \left[-\frac{\omega^2}{\beta} \bar{\eta}\, \eta^* + i\omega T \bar{\eta}\, \eta^* + \frac{1}{\rho_0} \nabla \bar{\eta}^T \nabla \eta^* \right] d\Omega \tag{33}$$

Define an adjoint equation for the performance measure of Eq. 28 by replacing p' in Eq. 29 by a virtual pressure $\bar{\eta}$ and equating the term to the sesquilinear forms as

$$b_u(\bar{\lambda}, \lambda) - \iint_{\Gamma^{as}} \bar{\eta} \lambda^{*T} n \, d\Gamma + d(\bar{\eta}, \eta) - \omega^2 \iint_{\Gamma^{as}} \eta^* \bar{\lambda}^T n \, d\Gamma = \iiint_{\Omega^a} \hat{\delta}(x - \hat{x}) \bar{\eta} \, d\Omega \tag{34}$$

which must hold for all kinematically admissible virtual states $\{\bar{\lambda}, \bar{\eta}\} \in Q$. Note that the solution of Eq. 34 is the complex conjugates $\{\lambda^*, \eta^*\}$. To take advantage of the adjoint equation, we may evaluate Eq. 34 at $\bar{\lambda}=z'$ and $\bar{\eta}=p'$, to obtain

$$b_u(z', \lambda) - \iint_{\Gamma^{as}} p' \lambda^{*T} n \, d\Gamma + d(p', \eta) - \omega^2 \iint_{\Gamma^{as}} \eta^* z'^T n \, d\Gamma = \iiint_{\Omega^a} \hat{\delta}(x - \hat{x}) p' \, d\Omega \tag{35}$$

which is the term on the right of Eq. 29 that we would like to write explicitly in terms of δu. Similarly, evaluate Eq. 23 at $\bar{z}^*=\lambda^*$ and $\bar{p}^*=\eta^*$ to obtain

$$b_u(z', \lambda) - \iint_{\Gamma^{as}} p' \lambda^{*T} n \, d\Gamma + d(p', \eta) - \omega^2 \iint_{\Gamma^{as}} \eta^* z'^T n \, d\Gamma = \ell'_{\delta u}(\lambda) - b'_{\delta u}(z, \lambda) \tag{36}$$

Since the left sides of Eqs. 35 and 36 are equal, the desired explicit design sensitivity expression can be obtained, from Eqs. 29, 35, and 36,

$$\begin{aligned} \psi_p' &= \ell'_{\delta u}(\lambda) - b'_{\delta u}(z, \lambda) \\ &= \iint_{\Omega^s} f_u^T \lambda^* \delta u \, d\Omega + \iint_{\Omega^s} \omega^2 m_u \lambda^{*T} z \delta u \, d\Omega - i\omega c'_{\delta u}(z, \lambda) - a'_{\delta u}(z, \lambda) \end{aligned} \tag{37}$$

Note that, to evaluate the design sensitivity of Eq. 37, the solution λ^* of Eq. 34, which is the complex conjugate of λ, must be obtained.

Another performance measure of the acousto-elastic built-up structure is the structural displacement at a point $\hat{x}$. For instance, the performance measure could be the vibration amplitude at a seat of the passenger vehicle, aircraft, or ship. The performance measure can be written as

$$\psi_{z_i} = \iint_{\Omega^s} \hat{\delta}(x - \hat{x}) z_i \, d\Omega, \qquad i=1,2,3 \tag{38}$$

The adjoint equation for this performance measure is defined as

$$b_u(\bar{\lambda}, \lambda) - \iint_{\Gamma^{as}} \bar{\eta}\lambda^{*T} n \, d\Gamma + d(\bar{\eta}, \eta) - \omega^2 \iint_{\Gamma^{as}} \eta^* \bar{\lambda}^T n \, d\Gamma = \iint_{\Omega^s} \hat{\delta}(x - \hat{x}) \bar{\lambda}_i \, d\Omega \tag{39}$$

which must hold for all kinematically admissible virtual states $\{\bar{\lambda}, \bar{\eta}\} \in Q$. Once the complex conjugate λ^* of the adjoint response is obtained from Eq. 39, the same design sensitivity expression of Eq. 37 can be used to obtain design sensitivity information. Another point to note is that, since sizing design u is defined only on the structural part, Eq. 37 requires only the structural response λ^* of adjoint Eqs. 34 or 39.

4. Design Components

To use the continuum DSA method, the first variations of the sesquilinear and semilinear forms must be derived for each design component so that these can be used to evaluate the design sensitivity of Eq. 37. In this paper, the built-up structure with structural damping is considered.

4.1 BEAM DESIGN COMPONENT

The sesquilinear form of the beam design component of length L and structural damping coefficient φ is [7]

$$b_u(z, \bar{z}) = -\int_0^L \omega^2 \left(\rho h \sum_{i=1}^{3} z_i \bar{z}_i^* + J z_4 \bar{z}_4^* \right) dx_1 + (1+i\varphi) \int_0^L (E h z_{1,1} \bar{z}_{1,1}^* + E I_3 z_{2,11} \bar{z}_{2,11}^* + E I_2 z_{3,11} \bar{z}_{3,11}^* + G J z_{4,1} \bar{z}_{4,1}^*) \, dx_1 \tag{40}$$

where z_1, z_2, z_3, and z_4, are the axial displacement, two orthogonal lateral displacements, and the angle of twist, respectively, and $z=[z_1, z_2, z_3, z_4]^T$. In Eq. 40, ρ is the mass density, E is Young's modulus, G is shear modulus, and h, I_2, I_3, and J are the cross-sectional area, two moments of inertia and the torsional moment of inertia, respectively. The semilinear form of external loads is

$$\ell_u(\bar{z}) = \int_0^L \left(\sum_{i=1}^{3} f_i \bar{z}_i^* + T_1 \bar{z}_4^* + M_2 \bar{z}_{3,1}^* + M_3 \bar{z}_{2,1}^* \right) dx_1 \tag{41}$$

where f_1, f_2, and f_3 are the axial and two orthogonal lateral harmonic loads, respectively. Also, T_1 is the harmonic torque and M_2 and M_3 are two harmonic moments.

The first variations of the sesquilinear and semilinear forms of Eqs. 40 and 41 can be obtained by taking the first variations of Eqs. 40 and 41 with respect to explicit dependency on design u as [7]

$$b'_{\delta u}(z, \bar{z}) = -\int_0^L \omega^2 \left(\rho h_{,u} \sum_{i=1}^{3} z_i \bar{z}_i^* + J_{,u} z_4 \bar{z}_4^* \right) \delta u \, dx_1 + (1+i\varphi) \int_0^L (E h_{,u} z_{1,1} \bar{z}_{1,1}^*$$

$$+ EI_{3,u} z_{2,11} \bar{z}_{2,11}^* + EI_{2,u} z_{3,11} \bar{z}_{3,11}^* + GJ_{,u} z_{4,1} \bar{z}_{4,1}^*) \, \delta u \, dx_1 \tag{42}$$

and

$$\ell'_{\delta u}(\bar{z}) = \int_0^L \left(\sum_{i=1}^{3} f_{i,u} \bar{z}_i^* + T_{1,u} \bar{z}_4^* + M_{2,u} \bar{z}_{3,1}^* + M_{3,u} \bar{z}_{2,1}^* \right) \delta u \, dx_1 \tag{43}$$

where the subscript u denotes the derivative of terms with respect to design u.

4.2 PLATE DESIGN COMPONENT

The sesquilinear form of the plate design component with structural damping coefficient φ is [7]

$$b_u(z, \bar{z}) = -\iint_\Omega \omega^2 \rho h \sum_{i=1}^{3} z_i \bar{z}_i^* \, d\Omega + (1+i\varphi) \iint_\Omega \left[h \sum_{i,j=1}^{2} \sigma^{ij}(v) \varepsilon^{ij}(\bar{v}^*) + \frac{h}{3} \sum_{i,j=1}^{2} \sigma^{ij}(z_3) \varepsilon^{ij}(\bar{z}_3^*) \right] d\Omega \tag{44}$$

where z_3 is the lateral displacement due to bending and $v=[z_1, z_2]^T$ is the in-plane displacement. For this design component, sizing design variable $u=h(x_1,x_2)$ is the thickness of the component. The semilinear form of external loads is

$$\ell_u(\bar{z}) = \iint_\Omega \sum_{i=1}^{3} f_i \bar{z}_i^* \, d\Omega + \int_{\Gamma^2} \sum_{i=1}^{2} T_i \bar{z}_i^* \, d\Gamma \tag{45}$$

where f_1 and f_2 are two in-plane harmonic loads; f_3 is the lateral harmonic load; and T_1 and T_2 are two in-plane harmonic traction loads applied at the traction boundary Γ^2.

The first variations of the sesquilinear and semilinear forms of Eqs. 44 and 45 can be obtained by taking the first variations of Eqs. 44 and 45 with respect to explicit dependency on design h as [7]

$$b'_{\delta u}(z, \bar{z}) = -\iint_\Omega \omega^2 \rho \sum_{i=1}^{3} z_i \bar{z}_i^* \, \delta h \, d\Omega$$

$$+ (1+i\varphi) \iint_\Omega \left[\sum_{i,j=1}^{2} \sigma^{ij}(v) \varepsilon^{ij}(\bar{v}^*) + \frac{1}{3} \sum_{i,j=1}^{2} \sigma^{ij}(z_3) \varepsilon^{ij}(\bar{z}_3^*) \right] \delta h \, d\Omega \tag{46}$$

and

$$\ell'_{\delta u}(\bar{z}) = \iint_\Omega \sum_{i=1}^{3} f_{i,h} \bar{z}_i^* \, \delta h \, d\Omega + \int_{\Gamma^2} \sum_{i=1}^{2} T_{i,h} \bar{z}_i^* \, \delta h \, d\Gamma \tag{47}$$

where the subscript h denotes the derivative of terms with respect to design h.

5. Numerical Examples

For the adjoint variable method, the adjoint load for each performance measure needs to be computed. For the displacement performance measure, a unit harmonic load is applied at the specified node of the structure in the direction of the displacement of which the design sensitivity information is to be computed. For the pressure performance measure, the adjoint load is the second time derivative of a unit volumetric strain at the point in the acoustic medium where the pressure performance measure is defined. The complex conjugates of the adjoint structural responses; i.e., solutions of Eqs. 34 and 39 can be obtained efficiently by the restart option of established FEA codes. Using the original response and complex conjugate of adjoint structural response, the design sensitivity information of Eq. 37 can be obtained by evaluating the integrands of Eq. 37 at Gauss points using the shape functions of the finite element and by carrying out numerical integration. Computational procedures for continuum design sensitivity analysis can be found in Refs. 7~9. If the direct differentiation method is used, the fictitious load on the right of Eq. 23 is computed using the shape functions of the finite element and numerical integration. An efficient solution of Eq. 23 can also be obtained using the restart option of FEA codes. Two vehicle systems, one with acoustic and the other without acoustic, are used to demonstrate feasibility of the continuum DSA method.

5.1 CONCEPT VEHICLE SYSTEM

Structural design criteria of passenger vehicle originate from the service load, safety consideration, and ride quality. The fuel economy compels engineers to design lighter body structure and favor a low RPM-high torque engine operating strategy. However, vehicle NVH problems occur at low RPM-high torque engine operation; i.e., the lugging or idling condition. Shuffling is an annoying fore and aft oscillation of the vehicle that is caused when engine roughness at rapid torque application excites a torsional resonance of the driveline. Shake is a high tactile vibration and boom is a noise measured by the sound pressure level. It is important at an early design stage to predict the vehicle's NVH characteristics, identify problems, and improve design.

The finite element model of a concept vehicle structure that consists of body-in-white, subframe, steering column, and suspension substructures is shown in Fig. 2. The large-scale finite element model consists of 618 beams, 732 springs, 2364 quadrilateral and 344 triangular plates, and 94 bar finite elements with 3324 grid points and 19944 degrees-of-freedom.

For this model, effects of rocker design on vibration amplitudes at driver and passenger seats are found using the DSA method. The vehicle rocker consists of two parts: the front part from Hinge pillar to B pillar, and the rear part from B pillar to C Pillar. Design variables on the rocker are two bending moments of inertia I_2, and I_3, and torsional moment of inertia J as shown in Fig. 3. The nominal values of I_2, I_3, and J are 1507948.5, 1511819.3, and 1380650.8 mm^4 at the front part and 1329718.7, 1410939.9, and 1309360.3 mm^4 at the rear part, respectively. The excitation force is the engine idling force that can be expressed as the function of frequency; i.e., the force magnitudes are $1.0(\omega)^0$ N, $62.8(\omega)^1$ N, and $39.4(\omega)^2$ N in the x, y, and z directions, respectively, where ω is the engine excitation frequency. Since the rocker neutral axis is offset from the neutral surface of floor pan, as can be seen in Fig. 3, the offset and shear effects are included in the DSA [11].

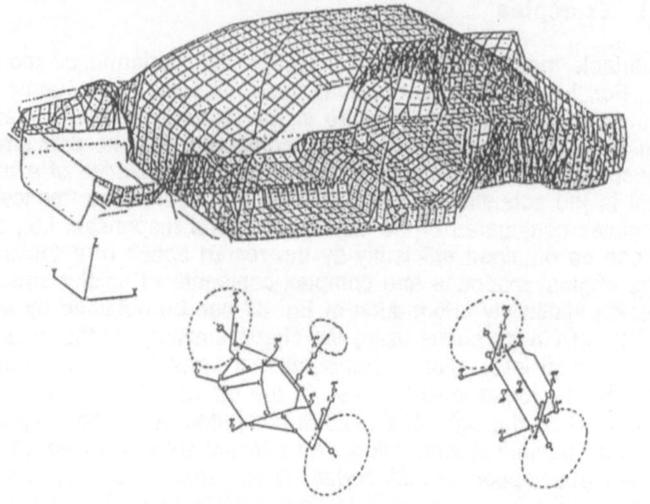

Figure 2 Finite Element Model of Concept Vehicle

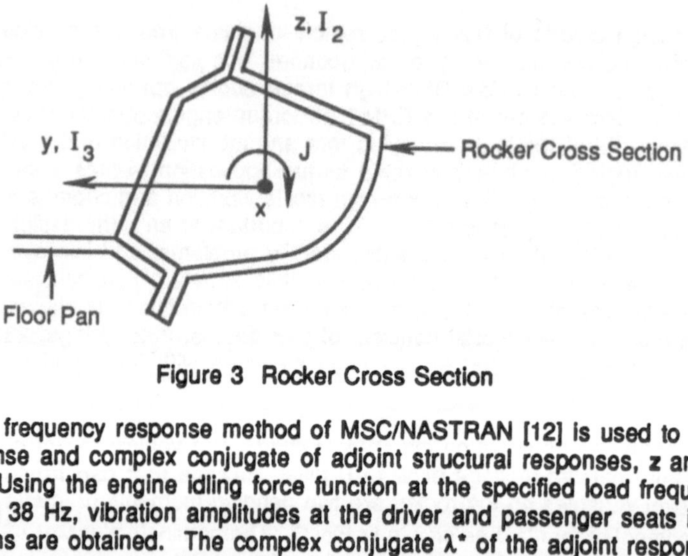

Figure 3 Rocker Cross Section

The direct frequency response method of MSC/NASTRAN [12] is used to obtain the original response and complex conjugate of adjoint structural responses, z and λ^*, respectively. Using the engine idling force function at the specified load frequencies of ω=24, 29, and 38 Hz, vibration amplitudes at the driver and passenger seats in x, y, and z directions are obtained. The complex conjugate λ^* of the adjoint response is then obtained by applying a unit harmonic excitation with the same frequency as the original load at the driver or passenger seat in the direction of the displacement. The multi-loading capability of MSC/NASTRAN is used to obtain the original and adjoint responses.

The accuracy of design sensitivity results have been checked using the finite difference results obtained from perturbed designs. In Tables 1~3, $\psi(u-\delta u)$ and $\psi(u+\delta u)$ are the values of the frequency responses at the driver seat of the perturbed designs u-δu and u+δu, respectively, where δu is the amount of variation in design. The central finite difference is denoted by Δψ=(ψ(u+δu)-ψ(u-δu))/2, and ψ' is the design sensitivity prediction. For the test, ±10% uniform perturbations of I_2, I_3, and J of the rocker are taken. Results in Tables 1~3 show good agreements between the design sensitivity predictions ψ' and finite differences Δψ except the x-direction vibration amplitude at 24 Hz in Table 1.

Table 1 Accuracy of Design Sensitivity of Vibration Amplitude at Driver Seat (Design Variable I_2)

Hz		ψ(u-δu)	ψ(u+δu)	Δψ	ψ'	ψ'/Δψ(%)
	x	0.6140D-01	0.7151D-01	0.5056D-02	0.1132D-01	223.8
24.0	y	0.4362D+00	0.4253D+00	-.5471D-02	-.5345D-02	97.7
	z	0.9291D+00	0.9280D+00	-.5266D-03	-.6191D-03	117.6
	x	0.6222D+00	0.5723D+00	-.2494D-01	-.2487D-01	99.7
29.0	y	0.3494D+00	0.3842D+00	0.1741D-01	0.1745D-01	100.2
	z	0.3958D+01	0.3983D+01	0.1288D-01	0.1218D-01	94.6
	x	0.2264D+00	0.2338D+00	0.3678D-02	0.3630D-02	98.7
38.0	y	0.2187D+01	0.2021D+01	-.8280D-01	-.8145D-01	98.4
	z	0.5907D+01	0.5724D+01	-.9157D-01	-.9426D-01	102.9

Table 2 Accuracy of Design Sensitivity of Vibration Amplitude at Driver Seat (Design Variable I_3)

Hz		ψ(u-δu)	ψ(u+δu)	Δψ	ψ'	ψ'/Δψ(%)
	x	0.4385D-01	0.5293D-01	0.4540D-02	0.4809D-02	105.9
24.0	y	0.4392D+00	0.4221D+00	-.8521D-02	-.8457D-02	99.3
	z	0.9361D+00	0.9218D+00	-.7141D-02	-.7029D-02	98.4
	x	0.6213D+00	0.5728D+00	-.2426D-01	-.2418D-01	99.7
29.0	y	0.3644D+00	0.3710D+00	0.3332D-02	0.3312D-02	99.4
	z	0.3949D+01	0.3990D+01	0.2061D-01	0.2084D-01	101.1
	x	0.2232D+00	0.2422D+00	0.9492D-02	0.1066D-01	112.3
38.0	y	0.2125D+01	0.2073D+01	-.2575D-01	-.2663D-01	103.4
	z	0.5802D+01	0.5819D+01	0.8584D-02	0.8721D-02	101.6

Table 3 Accuracy of Design Sensitivity of Vibration Amplitude at Driver Seat (Design Variable J)

Hz		ψ(u-δu)	ψ(u+δu)	Δψ	ψ'	ψ'/Δψ(%)
	x	0.4580D-01	0.4832D-01	0.1262D-02	0.1273D-02	100.9
24.0	y	0.4296D+00	0.4308D+00	0.5791D-03	0.5555D-03	95.9
	z	0.9268D+00	0.9302D+00	0.1724D-02	0.1678D-02	97.3
	x	0.6023D+00	0.5902D+00	-.6066D-02	-.6038D-02	99.5
29.0	y	0.3660D+00	0.3695D+00	0.1742D-02	0.1751D-02	100.5
	z	0.3969D+01	0.3972D+01	0.1678D-02	0.1601D-02	95.4
	x	0.2260D+00	0.2520D+00	0.1297D-01	0.1500D-01	115.7
38.0	y	0.2174D+01	0.2029D+01	-.7255D-01	-.7103D-01	97.9
	z	0.5914D+01	0.5720D+01	-.9694D-01	-.9615D-01	99.2

Design sensitivity of z-direction vibration amplitude at the driver seat with respect to the design variable J on the rocker are plotted in Fig. 4. In the figure, the negative sensitivity value indicates that the vibration amplitude at the driver seat decreases if the magnitude of the design J of the element is increased. The design sensitivity information are significant and negative from the Hinge pillar to B pillar and negligible between B pillar to C pillar.

To verify efficiency of the numerical method of continuum DSA, CPU time for design sensitivity computation is measured. CRAY X-MP and VAX 8650 computers are used for finite element and design sensitivity analyses, respectively. In this example, there are 144 design variables (I_2, I_3, and J on 48 rocker beam elements). Table 4 lists CPU times for the original and adjoint analyses on CRAY X-MP using multi-loading of the MSC/NASTRAN. CPU time for one adjoint load analysis is about 1% of the original analysis. The computation of design sensitivity for one performance measure with 144 design variables requires 55.0 seconds on the VAX 8650, which is estimated to be 3.3 seconds on the CRAY X-MP.

Table 4 CPU Time for Numerical Evaluation of Continuum DSA

	Computer	CPU Time
One Original Analysis	CRAY X-MP	420.0 sec
One Adjoint Analysis	CRAY X-MP	4.0 sec
DSA Computation for 144 DVs	VAX 8650	55.0 (3.3 sec)

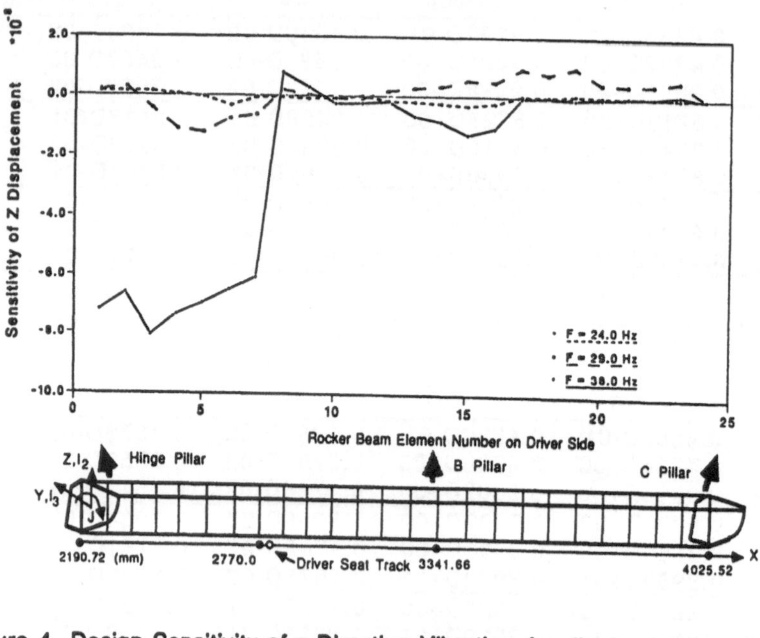

Figure 4 Design Sensitivity of z-Direction Vibration Amplitude at Driver Seat (Design Variable J)

5.2 ACOUSTO-ELASTIC VEHICLE SYSTEM

The low frequency noise in the passenger compartment of a vehicle occurs over a wide range of vehicle speed, and is known to be dominant at frequencies between 20 and 200 Hz. Considerable interaction between body structure motion and the acoustic medium vibration at critical frequencies has been observed, and it is suggested that the noise may be amplified by the acoustic medium resonance [13].

The simple box vehicle model which can be used to identify the system characteristics prior to a practical engineering model analysis of a vehicle system is shown in Fig. 5. The body structure is made of thin plates with uniform thickness that encloses the acoustic medium (air) and is mounted on the simplified suspension system consisting of springs and dampers. The mass density and Young's modulus of the plate are ρ=10 kg/m^3 E=0.5×10^9 N/m^2, respectively. The adiabatic bulk modulus and equilibrium density of the acoustic medium are β=141700 N/m^2 and ρ_0=1.20236 kg/m^3, respectively. The finite element model consists of 232 triangular shell elements for the body structure; 84 hexagonal and 8 tetrahedral elements for the acoustic medium; and 12 spring elements and 12 dampers for the suspension system. The body structure has 118 grid points and 708 degrees of freedom and the acoustic medium has 160 grid points with 160 degrees of freedom.

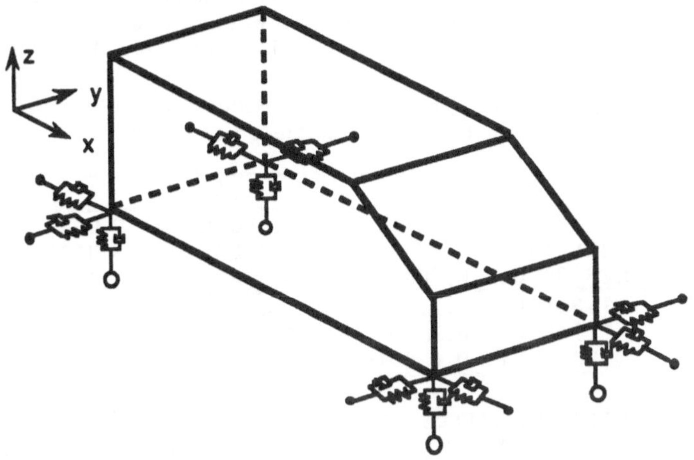

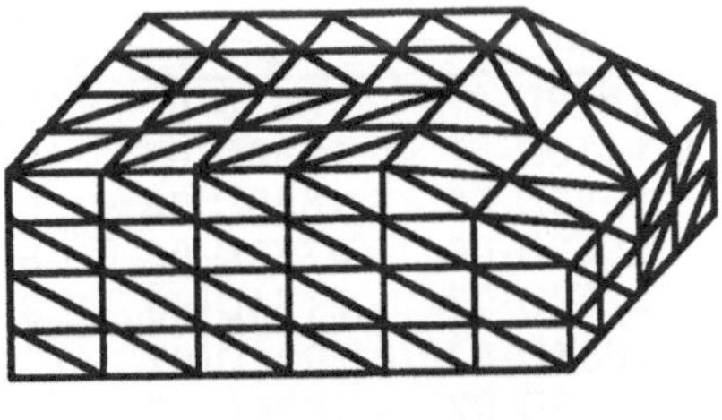

Body Structure

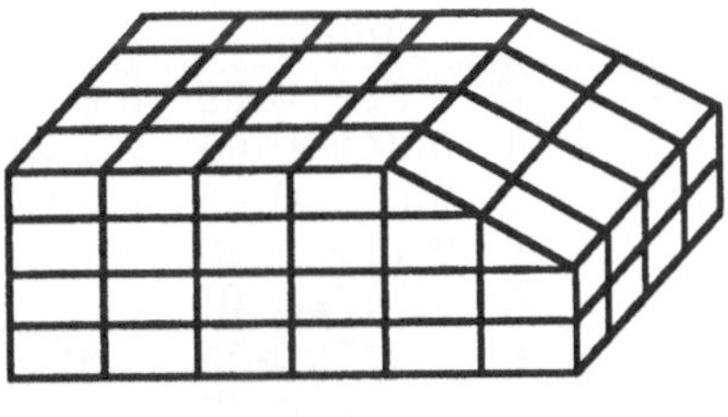

Acoustic Medium

Figure 5 Simple Box Vehicle System and Finite Element Models

For this model, effects of body structure design change on vibration amplitudes of acoustic pressure at the driver's and passenger's ear levels and the vibration amplitudes at the driver and passenger seats are investigated. Design variable is the thickness t of the body panels which is 0.009m at the current nominal design. Two rear suspension supports are harmonically excited in z-direction with an amplitude of 0.0001m and frequency of 60 Hz.

The direct frequency response method of ABAQUS [14] is used for analysis of the original and adjoint systems. The complex conjugates of the adjoint responses are obtained by solving the coupled system of equations with harmonic adjoint loads. For the acoustic pressure performance measure, the adjoint load is the second time derivative of a unit volumetric strain at the point of the performance measure. Also a unit force is applied for the displacement performance measure at the point where performance measure is defined.

Accuracy of design sensitivity results have been checked using the finite difference results as in the previous example. For the test, $\pm$0.01 % perturbations of the thickness of the body panels are taken. Table 5 shows accuracy of design sensitivity results for the acoustic pressure amplitudes. Pascal is the unit of pressure. Table 6 shows accuracy of design sensitivity results for the structural displacement amplitude in z-direction. Meter (m) is the unit of displacement. In Tables 5 and 6, R, I, and D are the real, imaginary, and magnitude of complex responses. Good agreements between the design sensitivity predictions φ' and finite differences $\Delta\varphi$ are obtained. In addition to the structural displacement amplitude, in Table 6, design sensitivities of the velocity and acceleration amplitudes, V and A, respectively, are presented.

Table 5 Accuracy of Design Sensitivity of Acoustic Frequency Responses at Driver's and Passenger's Ear Levels

Location		$\psi(u-\delta u)$	$\psi(u+\delta u)$	$\Delta\psi$	ψ'	$\psi'/\Delta\psi(\%)$
Driver	R	-.2316D-01	-.2299D-01	0.8600D-04	0.8944D-04	104.0
	I	-.1294D-01	-.1311D-01	-.8420D-04	-.8776D-04	104.2
	D	0.2653D-01	0.2647D-01	-.3349D-04	-.3547D-04	105.9
Passenger	R	0.1133D-01	0.1125D-01	-.4395D-04	-.4613D-04	105.0
	I	0.7174D-02	0.7265D-02	0.4549D-04	0.4831D-04	106.2
	D	0.1341D-01	0.1339D-01	-.1253D-04	-.1284D-04	102.5

Table 6 Accuracy of Design Sensitivity of Frequency Responses at Driver and Passenger Seats in z-Direction

Loaction		$\psi(u-\delta u)$	$\psi(u+\delta u)$	$\Delta\psi$	ψ'	$\psi'/\Delta\psi(\%)$
Driver	R	0.3681D-05	0.3650D-05	-.1579D-07	-.1667D-07	105.6
	I	0.3459D-05	0.3488D-05	0.1436D-07	0.1536D-07	107.0
	D	0.5052D-05	0.5048D-05	-.1583D-08	-.1530D-08	96.7
	V	0.1904D-02	0.1903D-02	-.5969D-06	-.5769D-06	96.7
	A	0.7179D+00	0.7175D+00	-.2250D-03	-.2174D-03	96.7
Passenger	R	-.3979D-05	-.3941D-05	0.1885D-07	0.1991D-07	105.6
	I	-.4586D-05	-.4620D-05	-.1708D-07	-.1829D-07	107.1
	D	0.6071D-05	0.6072D-05	0.6539D-09	0.8835D-09	135.1
	V	0.2289D-02	0.2289D-02	0.2465D-06	0.3330D-06	135.1
	A	0.8628D+00	0.8630D+00	0.9294D-04	0.1256D-03	135.1

6. Conclusion

A continuum sizing DSA method is developed for the dynamic frequency responses of acousto-elastic built-up structures using a variational approach with non-selfadjoint operators for complex variables. To derive a variational governing equation for the acousto-elastic built-up structure, interface conditions are identified and sesquilinear and semilinear forms are defined. Both the direct differentiation and adjoint variable methods are developed in which continuum formulations for the built-up structure and acoustic medium are retained throughout derivation of design sensitivity expressions. The method has been numerically implemented using analysis results from MSC/NASTRAN and ABAQUS FEA codes. Two vehicle systems, one without acoustic medium and the other with acoustic medium, are studied and good sensitivity results are obtained.

References

1. Crede, C., Shock and Vibration Concepts in Engineering Design, Prentice-Hall, Englewood Cliffs, N.J., 1965.
2. Fahy, F., Sound and Structural Vibration, Radiation, Transmission and Response, Academic Press, N.Y., 1985
3. Mroz, Z., "Optimal Design of Elastic Structures Subjected to Dynamic, Harmonically-varying Loads," Zeitschrift fur Angewande Mathematik und Mechanik, Vol. 50, 1970, pp. 303-309.
4. Lekszycki, T. and Olhoff, N., "Optimal Design of Viscoelastic Structures under Forced Steady-State Vibration," J. of Structural Mechanics, Vol. 9, No. 4, 1981, pp. 363-387.
5. Yoshimura, M., "Design Sensitivity Analysis of Frequency Response in Machine Structures," ASME, 83-DET-50.
6. Lekszycki, T. and Mroz, Z., "On Optimal Support Reaction in Viscoelastic Vibrating Structures," J. of Structural Mechanics, Vol. 11, No. 1, 1983, pp. 67-79.
7. Choi, K.K. and Lee, J.H., "Sizing Design Sensitivity Analysis of Dynamic Frequency Response of Vibrating Structures," ASME Journal of Mechanical Design, to appear, 1991.; also presented at the 15th ASME Design Automation Conference, September 1989.
8. Haug, E.J., Choi, K.K., and Komkov, V., Design Sensitivity Analysis of Structural Systems, Academic Press, New York, N.Y., 1986.
9. Choi, K.K., Santos, J.L.T., and Frederick, M.C., "Implementation of Design Sensitivity Analysis with Existing Finite Element Codes," ASME J. of Mechanisms, Transmissions, and Automation in Design, Vol. 109, No. 3, 1987, pp. 385-391.
10. Adams, R.A., Sobolev Spaces, Academic Press, N.Y., 1975.
11. Lee, J.H. and Choi, K.K., Sizing and Shape Design Sensitivity Analysis of Dynamic Frequency Response of Vibrating Structures, Technical Report R-59, Center for Simulation and Design Optimization, The University of Iowa, November 1989.
12. MSC/NASTRAN, MSC/NASTRAN User's Manual, Vol. I and II, The MacNeal-Schwendler Corporation, 815 Colorado Boulevard, Los Angeles, CA, 1988.
13. Kamal, M.M. and Wolf, J.A., Jr. ed., Modern Automotive Structural Analysis, Van Nostrand Reinhold Company, New York, 1982.
14. ABAQUS, ABAQUS Users' Manual, Hibbit, Karlsson & Sorensoen, Inc., 100 Medway St. Providence, RI, 1989.

SHAPE DESIGN SENSITIVITY ANALYSIS OF NONLINEAR NONCONSERVATIVE STRUCTURAL PPROBLEMS

JASBIR S. ARORA and TAE HEE LEE
Optimal Design Laboratory
College of Engineering
The University of Iowa
Iowa City, IA 52242
U.S.A.

ABSTRACT. The problem of design sensitivity analysis of *nonlinear nonconservative* systems is addressed using the *continuum formulation*. The principle of *virtual work* is used as the starting point and both the *adjoint and direct variations* methods are derived. Relative advantages and disadvantages of the two approaches for the *history dependent* problems are analyzed and discussed. The concept of reference domain is used to transform the problem to a fixed domain. This way, *shape and nonshape* design problems can be treated in a *unified* manner. The major effect of nonconservativeness of the problem is that the system matrices are *asymmetric* when the problem is discretized. Another major difference compared to other problems is that even for design sensitivity analysis, *iterations* are needed during solution of the sensitivity equations at the beginning of each load increment. Due to these reasons, the computational effort for such problems can be enormous.

1. Introduction

In many applications, it is important to include nonlinearities in the formulation of the problem to predict an accurate response of the structure to inputs. In some applications, the loads are dependent on the deformation. In addition, the materials can go into inelastic range and unloading can occur. Therefore there is internal dissipation of energy due to inelastic deformations. There is need to develop design sensitivity analysis of such problems so that they can be designed for their optimum performance. In this paper, a general formulation for analysis and design of such structures is presented. Some of the applications encompassed by the formulation are from aerospace, metal forming and tire manufacturing industries, and earthquake-resistant design of structures.

In the formulation, both *geometric and material nonlinearities*, and *nonconservative loads* are included. A slightly restricted form of the problem has been treated recently by Arora, Lee and Kumar (1992). That development is further generalized herein. To treat various material models, such as for plastic, viscoelastic and elasto-viscoplastic behavior, a general form of the constitutive equations based on the *internal variable theory* is introduced. A response functional representing limitation on stress, strain, displacement, and their rates is defined. The design sensitivity analysis for the functional is developed using the nonlinear equation of motion for the structural system based on Hamilton's principle. The total Lagrangian formulation based on the Green-Lagrange strain tensor and the second Piola-Kirchhoff stress tensor is used. Both the *direct variation method* and the *adjoint method* are derived, and the control volume/reference domain approach is used to *unify the shape and nonshape design sensitivity analyses* (Haber 1987; Cardoso and Arora 1988; Phelan and Haber 1989). The developed design sensitivity expressions can be specialized to viscoelastic, hyperelastic and linear elastic problems (Arora, Lee and Kumar 1992). It is assumed that the shape variations take place by adding or removing the material at the boundary of the structure. All the boundaries of the structure are allowed to change with the restriction that any change to the displacement specified boundary does not affect other boundaries (i.e., design sensitivity analysis of contact problems is not considered). Also, design sensitivity analysis of limit point and bifurcation point critical loads for conservative and nonconservative systems is not discussed here. For a comprehensive treatment of this topic, Komarakul-na-nakorn and Arora (1990) and various references cited there should be consulted.

G. I. N. Rozvany (ed.), Optimization of Large Structural Systems, Vol. I, 345–359.

Due to lack of space, no numerical examples are included in the paper. For numerical evaluations and discussion of the computer implementation aspects, the following references should be consulted: Lee et al (1991), Poldneff et al (1991), and Arora, Lee and Kumar (1992).

The problem of design sensitivity analysis of various classes of engineering problems has been addressed in the literature for the last about 25 years. Considerable body of *literature* now exists on the subject. A detailed review of the subject is beyond the scope of this paper; Haftka and Adelman (1989) can be consulted for review upto 1989.

Design sensitivity analysis and optimization of *nonlinear structures* is a relatively new area in which rapid developments have taken place in the last about 10 years. Initially, the finite dimensional models were treated (Ryu et al. 1985; Wang et al. 1985; Haririan et al. 1987; Wu and Arora 1987; Gopalakrishna and Greimann 1988) and more recently the continuum models have been treated (Haftka and Mroz 1986; Choi and Santos 1987; Cardoso and Arora 1988; Dems and Mròz 1989; Tsay and Arora 1990; Tortorelli et al. 1991). Both material derivative and control volume approaches have been developed. It has been recently shown that the two approaches are theoretically the same; i.e., one can recover the sensitivity expression of the one from the other (Arora, Lee and Cardoso 1992). Both the direct variation/differentiation and the adjoint methods have been addressed in the literature. It has been shown that the adjoint method is cumbersome and tedious to implement for problems with *history dependent* constitutive equations (Tsay and Arora 1990). This was realized earlier in the development of design sensitivity analysis for materially nonlinear problems by the adjoint method (Ryu et al. 1985). Methods of design sensitivity analysis for this class of problems are just beginning to be addressed. For example, Tsay and Arora (1989,1990), and Jao and Arora (1991) have developed and demonstrated the formulations for elastoplastic, viscoelastic and viscoplastic problems. Mukherjee and Chandra (1989) have used a boundary element formulation for design sensitivity analysis of geometrically linear but materially nonlinear elasto-viscoplastic problems.

NOMENCLATURE. The notation for analysis of nonlinear structural problems tends to be tedious and sometimes complex. It is important to clearly understand it and the symbols used for various quantities. To facilitate reading of the paper, the nomenclature used in the development of various concepts is summarized here. *Index notation* is used, so repeated indices imply summation. Left and right superscripts and subscripts have definite meaning. The *left superscript* is used to indicate the configuration in which the quantity occurs and a *left subscript* to indicate the reference configuration (Bathe 1982).

Left Superscripts. r: refers to the quantity measured over the fixed reference domain; t: refers to the quantity measured at time t; 0: refers to the quantity measured at time 0.

Left Subscripts. a: refers to the adjoint field; r: refers to the fixed reference domain; t: refers to the current configuration; 0: refers to the undeformed configuration.

Right Superscripts. D: deviator; e: elastic component; in: inelastic component; 0: prescribed quantity.

Right Subscripts. i,j,1,2,3: component of vector or tensor; u: displacement specified quantity; T: traction specified quantity; Γ: boundary.

List of Symbols

b_i	design variable vector
$C_{ijk\ell}$	elastic stress-strain material property tensor
${}^0dV$	differential volume in the undeformed configuration
${}^r dV$	differential volume in the fixed reference domain
$e_{ij}(\bullet,\bullet)$	a strain operator
${}^t_0 f_i$	body force per unit undeformed volume
${}^t_a f_i$	body force per unit volume for the adjoint structure
${}_0 f_i$, f_i	increment in the body force
g	integrand of the displacement specified boundary integral in the response functional

G	integrand of the volume integral in the response functional
$H^1(V)$	Sobolev function space of order 1 defined on the domain V; i.e., a collection of functions that are differentiable once, and the functions and their first derivatives are square integrable (belong to $L^2(V)$)
h	integrand of the traction specified boundary integral in the response functional
${}_0^tJ$, J	Jacobian of the transformation
${}^0n_k$	normal vector to the undeformed surface
${}_0S_{ij}$, S_{ij}	increment in the second Piola-Kirchhoff stress tensor
${}_0^tS_{ij}$	second Piola-Kirchhoff stress tensor at time t
${}_a^tS_{ij}$, ${}_a^tS_{ij}^I$	stress tensor for the adjoint structure and its initial value
t	time or load level for quasi-static problems
T_i^0, ${}_0T_i^0$	increment in the traction force vector
${}_0^tT_i^0$	fictitious surface traction vector in the undeformed configuration
${}_a^tT_i^0$	fictitious surface traction vector for the adjoint structure
${}^tu_i$, u_i	displacement field at time t and its increment
${}_0^tu_{i,j}$	$\partial^t u_i/\partial^0 x_j$, the displacement gradient
${}_a^tu_i$	displacement field for the adjoint structure
${}^0x_i$	coordinates in the undeformed configuration
${}^tx_i$	coordinates of the particles of the body at time t
${}_0^tx_{i,j}$	deformation gradient tensor
${}^0V$, ${}^tV$	volume in the undeformed configuration and the current configuration
${}^t\alpha_{ij}$, α	internal variables at time t, internal variable
δ_{ij}	Kronecker's delta function
${}_0^t\varepsilon_{ij}$	Green-Lagrange strain tensor
${}_a^t\varepsilon_{ij}$, ${}_a^t\varepsilon_{ij}^I$	strain tensor for the adjoint structure and its initial value
ϑ	integrand of performance functional to specify the terminal conditions
${}^0\Gamma$, ${}^t\Gamma$	surface in the undeformed configuration and the current configuration
${}^0\Gamma_T$, ${}^0\Gamma_u$	traction and displacement specified surfaces in the undeformed configuration
$\eta_{ij}(\bullet,\bullet)$	symmetric nonlinear strain operator
λ_{ij}	Lagrange multipliers associated with evolution equations
${}^t\sigma_{ij}$	Cauchy stress tensor at time t
τ	terminal time

Derivatives. The *'comma'* notation for partial derivatives is used for which the left subscript has the following meaning: it refers to the configuration for the independent variable; e.g., ${}_r^tu_{i,j}$ means $\partial^t u_i/\partial^r x_j$.

Design Variations. The following variational notation is used for design variation of various quantities: $\bar{\delta}(\,)$ *total* design variation of ( ); i.e., $\bar{\delta}(\,) = (d(\,)/db_k)\,\delta b_k$

$\bar{\bar{\delta}}(\,)$ — *explicit* design variation (*partial* derivative) of ( ); i.e., $\bar{\bar{\delta}}(\,) = (\partial(\,)/\partial b_k)\,\delta b_k$ for which state fields are frozen

$\tilde{\delta}(\,)$ — design variation of the fields that *implicitly* depend on the design variables, such as displacements, strains, strain rate and stress; also design variation of functionals with respect to the implicit state fields; for this variation, the explicit dependence on the design variables is frozen

2. Problem Formulation

Equation of Motion. Suppose that the particles of the body are labeled by their coordinates ${}^0x_i$ in an undeformed configuration at time $t = 0$. If at a later time t the particle ${}^0x_i$ has coordinates ${}^tx_i$, the motion of the body can be represented by the equations (note that the rectangular Cartesian coordinate system is used): ${}^tx_i = {}^tx_i({}^0x_k,t)$. Then the *deformation gradient* and the nonzero Jacobian of the transformation $({}^0x_k,t) \rightarrow ({}^tx_k,t)$ are defined as ${}^t_0x_{i,j} = (\partial^t x_i/\partial^0 x_j) = (\delta_{ij} + {}^t_0u_{i,j})$; ${}^t_0J = |{}^t_0x_{i,j}| > 0$, where $|\bullet|$ indicates the determinant of a matrix, ${}^tu_i$ is the displacement field at *time* t (or *load level* t) satisfying the appropriate smoothness requirements (e.g., ${}^t\mathbf{u} \in [H^1({}^tV)]^3$), and ${}^tx_i = {}^0x_i + {}^tu_i$, and ${}^t_0u_{i,j} = \partial^t u_i/\partial^0 x_j$ is the *displacement gradient*.

Let an *elasto-viscoplastic continuum* occupy a domain ${}^tV$ that is bounded by the surface ${}^t\Gamma$ at time t. Let ${}^t\Gamma_T$ represent the traction prescribed surface, and ${}^t\Gamma_u$ the displacement prescribed surface with ${}^tu_i^0$ as the specified displacement field. Since the control volume approach will be used for design sensitivity analysis, all the field variables and integrals need to be transformed to a fixed reference domain by introducing an independent variable transformation (Haber 1987; Cardoso and Arora 1988; Phelan and Haber 1989). Let ${}^rx_i$ be the coordinates in the fixed reference domain with volume ${}^rV$ and surface ${}^r\Gamma$ that remain unchanged during design variations and deformations. The differential volume and surface elements in the two coordinate systems are related as ${}^0dV = {}^0_rJ\,{}^rdV$; ${}^0d\Gamma = {}^0_rJ_\Gamma\,{}^rd\Gamma$, where ${}^0_rJ = |{}^0_rx_{i,j}|$, ${}^0_rx_{i,j} = \partial^0 x_i/\partial^r x_j$, ${}^0_rJ_\Gamma = {}^0_rJ\,\|{}^0_rx_{i,j}\,{}^rn_i\|$, and ${}^rn_i$ is the unit outward normal to the fixed reference boundary. Note that $\|\bullet\|$ denotes the *Euclidean norm* defined as $\|a_i\| = (a_ja_j)^{1/2}$ for a vector a_i. For simplicity of the notation, *left superscript and subscript on* ${}^0_rJ$ and ${}^0_rJ_\Gamma$ *will be dropped*. In the sequel, all the *equations* are written in the *reference domain*. Equations in the original domain can be recovered by letting ${}^0_rJ = 1$, ${}^0_rJ_\Gamma = 1$ and ${}^0_rx_{i,j} = {}^0_rx_{i,j} = \delta_{ij}$.

Using the *total Lagrangian* formulation to describe the motion of the continuum, the first law of thermodynamics under isothermal conditions gives the equation of motion at time t as

$$ {}^t_rA \equiv \int F J\,{}^rdV + \int R J_\Gamma\,{}^rd\Gamma = 0; \qquad F = -{}^0\rho\,{}^t\ddot{u}_i\,\delta^tu_i - {}^t_0S_{ij}\,\delta^t_0\varepsilon_{ij} + {}^t_0f_i\,\delta^tu_i \tag{1} $$

$$ R = {}^t_0x_{i,j}\,{}^t_0T_j\,\delta^tu_i = (\delta_{ij} + {}^t_0u_{i,j})\,{}^t_0T_j\,\delta^tu_i = (\delta_{ij} + {}^t_ru_{i,k}\,{}^r_0x_{k,j})\,{}^t_0T_j\,\delta^tu_i \tag{2} $$

where δ^tu_i is an arbitrary virtual displacement field satisfying the appropriate smoothness requirement (i.e., $\delta^tu_i \in H^1({}^tV)$), ${}^tu_i \in H^1({}^tV)$ is the displacement field, and ${}^0\rho$ is the mass density in the undeformed configuration. $\delta^t_0\varepsilon_{ij}$ is the virtual strain field that is compatible with δ^tu_i. The field ${}^t_0f_i$ denotes the body force in the current configuration per unit undeformed volume, and ${}^t_0T_i$ is a *pseudo-traction field* acting on the current surface per unit undeformed area and referred to the undeformed configuration. This pseudo-traction field is defined according to the equation ${}^tT_i\,{}^td\Gamma = {}^t_0x_{i,j}\,{}^t_0T_j\,{}^0d\Gamma$. The body force and the traction force are assumed to be *dependent on the deformation*; i.e., u_i and $\dot{u}_i$. ${}^t_0T_i^0$ is the traction specified on part of the surface Γ_T. The surface integral in Eq. (1) can be reduced to be only over the traction specified surface if δ^tu_i is taken as a kinematically admissible virtual displacement field.

The *Green-Lagrange strain tensor* and its arbitrary variation are given as

$$ {}^t_0\varepsilon_{ij} = \frac{1}{2}\left({}^r_0x_{m,j}\,{}^t_ru_{i,m} + {}^r_0x_{m,i}\,{}^t_ru_{j,m} + {}^r_0x_{m,i}\,{}^r_0x_{n,j}\,{}^t_ru_{k,m}\,{}^t_ru_{k,n} \right) \tag{3} $$

$$ \delta^t_0\varepsilon_{ij} = e_{ij}({}^t\mathbf{u}, \delta^t\mathbf{u}) \tag{4} $$

where e_{ij} is a strain operator that is defined for convenience of notation as (note that $e_{ij} = e_{ji}$)

$$ e_{ij}(\mathbf{a},\mathbf{b}) = \frac{1}{2}\left({}^r_0x_{m,j}\,{}_rb_{i,m} + {}^r_0x_{m,i}\,{}_rb_{j,m} + {}^r_0x_{m,i}\,{}^r_0x_{n,j}\,{}_ra_{k,m}\,{}_rb_{k,n} + {}^r_0x_{m,i}\,{}^r_0x_{n,j}\,{}_ra_{k,n}\,{}_rb_{k,m} \right) \tag{5} $$

Note ${}^r_0x_{i,j}$ *depends only explicitly* on design because this transformation matrix is independent of deformation. This makes ${}^t_0\varepsilon_{ij}$ and $e_{ij}({}^t\mathbf{u},\delta^t\mathbf{u})$ to be explicitly dependent on design as well. These facts will be used later in the derivation of design sensitivity expressions. The *second Piola-Kirchhoff stress tensor* referred to the configuration at time 0 (undeformed configuration) and the Cauchy formula are given as

$$ {}^t_0S_{ij} = {}^t_0J\,{}^0_tx_{i,k}\,{}^t\sigma_{k\ell}\,{}^0_tx_{j,\ell}; \quad {}^t_0T_i = {}^t_0S_{ij}\,{}^0n_j \text{ on } \Gamma \tag{6} $$

where ${}^t\sigma_{ij}$ is the Cauchy stress tensor at time t, ${}^0n_i$ is the unit outward normal to the undeformed boundary, and ${}^0_tx_{i,j}$ is the inverse of the deformation gradient, i.e., ${}^0_tx_{i,j} = [{}^t_0x_{i,j}]^{-1}$. Another way to treat the nonconservative surface traction is to use ${}^t_0\bar{T}_i = {}^t_0x_{i,j}\,{}^t_0T_j^0$ in Eq. (2). The Cauchy formula in Eq. (6) gives ${}^t_0\bar{T}_i = {}^t_0x_{i,j}\,{}^t_0S_{jk}\,{}^0n_k$. Note that ${}^t_0x_{i,j}\,{}^t_0S_{jk}$ is the first Piola-Kirchhoff stress which is in general not symmetric, but

${}_0^tS_{ij}$ is symmetric whenever ${}^t\sigma_{ij}$ is symmetric (nonpolar case). Both the formulations are equivalent, but they can give different expressions for numerical implementation.

The time integral of the functional ${}_r^tA$ given in Eq. (1) over a fixed time interval $T \in [0, \tau]$ when set to zero, gives the well known *Hamilton's principle* as

$$ {}_r^{\tau}H \equiv \int {}_r^tA\, dt = 0 \tag{7}$$

with the initial conditions for the displacement fields as ${}^0u_i$ and ${}^0\dot{u}_i$ at $t = 0$.

The equation of motion (1) can be solved for the displacement field by any of the well developed *implicit or explicit integration methods*. In reality, the physical and time domains will have to be discretized for a numerical solution of the complex problems. Incremental procedures will have to be used to solve the nonlinear equation (1). Therefore, it will be useful to obtain an *incremental form* of the equation of motion (1), or write the Hamilton's principle in Eq. (7) over a small time interval Δt. The incremental form of the equation of motion also gives some insights for the equations to be solved during the design sensitivity analysis phase. This will be seen later in the sections that contain design sensitivity analysis by the direct variation and adjoint methods.

To derive an incremental form of the equation of motion, assume that the motion is known at time t and it is desired at time t+Δt. To this end, decompose the state fields as

$$ {}^{t+\Delta t}u_i = {}^tu_i + u_i, \qquad {}_0^{t+\Delta t}S_{ij} = {}_0^tS_{ij} + {}_0S_{ij}, \qquad {}_0^{t+\Delta t}f_i = {}_0^tf_i + {}_0f_i, $$

$$ {}_0^{t+\Delta t}\varepsilon_{ij} = {}_0^t\varepsilon_{ij} + {}_0\varepsilon_{ij}, \qquad {}_0^{t+\Delta t}T_i^0 = {}_0^tT_i^0 + {}_0T_i^0 \tag{8}$$

where u_i, ${}_0S_{ij}$, ${}_0f_i$, ${}_0\varepsilon_{ij}$, and ${}_0T_i^0$ represent increments in the corresponding state fields. Note that in the following derivations *the left subscript on the various incremental fields will be dropped for simplicity.* Substituting displacement decomposition into Eq. (3), we get the incremental strain field as follows:

$$ \varepsilon_{ij} = e_{ij}({}^t\mathbf{u},\mathbf{u}) + \tfrac{1}{2}\eta_{ij}(\mathbf{u},\mathbf{u}) \tag{9}$$

where $\eta_{ij}(\mathbf{u},\mathbf{u})$ is a *nonlinear operator* that is defined using the following symmetric operator given in terms of **a** and **b**:

$$ \eta_{ij}(\mathbf{a},\mathbf{b}) = \tfrac{1}{2}\left({}_0^rx_{m,i}\, {}_0^rx_{n,j}\, {}_ra_{k,m}\, {}_rb_{k,n} + {}_0^rx_{m,i}\, {}_0^rx_{n,j}\, {}_ra_{k,n}\, {}_rb_{k,m} \right) \tag{10}$$

The arbitrary variation of the strain field at time t+Δt is given as

$$ \delta\, {}_0^{t+\Delta t}\varepsilon_{ij} = \delta\varepsilon_{ij} = e_{ij}({}^t\mathbf{u},\delta\mathbf{u}) + \eta_{ij}(\mathbf{u},\delta\mathbf{u}) \tag{11}$$

Now writing the equation of motion (1) at time t+Δt, substituting incremental decompositions of Eqs. (8), using the fact that Eq. (1) holds at time t, and neglecting the higher order term $S_{ij}\eta_{ij}(\mathbf{u},\delta\mathbf{u})$, we obtain the linearized incremental equation of motion and Hamilton's principle over the time interval Δt as

$$ A \equiv \int \left(-{}^0\rho\ddot{u}_i\delta u_i - S_{ij}e_{ij}({}^t\mathbf{u},\delta\mathbf{u}) - {}_0^tS_{ij}\eta_{ij}(\mathbf{u},\delta\mathbf{u}) + f_i\delta u_i \right) J\, {}^rdV $$

$$ + \int \left({}_0^tx_{i,j}\, T_j + {}_0u_{i,j}\, {}_0^{t+\Delta t}T_j \right) \delta u_i\, J_\Gamma\, {}^rd\Gamma = 0; \quad \text{and} \quad \int_t^{t+\Delta t} A\, dt = 0 \tag{12}$$

where the fact that δu_i is an *admissible field and* $\{\delta u_i, e_{ij}({}^t\mathbf{u},\delta\mathbf{u})\}$ is a *compatible set*, has been used while invoking Eq. (1) in order to obtain Eq. (12). Note that since the body force and surface traction are dependent on ${}^tu_i$ and ${}^t\dot{u}_i$, their increments are given as

$$ f_i = {}_0^tf_{i,{}^tu_j}\, u_j + {}_0^tf_{i,{}^t\dot{u}_j}\, \dot{u}_j; \quad T_i^0 = {}_0^tT^0_{i,{}^tu_j}\, u_j + {}_0^tT^0_{i,{}^t\dot{u}_j}\, \dot{u}_j \tag{13}$$

Substituting these into Eq. (12), the incremental displacements can be obtained and increments in other fields can be calculated using them. It is important to note that due to nonconservative loads, Eq. (12) has *asymmetric operators*. Therefore either an unsymmetric equation solver needs to be used, or the unsymmetric part needs to be transferred to the right hand side and iterations used to solve the linear system of equation. The latter procedure is preferred due to two reasons: (i) symmetric equation solvers can still be used, and (ii) iterations are needed in any case within the time/load increment to satisfy the constitutive evolution equations and to compensate for linearization of the incremental equation; i.e., for neglecting the term $S_{ij}\eta_{ij}(\mathbf{u},\delta\mathbf{u})$.

Constitutive Model. A number of constitutive equations for the elasto-viscoplastic (*rate-dependent* plastic-deformation) behavior of metallic materials have been developed recently (Lee, Arora and Kumar 1991a). These models consider inelastic deformation such as creep, stress relaxation, and plastic flow as a *unified* quantity. These constitutive equations can be modeled based on the thermodynamical concept which states that the current state of the material can be expressed in terms of the current values of the independent variables and a set of internal variables. In these theories, assuming small strains, the total strain is decomposed into elastic (recoverable) strains ${}^t_0\varepsilon^e_{ij}$ and inelastic (irrecoverable) strains ${}^t_0\varepsilon^{in}_{ij}$ as ${}^t_0\varepsilon_{ij} = {}^t_0\varepsilon^e_{ij} + {}^t_0\varepsilon^{in}_{ij}$. Further, the inelastic strains are assumed to be incompressible, i.e., ${}^t_0\varepsilon^{in}_{kk} = 0$. A similar decomposition can be assumed to hold for the case of large strains.

Now the problem is to find a proper *form* for ${}^t_0\varepsilon^{in}_{ij}$. In the *unified elasto-viscoplastic* constitutive equations, ${}^t_0\dot{\varepsilon}^{in}_{ij}$ are assumed to be a function of the stress ${}^t_0S_{ij}$, a set of internal variables ${}^t\alpha_{ij}$, and the time t. In general, they are governed by the following differential equations:

$$ {}^tp_{ij}({}^t_0\dot{S}_{k\ell}, {}^t_0S_{k\ell}, {}^t_0\dot{\varepsilon}^{in}_{k\ell}, {}^t_0\varepsilon^{in}_{k\ell}, {}^t\alpha_{k\ell}, t) = 0; \quad {}^tq_{ij}({}^t_0\dot{S}_{k\ell}, {}^t_0S_{k\ell}, {}^t_0\varepsilon^{in}_{k\ell}, {}^t\dot{\alpha}_{k\ell}, {}^t\alpha_{k\ell}, t) = 0 \tag{14} $$

where ${}^t\alpha_{ij}$ represent the internal variables at time t containing both the kinematic and isotropic hardening variables. Different constitutive models define specific forms for ${}^tp_{ij}$ and ${}^tq_{ij}$. Note that these differential equations are, in general, highly nonlinear and stiff (Kumar *et al* 1980). Their integration over the time increment gives the incremental inelastic strain $\varepsilon^{in}_{k\ell}$.

The stresses are given by the Hooke's law as

$$ {}^t_0S_{ij} = C_{ijk\ell}\,{}^t_0\varepsilon^e_{k\ell}; \quad C_{ijk\ell} = \lambda\,\delta_{ij}\delta_{k\ell} + \mu(\delta_{ik}\delta_{j\ell} + \delta_{i\ell}\delta_{jk}) \tag{15} $$

where the $C_{ijk\ell}$ are components of the initial elastic tangent modulus, and λ and μ are the Lamé constants. The incremental stresses are calculated as $S_{ij} = C_{ijk\ell}(\varepsilon_{k\ell} - \varepsilon^{in}_{k\ell})$.

For *time-independent (classical) plasticity*, any rate-type constitutive model can be used to describe the material behavior. The central concept in this case is the existence of a *yield surface* that encloses the zone of pure elastic response in the stress space. The yield surface is specified as a function of stress components and certain hardening variables, i.e., ${}^tF = {}^tF({}^t_0S_{ij}, {}^t\alpha_{ij}) \le 0$. The inelastic strain rate is replaced by the plastic strain rate ${}^t_0\dot{\varepsilon}^p_{ij}$ which is obtained by using the normality condition. This condition states that the plastic strain rate is proportional to the normal to the yield surface in the stress space as ${}^t_0\dot{\varepsilon}^p_{ij} = \lambda(\partial{}^tF/\partial{}^t_0S_{ij})$, where λ is the proportionality constant called the plastic multiplier. These equations can in general be written as ${}^tp_{ij}({}^t_0S_{k\ell}, {}^t_0\dot{\varepsilon}^p_{k\ell}, {}^t_0\varepsilon^p_{k\ell}, {}^t\alpha_{k\ell}, t) = 0$. The internal variables describing the kinematic and isotropic hardening behavior are represented as ${}^tq_{ij}({}^t_0S_{k\ell}, {}^t_0\dot{\varepsilon}^p_{k\ell}, {}^t_0\dot{\alpha}_{k\ell}, {}^t\alpha_{k\ell}, t) = 0$ Therefore we can describe the *time-independent plasticity* in a manner similar to that for the rate-dependent plasticity.

For *viscoelastic materials*, we cannot employ the stress-strain relation given above because there is no need for inelastic strain component. In that case, we can obtain stress-strain relation directly if we set ${}^tp_{ij}({}^t_0\dot{S}_{k\ell}, {}^t_0S_{k\ell}, {}^t_0\varepsilon_{k\ell}, t) = 0$, or ${}^tq_{ij}({}^t_0S_{k\ell}, {}^t_0\dot{\varepsilon}_{k\ell}, {}^t_0\varepsilon_{k\ell}, t) = 0$. For *nonlinear elastic material* behavior, the stresses are derivable from the strain energy density function ϕ as ${}^t_0S_{ij} = (\partial\phi/\partial{}^t_0\varepsilon_{ij})\,{}^t_0\varepsilon_{ij} \equiv {}^tC_{ijk\ell}\,{}^t_0\varepsilon_{k\ell}$, where ${}^tC_{ijk\ell}$ is the tangential stiffness modulus. This constitutive model has been integrated into design sensitivity analysis by Arora and Cardoso (1989), Dems and Mròz (1989) and Tsay and Arora (1990).

Based on the foregoing discussion, the stress-strain law in general can be written as

$$ {}^tr_{ij}({}^t_0\dot{S}_{k\ell}, {}^t_0S_{k\ell}, {}^t_0\dot{\varepsilon}_{k\ell}, {}^t_0\varepsilon_{k\ell}, {}^t_0\varepsilon^{in}_{k\ell}, t) = 0 \tag{16} $$

In Eqs. (14) to (16), t represents the real time for the dynamic case and pseudo-time in the quasistatic case.

Response Functional. Consider a general response functional (whose design sensitivity is desired) defined in the time-space domain [$T\times{}^0V$] and its boundary, and transformed to the reference domain, as

$$ \psi = \int\{\int G({}^t_0S_{ij}, {}^t_0\varepsilon_{ij}, {}^t_0\dot{\varepsilon}_{ij}, {}^tu_i, {}^t\dot{u}_i, {}^t\ddot{u}_i, b_i, t)\, J\, {}^rdV + \int g({}^tu_i, {}^t_0T_i, b_i, t)\, J_\Gamma\, {}^rd\Gamma_u $$

$$+\int h\left({}^t u_i, {}^t_0 T^0_i, b_i, t\right) J_\Gamma\, {}^r d\Gamma_T \Big\}\, dt + \int \vartheta\left({}^0 u_i, {}^0\dot{u}_i, {}^\tau u_i, {}^\tau \dot{u}_i, b_i, \tau\right) J\, {}^r dV \tag{17}$$

where G, g, h, and ϑ are continuous and differentiable functions with respect to their arguments and the domain ${}^0V$ is time independent. This response functional can represent limitations on stresses, strains, strain rates, displacements, velocities, accelerations, traction, and the initial and terminal conditions.

3. Design Sensitivity Analysis: Direct Variation Method

In the direct variation method (DVM), the total design variations of all the field variables are calculated by using the design variations of all the governing equations. Once these are known, design variation of any response functional, such as the one in Eq. (17), can be evaluated using the design variations of the field variables. Thus, this method can be derived without reference to any particular response functional.

For the derivations presented in the sequel, the following points should be noted: (i) since the displacement field ${}^t\mathbf{u}$ depends only implicitly on design, the notation $\bar\delta^t\mathbf{u}$ is used to denote the total design variation of ${}^t\mathbf{u}$ (instead of $\tilde\delta^t\mathbf{u}$), and (ii) the virtual displacement field is assumed to depend on design in an arbitrary manner; the form of this dependence is explored later.

The *total design variation of the strains and their arbitrary variations* are given as

$$\bar\delta^t_0\varepsilon_{ij} = \bar{\bar\delta}^t_0\varepsilon_{ij} + \tilde\delta^t_0\varepsilon_{ij}; \quad \tilde\delta^t_0\varepsilon_{ij} = e_{ij}({}^t u, \bar\delta^t u) \tag{18}$$

$$\bar{\bar\delta}^t_0\varepsilon_{ij} = \tfrac{1}{2}\big(\bar{\bar\delta}^r_0 x_{m,j}\, {}^t_r u_{i,m} + \bar{\bar\delta}^r_0 x_{m,i}\, {}^t_r u_{j,m} + \bar{\bar\delta}^r_0 x_{m,i}\, {}^r_0 x_{n,j}\, {}^t_r u_{k,m}\, {}^t_r u_{k,n} + {}^r_0 x_{m,i}\, \bar{\bar\delta}^r_0 x_{n,j}\, {}^t_r u_{k,m}\, {}^t_r u_{k,n}\big) \tag{19}$$

$$\bar\delta(\delta^t_0\varepsilon_{ij}) = e_{ij}({}^t u, \bar\delta(\delta^t u)) + \eta_{ij}(\bar\delta^t u, \delta^t u) + \bar{\bar\delta} e_{ij}({}^t u, \delta^t u) \tag{20}$$

The *total design variation of the constitutive equations* (14) to (16) gives

$$\bar\delta^t p_{ij} = 0, \quad \bar\delta^t q_{ij} = 0, \quad \bar\delta^t r_{ij} = 0 \tag{21}$$

$$\begin{aligned}\bar\delta^t p_{ij} = {}&{}^t p_{ij,{}^t_0\dot S_{k\ell}}\, \bar\delta^t_0\dot S_{k\ell} + {}^t p_{ij,{}^t_0 S_{k\ell}}\, \bar\delta^t_0 S_{k\ell} + {}^t p_{ij,{}^t_0\dot\varepsilon^{in}_{k\ell}}\, \bar\delta^t_0\dot\varepsilon^{in}_{k\ell} + {}^t p_{ij,{}^t_0\varepsilon^{in}_{k\ell}}\, \bar\delta^t_0\varepsilon^{in}_{k\ell}\\ &+ {}^t p_{ij,{}^t\alpha_{k\ell}}\, \bar\delta^t\alpha_{k\ell} + {}^t p_{ij,b_k}\, \delta b_k\end{aligned} \tag{22}$$

$$\begin{aligned}\bar\delta^t q_{ij} = {}&{}^t q_{ij,{}^t_0\dot S_{k\ell}}\, \bar\delta^t_0\dot S_{k\ell} + {}^t q_{ij,{}^t_0 S_{k\ell}}\, \bar\delta^t_0 S_{k\ell} + {}^t q_{ij,{}^t_0\varepsilon^{in}_{k\ell}}\, \bar\delta^t_0\varepsilon^{in}_{k\ell} + {}^t q_{ij,{}^t\dot\alpha_{k\ell}}\, \bar\delta^t\dot\alpha_{k\ell}\\ &+ {}^t q_{ij,{}^t\alpha_{k\ell}}\, \bar\delta^t\alpha_{k\ell} + {}^t q_{ij,b_k}\, \delta b_k\end{aligned} \tag{23}$$

$$\begin{aligned}\bar\delta^t r_{ij} = {}&{}^t r_{ij,{}^t_0\dot S_{k\ell}}\, \bar\delta^t_0\dot S_{k\ell} + {}^t r_{ij,{}^t_0 S_{k\ell}}\, \bar\delta^t_0 S_{k\ell} + {}^t r_{ij,{}^t_0\dot\varepsilon_{k\ell}}\, \bar\delta^t_0\dot\varepsilon_{k\ell} + {}^t r_{ij,{}^t_0\varepsilon_{k\ell}}\, \bar\delta^t_0\varepsilon_{k\ell}\\ &+ {}^t r_{ij,{}^t_0\varepsilon^{in}_{k\ell}}\, \bar\delta^t_0\varepsilon^{in}_{k\ell} + {}^t r_{ij,b_k}\, \delta b_k\end{aligned} \tag{24}$$

The design variation of the Hamilton's principle in Eq. (7) gives (note that $\delta^t_0\varepsilon_{ij} = e_{ij}({}^t u, \delta^t u)$ and Eq. (20) for $\bar\delta(\delta^t_0\varepsilon_{ij})$ are used; also $\delta^t u_i$ is now taken as kinematically admissible, i.e., $\delta^t u_i = 0$ on Γ_u):

$$\bar\delta^\tau_r H = \bar{\bar\delta}^\tau_r H + \tilde\delta^\tau_r H = 0 \tag{25}$$

$$\bar{\bar\delta}^\tau_r H = \int \bar{\bar\delta}^\tau_r A\, dt = \int\Big[\int \bar{\bar\delta}(FJ)\, {}^r dV + \int \bar{\bar\delta}(RJ_\Gamma)\, {}^r d\Gamma_T\Big]\, dt \tag{26}$$

$$\bar{\bar\delta}F = -\bar{\bar\delta}\,{}^0\rho\, {}^t\ddot u_i\, \delta^t u_i - \bar{\bar\delta}^t_0 S_{ij}\, e_{ij}({}^t u, \delta^t u) - {}^t_0 S_{ij}\, \bar{\bar\delta} e_{ij}({}^t u, \delta^t u) + \bar{\bar\delta}^t_0 f_i\, \delta^t u_i \tag{27}$$

$$\bar{\bar\delta}R = \bar{\bar\delta}({}^t_0 x_{i,j}\, {}^t_0 T^0_j)\, \delta^t u_i = ({}^t_r u_{i,k}\, \bar{\bar\delta}^r_0 x_{k,j}\, {}^t_0 T^0_j + {}^t_0 x_{i,j}\, \bar{\bar\delta}^t_0 T^0_j)\, \delta^t u_i \tag{28}$$

$$\tilde{\delta}^{\tau}_{r}H = \int\Big\{\int\big(-{}^{0}\rho\,\tilde{\delta}^{t}\ddot{u}_i\,\delta^{t}u_i - \tilde{\delta}^{t}_{0}S_{ij}\,e_{ij}({}^{t}u,\delta^{t}u) - {}^{t}_{0}S_{ij}\,\eta_{ij}(\tilde{\delta}^{t}u,\delta^{t}u) + \tilde{\delta}^{t}_{0}f_i\,\delta^{t}u_i\big)\,J^{r}dV$$
$$+\int\tilde{\delta}({}^{t}_{0}x_{i,j}\,{}^{t}_{0}T_j)\,\delta^{t}u_i\,J_{\Gamma}\,{}^{r}d\Gamma_T\Big\}\,dt + \int\Big\{\int\big(-{}^{0}\rho\,{}^{t}\ddot{u}_i\,\tilde{\delta}(\delta^{t}u_i)$$
$$-{}^{t}_{0}S_{ij}\,e_{ij}({}^{t}u,\tilde{\delta}(\delta^{t}u)) + {}^{t}_{0}f_i\,\delta(\tilde{\delta}^{t}u_i)\big)\,J^{r}dV + \int{}^{t}_{0}x_{i,j}\,{}^{t}_{0}T_j\,\tilde{\delta}(\delta^{t}u_i)\,J_{\Gamma}\,{}^{r}d\Gamma_T\Big\}\,dt \tag{29}$$

where Eq. (29) is obtained after rearranging and grouping certain terms. Let $\tilde{\delta}(\delta^{t}u_i)$ be specified as a kinematically admissible field that also satisfies appropriate smoothness requirements (e.g., $\tilde{\delta}(\delta^{t}u_i) \in H^1({}^{t}V)$ and $\tilde{\delta}(\delta^{t}u_i) = 0$ on Γ_u). Since $\{\tilde{\delta}(\delta^{t}u_i), e_{ij}({}^{t}u,\delta(\tilde{\delta}^{t}u)\}$ is a compatible set, the integrand of the second time integral in Eq. (29) satisfies the Hamilton's principle, and thus vanishes. If the virtual displacements $\delta^{t}u_i$ are assumed to be independent of design, then $\tilde{\delta}(\delta^{t}u_i) = 0$. In this case also, the terms in the second time integral in Eq. (29) vanish identically since $e_{ij}({}^{t}u,\delta(\tilde{\delta}^{t}u)) = 0$. Note, however, that even in this special case, $\tilde{\delta}^{t}_{0}\varepsilon_{ij} \neq 0$ and $\tilde{\delta}(\delta^{t}_{0}\varepsilon_{ij}) \equiv \tilde{\delta}e_{ij}({}^{t}u,\delta^{t}u) \neq 0$.

Assume that the velocity-dependent damping force and the displacement-dependent force are contained as part of the applied body forces ${}^{t}_{0}f_i$, and that the deformation-dependent loading (e.g., pressure) is considered in the surface traction ${}^{t}_{0}T^{0}_{i}$. Then implicit variations of these forces are written as

$$\tilde{\delta}^{t}_{0}f_i = {}^{t}_{0}f_{i,{}^{t}u_j}\,\tilde{\delta}^{t}u_j + {}^{t}_{0}f_{i,{}^{t}\dot{u}_j}\,\tilde{\delta}^{t}\dot{u}_j;\quad \tilde{\delta}^{t}_{0}T^{0}_{i} = {}^{t}_{0}T^{0}_{i,{}^{t}u_j}\,\tilde{\delta}^{t}u_j + {}^{t}_{0}T^{0}_{i,{}^{t}\dot{u}_j}\,\tilde{\delta}^{t}\dot{u}_j \tag{30}$$

Therefore substituting Eqs. (26), (29) and (30) into Eq. (25), the total design variation of the state equation becomes:

$$\bar{\delta}^{\tau}_{r}H = \int\Big\{\int\big[-{}^{0}\rho\,\tilde{\delta}^{t}\ddot{u}_i\,\delta^{t}u_i - \tilde{\delta}^{t}_{0}S_{ij}\,e_{ij}({}^{t}u,\delta^{t}u) - {}^{t}_{0}S_{ij}\,\eta_{ij}(\tilde{\delta}^{t}u,\delta^{t}u)$$
$$+({}^{t}_{0}f_{i,{}^{t}u_j}\,\tilde{\delta}^{t}u_j + {}^{t}_{0}f_{i,{}^{t}\dot{u}_j}\,\tilde{\delta}^{t}\dot{u}_j)\,\delta^{t}u_i\big]\,J^{r}dV + \int\big[{}^{t}_{0}x_{i,j}\,({}^{t}_{0}T^{0}_{j,{}^{t}u_k}\,\tilde{\delta}^{t}u_k$$
$$+{}^{t}_{0}T^{0}_{j,{}^{t}\dot{u}_k}\,\tilde{\delta}^{t}\dot{u}_k) + \tilde{\delta}^{t}_{r}u_{i,k}\,{}^{r}_{0}x_{k,j}\,{}^{t}_{0}T_j\,\delta^{t}u_i\big]\,J_{\Gamma}\,{}^{r}d\Gamma_T + \bar{\bar{\delta}}^{t}_{r}A\Big\}\,dt = 0 \tag{31}$$

The *DVM is now summarized* as follows: (i) solve the analysis problem, i.e., Eqs. (1), (14) to (16), (ii) form the response functional whose design sensitivity is desired, (iii) solve for the design variations of the response variables by simultaneously solving the problems in Eqs. (21) to (24) and (31), and (iv) substitute the design variations of the response variables into the design variation of the response functional. Note that the problem derived in Eq. (31) has same operators as in the incremental equation of motion (12) (compare these equations after substituting Eq. (13) into Eq. (12)); however, the pseudo-load depends on the solution variables $\tilde{\delta}^{t}_{0}\varepsilon^{in}_{ij}$. Therefore Eqs. (21) and (31) will have to be integrated simultaneously at a particular load/time level. These equations are linear in terms of the solution variables $\tilde{\delta}^{t}u_i$, $\tilde{\delta}^{t}_{0}\varepsilon^{in}_{ij}$ and $\tilde{\delta}^{t}\alpha_{ij}$ which can be solved using any technique for integration of differential equations accurately. To avoid dealing with the *asymmetric matrices* (in the discretized form) in Eq. (31), the asymmetric parts can be transferred to the right hand side of the equation, as in the analysis phase. Then *iterative procedures* can be used at a load/time level to solve for the sensitivities. For conservative problems, there is no asymmetry in the problem, so the foregoing procedure is not necessary. In any case, for *history dependent* problems, the sensitivities at any load/time level depend on their previous histories. This aspect is different from the path-independent problems where sensitivity calculation at any load/time level does not need sensitivities at any previous points.

It is important to note that the DVM method can be implemented in such a way that response calculations and sensitivity calculations can be done *simultaneously* during the incremental solution process. This is further explained later in the paper.

4. Design Sensitivity Analysis: Adjoint Method

The adjoint method (AM) uses certain *adjoint fields* to evaluate the design variations of a particular response functional. The total design variation of all the state fields is not computed directly. The explicit design variations of the primary state fields and the adjoint fields are used to obtain the total design variations for the response functional. Although the adjoint method apears to be tedious and cumbersome to implement for the history dependent problems in most cases (Tsay and Arora 1989), it is included here for completeness. Further work is needed to fully investigate usefulness of the adjoint method for practical applications.

In the adjoint structure method, we replace the arbitrary state fields in Eqs. (1) and (2) with the adjoint displacement field ${}^t_a u_i$ and adjoint strains ${}^t_a\varepsilon_{ij}$ as follows:

$$ {}^t_aA = \int {}_aF J\,{}^r dV + \int {}_aR_T J_\Gamma\,{}^r d\Gamma_T + \int {}_aR_u J_\Gamma\,{}^r d\Gamma_u = 0;\quad {}_aR_T = (\delta_{ij} + {}^t_r u_{i,k}\,{}^r_0 x_{k,j})\,{}^{t}_{0}T^{0\,t}_{j}\,{}_a u_i \tag{32} $$

$$ {}_aR_u = (\delta_{ij} + {}^t_r u_{i,k}\,{}^r_0 x_{k,j})\,{}^t_0 T_j\,{}^{t}_{a}u^{0}_{i};\quad {}_aF = -{}^0\rho\,{}^t\ddot{u}_i\,{}^t_a u_i - {}^t_0 S_{ij}\,{}^t_a\varepsilon_{ij} + {}^t_0 f_i\,{}^t_a u_i \tag{33} $$

Note that ${}^t_a u_i$ is an admissible displacement field, i.e., ${}^t_a u_i \in H^1({}^tV)$, but otherwise arbitrary at this stage. Later, we shall see that it must be the solution of an adjoint problem. Also note that since ${}^t_a\varepsilon_{ij}$ must be compatible with ${}^t_a u_i$, it is given, using Eq. (4), as

$$ {}^t_a\varepsilon_{ij} = e_{ij}({}^t u, {}^t_a u) \tag{34} $$

Note also that in Eq. (32), the surface integral has been split into traction specified and displacement specified surfaces. Inclusion of the surface integral over the displacement specified part allows one to calculate sensitivity of the reaction forces. Now the functional ${}^t_aH$ defined in Eq. (7) becomes

$$ {}^\tau_aH \equiv \int {}^t_aA\, dt = 0 \tag{35} $$

To develop the adjoint method using the *variational principle of design sensitivity analysis* (Arora and Cardoso 1991), we define an augmented functional as

$$ L \equiv \psi + {}^\tau_aH + M;\quad M \equiv \iint ({}^t\lambda^p_{ij}\,{}^t p_{ij} + {}^t\lambda^q_{ij}\,{}^t q_{ij} + {}^t\lambda^r_{ij}\,{}^t r_{ij})\, J\,{}^r dV\, dt \tag{36} $$

where ${}^t\lambda^p_{ij}(t)$, ${}^t\lambda^q_{ij}(t)$ and ${}^t\lambda^r_{ij}(t)$ are the adjoint variables (Lagrange multipliers) that will be determined later. ${}^t\lambda^p_{ij}(t)$ is the adjoint variable for the inelastic strain equation, ${}^t\lambda^q_{ij}(t)$ is the adjoint variable for the internal variable equation and ${}^t\lambda^r_{ij}(t)$ is the adjoint variable for the stress-strain equation. Substituting expressions for ψ, ${}^t_aH$ and M into the augmented functional L, we get

$$ L = \int\{\int \bar{G} J\,{}^r dV + \int \bar{g} J_\Gamma\,{}^r d\Gamma_u + \int \bar{h} J_\Gamma\,{}^r d\Gamma_T\}\, dt + \int \vartheta J\,{}^r dV \tag{37} $$

$$ \bar{G} = G + {}_aF + {}^t\lambda^p_{ij}\,{}^t p_{ij} + {}^t\lambda^q_{ij}\,{}^t q_{ij} + {}^t\lambda^r_{ij}\,{}^t r_{ij};\quad \bar{g} = g + {}_aR_u;\quad \bar{h} = h + {}_aR_T \tag{38} $$

According to the above mentioned variational principle, the *total design variation of ψ is equal to the explicit design variation of L*; i.e., $\bar{\delta}\psi = \bar{\bar{\delta}}L$, so $\bar{\delta}\psi$ and various quantities needed to compute it are given as

$$ \begin{aligned} \bar{\delta}\psi = \int\{ & \int \bar{G}\,\bar{\bar{\delta}}J\,{}^r dV + \int [\bar{\bar{\delta}}G + \bar{\bar{\delta}}_aF + {}^t\lambda^q_{ij}\,\bar{\bar{\delta}}{}^t p_{ij} + {}^t\lambda^q_{ij}\,\bar{\bar{\delta}}{}^t q_{ij} + {}^t\lambda^r_{ij}\,\bar{\bar{\delta}}{}^t r_{ij}]\, J\,{}^r dV \\ & + \int \bar{g}\,\bar{\bar{\delta}}J_\Gamma\,{}^r d\Gamma_u + \int [\bar{\bar{\delta}}g + \bar{\bar{\delta}}_aR_u]\, J_\Gamma\,{}^r d\Gamma_u + \int \bar{h}\,\bar{\bar{\delta}}J_\Gamma\,{}^r d\Gamma_T \\ & + \int [\bar{\bar{\delta}}h + \bar{\bar{\delta}}_aR_T]\, J_\Gamma\,{}^r d\Gamma_T\}\, dt + \int [\vartheta\,\bar{\bar{\delta}}J + \bar{\bar{\delta}}\vartheta J]\,{}^r dV \end{aligned} \tag{39} $$

$$ \bar{\bar{\delta}}G = G_{,{}^t_0S_{ij}}\,\bar{\bar{\delta}}{}^t_0S_{ij} + G_{,{}^t_0\varepsilon_{ij}}\,\bar{\bar{\delta}}{}^t_0\varepsilon_{ij} + (G_{,{}^t_0\dot{\varepsilon}_{ij}})\,\bar{\bar{\delta}}{}^t_0\dot{\varepsilon}_{ij} + G_{,b_i}\,\delta b_i \tag{40} $$

$$ \bar{\bar{\delta}}_aF = -\bar{\bar{\delta}}{}^0\rho\,{}^t\ddot{u}_i\,{}^t_a u_i - \bar{\bar{\delta}}{}^t_0S_{ij}\,{}^t_a\varepsilon_{ij} - {}^t_0S_{ij}\,\bar{\bar{\delta}}{}^t_a\varepsilon_{ij} + \bar{\bar{\delta}}{}^t_0f_i\,{}^t_a u_i \tag{41} $$

$$ \begin{aligned} \bar{\bar{\delta}}{}^t p_{ij} = {} & {}^t p_{ij,{}^t_0\dot{S}_{k\ell}}\,\bar{\bar{\delta}}{}^t_0\dot{S}_{k\ell} + {}^t p_{ij,{}^t_0S_{k\ell}}\,\bar{\bar{\delta}}{}^t_0S_{k\ell} + {}^t p_{ij,{}^t_0\dot{\varepsilon}^{in}_{k\ell}}\,\bar{\bar{\delta}}{}^t_0\dot{\varepsilon}^{in}_{k\ell} + {}^t p_{ij,{}^t_0\varepsilon^{in}_{k\ell}}\,\bar{\bar{\delta}}{}^t_0\varepsilon^{in}_{k\ell} \\ & + {}^t p_{ij,{}^t\alpha_{k\ell}}\,\bar{\bar{\delta}}{}^t\alpha_{k\ell} + {}^t p_{ij,b_k}\,\delta b_k \end{aligned} \tag{42} $$

$$ \begin{aligned} \bar{\bar{\delta}}{}^t q_{ij} = {} & {}^t q_{ij,{}^t_0\dot{S}_{k\ell}}\,\bar{\bar{\delta}}{}^t_0\dot{S}_{k\ell} + {}^t q_{ij,{}^t_0S_{k\ell}}\,\bar{\bar{\delta}}{}^t_0S_{k\ell} + {}^t q_{ij,{}^t_0\varepsilon^{in}_{k\ell}}\,\bar{\bar{\delta}}{}^t_0\varepsilon^{in}_{k\ell} + {}^t q_{ij,{}^t\dot{\alpha}_{k\ell}}\,\bar{\bar{\delta}}{}^t\dot{\alpha}_{k\ell} \\ & + {}^t q_{ij,{}^t\alpha_{k\ell}}\,\bar{\bar{\delta}}{}^t\alpha_{k\ell} + {}^t q_{ij,b_k}\,\delta b_k \end{aligned} \tag{43} $$

$$\bar{\bar{\delta}}{}^t r_{ij} = {}^t r_{ij,{}^t_0\dot{S}_{k\ell}}\, \bar{\bar{\delta}}{}^t_0\dot{S}_{k\ell} + {}^t r_{ij,{}^t_0 S_{k\ell}}\,\bar{\bar{\delta}}{}^t_0 S_{k\ell} + {}^t r_{ij,{}^t_0\dot{\varepsilon}_{k\ell}}\,\bar{\bar{\delta}}{}^t_0\dot{\varepsilon}_{k\ell} + {}^t r_{ij,{}^t_0\varepsilon_{k\ell}}\,\bar{\bar{\delta}}{}^t_0\varepsilon_{k\ell}$$
$$+ {}^t r_{ij,{}^t_0\varepsilon^{in}_{k\ell}}\,\bar{\bar{\delta}}{}^t_0\varepsilon^{in}_{k\ell} + {}^t r_{ij,b_k}\,\delta b_k \tag{44}$$

$$\bar{\bar{\delta}} g = g_{,{}^t u_i}\,\bar{\bar{\delta}}{}^t u_i^0 + g_{,{}^t_0 T_i}\,\bar{\bar{\delta}}{}^t_0 T_i + g_{,b_i}\,\delta b_i; \qquad \bar{\bar{\delta}} h = h_{,{}^t_0 T_i}\,\bar{\bar{\delta}}{}^t_0 T_i^0 + h_{,b_i}\,\delta b_i \tag{45}$$

$$\bar{\bar{\delta}}_a R_T = \left({}^t_r u_{i,k}\,\bar{\bar{\delta}}{}^r_0 x_{k,j}\,{}^t_0 T_j^0 + {}^t_0 x_{i,j}\,\bar{\bar{\delta}}{}^t_0 T_j^0\right){}^t_a u_i \tag{46}$$

$$\bar{\bar{\delta}}_a R_u = \left({}^t_r u_{i,k}\,\bar{\bar{\delta}}{}^r_0 x_{k,j}\,{}^t_0 T_j + {}^t_0 x_{i,j}\,\bar{\bar{\delta}}{}^t_0 T_j\right){}^t_a u_i^0 + {}^t_0 x_{i,j}\,{}^t_0 T_j\,\bar{\bar{\delta}}{}^t_a u_i^0 \tag{47}$$

$$\bar{\bar{\delta}}\vartheta = \vartheta_{,{}^0 u_i}\,\bar{\bar{\delta}}{}^0 u_i + (\vartheta_{,{}^0\dot{u}_i})\,\bar{\bar{\delta}}{}^0\dot{u}_i + \vartheta_{,b_i}\,\delta b_i \tag{48}$$

Note that in Eqs. (39), (42) to (44), $\bar{\bar{\delta}}{}^t_0\dot{\varepsilon}^{in}_{ij}$, $\bar{\bar{\delta}}{}^t_0\varepsilon^{in}_{ij}$, $\bar{\bar{\delta}}{}^t\dot{\alpha}_{ij}$ and $\bar{\bar{\delta}}{}^t\alpha_{ij}$ may be zero in most cases since they may not explicitly depend on the design variables.

To solve for the sensitivity coefficients from Eq. (39), the adjoint fields, and the Lagrange multipliers ${}^t\lambda^p_{ij}(t)$, ${}^t\lambda^q_{ij}(t)$ and ${}^t\lambda^r_{ij}(t)$ are needed. Equations that determine these quantities can be obtained by requiring $\tilde{\delta} L = 0$ (Arora and Cardoso 1991); i.e., by requiring L to be stationary with respect to the implicit state fields. This leads to the following equation using L defined in Eqs. (37) and (38):

$$\int\Big\{\int\big[\tilde{\delta}G + \tilde{\delta}_a F + {}^t\lambda^p_{ij}\,\tilde{\delta}{}^t p_{ij} + {}^t\lambda^q_{ij}\,\tilde{\delta}{}^t q_{ij} + {}^t\lambda^r_{ij}\,\tilde{\delta}{}^t r_{ij}\big]\,J\,{}^r dV$$
$$+ \int\big[\tilde{\delta}g + \tilde{\delta}_a R_u\big]\,J_\Gamma\,{}^r d\Gamma_u + \int\big[\tilde{\delta}h + \tilde{\delta}_a R_T\big]\,J_\Gamma\,{}^r d\Gamma_T\Big\}\,dt + \int\tilde{\delta}\vartheta\,J\,{}^r dV = 0 \tag{49}$$

$$\tilde{\delta}G = G_{,{}^t_0 S_{ij}}\,\tilde{\delta}{}^t_0 S_{ij} + G_{,{}^t_0\varepsilon_{ij}}\,\tilde{\delta}{}^t_0\varepsilon_{ij} + (G_{,{}^t_0\dot{\varepsilon}_{ij}})\,\tilde{\delta}{}^t_0\dot{\varepsilon}_{ij} + G_{,{}^t u_i}\,\tilde{\delta}{}^t u_i$$
$$+ (G_{,{}^t\dot{u}_i})\,\tilde{\delta}{}^t\dot{u}_i + (G_{,{}^t\ddot{u}_i})\,\tilde{\delta}{}^t\ddot{u}_i \tag{50}$$

$$\tilde{\delta}_a F = -{}^0\rho\,\tilde{\delta}({}^t\ddot{u}_i\,{}^t_a u_i) - \tilde{\delta}({}^t_0 S_{ij}\,{}^t_a\varepsilon_{ij}) + \tilde{\delta}({}^t_0 f_i\,{}^t_a u_i) \tag{51}$$

$$\tilde{\delta}{}^t p_{ij} = {}^t p_{ij,{}^t_0\dot{S}_{k\ell}}\,\tilde{\delta}{}^t_0\dot{S}_{k\ell} + {}^t p_{ij,{}^t_0 S_{k\ell}}\,\tilde{\delta}{}^t_0 S_{k\ell} + {}^t p_{ij,{}^t_0\dot{\varepsilon}^{in}_{k\ell}}\,\tilde{\delta}{}^t_0\dot{\varepsilon}^{in}_{k\ell} + {}^t p_{ij,{}^t_0\varepsilon^{in}_{k\ell}}\,\tilde{\delta}{}^t_0\varepsilon^{in}_{k\ell}$$
$$+ {}^t p_{ij,{}^t\alpha_{k\ell}}\,\tilde{\delta}{}^t\alpha_{k\ell} \tag{52}$$

$$\tilde{\delta}{}^t q_{ij} = {}^t q_{ij,{}^t_0\dot{S}_{k\ell}}\,\tilde{\delta}{}^t_0\dot{S}_{k\ell} + {}^t q_{ij,{}^t_0 S_{k\ell}}\,\tilde{\delta}{}^t_0 S_{k\ell} + {}^t q_{ij,{}^t_0\varepsilon^{in}_{k\ell}}\,\tilde{\delta}{}^t_0\varepsilon^{in}_{k\ell} + {}^t q_{ij,{}^t\dot{\alpha}_{k\ell}}\,\tilde{\delta}{}^t\dot{\alpha}_{k\ell}$$
$$+ {}^t q_{ij,{}^t\alpha_{k\ell}}\,\tilde{\delta}{}^t\alpha_{k\ell} \tag{53}$$

$$\tilde{\delta}{}^t r_{ij} = {}^t r_{ij,{}^t_0\dot{S}_{k\ell}}\,\tilde{\delta}{}^t_0\dot{S}_{k\ell} + {}^t r_{ij,{}^t_0 S_{k\ell}}\,\tilde{\delta}{}^t_0 S_{k\ell} + {}^t r_{ij,{}^t_0\dot{\varepsilon}_{k\ell}}\,\tilde{\delta}{}^t_0\dot{\varepsilon}_{k\ell} + {}^t r_{ij,{}^t_0\varepsilon_{k\ell}}\,\tilde{\delta}{}^t_0\varepsilon_{k\ell}$$
$$+ {}^t r_{ij,{}^t_0\varepsilon^{in}_{k\ell}}\,\tilde{\delta}{}^t_0\varepsilon^{in}_{k\ell} \tag{54}$$

$$\tilde{\delta}g = g_{,{}^t_0 T_i}\,\tilde{\delta}{}^t_0 T_i; \qquad \tilde{\delta}h = h_{,{}^t u_i}\,\tilde{\delta}{}^t u_i + h_{,{}^t_0 T_i^0}\,\tilde{\delta}{}^t_0 T_i^0 \tag{55}$$

$$\tilde{\delta}_a R_T = {}^t_0 x_{i,j}\,\tilde{\delta}{}^t_0 T_j^0\,{}^t_a u_i + \tilde{\delta}{}^t_r u_{i,k}\,{}^r_0 x_{k,j}\,{}^t_0 T_j^0\,{}^t_a u_i + {}^t_0 x_{i,j}\,{}^t_0 T_j^0\,\tilde{\delta}{}^t_a u_i \tag{56}$$

$$\tilde{\delta}_a R_u = {}^t_0 x_{i,j}\,\tilde{\delta}{}^t_0 T_j\,{}^t_a u_i^0 + \tilde{\delta}{}^t_r u_{i,k}\,{}^r_0 x_{k,j}\,{}^t_0 T_j\,{}^t_a u_i^0 + {}^t_0 x_{i,j}\,{}^t_0 T_j\,\tilde{\delta}{}^t_a u_i^0 \tag{57}$$

$$\tilde{\delta}\vartheta = \vartheta_{,{}^\tau u_i}\,\tilde{\delta}{}^\tau u_i + (\vartheta_{,{}^\tau\dot{u}_i})\,\tilde{\delta}{}^\tau\dot{u}_i \tag{58}$$

The implicit design variation of the adjoint strains from Eq. (34) is given as

$$\tilde{\delta}{}^t_a\varepsilon_{ij} = e_{ij}({}^t u, \tilde{\delta}{}^t_a u) + \eta_{ij}(\tilde{\delta}{}^t u, {}^t_a u) \tag{59}$$

The next step is to substitute Eqs. (50) to (59) into Eq. (49), collect terms and simplify the expression. In this process, the following equation will be used to eliminate certain terms:

$$\int\left(-{}^{0}\rho\,{}^{t}\ddot{u}_i\,\tilde{\delta}^{t}_{a}u_i - {}^{t}_{0}S_{ij}\,e_{ij}({}^{t}u,\tilde{\delta}^{t}_{a}u) + {}^{t}_{0}f_i\,\tilde{\delta}^{t}_{a}u_i\right) J\,{}^{r}dV + \int {}^{t}_{0}x_{i,j}\,{}^{t}_{0}T^{0}_{j}\,\tilde{\delta}^{t}_{a}u_i\,J_{\Gamma}\,{}^{r}d\Gamma_T = 0 \tag{60}$$

The above equation represents the equation of motion (1) for the primary structure since $\tilde{\delta}^{t}_{a}u_i$ can be specified as an arbitrary but kinematically admissible field satisfying the appropriate smoothness requirements (e.g., $\tilde{\delta}^{t}_{a}u_i \in H^1({}^{t}V)$ and $\tilde{\delta}^{t}_{a}u_i = 0$ on Γ_u) with $e_{ij}({}^{t}u,\tilde{\delta}^{t}_{a}u)$ compatible with it.

Now substitute Eqs. (30) and (50) to (60) into Eq. (49), carry out the integration by parts with respect to time for certain terms, use Eq. (18) for $\tilde{\delta}^{t}_{0}\varepsilon_{ij}$, collect various terms as coefficients of the implicit variation of the response variables, and require $\tilde{\delta}^{t}u_i$ to be arbitrary but admissible field ($\tilde{\delta}^{t}u_i \in H^1({}^{t}V)$) and $\tilde{\delta}^{t}_{0}S_{ij}$, $\tilde{\delta}^{t}_{0}\varepsilon^{in}_{ij}$ and $\tilde{\delta}^{t}\alpha_{ij}$ to be arbitrary on $L^2({}^{t}V)$. This leads to the governing equations of motion for the adjoint problems as

$$\int\Big\{\int\Big[-{}^{0}\rho\,{}^{t}_{a}\ddot{u}_i\,\tilde{\delta}^{t}u_i - {}^{t}_{a}S_{ij}\,e_{ij}({}^{t}u,\tilde{\delta}^{t}u) - {}^{t}_{0}S_{ij}\,\eta_{ij}(\tilde{\delta}^{t}u,{}^{t}_{a}u) + {}^{t}_{a}f_i\,\tilde{\delta}^{t}u_i\Big] J\,{}^{r}dV$$
$$+ \int \tilde{\delta}^{t}_{r}u_{i,k}\,{}^{r}_{0}x_{k,j}\,{}^{t}_{0}T_j\,{}^{t}_{a}u_i\,J_{\Gamma}\,{}^{r}d\Gamma + \int {}^{t}_{a}T^{0}_{i}\,\tilde{\delta}^{t}u_i\,J_{\Gamma}\,{}^{r}d\Gamma_T\Big\}\,dt = 0 \tag{61}$$

$$\left({}^{t}\lambda^{p}_{k\ell}\,{}^{t}p_{k\ell},{}^{t}_{0}\dot{\varepsilon}^{in}_{ij}\right)^{\cdot} - {}^{t}\lambda^{p}_{k\ell}\,{}^{t}p_{k\ell},{}^{t}_{0}\varepsilon^{in}_{ij} - {}^{t}\lambda^{q}_{k\ell}\,{}^{t}q_{k\ell},{}^{t}_{0}\varepsilon^{in}_{ij} - {}^{t}\lambda^{r}_{k\ell}\,{}^{t}r_{k\ell},{}^{t}_{0}\varepsilon^{in}_{ij} = 0 \tag{62}$$

$$\left({}^{t}\lambda^{q}_{k\ell}\,{}^{t}q_{k\ell},{}^{t}\dot{\alpha}_{ij}\right)^{\cdot} - {}^{t}\lambda^{q}_{k\ell}\,{}^{t}q_{k\ell},{}^{t}\alpha_{ij} - {}^{t}\lambda^{p}_{k\ell}\,{}^{t}p_{k\ell},{}^{t}\alpha_{ij} = 0 \tag{63}$$

$$\left({}^{t}\lambda^{p}_{ij}\,{}^{t}p_{ij},{}^{t}_{0}\dot{S}_{k\ell}\right)^{\cdot} - {}^{t}\lambda^{p}_{ij}\,{}^{t}p_{ij},{}^{t}_{0}S_{k\ell} + \left({}^{t}\lambda^{q}_{ij}\,{}^{t}q_{ij},{}^{t}_{0}\dot{S}_{k\ell}\right)^{\cdot} - {}^{t}\lambda^{q}_{ij}\,{}^{t}q_{ij},{}^{t}_{0}S_{k\ell} + \left({}^{t}\lambda^{r}_{ij}\,{}^{t}r_{ij},{}^{t}_{0}\dot{S}_{k\ell}\right)^{\cdot}$$
$$- {}^{t}\lambda^{r}_{ij}\,{}^{t}r_{ij},{}^{t}_{0}S_{k\ell} + {}^{t}_{a}\varepsilon_{k\ell} - {}^{t}_{a}\varepsilon^{I}_{k\ell} = 0 \tag{64}$$

During the process of integration by parts, certain terms at the terminal time $t = \tau$ are obtained. Annihilation of these terms gives the following terminal conditions for the adjoint problems:

$$\vartheta,{}_{{}^{\tau}\dot{u}_i} - {}^{0}\rho\,{}^{\tau}_{a}u_i + G,{}_{{}^{\tau}\ddot{u}_i} = 0 \quad \text{in } {}^{r}V \tag{65}$$

$$\vartheta,{}_{{}^{\tau}u_i} + {}^{0}\rho\,{}^{\tau}_{a}\dot{u}_i + {}^{\tau}_{a}u_j\,{}^{\tau}_{0}f_j,{}_{{}^{\tau}\dot{u}_i} + G,{}_{{}^{\tau}\dot{u}_i} - \left(G,{}_{{}^{\tau}\ddot{u}_i}\right)^{\cdot} = 0 \quad \text{in } {}^{r}V \tag{66}$$

$$G,{}_{{}^{\tau}_{0}\dot{\varepsilon}_{ij}} + {}^{\tau}\lambda^{r}_{k\ell}\,{}^{\tau}r_{k\ell},{}^{\tau}_{0}\dot{\varepsilon}_{ij} = 0 \quad \text{in } {}^{r}V \tag{67}$$

$${}^{\tau}_{0}x_{i,j}\,{}^{\tau}_{0}T^{0}_{j},{}_{{}^{\tau}\dot{u}_k}\,{}^{\tau}_{a}u_i + h,{}_{{}^{\tau}_{0}T^{0}_{i}}\,{}^{\tau}_{0}T^{0}_{i},{}_{{}^{\tau}\dot{u}_k} = 0 \quad \text{on } {}^{r}\Gamma_T \tag{68}$$

$${}^{\tau}\lambda^{p}_{ij}\,{}^{\tau}p_{ij},{}^{\tau}_{0}\dot{\varepsilon}^{in}_{k\ell} = 0 \quad \text{in } {}^{r}V \tag{69}$$

$${}^{\tau}\lambda^{q}_{ij}\,{}^{\tau}q_{ij},{}^{\tau}\dot{\alpha}_{k\ell} = 0 \quad \text{in } {}^{r}V \tag{70}$$

$${}^{\tau}\lambda^{p}_{ij}\,{}^{\tau}p_{ij},{}^{\tau}_{0}\dot{S}_{k\ell} + {}^{\tau}\lambda^{q}_{ij}\,{}^{\tau}q_{ij},{}^{\tau}_{0}\dot{S}_{k\ell} + {}^{\tau}\lambda^{r}_{ij}\,{}^{\tau}r_{ij},{}^{\tau}_{0}\dot{S}_{k\ell} = 0 \quad \text{in } {}^{r}V \tag{71}$$

While collecting various terms in the foregoing derivation, the following quantities for the adjoint system appear. For the convenience notation, they are defined as follows:

Initial strain: $${}^{t}_{a}\varepsilon^{I}_{ij} = G,{}_{{}^{t}_{0}S_{ij}} \quad \text{in } {}^{r}V \tag{72}$$

Initial stress: $${}^{t}_{a}S^{I}_{ij} = -G,{}_{{}^{t}_{0}\varepsilon_{ij}} + \left(G,{}_{{}^{t}_{0}\dot{\varepsilon}_{ij}}\right)^{\cdot} \quad \text{in } {}^{r}V \tag{73}$$

Body force: $${}^{t}_{a}f_i = G,{}_{{}^{t}u_i} - \left(G,{}_{{}^{t}\dot{u}_i}\right)^{\cdot} + \left(G,{}_{{}^{t}\ddot{u}_i}\right)^{\cdot\cdot} + {}^{t}_{a}u_j\,{}^{t}_{0}f_j,{}_{{}^{t}u_i} - \left({}^{t}_{a}u_j\,{}^{t}_{0}f_j,{}_{{}^{t}\dot{u}_i}\right)^{\cdot} \quad \text{in } {}^{r}V \tag{74}$$

Boundary displacement: $$ {}^{t}_{a}u^{0}_{i} = -{}_{t}x^{0}_{j,i}\; g_{,{}^{t}_{0}T_{j}} \quad \text{on } {}^{r}\Gamma_{u} \tag{75}$$

Specified traction: $$ {}^{t}_{a}T^{0}_{i} = h_{,{}^{t}u_{i}} + h_{,{}^{t}_{0}T^{0}_{j}}\, {}^{t}_{0}T^{0}_{j,{}^{t}u_{i}} - \left(h_{,{}^{t}_{0}T^{0}_{j}}\, {}^{t}_{0}T^{0}_{j,{}^{t}\dot{u}_{i}}\right)^{\cdot} + {}^{t}_{0}x_{i,j}\, {}^{t}_{0}T^{0}_{j,{}^{t}u_{k}}\, {}^{t}_{a}u_{k}$$
$$ - \left({}^{t}_{0}x_{i,j}\, {}^{t}_{0}T^{0}_{j,{}^{t}\dot{u}_{k}}\, {}^{t}_{a}u_{k}\right)^{\cdot} \quad \text{on } {}^{r}\Gamma_{T} \tag{76}$$

In addition, collection of various terms leads to the following definition of the stress-strain law for the adjoint structure:

$$ {}^{t}_{a}S_{ij} = \left({}^{t}\lambda^{r}_{k\ell}\, {}^{t}r_{k\ell,{}^{t}_{0}\dot{\varepsilon}_{ij}}\right)^{\cdot} - {}^{t}\lambda^{r}_{k\ell}\, {}^{t}r_{k\ell,{}^{t}_{0}\varepsilon_{ij}} + {}^{t}_{a}S^{I}_{ij} \tag{77}$$

Also during the process of integration by parts, certain terms at the initial time t = 0 involving design variations of the state fields are obtained. These terms can be collected and written as

$$\begin{aligned}
\bar{\bar{\delta}}B_{0} = &- \int \left[{}^{0}\rho\, {}^{t}_{a}\dot{u}_{i} + {}^{t}_{a}u_{j}\, {}^{t}_{0}f_{j,{}^{t}\dot{u}_{i}} + G_{,{}^{t}\dot{u}_{i}} - \left(G_{,{}^{t}\ddot{u}_{i}}\right)^{\cdot}\right]_{t=0} \bar{\bar{\delta}}\,{}^{0}u_{i}\, J\, {}^{r}dV \\
&- \int \left[\left(G_{,{}^{t}_{0}\dot{\varepsilon}_{ij}}\right) \bar{\bar{\delta}}\,{}^{t}_{0}\varepsilon_{ij}\right]_{t=0} J\, {}^{r}dV + \int \left[{}^{0}\rho\, {}^{t}_{a}u_{i} - G_{,{}^{t}\ddot{u}_{i}}\right]_{t=0} \bar{\bar{\delta}}\,{}^{0}\dot{u}_{i}\, J\, {}^{r}dV \\
&- \int \left[{}^{t}_{0}x_{i,j}\, {}^{t}_{0}T^{0}_{j,{}^{t}\dot{u}_{k}}\, {}^{t}_{a}u_{i} + h_{,{}^{t}_{0}T^{0}_{i}}\, {}^{t}_{0}T^{0}_{i,{}^{t}\dot{u}_{k}}\right]_{t=0} \bar{\bar{\delta}}\,{}^{0}u_{k}\, J_{\Gamma}\, {}^{r}d\Gamma_{T} \\
&- \int \left[{}^{t}\lambda^{p}_{ij}\, {}^{t}p_{ij,{}^{t}_{0}\dot{\varepsilon}^{in}_{k\ell}}\, \bar{\bar{\delta}}\,{}^{t}_{0}\varepsilon^{in}_{k\ell} + {}^{t}\lambda^{q}_{ij}\, {}^{t}q_{ij,{}^{t}\dot{\alpha}_{k\ell}}\, \bar{\bar{\delta}}\,{}^{t}\alpha_{k\ell}\right]_{t=0} J\, {}^{r}dV \\
&- \int \left[\left({}^{t}\lambda^{p}_{ij}\, {}^{t}p_{ij,{}^{t}_{0}\dot{S}_{k\ell}} + {}^{t}\lambda^{q}_{ij}\, {}^{t}q_{ij,{}^{t}_{0}\dot{S}_{k\ell}} + {}^{t}\lambda^{r}_{ij}\, {}^{t}r_{ij,{}^{t}_{0}\dot{S}_{k\ell}}\right) \tilde{\delta}\,{}^{t}_{0}S_{k\ell}\right]_{t=0} J\, {}^{r}dV \\
&- \int \left[{}^{t}\lambda^{r}_{ij}\, {}^{t}r_{ij,{}^{t}_{0}\dot{\varepsilon}_{k\ell}}\, \tilde{\delta}\,{}^{t}_{0}\varepsilon_{k\ell}\right]_{t=0} J\, {}^{r}dV
\end{aligned} \tag{78}$$

The terms in Eq. (78) are either zero if the state fields are specified at t = 0, or they are explicit design variations if they are treated as design dependent (this is the reason for replacing δ by $\bar{\bar{\delta}}$ in Eq. (78)). Therefore, Eq. (78) can be merged with Eq. (39) to obtain design derivatives with respect to the initial values of the state fields.

The *adjoint method is now summarized* as follows: (i) solve the analysis problem for the primary structure, (ii) form the performance functional whose design sensitivity is desired, (iii) define the adjoint problems given in Eqs. (61) to (64) and solve them for the adjoint fields using the terminal conditions given in Eqs. (65) to (71), and (iv) assemble the final sensitivity coefficients using the adjoint fields and the explicit design variations (Eq. 39).

Note that the adjoint problems in Eqs. (61) to (64) are *terminal value problems*. Therefore these problems cannot be solved until the analysis problem has been solved for the entire time interval, and a point along the loading path has been determined where the sensitivity is desired. This can be a major drawback of the adjoint method because it either requires massive storage of information generated during the analysis phase, or regeneration of the same information for solution of the adjoint problem. Moreover, Eqs. (61) to (64) should be solved for the adjoint variables simultaneously because they are coupled to each other. They are, however, linear in terms of the adjoint displacement field and Lagrange multipliers. Equation (61) has asymmetric operators, as in the analysis problem. As in the direct variation method, iterations will have to be used at a particular load/time level to avoid dealing with the asymmetric set of equations.

5. Computational Aspects

To solve practical problems for response analysis and sensitivity analysis, it is obvious that *numerical methods* will have to be used. The physical domain as well as the time domain will have to be *discretized*. The finite element method is the usual approach for solving complex structural problems. In the time domain, explicit or implicit methods can be used to integrate the discretized equations of motion. It can be seen that the equation of motion (1) is coupled with the constitutive equations (14) to (16). Therefore these equations must be solved simultaneously. To complicate the matters further, these equations are highly nonlinear. Also the constitutive equations can be quite stiff, requiring special procedures for their integration (Lee *et al* 1991). In the analysis phase the usual solution procedure will be to integrate the evolution equations (14) and determine the stresses using Eq. (16). Knowing the inelastic strain, its contribution to the effective load for the structure can be calculated. Thus, the equation of motion can be integrated for the displacement increment. Then the stresses and total strains can be updated. Using the updated values, the evolution Eqs. (14) can be integrated for the new values of the inelastic strains and the internal variables. This iteration process can be continued within the load/time increment until a convergence criterion is satisfied.

In the *direct variation method* of sensitivity analysis, once the equilibrium is obtained at a load level/time, design sensitivities of the state variables can be calculated. At the current time, all the response quantities are available, so using them all the system matrices can be updated. These can be then used for design sensitivity analysis at the current time as well as response analysis for the next load/time increment. The numerical procedure for design sensitivity analysis is the same as for the analysis phase, i.e., the varied form of the evolution equations (22) to (24) and the sensitivity equation (31) need to be solved simultaneously. The same integration routines as for the analysis phase can be used here as well. Note that iterations may be needed here also to compute increments in the design sensitivities. This is due to the reason that either self adaptive explicit or implicit time integration methods will have to be used for stability and accuracy of the numerical solution process. The process also depends very much on the time increments used in the numerical calculations.

The *adjoint method* of sensitivity analysis appears to be more complicated to implement than the direct variation method. First, the adjoint problems are terminal value problems, so one needs to know the time point at which the response functional has critical value. Then the sensitivity at that point can be evaluated. Second, the definition of the adjoint problems depends on the response functional that needs sensitivity analysis. In the design process, the response functionals needing sensitivity calculations are not known until the entire response history has been generated. Assuming that this has been done, the numerical process of integrating the adjoint problems backward in time is quite tedious. It requires either massive storage of all the system matrices and response quantities along the entire loading path, or they need to be regenerated. Also in the self adaptive explicit or implicit time integration method, the response may be needed at the points other than the time grid points. This requires some interpolation scheme for the variables. Thus, this method can be quite tedious and cumbersome to implement, and may be even inefficient.

6. Discussion and Conclusions

Design sensitivity analysis of a structural system that needs to describe the *inelastic behavior of materials* at the elevated temperature and high strain rate is developed. The feasibility of using *viscoplastic models* to perform the design sensitivity analysis is investigated. To describe motion of the *continuum*, the Hamilton's principle is used. *Geometric nonlinearities* are included in the formulation by using the Green-Lagrange strain and the second Piola-Kirchhoff stress that are referred to the undeformed configuration. Deformation dependent loads are treated in the formulation. The *control volume concept* that can be translated into the isoparametric finite elements for discretization of the design sensitivity equations is employed. This way, the shape and nonshape design sensitivity analysis problems are treated simultaneously and consistently.

A *general response functional* that needs design sensitivity calculations is defined. It can be used to impose constraints on displacements, velocities, accelerations, stresses, strains, and strain rates at any time and at any point in the structure as well as on the initial and terminal conditions. A *variational principle of design sensitivity analysis* is used to obtain the sensitivity expression for this functional in the adjoint method. The direct variation method is also derived.

In the analysis phase, the equation of motion and the constitutive evolution equations need to be *integrated simultaneously*. During each load/time increment, *iteration* is necessary to treat nonlinearities and the *load-correction stiffness* matrix due to nonconservative loading. The basic equation to be solved during sensitivity analysis is quite similar to the incremental equation of motion for the analysis phase; however, it contains asymmetric terms as for the analysis problem. Due to this, and accuracy and stability consideration of the numerical process, iterations are also needed at each load/time increment to determine the increments in design sensitivities of the state fields, or adjoint fields in the adjoint method. Design variations of the constitutive evolution equations, or the associated adjoint equations must be integrated simultaneously with other equations. Note that all the sensitivity equations are linear in terms of the solution variables. It is also important to note that the design sensitivities at a particular time are dependent on their previous history. This feature of design sensitivity analysis is different from that for history independent problems where the sensitivity analysis is not history dependent.

It is found that the *adjoint problems* associated with the formulation are terminal value problems, so they must be integrated backward in time/load. In addition, loading for the problems depends on the response functional whose sensitivity is desired. Therefore, the adjoint problems cannot be defined until the entire response history has been generated. This is the usual feature of the adjoint methods. What makes the method different for the present formulation is the presence of the nonlinearities, the history-dependent effects and the deformation dependent loading. Due to this, most of the data generated during the analysis phase needs to be either stored or re-generated during the solution process for the adjoint problems. This makes the method tedious to implement and perhaps inefficient compared to the direct variation method. Nevertheless, the method needs to be investigated for its applicability to practical problems.

The *direct variation method* is relatively straightforward to implement. Its implementation does not depend on a particular response functional whose sensitivity is desired. Essentially, the response analysis problems and sensitivity analysis problems can be solved simultaneously, and the sensitivity of any response functional can be computed at any point along the loading path even after the final load level has been reached. Thus, the method is *recommended* for this class of problems. The method has been implemented for planar quadrilateral elements, and demonstrated on two quasistatic problems. Several viscoplastic models have been implemented (Arora, Lee and Kumar 1992; Lee *et al* 1991). It has been shown that quite accurate sensitivities are obtained with respect to the shape design variables.

In another implementation (Poldneff et al. 1991), the effect of *nonconservative follower forces* on the accuracy of design sensitivity analysis has been studied. There also, the direct variation method was implemented with the semianalytical approach where the partial derivatives with respect to the design variables at the element level were calculated using the central difference approach. It was shown that the sensitivities improve substantially when the effect of load-stiffness correction matrix was included in analysis as well as design sensitivity analysis.

Recently, an *alternate treatment of the viscoplastic problems* has also been developed and presented (Jao and Arora 1991). There, a rate-type equation of motion is used along with a unified endochronic constitutive model. The direct variation method is used there also, and quite accurate sensitivities are obtained.

Finally, it is *concluded* that shape design sensitivity analysis and optimum design of viscoplastic structures is possible. However, we must be willing to bear substantial cost for implementation of the theory and the computational costs for response analysis, design sensitivity analysis, and optimization.

7. References

Arora, J.S. and Cardoso, J.B. (1991), "A Variational Principle for Shape Design Sensitivity Analysis," *32nd AIAA/ASME/ASCE/AHS/ASC Structures, Structural Dynamics and Materials Conference*, Paper No. 91-1213, Baltimore, MD, April 8-10; to appear in *AIAA Journal*.

Arora, J.S., Lee, T.H. and Cardoso, J.B. (1992), "Structural Shape Sensitivity Analysis: Relationship between Material Derivative and Control Volume Approaches," *AIAA Journal*, to appear.

Arora, J.S., Lee, T.H. and Kumar, V. (1992), "Design Sensitivity Analysis of Nonlinear Structures - III: Shape Variation of Viscoplastic Structures," in AIAA Series on Progress in Aeronautics and Astronautics, *Structural Optimization: Status & Promise*, M.P. Kamat (ed.), to be published by AIAA.

Bathe, K.J. (1982), *Finite Element Procedures in Engineering Analysis*, Prentice-Hall, Englewood Cliffs, New Jersey.

Cardoso, J.B. and Arora, J.S. (1988), "Variational Method for Design Sensitivity Analysis in Nonlinear Structural Mechanics," *AIAA Journal*, Vol. 26, pp. 595-603.

Choi, K.K. and Santos, J.L.T. (1987), "Design Sensitivity Analysis of Nonlinear Structural Systems Part 1: General Theory," *International J. for Numerical Methods in Engineering*, Vol. 24, pp. 2039-2055.

Dems, K. and Mròz, Z. (1989), "Shape Sensitivity Analysis and Optimal Design of Physically Nonlinear Plates," *Archives of Mechanics*, Vol. 41, No. 4, pp. 481-501.

Gopalakrishna, H.S. and Greimann, L.F. (1988), "Newton-Raphson Procedure for the Sensitivity Analysis of Nonlinear Structural Behavior," *Computers & Structures*, Vol. 30, No. 6, pp. 1263-1273.

Haber, R.B. (1987), "A New Variational Approach to Structural Shape Design Sensitivity Analysis," *Computer Aided Optimal Design: Structural and Mechanical Systems*, Edited by Mota Soares, C.A., NATO ASI Series, Vol. F27, Springer-Verlag Berlin Heidelberg, pp. 573-587.
Haftka, R.T. and Adelman, H.M. (1989), "Recent Developments in Structural Sensitivity Analysis," *Structural Optimization*, Vol. 1, pp. 137-151.
Haftka, R.T. and Mròz, Z. (1986), "First- and Second-Order Sensitivity Analysis of Linear and Nonlinear Structures," *AIAA Journal*, Vol. 24, pp 1187-1192.
Haririan, M., Cardoso, J.E.B. and Arora, J.S. (1987), "Use of ADINA for Design Optimization of Nonlinear Structures," *Computers and Structures*, Vol. 26, Nos. 1/2, pp. 123-134.
Jao, S.Y. and Arora, J.S. (1991a), "Design Sensitivity Analysis of Nonlinear Structures Using Endochronic Constitutive Model," *Computational Mechanics*, to appear, 1991.
Jao, S.Y. and Arora, J.S. (1991b), "Design Optimization of Nonlinear Structures with Rate-Dependent and Rate-Independent Constitutive Models," *International J. for Numerical Methods in Engineering*, submitted.
Komarakul-na-nakorn, A. and Arora, J.S. (1990), *A Study of Nonlinear Stability Limit and Its Design Sensitivity Analysis*, Technical Report No. ODL-90.10, Optimal Design Laboratory, College of Engineering, The University of Iowa, Iowa City, IA 52242 U.S.A.
Kumar, V., Morjaria, M., and Mukherjee, S. (1980), "Numerical Integration of Some Stiff Constitutive Models of Inelastic Deformation," *J. of Engrg. Materials and Technology*, ASME, Vol. 102, pp. 92-96.
Lee, T.H., Arora, J.S. and Kumar, V. (1991a), "Design Sensitivity Analysis of Viscoplastic Structures 1: Constitutive Models," *Computer Methods in Applied Mechanics and Engineering*, submitted.
Lee, T.H., Arora, J.S. and Kumar, V. (1991b), "Design Sensitivity Analysis of Viscoplastic Structures 2: Shape Variations," *Computer Methods in Applied Mechanics and Engineering*, submitted.
Lee, T.H., Arora, J.S. and Rim, K. (1991), "Design Sensitivity Analysis of Thermoviscoplastic Mechanical and Structural Systems," *Thecnical Report No. ODL-91.19, Optimal Design Laboratory*, College of Engineering, The University of Iowa, Iowa City, Iowa 52242 U.S.A.
Mukherjee, S. and Chandra, A. (1989), "A Boundary Element Formulation for Design Sensitivities in Materially Nonlinear Problems," *Acta Mechanica*, Vol. 78, pp. 243-253.
Phelan, D.G. and Haber, R.B. (1989), "Sensitivity Analysis of Linear Elastic Systems Using Domain Parameterization and A Mixed Mutual Energy Principle," *Computer Methods in Applied Mechanics and Engineering*, Vol. 77, pp. 31-59.
Poldneff, M.J., Rai, I.S. and Arora, J.S. (1991), "Design Variations of Nonlinear Elastic Structures Subjected to Follower Forces," *Computer Methods in Applied Mechanics and Engineering*, submitted.
Ryu, Y.S., Haririan, M., Wu, C.C. and Arora, J.S. (1985), "Structural Design Sensitivity Analysis of Nonlinear Response," *Computers and Structures*, Vol. 21, No. 1/2, pp. 245-255.
Tortorelli, D.A., Haber, R., and Lu, S.C.Y. (1991), "Adjoint Sensitivity Analysis for Nonlinear Dynamic Thermoelastic Systems," *AIAA Journal*, Vol. 29, pp. 253-263.
Tsay, J.J. and Arora, J.S. (1989), "Optimum Design of Nonlinear Structures with Path Dependent Response," *Structural Optimization*, Vol. 1, No. 4, pp. 203-214.
Tsay, J.J. and Arora, J.S. (1990), "Nonlinear Structural Design Sensitivity for Path Dependent Problems. Part 1: General Theory," *Computer Methods in Applied Mechanics and Engrg.*, Vol. 81, pp. 183-208.
Wang, S-Y., Sun, Y. and Gallagher, R.H. (1985), "Sensitivity Analysis in Shape Optimization of Continuum Structures," *Computers & Structures*, Vol. 20, No. 5, pp. 855-867.
Wu, C.C. and Arora, J.S. (1987), "Design Sensitivity Analysis and Optimization of Nonlinear Structural Response Using Incremental Procedure," *AIAA Journal*, Vol. 25, No. 8, pp. 1118-1125.

Acknowledgement

The work reported in this paper is based on a part of the research sponsored by the U.S. National Science Foundation under Grant No. MSM 89-13218 entitled, "Design Sensitivity Analysis and Optimization of Nonlinear Systems."

NEW METHOD OF ERROR ANALYSIS AND DETECTION IN SEMI-ANALYTICAL SENSITIVITY ANALYSIS

GENGDONG CHENG* and NIELS OLHOFF
Institute of Mechanical Engineering
Aalborg University, DK-9220 Aalborg, Denmark

ABSTRACT. This paper introduces a new method, the rigid-body motion test, for analysis and detection of possible abnormal errors of shape design sensitivities determined by the semi-analytical method. It is found that rigid-body rotation of an individual finite element of given type gives rise to non-vanishing semi-analytical pseudo-loads, the resultant moment of which can be applied to predict the abnormal error behaviour that has been reported for cases of semi-analytical sensitivity analysis in the literature. The proposed method is illustrated for a number of different types of elements in this paper.

1. Introduction

An efficient and reliable method of sensitivity analysis is a necessity in areas like structural optimization, structural identification and optimal control. For structural size design variables such as cross-sectional area of bars and beams , thickness of plate elements, etc., the method of analytical sensitivity analysis for discretized structures is well known and has been applied successfully for many years. For structural shape design variables, i.e., variables defining the form or shape of structures to be optimized or redesigned, the development and implementation of analytical sensitivity analysis is quite different and more complex. For the latter type of problem, an alternative method, the semi-analytical method of sensitivity analysis, was applied as early as in 1973 by Zienkiewicz *et al*, see Ref.[1]. Due to its high efficiency, quite simple programming and reasonable accuracy, this method has become increasingly popular during recent years. A number of computer aided structural optimization programs are based on the semi-analytical design sensitivity technique and have been applied for optimization of practical, complicated structures. The efficiency, accuracy and possible ways of improvement of the semi-analytical method have been the subjects of many recent studies, see Refs.[2-10].

The basic idea of the semi-analytical method of sensitivity analysis is straight-forward, i.e., we replace the partial derivatives of the finite element stiffness matrix with respect to design variables by corresponding finite difference approximations. If this is done at the element level,

*) Guest professor during the period 1 September - 1 December 1990. Permanent address: Dalian University of Technology, Dalian 116024, Liaoning, P.R. China.

G. I. N. Rozvany (ed.), Optimization of Large Structural Systems, Vol. I, 361–383.

the implementation of the semi-analytical method only demands limited knowledge of the data structure, element stiffness matrix formulation and computer code of existing FEM packages. Because of the complexity of the analytical formulas for the sensitivities of element stiffness matrices with respect to shape design variables, the computational efficiency of the semi-analytical method is competitive with the analytical method. Also, implementation of the latter method requires a considerable amount of analytical work and programming. For example, for an existing commercial FEM package such as ANSYS, whose element library includes nearly 100 different types of elements, it would be a formidable task to identify the various mathematical formulations for all the elements, and to carry out the analytical derivation and programming of sensitivities with respect to varieties of types of design variables.

Nevertheless, a problem which needs further thorough understanding is the accuracy of sensitivities calculated by means of the semi-analytical method. The accuracy problem is caused by the approximation introduced in the semi-analytical method when partial derivatives of the element stiffness matrix are replaced by their finite difference approximations. Two types of abnormal errors have been noticed by Barthelemy and Haftka [7]. For structures modeled by beam or plate elements, the numerical errors of displacement sensitivities with respect to shape design variables such as length of a beam or plate increases inversely proportional to the square of the size of the perturbed elements. Another type of error appears when long span beam-like structures are modeled by truss elements, solid elements or plane stress elements of fixed size and are subjected to bending. Here, the errors of sensitivities of displacements with respect to beam length increase rapidly with the beam aspect ratio or the length of the beam.

A number of recent papers have been devoted to semi-analytical error behaviour. Pedersen, Cheng and Rasmussen [8] studied the model problem of a cantilever beam subjected to a moment at the free end and observed that the 16 components of the beam element stiffness matrix depend on the design variable considered, i.e., the element length ℓ , in three different powers. Thus, when a first-order finite difference scheme is used, the approximate derivatives of the components of different powers are of different accuracy, and these uneven truncation errors result in relative errors of the pseudo-forces that are different from those of the pseudo-moments, and gives rise to the first type of abnormal error behaviour. Olhoff and Rasmussen [9] extended the error analysis in [8] by deriving the analytical solution to the global set of finite element equations for the semi-analytic design sensitivity analysis problem for any degree of discretization, which enabled the authors to precisely identify and explain the source of the numerical inaccuracy problem. Barthelemy and Haftka [7] have suggested that mismatch between the derivatives of the displacements and the rotations is responsible for the error behaviour, and Haftka [11] has studied the condition number of global stiffness matrices and suggested an error magnification index for estimation of error. Cheng *et al* [10] carried out studies of three- and six-node triangular plane stress elements and various tests of beam, plate, plane stress and solid elements. By studies of the resultants of the pseudo-loads at the element level, which will be further developed in this paper, it was found in [10] that the first type of abnormal errors were only exhibited by finite elements that have different types of nodal degrees of freedom.

Different approaches have been suggested for improvement of inaccurate semi-analytical design sensitivities. Haftka and Adelman have advocated use of central difference instead of forward difference approximations which implies additional computational cost. Taking into consideration that computational efficiency is desirable and that severe abnormal error only manifests itself when the number of perturbed elements is large, Cheng *et al* [10] proposed an alternate forward/backward finite difference scheme which is now implemented in the computer aided optimization software MCADS [13]. In procedures of structural reanalysis, the second order correction method [13] (or even higher order corrections) is also an alternative to error reduction, but as is pointed out in [10] and [13], all these methods cannot completely eliminate the errors. The reader is referred to [14] for a method by which both error associated with mesh refinement and design variable pertubation can be eliminated for a class of problems that comprises the model problem considered in Refs.[8] and [9].

2. Preliminary Considerations

Using the method of semi-analytical sensitivity analysis, we compute the sensitivity of the displacement U with respect to a design variable d by means of the following formula

$$\frac{\partial U}{\partial d} \doteq \bar{s} = -K^{-1}\left(\left[\frac{K(d+\Delta d) - K(d)}{\Delta d}\right] U\right) \tag{1}$$

which is obtained by replacing in the exact formula,

$$\frac{\partial U}{\partial d} = s = -K^{-1}\left(\frac{\partial K}{\partial d} U\right) \tag{2}$$

the partial derivative of the stiffness matrix K by its corresponding finite difference approximation.

As is mentioned in the Introduction, two types of abnormal errors have been observed when the semi-analytical approach is applied, and both manifest themselves when we increase the number of finite elements used in the computation. Since the number of elements directly determines the size of the global stiffness matrix, it has been suspected that the difficulties are associated with the size or the properties of the global stiffness matrix K, see Ref.[11]. However, it is known from the literature [10] that the abnormal error of sensitivities appears even if we only use 16 elements to model a cantilever beam, if we choose a large relative perturbation $\Delta d/d = 0.1$ of the design variable. For a structure modeled by 16 elements, the finite element solution process is simple and well-behaved. Hence, the cause of the problem must be sought and studied differently. In this respect, let us note the fact that the accurate sensitivity s and approximate sensitivity $\bar{s}$ are obtained from equations with the same stiffness matrix K for a given number of elements, but that the equations have different right hand terms, i.e.,

$$Q = -\frac{\partial K}{\partial d} U \tag{3}$$

$$\bar{Q} = -\left[\frac{K(d+\Delta d) - K(d)}{\Delta d}\right] U \tag{4}$$

where Q and $\bar{Q}$, respectively, are termed the accurate structural pseud-load and the approximate structural pseudo-load. Thus, the abnormal behaviour of the error $\epsilon_s = \bar{s} - s$ must be expected to result from the error $\epsilon_q = \bar{Q} - Q$ of the pseudo-load. The structural pseudo-loads above are equal to sums of contributions from individual finite elements of the accurate and approximate element pseudo-loads Q_i and $\bar{Q}_i$, respectively,

$$Q = -\frac{\partial K}{\partial d} U = -\sum_i \frac{\partial K_i}{\partial d} U_i = \sum_i Q_i \tag{5}$$

$$\bar{Q} = -\left[\frac{K(d+\Delta d) - K(d)}{\Delta d}\right] U = -\frac{\sum_i \left(K_i(d+\Delta d) - K_i(d)\right) U_i}{\Delta d} = \sum_i \bar{Q}_i \tag{6}$$

Here the summations are carried out over the perturbed elements. Instead of considering the error ϵ_q of the structural pseudo-load $\bar{Q}$, we shall rather direct our attention to the error ϵ_{qi} of the element pseudo-load $\bar{Q}_i$

$$\epsilon_{qi} = \bar{Q}_i - Q_i \tag{7}$$

Moreover, recalling the fact that in the finite element theory the nodal forces can be obtained in either a consistent way or via ad hoc lumped methods, and that consistent and lumped methods become identical as a mesh is repeatedly subdivided [15], and also considering the fact that the abnormal errors observed so far have only manifested themselves when the size of perturbed elements is small in comparison with the structural size, it is interesting to study the resultants $R(\epsilon_{qi})$ of the errors rather than the entire error vectors ϵ_{qi} themselves. It is our expectation that we will keep track of the dominating abnormal behaviour by studying the resultants of the errors. By decomposing the element displacement vector into the rigid body translation, the rigid body rotation and the element deformation and studying the resultants of the pseudo-load errors associated with those three displacements we indeed observed [10] that in shape optimization for elements whose nodal degrees of freedom have the same dimension the resultants of the errors associated with the rigid rotation and the deformation have the same order, and for elements whose nodal degrees of freedom have different dimension the resultants of the errors assosiated with the rigid rotation and the deformation can have different order. This explains the first type of abnormal behaviour very well. The basic idea in [10] is further pursued in this paper.

3. New Rigid-Body Motion Test

As is well known in finite element theory, a proper finite element formulation should include the rigid-body motion capability, i.e.,

$$KU_t = 0 \quad , \quad KU_r = 0 \tag{8}$$

Here U_t and U_r , respectively, denote a unit rigid-body translation and rotation of the finite element, and K is the element stiffness matrix. By differentiating Eqs.(8) with respect to a design variable d, we get

$$\frac{\partial K}{\partial d} U_t = -K \frac{\partial U_t}{\partial d} \quad , \quad \frac{\partial K}{\partial d} U_r = -K \frac{\partial U_r}{\partial d} \tag{9}$$

For size design variables such as bar cross-sectional area, plate thickness, etc., U_t and U_r do not depend on d, so we have

$$\frac{\partial K}{\partial d} U_t = 0 \quad , \quad \frac{\partial K}{\partial d} U_r = 0 \tag{10}$$

For shape design variables, such as length of bar or beam, or coordinates of a nodal point of a plane-stress element, U_t is independent of d, whereas U_r depends on d, so in general we have

$$\frac{\partial K}{\partial d} U_t = 0 \quad , \quad \frac{\partial K}{\partial d} U_r = -K \frac{\partial U_r}{\partial d} \neq 0 \tag{11}$$

In structural optimization, the terms

$$-\frac{\partial K}{\partial d} U_t \quad , \quad -\frac{\partial K}{\partial d} U_r$$

are identified as the pseudo-load contributed from the element due to the rigid body translation and rigid body rotation, respectively, and we will use the notation

$$Q_t = -\frac{\partial K}{\partial d} U_t \quad , \quad Q_r = -\frac{\partial K}{\partial d} U_r \tag{12}$$

for these rigid-body based pseudo-loads.

Now let us study the components F_{tt}, F_{tr} of the resultant force and the components F_{rt}, F_{rr} of the resultant moment which are defined as follows

$$\begin{aligned} F_{tt} &= U_t^T Q_t = -U_t^T \frac{\partial K}{\partial d} U_t \quad , \quad F_{rt} = U_r^T Q_t = -U_r^T \frac{\partial K}{\partial d} U_t \\ F_{tr} &= U_t^T Q_r = -U_t^T \frac{\partial K}{\partial d} U_r \quad , \quad F_{rr} = U_r^T Q_r = -U_r^T \frac{\partial K}{\partial d} U_r \end{aligned} \tag{13}$$

From Eqs. (10) and (11) it is obvious that

$$F_{tt} = F_{tr} = F_{rt} = F_{rr} = 0, \qquad \text{for size design variables} \tag{14}$$

$$F_{tt} = F_{tr} = F_{rt} = 0, \qquad \text{for shape design variables} \tag{15}$$

For cases where d denotes a shape design variable we see that also the term F_{rr} vanishes since, in view of Eq. (9) (assuming rigid-body capability),

$$F_{rr} = -U_r^T \frac{\partial K}{\partial d} U_r = U_r^T K \frac{\partial U_r}{\partial d} = 0 \tag{16}$$

So we have shown that the components of the resultant force and moment of the pseudo-loads associated with rigid-body motions are all zero if the differentiation of the components of the element stiffness matrix is exact, and the finite element possesses the rigid-body rotation capability.

Let us now consider the approximate pseudo-loads $\bar{Q}_t$ and $\bar{Q}_r$ associated with unit valued rigid-body translation and rotation, respectively, that will be obtained by use of the semi-analytical method of sensitivity analysis,

$$\begin{aligned} \bar{Q}_t &= -\frac{\Delta K}{\Delta d} U_t = -\frac{K(d+\Delta d) - K(d)}{\Delta d} U_t \\ \bar{Q}_r &= -\frac{\Delta K}{\Delta d} U_r = -\frac{K(d+\Delta d) - K(d)}{\Delta d} U_r \end{aligned} \tag{17}$$

Assuming that the applied finite element fulfils the rigid-body requirement implied by Eqs. (8), the expressions (17) for the approximate pseudo-loads can be simplified to

$$\bar{Q}_t = -\frac{K(d+\Delta d)}{\Delta d} U_t \quad , \quad \bar{Q}_r = -\frac{K(d+\Delta d)}{\Delta d} U_r \tag{18}$$

In cases where d identifies a size design variable, we again have

$$\bar{Q}_t = \bar{Q}_r = 0 \tag{19}$$

because the vectors U_t and U_r for rigid-body motions of the element are both independent of d. Thus, Eqs. (19) imply

$$\bar{F}_{tt} = \bar{F}_{tr} = \bar{F}_{rt} = \bar{F}_{rr} = 0 \qquad \text{for size design variables} \tag{20}$$

The rigid-body translation U_t is also independent of d when d denotes a shape design variable, so we still have $\bar{Q}_t = 0$ which leads to $\bar{F}_{tt} = \bar{F}_{tr} = \bar{F}_{rt} = 0$, but $\bar{Q}_r \neq 0$ in general.

Furthermore, it has been shown for many types of elements studied up to now that

$$\bar{F}_{rr} = U_r^T \bar{Q}_r = -U_r^T \frac{\Delta K}{\Delta d} U_r \neq 0 \tag{21}$$

in general. By comparing Eqs. (21) with Eqs. (16), it is obvious that the semi-analytical method introduces an extra moment in addition to the accurate pseudo-load, and that this moment is the resultant of the approximate pseudo-loads associated with the rigid-body rotation of the element. In the sequel, the behaviour of this extra moment $\bar{F}_{rr}$ will be shown to provide excellent a priori indication of whether a given type of finite element is prone to exhibit abnormal design sensitivity error if the semi-analytical method is applied.

The extra moment $\bar{F}_{rr}$ is very easily determined. All that is needed is to calculate a simple element matrix and the possible rigid-body rotations. Computation for different element sizes, different perturbations and different orientations of the element then provide desired information about the approximate sensitivity. In fact, we propose these simple calculations as a new rigid-body motion test against inaccurate semi-analytical sensitivity computation.

4. Accuracy Problem for Beam-like Structure Modeled by Bar Elements

Barthelemy and Haftka [7] have analyzed a truss beam composed of square cells that each contain 5 bar elements.

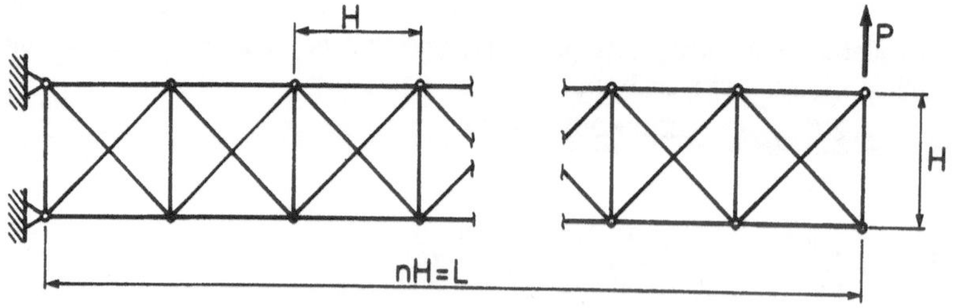

Fig. 1. Truss beam modeled by bar elements.

The number of cells was varied from 1 to 20, changing the beam aspect ratio from 1 to 20, and it is found in the numerical study in Ref.[7] that the errors of the semi-analytical design sensitivities of the tip displacement with respect to the total length L increase drastically with the beam aspect ratio. The results reported in Ref.[7] indicate that the error is proportional to the square of the aspect ratio. In the following, we present an explanation of this abnormal error behaviour.

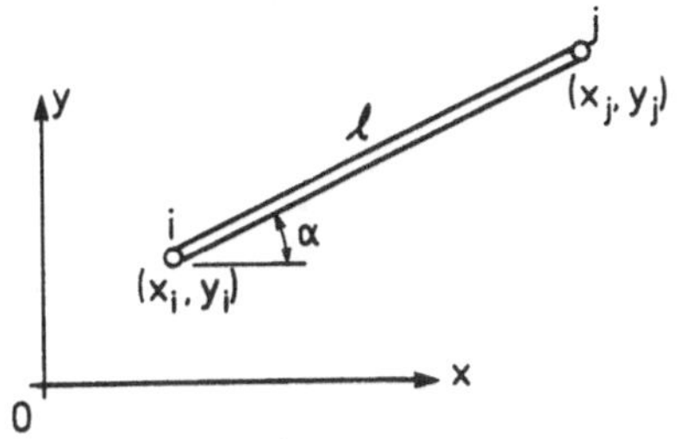

Fig. 2. Bar element.

Firstly, we shall estimate the extra moment $\bar{F}_{rr}$ by making use of Eq.(21). To this end, consider an arbitrarily oriented bar element as shown in Fig. 2. The element stiffness matrix $\boldsymbol{K}$ is given by

$$\boldsymbol{K} = \begin{bmatrix} K_{11} & K_{12} & -K_{11} & -K_{12} \\ K_{12} & K_{22} & -K_{12} & -K_{22} \\ -K_{11} & -K_{12} & K_{11} & K_{12} \\ -K_{12} & -K_{22} & K_{12} & K_{22} \end{bmatrix} \tag{22}$$

where

$$K_{11} = \frac{AE}{\ell} c^2 \quad , \quad K_{12} = \frac{AE}{\ell} cs \quad , \quad K_{22} = \frac{AE}{\ell} s^2$$
$$c = (x_j - x_i)/\ell, \quad s = (y_j - y_i)/\ell, \quad \ell = \sqrt{(x_j-x_i)^2 + (y_j-y_i)^2} \tag{23}$$

We have carried out the detailed calculation and checked that Eq. (16) is fulfilled. The element rigid-body rotation of unit angle is given by

$$\boldsymbol{U}_r = \ell \, \{0 \quad 0 \quad -s \quad c\}^T \tag{24}$$

In the subsequent calculations, we assume that the coordinate x_j of the nodal point j of the bar element is perturbed by Δx_j implying that the total length L of the truss beam is perturbed by $\Delta L = n \cdot \Delta x_j$.

The approximate pseudo-load vector $\bar{\boldsymbol{Q}}_r$, see Eq.(18), that corresponds to the rigid-body rotation (24) is given by

$$\bar{Q}_r = \{\bar{Q}_{r1}, \bar{Q}_{r2}, \bar{Q}_{r3}, \bar{Q}_{r4}\}^T$$

$$\bar{Q}_{r1} = -\left[\frac{AE}{\ell_1} c_1^2\, s\ell - \frac{AE}{\ell_1} c_1 s_1 c\ell\right] \frac{1}{\Delta L}$$

$$\bar{Q}_{r2} = -\left[\frac{AE}{\ell_1} c_1 s_1 s\ell - \frac{AE}{\ell_1} s_1^2\, c\ell\right] \frac{1}{\Delta L} \tag{25}$$

$$\bar{Q}_{r3} = \left[\frac{AE}{\ell_1} c_1^2\, s\ell - \frac{AE}{\ell_1} c_1 s_1 c\ell\right] \frac{1}{\Delta L}$$

$$\bar{Q}_{r4} = \left[\frac{AE}{\ell_1} c_1 s_1 s\ell - \frac{AE}{\ell_1} s_1^2 c\ell\right] \frac{1}{\Delta L}$$

where

$$\ell_1 = \sqrt{(x_j+\Delta x_j-x_i)^2 + (y_j-y_i)^2} \;, \quad c_1 = \frac{(x_j+\Delta x_j-x_i)}{\ell_1}$$

$$s_1 = \frac{(y_j - y_i)}{\ell_1} \tag{26}$$

The extra moment $\bar{F}_{rr}$ is now obtained by multiplying $\bar{Q}_r$ by the unit rigid-body rotation vector in Eq. (24), and we get

$$\bar{F}_{rr} = \frac{-AE}{\ell_1} \ell^2 \sin^2(\alpha-\alpha_1) \frac{1}{\Delta L} \tag{27}$$

where

$$\sin\alpha = s \;, \quad \cos\alpha = c \;, \quad \sin\alpha_1 = s_1 \;, \quad \cos\alpha_1 = c_1 \tag{28}$$

Let us now apply the result (27) to the structure shown in Fig. 1. It is well known that the behaviour of a discrete, long-span beam-like structure can be obtained if we model the discrete structure as a continuum beam as shown in Fig.3 with proper cross-sectional area A and area moment of inertia I. The actual values of A and I, however, are not important for the present study. For the equivalent beam in Fig.3, the tip deflection is

$$U_L = \frac{P}{3EI} (nH)^3 \tag{29}$$

and the deflection at any point x is given by

$$U(x) = \frac{Px^2}{2EI}\left[nH - \frac{x}{3}\right] \tag{30}$$

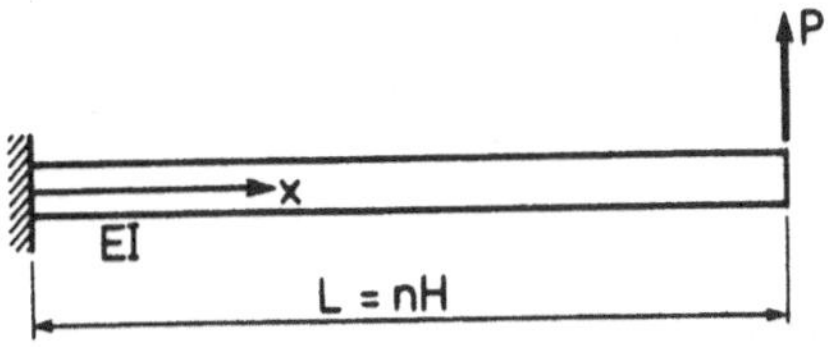

Fig. 3. Equivalent Bernoulli-Euler beam.

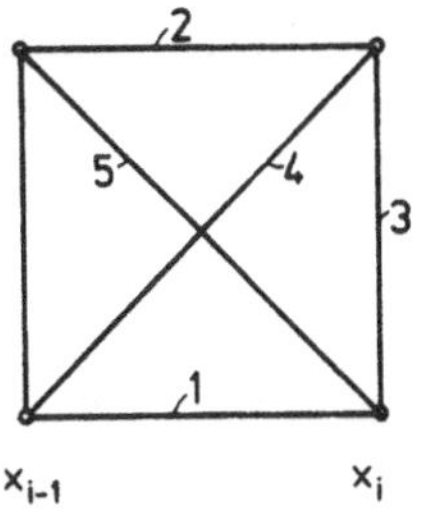

Fig. 4. The i-th cell of the truss beam in Fig.1, with x_{i-1} = (*i*-1)H and x_i = *iH*.

The sensitivity of the tip displacement U_L with respect to the design variable *L*, is

$$\frac{\partial U_L}{\partial L} = \frac{P(nH)^2}{EI} \tag{31}$$

and the rotation at any point *x* is given by

$$\theta = \frac{\partial U}{\partial x} = \frac{PnHx}{EI} - \frac{Px^2}{2EI} \tag{32}$$

Now we consider the *i*-th cell depicted in Fig.4 of the structure shown in Fig.1. As the total length *L* of the structure is perturbed by ΔL, the length of an individual cell is perturbed by $\Delta x_i = \Delta L/n$. Note that for the truss members 1 , 2 and 3 , the perturbation does not change the angle between the bar axis and the *x* axis, i.e., α-α_1 in Eq.(27) is equal to zero. Hence, the only contributions to the aforementioned extra moment come from the members 4 and 5. It is interesting to notice that in Ref.[7], the cross-sectional areas of the bars 4 and 5 were taken to be 125 times the areas of the bars 1 and 2 which clearly amplifies the extra moment, and thereby the magnitude of the abnormal sensitivity error. This large difference between the bar areas at the same time justifies the use of the equivalent Bernoulli-Euler beam in Fig.3 because the bars 4 and 5 of the actual truss beam provide large shear stiffness. Now, the rotation of the bars 4 and 5 due to the given tip load on the truss beam can be approximated by the value of $\theta(x)$ at $x = iH$

$$\theta_i = \frac{PH^2}{EI}\left(in - \frac{i^2}{2}\right) \tag{33}$$

The sum of the extra moment m_i from bar 4 and 5 corresponding to the angle of rotation θ_i is then given by

$$m_i = 2\bar{F}_{rr}\theta_i = \frac{-2AE\ell^2\theta_i}{\Delta x_i\,\ell_1\, n}\sin^2(\alpha - \alpha_1) = -\beta\left[i - \frac{i^2}{2n}\right] \tag{34}$$

where

$$\beta = \frac{2APH^2\ell^2}{\Delta x_i I\ell_1}\sin^2(\alpha - \alpha_1) \tag{35}$$

This extra moment m_i acts at the middle of the cell. Since we are only interested in an estimation, we assume for simplicity that the moment acts at the point $x = iH$. The tip displacement due to m_i is

$$\delta_i = m_i \cdot iH\left[nH - \frac{iH}{2}\right]\frac{1}{EI} \tag{36}$$

which is obtained by the simple beam theory. The sum of the contributions to the tip displacement from the number of n extra moments m_i is thus

$$\delta = \sum_{i=1}^{n}\delta_i = -\frac{\beta H^2}{EI}\left\{ n\sum_1^n i^2 - \sum_1^n i^3 + \frac{1}{4n}\sum_1^n i^4\right\} \tag{37}$$

Making use of the well-known formulas for $\sum_1^n i^2, \sum_1^n i^3, \sum_1^n i^4$, we obtain the following estimate of the total tip displacement caused by all the extra moments:

$$\delta = -\frac{\beta H^2}{EI}\frac{\left(16n^4 + 15n^3 - 1\right)}{120} \tag{38}$$

This result represents the estimation of the error, contributed from the moments associated with the rigid body rotations, of the sensitivity of the tip displacement with respect to the length L of the structure, when semi-analytical sensitivity analysis is applied.

The relative error is given by

$$\delta\Big/\left(\frac{\partial U_L}{\partial L}\right) = \frac{-\beta}{120P}\left[16n^2 + 15n - \frac{1}{n^2}\right] = O\left(n^2\right) \tag{39}$$

This behaviour of the error coincides with that obtained numerically in Ref.[7].

In Sections 5 and 6 we shall show that the extra moments for triangular plane elements and Bernoulli-Euler beam elements exhibit similar features as the truss element considered here. This explains the similar error behaviour of these elements that was observed in Ref.[7].

5. Accuracy Problem for Beam-like Structure Modeled by Three-node Triangular Constant Stress Elements

The study in this section follows the lines of Section 4. The constant stress triangular element, see Fig.5, has an explicit stiffness matrix $\boldsymbol{K}$ given by

$$\boldsymbol{K} = tA\boldsymbol{B}^T\boldsymbol{E}\boldsymbol{B} \tag{40}$$

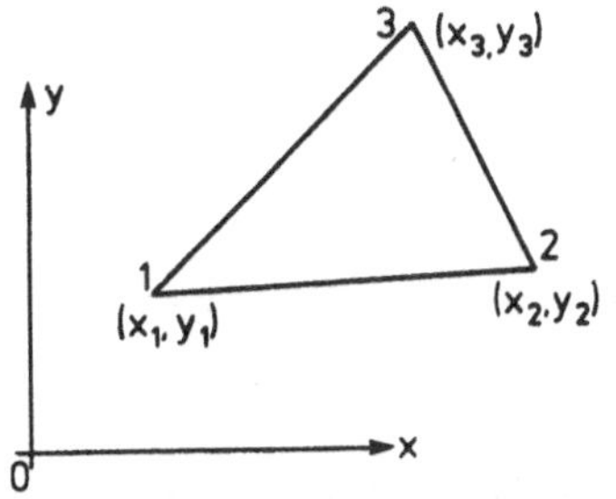

Fig. 5. Three-node triangular finite element.

where t and A, respectively, denote thickness and area of the element, and the constitutive matrix $\boldsymbol{E}$ and the strain-displacement matrix $\boldsymbol{B}$ are given by

$$\boldsymbol{E} = \frac{E}{(1+\nu)(1-2\nu)} \begin{bmatrix} 1-\nu & \nu & 0 \\ \nu & 1-\nu & 0 \\ 0 & 0 & \frac{1-2\nu}{2} \end{bmatrix} \quad \text{for plane strain} \tag{41}$$

$$\boldsymbol{E} = \frac{E}{1-\nu^2} \begin{bmatrix} 1 & \nu & 0 \\ \nu & 1 & 0 \\ 0 & 0 & \frac{(1-\nu)}{2} \end{bmatrix} \quad \text{for plane stress} \tag{42}$$

$$\boldsymbol{B} = \frac{1}{2A}\bar{\boldsymbol{B}}, \qquad \bar{\boldsymbol{B}} = \begin{bmatrix} y_{23} & 0 & y_{31} & 0 & y_{12} & 0 \\ 0 & x_{32} & 0 & x_{13} & 0 & x_{21} \\ x_{32} & y_{23} & x_{13} & y_{31} & x_{21} & y_{12} \end{bmatrix} \tag{43}$$

$$x_{ij} = x_i - x_j \ , \quad y_{ij} = y_i - y_j \tag{44}$$

With the above definition of the matrix $\bar{\boldsymbol{B}}$ we can rewrite the stiffness matrix $\boldsymbol{K}$ as

$$\boldsymbol{K} = \frac{t}{4A}\bar{\boldsymbol{B}}^T\boldsymbol{E}\bar{\boldsymbol{B}} \tag{45}$$

which is a form suitable for our subsequent derivations. Now we consider the most general perturbation of the nodal coordinates, i.e., we perturb the coordinate vector $\boldsymbol{x} = (x_1 \ y_1 \ x_2 \ y_2 \ x_3 \ y_3)^T$ by $\Delta\boldsymbol{x} = (\Delta x_1 \ \Delta y_1 \ \Delta x_2 \ \Delta y_2 \ \Delta x_3 \ \Delta y_3)^T$ to the new coordinate vector $\boldsymbol{x} + \Delta\boldsymbol{x}$.

Let us denote by $\boldsymbol{K}(\boldsymbol{x} + \Delta\boldsymbol{x})$, $\bar{\boldsymbol{B}}(\boldsymbol{x} + \Delta\boldsymbol{x})$ and $A(\boldsymbol{x} + \Delta\boldsymbol{x})$ the stiffness matrix, the strain-displacement matrix and the element area that correspond to the new triangle with the nodal coordinates $(x_1 + \Delta x_1 \ y_1 + \Delta y_1 \ x_2 + \Delta x_2 \ y_2 + \Delta y_2 \ x_3 + \Delta x_3 \ y_3 + \Delta y_3)$.

The stiffness matrix associated with the perturbed form of the triangular element is

$$K\,(x + \Delta x) = \frac{t}{4A(x+\Delta x)}\,\bar{B}\,(x+\Delta x)^T E \bar{B}\,(x+\Delta x) \tag{46}$$

where

$$\bar{B}(x + \Delta x) = \bar{B}(x) + \begin{bmatrix} \Delta y_{23} & 0 & \Delta y_{31} & 0 & \Delta y_{12} & 0 \\ 0 & \Delta x_{32} & 0 & \Delta x_{13} & 0 & \Delta x_{21} \\ \Delta x_{32} & \Delta y_{23} & \Delta x_{13} & \Delta y_{31} & \Delta x_{21} & \Delta y_{12} \end{bmatrix} \tag{47}$$

$$= \bar{B}(x) + \Delta\bar{B}$$

$$\Delta x_{ij} = \Delta x_i - \Delta x_j \ , \ \Delta y_{ij} = \Delta y_i - \Delta y_j \tag{48}$$

Here, the definition of $\Delta\bar{B}$ is obvious. Let us now consider the extra moment F_{rr} due to the pseudo-load associated with rigid-body rotation of unit angle. The rigid-body rotation of unit angle is

$$U_r = (\,0 \quad 0 \quad -y_{21} \quad x_{21} \quad -y_{31} \quad x_{31}\,)^T \tag{49}$$

and the corresponding extra moment becomes

$$\bar{F}_{rr} = -\frac{1}{\Delta d}\,U_r^T \Delta K U_r = \frac{-t}{4A(x + \Delta x)\cdot\Delta d}\,[\;U_r^T \bar{B}^{\,T}(x) E \bar{B}(x) U_r +$$

$$U_r^T (\Delta\bar{B})^T E \bar{B}(x)\, U_r + U_r^T \bar{B}^{\,T}(x) E \Delta\bar{B}\, U_r + U_r^T (\Delta\bar{B})^T E (\Delta\bar{B})\, U_r] \tag{50}$$

Here, the first term in the parenthesis is obviously equal to zero because the original stiffness matrix possesses rigid-body motion capability. The second term and third term are also zero since

$$\bar{B}(x)\, U_r = 0 \tag{51}$$

The fourth term is $U_r^T (\Delta\bar{B})^T E (\Delta\bar{B})\, U_r$ where we have

$$(\Delta\bar{B})\, U_r = \begin{Bmatrix} -y_{21}\Delta y_{31} - y_{31}\Delta y_{12} \\ x_{21}\Delta x_{13} + x_{31}\Delta x_{21} \\ -y_{21}\Delta x_{13} + x_{21}\Delta y_{31} - y_{31}\Delta x_{21} + x_{31}\Delta y_{12} \end{Bmatrix} \tag{52}$$

For our purpose we now consider two cases.

<u>Case a:</u> The element shown in Fig.6a with $y_1 = y_2$, $x_1 = x_3$ is perturbed by Δx_2 while $\Delta x_1 = \Delta x_3 = 0$, $\Delta y_1 = \Delta y_2 = \Delta y_3 = 0$.

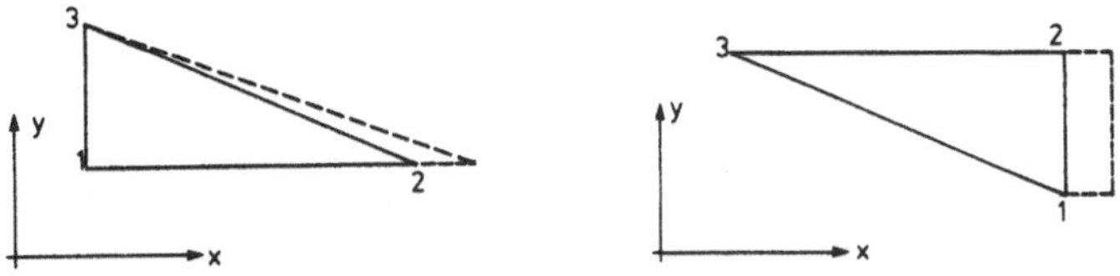

(a) (b)

Fig. 6. Two cases.

We get:

$$(\Delta\bar{B})U_r = \{0 \quad 0 \quad -y_{31}\Delta x_2\}^T$$

$$\bar{F}_{rr} = \frac{-t\Delta x_2^2 E y_{31}^2}{8\Delta d(1+\nu)A(x+\Delta x)} \qquad \text{for plane strain} \tag{53}$$

The same result is obtained for plane stress.

<u>Case b:</u> Let $y_2 = y_3, x_2 = x_1$, cf. Fig.6b, and consider a pertubation given by $\Delta x_1 = \Delta x_2 \neq 0$, $\Delta x_3 = \Delta y_1 = \Delta y_2 = \Delta y_3 = 0$.
We get:

$$(\Delta\bar{B})\,U_r = \{0 \quad 0 \quad -y_{21}\Delta x_2\}^T$$

$$\bar{F}_{rr} = \frac{-t\;\Delta x_2^2\;E\;y_{21}^2}{8(1+\nu)\;A(x+\Delta x)\Delta d} \qquad \text{for plane strain} \tag{54}$$

The same result is obtained for plane stress.

Now let us consider the problem of a cantilever beam modeled by plane triangular elements. The problem was studied in [7] and is depicted in Fig.7. The total length L of the beam is perturbed by ΔL, $\Delta L/L = \eta$, where the relative increment η is fixed. The i-th section may be viewed as being perturbed as indicated in Fig.8.

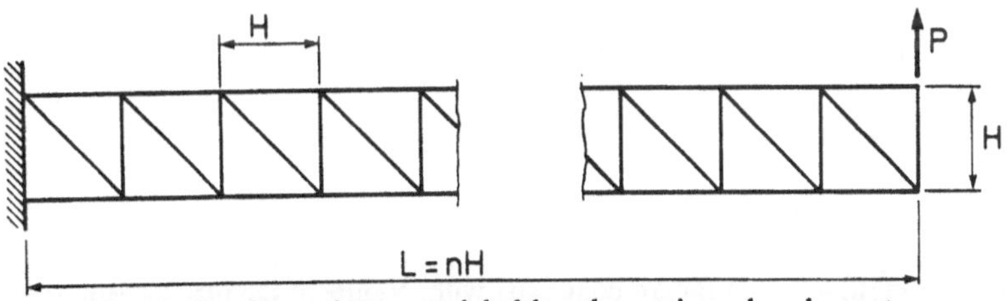

Fig. 7. Cantilever beam modeled by plane triangular elements.

By comparing Figs. 6(a) and 6(b) with Fig.7, the connection is immediately seen. According to Eqs. (53) and (54), the extra moment m_i associated with a rigid-body rotation of angle θ can be written as

$$m_i = \bar{F}_{rr}\,\theta = -\frac{t(\Delta H)^2 E H^2}{8n\Delta H(1+\nu)\,A(x+\Delta x)}\,\theta = -3\frac{\eta}{1+\eta}\,\frac{EI}{nH^2(1+\nu)}\,\theta \tag{55}$$

Here, we have made use of $tH^3/12 = I$ and employed the relation $A(x+\Delta x) = H^2(1+\eta)/2$ which holds good for both Case a and Case b.

Consider now the i-th beam section shown in Fig.8. Assuming that the extra moment m_i acts at the center of gravity of the finite element, the point $x = x_i$ of action for the "Case a" element of the i-th section is

$$x_i = H\left(i - \frac{2}{3}\right) \tag{56}$$

and the extra moment of the "Case b" element of the i-th section acts at the point

$$x_i = H\left(i - \frac{1}{3}\right) \tag{57}$$

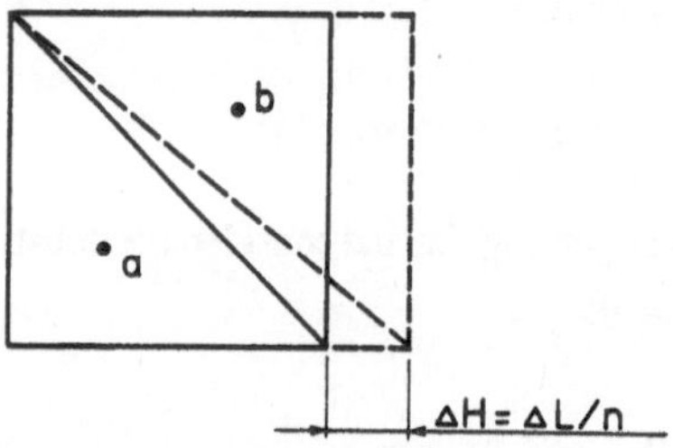

Fig. 8. Pertubation of section of beam in Fig. 7.

In what follows, we adopt similar expressions for the Bernoulli-Euler beam deflections and rotations as in Section 4, i.e.,

$$U(x) = \frac{Px^2}{2E'I}\left(nH - \frac{x}{3}\right), \quad U_L = \frac{P(nH)^3}{3E'I}$$
$$\frac{\partial U_L}{\partial L} = \frac{P(nH)^2}{E'I}, \quad \theta = \frac{\partial U}{\partial x} = \frac{PnHx}{E'I} - \frac{Px^2}{2E'I} \tag{58}$$

but employ a "modulus of elasticity" E' given by

$$E' = \frac{E}{1-\nu^2} \quad \text{for plane strain}$$
$$E' = E \quad \text{for plane stress} \tag{59}$$

when the results (58) are used for, or compared with, results of the plane strain or plane stress finite element model.

In view of Eqs. (56)-(58), the values of the rotations at the points of action of extra moments become

$$\theta_i = \frac{PH^2}{E'I}\left(i-\frac{2}{3}\right)\left[n-\frac{1}{2}\left(i-\frac{2}{3}\right)\right] \qquad \text{for "Case a"}$$
(60)
$$\theta_i = \frac{PH^2}{E'I}\left(i-\frac{1}{3}\right)\left[n-\frac{1}{2}\left(i-\frac{1}{3}\right)\right] \qquad \text{for "Case b"}$$

The tip displacement δ_i due to the action of a moment m_i at the point x_i is

$$\delta_i = \frac{m_i x_i}{E'I}\left(nH-\frac{x_i}{2}\right) \tag{61}$$

so the contribution from the "Case a" element of the i-th section is

$$\delta_i' = -3\,\frac{\eta}{(1+\eta)n}\cdot\frac{PEH^2}{E'^2I(1+\nu)}\left(i-\frac{2}{3}\right)^2\left[n-\frac{1}{2}\left(i-\frac{2}{3}\right)\right]^2 \tag{62}$$

and the contribution from the "Case b" element of the i-th section is

$$\delta_i'' = -3\,\frac{\eta}{(1+\eta)n}\cdot\frac{PEH^2}{E'^2I(1+\nu)}\left(i-\frac{1}{3}\right)^2\left[n-\frac{1}{2}\left(i-\frac{1}{3}\right)\right]^2 \tag{63}$$

After some algebra the contribution from the i-th section is found to be

$$\delta_i = \delta_i' + \delta_i'' = -3\,\frac{\eta}{(1+\eta)n}\cdot\frac{PEH^2}{E'^2I(1+\nu)}\left[\frac{1}{2}i^4 - (1+2n)i^3 + \left(\frac{5}{6}+3n+2n^2\right)i^2 - \left(\frac{1}{3}+\frac{5}{3}n+2n^2\right)i + \frac{17}{324}+\frac{1}{3}n+\frac{5}{9}n^2\right] \tag{64}$$

Summing up the contributions from all the beam sections, we find that the error of the sensitivity of the tip displacement will be

$$\delta = \sum_{n=1}^{n}\delta_i = -3\,\frac{\eta}{(1+\eta)}\cdot\frac{PEH^2}{E'^2I(1+\nu)}\left\{\frac{4}{15}n^4 + \frac{13}{1620}\right\} \tag{65}$$

and considering the analytical solution for $\partial U_L/\partial L$ in Eq. (58) with E' defined by (59), we find the following expression for the relative error of the design sensitivity of the displacement at the tip of the beam,

$$\delta/\left(\partial U_L/\partial L\right) \propto -c\cdot\eta n^2 , \quad |\eta| \ll 1,\ n \gg 1,$$
$$c = \frac{4(1-\nu)}{5} \qquad \text{for plane strain} \tag{66}$$
$$c = \frac{4}{5(1+\nu)} \qquad \text{for plane stress}$$

We again notice the abnormal behaviour that the relative error of the beam sensitivity increases quadratically with the number of finite elements. However, as expected [9], the relative error is linearly proportional with the relative pertubation η of the design variable.

6. Cantilever Modeled by Bernoulli-Euler Beam Elements

This section is devoted to a model problem which has been studied in several papers [7,8,9,10] in order to gain understanding of the semi-analytical inaccuracy problem. Here, we shall apply the new concept of the extra moment.

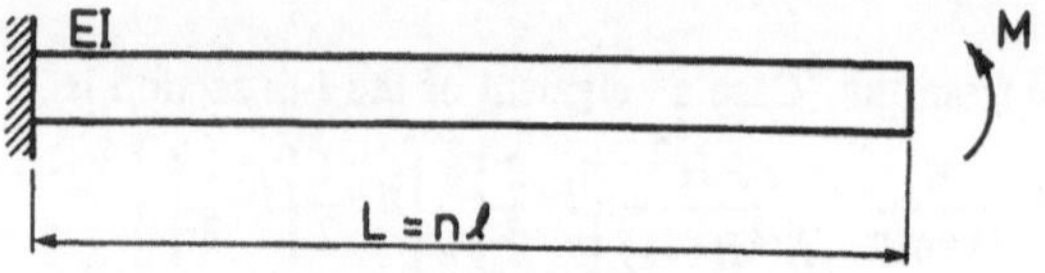

Fig. 9. Cantilever loaded by tip moment.

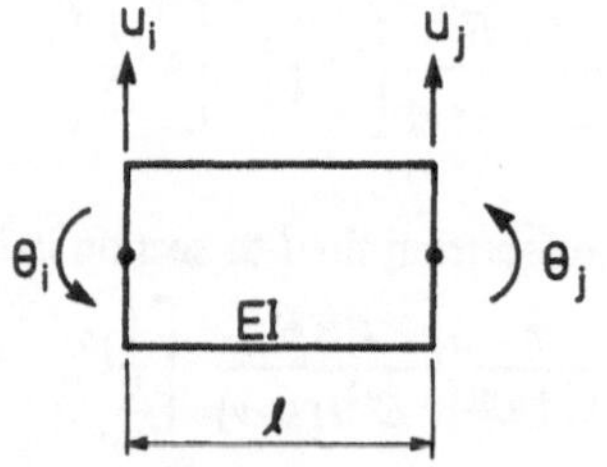

Fig. 10. Bernoulli-Euler beam element.

The problem concerns a cantilever beam of constant bending stiffness EI and variable length L that is loaded at its tip by a given, constant moment M, see Fig.9. The beam is discretized into a total number of n Bernoulli-Euler beam elements of equal length ℓ , see Fig.10, and the length of the element is given by $\ell = L/n$. It is our objective to determine the dominant term of the error of the sensitivity $\Delta U_L/\Delta L$, where U_L is the displacement of the tip of the beam.

The Bernoulli-Euler beam element stiffness matrix is

$$\boldsymbol{K} = \begin{bmatrix} K_{11} & K_{12} & -K_{11} & K_{12} \\ K_{12} & K_{22} & -K_{12} & \frac{1}{2}K_{22} \\ -K_{11} & -K_{12} & K_{11} & -K_{12} \\ K_{12} & \frac{1}{2}K_{22} & -K_{12} & K_{22} \end{bmatrix} \tag{67}$$

where

$$K_{11} = \frac{12EI}{\ell^3} \quad , K_{12} = \frac{6EI}{\ell^2} \quad , K_{22} = \frac{4EI}{\ell} \tag{68}$$

The unit rigid-body rotation of the beam element is

$$U_r = \left(U_i \quad \theta_i \quad U_j \quad \theta_j\right)^T = (0 \quad 1 \quad \ell \quad 1)^T \tag{69}$$

Using the notation of Ref.[8], the forward finite difference quotients of the typical components of the stiffness matrix can be expressed as follows in terms of the corresponding exact derivatives:

$$\frac{\Delta K_{11}}{\Delta \ell} = \frac{\partial K_{11}}{\partial \ell}\left(1+\alpha_1\right) \quad , \quad \frac{\Delta K_{12}}{\Delta \ell} = \frac{\partial K_{12}}{\partial \ell}\left(1+\alpha_2\right) \quad , \quad \frac{\Delta K_{22}}{\Delta \ell} = \frac{\partial K_{22}}{\partial \ell}\left(1+\alpha_3\right) \tag{70}$$

where

$$\alpha_1 = -\eta\frac{6+8\eta+3\eta^2}{3(1+\eta)^3} \quad , \quad \alpha_2 = -\eta\frac{3+2\eta}{2(1+\eta)^2}$$
$$\alpha_3 = -\eta\frac{1}{1+\eta} \quad , \quad \eta = \Delta\ell/\ell \tag{71}$$

The pseudo-load vector $\bar{Q}_r$ in Eq.(18) associated with the rigid-body rotation of unit angle is

$$\bar{Q}_r = \left(\bar{Q}_{r1} \quad \bar{Q}_{r2} \quad \bar{Q}_{r3} \quad \bar{Q}_{r4}\right)^T$$

$$\bar{Q}_{r1} = -\left[2\frac{\partial K_{12}}{n\partial \ell}\,\alpha_2 - \frac{\partial K_{11}}{n\partial \ell}\,\alpha_1\ell\right] = -\frac{12EI}{n\ell^3}\left(3\alpha_1 - 2\alpha_2\right)$$

$$\bar{Q}_{r2} = -\left[\frac{3}{2}\,\frac{\partial K_{22}}{n\partial \ell}\,\alpha_3 - \frac{\partial K_{12}}{n\partial \ell}\,\alpha_2\ell\right] = -\frac{6EI}{n\ell^2}\left(2\alpha_2 - \alpha_3\right) \tag{72}$$

$$\bar{Q}_{r3} = -\left[-2\frac{\partial K_{12}}{n\partial \ell}\,\alpha_2 + \frac{\partial K_{11}}{n\partial \ell}\,\alpha_1\ell\right] = \frac{12EI}{n\ell^3}\left(3\alpha_1 - 2\alpha_2\right)$$

$$\bar{Q}_{r4} = -\left[\frac{3}{2}\,\frac{\partial K_{22}}{n\partial \ell}\,\alpha_3 - \frac{\partial K_{12}}{n\partial \ell}\,\alpha_2\ell\right] = -\frac{6EI}{n\ell^2}\left(2\alpha_2 - \alpha_3\right)$$

The resultant moment $\bar{F}_{rr}$ defined in Eq. (21) of the above pseudo-load vector components corresponds to unit angle of rigid-body rotation. The moment m_i associated with a rigid-body rotation of angle θ is then given by

$$m_i = \bar{F}_{rr}\,\theta = \left(\left(\bar{Q}_{r2} + \bar{Q}_{r4}\right) + \ell\bar{Q}_{r3}\right)\theta = \frac{12EI\theta}{n\ell^2}\left(3\alpha_1 - 4\alpha_2 + \alpha_3\right) \tag{73}$$

Here, the rigid-body rotation θ of each individual element may be taken from the well-known analytical solution of the problem:

$$U(x) = \frac{Mx^2}{2EI} \ , \ \theta(x) = \frac{Mx}{EI} \tag{74}$$

The extra moment m_i in Eq.(73) is acting at the mid-point of the i-th beam element, so we evaluate the angle $\theta(x)$ of the rigid-body rotation angle of the element at this point

$$x = (i-\frac{1}{2})\ell = (i-\frac{1}{2})L/n$$

whereby

$$\theta_i = \frac{M\ell}{EI}\left[i-\frac{1}{2}\right] \tag{75}$$

so the extra moment is given by

$$m_i = \frac{12M}{\ell}(3\alpha_1 - 4\alpha_2 + \alpha_3)\left(i - \frac{1}{2}\right) \tag{76}$$

The displacement δ_i of the tip of the beam caused by m_i can be determined by making use of the simple beam theory, and we easily find

$$\begin{aligned}\delta_i &= m_i\,(i-\frac{1}{2})\ell\left[n\ell - \frac{\ell}{2}\left[i-\frac{1}{2}\right]\right]\frac{1}{EI}\\ &= \frac{12M\ell}{EIn}(3\alpha_1 - 4\alpha_2 + \alpha_3)\left(i-\frac{1}{2}\right)^2\left(n-\frac{1}{2}\left(i-\frac{1}{2}\right)\right)\\ &= \frac{6M\ell}{EIn}(3\alpha_1 - 4\alpha_2 + \alpha_3)\left[-i^3 + \left(2n+\frac{3}{2}\right)i^2 - \left(2n+\frac{3}{4}\right)i + \frac{1}{2}n + \frac{1}{8}\right]\end{aligned} \tag{77}$$

The total displacement of the tip caused by the extra moments of all the elements comprises the dominant part of the error of the tip displacement sensitivity, and we find

$$\delta = \sum_{i=1}^{n}\delta_i = \frac{5ML}{2EI}(3\alpha_1 - 4\alpha_2 + \alpha_3)\left(n^2 + \frac{1}{10}\right) \tag{78}$$

The exact sensitivity of the tip displacement with respect to the length L of the beam is

$$\frac{\partial U_L}{\partial L} = \frac{ML}{EI} \tag{79}$$

so the relative error of the sensitivity is given by

$$\delta/(\partial U_L/\partial L) \propto \frac{5}{2}(3\alpha_1 - 4\alpha_2 + \alpha_3)\,n^2 \ , \quad n \gg 1 \tag{80}$$

This result precisely identifies the critical term that causes the abnormal sensitivity error. The same result has recently been obtained by Olhoff and Rasmussen in Ref.[9] via a different path. In Ref.[9], a complete, exact error analysis is carried out for the current sensitivity error, in which use is made of the exact displacements and rotations, and the term in Eq.(80) is found to be the leading one in the complete, exact expression for the relative error of the sensitivity derived in Ref. [9]. Notice that in the present paper basically we only work with the extra

TABLE 1
Displacement design sensitivities vs. number n of finite elements used.

Number n of elements	Exact sensitivities	Semi-Analytical sensitivities	Corrected sensitivities
1	1	.9997	.9999
2	1	.9989	.9999
3	1	.9977	.9999
4	1	.9959	.9999
5	1	.9937	.9999
6	1	.9909	.9999
7	1	.9877	.9999
8	1	.9839	.9999
9	1	.9797	.9999
10	1	.9749	.9999
12	1	.9639	.9999
14	1	.9509	.9999
16	1	.9359	.9999
18	1	.9189	.9999
20	1	.8999	.9999
22	1	.8789	.9999
24	1	.8559	.9999
26	1	.8310	.9999
28	1	.8040	.9999
30	1	.7750	.9999
32	1	.7440	.9999
34	1	.7110	.9999
36	1	.6760	.9999
38	1	.6390	.9999
40	1	.6000	.9999
42	1	.5590	.9999
44	1	.5160	.9999
46	1	.4711	.9999
48	1	.4241	.9999
50	1	.3751	.9999
52	1	.3241	.9999
54	1	.2711	.9999
56	1	.2161	.9999
58	1	.1592	.9999
60	1	.1002	.9999
62	1	.0392	.9999
64	1	-.0238	.9999
66	1	-.0888	.9999
68	1	-.1558	.9999
70	1	-.2247	.9999
72	1	-.2957	.9999
74	1	-.3687	.9999
76	1	-.4437	.9999
78	1	-.5206	.9999
80	1	-.5996	.9999
82	1	-.6806	.9999
84	1	-.7636	.9999
86	1	-.8485	.9999
88	1	-.9355	.9999
90	1	-1.0245	.9999
92	1	-1.1155	.9999
94	1	-1.2084	.9999
96	1	-1.3034	.9999
98	1	-1.4004	.9999
100	1	-1.4994	.9999

Multiplying factor: ML/EI

moment of the pseudo-load associated with a simple rigid-body rotation of a single finite element, and that this directly leads us to the dominant part of the error. For example, in the present case we can remove the dominant, abnormal part of the error and obtain much more accurate sensitivity results by using the result in Eq. (80). From Eqs. (71) we find that $3\alpha_1 - 4\alpha_2 + \alpha_3 = -\eta/(1+\eta)^3$ where η is the relative pertubation of the design variable, so Eq. (80) may be written as

$$\delta/(\partial U_L/\partial L) \propto -\frac{5}{2}\eta n^2 \quad , \quad |\eta| \ll 1,\ n \gg 1 \tag{81}$$

If we remove this dominant part of the error of the semi-analytical sensitivity of the cantilever beam example considered in this section, we obtain results as presented in Table 1. Here, the first column contains the values taken for the number n of finite elements used to model the beam. The second and third column contain the corresponding exact sensitivity and semi-analytically computed sensitivity, respectively, of the beam tip displacement with respect to change of the total length L of the beam. The semi-analytical sensitivities are all computed on the basis of the value $\Delta\ell/\ell = \eta = 0.0001$ of the relative pertubation of the length of the beam, and the fourth column of Table 1 presents the values of the semi-analytical sensitivities obtained upon correction by subtraction of the dominant error term in Eq.(81). We notice that considerable improvement is achieved.

7. Tests on Various Elements

As was discussed in the preceding section, the extra moment of the approximate pseudo-load associated with element rigid-body rotation is an excellent indicator for the error behaviour of sensitivities calculated by the semi-analytical method. For the purpose of further demonstration of this point, we have carried out a number of numerical tests for different types of finite elements contained in the element library of the commercially available ANSYS code. Although the formula (21) for the extra moment is very simple, the implementation of this formula for particular finite elements requires calculation of the element stiffness matrix, finite difference approximations of the derivatives of components influenced by pertubation of an element shape variable, and calculation of the pseudo-load associated with rigid-body rotation. Notice that the implementation of the formula (21) in a finite element computer code like ANSYS would be an important initial step towards introduction of the semi-analytical method of sensitivity analysis in the code.

For an ordinary user, ANSYS is a black box. During an ANSYS run, the program generates a set of intermediate files, most of which are binary blocked files. However, a special ANSYS utility function is available which can dump desired binary blocked files into readable files. We take advantage of this utility and develop a dedicated program to scan the readable file, pick the necessary stiffness matrix, carry out the perturbation and computation of finite difference based stiffness sensitivities associated with element shape variables, and perform the computation. The entire operation and the required data input is organized in a command file in such a way that tests of different types of finite elements and different types of element shape pertubations can be relatively easily carried out.

At this stage we have tested the following elements: (a) inclined bar element, (b) horizontal beam element, (c) inclined frame element, (d) four-node isoparametric plane stress element, (e) six-node triangular plane stress element, (f) four-node plate bending element, and (g) three-node triangular plate bending element. For each type of element we have calculated the extra moment $\bar{F}_{rr}$ caused by a unit rigid-body rotation U_r,

$$\bar{F}_{rr} = -U_r^T \frac{\Delta K}{\Delta d} U_r \tag{82}$$

for two elements of different sizes. The large element has linear in-plane dimensions that are 10 times larger than those of the small element. The design variable d is chosen to be a characteristic linear dimension of the element such as the length of a bar, beam or frame element, or the length of one of the sides of a plane stress element or a plate bending element. The perturbation Δd is taken to be 1% of the value of the design variable. It is very important to remember the difference between d and actual structural design variables adopted for structural optimization. For example, let us consider optimization of a beam and take the length L of the beam as the structural design variable. If a number of n beam elements are used to model the beam, the extra moment associated with a rigid-body rotation of the beam of magnitude θ is

$$m_\theta = \bar{F}_{rr}\theta = -U_r^T \frac{\Delta K}{\Delta L} U_r\theta = -U_r^T \frac{\Delta K}{\Delta d \cdot n} U_r\theta \tag{83}$$

and we note that n is large for small elements. Now, since absolute values of m_θ are less interesting in the present context, we rather study the ratio ρ defined as

$$\rho = \frac{m_\theta \text{ for large element}}{m_\theta \text{ for small element}} \tag{84}$$

Values of the ratio ρ obtained for the types of elements (a) - (g) mentioned above are listed in Table 2.

TABLE 2
Ratios ρ for different types of elements

Case	ρ	Element type
a	10	inclined bar
b	0.1	horizontal beam
c	1.2	inclined frame
d	100	plane stress (d)
e	100	plane stress (e)
f	1	plate bending (f)
g	1	plate bending (g)

Note that in cases f and g, the rigid-body rotations and the resultant moments are both vectors, i.e., they both have in-plane and out-of-plane components. The results shown in Table 2 for the plate bending elements (f) and (g) refer to the in-plane component of the resultant moment and thus corresponds to out-of-plane bending.

Let us consider a beam example in order to illustrate the importance of the values of the above ratios. Suppose that a short section of the beam is discretized into 1 and 10 elements, respectively. Denote the moments for each of the corresponding elements by m_1 and m_{10}, respectively. Since the considered section of the beam is short, the effect of 10 extra moments m_{10} contributed from 10 small elements should be nearly equal to the effect of their resultant moment $10 \cdot m_{10}$. Obviously, the effect of the extra moment will diminish if the mesh is refined, if $10 \cdot m_{10} << m_1$. On the other hand, the effect of the extra moment will increase as the mesh is refined, if $10 \cdot m_{10} > m_1$. In the latter case, we will be confronted with type 1 abnormal error behaviour, i.e., the error of the sensitivity increases as the mesh is refined.

From Table 2 we can see that the bar and the plane stress elements have large ratios ρ which implies that they do not exhibit type 1 abnormal error. In contrast to this, beam, frame, and plate bending elements have small values of the ratio and thus reveal that they will be associated with type 1 abnormal error. We note that these findings are in perfect agreement with the results presented in Ref.[10].

8. Concluding Remarks

Two types of error may manifest themselves in finite element based semi-analytical sensitivity analysis. The source of both error types is the numerical differentiation of the components of the stiffness matrix which is the characteristic of the method.

The first type of error appears when elements with different types of nodal degrees of freedom are applied, for example beam, frame, and plate elements. Here, the stiffness components will depend on a given in-plane element design variable in different powers which gives rise to uneven truncation errors that become seriously magnified by the finite element based solution procedure for the sensitivities, cf. Sections 5 and 6. Remedies for reduction and even removal of this first type of error are already discussed in the Introduction.

Bar, truss and plane stress or strain elements are not prone to this type of error, but may display a second type of semi-analytical sensitivity error if a long span beam-like structure is modeled by such elements. The error can be traced back to an uneven truncation error connected with numerical differentiation of the coordinate transformation matrix, cf. Section 4. A viable path for removal of this second type of error is analytical representation and differentiation of the element coordinate transformation matrix. In addition, it should be borne in mind that a long-span cantilever beam-like structure behaves differently from most other structures. Hence, even the structural analysis of a long-span cantilever beam-like structure needs special attention concerning numerical error because any small rotation near the root of the cantilever will cause large displacement at the tip, if the beam is long enough.

It is shown in this paper that the dominant term of both types of error may be identified at the element level as an extra moment associated with rigid-body rotation, and a new finite element test is developed by means of which it can be revealed a priori whether a given type of element will exhibit semi-analytical sensitivity error behaviour.

Acknowledgement - Fruitful discussions with prof. John Rasmussen are gratefully acknowledged.

References

[1] O.C. Zienkiewicz & J.S. Campbell (1973) 'Shape Optimization and Sequential Linear Programming' in: Optimum Structural Design, Theory and Applications (Eds. R.H. Gallagher and O.C.Zienkiewicz), Wiley and Sons,London, pp. 109-126.

[2] B.J.D. Esping (1983) 'Minimum Weight Design of Membrane Structures', Ph.D. Thesis, Dept. Aeronautical Structures and Materials, The Royal Institute of Technology, Stockholm, Report 83-1.

[3] G. Cheng & Y. Liu (1987) 'A New Computation Scheme for Sensitivity Analysis', Eng. Opt. 12, pp. 219-235.

[4] R.T. Haftka & H.M. Adelman (1989) 'Recent Developments in Structural Sensitivity Analysis', Structural Optimization 1, pp. 137-151.

[5] B. Barthelemy, C.T. Chon & R.T. Haftka (1988) 'Sensitivity Approximation of Static Structural Response', Finite Elements in Analysis and Design 4, pp. 249-265.

[6] K.K. Choi & S.-L. Twu 'On Equivalence of Continuum and Discrete Methods of Shape Sensitivity Analysis', AIAA J. (to appear).

[7] B. Barthelemy & R.T. Haftka (1988) 'Accuracy Analysis of the Semi-Analy tical Method for Shape Sensitivity Calculation', AIAA Paper 88-2284, Proc. AIAA/ASME/ASCE/ASC 29th Structures, Structural Dynamics and Materials Conf., (held in Williamsburg, Va., April 18-20, 1988), Part 1, pp. 562-581 Also: Mech. Struct. Mach. 18, pp. 407-432.

[8] P. Pedersen, G. Cheng & J. Rasmussen (1989) 'On Accuracy Problems for Semi-Analytical Sensitivity Analyses', Mech. Struct. Mach. 17, pp. 373-384.

[9] N. Olhoff & J. Rasmussen (1990) 'Study of Inaccuracy in Semi-Analytical Sensitivity Analysis - A Model Problem', Report No. 28, Institute of Mechanical Engineering, University of Aalborg, Denmark, Structural Optimization (to appear).

[10] G. Cheng, Y. Gu & Y. Zhou (1989) 'Accuracy of Semi-Analytic Sensitivity Analysis', Finite Elements in Analysis and Design 6, pp. 113-128.

[11] R.T. Haftka (1990) 'Stiffness-Matrix Condition Number and Shape Sensitivity Errors', AIAA J. 28, pp. 1322-1324.

[12] Y. Gu & C. Cheng (1990) 'Structural Shape Optimization Integrated with CAD Environment', Structural Optimization 2, pp. 23-28.

[13] G. Cheng, Y. Gu & X. Wang (1991) 'Improvement of Semi-Analytic Sensitivity Analysis and MCADS' in: Engineering Optimization in Design Processes (Eds. H.A. Eschenauer, C. Mattheck and N. Olhoff), pp. 211-223. Springer-Verlag, Berlin.

[14] N. Olhoff & J. Rasmussen (1991) 'Method of Error Elimination for a Class of Semi-Analytical Sensitivity Analysis Problems', these proceedings.

[15] R.D. Cook, D.S. Malkus & M.E. Plesha (1989) 'Concepts and Applications of Finite Element Analysis', Third Edition, John Wiley & Sons, Inc.

METHOD OF ERROR ELIMINATION FOR A CLASS OF SEMI-ANALYTICAL SENSITIVITY ANALYSIS PROBLEMS

NIELS OLHOFF and JOHN RASMUSSEN
Institute of Mechanical Engineering
Aalborg University, DK-9220 Aalborg, Denmark

ABSTRACT: The semi-analytical method of sensitivity analysis [1-3] of finite element discretized structures is indispensable in a computer aided engineering environment for interactive design and optimization. However, it has been shown [3-10] that the method may exhibit serious inaccuracies when applied to structures modeled by beam, plate, shell, and Hermite elements. The inaccuracy of primary concern is associated with the dependence of design sensitivity error on finite element mesh refinement [3-10], but also errors subject to the pertubation of design variables may manifest themselves. Truncation errors due to conditioning of algebra and limited computer precision will not be considered here. In this paper we present a new method developed in [10] for elimination of inaccuracy in semi-analytical sensitivity analysis for a class of problems. The method is advantageous from the point of view that problem dependent, exact error analysis is not required, and that it both eliminates the dependence of the error of the sensitivity on finite element mesh refinement and on design variable pertubation. Also, the method is computationally inexpensive because the differentation of the stiffness components can be exclucively carried out via a forward difference scheme, provided that a set of simple correction factors has been computed. The correction factors may be determined once and for all for a given type of finite element, or as an initial step of the procedure.

1. INTRODUCTION

Among the different methods available for sensitivity analysis of a finite element discretized structure, i.e., the overall finite difference technique, the analytical technique, and the semi-analytical technique [1-4], the latter is preferable for a broad class of problems.

The method is based on the global equilibrium equations for a finite element discretized problem

G. I. N. Rozvany (ed.), Optimization of Large Structural Systems, Vol. I, 385–396.

$$[S]\{D\} = \{F\} \ , \tag{1}$$

where {F} is the vector of external loading, [S] the stiffness matrix, and {D} the resulting displacement vector. In a design problem, [S] and {D} depend on a vector $\{a\}$ of design variables a_j, j = 1,..,J. We shall assume that the external loads are independent of design so that $\partial\{F\}/\partial a_j = \{0\}$, j = 1,..,J.

The primary goal of design sensitivity analysis is to determine the sensitivities $\partial\{D\}/\partial a_j$ of the nodal displacements with respect to design. To this end, (1) is differentiated with respect to a_j, j = 1,..,J, and with design independent external loads, we obtain

$$[S(\{a\})] \ \frac{\partial\{D\}}{\partial a_j} = \{\tilde{F}\}_j \qquad j = 1, \ldots, J \ , \tag{2}$$

where

$$\{\tilde{F}\}_j = - \ \frac{\partial[S(\{a\})]}{\partial a_j} \ \{D\} \qquad j = 1, \ldots, J \ , \tag{3}$$

is the so-called *pseudo load vector* associated with the design variable a_j.

The sensitivities $\partial\{D\}/\partial a_j$ can now be solved from (2) using the same factorization of the global stiffness matrix [S] as is employed in the initial solution of the finite element equilibrium equations (1) for the nodal displacements {D} in a given step of redesign.

With {D} obtained from (1), the determination of $\partial\{D\}/\partial a_j$ from (2) only requires knowledge of the pseudo loads $\{\tilde{F}\}_j$ from (3), where the design sensitivities $\partial[S]/\partial a_j$ of the stiffness matrix must be available. If the latter sensitivities are determined analytically, the above approach is called the method of *Analytical* sensitivity analysis, and if they are determined by numerical differentation, the term *Semi-analytical* sensitivity analysis is used.
In the recent papers [3-10] it has been demonstrated that the method of semi-analytical sensitivity analysis may suffer serious accuracy drawbacks when applied to finite element discretized structures modeled by beam, plate, and Hermite elements. Thus, in the papers [8,10], error analyses were carried out for a model problem of a finite element discretized beam, whose length was taken as a design variable, i.e., a simplified type of shape optimization problem was studied. Along with the expected and acceptable feature that the sensitivity error is proportional to the relative pertubation of the design variable, it was also found in [8,10] that, unfortunately, the sensitivity error is at the same time proportional to the square of the number of finite elements used to model the beam.

The source of the latter severe inaccuracy problem was found to be two-fold in [10]. Firstly, the components of the stiffness matrix of the finite element used, depend on the design variable in three different powers because the element both possesses translational and rotational degrees of freedom. Secondly, given this fact, the order of approximation behind a standard forward finite difference scheme (or, for that matter, a central finite difference scheme) for numerical differentiation of the stiffness components, is insufficient to make associated stiffness errors equal (preferably to zero), which was found to be a requirement for elimination of error dependence on mesh refinement in [10].

We shall now consider a method [10] for an extended class of problems, that does not only eliminate the accuracy problem pertaining to the number of finite elements used in the discretization, but also removes the error subject to the pertubation of design variables.

2. CLASS OF PROBLEMS AND METHOD OF ERROR ELIMINATION

Suppose that a global or local finite element stiffness matrix S({a}) is to be differentiated numerically with respect to a design variable a_j, j = 1,.., J, and assume that the typical a_j-dependent stiffness components s_r, r = 1,.., R, contain a_j in different negative integer powers and have the form

$$s_r(l) = p_r + q_r l^{-r} \quad , \quad r = 1,..,R, \tag{4}$$

where $l = a_j$, and R ≥ 1. Eq. (4) implies that l will typically be a characteristic element length or dimension in the plane of the finite element. The terms p_r and coefficients q_r, r = 1,.., R, depend, in general, on the remaining design variables, i.e.,

$$\begin{aligned} p_r &= p_r(a_1,.., a_{j-1}, a_{j+1},.., a_J) \ , \\ q_r &= q_r(a_1,.., a_{j-1}, a_{j+1},.., a_J) \ , \quad r = 1,..,R \ . \end{aligned} \tag{5}$$

If s_r, r = 1,.., R, have the form (4), we introduce a substitution of variable such that a_j is represented by the *reciprocal* variable z,

$$z = l^{-1} \ , \tag{6}$$

whereby (4) can be written in the form $s_r^*(z) = p_r + q_r z^r$, $r = 1,..R$, such that we have $s_r^*(z) = s_r^*(l^{-1}) = s_r(l) = s_r(z^{-1})$. As is obvious from the latter relationships, the introduction of a_j = z rather than $a_j = l$ as a design variable presents no barring for practical application, and can be easily implemented. Let us refrain from applying asterisks as indicators of stiffness components given as functions of z, and just write

$$s_r(z) = p_r + q_r z^r \quad , \quad r = 1,.., R \ . \tag{7}$$

We shall now assume that the numerical differentiation with respect to z of a given stiffness component s_r given by (7) is performed by means of a standard finite difference operator $d_z^{(m)}$, where m designates the order of the polynominal approximation of $s_r(z)$ that constitutes the basis for computation of the finite difference approximation $d_z^{(m)} s_r$ to the exact first derivative $\partial s_r/\partial z$.

For, e.g., first through fourth order approximation of s_r, we have the following well-known formulas for computation of the first derivative:

$$(m = 1) \quad d_z^{(1)} s_r = \frac{\Delta s_r}{\Delta z} = \frac{1}{\Delta z}\left[s_r(z+\Delta z) - s_r(z)\right] \tag{8a}$$

$$(m = 2) \quad d_z^{(2)} s_r = \frac{1}{2\Delta z}\left[s_r(z+\Delta z) - s_r(z-\Delta z)\right] \tag{8b}$$

$$(m = 3) \quad d_z^{(3)} s_r = \frac{1}{6\Delta z}\left[-s_r(z+2\Delta z) + 6s_r(z+\Delta z) - 3s_r(z) - 2s_r(z-\Delta z)\right] \tag{8c}$$

$$(m = 4) \quad d_z^{(4)} s_r = \frac{1}{12\Delta z}\left[-s_r(z+2\Delta z) + 8s_r(z+\Delta z) - 8s_r(z-\Delta z) + s_r(z-2\Delta z)\right] \tag{8d}$$

The fact that the computational cost increases with the number of incremented values of z at which the stiffness components s_r have to be evaluated, must naturally be taken into account in the computational procedure.

Let us denote by η_z the relative increment (pertubation) of the design variable z, i.e.,

$$\eta_z = \frac{\Delta z}{z} \quad . \tag{9}$$

For a given order m of approximation, we now express the *finite difference approximations* $d_z^{(m)} s_r$ in terms of the *exact first derivatives* $\partial s_r/\partial z$ and the *relative error factors* $\alpha_r^{(m)}$:

$$d_z^{(m)} s_r = \frac{\partial s_r}{\partial z}\left(1 + \alpha_r^{(m)}\right) \quad , r = 1,.., R \quad . \tag{10}$$

It is easily verified that, due to the form of Eqs. (7), (8) and (10), *the relative error factors $\alpha_r^{(m)}$ will be independent of the actual value of the design variable z. Thus, if a particular $\alpha_r^{(m)}$ is non-vanishing, it will only depend on the relative pertubation* η_z as defined by (9). The crucial point is that *when written in the form of (7), the stiffness components* s_r, r = 1,.., R, *depend on the design variable* a_j = z *in non-negative integer powers* r = 1,.., R. *This implies that numerical differentiation of these stiffness components by means of a formula from among Eqs. (8), all of which are based on standard polynomial approximations, will furnish exact derivatives of all the stiffnesses* s_r, r = 1,.., R, *provided that the order m of the polynomial approximation behind the applied formula is equal to or larger than R, i.e.,* m ≥ R.

Similar advantage is not achieved if the stiffness components s_r are considered functions of a design variable appearing in *negative powers* as in (4), because such a form of the stiffness components *cannot* be represented exactly by *any* standard polynomium of the design variable in question.

Thus, only when using a_j = z as a design variable, can we make the finite difference based first derivatives coincide with the exact first derivatives of the stiffness components. This requires that we take m $\geq$ R, and can be expressed as

$$d_z^{(m)} s_r = \frac{\partial s_r}{\partial z} \quad \Leftrightarrow \quad \alpha_r^{(m)} = 0 \ , \quad r = 1,.., R, \quad \text{if } m \geq R \ . \tag{11}$$

The fact that the relative error factors $\alpha_r^{(m)}$ of the derivatives of the a_j-dependent stiffnesses s_r, r = 1,.., R, become equal for m $\geq$ R, implies that *the contribution from the design variable* $a_j = z$ *to the sensitivity error associated with finite element mesh refinement is eliminated if we take* m $\geq$ R. Moreover, the vanishing of the error factors associated with m $\geq$ R implies that they have become independent of the value of the relative pertubation η_z of the design variable a_j = z. This means that also the numerically computed derivatives of the stiffness components become independent of η_z if m $\geq$ R. From this we may conclude that *even the contribution to the sensitivity error from the pertubation of the design variable* $a_j = z$ *is eliminated if we take* $m \geq R$.

We shall now implement the above results in an efficient computational procedure of low cost. This means that use of higher order formulas in (8) is to be limited as much as possible, because, as we have already discussed, the computational cost increases with m.

Write now (11) for the smallest order m of approximation which yields the *exact* derivatives $\partial s_r/\partial z$ of all the z-dependent stiffness derivatives, i.e., m = R:

$$d_z^{(R)} s_r = \frac{\partial s_r}{\partial z} \ , \quad r = 1,.., R \ . \tag{12}$$

Since the derivative at the right hand side is independent of η_z (but depends on z), the same holds true for the derivative at the left hand side.

Next, consider (10) in the case of first order approximation, m = 1, where $d_z^{(1)} s_r$ is the simple first order derivative $\Delta s_r/\Delta z$, cf. (8a), so that we may write

$$c_r \frac{\Delta s_r}{\Delta z} = \frac{\partial s_r}{\partial z} \ ; \quad c_r = \frac{1}{1 + \alpha_r^{(1)}} \ , \quad r = 1,..,R \ , \tag{13}$$

where the dimensionless factors c_r, r = 1,.., R, will be termed *correction factors* in the sequel. The correction factors c_r must follow the error factors $\alpha_r^{(1)}$ in being independent of the actual value of the design variable z and only dependent on the value of the relative increment $\eta_z = \Delta z/z$. It is characteristic, though, that we will always have $c_1 = 1$, cf. (13) with r = 1, since the first order formula (8a) yields the exact result for the linear stiffness component s_1.

Obtain now the following expression for the correction factors c_r by combination of (12) and (13):

$$c_1 = 1; \quad c_r = \frac{d_z^{(R)} s_r}{\dfrac{\Delta s_r}{\Delta z}}, \quad r = 2,.., R. \tag{14}$$

Define then the corrected first derivatives of the stiffness components $(\Delta s_r/\Delta z)_{corr}$, $r = 1,..,R$, as

$$\left(\frac{\Delta s_r}{\Delta z}\right)_{corr} = c_r \frac{\Delta s_r}{\Delta z}, \quad r = 1,.., R. \tag{15}$$

It is the primary objective of the new computational procedure to determine these *corrected* derivatives, and we see by means of (13) that they correspond to the *exact* derivatives,

$$\left(\frac{\Delta s_r}{\Delta z}\right)_{corr} = \frac{\partial s_r}{\partial z}, \quad r = 1,.., R. \tag{16}$$

Let us now set up the *computational procedure* pertaining to a given design variable $a_j = z$:

(0) Choose an appropriate value of the relative increment $\eta_z = \Delta z/z$ to be used throughout.

(I) As an initial step, determine for all subsequent computations, the correction factors c_r, $r = 1,.., R$, by means of Eq. (14) with the denominator given by (8a) and the numerator given by (8b) if $R = 2$, (8c) if $R = 3$, etc. The values of s_r to be used are obtained from (7) on the basis of an arbitrarily chosen value of z and the above value of η_z, which furnishes $\Delta z = \eta_z z$.

(II) In all subsequent computations where it is required to determine the derivatives of the stiffness components subject to a specified value of the design variable z, apply Eq. (15) with the values of c_r, $r = 1,.., R$, obtained in step (I). The determination of $\Delta s_r/\Delta z$, $r = 1,.., R$, on the right hand side of (15) implies a simple, first order forward difference approach based on formula (8a), and approximate derivatives are then corrected by means of the factors c_r such that we obtain $(\Delta s_r/\Delta z)_{corr}$, $r = 1,.., R$, which correspond to the exact derivatives, cf. (16).

(III) The corrected values of the first derivatives of the stiffness components with respect to the design variable $a_j = z$ can now be assembled in $\partial[S(\{a\})]/\partial a_j$ and substituted into (3) along with the displacement vector {D} obtained from the analysis problem (1). The pseudo load vector determined from (3) is then substituted into (2), which we can solve for the desired displacement derivatives $\partial\{D\}/\partial a_j$.

Since the result $(\Delta s_r/\Delta z)_{corr}$, $r = 1,.., R$, obtained in step (II) subject to any specified value of z, is *independent* of the relative pertubation η_z, *it is not required to assign* η_z *a small value* in step (0). In fact, values of η_z taken in the range between 10^{-1} and 1 have a beneficial effect on reducing truncation errors due to the conditioning of the algebra and the computational accuracy of the computer.

It should be observed that the initial steps (0) and (I) may actually be executed once and for all for the type of element and the element design variable a_j in question. Thus, *precomputed values of the correction factors* c_r *for a given value of* η_z *will be applicable for all future sensitivity analysis problems involving the given type of element and design variable* a_j, *provided that the original value of* η_z *is used.*

It is a notable feature of the computational procedure that, *after the initial steps* (0) and (I), *it is only required to perform the numerical differentation of the stiffness components by means of simple, computationally inexpensive, first order finite differences, and yet exact stiffness derivatives are obtained.*

The above results are valid, and the computational procedure is applicable, for any of the design variables a_j, and for any set of values assigned to them. Attention should also be drawn to the fact that the correction factors c_r, cf. (14), and the computational procedure as such, are *independent* of actual values of the coefficients p_r and q_r of the typical stiffness components, s_r, $r = 1,..,R$, in Eq.(7). This implies that *although the finite element mesh alterations inherent in shape optimization problems imply changes of the coefficients* p_r *and* q_r *through changes of the values of the design variables*, see (5), *these changes will not affect the computational procedure and the applicability of the initially determined values of the correction factors* c_r.

Thus, within the class of stiffness matrices considered, our development may be said to represent a general, efficient, and cost competitive method for elimination of inaccuracy subject to both finite element mesh refinement and pertubation of design variables in semi-analytical sensitivity analyses. Here, we tacitly assume that an accurate and efficient solution procedure for linear equations, is available for solution of the pure analysis problems (1) and (2).

3. EXAMPLE

We consider an example that has been adopted for display and numerical investigation of the semi-analytical sensitivity inaccuracy problem in [5,6] and error analyses in [8,10]. The example pertains to a finite element modeled uniform Bernoulli-Euler beam of constant bending stiffness EI and variable length L, see Fig. 1. The beam is loaded by a given, concentrated bending moment M at the free end, i.e., the external nodal load vector {F} and the associated nodal displacement vector {D} of the *analysis problem* (1) are

$$\{F\}^T = \{\, 0,\, 0,\, .\, .\, .\, .\, .\, ,\, M\}^T \;, \qquad \{D\}^T = \{u_1,\, \theta_1, \ldots,\, u_n,\, \theta_n\}^T \;. \qquad (17)$$

As in [5,6,8,10], the study will be devoted to the sensitivity of the transverse deflection $u_n(L)$

= $ML^2/2EI$ at the free end with respect to a change of the length L of the beam, so the *exact* result within Bernoulli-Euler theory is

$$\frac{\partial u_n}{\partial L} = \frac{ML}{EI} \quad . \tag{18}$$

The example only involves *one* design variable $a_j = a$ since only the total beam length L may vary. Let us choose to discretize the beam into a total number of n finite elements of equal length l = L/n, see Fig.1. In order to investigate the inaccuracy problem associated with finite element mesh refinement, the choice of design variable must both reflect the discretization and the beam length L, and two alternate choices of the design variable a will be considered, namely

$$a = l \ \text{ and } \ a = z = l^{-1} \ , \quad \text{where } l = L/n \quad . \tag{19}$$

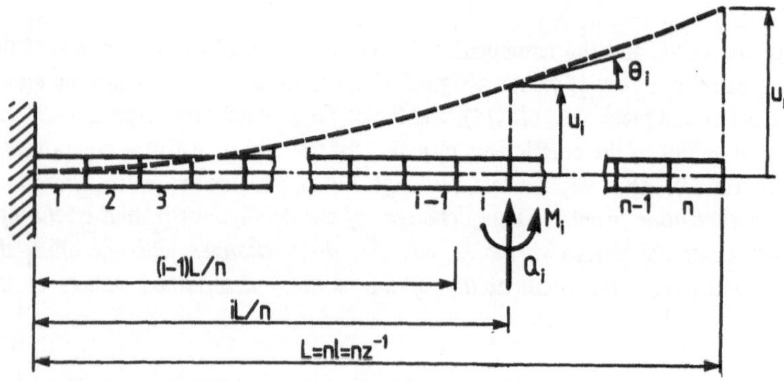

Fig. 1. Global finite element model.

We adopt a beam finite element which is exact within Bernoulli-Euler theory, whereby the element stiffness matrix is (see, e.g., [11]):

$$\begin{bmatrix} s_{11} & s_{12} & -s_{11} & s_{12} \\ s_{12} & s_{22} & -s_{12} & \frac{1}{2}s_{22} \\ -s_{11} & -s_{12} & s_{11} & -s_{12} \\ s_{12} & \frac{1}{2}s_{22} & -s_{12} & s_{22} \end{bmatrix}$$

with stiffness components given by

$$s_{11} = 12\,\frac{EI}{l^3}\ ,\quad s_{12} = 6\,\frac{EI}{l^2}\ ,\quad s_{22} = 4\,\frac{EI}{l}\ . \tag{20}$$

TABLE

Computed displacement design sensitivities vs. number n of finite elements used in beam model

n	$\frac{\partial u_n}{\partial L}$	$\left(\frac{\Delta u_n}{\Delta L}\right)_l$	ε_l	$\left(\frac{\Delta u_n}{\Delta L}\right)_z$	ε_z	$\left(\frac{\Delta u_n}{\Delta L}\right)_{corr}$	ε
1	1.000	1.000	.000	1.000	.000	1.000	0.3E-14
2	1.000	.999	-.001	1.001	.001	1.000	0.9E-14
3	1.000	.998	-.002	1.002	.002	1.000	0.3E-13
4	1.000	.996	-.004	1.004	.004	1.000	0.3E-13
5	1.000	.994	-.006	1.006	.006	1.000	-0.1E-13
6	1.000	.991	-.009	1.009	.009	1.000	0.2E-12
7	1.000	.988	-.012	1.012	.012	1.000	-0.2E-12
8	1.000	.984	-.016	1.016	.016	1.000	0.1E-12
9	1.000	.980	-.020	1.020	.020	1.000	0.6E-13
10	1.000	.975	-.025	1.025	.025	1.000	0.1E-13
12	1.000	.964	-.036	1.036	.036	1.000	0.3E-12
14	1.000	.951	-.049	1.049	.049	1.000	-0.1E-11
16	1.000	.936	-.064	1.064	.064	1.000	0.7E-12
18	1.000	.919	-.081	1.081	.081	1.000	0.4E-12
20	1.000	.900	-.100	1.100	.100	1.000	0.2E-12
22	1.000	.879	-.121	1.121	.121	1.000	-0.8E-12
24	1.000	.856	-.144	1.144	.144	1.000	0.2E-11
26	1.000	.831	-.169	1.169	.169	1.000	0.2E-11
28	1.000	.804	-.196	1.196	.196	1.000	-0.4E-11
30	1.000	.775	-.225	1.225	.225	1.000	0.1E-11
32	1.000	.744	-.256	1.256	.256	1.000	0.3E-11
34	1.000	.711	-.289	1.289	.289	1.000	-0.3E-11
36	1.000	.676	-.324	1.324	.324	1.000	0.8E-12
38	1.000	.639	-.361	1.361	.361	1.000	-0.4E-11
40	1.000	.600	-.400	1.400	.400	1.000	-0.1E-11
42	1.000	.559	-.441	1.441	.441	1.000	0.9E-11
44	1.000	.516	-.484	1.484	.484	1.000	0.3E-12
46	1.000	.471	-.529	1.529	.529	1.000	0.4E-11
48	1.000	.424	-.576	1.576	.576	1.000	0.5E-11
50	1.000	.375	-.625	1.625	.625	1.000	0.1E-11
52	1.000	.324	-.676	1.676	.676	1.000	0.9E-11
54	1.000	.271	-.729	1.729	.729	1.000	0.7E-11
56	1.000	.216	-.784	1.784	.784	1.000	-0.1E-10
58	1.000	.159	-.841	1.841	.841	1.000	0.7E-11
60	1.000	.100	-.900	1.900	.900	1.000	-0.4E-11
62	1.000	.039	-.961	1.961	.961	1.000	-0.1E-10
64	1.000	-.024	-1.024	2.024	1.024	1.000	0.2E-10
66	1.000	-.089	-1.089	2.089	1.089	1.000	-0.5E-11
68	1.000	-.156	-1.156	2.156	1.156	1.000	-0.2E-10
70	1.000	-.225	-1.225	2.225	1.225	1.000	-0.3E-11
72	1.000	-.296	-1.296	2.296	1.296	1.000	0.1E-10
74	1.000	-.369	-1.369	2.369	1.369	1.000	0.4E-11
76	1.000	-.444	-1.444	2.444	1.444	1.000	-0.1E-10
78	1.000	-.521	-1.521	2.521	1.521	1.000	0.3E-11
80	1.000	-.600	-1.600	2.600	1.600	1.000	0.4E-11
82	1.000	-.681	-1.681	2.681	1.681	1.000	0.6E-11
84	1.000	-.764	-1.764	2.764	1.764	1.000	0.4E-10
86	1.000	-.849	-1.849	2.849	1.849	1.000	0.6E-10
88	1.000	-.936	-1.936	2.936	1.936	1.000	0.4E-11
90	1.000	-1.024	-2.024	3.025	2.025	1.000	0.8E-12
92	1.000	-1.115	-2.115	3.116	2.116	1.000	0.4E-11
94	1.000	-1.208	-2.208	3.209	2.209	1.000	0.2E-10
96	1.000	-1.303	-2.303	3.304	2.304	1.000	0.2E-10
98	1.000	-1.400	-2.400	3.401	2.401	1.000	-0.4E-10
100	1.000	-1.499	-2.499	3.500	2.500	1.000	0.8E-11

Multiplier : ML/EI

In the Table, we now present numerically computed values of the subject design sensitivity for a series of values of the number n of elements used in the finite element modeling of the beam, see Fig. 1. The unit values in the second column of the Table serve as the exact result to be compared with by sensitivities determined by the semi-analytical method.

Firstly, we present some results which we obtain [10], if we apply a standard approach of semi-analytical sensitivity analysis, where the current method for error elimination is *not implemented*.

Thus, the values of the semi-analytical sensitivity $(\Delta u_n/\Delta L)_l$ in the third column are based on a standard first order forward difference approximation of stiffness derivatives with the use of l as a design variable [10], and the relative increment is taken to be $\eta_l = \Delta l/l = 10^{-4}$. The results show that the semi-analytical sensitivities become increasingly inaccurate with increasing n, and that even the sign becomes wrong when we use more than 63 finite elements. The subsequent column displays the relative error ϵ_l, and clearly illustrates the n^2-dependence of ϵ_l. The results [10] for $(\Delta u_n/\Delta L)_z$ in the fifth column are based on application of the reciprocal design variable $z = l^{-1}$ and the value $\eta_z = \Delta z/z = 10^{-4}$ of the relative increment, i.e., the same value as was used for $\eta_l = \Delta l/l$, and again standard semi-analytical sensitivity analysis is carried out. Except for having the correct sign for all values of n, the sensitivities are seen to be no more accurate than when l is used as a design variable, and we notice again that the relative error ϵ_z exhibits servere n^2-dependence. The sources of these problems are revealed in [10].

Let us now adopt the present method of error elimination to help this unfortunate state of affairs. Then z is required to be the design variable, i.e., $a_j = a = z$, and from (7) and (20) we identify $R = 3$ and the typical z-dependent stiffness components to be

$$s_1 = s_{22} = 4EIz, \quad s_2 = s_{12} = 6EIz^2, \quad s_3 = s_{11} = 12EIz^3 \, , \tag{21}$$

where $z = n/L$, cf. (19). Since we seek to determine $\Delta u_n/\Delta L$ and know that ML/EI factors out from the result, we assign M, L, and EI unit values in the subsequent computations, whereby we simply have $z = n$.

In step (0) of the computational procedure, we select η_z as $\eta_z = 10^{-1}$. In the initial step (I), based on the arbitrarily chosen value $z = 200$ and thus $\Delta z = 20$, from the expressions

$$c_1 = 1 \; ; \; c_r = \frac{1}{6} \, \frac{-s_r(z+2\Delta z) + 6s_r(z+\Delta z) - 3s_r(z) - 2s_r(z-\Delta z)}{s_r(z+\Delta z) - s_r(z)} \, , \; r = 2,3, \tag{22}$$

we determine the correction factors as $c_1 = 1$, $c_2 = 0.952381..$, $c_3 = 0.906344..$.

Based on these values, we now in step (II) apply the formula

$$\left[\frac{\Delta s_r}{\Delta z}\right]_{corr} = c_r \frac{1}{\Delta z}\left[s_r(z+\Delta z) - s_r(z)\right] , \quad r = 1,2,3 \ , \tag{23}$$

cf. (15) and (8a), for computation of the corrected derivatives of the stiffness components $(\Delta s_r/\Delta z)_{corr}$, r = 1, 2, 3, for all the values of n = z required in the Table. For each of these values of n = z, we then follow the scheme of step (III) of the computational procedure and obtain the desired displacement design sensitivity. The results are denoted by $(\Delta u_n/\Delta L)_{corr}$ and are listed in the second-last column of the Table, with associated errors ϵ given in the last column. The errors ϵ are simply computed as the difference between $(\Delta u_n/\Delta L)$ and unity.

The results illustrate most convincingly that the types of error considered in this paper can be completely eliminated by means of the proposed method. The errors are clearly seen to be small truncation errors due to the conditioning of the algebra and limited precision of the computer. Except for this type of error, the values of the semi-analytical sensitivities cannot be distinguished from the exact ones.

ACKNOWLEDGMENTS - Authors gratefully acknowledge stimulating discussions with prof. Gengdong Cheng. This work received support from the Danish Technical Research Council (Programme of Research on Computer Aided Design) and from the Research Council (Konsistoriums forskningsudvalg) of the University of Aalborg.

REFERENCES

[1] *O.C. Zienkiewicz & J.S. Campbell* (1973) Shape Optimization and Sequential Linear Programming, in: Optimum Structural Design, Theory and Applications (Eds. R.H. Gallagher and O.C.Zienkiewicz), Wiley and Sons, London, pp. 109-126.

[2] *B.J.D. Esping* (1983) Minimum Weight Design of Membrane Structures, Ph.D. Thesis, Dept. Aeronautical Structures and Materials, The Royal Institute of Technology, Stockholm, Report 83-1.

[3] *G. Cheng & Y. Liu* (1987) A New Computation Scheme for Sensitivity Analysis, Eng. Opt. 12, pp. 219-235.

[4] *R.T. Haftka & H.M. Adelman* (1989) Recent Developments in Structural Sensitivity Analysis, Structural Optimization 1, pp. 137-151.

[5] *B. Barthelemy, C.T. Chon & R.T. Haftka* (1988) Sensitivity Approximation of Static Structural Response, Finite Elements in Analysis and Design 4, pp. 249-265.

[6] *B. Barthelemy & R.T. Haftka* (1988) Accuracy Analysis of the Semi-Analytical Method for Shape Sensitivity Calculation, AIAA Paper 88-2284, Proc. AIAA/ASME/ASCE/ASC 29th Structures, Structural Dynamics and Materials Conf., (held in Williamsburg, Va., April 18-20, 1988), Part 1, pp. 562-581.

[7] *K.K. Choi & S.-L. Twu* On Equivalence of Continuum and Discrete Methods of Shape Sensitivity Analysis, AIAA J. (to appear).

[8] *P. Pedersen, G. Cheng & J. Rasmussen* (1989) On Accuracy Problems for Semi-Analytical Sensitivity Analyses, Mech. Struct. Mach. 17, pp. 373-384.

[9] *G. Cheng, Y. Gu & Y. Zhou* (1989) Accuracy of Semi-Analytic Sensitivity Analysis, Finite Elements in Analysis and design 6, pp. 113-128.

[10] *N. Olhoff & J. Rasmussen* (1990) On Elimination of Inaccuracy in Semi-Analytical Sensitivity Analysis, Report No. 28, Institute of Mechanical Engineering, University of Aalborg, Denmark. Structural Optimization (to appear).

[11] *R.D. Cook, D.S. Malkus & M.E. Plesha* (1989) Concepts and Applications of Finite Element Analysis, 3rd Edition Wiley & Sons, New York.

INTRODUCTION TO SHAPE SENSITIVITY
THREE - DIMENSIONAL AND SURFACE SYSTEMS

B. Rousselet
Laboratoire de Mathématiques
Université de Nice 06034 Nice cedex
France

ABSTRACT. Basic material for shape optimization of stuctures; continuous approach.

1 Introduction

In these lectures we present some basic material for the shape optimization of structures. We emphasise the so - called continuous approach with few results on numerical approximation with finite elements or boundary integrals; this approach is traditional in mathematics and theoretical mechanics, whereas in mechanical engineering the tendency is to first approximate the behaviour of the structure with finite elements and afterwards to tackle optimization.

The choice of one of these approaches depends on the habits of thought; in many cases, discretisation in the first or second step yields the same results; this has been proved when one uses <u>conformal</u> finite elements (S. Moriano 1988 , M. Masmoudi 1987). If one is interested in deriving necessary optimality conditions and finding explicit solutions, then the continuous approach is necessary; this is the route followed by the Pragerian school.

G. I. N. Rozvany (ed.), Optimization of Large Structural Systems, Vol. I, 397–432.

However in connection with finite elements, the continuous approach is quite versatile: it enables the addition of design sensitivity to a commercial finite element code (Melao Barros & Mota soares 1987, Chenais & Knopf-Lenoir 1988); but it also enables the inclusion of design sensitivity in an open finite element library such as Modulef (1985) and makes good use of existing software (Mehrez-Palma-Rousselet 1991 to appear).

Moreover, formulae obtained with the continuous approach can be implemented with boundary elements (Masmoudi (1987), Mota soares, Rodrigues Choi (1984)).

It should also be pointed out that these techniques may be used and are used in other fields of application; for example in acoustic (Masmoudi 1987) and in fluid mechanics (Pironneau 1984).

However what is shape optimization ? It is an optimal design problem where the design variable is the shape of the domain Ω occupied by the physical system; the best shape of a fillet in a tension bar will provide a classsical engineering example (Haug,Choi-Komkov 1986):
we want to find the best shape of Γ_0 to minimize volume with constraints on Von-Mises yield stress.

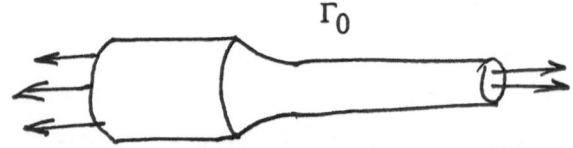

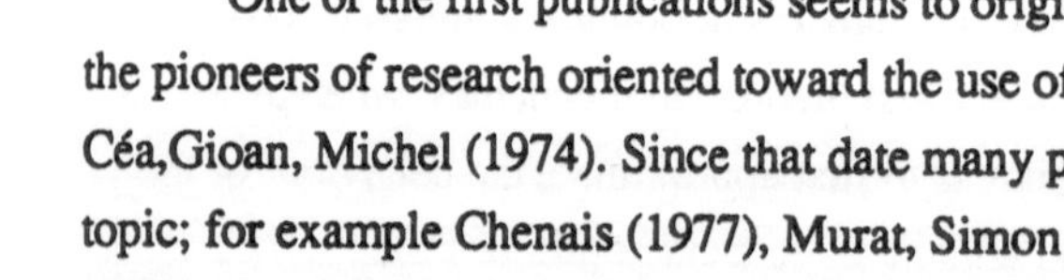

One of the first publications seems to originate with Hadamard (1908) but the pioneers of research oriented toward the use of computers seems to be Céa,Gioan, Michel (1974). Since that date many papers have been devoted to this topic; for example Chenais (1977), Murat, Simon (1976), Rousselet (1976,1977,1982), Dems,Mroz (1984), Pironeau (1984). INRIA schools devoted to shape optimization have been organized by Pironeau (1982) and Céa, Rousselet (1983).

2 shape optimization and continuum mechanics

x ω_0

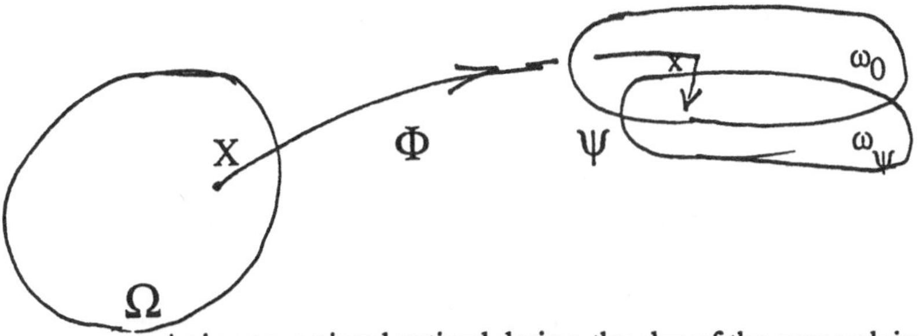

As in conventional optimal design, the clue of the approach is to obtain first - order estimates of the variation of a functional of the state of the system; but for shape optimization one soon realizes that the set of possible domains has no standard vector space structure, so that it seems that classical differential calculus and calculus of variations cannot apply here.

Indeed, these techniques can be used if one realizes that for a given topology and regularity of the boundary, it is natural to look for domains as mappings of a given domain Ω; we shall denote

$$\omega_{\Psi} = (\Phi + \tilde{\Psi})(\Omega) = \{ x \in E \,/\, x = \Phi(X) + \tilde{\Psi}(X) \ \forall X \in \Omega \}$$

where E is the usual Euclidean space (in one,two or three dimensions); Ψ is an element of a vector space of functions; it will enable to define variations of $\omega = \Phi(\Omega)$.

Anyone familiar with the foundations of continuum mechanics should realise that this is what we are doing when we are using a Lagrangian representation of the movement of a continuous medium; this is the usual representation in solid mechanics. For the implementation of the analysis of large deformations of solids, it is usual to use an updated Lagrangian formulation: this amounts to linearizing the behaviour of the solid around a configuration obtained with a fraction of the load.

Here we are going to linearize around the given domain ω, but we should keep in mind that in the overall process of optimization we shall update the domain ω around which we linearize the cost functional and the constraints.

Basic tools for this linearization are well - known in continuum mechanics, but were derived independently for shape optimization by several authors including Dervieux-Palmerio (1975), Murat-Simon (1976), Rousselet (1976). These tools are recalled in the next two sections.

3 Differential calculus and linearization around a given domain

To join domain sensitivity and surface sensitivity, we recall some basic notations of curvilinear coordinates; in fact the mapping $\Phi: \Omega \longrightarrow \omega$, $X \longmapsto x = \Phi(X)$ defines curvilinear coordinates in ω ; we denote by $g_i(x) = \dfrac{\partial \Phi}{\partial X_i}$ the local basis; generally it is not orthonormal so that it is convenient to use the dual basis g^i defined by $g^i \cdot g_j = \delta^i_j$.

With these notations the matrix of $\dfrac{\partial \Phi}{\partial X}$ is $(g_1\, g_2\, g_3)$.

Let $f : \omega \longrightarrow R$ be a scalar funtion; if we set $f_{,i} = \dfrac{\partial f(\Phi(X))}{\partial X_i}$ the chain rule yields $\dfrac{\partial f}{\partial x} = f_{,i}\, g^i$; it is usual to set $\text{grad} f = g^{ij} f_{,i}\, g_j$ where $g^{ij} = g^i \cdot g^j$.

For future reference we recall that

$$\int_\omega f(x)\, dx = \int_\Omega f(\Phi(X))\, |D\Phi|\, dX$$

with $|D\Phi| = \det(g_1\, g_2\, g_3) = \sqrt{g}$

where $g = \det(g_{ij})$ with $g_{ij} = g_i \cdot g_j$

Note. In the following repeated latin indeces are summed from 1 to 3 and greek indeces from 1 to 2.

For a vector field v defined in Ω_Φ , the chain rule also yields :

$$\frac{\partial v}{\partial x} = v_{,i}\, g^i \qquad \text{where} \qquad v_{,i} = \frac{\partial v(\Phi(X))}{\partial X_i}$$

but if we express v in the local basis g_i :

and wish to express $\frac{\partial v}{\partial x}$ with these components it is classsical to introduce Christoffel symbols $\Gamma^i_{kj} = g^i \cdot g_{k,j}$ so that

$$\frac{\partial v}{\partial x} = v^i|_j \; g_i \otimes g^j \quad \text{where} \quad v^i|_j = v^i_{,j} + \Gamma^i_{jk}\, v^k$$ and $g_i \otimes g^j$ is the linear maping defined by $(g_i \otimes g^j) (h) = g_i\, h^j$

The divergence operator is well-known in continuum mechanics; we recall here some formulae which have similar features when applied to surfaces. We first consider as a definition the following equality which should hold for any continuously differentiable function f with compact support in ω :

$$\int_\omega f \operatorname{div} v \, dx = - \int_\omega \frac{\partial f}{\partial x} v \, dx$$

where component-wise $\frac{\partial f}{\partial x} v = f_{,i} (g^i v) = f_{,i}\, v^i$

to obtain an expression of div v in local basis it is usefull to state

Lemma 3.1

(i) $\frac{\partial g}{\partial g_{ij}} = g\, g^{ij}$

(ii) $\frac{\partial g}{\partial X_i} = 2\, g\, \Gamma^j_{i\,j}$ or $\frac{\partial \sqrt{g}}{\partial X_i} = \sqrt{g}\, \Gamma^j_{i\,j}$

Proof.

(i) Comes from $g_{ij}\, g^{jk} = \delta_i^k$

(ii) The proof uses (i) and some manipulations.

Proposition 3.2 Let $v = v^i\, g_i$

(i) The following expressions hold

$$\operatorname{div} v = \frac{1}{\sqrt{g}}(v^i \sqrt{g})_{,i} = v^i{}_{|i} = g^i \cdot v_{,i}$$

(ii) The following identity holds

$$\operatorname{div}(f\, v) = f \operatorname{div} v + \frac{\partial f}{\partial x} v$$

Proof.

(i) The first identity comes from the definition (3,11) and (3,5), (3,6), (3,7).

(ii) Is straightforward in components.

The formula which provides **the first-order change of an integral over a domain ω with respect to changes of its shape** is well-known in continuum mechanics (see for example Germain 1979) and is now widely used in shape optimal design (see e.g. Céa (1975, 1986), Masmoudi (1987)). Here we try to provide a presentation which is introductory to the more complex case of surface variation.

We recall from § 2 that $\omega_\Psi = (\Phi + \tilde{\Psi})(\Omega) = (I + \Psi)(\omega)$ with $\omega = \Phi(\Omega)$ and $\tilde{\Psi} = \Psi o \Phi$

To make precise the first variation of a function f_Ψ defined on a variable domain ω_Ψ we set the

Definition 3.3.

Let f_Ψ be a function defined on ω_Ψ ; this function may depend explicitly on the vector field Ψ and implicitly through the position of the point
$x = x + \Psi(x)$ where it is evaluated; we set

$$\tilde{f}_\Psi(X) = f_\Psi(\Phi(X) + \Psi o \Phi(X)) \qquad \text{and}$$

$$\delta f(x) = \lim_{t \to 0} \frac{f_{t\Psi}(x + t\,\Psi(x)) - f(x)}{t} =$$

$$\frac{d}{dt} f_{t\Psi} \circ (\Phi + t\tilde{\Psi}) \Big|_{t=0} = \frac{\partial}{\partial t} \tilde{f}_{t\Psi}(X) \Big|_{t=0}$$

Remark 1.

We note that δf is a function defined on ω ; it is linear with respect to ψ; if f does not depend explicitly on Ψ, the chain rule yields $\delta f = \frac{\partial f}{\partial x} \Psi$; on the other hand if $\Psi(x) = 0$, $\delta f(x)$ is just the partial derivative with respect to Ψ . Moreover if f does not depend on x, δf is just the directional derivative with respect to Ψ.

Remark 2.

The usual rules for computing derivatives of a sum or of a product of functions hold for the operator δ .

Remark 3.

In continuum mechanics, when a flow $t \longmapsto x(t)$ is defined on ω, the <u>material derivative</u> of $f(t, x(t))$ is \o(lim,\s\do8(δt-->0)) \f(f(t, x(t +δt))- f(t, x(t)),δt)
we note that δf is a particular case when $x(t) = X + t\Psi(X)$; this simple flow is what is needed to define first order variation of ω . Herein we call <u>material derivative</u> the δ operator .

To compute the variation of the integrals we need the following lemma.

<u>Lemma 3.4.</u>

Let $g = \det(g_{ij})$ then
$\delta\sqrt{g} = \sqrt{g}\,\mathrm{div}\Psi$

<u>Proof</u>.

We have $\delta g = \frac{\partial g}{\partial g_{ij}} \delta g_{ij}$; as $g_{ij} = g_i \cdot g_j$ we obtain

$\delta g_{ij} = \tilde{\Psi}_i \cdot g_j + g_i \cdot \tilde{\Psi}_j$; then using Lemma 3.1 we obtain

$\delta g = 2\, g\, g^j \cdot \tilde{\Psi}_j = 2g\,\mathrm{div}\,\Psi$

<u>Proposition 3.5.</u>

$$\delta \int_\omega u \, d\omega = \int_\omega \delta u \, d\omega + \int_\omega u \, \mathrm{div}\Psi \, d\omega$$

Proof.

We obtain from (3.5), (3,6): $\int_\omega u \, d\omega = \int_\Omega \tilde{u} \sqrt{g} \, dX$

The result then comes from the definition and Lemma 3.4.

Example.

$$\mathrm{vol}(\omega) = \int_\omega d\omega \quad \text{yields}$$

$$\delta \, \mathrm{vol}(\omega) = \int_\omega \mathrm{div}\Psi \, d\omega = \int_{\partial\omega} \Psi \, . \, \nu \, d\sigma$$

as could be expected.

We now state the **variation of a derivative**.

Proposition 3.6.

Let u be a function defined in ω_Ψ ; we have

$$\delta \frac{\partial u}{\partial x} = \frac{\partial}{\partial x} \delta u - \frac{\partial u}{\partial x} . \frac{\partial \Psi}{\partial x}$$

Proof.

We note that ; $\dfrac{\partial u_{t\Psi}}{\partial x} = \dfrac{\partial \tilde{u}_{t\Psi}}{\partial X} \left(\dfrac{\partial (\Phi + t \tilde{\Psi})}{\partial X} \right)^{-1}$ We use the definition

of δ: $$\delta \frac{\partial u_{t\Psi}}{\partial x} = \frac{\partial}{\partial t} \left(\frac{\partial \tilde{u}_{t\Psi}}{\partial X} \right)_{|t=0} \left(\frac{\partial \Phi}{\partial X} \right)^{-1} +$$

$$\frac{\partial \tilde{u}_{t\Psi}}{\partial X}_{|t=0} \frac{\partial}{\partial t} \left(\frac{\partial (\Phi + t \tilde{\Psi})}{\partial X} \right)^{-1}_{|t=0}$$

$$= \frac{\partial}{\partial X}\left(\frac{\partial \tilde{u}_{t\Psi}}{\partial t}\right)\Big|_{t=0} \left(\frac{\partial \Phi}{\partial X}\right)^{-1} - \frac{\partial \tilde{u}_0}{\partial X}\left(\frac{\partial \Phi}{\partial X}\right)^{-1} \frac{\partial \tilde{\Psi}}{\partial X}\left(\frac{\partial \Phi}{\partial X}\right)^{-1}$$

$$= \frac{\partial}{\partial x}\delta u - \frac{\partial u}{\partial x} \cdot \frac{\partial \Psi}{\partial x}$$

which proves the result.

4. Shape sensitivity for a model system.

We apply the previous results to shape sensitivity of the simplest example: a membrane prestressed with an inplane tension T and submitted to a normal density of force f ; the normal deflection is the solution of:

$$(4,1)\quad \begin{cases} -T\,\Delta u = f & \text{in } \omega \\ u = 0 \text{ on the part } \gamma_1 & \text{of the boundary } \partial\omega \text{ where it is fixed.} \\ T\dfrac{\partial u}{\partial n} = 0 & \text{on } \gamma_2 \text{ where it is free} \end{cases}$$

We denote by V the space of kinematically admissible displacements; the principle of virtual work states that

(4,2) $\forall v \in V \quad a(u, v) = l(v)$ where

(4,3) $a(u, v) = \int_\omega T \frac{\partial u}{\partial x} \overline{\frac{\partial v}{\partial x}}\, dx$ and $l(v) = \int_\omega f\, v\, dx$

Note. The overbar denotes the vector associated to a linear form and vice versa: $\overline{\frac{\partial v}{\partial x}}$ *is the grdient of v .*

The variation of the solution is itself the solution of an equation as stated below.

Proposition4.1

Let u be the solution of (4.2), then its first-order variation δu satisfies

(4,4) $\forall v \in V \quad a(\delta u, v) = -(\delta a)(u, v) + (\delta l)(v)$

where δa and $\delta 1$ are variations of a and 1 *for fixed u and v :*

(4,5) $$(\delta a)(u, v) = \int_{\omega} T\left(\frac{\partial u}{\partial x} \cdot \frac{\partial \psi}{\partial x} \cdot \overline{\frac{\partial v}{\partial x}}\, dx\right) + \frac{\partial v}{\partial x} \cdot \frac{\partial \psi}{\partial x} \cdot \overline{\frac{\partial u}{\partial x}}\, dx) + \int_{\omega} T \frac{\partial u}{\partial x} \cdot \overline{\frac{\partial v}{\partial x}}\, \mathrm{div}\, \psi\, dx$$

(4,6) $$(\delta 1)(v) = \int_{\Omega} v\, \delta f\, dx + \int_{\Omega} f\, v\, \mathrm{div}\, \Psi\, dx$$

The <u>proof</u> is a direct application of Propositions 3.4 and 3.5 .

<u>Shape sensitivity of a functional .</u>

We consider the simplest case

(4,7) $$J = \int_{\omega} \alpha(u)\, dx$$

The proposition 3.5 yields

(4,8) $$\delta J = \int_{\omega} \alpha'(u)\, \delta u\, dx + \int_{\omega} \alpha(u)\, \mathrm{div}\, \Psi\, dx$$

As in conventional design sensitivity this expression is not explicit with respect to Ψ: δu is defined through equation 4.4; but this expression may be transformed.

<u>Proposition 4.2</u>

Let $L(u, v) = J(u) + a(u, v) - l(v)$

and set p the solution of

(4,9) $$\forall w \in V \quad \frac{\partial L}{\partial u} w = 0 \qquad \text{or} \qquad a(w, p) = -\frac{\partial J}{\partial u} w$$

then $\delta J = (\delta L)(u,p)$ where the variation of L is computed at u and p fixed; or more precisely:

(4,10) $$\delta J = \int_{\Omega} \alpha(u)\, \mathrm{div}\, \psi\, dx + (\delta a)(u, p) - (\delta 1)(p)$$

with δa and δl given in the previous Proposition.

5 Surface differential calculus.

We consider now a surface S imbedded in a 3D space E^3, parametrized by a single-valuedΦ from a reference open domain Ω of a 2D space E^2. The striking difference with section 3 is that Φ is a mapping from a 2D space to a 3D space. With simplifications all the material presented would be adequate for plane curves, although the use of arc length would simplify some formulae.

To emphasize that Φ stems from a two - dimensional space, we denote by ξ the variable in Ω and greek indeces are implicitely running from 1 to 2 ; repeated indeces mean summation, from 1 to 2.

The local basis is noted

(5,1) $a_\alpha = \dfrac{\partial \Phi}{\partial \xi_\alpha}$; it is a basis of the tangent space to S at the point $m = \Phi(\xi)$.

The dual basis is defined by $a^\alpha . a_\beta = \delta^\alpha_\beta$

where the dot means the usual scalar product of E^3.

So $(a_1 \ a_2)$ is the matrix of $\dfrac{\partial \Phi}{\partial \xi} = a_\alpha \otimes e^\alpha$ where $e^\alpha = e_\alpha$ is the standard basis of E^2. We note that

$$\begin{pmatrix} {}^t a^1 \\ {}^t a^2 \end{pmatrix} \begin{pmatrix} a_1 & a_2 \end{pmatrix} = I_{\mathbf{R}^2}$$

or in tensor notations $(e_\alpha \otimes a^\alpha) . (a_\alpha \otimes e^\alpha) = e_\alpha \otimes e^\alpha$

but $$\begin{pmatrix} a_1 & a_2 \end{pmatrix} \begin{pmatrix} {}^t a^1 \\ {}^t a^2 \end{pmatrix} = a_1 \otimes a^1 + a_2 \otimes a^2$$

is the matrix of Π, the orthogonal projection onto the tangent plane.

Let now f: S ---------> **R** be a real function defined on S. If we set

(5,2) $$\frac{\partial f}{\partial m} = \frac{\partial \tilde{f}}{\partial \xi^{\alpha}} a^{\alpha}$$ it is easy to check that this linear mapping from the tangent plane to **R** is independent of the parametrization; we also have $\frac{\partial f}{\partial m} a_{\alpha}$ $= \frac{\partial \tilde{f}}{\partial \xi_{\alpha}}$ or $\frac{\partial f}{\partial m} \frac{\partial \Phi}{\partial \xi} = \frac{\partial \tilde{f}}{\partial \xi}$

In the following all the notions introduced are independent of the parametrization with the exception of the Christoffel symbols.

The integral over the surface may be written with a parametrization:

(5,3) $$\int_S f(m)\, dS = \int_{\Omega} f(\Phi(\xi))\, \sqrt{a}\, d\xi$$

where

(5,4) $a = \det(a_{\alpha\beta})$ with $a_{\alpha\beta} = a_{\alpha} \cdot a_{\beta}$ or

$\sqrt{a} = || a_{\alpha} \times a_{\beta} ||$ (area element).

<u>**The differentiation of a vector field**</u> is here more intricate; this is intuitively obvious with a circle: let $T(\theta)$ be a unitary tangent vector field. It is clear that when $T(\theta)$ is near $T(\theta_0)$, the first-order change is not tangent but rather orthogonal to the circle; thus we need to introduce the orthogonal projection Π onto the tangent plane; Note that its matrix is

$$a_1 \otimes a^1 + a_2 \otimes a^2 \qquad \text{we set } a_{\alpha,\beta} = \frac{\partial a_{\alpha}}{\partial \xi^{\beta}}$$

where $\Pi a_{\alpha,\beta}$ is a tangent vector. Its decomposition in the local basis is classically expressed with Christoffel symbols (they <u>do</u> depend on the parametrization!):

(5,5) $$\Pi a_{\alpha,\beta} = \Gamma^{\lambda}_{\alpha\beta} a_{\lambda}$$

Note that $\Gamma^{\lambda}_{\alpha\beta} = \Gamma^{\lambda}_{\beta\alpha}$ as $a_{\alpha,\beta} = a_{\beta,\alpha}$

Similarly for a tangent vector field $v_t = v^\alpha a_\alpha$ as $\dfrac{\partial v_t}{\partial \xi^\alpha}$ is not tangent,

we consider

(5,6) $\Pi \dfrac{\partial v_t}{\partial m} = v^\alpha_{\ |\beta}\, a_\alpha \otimes a^\beta$ with

(5,7) $v^\alpha_{\ |\beta} = v^\alpha_{\ ,\beta} + \Gamma^\alpha_{\lambda\beta}\, v^\lambda$

We turn now to the <u>**divergence of a tangent vector field**</u> also defined by an integration by parts.
For any continuously differentiable function f with compact support in S:

(5,8) $\displaystyle\int_S f \operatorname{div} v_t \, dS = - \int_S \frac{\partial f}{\partial m} v_t \, dS$

To obtain an expression in the local basis it is convenient to note:

<u>Lemma 5.1</u>

(i) $\dfrac{\partial a}{\partial a_{\alpha\beta}} = a\, a^{\alpha\beta}$

(ii) $\dfrac{\partial a}{\partial \xi^\alpha} = 2\, a\, \Gamma^\lambda_{\alpha\lambda}$ or $\dfrac{\partial \sqrt{a}}{\partial \xi^\alpha} = \sqrt{a}\, \Gamma^\lambda_{\alpha\lambda}$

<u>Proof</u>:
It is based on $(a^{\alpha\beta})(a_{\beta\gamma}) = \delta^\alpha_\gamma$ and

$a_{\lambda\mu,\alpha} = \Gamma^\gamma_{\lambda\alpha}\, a_{\gamma\mu} + \Gamma^\gamma_{\mu\gamma}\, a^{\lambda\gamma}$

which comes directly from the definition of Christoffel symbols.

<u>Proposition 5.2</u>

Let $v_t = v^\alpha a_\alpha$ be a tangent vector field, we then have the following expressions of the divergence

(5,9) $\operatorname{div} v_t = \frac{1}{\sqrt{a}}(\sqrt{a}\ v^{\alpha})_{,\alpha} = v^{\alpha}_{\ |\alpha} = a^{\alpha}.\Pi\ \frac{\partial \tilde{v}_t}{\partial \xi^{\alpha}} = a^{\alpha}.\Pi \frac{\partial v_t}{\partial m}\ a_{\alpha}$

Proof.

The first equality comes from the definition and (5,3); then we obtain

$$\operatorname{div} v_t = v^{\alpha}_{\ ,\alpha} + v^{\alpha}\frac{(\sqrt{a})_{,\alpha}}{\sqrt{a}}$$ and with Lemma5.1:

$$\operatorname{div} v_t = v^{\alpha}_{\ ,\alpha} + v^{\alpha}\Gamma^{\lambda}_{\alpha\lambda} = v^{\alpha}_{\ |\alpha}$$; (5,6) now gives

$$v^{\alpha}_{\ |\alpha} = a^{\alpha}.\Pi\frac{\partial v_t}{\partial m}\ a_{\alpha} = a^{\alpha}.\Pi\ \frac{\partial \tilde{v}_t}{\partial \xi^{\alpha}}$$

Because we are interested in variation of S, we shall have to consider vector fields Ψ which are transverse to S ; so now we recall how to compute derivatives of transverse vector fields.

It is usual to introduce a unitary normal vector

(5,10) $$a_3 = \frac{|| a_1 \times a_2 ||}{|| a_1 \times a_2 ||}$$

As $a^3 . a^3 = 1$ we have $\frac{\partial a_3}{\partial m}.a_3 = 0$ so that $\frac{\partial a_3}{\partial m}$

may be considered as an operator of the tangent plane. Its expression in the local basis is usually written:

(5,11) $\frac{\partial a_3}{\partial m} = - b^{\alpha}_{\beta}\ a_{\alpha} \otimes a^{\beta}$ so that

$$b^{\alpha}_{\beta} = - a^{\alpha}.\frac{\partial a_3}{\partial m}.a_{\beta}$$

we also set $b_{\alpha\beta} = - a_\alpha \cdot \frac{\partial a_3}{\partial m} \cdot a_\beta$

Note that the lowering of indeces is performed systematiclly with the metric tensor $a_{\alpha\beta}$:

$$b_{\alpha\beta} = a_{\alpha\lambda}\, b^{\lambda}_{\beta}.$$

The derivative of a tangent vector a_α may be written:

(5,12) $$a_{\alpha,\beta} = \Gamma^{\lambda}_{\alpha\beta}\, a_\lambda + b_{\alpha\beta}\, a_3$$

Now let Ψ be a transversed vector field:

(5,13) $$\psi = \psi_\lambda\, a^\lambda + \psi_3\, a_3 = \psi^\lambda\, a_\lambda + \psi_3\, a_3$$

from the previous formula we can obtain:

(5,14) $$\frac{\partial \tilde{\psi}}{\partial \xi^\beta} = (\psi_{\lambda|\beta} - b_{\lambda\beta}\, \psi_3)\, a^\lambda + (\psi_{3,\beta} + b^{\lambda}_{\beta}\, \psi_\lambda)\, a^3$$

or

(5,15) $$\frac{\partial \tilde{\psi}}{\partial \xi^\beta} = (\psi^{\lambda}_{|\beta} - b^{\lambda}_{\beta}\, \psi_3)\, a_\lambda + (\psi_{3,\beta} + b_{\lambda\beta}\, \psi^\lambda)\, a_3$$

from which we obtain

(5,16) $$\frac{\partial \psi}{\partial m} = \psi^{\lambda}_{||\beta}\; a_\lambda \otimes a^\beta + \psi^{3}_{||\beta}\; a_3 \otimes a^\beta$$ with

$$\psi^{\lambda}_{||\beta} = (\psi^{\lambda}_{|\beta} - b^{\lambda}_{\beta}\, \psi_3)$$

$$\psi^{3}_{||\beta} = (\psi_{3,\beta} + b_{\lambda\beta}\, \psi^\lambda)$$

An important operator for surface variation is the **tangential divergence of a vector field:**

(5,17) $$\mathrm{div}_S\, \psi = a^\beta \cdot \frac{\partial \psi}{\partial m}\, a_\beta = (\psi^{\beta}_{|\beta} - b^{\beta}_{\beta}\, \psi_3) = \psi^{\beta}_{||\beta}$$

We recognize that $\psi^{\beta}_{|\beta} = \mathrm{div}\,\Pi\,\psi$ and it is usual to set $H = -\,b^{\beta}_{\beta}$

(mean curvature of S), so that we can also write:

(5,18) $\mathrm{div}_S\,\psi = \mathrm{div}\,\Pi\,\psi + H\,\psi^3$

We note that $\mathrm{div}_S\; a^3 = H$.

Now we are to provide some basic formulas for surface variation.

The material derivative operator δ is defined in the same way as for domain variation

(5,19) $\delta f(m) = \frac{d}{dt} f_{t\Psi} \circ (\Phi + t\tilde{\Psi})(\xi)\,|_{t=0} = \frac{\partial}{\partial t}\tilde{f}_{t\Psi}(\xi)\,|_{t=0}$

We should emphasize some differences:
S is a surface; Ψ is a transverse vector field to S ;

(5,20) $S_{t\psi} = \{ m \mid \forall M \in S \quad m = M + t\,\tilde{\psi}(M) \}$

$f_{t\psi}$ is defined on $S_{t\psi}$.

First-order variation of integrals will be obtained with the following lemma.

Lemma.5.3

Let $a = \det(a_{\alpha\beta})$ we have

(5,21) $\delta\sqrt{a} = \sqrt{a}\,\mathrm{div}_S\,\psi$

From which we obtain

Proposition 5.4

(5,22) $\delta \int_S f_S\, dS = \int_S \delta f\, dS + \int_S f_S\, \mathrm{div}_S\,\psi\, dS$

Example.

area (S) = $\int_S dS$ implies $\delta\,\mathrm{area}(S) = \int_S \mathrm{div}\Pi\,\psi\, dS + \int_S H\,\psi^3\, dS$

now a Green's formula yields

$$\int_S \operatorname{div}\Pi\, \psi\, dS = \int_{\partial S} (\Pi\, \psi) . n \;\; d\sigma$$

n being normal to S in the tangent plane, the interpretation is obvious.

The second term $\int_S H\, \psi^3\, dS$ means that for a given area the change depends on H ; this is obvious for the one-dimensional example of the circle: a given area means $R\alpha = \text{cste}$;

$$\delta \text{ length } \int_S \frac{1}{R} dR\, R\, d\theta = \delta R\, \alpha = \delta R \text{ cst} / R$$

in the limit if R -------> + ∞ (rectilinear segment), the variation of length is zero as it should be for a normal vector field to a rectilinear segment.

Now we study the <u>variation of a derivative. This is more difficult than in the volumic case.</u> We first state and "prove" a simple but <u>*wrong*</u> result .

$$(5,23) \qquad \delta\frac{\partial u}{\partial m} = \frac{\partial \delta u}{\partial m} - \frac{\partial u}{\partial m}\Pi\frac{\partial \Psi}{\partial m}$$

The natural but <u>**wrong**</u> proof is as follows:

$$\frac{\partial u}{\partial m} = \frac{\partial \tilde{u}}{\partial \xi}\left(\frac{\partial \Phi}{\partial \xi}\right)^{-1} \qquad \text{now if} \qquad \Phi(\xi) = \Phi_0(\xi) + \tilde{\psi}(\xi)$$

$$\frac{\partial \Phi}{\partial \xi} = \frac{\partial \Phi_0}{\partial \xi} + \frac{\partial \tilde{\Psi}}{\partial \xi} = \frac{\partial \Phi_0}{\partial \xi}\left(I + \left(\frac{\partial \Phi_0}{\partial \xi}\right)^{-1} \frac{\partial \tilde{\Psi}}{\partial \xi} \right)$$

so that expanding up to first-order

$$\left(\frac{\partial \Phi}{\partial \xi}\right)^{-1} = \left(\frac{\partial \Phi_0}{\partial \xi}\right)^{-1} - \left(\frac{\partial \Phi_0}{\partial \xi}\right)^{-1} \frac{\partial \tilde{\Psi}}{\partial \xi} \left(\frac{\partial \Phi_0}{\partial \xi}\right)^{-1}$$

as in the volumic case; then expanding $u = u_0 + \delta u + \ldots$

$$\frac{\partial \tilde{u}}{\partial \xi}\left(\frac{\partial \Phi_0}{\partial \xi}\right)^{-1} = \frac{\partial u}{\partial m_0} = \frac{\partial u_0}{\partial m_0} + \frac{\partial \delta u}{\partial m_0} + \ldots \qquad \text{and}$$

$$\frac{\partial \tilde{u}}{\partial \xi}\left(\frac{\partial \Phi_0'}{\partial \xi}\right)^{-1}\frac{\partial \tilde{\Psi}}{\partial \xi}\left(\frac{\partial \Phi_0}{\partial \xi}\right)^{-1} = \frac{\partial u}{\partial m_0}\Pi\frac{\partial \psi}{\partial m_0}$$ so that up to first-order

$$\frac{\partial u}{\partial m_0} - \frac{\partial u_0}{\partial m_0} = \frac{\partial \delta u}{\partial m_0} - \frac{\partial u_0}{\partial m_0}\Pi\frac{\partial \psi}{\partial m_0} + \ldots$$ which is equivalent to (5,23)

What is wrong? The crucial point is that $\frac{\partial \Phi_0}{\partial \xi}\left(\frac{\partial \Phi_0}{\partial \xi}^{-1}\right)$ cannot be the identity! $\frac{\partial \Phi_0}{\partial \xi}$ is not on-to so it cannot have a right inverse (its image is a 2D tangent plane); it has a left inverse $B = e_\alpha \otimes a^\alpha$, the matrix of which is $\begin{pmatrix} {}^t a^1 \\ {}^t a^2 \end{pmatrix}$, where a^α means the components of a^α in an orthonormal basis of E^2. Indeed, we have seen that $\frac{\partial \Phi_0}{\partial \xi} B$ is the orthogonal projection onto the tangent plane; this means

$$I_{E^3} = \frac{\partial \Phi_0}{\partial \xi} B + a^3 \otimes a^3$$

Rather than modifying the previous proof, we are going to use more directly the basis vectors, but first of all the **right** proposition is as following.

Proposition 5.5:

$$\delta\frac{\partial u}{\partial m} = \frac{\partial \delta u}{\partial m} - \frac{\partial u}{\partial m}\Pi\frac{\partial \Psi}{\partial m} + a^3\left(a_3 \cdot \frac{\partial \Psi}{\partial m} \cdot \frac{\partial u}{\partial m}\right)$$ or component-wise

$$\delta\frac{\partial u}{\partial m} = (\delta \tilde{u})_{,\alpha}\, a^\alpha - \tilde{u}_{,\alpha}\,(a^\alpha \cdot \tilde{\psi}_{,\mu})\, a^\mu + \tilde{u}_{,\alpha}\, a^{\alpha\mu}\,(a^3 \cdot \tilde{\psi}_{,\mu})\, a^3$$

The proof rests on the following lemma.

Lemma 5.6:

(i) $\delta\, a_\alpha = \frac{\partial \tilde{\psi}}{\partial \xi^\alpha}$ or $\delta\frac{\partial \Phi}{\partial \xi} = \frac{\partial \tilde{\psi}}{\partial \xi}$

(ii) $\delta a^3 = -(a^3 . \frac{\partial \tilde{\psi}}{\partial \xi^\mu}) a^\mu = -a^3 . \frac{\partial \Psi}{\partial m} = -\psi^3_{||\beta} a^\beta$

(iii) $\delta a^\alpha = -(a^\alpha . \frac{\partial \tilde{\psi}}{\partial \xi^\mu}) a^\mu + a^3 a^{\alpha\mu} (a^3 . \frac{\partial \tilde{\psi}}{\partial \xi^\mu})$

or $\delta B = -B \frac{\partial \Psi}{\partial m} \Pi + B\, \overline{(a^3 . \frac{\partial \psi}{\partial m})}\, a^3$

(overbar means transposition; see section 4)

Proof:

We set $\Phi = \Phi_0 + \Psi$ we have $a_\alpha = \frac{\partial \Phi}{\partial \xi^\alpha}$

(i) so that

$a_\alpha = \overset{0}{a}_\alpha + \frac{\partial \tilde{\psi}}{\partial \xi^\alpha}$ which gives $\delta a_\alpha = \frac{\partial \tilde{\psi}}{\partial \xi^\alpha}$

(ii) We use $a^3 . a_\mu = 0$ so that $(\delta a^3) . a_\mu = -a^3 . \delta a_\mu$

moreover $a^3 . a^3 = 1$ gives $(\delta a^3) . a^3 = 0$

then as $\delta a^3 = (\delta a^3 . a_\mu) a^\mu = -(a^3 . \delta a_\mu) a^\mu$

we obtain $\delta a^3 = -(a^3 . \frac{\partial \tilde{\psi}}{\partial \xi^\mu}) a^\mu$

then as $\frac{\partial \psi}{\partial m} = \frac{\partial \tilde{\psi}}{\partial \xi^\mu} a^\mu$

we have obtained the second equality; the third one comes just from the notation (5,16)

$$\frac{\partial \psi}{\partial m} = \psi^\lambda_{||\beta}\, a_\lambda \otimes a^\beta + \psi^3_{||\beta}\, a_3 \otimes a^\beta$$

(iii) $a^\alpha . a^3 = 0$ implies $(\delta a^\alpha) . a^3 = -(a^\alpha . \delta a^3)$

and $a^\alpha . a_\lambda = \delta^\alpha_\lambda$ gives $\delta a^\alpha . a_\lambda = a^\alpha . \delta a_\lambda$

so that

$$\delta a^\alpha = -(a^\alpha . \delta a_\lambda) a^\lambda + (a^\alpha . \delta a^3) a^3$$

$$= - (a^{\alpha} . \frac{\partial \tilde{\psi}}{\partial \xi^{\lambda}}) a^{\lambda} + a^{\alpha} . a^{\mu} (a^3 . \frac{\partial \tilde{\psi}}{\partial \xi^{\mu}}) a^3$$

On the other hand $\quad B\,h = e_{\alpha} (a^{\alpha} . h)$

so that

$$\delta B\,h = e_{\alpha} (\delta a^{\alpha} . h) = - e_{\alpha} (a^{\alpha} . \frac{\partial \tilde{\psi}}{\partial \xi^{\lambda}}) (a^{\lambda} . h) + e_{\alpha} (a^{\alpha} . a^{\mu} (a^3 . \frac{\partial \tilde{\psi}}{\partial \xi^{\mu}})$$

$)(a^3 . h)$

<u>Proof of the Proposition5.5:</u>

$$\frac{\partial u}{\partial m} = \tilde{u}_{,\alpha} a^{\alpha} \qquad \text{so that} \qquad \delta \frac{\partial u}{\partial m} = (\delta \tilde{u})_{,\alpha} a^{\alpha} + \tilde{u}_{,\alpha} \delta$$

a^{α}

we note that $\tilde{u}$ is computed at a fixed point so that δ and $\frac{\partial}{\partial \xi^{\alpha}}$ commute;

secondly we use (iii) of the previous Lemma:

$$\delta \frac{\partial u}{\partial m} = (\delta \tilde{u})_{,\alpha} a^{\alpha} + \tilde{u}_{,\alpha} (a^{\alpha} . \frac{\partial \tilde{\psi}}{\partial \xi^{\lambda}}) a^{\lambda} + \tilde{u}_{,\alpha} a^{\alpha \mu} (a^3 . \frac{\partial \tilde{\psi}}{\partial \xi^{\mu}}) a^3$$

this is the component-wise formula of the proposition; the intrinsic formula stems from

$$\frac{\partial u}{\partial m} = \tilde{u}_{,\alpha} a^{\alpha} \quad \text{and} \quad \frac{\partial \psi}{\partial m} = \tilde{\psi}_{,\lambda} a^{\lambda}$$

6 Sensitivity analysis for surface heat equation.

We still consider a simple surface system; i.e. a stationnary surface heat conduction equation; we set:

f surface density of heat source,
g line density of heat source,
q heat flux vector,
u deviation of temperature from the natural state.

We assume Fourier-law for an isotropic homogeneous medium:

$$q = -c \frac{\partial u}{\partial x}$$

the conservation of heat gives: div $q = f$.

We assume prescribed zero deviation of the temperature on γ_1 ; $u = 0$;

prescibed heat flux on γ_2 : $q.\, n = -g$; note that the minus sign is a convention, q is pointing toward the cold subset, g is positive when heat is received and n is outward).

Finally we have

$$\begin{cases} \text{div } q = f & \text{in } S \\ u = 0 & \text{on } \gamma_1 \\ q.n = -g & \text{on } \gamma_2 \\ q = -\overline{\dfrac{\partial u}{\partial m}} & \end{cases} \quad \text{or component-wise} \quad \begin{cases} q^{\alpha}_{\ |\alpha} = f & \text{in } S \\ u^{\alpha} = 0 & \text{on } \gamma_1 \\ q^{\alpha} .\, n_{\alpha} = -g \text{ in } S & \\ q^{\alpha} = -c\, a^{\alpha\beta} u_{,\beta} & \end{cases}$$

In a standard way we consider the variational formulation:

$\forall v \in V \;\; a(u, v) = l(v)$ where

$$V = \left\{ v \in H^1(S) \;|\; v|_{\gamma_1} = 0 \right\}$$

$$a(u, v) = \int_S c \frac{\partial u}{\partial m} \cdot \overline{\frac{\partial v}{\partial m}} \, dS$$

$$l(v) = \int_S f\, v\, dS + \int_{\gamma_2} g\, v\, d\sigma$$

As in the volumic case δu is the solution of an equation with the same bilinear form a.

Proposition 6.1

Let u be the solution of (6.1), then its variation du satisfies

$$\forall v \in V \quad a(du, v) = -(\delta a)(u, v) + (\delta l)(v)$$

where δa and δl are the variations of a and l *for fixed u and v :*

$$(\delta a)(u,v) = -\int_S c\{ \frac{\partial u}{\partial m} \Pi \frac{\partial \psi}{\partial m} \cdot \overline{\frac{\partial v}{\partial m}} + \frac{\partial v}{\partial m} \Pi \frac{\partial \psi}{\partial m} \cdot \overline{\frac{\partial u}{\partial m}} \} \, dS$$

$$+ \int_S c \frac{\partial u}{\partial m} \cdot \overline{\frac{\partial v}{\partial m}} \, \mathrm{div}_S \, \psi \, dS$$

$$(\delta l)(v) = \int_S v \, \delta f \, dS + \int_S f \, v \, \mathrm{div}_S \, \psi \, dS + \int_{\gamma_2} v \, \delta g \, d\sigma +$$

$$\int_{\gamma_2} v \, g \, \mathrm{div}_{\gamma_2} \, \psi \, d\sigma$$

As in the volumic case, the proof is simple if one uses the previous results: Propositions 5.4 and 5.5. As a touch of "humour" we note that the "wrong" Proposition 5.5 would give here the same Proposition 6.1. This is because the term $(a_3 \cdot \frac{\partial \Psi}{\partial m} \cdot \frac{\partial u}{\partial m}) a_3$ has a zero scalar product with $\frac{\partial v}{\partial m}$, which is a tangential vector.

Surface sensitivity of a functional.

The simplest case is $J(u) = \int_S \alpha(u) \, dS$

With proposition 5.4 we obtain $\delta J = \int_S \alpha'(u)\, \delta u \, dS \qquad \int_S \alpha(u)\, \mathrm{div}_S \psi \, dS$

To make this expression explicit with respect to Ψ we use the same Proposition 4.3 to obtain

$\delta J = (\delta L)(u, p)$ where $L = J + a(u, p) - l(p)$ and δL is computed at fixed u and p:

$$\delta J = \int_S \alpha(u)\, \mathrm{div}_S\, \psi \, dS + \delta a(u, p) - \delta l(p)$$

where δa and δl are given in Proposition 6.1 .

7 Boundary expression of shape sensitivity

We turn here to the volumic case of section 4. It is possible to obtain a different expression of the shape sensitivity of a functional: formula (4,10) may be tranformed to a formula which involves boundary integrals. We need some auxiliary lemmas.

Lemma 7.1

The solution u of (4,1) satisfies

$$(7,1)\quad \forall\, w \in \omega \quad a(u, w) = l(w) + l_{\gamma_1}(w)$$

where W is the space of virtual displacements which do not necessarily satisfy $w = 0$ on γ_1 and

$$(7,2)\ \text{and}\quad l_{\gamma_1}(w) = \int_{\gamma_1} Tw\frac{\partial u}{\partial n}\, d\sigma$$

the proof just uses the Stockes formula:

$$\int_{\omega} -T\Delta u\, w\, dx = \int_{\omega} T\frac{\partial u}{\partial x}\frac{\partial \overline{w}}{\partial x}\, dx - \int_{\partial\omega} Tw\frac{\partial u}{\partial n}\, d\sigma$$

which gives Lemma 7.1 as $T\frac{\partial u}{\partial n} = 0$ on γ_2

Lemma 7.2

$$\text{(i)}\quad \frac{\partial u}{\partial x}\frac{\partial \psi}{\partial x}\frac{\partial \overline{p}}{\partial x} = \frac{\partial}{\partial x}\left(\frac{\partial u}{\partial x}\psi\right)\frac{\partial \overline{p}}{\partial x} - \overline{\psi}\frac{\partial^2 u}{\partial x^2}\frac{\partial \overline{p}}{\partial x}$$

$$\text{(ii)}\quad \frac{\partial p}{\partial x}\frac{\partial \psi}{\partial x}\frac{\partial \overline{u}}{\partial x} = \frac{\partial}{\partial x}\left(\frac{\partial p}{\partial x}\psi\right)\frac{\partial \overline{u}}{\partial x} - \overline{\psi}\frac{\partial^2 p}{\partial x^2}\frac{\partial \overline{u}}{\partial x}$$

$$\text{(iii)}\quad \frac{\partial u}{\partial x}\frac{\partial \psi}{\partial x}\frac{\partial \overline{p}}{\partial x} + \frac{\partial p}{\partial x}\frac{\partial \psi}{\partial x}\frac{\partial \overline{u}}{\partial x}$$

$$= \frac{\partial}{\partial x}\left(\frac{\partial u}{\partial x}\psi\right)\frac{\partial \overline{p}}{\partial x} + \frac{\partial}{\partial x}\left(\frac{\partial p}{\partial x}\psi\right)\frac{\partial \overline{u}}{\partial x} - \frac{\partial}{\partial x}\left(\frac{\partial u}{\partial x}\frac{\partial \overline{p}}{\partial x}\right)\psi$$

(iv) $\frac{\partial u}{\partial x}\frac{\partial \overline{p}}{\partial x}\operatorname{div}\psi = \operatorname{div}\left(\left(\frac{\partial u}{\partial x}\frac{\partial \overline{p}}{\partial x}\right)\psi\right) - \frac{\partial}{\partial x}\left(\frac{\partial u}{\partial x}\frac{\partial \overline{p}}{\partial x}\right)\psi$

(v) $-\frac{\partial u}{\partial x}\frac{\partial \psi}{\partial x}\frac{\partial \overline{p}}{\partial x} - \frac{\partial p}{\partial x}\frac{\partial \psi}{\partial x}\frac{\partial \overline{u}}{\partial x} + \frac{\partial u}{\partial x}\frac{\partial \overline{p}}{\partial x}\operatorname{div}\psi =$

$$-\frac{\partial}{\partial x}\left(\frac{\partial u}{\partial x}\psi\right)\frac{\partial \overline{p}}{\partial x} - \frac{\partial}{\partial x}\left(\frac{\partial p}{\partial x}\psi\right)\frac{\partial \overline{u}}{\partial x} + \operatorname{div}\left(\left(\frac{\partial u}{\partial x}\frac{\partial \overline{p}}{\partial x}\right)\psi\right)$$

(vi) $p\,\delta f + p f \operatorname{div}\psi = \operatorname{div}(p f\,\psi) - f\frac{\partial p}{\partial x}\psi + p\left(\delta f - \frac{\partial f}{\partial x}\psi\right)$

and if f does not depend explicitly on ω :

$$p\,\delta f + p f \operatorname{div}\psi = \operatorname{div}(p f\,\psi) - f\frac{\partial p}{\partial x}\psi$$

The proof uses the definitions and is left to the reader.

Proposition 7.3

Let u be the solution of the model system (4.1) and $J_\omega(u)$ be the functional (4.7), then its variation given by (4.10) may be also expressed as

(7,3) $$\delta J = -T\int_{\gamma_1}\frac{\partial u}{\partial n}\frac{\partial p}{\partial n}(\psi\,.\,n)\,d\sigma +$$

$$\int_{\gamma_2}\alpha(u)\,(\psi.\,n)\,d\sigma + \int_{\gamma_2}\left(T\frac{\partial u}{\partial \sigma}\frac{\partial \overline{p}}{\partial \sigma} - pf\right)(\psi\,.\,n)\,d\sigma$$

Proof.

Lemma 7.2 (v) enables us to derive from (4.5):

$$(\delta a)(u,p) = -$$

$$T\int_{\omega} \left(\frac{\partial}{\partial x}\left(\frac{\partial u}{\partial x}\, \psi\right) \frac{\partial \overline{p}}{\partial x} + \frac{\partial}{\partial x}\left(\frac{\partial p}{\partial x}\, \psi\right) \frac{\partial \overline{u}}{\partial x} \right) dx +$$

$$(7,4) \qquad T\int_{\omega} \operatorname{div}\left(\left(\frac{\partial u}{\partial x}\, \frac{\partial \overline{p}}{\partial x}\right) \psi \right) dx$$

Next we use Lemma 7.2 (vi) with (4,6) and assume for simplicity that f does not depend explicitly on ω : $\delta f = \frac{\partial f}{\partial x} . \psi$

$$(7.5) \quad (\delta\, 1)(f) = \int_{\omega} \operatorname{div}(p\, f\, \psi)\, dx - \int_{\omega} f\, \frac{\partial p}{\partial x}\, \psi\, dx$$

Then we note that Lemma 7.1 for the adjoint state (4,9) gives

$$\forall\, w \in W \quad a(p, w) = -\left\langle \frac{\partial J}{\partial u}, w \right\rangle + \int_{\gamma_1} T\, w\, \frac{\partial p}{\partial n}\, d\sigma$$

so that:

$$(7,6) \quad -T\int_{\omega} \frac{\partial}{\partial x}\left(\frac{\partial u}{\partial x}\, \psi\right) \frac{\partial \overline{p}}{\partial x} =$$

$$\left\langle \frac{\partial J}{\partial u}, \frac{\partial u}{\partial x} w \right\rangle - \int_{\gamma_1} T\left(\frac{\partial u}{\partial x}\, \psi\right) \frac{\partial p}{\partial n}\, d\sigma$$

Similarly as the state u of the system sartisfies:

$$\forall\, w \in W \quad a(u\,,w) = \int_{\omega} f\, w\, dx \;+\; \int_{\gamma_1} T \;\; w\, \frac{\partial u}{\partial n}\, d\sigma$$

we have

$$(7{,}7)\ -T \int_{\omega} \frac{\partial}{\partial x}\left(\frac{\partial p}{\partial x}\;\; \psi\right) \;\; \frac{\partial\, \overline{u}}{\partial x} \;=$$

$$\int_{\omega} \frac{\partial p}{\partial x}\, \psi\, dx \;+\; \int_{\gamma_1} T \;\left(\frac{\partial p}{\partial x}\;\; \psi\right) \frac{\partial\, u}{\partial n}\, d\sigma$$

Finally we recall from Proposition 4.3 that if

$$(7{,}8)\quad J_{\omega}(u) = \int_{\omega} \alpha\,(u)\; dx$$

$$\delta\, J = \int_{\omega} \alpha\,(u)\, \mathrm{div}(\,\psi\,)\, dx \;+\; \delta\, a(\,u\,,\,p\,) - \delta\, l(\,p)$$

so that (7,4), (7,5), (7,6), (7,7) provide

$$(7{,}9)\quad \delta\, J = \int_{\omega} \alpha\,(u)\; \mathrm{div}(\,\psi\,)\, dx \;+\; < \frac{\partial J}{\partial u}, \frac{\partial u}{\partial x}\, w > -$$

$$\int_{\gamma_1} T\,\left(\frac{\partial\, u}{\partial x}\;\; \psi\right) \frac{\partial\, p}{\partial n}\, d\,\sigma \;-\; \int_{\gamma_1} T \;\left(\frac{\partial p}{\partial x}\;\; \psi\right) \frac{\partial\, u}{\partial n}\, d\sigma \;+$$

$$T \int_{\omega} \mathrm{div}\;\left(\left(\frac{\partial\, u}{\partial x}\;\; \frac{\partial\, \overline{p}}{\partial x} \right) \psi\,\right) dx \;-\; \int_{\omega} \mathrm{div}(p\;\, f\;\, \psi\,)\; dx$$

Then we note $<\frac{\partial J}{\partial u}, \frac{\partial u}{\partial x}\psi> = \int_{\omega} \alpha'(u) \frac{\partial u}{\partial x} \cdot \psi \, dx$, so that the first two terms of (7,9) provide:

$$(7,10) \quad \int_{\omega} \alpha(u) \operatorname{div}(\psi) \, dx + \int_{\omega} \alpha'(u) \frac{\partial u}{\partial x} \cdot \psi \, dx =$$

$$\int_{\omega} \operatorname{div}(\alpha(u) \, \psi) dx = \int_{\gamma_2} \alpha(u) \, \psi \, . \, n \, d\sigma$$

Using Green's formula in (7,9) we obtain:

$$(7,11) \; \delta J = \int_{\gamma_2} \alpha(u) \, \psi \, . \, n \, d\sigma - \int_{\gamma_1} T\left(\frac{\partial u}{\partial x} \, \psi\right) \frac{\partial p}{\partial n} \, d\sigma$$

$$- \int_{\gamma_1} T \left(\frac{\partial p}{\partial x} \, \psi\right) \frac{\partial u}{\partial n} d\sigma +$$

$$T \int_{\partial\omega} \left(\frac{\partial u}{\partial x} \frac{\partial \overline{p}}{\partial x}\right) (\psi . n) \, d\sigma - \int_{\partial\omega} (p \, f \, \psi . n) \, d\sigma$$

Then we note that on γ_1 $u = 0$ and $p = 0$ so that $\frac{\partial u}{\partial x} = \frac{\partial u}{\partial n} n$ and $\frac{\partial p}{\partial x} = \frac{\partial p}{\partial n} n$ and on γ_2 $\frac{\partial u}{\partial x} = \frac{\partial u}{\partial \sigma^{\alpha}} a^{\alpha}$ so that

$$\delta J = - \int_{\gamma_1} T \frac{\partial u}{\partial n} \frac{\partial p}{\partial n} (\psi . n) d\sigma + \int_{\gamma_2} \alpha(u) \, \psi \, . \, n \, d\sigma +$$

$$+ \int_{\gamma_2} (T \frac{\partial u}{\partial \sigma} \frac{\partial p}{\partial \sigma} - p f) (\psi . n) d \sigma$$

Remark.
When one uses finite elements to solve (4,1), it has been observed that the boundary expression (7,3) is not very accurate; theoretical support of this fact will be given in section 8.

8 Use of finite elements and boundary integrals

We consider the model system of section 4 and to make things simpler we assume $u = 0$ is the only boundary condition. We shall give error estimates of δJ when we replace u and p by finite elements approximation; many results of this type may be found in Masmoudi (1987).

Firstly we consider the boundary expression (7,3) which in the case of $u = 0$ on $\partial\omega$ is

$$(8,1) \quad \delta J^B = -T \int_{\partial \omega} \frac{\partial u}{\partial n} \frac{\partial p}{\partial n} (\psi . n) d\sigma$$

We denote u_h and p_h finite elements approximations of u and p ; h denotes the mesh size. We set

$$(8,2) \quad \delta J_h^B = -T \int_{\partial \omega} \frac{\partial u_h}{\partial n} \frac{\partial p_h}{\partial n} (\psi . n) d\sigma$$

For simplification we assume that there is no error in the approximation of the geometry; we need an error estimate of the normal derivative of u and p on $\partial\omega$; an accurate estimation may be based on Rannacher & Scott (1982):

$$(8,3) \quad || u - u_h ||_{1,\infty;\Omega} \leq c h^k || u ||_{k+1,\infty;\Omega}$$

where $||u||_{k,\infty;\Omega} = ||u||_{L^\infty(\Omega)} + \sum_{l=0}^{l=k} ||\frac{\partial^l u}{\partial x^l}||_{L^\infty(\Omega)}$

and k is the degree of polynomials used in the finite element approximation. We note that for $k = 1$, the second derivatives of the solution u should be essentially bounded ($\frac{\partial^2 u}{\partial x^2} \in L^\infty(\Omega)$); this is an assumption which, for example, excludes reintrant corners in $\partial\Omega$ (Grisvard 1985).

Proposition 8.1.

If datas are smooth enough such that (8.3) holds then

$$|\delta J^B - \delta J_h^B| \leq$$

$$c h^k ||u||_{k+1,\infty;\Omega} ||p||_{k+1,\infty;\Omega} ||\psi.n||_{0,\infty;\Gamma}$$

Proof.

$$|\delta J^B - \delta J_h^B| \leq T \int_{\partial\omega} |\frac{\partial u}{\partial n} - \frac{\partial u_h}{\partial n}| \, |\frac{\partial p}{\partial n}| \, |(\psi.n)| d\sigma +$$

$$T \int_{\partial\omega} |\frac{\partial u}{\partial n} - \frac{\partial u_h}{\partial n}| \, |\frac{\partial p}{\partial n} - \frac{\partial p_h}{\partial n}| \, |(\psi.n)| d\sigma +$$

$$T \int_{\partial\omega} |\frac{\partial u}{\partial n}| \, |\frac{\partial p}{\partial n} - \frac{\partial p_h}{\partial n}| \, |(\psi.n)| d\sigma$$

fom which we obtain:

$$|\delta J^B - \delta J_h^B| \leq T(||\frac{\partial u}{\partial n} - \frac{\partial u_h}{\partial n}||_{0,\infty,\gamma} ||\frac{\partial p}{\partial n}||_{0,\infty,\gamma} +$$

$$\left|\left|\frac{\partial u}{\partial n} - \frac{\partial u_h}{\partial n}\right|\right|_{0,\infty,\gamma} \left|\left|\frac{\partial p}{\partial n} - \frac{\partial p_h}{\partial n}\right|\right|_{0,\infty,\gamma} +$$

$$\left|\left|\frac{\partial u}{\partial n}\right|\right|_{0,\infty\,\gamma} \left|\left|\frac{\partial p}{\partial n} - \frac{\partial p_h}{\partial n}\right|\right|_{0,\infty;\,\gamma}) \,||\,\psi.\,n\,||_{0,\infty;\,\gamma}$$

We note that (8.3) implies:

$$\left|\left|\frac{\partial u}{\partial n} - \frac{\partial u_h}{\partial n}\right|\right|_{1,\infty;\,\gamma} \leq c\,h^k\,||\,u\,||_{k+1,\infty;\,\omega}$$

and equivalently for p from which we obtain the proposition.

Secondly we consider the domain expression.

We only state the result; we set δJ^D the expression given by (4,10); δJ^D_h means that u and p are replaced by a finite element approximation u_h and p_h ; k stands for the degee of polynomial approximation and k' the order of derivatives of ψ which are essentially bounded.

Proposition8.2.

If the datas are smoth enough then

$$|\delta J^D - \delta J^D_h| \leq c(u)\,h^{k+k'}(\,||p||_{K+1;\Omega} + 1\,)\,||\psi||_{k'+1,\infty;\Omega}$$

The proof is technical but the result may be understood directly. It means that if the vector field ψ has essentially bounded second derivatives, the error estimate is in h^{k+1} and, moreover, if the second derivatives are small it will be multiplied by a small constant; this error estimate is to be compared with h^k of Proposition 8.1 ; numerical evidence of this result may be found in Rochette (1990).

Thirdly we consider the use of boundary integrals to solve the state equation. For simplification, we consider a membrane with no density of force and prescribed constant displacement on the two pieces of the boundary.

$$(8,4)\quad \begin{cases} -T\Delta u = 0 & \text{in } \omega \\ u = b & \text{on } \gamma_1 \\ u = 0 & \text{on } \gamma_0 \end{cases}$$

The transformation of this boundary value problem into a boundary integral equation is performed by introducing the classical elementaty solution of the Laplacian Δ :

$$(8,5)\quad E(x,z) = -\frac{1}{2\pi} \log | x - z | \qquad \text{(in two dimensions);}$$

classically $\Delta_z E(x, z) = \delta(x - z)$ holds; and if we consider

$$(8,6)\quad v(x) = \int_\gamma E(x, y)\, q(y)\, d\gamma_y \qquad \text{where } \gamma \text{ is the boundary of } \Omega,$$

this funtion satisfies

$$(8,7)\quad \begin{cases} \Delta v = 0 & \text{outside } \gamma \\ v & \text{is continuous in } \mathbf{R}^2 \\ \dfrac{\partial v^i}{\partial n} - \dfrac{\partial v^e}{\partial n} = q & \text{on } \gamma \end{cases}$$

where $v^i = v|_\omega$ and $v^e = v|_{R^2 - \omega}$

So, the solution of (8,4) is given by $u(x) = \int_\gamma E(x,y)q(y)\, d\gamma_y$

where q is solution of

$$(8,8)\quad \begin{cases} \displaystyle\int_\gamma E(x,y)q(y)\, d\gamma_y = b(x) & \forall\, x \in \gamma_1 \\ \displaystyle\int_\gamma E(x,y)q(y)\, d\gamma_y = 0 & \forall\, x \in \gamma_0 \end{cases}$$

Now, we note $u = 0$ outside of γ_0 and $u = b$ inside of γ_1 are the solution of (8,4) with the right boundary conditions, so that using (8,7) we obtain

(8,9) $\qquad \frac{\partial u}{\partial n} = q \quad$ on γ

If we consider now the simple functional $J_\omega = \int_\omega |\frac{\partial u}{\partial x}|^2 dx$ we

note that $\qquad \frac{\partial J}{\partial u} \cdot w = \int_\omega \frac{\partial u}{\partial x} \frac{\partial w}{\partial x} dx$

so that the adjoint state is $p = -u$ and

$\delta J^B = T \int_\gamma |\frac{\partial u}{\partial n}|^2 (\psi . n) \, d\sigma = T \int_\gamma q^2 \, (\psi \, . \, n) \, d\sigma$ is quite easy to

compute. Moreover it is easy to compute with Green's formula.

$$J_\omega(u) = \int_{\gamma_1} T \; b \; q \; d\sigma$$

The case of an equation with a non zero right-hand side may also be computed with integral equations but we need to use the second Green formula; we skip the details and refer to Masmoudi (1987). Before stating the error estimate of δJ, we should recall that many error estimates for the solution of boundary integral equations have been obtained by Nedelec (1976,1977) and the proof is still in Masmoudi (1987).

Proposition8.3

For smooth enough datas

$| \delta J^I - \delta J^I_h | \leq c \, h^{\min(2k+1,l+1)} \, || q ||_{k+1,\gamma} \, || \Pi ||_{k+1,\gamma} ||$

$\Psi.n ||$

where:

- l means the degree of polynomials used to approximate the boundary;
- k the degree of finite elements approximation
- q is the solution of an equation of the type of (8,9) set on γ;
- Π is the solution of a similar equation for the adjoint state.

Remark. We note that for $k = 2$ we should use $l = 4$ so that the error estimate is in h^5; finally we should note that with a smooth approximation of the boundary the error estimate of δJ is quite good; this is in contrast with what we obtained with finite elements.

REFERENCES

Céa J. , Conception optimale ou identification de domaines: calcul rapide de la derivée directionnelle de la fonction cout, Math. modeling & numerical analysis, vol. 20, 1986, pp 371-402 .

Céa J., Une méthode numérique pour la recherche d'un domaine optimal, Publications IMAN-P2 Nice , 1975.

Céa J, Gioan A., Michel J.,Quelques résultats sur l'identification de domaines, Calcolo, III-IV, 1973.

Céa J. & B. Rousselet B.,Ed., Conception optimale de formes, Ecole INRIA (26-30 septembre 1983, Université de Nice).

Chenais D., Sur une famille de variétés à bord lipschtziennes: une application à un problème d'identification de domaines, Annales de l'Institut fourier (1977), 27, pp 201-231.

Chenais D. Knopf-Lenoir C., Sur la communication entre logiciels éléments finis et optimiseur en controle distribué; Calcul des structures et intelligence artificelle, publié par Fouet J.M. Ladevèze P. Ohayon R., Pluralis 1988.

Germain P., Cours de mécanique, Ecole Polytechnique, Palaiseau, France, 1980.

Grisvard P., Elliptic problems in non smooth domains , Pitman 1985.

Hadamard J., Mémoire sur le problème d'analyse relatif à l'équilibre des plaques elastiques encastrées (1908), Oeuvre de J. Hadamard, CNRS, Paris 1968.

Haug E. 1 Choi K. & Komkov V., Design sensitivity analysis structural systems, Academic Press, New York (1986).

Masmoudi M., Outils pour la conception optimale de formes, thèse d'Etat, Université de Nice, 1987.

Modulef, Une bibliothèque modulaire d'éléments finis, INRIA (1985), Rocquencourt, France.

Mota Soares C.A., Rodrigues H.C. & Choi K.K., Shape optimal structural design using boundary elements and minimum compliance techniques, ASME Journal of mechanisms, transmissions & automation in design, vol 106, pp518-523, 1984.

Murat F. & Simon J., Etude de problèmes d'optimum design, Proceedings of the 7th IFIP Conf., Springer Verlag, Lecture Notes in Computer Sciences, n° 41, 1976, pp 54-62.

Nédelec J.C., Curved finite element methods for a solution of singular integral equation on surfaces in $\mathbf{R}^3$, Comp. meth. applic. mech. eng., 1976.

Nédelec J.C., Approximation des équations intégrales en mécanique et en physique, cours EDF-CEA-INRIA, Centre de mathématiques appliquées, Ecole Polytechnique, Palaiseau, 1977.

Pironneau O., Ed., Optimisation de forme dans les systèmes à paramètres distribués: résolution numérique et applications, Ecole INRIA, Rocquencourt, France (8-10 novembre 1982).

Pironneau O., Optimal shape design for elliptic systems, Springer series in computational physics, Springer Verlag 1984.

Rannacher R. & Scott, Some optimal error estimates for piecewise linear finite element approximation, Math. of comp., vol. 38, pp 437-445.

Rochette M., Conception optimale de formes appliquée aux résistances ajustables, thèse, Université de Nice Sophia-Antipolis, 1990.

Rousselet B., Etude de la régularité desz valeurs propres par rapport à des déformations bilipschitziennes du domaine, CRAS 283, série A, 1976, p 507.

Rousselet B., Problèmes inverses de valeurs propres, pp 77-85, "Optimization techniques", Céa ed., Lect. notes in computer sciences,Springer-Verlag, 1976.

Rousselet B., Shape design sensitivity of a membrane, pp 595-623, JOTA, 1983;

Rousselet B., Quelques résultats en optimisation de domaines, thèse d'état, Université de Nice, 1982.

Rousselet B. Shape design sensitivity, from partial differential equation to implementation. Eng. Opt.,1987,vol 11,pp151-171.

Rousselet B., Shape optimization of structures with state constraints, pp 255-264 dans "Control of partial differential equations", A. Bermudez ed; Lect. notes in control and information sciences, vol. 114, Springer Verlag.

A FINITE STRAIN ROD MODEL AND ITS DESIGN SENSITIVITY

B. ROUSSELET
Laboratoire de Mathématiques
Université de Nice Sophia Antipolis
06034 Nice CEDEX

Lecture notes to NATO/DFG Advanced Study Institute

Optimization of large structural systems
(Berchtesgaden, Germany, Sept 23-Oct4, 1991)

ABSTRACT. Basic equtions of a model of beam in large strain and design sensitivity with respect to midline shape.

1 Introduction

The model considered here includes finite rotation (bending and torsion) and shearing. Introduction of this director type formulation seems due to Antman (1972) ; it is an extension of the classical Kirchoff-Love model (1927) ; several theoretical and numerical papers have been devoted to its study ; among others Antman & Kenney (1981), Simo (1985) Simo & Vu-Quoc (1986), Bourgat & Le Tallec & Mani (1988).

We present in this lecture a basic description of the equations of the rod following the aforementioned papers. In section 5, design sensitivity with respect to midline shape will be adressed.

2 Kinematic description. Stresses

2.1. KINEMATIC DESCRIPTION

Here we present the set of admissible configurations; time is not included as the case of static loading will be only considered later.

The reference configuration (before deformation) is defined by a family of cross sections, the centroïds of which are connected by a curve refered to, as the line of centroïds.

On the contrary of previous presentations we do not assume the reference line of centroïds to be straight and we

G. I. N. Rozvany (ed.), Optimization of Large Structural Systems, Vol. I, 433–453.

do not assume either this line to be parametrized by its curvilinear obscissa ; in other words

$\xi \to \varphi_o(\xi)$ is a map $R \to R^3$ with $||\varphi'_o(\xi)||$ not necessarily equal to one ; the arc length element is $ds = ||\varphi'_o(\xi)|| d\xi$.

The possibility of considering the reference line in such a way is quite important for stating the design sensitivity of the next lecture.

The <u>reference cross-section</u> is defined by a moving <u>orthonormal</u> frame attached to it : $E_1(\xi)$, $E_2(\xi)$, $E_3(\xi)$; the last vector is normal to the cross section ; we assume the reference cross-section is normal to the line of centriods so that

$$\varphi'_o(\xi) = ||\varphi'_o(\xi)|| E_3.$$

<u>Before deformation</u>, the rod may be described by

$$\overline{\Omega}_o = \Big\{ X \in R^3,\ X = \varphi_o(\xi) + X_1 E_1(\xi) + X_2 E_2(\xi),\ 0 \le \xi \le a,\ (X_1, X_2) \in \varpi(\xi) \Big\}$$

where $\varpi(\xi)$ is a compact subset of R^2 ; it gives the stape of the cross-section ; as a simple example it may be a circle, the radius of which depends (linearly) on the arc length S.

Similarly an <u>admissible configuration</u> is defined by the position of the line of centroïds and the cross-section :

$$\overline{\Omega} = \Big\{ x \in R^3,\ x = \varphi(\xi) + X_1 t_1(\xi) + X_2 t_2(\xi),\ 0 \le \xi \le a,\ (X_1, X_2) \in \varpi(\xi) \Big\}$$

where $t_i(\xi)$ $i = 1,2,3$ is an orthonormal frame with t_3 normal to the cross-section ; the parameters $(X_1, X_2) \in \varpi(\xi)$ are the same as those used before deformation ; it is so because the cross-section is assumed to be a rigid body.

Note that <u>shearing</u> is taken into account as $\varphi'(\xi)$ is <u>not</u> assumed to be equal to $||\varphi'(\xi)|| t_3$.

2.2. STRESS RESULTANT AND STRESS COUPLE

We denote by T the first tensor of Piola-Kirchoff (also called Boussinesq or Piola-Lagrange tensor) and by σ the Cauchy tensor. For our purpose we need only to recall the relation

$$T(X)\, N(X)\, dA = \sigma(x)\, \upsilon(x)\, da$$

where dA is the area element in the undeformed configuration at the point X ; da is the corresponding area element in the deformed configuration at the corresponding point x ; N is a unitary normal to the undeformed area element at the point X; υ is a corresponding unitary normal at the point x.

So that the stress resultant (per unit of reference arc length) on section :

$$\overline{\Omega}(\xi) = \left\{ x \in R^3,\ x = \varphi(\xi) + X_\alpha t_\alpha(\xi)\ (X_1, X_2) \in \varpi(\xi) \right\}$$

is given by

$$(2.1)\quad n(\xi) = \int_{\varpi(\xi)} T(X_1, X_2, \xi)\, N(\xi)\, dX_1\, dX_2$$

and the stress couple acting on the same section is

$$(2.2)\quad m(\xi) = \int_{\varpi(\xi)} (x(X_1, X_2, \xi) - \varphi(\xi)) \times (T(X_1, X_2, \xi)\, N(\xi))\, dX_1 dX_2$$

Starting from three dimensional balance equation, reduced balance equation for n and m may be oltained (see for example Simo (1985) :

$$(2.3)\begin{cases} \dfrac{dn}{ds_o} + \overline{n} = 0 \\ \dfrac{dm}{ds_o} + \dfrac{d\varphi_o}{ds_o} \times n + \overline{m} = 0 \end{cases}$$

where $\dfrac{d}{ds_o} = \dfrac{1}{||\varphi'_o(\xi)||}\dfrac{d}{d\xi}$, $\overline{n}$ is the resultant applied load and $\overline{m}$ is the applied couple (per unit of reference arc length).

In the sequel, it will be convenient to consider vectors N and M expressed in a fixed frame associated to n and m ; componentwise :

$$(2.4)\quad \begin{cases} n = N^i\, t_i \\ m = M^i\, t_i \end{cases} \qquad \begin{cases} N = N^i\, E_i \\ M = M^i\, E_i \end{cases}$$

Here t_i and E_i are the frames defined in § 2.1 and repeated latin indexes mean numeration from 1 to 3.

2.3. ADMISSIBLE VARIATIONS

Due to the constraint of rigid body motion of the cross section, the admissible configurations are not a vector space but rather a (sub)mani fold ; in pratice this means that the variations δt_i are not independant but satisfy $\delta t_i \cdot t_j + t_i \cdot \delta t_j = 0$; it is well known in differential geometry and in rigid body dynamics that these conditions imply the existence of a vector θ such that $\delta t_i = \theta \times t_i$.

For our distributed model it has been proved in a functional setting by Bourgat & Le Tallec & Mani (1988) that if we consider the set of kinematic admissible configurations to be

$$K=\Big\{ (\varphi , t_i) \in H^1(0,a ;R^{3*4})/$$

$$\varphi(0) = \varphi_0 ,\ \forall\xi \in [0,a] \quad t_i . t_j = \delta_{ij}$$

$$\det(t_1,t_2,t_3) >0,\ \varphi(a) = \varphi_L ,\ t_2(a) = t_{2a}\Big\}$$

then the tangent space $\delta K(\varphi,t_i)$ to K in (φ,t_i) is

$$\delta K(\varphi,t_i)=\Big\{ (\delta\varphi,\delta t_i) \in H^1 (0,a;R^{3*4})/$$

$$\delta\varphi(0) = 0\ \delta\varphi(a) = 0\ \delta t_2(a) = 0$$

$$\exists\, \theta \in H^1(0,a;R^3)\ \forall\xi \in [0,a]\ \delta t_i= \theta \times t_i\Big\}$$

It should be pointed out that the boundary conditions considered here are just given as an example (free rotation at φ_o and articulation around t_{2a} at φ_L); other cases are in the afore mentioned paper.

In the sequel δ will denote the variation, or differentiation on K with respect to these tangent vectors ; $\hat{\delta}$ will be the corotational variation (relative to an observation in the moving frame ; the effect of the "spin" θ is substracted) ; in particular we get easily:

$$\begin{cases} \delta N = \hat{\delta} n = \delta n - \theta \times n \\ \delta M = \hat{\delta} m = \delta m - \theta \times m \end{cases}$$

3 Weak form of balance equation ; strain measures

Starting from the reduced balance equation (2.3) we can obtain its weak form ; find $(\varphi, t_i) \in K$,

$$\forall(\delta\varphi,\delta t_i)\in \delta K(\varphi,t_i) \qquad G(\varphi,t_i ; \delta\varphi,\delta t_i) = 0$$

with

$$(3.1)\quad G(\varphi,t_i ; \delta\varphi,\delta t_i) = \int_0^L [n \cdot (\frac{d\delta\varphi}{ds_o} - \theta x \frac{d\varphi}{ds_o}) +$$

$$m \cdot \frac{d\varphi}{ds}]ds_o + \int_0^L [\bar{n} \cdot \delta\varphi + \bar{m} \cdot \theta]\, ds_o$$

where still, $ds_o = ||\varphi'_o(\xi)||\, d\xi$.

This weak form of balance equation may be reset using variations of strain measures that will be associated to the reduced stresses n and m.

The elongation and shear are measured by

$$(3.2) \quad \gamma = \frac{d\varphi}{ds_o} - t_3$$

Then we denote by ω_0 the unique vector field ω_0 which relates $\frac{dE_i}{ds_o}$ and E_i :

$$(3.3) \quad \frac{dE_i}{ds_o} = \omega_0 \times E_i$$

In the case where E_i is the Frenet frame of the undeformed line of centroïds, this last formula is a vector form of the classical Serre-Frenet formulas. We have $\omega_0 = \omega_0^i \, E_i$ with

$$\omega_0^1 = 0 \qquad \omega_0^2 = \frac{1}{\rho_0}$$

$$\omega_0^3 = \frac{1}{\tau_0}$$

. It is then natural to use as a measure of change of curvature and teorsion the unique vector field ω such that

$$(3.4) \quad \frac{dt_i}{ds_o} = (\omega + \omega_0) \times t_i$$

As for the stresses, it is convenient to associate vectors Γ and χ expressed int the reference frame :

$$(3.5) \text{ we set } \begin{cases} \gamma = \Gamma^i \, t_i \\ \omega = \chi^i \, t_i \end{cases} \text{ and we define } \begin{cases} \Gamma = \Gamma^i \, E_i \\ \chi = \chi^i \, E_i \end{cases}$$

Using the variation δ and corotational variation $\hat{\delta}$ defined in the § 2.3, the following lemma may be proved :

<u>lemma</u> :

$$(3.6)\quad \begin{cases} \hat{\delta}\gamma = \dfrac{d}{ds_o}\delta\varphi - \theta \times \dfrac{d\varphi}{ds_o} \\ \\ \hat{\delta}\omega = \dfrac{d\theta}{ds_o} \end{cases}$$

and

$$(3.7)\quad \begin{cases} t_i \,.\, \hat{\delta}\gamma = E_i \,.\, \delta\Gamma \\ \\ t_i \,.\, \hat{\delta}\omega = E_i \,.\, \delta\chi \end{cases}$$

With this lemma, it is then obvious that the weak form of the reduced balance equation may be written

$$(3.8)\, G(\varphi, t_i; \delta\varphi, \delta t_i) = \int_0^L (n \,.\, \hat{\delta}\gamma + m \,.\, \hat{\delta}\omega) ds_o \; + \int_0^L (\bar{n} \,.\, \delta\varphi + \bar{m} \,.\, \theta) ds_o$$

$$= \int_0^L (N \,.\, \delta\Gamma + M \,.\, \delta\chi) ds_o + \int_0^L (\bar{n} \,.\, \delta\varphi + \bar{m} \,.\, \theta) ds_o$$

An existence proof of a solution of this weak equation is given in Bourgat-Le Tallec- Mani (1988) cander the assumptions of dead loading and a quadratic elastic energy density .

To close our model we define now the <u>stress-strain relations</u>.

We assume here the material is hyperelastic with an elastic potential W (S_o, Γ, χ) ; so

$$(3.9)\quad N = \frac{\partial W}{\partial \Gamma} \quad M = \frac{\partial W}{\partial \chi}$$ are the stress-strain relation.

The simplest case is the one where W is quadratic (St Venant - Kirchoff material) :

$$(3.10)\quad W(S_o, \Gamma, \chi)=\frac{GA_1}{2}(\Gamma_1)^2 + \frac{GA_2}{2}(\Gamma_2)^2 + \frac{EA}{2}(\Gamma_3)^2 + \frac{EI_1}{2}(\chi_1)^2 + \frac{EI_2}{2}(\chi_2)^2 + \frac{GJ}{2}(\chi_3)^2$$

With these stress-strain relations, we may introduce the potential energy :

$$(3.11)\quad \pi(\varphi,t_i) = \int_0^L W\,(S_o,\Gamma,\chi)ds_o - \int_0^L (\bar{n}\cdot\varphi + \bar{m}\,.\,\theta)ds_o$$

Under dead loading assumptions, the reduced balance equation in weak form may be written : for any $(\delta\varphi,\delta t_i) \in \delta K(\varphi,t_i)$

$$(3.12)\quad \frac{\partial\pi}{\partial\varphi}\delta\varphi + \frac{\partial\pi}{\partial t_i}\quad \delta t_i = 0$$

4 Principles and notations of design sensitivity

The weak from of the balance equation stated in § 3 will be written here : find $(\varphi,t_i) \in K$ such that

$$(4.1)\quad \forall(\delta\varphi,\delta t_i) \in TK(\varphi,t_i) \quad G\,(\varphi_o;\,\varphi,\,t_i,\,\delta\varphi,\delta t_i) = 0$$

where we have added φ_o as a first argument ; it stands for a parametrization of the reference line of centroïds ; in fact this paragraph is devoted to general results so that here it can stand for any design variable as for example the area of the cross section ; it should be pointed out that the virtual displacements $(\delta\varphi,\delta t_i)$ appear linearly in this equation ;

moreover the tangent space $TK(\varphi, t_i)$ does not depend on φ_o as it is tangent to the curent configuration.

As usual we consider a functional J of the state of the structure :
$J(\varphi_o; \varphi, t_i)$ and we set

$$(4.2) \quad j(\varphi_o) = J(\varphi_o; \varphi, t_i)$$

Here (φ, t_i) stands for the solution of (4.1) ; it depends implicitely on φ_o ; $j(\varphi_o)$ takes into account the explicit and the implicit dependence on φ_o.

As a simple example we consider

$$(4.3) \quad J(\varphi_o; \varphi, t_i) = \int_0^L (\varphi - \varphi_o)^2 \, ds_o + \int_0^L (t_i - E_i)^2 \, ds_o$$

explicit dependence on φ_o appears in the first integral but also in the second as E_i is bound to the reference line of centroïds ; in § 3 we chose E_i to be the Frenet frame of this line it appears also in the arc length ds_o.

A second example is the potential energy π.

In the general case we assume the dependence with respect to φ_o to be smooth so that the chain rule gives :

$$(4.4) \quad \frac{dj}{d\varphi_o} \Delta\varphi_o = \frac{\partial J}{\partial \varphi_o} \Delta\varphi_o + \frac{\partial J}{\partial \varphi} \frac{\partial \varphi}{\partial \varphi_o} \Delta\varphi_o + \frac{\partial J}{\partial t_i} \frac{\partial t_i}{\partial \varphi_o} \Delta\varphi_o$$

To be valuable we need a formula where $\Delta\varphi_o$ appears explicitely whereas here $\frac{\partial \varphi}{\partial \varphi_o} \Delta\varphi_o$ and $\frac{\partial t_i}{\partial \varphi_o} \Delta\varphi_o$ are not known explicitely.

A standard practice is to differentiate the balance equation with respect to the design variable :

(4.5) $$\frac{\partial G}{\partial \varphi_o}\Delta\varphi_o + \frac{\partial G}{\partial \varphi}\frac{\partial \varphi}{\partial \varphi_o}\Delta\varphi_o + \frac{\partial G}{\partial t_i}\frac{\partial t_i}{\partial \varphi_o}\Delta\varphi_o = 0$$

to connect this differentiated balance equation with formula (4.4) we introduce a so called <u>adjoint equation</u> :

find $(d\psi, d\tau_i) \in TK(\varphi, t_i)$ such that

$\forall (d\varphi, dt_i) \in TK(\varphi, t_i)$

(4.6) $$\frac{\partial G}{\partial \varphi}(\varphi_o;\, \varphi,\, t_i,\, d\psi, d\tau_i)\, d\varphi + \frac{\partial G}{\partial t_i}(\varphi_o;\, \varphi,\, t_i,\, d\psi, d\tau_i)\, dt_i$$
$$= -\frac{\partial J}{\partial \varphi}(\varphi_o;\, \varphi,\, t_i)\, d\varphi - \frac{\partial J}{\partial t_i}(\varphi_o;\, \varphi,\, t_i)\, dt_i$$

It should be pointed out that the unknown $(d\psi, d\tau_i)$ appears linearly in this equation and it lies in the vector space $TK(\varphi, t_i)$; then we set in this adjoint equation $d\varphi = \frac{\partial \varphi}{\partial \varphi_o}\Delta\varphi_o$ and $dt_i = \frac{\partial t_i}{\partial \varphi_o}\Delta\varphi_o$ and in the differentiated balance equation we set $\delta\varphi = d\psi$ and $\delta t_i = d\tau_i$ substracting these two equations we obtain:

(4.7) $$\frac{\partial G}{\partial \varphi_o}(\varphi_o;\, \varphi,\, t_i,\, d\psi, d\tau_i) = \frac{\partial J}{\partial \varphi}(\varphi_o;\, \varphi,\, t_i)\frac{\partial \varphi}{\partial \varphi_o}\Delta\varphi_o + \frac{\partial J}{\partial t_i}(\varphi_o;\, \varphi,\, t_i)\frac{\partial t_i}{\partial \varphi_o}\Delta\varphi_o$$

so that we have obtained

(4.8) $$\frac{dj}{d\varphi_o}\Delta\varphi_o = \frac{\partial J}{\partial \varphi_o}\Delta\varphi_o + \frac{\partial G}{\partial \varphi_o}(\varphi_o;\, \varphi,\, t_i,\, d\psi, d\tau_i)\,\Delta\varphi_o$$

where (φ, t_i) is solution of the balance equation (4.1) and $(d\psi, d\tau_i)$ is solution of the adjoint equation (4.6).

The derivation is only formal ; a precise proof resting on the implicit function theorem should be written carefully ; the case of buckling deserves special attention ; in particular the adjoint equation which is linear could be non well posed in this case ; here we simply <u>assume</u> the adjoint equation to be well posed.

In the case where J is the potential energy (3.11)

(4.9) $j(\varphi_o) = \pi(\varphi_o; \varphi, t_i)$ using the balance equation we obtain (the adjoint state is zero)

$$(4.10) \quad \frac{\partial j}{\partial \varphi_o} \Delta\varphi_o = \frac{\partial \pi}{\partial \varphi_o} \Delta\varphi_o$$

We notice that this formula is totally identical to the one of the case of a linear balance equation.

Now, we <u>summarize</u> the <u>basic steps</u> of design sensitivity.

(i) solve the balance equation (4.1) :
find $(\varphi, t_i) \in K$ such that

$$\forall(\delta\varphi,\delta t_i) \in TK(\varphi,t_i) \quad G(\varphi_o; \varphi, t_i, \delta\varphi,\delta t_i) = 0$$

(ii) compute the functional (4.2)
$j(\varphi_o) = J(\varphi_o; \varphi, t_i)$

(iii) solve the adjoint equation (4.6) : find
$(d\psi,d\tau_i) \in TK(\varphi,t_i)$ such that

$\forall(d\varphi,dt_i) \in TK(\varphi,t_i)$

$$\frac{\partial G}{\partial \varphi}(\varphi_o; \varphi, t_i, d\psi,d\tau_i)\, d\varphi + \frac{\partial G}{\partial t_i}(\varphi_o; \varphi, t_i, d\psi,d\tau_i)\, dt_i$$
$$= - \frac{\partial J}{\partial \varphi}(\varphi_o; \varphi, t_i)\, d\varphi - \frac{\partial J}{\partial t_i}(\varphi_o; \varphi, t_i)\, dt_i$$

(i υ) the design sensitivity is given by (4.8) :

$$\frac{dj}{d\varphi_o} \Delta\varphi_o = \frac{\partial J}{\partial \varphi_o} \Delta\varphi_o + \frac{\partial G}{\partial \varphi_o}(\varphi_o; \varphi, t_i, d\psi,d\tau_i)\, \Delta\varphi_o$$

Remark that in the particular case of potential energy indicated before the summary, the solution of the adjoint equation is zero.

From this summary we can draw the following pratical conclusions.

In order to perform design sensitivity obviously one should first solve the balance equation (4.1) ; then he should solve the adjoint equation; under the assumption of well posedness this is an easy task because the left hand side is a linearisation of equation (4.1) ; if one solves equation (4.1) with a Newton type method, all the matrices needed to set up a linearized equation are already written and implemented ; the only new task is the right hand side. To get design sensitivity we need to compute the differential of the balance equation with respect to the design variable φ_o ; this differential is only to be taken for the explicit dependence in φ_o.

5 Shape derivative of the balance equation

5.1. INTRODUCTION

To make things precise, we assume the material to be hyperlastic ; we denote W $(\varphi_o; S_o, \tau, \chi)$ the elastic potential ; we have added the shape φ_o as a first argument.

From § 3, the balance equation may be written

$$\forall(\delta\varphi,\delta t_i) \in TK(\varphi,t_i) \quad G\,(\varphi_o;\, \varphi,\, t_i,\, \delta\varphi,\delta t_i) = 0$$

with

$$G\,(\varphi_o;\, \varphi,\, t_i,\, \delta\varphi,\delta t_i) = \frac{\partial\pi}{\partial\varphi}\,(\varphi_o;\, \varphi,\, t_i)\,\delta\varphi + \frac{\partial\pi}{\partial t_i}\,(\varphi_o;\, \varphi,\, t_i)\,\delta t_i$$

where the potential energy is given by

$$\pi(\varphi_o;\varphi, t_i) = \int_0^L W(\varphi_o;\ S_o, \Gamma, \chi)\, ds_o -$$

$$\int_0^L (\bar{n} \cdot \varphi + \bar{f}_\alpha \cdot t_\alpha) ds_o$$

so that

$$G\,(\varphi_o;\ \varphi,\ t_i,\ \delta\varphi, \delta t_i) = \int_0^L (N \cdot \delta\Gamma + M \cdot \delta\chi)\, ds_o \quad -$$

$$\int_0^L (\bar{n} \cdot \delta\varphi + \bar{m} \cdot \theta)\, ds_o$$

with $N = \dfrac{\partial W}{\partial \Gamma}$ $M = \dfrac{\partial W}{\partial \chi}$ and $\bar{m} = t_\alpha \times f_\alpha$

In the case of a quadratic elastic potential we get

$$G\,(\varphi_o;\ \varphi,\ t_i;\ \delta\varphi, \delta t_i) = \int_0^L [\,G\,A\,\alpha\ \Gamma_\alpha\ \delta\Gamma_\alpha + E\,A\ \ \Gamma_3\ \delta\Gamma_3$$

$$+ EI_\alpha\ \chi_\alpha\ \delta\chi_\alpha + GJ\ \chi_3\ \delta\chi_3]\ ds_o$$

In any case we have the following relations for the <u>strain measures</u>

$$\Gamma_\alpha = t_\alpha \cdot \frac{d\varphi}{ds_o} \qquad\qquad \Gamma_3 = t_3 \cdot \frac{d\varphi}{ds_o} - 1$$

$$\chi_j = \frac{1}{2}\ e_{jkl} \left(\frac{dt_k}{ds_o} \cdot t_l - \frac{dE_k}{ds_o} \cdot E_l \right)$$

or more precisely using the Serre-Frenet formulas (see §3)

$$\chi 1 = \frac{1}{2}\ e\ 1\ k\ l\ \frac{dt_k}{ds_o}\ .\ t_l$$

$$\chi 2 = \frac{1}{2}\ e\ 2\ k\ l\ \frac{dt_k}{ds_o}\ .\ t_l\ -\ \frac{1}{\rho_0}$$

$$\chi 3 = \frac{1}{2}\ e\ 3k\ l\ \frac{dt_k}{ds_o}\ .\ t_l\ -\ \frac{1}{\tau_0}$$

here $e\ j\ k\ l$ is the orientation tensor :

$$E_j \times E_k = e\ j\ k\ l\ \ E_l$$

From (3.6), (3,7) we have

$$\delta\Gamma_k = t_k\ .\ \hat{\delta}\,\gamma\ = t_k\ .\ \frac{d\delta\varphi}{ds_o}\ -\ t_k\ .\ (\theta \times \frac{d\varphi}{ds_o})$$

$$\equiv t_k\ .\ \frac{d\delta\varphi}{ds_o} + \frac{d\varphi}{ds_o}\ .\ (\theta \times\ t_k)$$

and

$$\delta\chi_k = t_k\ .\ \hat{\delta}\,\omega = t_k\ .\ \frac{d\theta}{ds}$$

5.2. PERTUBATION OF STRAIN MEASURES

We notice that the undeformed line of centroïds that we consider as design variable appears in Γ^k and χ^k in two ways : firstly with derivatives with respect to the curivilinear abscissa s_o , secondly by the Frenet frame of this line.

The first case is adressed in the following elementary.

<u>Lemma</u> 5.1

If the curve $\xi \rightarrow \varphi_o\ (\xi)$ is changed in $\xi \rightarrow \varphi_o\ \ (\xi) + \Delta\varphi_o\ (\xi)$ then the variation of the length of its tangent vector $\varphi'\ (\xi_o)$ is given by

$$\Delta \left\| \frac{d\varphi_o}{d\xi} \right\| =$$

$$\left\| \frac{d\varphi_o}{d\xi} \right\| \frac{d\varphi_o}{ds_o} \cdot \frac{d\Delta\varphi_o}{ds_o} \equiv \left\| \frac{d\varphi_o}{d\xi} \right\| E3 \cdot \frac{d\Delta\varphi_o}{ds_o}$$

and

$$\Delta \left\| \frac{d\varphi_o}{d\xi} \right\|^{-1} = - \left\| \frac{d\varphi_o}{d\xi} \right\|^{-1} \frac{d\varphi_o}{ds_o} \cdot \frac{d\Delta\varphi_o}{ds_o}$$

$$\equiv - \left\| \frac{d\varphi_o}{d\xi} \right\|^{-1} E3 \cdot \frac{d\Delta\varphi_o}{ds_o}$$

or more concisely :

$$\Delta\, ds_o = \left(\frac{d\varphi_o}{ds_o} \cdot \frac{d\Delta\varphi_o}{ds_o}\right) ds_o \qquad \Delta \frac{d}{ds_o} = - \left(\frac{d\varphi_o}{ds_o} \cdot \frac{d\Delta\varphi_o}{ds_o}\right) \frac{d}{ds_o}$$

$$\equiv \left(E3 \cdot \frac{d\Delta\varphi_o}{ds_o}\right) ds_o \qquad \equiv - \left(E3 \cdot \frac{d\Delta\varphi_o}{ds_o}\right) \frac{d}{ds_o}$$

The <u>proof</u> is elementary and is based on

$$\Delta \left(\frac{d\varphi_o}{d\xi} \cdot \frac{d\varphi_o}{d\xi}\right) = 2 \frac{d\varphi_o}{d\xi} \cdot \Delta \frac{d\varphi_o}{d\xi} = 2 \frac{d\varphi_o}{d\xi} \cdot \frac{d\Delta\varphi_o}{d\xi}$$

the consise formulas use the following relations :

$$ds_o = \left\| \frac{d\varphi_o}{d\xi} \right\| d\xi \quad \text{and} \quad \frac{d}{ds_o} = \left\| \frac{d\varphi_o}{d\xi} \right\|^{-1} \frac{d}{d\xi}$$

We get easily the following corollary.

Corollary 5.2

(i)

$$\Delta\frac{d\varphi}{ds_o} = -(\frac{d\varphi_o}{ds_o} \cdot \frac{d\Delta\varphi_o}{ds_o})\frac{d\varphi}{ds_o} \equiv -(E_3 \cdot \frac{d\Delta\varphi_o}{ds_o})\frac{d\varphi}{ds_o}$$

(ii)

$$\Delta\frac{dt_k}{ds_o} = -(\frac{d\varphi_o}{ds_o} \cdot \frac{d\Delta\varphi_o}{ds_o})\frac{dt_k}{ds_o} \equiv -(E_3 \cdot \frac{d\Delta\varphi_o}{ds_o})\frac{dt_k}{ds_o}$$

(iii) $\Delta\Gamma_k = t_k \cdot \Delta\dfrac{d\varphi}{ds_o}$ $\qquad 1 \le k \le 3$ or

(i υ) $\Delta\Gamma_\alpha = -(E_3 \cdot \dfrac{d\Delta\varphi_o}{ds_o})\Gamma_\alpha \qquad 1 \le \alpha \le 2$

(υ) $\Delta\Gamma_3 = -(E_3 \cdot \dfrac{d\Delta\varphi_o}{ds_o})(1 + \Gamma_3)$

To obtain the perturbation of χ_j with respect to φ_o we need also the perturbation of the Frenet frame E_k; the following lemma deals with the unit tangent vector E_3 :

Lemma 5.3

(i) $\Delta E_3 = \dfrac{d\Delta\varphi_o}{ds_o} - (E_3 \cdot \dfrac{d\Delta\varphi_o}{ds_o})E_3 = (E_\alpha \cdot \dfrac{d\Delta\varphi_o}{ds_o})E_\alpha$

(ii)

$$\Delta\frac{dE_3}{ds_o} = \frac{d}{ds_o}\Delta E_3 - (E_3 \cdot \frac{d\Delta\varphi_o}{ds_o})\frac{dE_3}{ds_o}$$

$$\equiv -\frac{2}{\rho_o}(E_3 \cdot \frac{d\Delta\varphi_o}{ds_o})E_1 - \frac{1}{\rho_o}(E_1 \cdot \frac{d\Delta\varphi_o}{ds_o})E_3$$

$$+ (E_\alpha \cdot \frac{d^2\Delta\varphi_o}{ds_o^2})E_\alpha$$

To start the proof we notice that

$$E_3 = \frac{d\varphi_o}{ds_o} = \left|\left|\frac{d\varphi_o}{d\xi}\right|\right|^{-1}\frac{d\varphi_o}{d\xi}$$

so that

$$\Delta E_3 = (\Delta \left\| \frac{d\varphi_o}{d\xi} \right\|^{-1}) \frac{d\varphi_o}{d\xi} + \left\| \frac{d\varphi_o}{d\xi} \right\|^{-1} \Delta \frac{d\varphi_o}{d\xi}$$

from which the first equality of (i) is deduced ; for the second equality we notice

$$E_\alpha \cdot \Delta E_\alpha = E_\alpha \cdot \frac{d\Delta\varphi_o}{ds_o} \text{ and } E_3 \cdot \Delta E_3 = 0$$

The same method gives the first equality of (ii) ; then as

$$\frac{dE_1}{ds_o} = -\frac{1}{\rho_o} E_3 + \frac{1}{\tau_o} E_2 \quad \text{and}$$
$$\frac{dE_2}{ds_o} = -\frac{1}{\tau_o} E_1$$

we get from the second formula of (i)

$$\frac{d\Delta E_3}{ds_o} = [(-\frac{1}{\rho_o} E_3 + \frac{1}{\tau_o} E_2) \cdot \frac{d\Delta\varphi_o}{ds_o}] E_1$$

$$+ (E_1 \cdot \frac{d^2\Delta\varphi_o}{ds_o^2}) E_1 + (E_1 \cdot \frac{d\Delta\varphi_o}{ds_o})(\frac{-1}{\rho_o} E_3 + \frac{1}{\tau_o} E_2)$$

$$+ (-\frac{1}{\tau_o} E_1 \cdot \frac{d\Delta\varphi_o}{ds_o}) E_2 + (E_2 \cdot \frac{d^2\Delta\varphi_o}{ds_o^2}) E_2$$

$$- (E_2 \cdot \frac{d\Delta\varphi_o}{ds_o}) \frac{1}{\tau_o} E_1$$

$$\frac{d\Delta E_3}{ds_o} = \frac{-1}{\rho_o} (E_3 \cdot \frac{d\Delta\varphi_o}{ds_o}) E_1 - \frac{1}{\rho_o} (E_1 \cdot \frac{d\Delta\varphi_o}{ds_o}) E_3$$

$$+ (E_\alpha \cdot \frac{d^2\Delta\varphi_o}{ds_o^2}) E_\alpha$$

this is the second equality of (ii).

Now we can obtain variation of curvature

Corollary 5.4

$$\Delta\chi_2 = \Delta\frac{1}{\rho_o} = E_1 \cdot \frac{d\Delta E_3}{ds_o} - \frac{1}{\rho_o}(E_3 \cdot \frac{d\Delta\varphi_o}{ds_o})$$

$$= E_1 \cdot \frac{d^2\Delta\phi_0}{ds_o{}^2} - \frac{2}{\rho_o}(E_3 \cdot \frac{d\Delta\varphi_o}{ds_o})$$

The proof is based on previous lemma.

Now we can state the perturbations of E_1, E_2.

Lemma 5.5

(i) $\Delta E_1 = \rho_o(E_2 \cdot \frac{d^2\Delta\phi_0}{dS_o{}^2})E_2 + (E_1 \cdot \frac{d\Delta\varphi_o}{ds_o}) E_3$

(ii) $\Delta E_2 = -\rho_o(E_2 \cdot \frac{d^2\Delta\phi_0}{ds_o{}^2})E_1 - (E_2 \cdot \frac{d\Delta\varphi_o}{ds_o}) E_3$

(iii) $\Delta\frac{dE_2}{ds_o} = \frac{1}{\tau_o}(E_3 \cdot \frac{d\Delta\varphi_o}{ds_o}) E_1 + \frac{d\Delta E_2}{ds_o}$

Corollary 5.6

(i) $\Delta\chi_3 = \Delta\frac{1}{\tau_o} = -\tau_o \frac{dE_2}{ds_o} \cdot \Delta\frac{dE_2}{ds_o}$

The proof of this lemma and corollary are based on previous results and on the orthonormality of the basis E_i.

To end this paragraph, it remains to compute perturbation with respect to the reference line of centroids of virtual variations of strain measures $\delta\Gamma_k$ and $\delta\chi_k$; from formulas of §5.1 it appears that the reference line of centroids appears only through the derivatives with respect to S_0 ; so a straightforward use of lemma 5.1 gives

Corollary 5.7

(i) $$\Delta\,\delta\,\Gamma_k = -\left(\left(t_k \cdot \frac{d\delta\varphi}{ds_o} \right) + \frac{d\varphi}{ds_o} \cdot (\theta \times t_k) \right) \left(E_3 \cdot \frac{d\Delta\varphi_o}{ds_o}\right)$$

(i) $$\Delta\,\delta\,\chi_k = -\left(t_k \cdot \frac{d\theta}{ds_o} \right)\left(E_3 \cdot \frac{d\Delta\varphi_o}{ds_o}\right)$$

5.3 COMPUTATION OF THE SHAPE DERIVATIVE OF THE BALANCE EQUATION.

After having obtained the perturbations of strain measures we turn now to what appeared to be the main step of design sensitivity at the end of §4: the derivative of the left hand side of the balance equation: $\frac{\partial G}{\partial \varphi_o}(\varphi_o; \varphi, t_i, \delta\varphi, \delta t_i)\, \Delta\varphi_o$.

After the technical results of §5.2 it is now an easy task.In the case of an hyperelastic material the perturbation of stress resultants and couples are straightforward:

$$N = \frac{\partial W}{\partial \Gamma} \quad M = \frac{\partial W}{\partial \chi} \quad \text{gives}$$

$$\Delta N = \frac{\partial^2 W}{\partial \Gamma^2} \Delta\Gamma + \frac{\partial^2 W}{\partial \Gamma \partial \chi} \Delta\chi$$

$$\Delta M = \frac{\partial^2 W}{\partial \chi \partial \Gamma} \Delta\Gamma + \frac{\partial^2 W}{\partial \chi^2} \Delta\chi$$

In the particular case of a quadratic elastic potential as

$$N_\alpha = G\, A_\alpha\, \Gamma_\alpha \quad N_3 = E\, A\Gamma_3$$

$$M_\alpha = E\, I_\alpha\, \chi_\alpha \quad M = G\, J\, \chi_3$$

we get simply

$$\Delta N_\alpha = \frac{\partial^2 W}{\partial \Gamma_\alpha^2} = G A_\alpha \, \Delta \Gamma_\alpha \qquad \Delta N_3 = \frac{\partial^2 W}{\partial \Gamma_3^2} = E A \, \Delta \Gamma_3$$

$$\Delta M_\alpha = E I_\alpha \, \Delta \chi_\alpha \qquad \Delta M_3 = G J \, \Delta \chi_3$$

where $\Delta \Gamma_k$ is given in lemma 5.2 and $\Delta \chi_k$ in corollary 5.4 and 5.5 ; so we have obtained the following result.

Proposition 5.8

If the datas are smooth enough we have

$$\frac{\partial G}{\partial \varphi_o}(\varphi_o; \varphi, t_i, \delta\varphi, \delta t_i)\, \Delta\varphi_o =$$

$$\int_0^L (\Delta N \,.\, \delta\Gamma + \Delta M \,.\, \delta\chi + N \,.\, \Delta\delta\Gamma + M \,.\, \Delta\delta\chi)\, ds_o \;+$$

$$\int_0^L (N \,.\, \delta\Gamma + M \,.\, \delta\chi)\, \Delta ds_o$$

$$- \int_0^L (\Delta \bar{n} \,.\, \delta\varphi + \Delta \bar{m} \,.\, \theta)\, ds_o \;-\; \int_0^L (\bar{n} \,.\, \delta\varphi + \bar{m} \,.\, \theta)\, \Delta ds_o$$

where ΔN and ΔM are expressed above and the variations of $\Gamma, \chi, \delta\Gamma$ and $\delta\chi$ are given in lemmas and corollary of §5.2 ; but the variations of the applied loads $\Delta \bar{n}$ and $\Delta \bar{m}$ should be considered as problem dependent datas.

6. Conclusion

If one uses finite elements to solve the balance equation of this finite strain rod model, the performance of the basic steps of design sensitivity described at the end of §4 could be implemented in a similar way as in the linear case; see e.g. Rousselet (1987).

If we compare with the design sensitivity of a plane arch with linear elastic assumptions (Chenais- Rousselet -Benedict 1988), it appears that the method is nearly as complex as for this highly non linear model. I believe that this is due to the fact that this model is well expressed in a continuous setting so that it seems quite attractive for shape sensitivity. Design sensitivity of an engineering model where the finite element approximation is mixed up with the numerical method which solves the balance equation, would lead to a more cumbersome approach.

Theoretical foundations and numerical implementation of this approach are under study.

REFERENCES

Antman, The theory of rod, Handbuch der Physic, VIa/2, Springer Verlag, 1972.

Antman & Kenney Large buckled states of nonlinearly elastic rods under torsion, thrust and gravity, pp289-238, Arch. Rat. Mech. Anal.,1981.

Bourgat & Le Tallec & Mani Modélisation et calcul des grands déplacements de tuyaux élastiques en flexion-torsion, pp379-408, J. Méca. Théor. Appli., 1988.

Chenais & Rousselet & Benedict, Design sensitivity for arch structures with respect to midsurface shape under static loading, pp 225-239 ,J. Opt. th. & appl., 1988.

Love, The mathematical theory of elasticity, Dover, New York, 1944.

Rousselet, Shape design sensitivity , from partial differential equation to implementation, pp151-171, Eng. Opt., 1987.

Simo A finite strain beam formulation. the three dimensional dynamic problem. Part I, pp55-70, Comp. Meth. Appl. Mech. Engrg., 1985.

Simo & Vu-Quoc A three dimensional finite-strain rod model. PartII: computational aspects, pp 79-116, (1986).

Design sensitivity of critical loads and vibration frequencies of nonlinear structures

Zenon Mróz

Institute of Fundamental Technological Research,
Polish Academy of Sciences, Warsaw, Poland

ABSTRACT: General design sensitivity formalism is introduced for non linear elastic structures reaching critical equilibrium states. The discretized formulation is first considered and the sensitivity analysis is discussed for regular and critical states including analysis of post-critical behaviour. Next, the variational approach is presented for beam and surface structures. The design sensitivity of critical loads and vibration frequencies is derived. Some illustrative examples are presented.

1 INTRODUCTION

The present article summarizes recent research concerned with sensitivity analysis of non-linear structures in regular and critical states. Usually a non-linear elastic structure first passes through a regular deformation range associated with stable and unique response. When at some load value the critical state (limit of bifurcation point) is reached, the typical problem is to determine the critical load and mode, study post-critical behaviour and also the variation of critical load with structural parameter or imperfection variation. For some cases, the character of critical state is preserved when small design variation occurs. In other cases, however, the design variation may induce disappearance of critical point or variation of its character. We shall call such cases regular or singular design modifications. Our analysis will follow the previous work by Mróz (1987), Mróz and Haftka (1988,1992) Szefer, Mróz and Demkowicz (1988), Cohen and Haftka (1989), Haftka, Cohen and Mróz (1990), Dems and Mróz (1990) on sensitivity of buckling loads and vibrations frequencies of plates and shells with respect to variation of stiffness parameters and shape.

2 SENSITIVITY ANALYSIS FOR REGULAR STATES

Consider an elastic discretized structure whose deformation is described by a set of generalized coordinates q_i and whose potential energy has the form

G. I. N. Rozvany (ed.), Optimization of Large Structural Systems, Vol. I, 455–476.

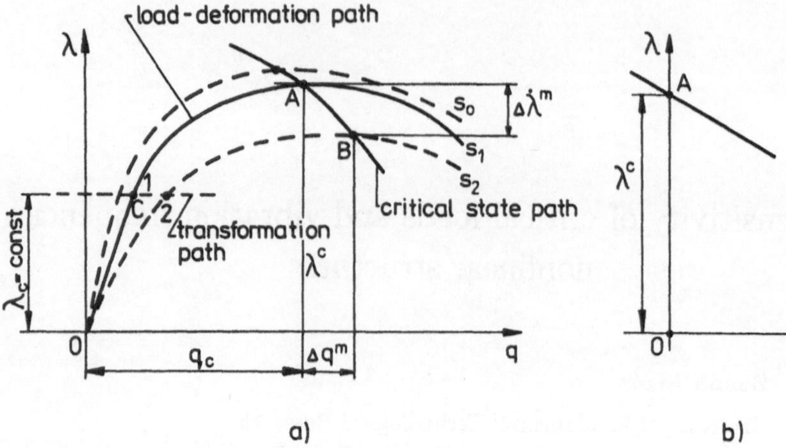

Figure 1: **a) Load-deformation path of structure passing through limit point critical state path and transformation path, b) Sensitivity diagram**

$$V = V(q_k, \lambda, s) = 0 \quad , \qquad i = 1, 2, ...n \tag{1}$$

where λ denotes the loading parameter, s denotes the design or imperfection parameter. The equilibrium equations are generated from (1)

$$\frac{\partial V}{\partial q_i} = V_i(q_k, \lambda, s) = 0 \qquad , \qquad i = 1, 2, ...n \tag{2}$$

Consider the equilibrium path in the $n+1$ load-configuration space, Fig. 1. Denote the progression parameter along equilibrium state by η, thus $\mathbf{q} = \mathbf{q}(\eta)$. Then, at any equilibrium state $q^0(\eta_0)$, $\lambda^0(\eta^0)$ we can write

$$\begin{aligned} q_i &= q_i^0 + \dot{q}_i \Delta\eta + \frac{1}{2}\ddot{q}_i \Delta\eta^2 + \cdots \\ \lambda &= \lambda^0 + \dot{\lambda}\Delta\eta + \frac{1}{2}\ddot{\lambda}\Delta\eta^2 + \cdots \end{aligned} \tag{3}$$

where $\dot{q}_i, \ddot{q}_i, ...\dot{\lambda}, \ddot{\lambda}, ..$ denote derivatives of q_i and λ with respect to η at $\mathbf{q} = \mathbf{q^0}$, $\lambda = \lambda^0$, $\eta = \eta^0$ and $\Delta\eta = \eta - \eta^0$. Equations (3) specify the load-deformation process in the vicinity of a considered equilibrium state q_i^0,η_0. On the other hand, for varying s and constant λ we have the structure transformation process. It can therefore be assumed

$$s = s^0 + \dot{s}\Delta\eta + \frac{1}{2}\ddot{s}\Delta\eta^2 + \cdots \tag{4}$$

where $\dot{s}, \ddot{s}, \ldots$ denote derivatives of s with respect to η at $\eta = \eta_0$. Differentiating (2) one obtains a set of identity relations expressing equilibrium conditions associated with variation of configuration, loading and of structure parameters, namely

$$V_{ij}\dot{q}_j + V_{i\lambda}\dot{\lambda} + V_{is}\dot{s} = 0 \tag{5}$$

$$\begin{aligned} &V_{ijk}\dot{q}_j\dot{q}_k + 2V_{ij\lambda}\dot{\lambda}\dot{q}_j + 2V_{ijs}\dot{s}\dot{q}_j + 2V_{is\lambda}\dot{s}\dot{\lambda} + V_{i\lambda\lambda}\dot{\lambda}^2 + V_{iss}\dot{s}^2 \\ &+ V_{ij}\ddot{q}_j + V_{i\lambda}\ddot{\lambda} + V_{is}\ddot{s} = 0 \end{aligned} \tag{6}$$

and the third order perturbation equation

$$\begin{aligned} &V_{ijkl}\dot{q}_j\dot{q}_k\dot{q}_l + 3V_{ijk}\dot{q}_j\ddot{q}_k + V_{ij}\dddot{q}_j + 3V_{ijk\lambda}\dot{q}_j\dot{q}_k\dot{\lambda} \\ &3V_{ijks}\dot{q}_j\dot{q}_k\dot{s} + 3V_{ij\lambda}\ddot{q}_j\dot{\lambda} + 3V_{ijs}\ddot{q}_j\dot{s} + 3V_{ij\lambda}\dot{q}_j\ddot{\lambda} + \\ &3V_{ijs}\dot{q}_j\ddot{s} + V_{i\lambda}\dddot{\lambda} + V_{is}\dddot{s} + 3V_{ij\lambda\lambda}\dot{q}_j\dot{\lambda}^2 + 3V_{ijss}\dot{q}_j\dot{s}^2 + \\ &6V_{ij\lambda s}\dot{q}_j\dot{\lambda}\dot{s} + 3V_{i\lambda\lambda}\dot{\lambda}\ddot{\lambda} + 3V_{iss}\dot{s}\ddot{s} + 3V_{i\lambda s}\ddot{\lambda}\dot{s} + 3V_{i\lambda s}\dot{\lambda}\ddot{s} + \\ &3V_{is\lambda\lambda}\dot{\lambda}^2\dot{s} + 3V_{iss\lambda}\dot{\lambda}\dot{s}^2 + V_{i\lambda\lambda\lambda}\dot{\lambda}^3 + V_{isss}\dot{s}^3 = 0 \end{aligned} \tag{7}$$

This set of perturbation equations allows for study of both deformations response, identification of critical points and also of sensitivity to imperfection and design variations.

2.1 Incremental load-deformation response

Let us first discuss the regular case and assume that

$$V_{ij}\dot{q}_j \neq 0 \qquad or \qquad det\,[V_{ij}] \neq 0 \tag{8}$$

With $\dot{s} = 0$, Eq. (5) provides the incremental equilibrium conditions along the load path

$$V_{ij}q'_j + V_{i\lambda}\lambda' = 0 \tag{9}$$

where $V_{ij}(q_k^0, \lambda^0, s^0)$ is the tangent stiffness matrix, and the derivatives with respect to loading parameters are denoted with primes instead of dots. When V_{ij} is positive definite, the incremental problem (9) is associated with the minimum principle of the incremental potential energy function

$$W(q'_i, \lambda') = \frac{1}{2}V_{ij}\,q'_i\,q'_j + V_{i\lambda}\,q'_i\,\lambda' \tag{10}$$

In fact, the stationary condition for W leads to (9), that is

$$\delta W = (V_{ij}\,q'_j + V_{i\lambda})\,\delta q'_i = 0 \tag{11}$$

where $\delta q_i'$ denotes a kinematically admissible variation. The absolute minimum of W occurs in the class of kinematically admissible q_i^k, thus in view of (11)

$$W(q_i'^k, \lambda') - W(q_i', \lambda') = \frac{1}{2} V_{ij} q_i'^k q_j'^k - V_{ij} q_i' q_j' + V_{i\lambda}(q_i'^k - q_i') =$$
$$V_{ij}(q_i'^k - q_i')(q_j'^k - q_j') \geq 0 \quad (12)$$

2.2 Incremental transformation response. Sensitivity analysis

Consider now the transformation path occurring at fixed λ but with varying design or imperfection parameter s. We denote the derivative of $\mathbf{q}$ with respect to s by $\mathbf{q_s}$, and then from (5) it follows that

$$V_{ij} q_{sj} + V_{is} = 0 \quad (13)$$

The solution of (13) provides the sensitivity $\mathbf{q_s}$. Consider now an analytical function

$$G = G(\mathbf{q}, s) = G(\mathbf{q}(s), s) \quad (14)$$

The variation of G along the transformation path can be presented as follows:

$$G = G_O + \dot{G}\Delta s + \frac{1}{2}\ddot{G}\Delta s^2 + \frac{1}{2}\dddot{G}\Delta s^3 + \ldots \quad (15)$$

where

$$\dot{G} = G_i q_{si} + G_s \quad , \quad \ddot{G} = G_{ij} q_{si} q_{sj} + 2G_{is} q_{si} + G_{ss} + G_i q_{ssi} + \ldots \quad (16)$$

and q_{ssi}, the second derivative of the displacement with respect to s, can be determined from (5), specialized to the case of $\eta = s$

$$V_{ij} q_{ssj} + V_{ijk} q_{sj} q_{sk} + 2V_{ijs} q_{sj} + V_{iss} = 0 \quad (17)$$

Consider now the adjoint method. We derive the variation of G due to variation of s subject to the equilibrium constraint (2). Assuming λ as constant, the augmented function is

$$\bar{G} = G(q_j, s) - \mu_i V_i(q_j, s) \quad (18)$$

where $\boldsymbol{\mu}$ is a Lagrange multiplier vector. The variation of G is expressed as follows:

$$\dot{\bar{G}} = G_j q_{sj} + G_s - \mu_i V_{ij} q_{sj} - \mu_i V_{is} = (-V_{ij}\mu_i + G_j) q_{sj} + G_s - \mu_i V_{is} \quad (19)$$

In order to eliminate the (computationally expensive) term q_{sj} in (19) we require the adjoint structure to satisfy

$$V_{ij}\mu_i - G_j = 0 \quad (20)$$

which requires a solution with the tangential stiffness matrix V_{ij}. Now the first order sensitivity G can be expressed as

$$\dot{G} = \dot{\tilde{G}} = G_s - \mu_i V_{is} \tag{21}$$

Instead of direct determination of q_{sj} from (13) followed by calculation of G_1 from (16), we may use the adjoint state μ_i from (20) and calculate G_1 from (21). The adjoint method is efficient when we denote the derivative of G with respect to many variables, since only one adjoint solution is needed. The expression for $\ddot{G}$ in (16) requires the calculation of the second derivative field q_{ss}. Using the adjoint method we can eliminate this term. We start by specializing (6) to the transformation path and multiply by μ_i to obtain

$$V_{ijk}\mu_i q_{sj} q_{sk} + 2V_{ijs}\mu_i q_{sj} + V_{iss}\mu_i + V_{ij}\mu_i q_{ssj} = 0 \tag{22}$$

Using (20), the last term in (22) is equal to $G_j q_{ssj}$, and then $\ddot{G}$ from (16) may be written as

$$\ddot{G} = (-V_{ijk}\mu_i q_{sk} - 2V_{ijs}\mu_i + G_{ij} q_{si} + G_{js}) q_{sj} + G_{ss} - V_{iss}\mu_i \tag{23}$$

3 SENSITIVITY ANALYSIS FOR CRITICAL STATES

Consider now critical state satisfying the condition

$$V_{ij} q_{1j} = 0 \qquad or \qquad det[V_{ij}] = 0 \tag{24}$$

where q_{1j} is the eigenvector of V_{ij} associated with the zero eigenvalue (assumed to be a simple eigenvalue). Multiplying (9) by q_{1i} we obtain

$$q_{1j} V_{i\lambda} \lambda' = 0 \tag{25}$$

so that either $V_{i\lambda} q_{1i} = 0$, or $\lambda' = 0$, at the critical state. The first case corresponds to bifurcation and the second to limit point.

3.1 Limit point: post-critical response and sensitivity

For a limit point we have $\lambda' = 0$ as well as (24). Also for this case $q_{1i} = q_i'$. Setting $\dot{s} = 0$ in (6) and using primes to denote load path derivatives we get

$$V_{ijk} q_j' q_k' + 2V_{ij\lambda}\lambda' q_j' + V_{i\lambda\lambda}\lambda'^2 + V_{ij} q_j'' + V_{i\lambda}\lambda'' = 0 \tag{26}$$

Multiplying (26) by q_{1i} and evaluating at the limit point ($\lambda = \lambda_c, \lambda' = 0, q_i' = q_{1i}$) we obtain in view of (24)

$$\lambda_c'' = -\frac{V_{ijk} q_{1i} q_{1j} q_{1k}}{V_{i\lambda} q_{1i}} \tag{27}$$

Design or imperfection sensitivity can be calculated from (5) and (6) by considering the *critical – statepath* following limit point. Along that path we can have $s = \eta$ so that $\dot{s} = 1$, $\ddot{s} = 0$, etc. However, now both λ and s change simultaneously. Multiplying (5) by q_{1i} we can solve for $\dot{\lambda}$ which is the sensitivity of λ_c with respect to s

$$\lambda_{cs} = \dot{\lambda} = -\frac{V_{is}q_{1i}}{V_{i\lambda}q_{1i}} \tag{28}$$

Similarly, by multiplying (6) by q_{1i} one obtains

$$\lambda_{css} = \ddot{\lambda} = -\frac{(V_{ijk}\dot{q}_j\dot{q}_k + 2V_{ij\lambda}\dot{q}_j + 2V_{ijs}\dot{q}_j + 2V_{i\lambda s}\dot{\lambda} + V_{i\lambda\lambda}\dot{\lambda}^2 + V_{iss})q_{1i}}{V_{i\lambda}q_{1i}} \tag{29}$$

To evaluate $\dot{\mathbf{q}}$ appearing in (29) we need to solve Eq. (5) which is singular at $\lambda = \lambda_c$. Equation (28) provides the consistency condition guaranteeing that (5) has a solution at $\lambda = \lambda_c$. However, we need one additional equation to make the solution of (5) unique. This extra conditions is provided by differentiating (24) along the critical path to obtain

$$V_{ijk}q_{1j}\dot{q}_k + V_{ij\lambda}q_{1j}\dot{\lambda} + V_{ijs}q_{1j} + V_{ij}\dot{q}_{1j} = 0 \tag{30}$$

Multiplying by q_{1i} we have

$$V_{ijk}q_{1i}q_{1j}\dot{q}_k + V_{ij\lambda}q_{1i}q_{1j}\dot{\lambda} + V_{ijs}q_{1i}q_{1j} = 0 \tag{31}$$

which together with (5) provides the solution for $\dot{\mathbf{q}}$. Equation (30) can also be used to find $\dot{q}_{1j}$, the derivative of the limit-load eigenvector. However, this equation is also singular and must be supplemented by another relation derived from the normalized condition for the eigenvector, namely

$$T_{ij}q_{1i}q_{1j} = 1 \tag{32}$$

where T_{ij} is a positive definite matrix, then by differentiating (32) we obtain

$$T_{ij}q_{1j}\dot{q}_{1i} = 0 \tag{33}$$

which is an orthogonality condition on $\dot{\mathbf{q}}_1$.

3.2 Bifurcation point: post-critical behavior

Consider now the bifurcation point for which the following conditions are satisfied

$$V_{ij}q_{1j} = 0, \qquad V_{i\lambda}q_{1i} = 0 \tag{34}$$

We denote the generalized displacements along the fundamental loading path by $\mathbf{q}_0$, and those along the post critical path by $\mathbf{q}$, so that after bifurcation

$$q_j = q_{0j} + \Delta\eta q_{1j} + \frac{1}{2}\Delta\eta^2 q_{2j} + \frac{1}{6}\Delta\eta^3 q_{3j} + \cdots \tag{35}$$

where η is the post-critical path parameter. We assume that q_{0j} is evaluated at the same load as q_j which brings in an indirect dependence of q_{0j} on η. This is

$$\begin{aligned} \dot{q}_j &= q'_{0j}\dot{\lambda} + q_{1j} + \Delta\eta q_{2j} + \tfrac{1}{2}\Delta\eta^2 q_{3j} + \cdots \\ \ddot{q}_j &= q''_{0j}\dot{\lambda}^2 + q_{0j}\ddot{\lambda} + q_{2j} + \Delta\eta q_{3j} + \cdots \end{aligned} \tag{36}$$

where primes denote derivatives of the prebuckling state with respect to the load. The equations to obtain q'_{0j} and q''_{0j} are obtained from (5) and (6) by setting the path parameter to be λ

$$V^c_{ij}q'_{0j} + V^c_{i\lambda} = 0 \tag{37}$$

$$V^c_{ijk}q'_{0j}q'_{0k} + 2V^c_{ij\lambda}q'_{0j} + V^c_{i\lambda\lambda} + V^c_{ij}q''_{0j} = 0 \tag{38}$$

where the superscript c denotes the bifurcation (critical) point. Since V^c_{ij} is singular q'_{0j} cannot be completely evaluated from Eq. (37), and an additional condition is required. This is obtained by multiplying Eq. (38) by q_{1i}

$$V^c_{ijk}q_{1i}q'_{0j}q'_{0k} + 2V^c_{ij\lambda}q_{1i}q'_{0j} + V^c_{i\lambda\lambda}q_{1i} = 0 \tag{39}$$

We can also write (6) for the post-critical path parameter, setting $\dot{s} = \ddot{s} = 0$ and using (36) with $\Delta\eta = 0$ to obtain

$$\begin{aligned} &V^c_{ijk}(\dot{\lambda}q'_{0j} + q_{1j})(\dot{\lambda}q'_{0k} + q_{1k}) + 2V^c_{ij\lambda}(\dot{\lambda}q'_{0j} + q_{1j})\dot{\lambda} + \\ &V_{i\lambda\lambda}\dot{\lambda}^2 + V^c_{ij}(\dot{\lambda}^2 q''_{0j} + \ddot{\lambda}q'_{0j} + q_{2j}) + V^c_{i\lambda}\ddot{\lambda} = 0 \end{aligned} \tag{40}$$

We now substract $\dot{\lambda}_2$ times (38) from (40) and use (37) to obtain

$$V^c_{ijk}q_{1j}q_{1k} + 2\dot{\lambda}V^c_{ijk}q'_{0j}q_{1k} + 2\dot{\lambda}V^c_{ij\lambda}q_{1j} + V^c_{ij}q_{2j} = 0 \tag{41}$$

where we have made use of the symmetry $V_{ijk} = Vikj$. Multiplying Eq. (41) by q_{1i} and solving for $\dot{\lambda}$ we obtain the load variation for asymmetric bifurcation

$$\dot{\lambda} = -\frac{V^c_{ijk}q_{1i}q_{1j}q_{1k}}{2V^c_{ijk}q_{1i}q'_{0j}q_{1k} + 2V^c_{ijk}q_{1i}q_{1j}} \tag{42}$$

3.3 Bifurcation point: sensitivity analysis

Assume the regular sensitivity case presenting the bifurcation point. We again consider a critical-state path that connects bifurcation points as the design is changed (see Fig. 2), with $s = \eta$. We cannot use the same procedure as we used for the limit load, because now both the numerator and denominator of (28) are zero. Instead we define an energy function

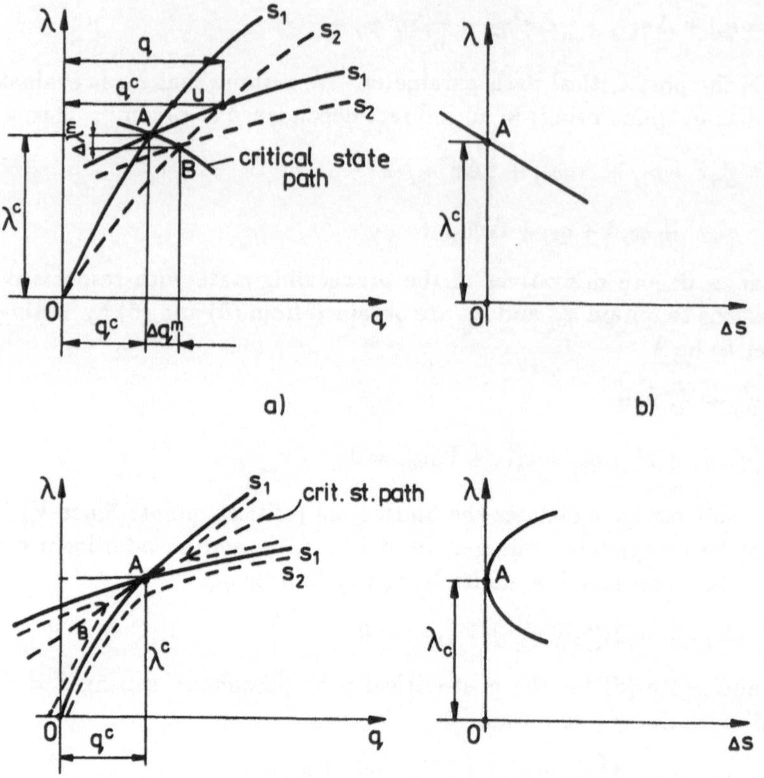

Figure 2: **a) Load deformation path of structure passing through bifurcation point and critical state path for regular case, b) Sensitivity diagram for regular case, c) Load deformation path of structure and critical state path for singular case, d) Sensitivity diagram for singular case**

$$E = V_{ij}q_{1i}q_{1j} + \mu_i V_i \tag{43}$$

where $\boldsymbol{\mu}$ is a Lagrange multiplier vector used to enforce prebuckling equilibrium. Differentiating (43) with respect to the critical path parameter we have

$$V^c_{ijk}\dot{q}_{0k}q_{1i}q_{ij} + \lambda_{cs}V^c_{ij\lambda}q_{1i}q_{1j} + V^c_{ijs}q_{1i}q_{1j} + \mu_i(V^c_{ij}\dot{q}_{0j} + \lambda_{cs}V^c_{i\lambda} + V^c_{is}) = 0 \tag{44}$$

Along the critical path both the load and the stiffness vary simultaneously so that

$$\dot{q}_{0k} = \lambda_{cs}q'_{0k} + q_{0sk} \tag{45}$$

so that (44) may be written as

$$\begin{aligned}(V^c_{ijk}q_{1i}q_{1j} + \mu_i V^c_{ik})q_{0sk} + \lambda_{cs}[(V^c_{ijk}q'_{0k} + V^c_{ij\lambda})q_{1i}q_{1j} \\ + \mu_i(V^c_{ij}q'_{0j} + V^c_{i\lambda})] + V^c_{ijs}q_{1i}q_{1j} + \mu_i V^c_{is} = 0\end{aligned} \tag{46}$$

We now have two options for calculating λ_{cs} from (46). We can set μ to zero and obtain

$$\lambda_{cs} = -\frac{V_{ijk}q_{1i}q_{1j}q_{0sk} + V^c_{ijs}q_{1i}q_{1j}}{(V^c_{ijk}q'_{0k} + V^c_{ij\lambda})q_{1i}q_{1j}} \tag{47}$$

This form requires the calculation of the prebuckling sensitivity $\mathbf{q}_{os}$. Instead we can define the adjoint field $\boldsymbol{\mu}$ eliminate the coefficient of q_{osk} in (46)

$$V^c_{ik}\mu_i + V^c_{ijk}q_{1i}q_{1j} = 0 \tag{48}$$

and then, using (37), we obtain from (46)

$$\lambda_{cs} = -\frac{V^c_{ijs}q_{1i}q_{1j} + \mu_i V^c_{is}}{(V^c_{ijk}q'_{0k} + V^c_{ij\lambda})q_{1i}q_{1j}} \tag{49}$$

4 SENSITIVITY OF FREQUENCIES AND EIGENMODES OF SMALL HARMONIC VIBRATIONS

Consider now small harmonic oscillations superposed on initial equilibrium state specified by q^0_i, λ^0, such that

$$V_i(q^0_i, \lambda^0, s) = 0 \tag{50}$$

and $q_i = q^0_i + u_i cos\omega t$, where u_i and ω denote vibration amplitude mode and vibration frequency. The equations of motions now take the form

$$V_{ij}u_j - \omega^2 M_{ij}u_j = 0 \tag{51}$$

where M_{ij} denotes the mass matrix. Note that $V_{ij} = V_{ij}(q^0_k, \lambda^0, s)$, $M_{ij} = M_{ij}(s)$. Multiplying (51) by u^i one obtains the Rayleigh quotient

$$\omega^2 = \frac{V_{ij}u_iu_j}{M_{ij}u_iu_j} \tag{52}$$

where u_i is the eigenvector of the matrix $V_{ij} - \omega^2 M_{ij}$. Equations (51) and (52) occur for a discrete set of eigenvalues and eigenvectors $(\omega_1, u_i^{(1)}),(\omega_2, u_i^{(2)}),...$ Consider first the variation of frequency due to load variation. From (50) and (51), we obtain

$$V_{ik}\dot{q}_k^0 + V_{i\lambda}\dot{\lambda} = 0 \quad or \quad \dot{q}_k^0 = -V_{ik}^{-1}V_{i\lambda}\dot{\lambda} = t_k^0\dot{\lambda} \tag{53}$$

$$V_{ij}\dot{u}_j + V_{ijk}u_j\dot{q}_k^0 + V_{ij\lambda}u_j\dot{\lambda} - \omega^2 M_{ij}\dot{u}_j - (\omega^2)^{\cdot}_{\lambda}M_{ij}u_j = 0 \tag{54}$$

where $(\omega^2)^{\cdot}_{\lambda} = (\omega^2)_{\lambda}\dot{\lambda}$ and $(\omega^2)_{\lambda}$ is the derivative of ω^2 with respect to λ. Multiplying (54) by u_i and accounting for (51), we have

$$(\omega^2)_{\lambda} = \frac{V_{ijk}u_iu_k\dot{q}_k^0 + V_{ij\lambda}u_iu_j}{M_{ij}u_iu_j} = \frac{V_{ijk}u_iu_jt_k^0 + V_{ij\lambda}u_iu_j}{M_{ij}u_iu_j} \tag{55}$$

Consider now the structure transformation process under constant load. Similarly to (53) and (54), we have

$$V_{ik}\dot{q}_k^s + V_{is}\dot{s} = 0 \tag{56}$$

$$V_{ij}\dot{u}_j^s + V_{ijk}u_j\dot{q}_k^{0s} + V_{ijs}u_j\dot{s} - \omega^2 M_{ij}\dot{u}_j^s - (\omega^2)^{\cdot}_s M_{ij}u_j - \omega^2 M_{ijs}u_j\dot{s} = 0 \tag{57}$$

Multiplying Eq. (57) by u_i, in view of (51), we obtain

$$(\omega^2)^{\cdot}_s = \frac{V_{ijk}u_iu_j\dot{q}_k^{0s} + V_{ijs}u_iu_j\dot{s} - \omega^2 M_{ijs}u_iu_j\dot{s}}{M_{ij}u_iu_j} \tag{58}$$

where $\dot{q}_k^{0s} = -V_{ik}^{-1}V_{is}\dot{s}$ is specified from (56). Instead of solution for $\dot{q}_k^{0s}$ from (56), the components $\dot{q}_k^{0s}$ can be eliminated by using the adjoint solution. In fact, regarding (56) as constraint equations imposed on (57) and the Lagrange multiplier vector μ_i, we can write

$$(-V_{ik}\mu_i + V_{ijk}u_iu_j)\dot{q}_k^{0s} + V_{ijs}u_iu_j\dot{s} - \mu_iV_{is}\dot{s} - \omega^2\dot{M}_{ij}u_iu_j = (\omega^2)^{\cdot}_s M_{ij}u_iu_j \tag{59}$$

Similarly as previously, introducing the adjoint solution

$$V_{ik}\mu_i - V_{ijk}u_iu_j = 0 \tag{60}$$

the frequency derivative with respect to s is expressed as follows

$$(\omega^2)_s = \frac{V_{ijs}u_iu_j - \mu_iV_{is} - \omega^2 M_{ijs}u_iu_j}{M_{ij}u_iu_j} \tag{61}$$

where $(\omega^2)^{\cdot}_s = (\omega^2)_s\dot{s}$, $\dot{M}_{ij} = M_{ijs}\dot{s}$. The critical state is specified by the condition

$$\omega^2(\lambda^c, s) = 0 \tag{62}$$

from which it follows that

$$(\omega^2)_\lambda \lambda_s^c + (\omega^2)_s = 0 \tag{63}$$

Hence in view of (55) and (61), we obtain

$$\frac{1}{s}\dot{\lambda}^m = \lambda_s^c = -\frac{(\omega^2)_s}{(\omega^2)_\lambda} = -\frac{V_{ijs}u_i u_j - \mu_i V_{is}}{V_{ijk}u_i u_j t_k^0 + V_{ij\lambda}u_i u_j} \tag{64}$$

Consider now the eigenmode variation due to load or design parameter variation. Assume that the set of eigenfrequencies $\omega_1, \omega_2, \ldots$ and eigenmodes $u_1^{(1)}, u_1^{(2)}, \ldots$ has been specified and Eq. (51) occurs for any pair $\omega_k, u_i^{(k)}$ thus

$$V_{ij}u_j^{(k)} - \omega^2 M_{ij}u_j^{(k)} = 0, \quad V_{ij}u_j^{(1)}u_j^{(k)} = M_{ij}u_j^{(1)}u_j^{(k)} = 0, \quad k \neq l \tag{65}$$

Consider, for instance, variation of the first mode and use the modal expansions

$$\dot{u}_j^{(1)\lambda} = C_2^\lambda u_j^{(2)} + C_3^\lambda u_j^{(3)} + \cdots, \qquad \dot{u}_j^{(1)s} = C_2^s u_j^{(2)} + C_3^s u_j^{(3)} + \cdots \tag{66}$$

where $\dot{u}_j^{(1)\lambda}$ and $\dot{u}_j^{(1)s}$ denote mode variations due to load and design parameter changes. Substituting (66) into (65), we obtain

$$C_m^\lambda = -\frac{V_{ijk}u_i^{(m)}u_j^{(1)}\dot{q}_k^0 + V_{ijs}u_i^{(m)}u_j^{(1)}\dot{s} - \omega^2 \dot{M}_{ij}u_i^{(m)}u_j^{(1)}}{M_{ij}u_i^{(m)}u_j^{(1)}}, \quad m = 2, 3, 4, \ldots \tag{67}$$

and the respective expressions for the expansion coefficients in the case of design parameter variation

$$C_m^s = -\frac{V_{ijk}u_i^{(m)}u_j^{(1)}\dot{q}_k^{0s} + V_{ijs}u_i^{(m)}u_j^{(1)}\dot{s} - \omega^2 \dot{M}_{ij}u_i^{(m)}u_j^{(1)}}{M_{ij}u_i^{(m)}u_j^{(m)}}, \quad m = 2, 3, 4, \ldots \tag{68}$$

or using the adjoint system solution

$$C_m^s = -\frac{V_{ijs}u_i^{(m)}u_j^{(1)}\dot{s} - \mu_i^{(m)}V_{is}\dot{s} - \omega_m^2 \dot{M}_{ij}u_i^{(m)}u_j^{(1)}}{M_{ij}u_i^{(m)}u_j^{(m)}} \tag{69}$$

where $\mu_i^{(m)}$ satisfies the equation

$$V_{ik}\mu_i^{(m)} - V_{ijk}u_i^{(m)}u_j^{(1)} = 0, \qquad m = 2, 3, \ldots \tag{70}$$

5 SENSITIVITY OF BUCKLING LOADS OF TRUSS AND BEAM STRUCTURES

5.1 Discretized formulation

Consider a discretized beam structure at the state of buckling. The critical state condition can now be specified as follows

$$[w]^T[K][w] + \lambda\,[w]^T[H][w] = 0 \tag{71}$$

where $[K]$ denotes the global stiffness matrix, $[H]$ is the global geometric stiffness matrix and λ is the load factor. Here w denotes the buckling mode and $\lambda = \lambda_c$ specifies the critical load value. The eigenvalue problem associated with (71) is of the form

$$([K] + \lambda_k[H])\,[w^{(k)}] = [0] \tag{72}$$

where $\lambda_k = \lambda_1, \lambda_2, \lambda_3, ...$ denote the subsequent eigenvalues and $w^{(1)}, w^{(2)}, w^{(3)}, ...$ are the corresponding eigenmodes (eigenvectors). In the finite element discretization, Eq. (71) is written as follows

$$\sum_i [w_i]^T[K_i][w_i] + \lambda \sum_i N_i[w_i]^T[H_i][w_i] = 0 \tag{73}$$

where $[H_i]$ denotes the geometric stiffness matrix of the i-th element for a unit axial force, expressed in a global reference system and N_i denotes the axial force in this element. When variation of the design parameter s occurs, the perturbation of (73) provides

$$\delta\lambda = -\frac{\sum_i [w_i]^T[\delta K_i][w_i] + \lambda_c \sum_i N_i[w_i]^T[\delta H_i][w_i] + \lambda_c \sum_i \delta N_i[w_i]^T[H_i][w_i]}{\sum_i [w_i]^T[H_i][w_i]} \tag{74}$$

In order to eliminate variations δN_i resulting from axial force redistribution due to design variation, let us introduce the linear adjoint system subjected to initial distortions

$$u^a_{Di} = [w_i]^T[H_i][w_i] \tag{75}$$

in each element. The adjoint solution $[w^a_i]$ of a linear problem permits the elimination of δN_i from (74). In fact, we have

$$\lambda \sum_i \delta N_i[w_i]^T[H_i][w_i] = \lambda \sum_i \delta N_i u^a_{Di} = \lambda \sum_i [w^a_i]^T[\delta K_i][w_i] \tag{76}$$

and (174) becomes

$$\delta\lambda = -\frac{\sum_i [w_i]^T\left([\delta K_i] + \lambda_c N_i[\delta H_i]\right)[w_i] + \lambda_c \sum_i [w^a_i]^T[\delta K_i][w_i]}{\sum_i [w_i]^T[H_i][w_i]} \tag{77}$$

In order to determine eigenvector variation, assume the modal representation

$$\left[\delta w^{(1)}\right] = \delta C_{12}\left[w^{(2)}\right] + \delta C_{13}\left[w^{(3)}\right] + \cdots \tag{78}$$

Following the general derivation of the previous section, we obtain

$$\delta C_{1k} = -\frac{\sum_i [w_i^{(k)}]^T \left([\delta K_i] + \lambda_1 N_i [\delta H_i]\right) [w_i^{(1)}] + \lambda_1 \sum_i \delta N_i [w_i^{(k)}]^T [H_i][w_i^{(1)}]}{(\lambda_1 - \lambda_2) \sum_i [w_i^{(k)}]^T [H_i][w_i^{(k)}]} \tag{79}$$

where $k = 2, 3 \ldots$. The variations δN_i can be eliminated by solving linear adjoint problems with initial distortions specified by (75).
The sensitivity formulae contain variations of stiffness, geometric stiffness and mass matrices expressed in the global coordinate system. The relations between matrices in both systems are

$$[K_i] = [T_i]^T [k_i][T_i], \quad [H_i] = [T_i]^T [h_i][T_i], \quad [M_i] = [T_i]^T [m_i][T_i] \tag{80}$$

where $[k_i], [h_i], [m_i]$ are the matrices expressed in the local reference system and $[T_i]$ denote the transformation matrices. Variations of (80) are

$$[\delta K_i] = [T_i]^T [\delta k_i][T_i] + [\delta T_i]^T [k_i][T_i] + [T_i]^T [k_i][\delta T_i] \tag{81}$$

and similar expressions for $[\delta H_i]$ and $[\delta M_i]$. The first type of structure variation involves only variation of local matrices with transformation matrices unchanged. This case occurs when cross sectional properties or material stiffness moduli are varied. On the other hand, when the structure configuration is modified, for instance, by varying node positions of trusses or frames, the transformation matrices also undergo variation. For instance, the stiffness matrix of the i-th bar of truss has the form

$$[k_i] = \frac{E_i A_i}{l_i} \begin{bmatrix} 1 & -1 \\ -1 & 1 \end{bmatrix} \tag{82}$$

and when its cross sectional area is modified, there is

$$[\delta k_i] = \frac{E_i}{l_i} \begin{bmatrix} 1 & -1 \\ -1 & 1 \end{bmatrix} \delta A_i \tag{83}$$

Consider now the variation of node position of a truss by specifying the translation vector δs inclined at the angle β to the $x_1 - axis$; Fig. 3. As the components of δs are $\delta x_1 = \delta s \cos\beta$, $\delta x_2 = \delta s \sin\beta$, we have

$$\begin{aligned} \delta l_i &= -a_i \cos\alpha_i \, \delta x_1 & \delta l_i &= -a_i \sin\alpha_i \, \delta x_2 \\ \delta(\cos\alpha_i) &= -\frac{a_i}{l_i} \sin^2\alpha_i \, \delta x_1 & \delta(\cos\alpha_i) &= \frac{a_i}{l_i} \sin\alpha_i \cos\alpha_i \, \delta x_2 \\ \delta(\sin\alpha_i) &= -\frac{a_i}{l_i} \sin\alpha_i \cos\alpha_i \, \delta x_1 & \delta(\sin\alpha_i) &= -\frac{a_i}{l_i} \cos^2\alpha_i \, \delta x_2 \end{aligned} \tag{84}$$

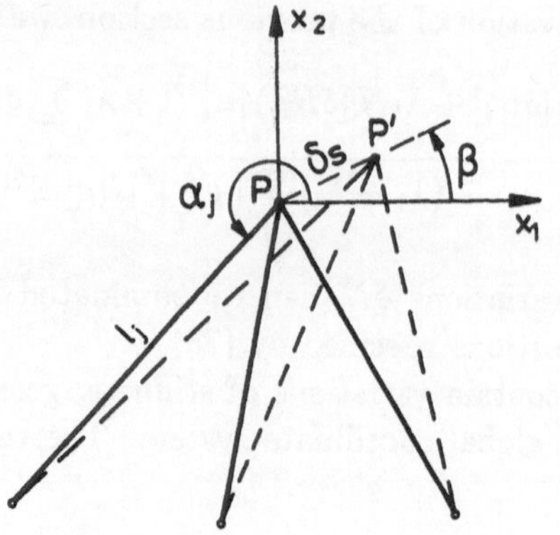

Figure 3: Translation of truss node from P to P'

where $a_i = l_i/L_i$ denotes the ratio of length of the $i-th$ element to the total length of the member. In the case of trusses, there is $a_i = 1$. The angles α_i specify member orientation with respect to the $x_1 - axis$, Fig. 3. Consider, for instance, translation of the truss node along the $x_2 - axis$. The local stiffness matrix of each bar is expressed by (82) and the transformation matrix has the form

$$[T_i] = \begin{bmatrix} \cos\alpha_i & \sin\alpha_i & 0 & 0 \\ 0 & 0 & \cos\alpha_i & \sin\alpha_i \end{bmatrix} \tag{85}$$

In view of (84), the variations of $[h_i]$ and $[T_i]$ are

$$[\delta h_i] = \frac{E_i A_i \sin\alpha_i}{l_i^2} \begin{bmatrix} 1 & -1 \\ -1 & 1 \end{bmatrix} \delta x_2,$$

$$[\delta T_i] = \frac{\cos\alpha_i}{l_i} \begin{bmatrix} \sin\alpha_i & -\cos\alpha_i & 0 & 0 \\ 0 & 0 & \sin\alpha_i & -\cos\alpha_i \end{bmatrix} \delta x_2 \tag{86}$$

Similar expressions are obtained for the variation δx_1 of node position. For an arbitrary variation δs, the respective variations of $[K_i]$, $[H_i]$ and $[M_i]$ are generated by (81) by superposing variations δx_1, δx_2 of each node. For a frame element (with account for axial forces) the stiffness and transformation matrices are

$$[k_i] = \frac{E_i A_i}{l_i^3}\begin{bmatrix} \frac{l_i^2}{i_i^2} & 0 & 0 & -\frac{l_i^2}{i_i^2} & 0 & 0 \\ 0 & 12 & 6l_i & 0 & -12 & 6l_i \\ 0 & 6l_i & 4l_i^2 & 0 & -6l_i & 2l_i^2 \\ -\frac{l_i^2}{i_i^2} & 0 & 0 & \frac{l_i^2}{i_i^2} & 0 & 0 \\ 0 & -12 & -6l_i & 0 & 12 & -6l_i \\ 0 & 6l_i & 2l_i^2 & 0 & -6l_i & 4l_i^2 \end{bmatrix},$$

$$[T_i] = \begin{bmatrix} \cos\alpha_i & \sin\alpha_i & 0 & 0 & 0 & 0 \\ -\sin\alpha_i & \cos\alpha_i & 0 & 0 & 0 & 0 \\ 0 & 0 & 1 & 0 & 0 & 0 \\ 0 & 0 & 0 & \cos\alpha_i & \sin\alpha_i & 0 \\ 0 & 0 & 0 & -\sin\alpha_i & \cos\alpha_i & 0 \\ 0 & 0 & 0 & 0 & 0 & 1 \end{bmatrix} \tag{87}$$

Their variations associated with node translation through δx_2 are expressed as follows

$$[\delta k_i] = \frac{a_i E_i \sin\alpha_i}{l_i^4}\begin{bmatrix} \frac{l_i^2}{i_i^2} & 0 & 0 & -\frac{l_i^2}{i_i^2} & 0 & 0 \\ 0 & 36 & 12l_i & 0 & -36 & 12l_i \\ 0 & 12l_i & 4l_i^2 & 0 & -12l_i & 2l_i^2 \\ -\frac{l_i^2}{i_i^2} & 0 & 0 & \frac{l_i^2}{i_i^2} & 0 & 0 \\ 0 & -36 & -12l_i & 0 & 36 & -12l_i \\ 0 & 12l_i & 2l_i^2 & 0 & -12l_i & 4l_i^2 \end{bmatrix}\delta x_2,$$

$$[\delta T_i] = -\frac{a_i \cos\alpha_i}{l_i}\frac{d[T_i]}{d\alpha}\delta x_2 \tag{88}$$

where i_i denotes the radius of inertia of the $i-th$ cross section.

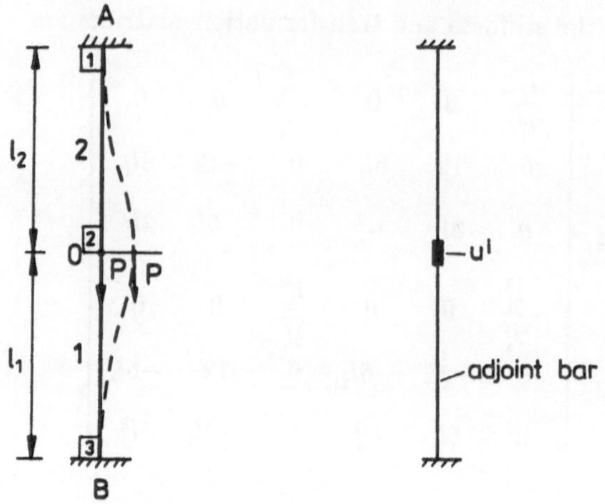

Figure 4: **Built-in column loaded at middle cross section**

Example. Sensitivity of buckling load for a clamped column loaded at the middle cross section.

Consider a column of Fig. 4 loaded axially at the mid point A and clamped at its ends. Let us derive the buckling load and mode variation due to length variation of two beam spans l_1 and l_2. Assume the initial design to have $l_1 = l_2$, and the same stiffnesses EA and EI of the elements 1 and 2. Consider first the buckling problem. The buckling mode vector is specified by the horizontal displacement and the angle of rotation of node 2, so that

$$[w^*]^T = [w_{x2}, \theta] \tag{89}$$

The respective stiffness matrices of the elements 1 and 2 are

$$[K_1^*] = \frac{EI}{l^3}\begin{bmatrix} 12 & -6l \\ -6l & 4l^2 \end{bmatrix}, \qquad [K_2^*] = \frac{EI}{l^3}\begin{bmatrix} 12 & 6l \\ 6l & 4l^2 \end{bmatrix} \tag{90}$$

and the geometric stiffness matrices are

$$[H_1] = \frac{1}{30l}\begin{bmatrix} 36 & -3l \\ -3l & 4l^2 \end{bmatrix}, \qquad [H_2] = \frac{1}{30l}\begin{bmatrix} 36 & 3l \\ 3l & 4l^2 \end{bmatrix} \tag{91}$$

The eigenvalue problem specifying the critical load factor λ and the eigenvector is expressed as follows

$$\left(\begin{bmatrix} 24 & 0 \\ 0 & 8l^2 \end{bmatrix} + \frac{\lambda n_1 l^2}{30EI}\begin{bmatrix} 0 & 6l \\ 6l & 0 \end{bmatrix}\right)\begin{bmatrix} w_{x2} \\ \theta \end{bmatrix} = \begin{bmatrix} 0 \\ 0 \end{bmatrix} \tag{92}$$

where n_1 denotes the axial force due to unit load at A. The first eigenvalue is

$$\lambda_1 = 277.128 \frac{EI}{l^2} n_1 \tag{93}$$

To determine the force n_1, apply the unit load at A and consider a linear problem for a redundant structure. The stiffness matrices of the elements 1 and 2 are

$$[K_1] = [K_2] = \frac{EA}{l}\begin{bmatrix} 1 & -1 \\ -1 & 1 \end{bmatrix} \tag{94}$$

The displacement state, and the force n_1 are now expressed as follows

$$[w_1] = \frac{l}{2EA}\begin{bmatrix} 0 \\ -1 \end{bmatrix}, \quad [w_2] = \frac{l}{2EA}\begin{bmatrix} -1 \\ 0 \end{bmatrix}, \quad n_1 = -\frac{1}{2} \tag{95}$$

In view of (7) the buckling load is

$$\lambda_1 = 138.564 \frac{EI}{l^2} \tag{96}$$

Assuming the normalization condition in the form

$$[\overset{(i)}{w^*}]^T [H] [\overset{(i)}{w^*}] = \frac{n_1 l}{30} \tag{97}$$

the eigenvector associated with the fundamental eigenvalue λ_1 equals

$$[\overset{(1)}{w^*}] = [\overset{(1)}{w_1^*}] = [\overset{(1)}{w_2^*}] = \begin{bmatrix} 0.219345l \\ -0.37991 \end{bmatrix} \tag{98}$$

The variation of the buckling load is expressed as follows

$$\delta\lambda_1 = -\frac{\sum_{i=1}^{2} [\overset{(1)}{w_i^*}]^T \left([\delta K_i^*] + \lambda_1 n_i [\delta H_i]\right) [\overset{(1)}{w_i^*}] + \lambda_1 \sum_{i=1}^{2} [w_i^a]^T [\delta K_i][w_i]}{[\overset{(1)}{w_i^*}]^T [H_i] [\overset{(1)}{w_i^*}]} \tag{99}$$

where $[w_i^a]$ denotes the solution for the adjoint problem. The adjoint structure is assumed in the form shown in Fig. 4 with an initial axial distortion

$$u_D^a = [\overset{(1)}{w_1^*}]^T [H_1] [\overset{(1)}{w_1^*}] = \frac{l}{30} \tag{100}$$

The adjoint solution is obtained for a linear system, so that

$$[w_1^a] = \frac{l}{180}\begin{bmatrix} 0 \\ 13.8 \end{bmatrix}, \qquad [w_2^a] = \frac{l}{180}\begin{bmatrix} -13.8 \\ 0 \end{bmatrix} \tag{101}$$

The variations of stiffness matrices now are

$$\begin{aligned} [\delta K_i^*] &= [T_i^*]^T[\delta k_i^*][T_i^*], \qquad i = 1,2 \\ [\delta K_i] &= [T_i]^T[\delta k_i][T_i] \\ [\delta H_i] &= [T_i^*]^T[\delta h_i][T_i] \end{aligned} \tag{102}$$

and we obtain respectively

$$[\delta K_1^*] = \frac{EI}{l^4}\begin{bmatrix} -36 & 12l \\ 12l & -4l^2 \end{bmatrix}\delta l_1, \qquad [\delta K_2^*] = \frac{EI}{l^4}\begin{bmatrix} -36 & -12l \\ -12l & -4l^2 \end{bmatrix}$$

$$[\delta K_1] = -\frac{EA}{l^2}\begin{bmatrix} 1 & -1 \\ -1 & 1 \end{bmatrix}\delta l_1, \qquad [\delta K_2] = -\frac{EA}{l^2}\begin{bmatrix} 1 & -1 \\ -1 & 1 \end{bmatrix}$$

$$[\delta H_1] = \frac{2}{15l^2}\begin{bmatrix} -9 & 0 \\ 0 & l^2 \end{bmatrix}\delta l_1, \qquad [\delta H_2] = \frac{2}{15l^2}\begin{bmatrix} -9 & 0 \\ 0 & l^2 \end{bmatrix} \tag{103}$$

Moreover, we have

$$[\overset{(1)}{w^*}]^T[H][\overset{(1)}{w^*}] = -\frac{l}{60} \tag{104}$$

and from (11) it follows that

$$\delta\lambda_1 = -98.564\frac{EI}{l^3}\delta l_1 + 320\frac{EI}{l^3}\delta l_1 \tag{105}$$

where first term corresponds to buckling load variation with neglect of the axial force redistribution and the second represents the effect of force variation expressed in terms of the adjoint solution. It is seen that the effect of axial force variation with redesign is very significant in this case and cannot be neglected.

6 VARIATIONAL FORMULATION OF DESIGN SENSITIVITY

Analogously to discrete formulation, consider now any surface or beam structure for which the state is represented by generalized stress $\boldsymbol{\sigma}$, strain $\boldsymbol{\varepsilon}$, and displacement $\boldsymbol{u}$. Following Budiansky notation the strain-displacement relation is

$$\boldsymbol{\varepsilon} = \mathcal{L}_1(\boldsymbol{u}) + \frac{1}{2}\mathcal{L}_2(\boldsymbol{u}) \tag{106}$$

where $\mathcal{L}_1$ and $\mathcal{L}_2$ are first and second order homogeneous operators. The variation of strain is

$$\delta\boldsymbol{\varepsilon} = \mathcal{L}_1(\delta\boldsymbol{u}) + \mathcal{L}_{11}(\boldsymbol{u}, \delta\boldsymbol{u}) \tag{107}$$

where $\mathcal{L}_{11}$ is a symmetric bilinear operator, that is $\mathcal{L}_{11}(\boldsymbol{u}, \boldsymbol{v}) = \mathcal{L}_{11}(\boldsymbol{v}, \boldsymbol{u})$ and

$$\mathcal{L}_2(\boldsymbol{u} + \boldsymbol{v}) = \mathcal{L}_2(\boldsymbol{u}) + \mathcal{L}_2(\boldsymbol{v}) + 2\mathcal{L}_{11}(\boldsymbol{u}, \boldsymbol{v}) \tag{108}$$

so that $\mathcal{L}_{11}(\boldsymbol{u}, \boldsymbol{u}) = \mathcal{L}_2(\boldsymbol{u})$. The linear stress strain law is written as

$$\boldsymbol{\sigma} = \mathcal{D}\boldsymbol{\varepsilon} \tag{109}$$

where $\mathcal{D}$ is the stiffness operator. Assuming structure to be loaded by the deformation independent load $\lambda\boldsymbol{f}$, the equilibrium equations are expressed by the principle of virtual work

$$\boldsymbol{\sigma} \cdot \delta\boldsymbol{\varepsilon} = \lambda\boldsymbol{f} \cdot \delta\boldsymbol{u} \tag{110}$$

where *dot* denotes the scalar product integrated over the structural domain. Considering variation of $\mathcal{D}$ dependent on a design parameter s, we can write along the transformation path

$$\begin{aligned}
&\dot{\boldsymbol{\varepsilon}} = \mathcal{L}_1(\dot{\boldsymbol{u}}) + \mathcal{L}_{11}(\boldsymbol{u}, \dot{\boldsymbol{u}}) \\
&\dot{\boldsymbol{\sigma}} = \mathcal{D}\dot{\boldsymbol{\varepsilon}} + \dot{\mathcal{D}}\boldsymbol{\varepsilon} \\
&\dot{\boldsymbol{\sigma}} \cdot \delta\boldsymbol{\varepsilon} + \boldsymbol{\sigma} \cdot \mathcal{L}_{11}(\dot{\boldsymbol{u}}, \delta\boldsymbol{u}) = \dot{\lambda}\boldsymbol{f} \cdot \delta\boldsymbol{u}
\end{aligned} \tag{111}$$

At the limit point the derivative of the load factor with respect to s is, cf. *Mróz and Haftka (1992)*

$$\lambda_{cs} = \frac{\mathcal{D}_{,s}\boldsymbol{\varepsilon}^c \cdot \boldsymbol{\varepsilon}_1}{\boldsymbol{f} \cdot \boldsymbol{u}_1} \tag{112}$$

At the bifurcation point, the respective formula is expressed as follows

$$\lambda_{cs} = \frac{\mathcal{D}_{,s}\boldsymbol{\varepsilon}^c \cdot \boldsymbol{\varepsilon}^a - \mathcal{D}_{,s}\boldsymbol{\varepsilon}_1 \cdot \boldsymbol{\varepsilon}_1}{2\boldsymbol{\sigma}_1 \cdot \mathcal{L}_{11}(\boldsymbol{u}^{c'}, \boldsymbol{u}_1) + \boldsymbol{\sigma}^{c'} \cdot \mathcal{L}_2(\boldsymbol{u}_1)} \tag{113}$$

Where $\boldsymbol{\varepsilon}^a$ denotes the adjoint strain specified by the relations

$$\begin{aligned}
&\boldsymbol{\varepsilon}^a = \mathcal{L}_1(\boldsymbol{u}^a) + \mathcal{L}_{11}(\boldsymbol{u}^c, \boldsymbol{u}^a) - \frac{1}{2}\mathcal{L}_2(\boldsymbol{u}_1) \\
&\boldsymbol{\sigma}^a = \mathcal{D}\boldsymbol{\varepsilon}^a \\
&\boldsymbol{\sigma}^a \cdot \delta\boldsymbol{\varepsilon} - 2\boldsymbol{\sigma}_1 \cdot \mathcal{L}_{11}(\boldsymbol{u}_1, \delta\boldsymbol{u}) + \boldsymbol{\sigma}^c \cdot \mathcal{L}_{11}(\boldsymbol{u}^a, \delta\boldsymbol{u}) = 0
\end{aligned} \tag{114}$$

Here $\boldsymbol{\sigma}_1, \boldsymbol{\varepsilon}_1, \boldsymbol{u}_1$ denotes the eigenmode and $\boldsymbol{\sigma}^c, \boldsymbol{\varepsilon}^c, \boldsymbol{u}^c$ is the critical state.
In the case of bar system, the critical state condition has the form

$$\sum_i \int_0^{l_i} \frac{1}{2} EIk^2 dx_i + \sum_i \int_0^{l_i} N(w')^2 dx_i = 0 \tag{115}$$

where k denotes the curvature, N is the axial force, w is the buckling mode and w' denotes its slope. The initial strain introduced into the adjoint system equals

$$\varepsilon^i = \frac{1}{2}(w')^2 \tag{116}$$

The variation of the critical load due to variation of inertia and cross section area now is

$$\delta\lambda = \frac{\sum_i (2\int_0^{l_i} kk^r E dx_i - \int_0^{l_i} k^2 E dx_i)\delta I_i + 2\sum_i \int_0^{l_i} \varepsilon \cdot \varepsilon^r dx_i \delta A_i}{\sum_i \int_0^{l_i} N(w')^2 dx_i} \tag{117}$$

where $\varepsilon^r = \varepsilon^a - \varepsilon^i = \dfrac{N^a}{EA}$, $k^r = \dfrac{M^a}{EI}$.

<u>Example</u>. Two bar truss

For a two bar truss loaded by a rigid plate, Fig. 5, the bar forces are

$$N_1 = -P\frac{A_1}{A_1 + A_2}, \qquad N_2 = -P\frac{A_2}{A_1 + A_2} \tag{118}$$

The critical forces of bars 1 and 2 respectively, are

$$P_{c1} = \frac{\pi^2 EI_1}{l^2}\frac{A_1 + A_2}{A_1}, \qquad P_{c2} = \frac{\pi^2 EI_2}{l^2}\frac{A_1 + A_2}{A_2} \tag{119}$$

and the buckling mode is

$$w = B\sin\frac{\pi x}{l} \tag{120}$$

When the bar 1 buckles first the adjoint truss is submitted to the initial strain in first bar

$$u^i = \frac{1}{2}\int_0^l (w')^2 dx = \frac{B^2\pi^2}{4l} \tag{121}$$

and the forces in bars of the adjoint system are

$$N_2^a = -N_1^a = \frac{u^i EA_1A_2}{l(A_1 + A_2)} \tag{122}$$

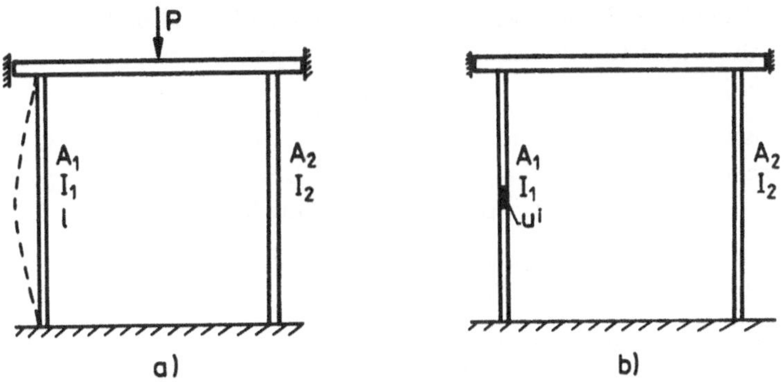

Figure 5: a) Two bar system loaded by a rigid plate, b) Adjoint system

The formula (117) now provides

$$\delta\lambda = \frac{-\int_0^l k^2 E dx_1 \delta I_1 + 2\int_0^l \varepsilon \cdot \varepsilon^r dx_1 \delta A_1 + 2\int_0^l \varepsilon \cdot \varepsilon^r dx_2 \delta A_2}{\int_0^l N_1^0 (w')^2 dx} \tag{123}$$

and since

$$\int_0^l N_1^0 (w')^2 dx_1 = -\frac{2A_1}{A_1 + A_2} u^i, \quad \int_0^l k^2 E dx_1 = \frac{N_1^2}{EI_1^2} \int_0^l w^2 dx_1 = \\ = \frac{P_c^2 B^2 A_1^2 l}{2EI_1 (A_1 + A_2)^2}, \quad \int_0^l \varepsilon\varepsilon^r E dx_1 = \frac{N_1 N_1^a l}{EA_1^2}, \quad \int_0^l \varepsilon\varepsilon^r E dx_2 = \frac{N_2 N_2^a l}{EA_2^2} \tag{124}$$

from (123) it follows that

$$\delta\lambda = \frac{\pi^2 E}{l^2} \frac{A_1 + A_2}{A_1} \delta I_1 - \frac{\pi^2 E I_1}{l^2} \frac{A_2}{A_1^2} dA_1 + \frac{\pi^2 E I_1}{l^2} \frac{1}{A_1} dA_2 \tag{125}$$

REFERENCES

Cohen G. A. and Haftka R. T., Sensitivity of buckling loads of anisotropic shells of revolution to geometric imperfection and design changes, *Computers and Structures*, 31, 985-995, (1989)

Dems, K. and Mróz, Z., Sensitivity of buckling loads and vibration frequency with respect to shape of stiffened and unstiffened plates, *Journ. Mech. Struct. Machines*, 17, 4, 431-457, (1989)

Haftka, R. T., Cohen, G. A., and Mróz, Z., Derivatives of buckling loads and vibration frequencies with respect to stiffness and initial strain parameters, *Journ. Appl. Mech.*, 57, 18-24, (1990)

Mróz, Z., Sensitivity analysis and optimal design with account for varying shape and support conditions, *Computer Aided Optimal Design, Structural and Mechanical Systems*, C. A. Mota Soares, Ed. Springer Verl., New York, 407-438, (1987)

Mróz, Z., and Haftka, R. T., Sensitivity of buckling loads and vibrations frequencies of plates, *Josef Singer Ann. Vol.*, Springer Verl., New York, (1988)

Mróz, Z., Kamat, M. P. and Plaut, R. H., Sensitivity analysis and optimal design of non-linear beams and plates, *Journ. Struct. Mech.*, 13, 245-266, (1985)

Szefer, G., Mróz, Z. and Demkowicz, L., Variational approach to sensitivity analysis in non-linear elasticity, *Arch. Mech.*, 39, 247-259, (1987)

Thompson, J. H. T. and Hunt, G. W., A general theory of elastic stability, J. Wiley Sons, (1978)

Mróz, Z., and Haftka, R. T., Design sensitivity analysis of non-linear structures in regular and critical states, *Int. J. Struct.*, (submitted for publ.), (1992)

Bojczuk, D., and Mróz, Z., Sensitivity analysis of truss and frame structures, (in preparation), (1992)

DIRECT AND ADJOINT APPROACH TO FIRST- AND SECOND-ORDER SHAPE SENSITIVITY AND OPTIMAL DESIGN OF STRUCTURE

K. DEMS
Łódź Technical University I-26
ul. Żwirki 36
90-924 Łódź
Poland

ABSTRACT. *For an arbitrary stress, strain and displacement functional, its first- and second-order sensitivities with respect to varying structural shape are discussed. It is assumed that the shape modification is described by a set of shape design parameters. The first derivatives of a functional with respect to these parameters are derived using both the direct and adjoint approaches. Next the second derivatives are obtained using the mixed approach in which both the direct and adjoint first-order solutions are used. Some simple example illustrates the presented theory.*

1. Introduction

There are numerous problems of mechanics, where we need to assess the change of any local or global structural response characteristic due to variation of structural shape. For instance, in redesign or optimization problems the variations of state fields and of an objective functional are evaluated at each redesign step. Generally, if G denotes any scalar functional depending on state fields of structure and **b** constitute a set of shape parameters of structural boundary, the sensitivity analysis is aimed to expressing the variation of G in terms of **b**, for instance in a form of polynomial expansion $\Delta G = \mathbf{S}\cdot\delta\mathbf{b} + \frac{1}{2}\delta\mathbf{b}^T\cdot\mathbf{R}\cdot\delta\mathbf{b} + .. = \delta G + \frac{1}{2}\delta^2 G + ..,$ where **S** and **R** are the first-order sensitivity vector and the second-order sensitivity matrix.

This class of problems have been illustrated in an extensive literature. The first-order sensitivity analysis was generally discussed, for instance, by Haug and Aurora(1978), Dems and Mróz(1984), Dems and Haftka(1989) and others. The second-order sensitivity analysis with respect to stiffness properties was presented by Haug(1981), Haftka(1982), Dems and Mróz(1985) and Haftka and Mróz(1986). The first- and

G. I. N. Rozvany (ed.), Optimization of Large Structural Systems, Vol. I, 477–491.

second-order shape sensitivities was considered by Fujii-(1986). The present paper extends the analysis presented by Dems and Mróz(1984) and Dems and Haftka(1989) and discusses the general equations for calculation of first- and second-order sensitivities for physically nonlinear elastic structures. As in Dems and Haftka(1989), the concept of total material derivatives with respect to shape parameters b_p is used. The detailed analysis of this problem can be found in Dems (1991). The analysis is limited to shape modification of traction-free boundary part of structure, although the obtained results can be easily extended to the case of loaded and/or supported boundary parts modification.

2. First-order sensitivities for varying free boundary

Similarly is in Dems and Haftka(1989), let us consider a primary structure occupying a domain V with an external boundary S, Fig.1. The structure is subjected to body forces f within its domain V, surface traction T^o on a part S_T of

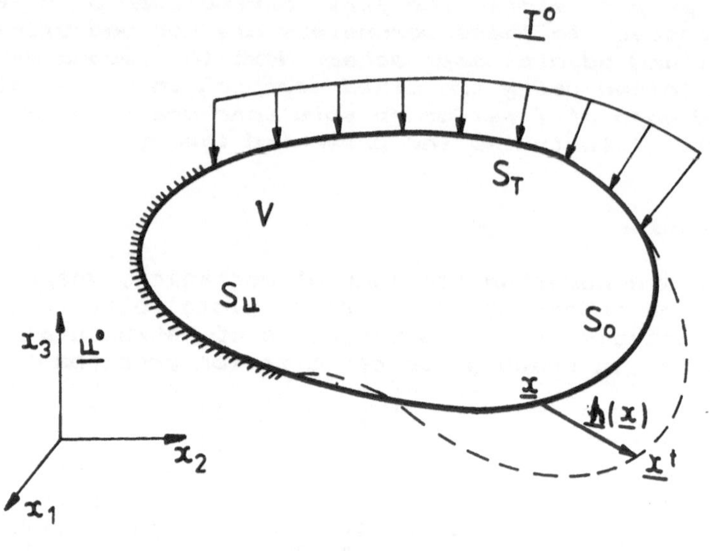

Fig. 1 Structure with varying free boundary part S .

boundary S and prescribed displacement u^o on a part S_u. The remaining portion S_o of external boundary is traction-free. The stress, strain and displacement fields s, e and u satisfy equilibrium, compatibility and boundary conditions given in the form

$$\text{div}\, s + f = 0 \quad , \quad e = \frac{1}{2}(\nabla u + \nabla^T u) \qquad \text{within } V$$
$$u = u^o \text{ on } S_u \quad , \quad s\cdot n = T^o \text{ on } S_T \quad , \quad s\cdot n = 0 \text{ on } S_o \tag{1}$$

Thus our analysis will be confined to small strain theory but a nonlinear stress-strain relation, given in a general form

$$s = E(e) \tag{2}$$

where $E(e)$ is generated by a potential rule associated with the specific strain energy $U(e)$ as $E = \partial U/\partial e$.

Besides a *deformation process* due to applied load, consider a *transformation process* $x^t = x + h(x,b)$ (cf. Fig.1), where h can be considered as a given transformation field depending on a set of design parameters b_p , $p=1,2,\dots,P$. This field modifies the structural shape satisfying the condition $h(x,b) = 0$ on $S_T \cup S_u$. Thus, only the traction-free portion S_o of external boundary S is subject to shape modification in transformation process. Consider now the infinitesimal transformation δh from initial configuration of a structure caused by the infinitesimal variation δb of design parameters. The variation δh can then be expressed as

$$\delta h_k = \frac{\partial h_k}{\partial b_p}\,\delta b_p = v_k^p \delta b_p \tag{3}$$

where h_k (k=1,2,3) denotes the k-th component of the transformation vector h with respect to fixed Cartesian coordinate system, and v_k^p is the k-th component of the transformation velocity field associated with shape parameter b_p, treated as time-like parameter. The transformation velocity field should satisfy the condition

$$v^p(x,b) = 0 \qquad \text{for } x \in S_T \cup S_u \tag{4}$$

Consider now an arbitrary functional G of the form

$$G = \int_{V(h)} \Psi(s,e,u)\,dV + \int_{S_T\cup S_u} h(u,T)\,dS \tag{5}$$

where Ψ and h are continuous and differentiable functions of their arguments. The first variation of G with respect to variations of shape parameters b can be generally expressed by

$$\delta G = S\cdot\delta b = \frac{DG}{Db_p}\delta b_p =$$

$$\left\{\int\left[\frac{D\Psi}{Db_p}dV + \Psi\frac{D(dV)}{Db_p}\right] + \int\left[\frac{Dh}{Db_p}dS + h\frac{D(dS)}{Db_p}\right]\right\}\delta b_p \tag{6}$$

where $S = DG/Db$ is the first-order sensitivity vector. The derivation of components of this vector will be briefly discussed in this Section by applying both the direct and adjoint approaches. The obtained results will be next used in Section 3 in order to determine the second-order sensitivity matrix of functional (5).

Since the surface tractions T^o on fixed boundary portion S_T and prescribed displacements u^o on fixed boundary portion S_u are independent on shape parameters b , we can write

$$\frac{DT^o}{Db_p} = 0 \quad \text{on} \quad S_T \quad , \quad \frac{Du^o}{Db_p} = 0 \quad \text{on} \quad S_u \tag{7}$$

Recalling now the results of Dems and Haftka (1989) and taking into account conditions (4) and (7), the sensitivities of G following from (6) can be written in the form

$$\frac{DG}{Db_p} = \int(\Psi_{,s}\cdot s_{,p} + \Psi_{,e}\cdot e_{,p} + \Psi_{,u}\cdot u_{,p})dV + \int h_{,u}\cdot u_{,p}dS_T +$$

$$+ \int h_{,T}\cdot s_{,p}\cdot n dS_u + \int \Psi n\cdot v^p dS_o \tag{8}$$

where $\Psi_{,s}$, $\Psi_{,e}$, $\Psi_{,u}$ and $h_{,u}$, $h_{,T}$ denote the partial derivatives of integrands of (5) with respect to state variables and n is the normal unit vector to boundary S. Furthermore, the quantities $s_{,p}$, $e_{,p}$ and $u_{,p}$ in (8) are the local derivatives of primary state fields with respect to shape parameter b_p (cf. Dems and Haftka (1989)).

The equation (8) constitutes the basic expression in first-order sensitivity analysis of any functional, to which the direct or adjoint approach can be next applied. When the direct sensiti vity method is used, it requires an additional solution of boundary-value problem for each particular variation of design parameter b_p. Thus, for P design parameters we need P additional sensitivity solutions. On the other hand, the adjoint state method requires only one additional solution of an adjoint problem for specified functional independently of design variations. The two approaches have received considerable attention in the literature, cf. Choi and Haug (1983), Dems and Mróz (1984), Choi and Seong (1986), Ha ber (1986), Dems and Haftka (1989) and others. The choice between these two approaches depends on number of objective functionals and design parameters, and also on the relative difficulty of obtaining adjoint or direct solutions. As it will be shown in Section 3, when the second-order sensitivities are considered, both above method should be used to generate the adjoint and sensitivity state fields.

Applying the direct approach in sensitivity analysis, we should derive the sensitivity state fields which are the local derivatives of primary state fields appearing in (8), associated with each shape parameter b_p. To do this, we should solve the additional boundary-value problems associated with particular variation of each design parameter. Equations describing these problems are obtained by differentiation with respect to each b_p the set (1) and (2) describing the primary boundary-value problem. Introducing the notation

$$\hat{u}^p = u_{,p} \quad , \quad \hat{e}^p = e_{,p} \quad , \quad \hat{s}^p = s_{,p} \quad , \quad \hat{T}^p = s_{,p} \cdot n \qquad (9)$$

and assuming that the body forces f are design insensitive, the equations of p th additional boundary-value problem have the form (cf. Dems and Haftka (1989))

$$\begin{aligned} &\operatorname{div} \hat{s}^p = 0 \quad , \quad \hat{e}^p = \tfrac{1}{2}(\nabla \hat{u}^p + \nabla^T \hat{u}^p) \quad \text{within } V \\ &\hat{u}^p = 0 \quad \text{on } S_u \quad , \quad \hat{T}^p = \hat{s}^p \cdot n = 0 \quad \text{on } S_T \\ &\hat{T}^p = \hat{s}^p \cdot n = -s_{,k} \cdot n v_k^p + s \cdot \nabla v_k^p n_k \quad \text{on } S_o \\ &\hat{s}^p = D \cdot \hat{e}^p \\ &p = 1, 2, \ldots, P \end{aligned} \qquad (10)$$

where $D = \partial E/\partial e$ can be regarded as a tangential stiffness matrix at the equilibrium solution s, e of the primary structure. For a stable elastic material D is symmetric and positive definite. Thus, though the primary problem can be physically nonlinear, all the additional problems are always linear.

Having P design parameters, P+1 boundary-value problems (1), (2) and (10) have to be solved in order to calculate the first-order sensitivity vector of any functional G, independently on the number of functionals. Using the solutions of problems (10), the sensitivities of G can be finally expressed in the form

$$\frac{DG}{Db_p} = \int(\psi_{,s}\cdot\hat{s}^p + \Psi_{,e}\cdot\hat{e}^p + \Psi_{,u}\cdot\hat{u}^p)\,dV + \int h_{,u}\cdot\hat{u}^p dS_T + \int h_{,T}\cdot\hat{T}^p dS_u + \int\Psi v_n^p dS_o \qquad (11)$$

$$p = 1,2,..,P$$

where $\hat{s}^p$, $\hat{e}^p$ and $\hat{u}^p$ are the local sensitivities of primary state fields with respect to design parameter b_p and v_n^p denotes the normal component of transformation velocity field associated with b_p on varying boundary portion S_o.

The alternate method of determining sensitivity vector of G consists in introducing an adjoint structure with an imposed fields of body forces and initial strains and stresses

$$f^a = \Psi_{,u} \quad , \quad e^{ai} = \Psi_{,s} \quad , \quad s^{ai} = \Psi_{,e} \qquad (12)$$

and subjected to the following set of boundary conditions

$$u^{ao} = -h_{,T} \quad \text{on } S_u \quad , \quad s^a\cdot n = T^{ao} = h_{,u} \quad \text{on } S_T \qquad (13)$$

The adjoint state equations can be written in the form (cf. Dems and Mróz (1984))

$$\operatorname{div} s^a + f^a = 0 \quad , \quad e^a = \tfrac{1}{2}(\nabla u^a + \nabla^T u^a) \quad \text{within } V$$
$$s^a = D\cdot(e^a - e^{ai}) - s^{ai} \qquad (14)$$

where s^a, e^a and u^a are the state fields of adjoint structure. Using now (12), (13) and constitutive laws of additional and adjoint boundary-value problems (10) and (14) as well as the boundary conditions of p-th additional structure, the first three integrals on the right-hand side of sensitivity expression (8) can be rewritten in the form

$$\int(e^{ai}\cdot s_{,p} + s^{ai}\cdot e_{,p} + f^a\cdot u_{,p})\,dV + \int T^{ao}\cdot u_{,p}dS_T - \int u^{ao}\cdot s_{,p}\cdot n dS_u = \left\{\int e^a\cdot s_{,p}dV - \int u^{ao}\cdot s_{,p}\cdot n dS_u\right\} -$$

$$\left\{\int(s^a\cdot e_{,p}-f^a\cdot u_{,p})dV-\int T^{ao}\cdot u_{,p}dS_T\right\}=\int u^a\cdot s_{,p}\cdot ndS_o= \tag{15}$$

$$\int(u^a\cdot s_k\cdot\nabla v_k^p n_k-u^a\cdot s_{,k}\cdot nv_k^p)dS_o$$

Substituting (15) into (8), after some transformations, the first order sensitivities of any functional G, expressed in terms of primary and adjoint solutions, have the final form (cf. Dems and Haftka (1989))

$$\frac{DG}{Db_p}=\int(\Psi-s\cdot e^a+f\cdot u^a)v_n^p dS_o \tag{16}$$

and are obtained as the results of solutions of primary problem and one additional adjoint problem for each functional to be considered, independently on the number of design parameters.

Thus, when we limit ourselves to the first-order sensitivity analysis and there is M functionals and P design parameters, then the direct approach would still require P+1 solutions in order to generate the sensitivities of M functionals, whereas the adjoint approach would need M+1 solutions.

3. Second-order sensitivities for varying free boundary

Let us consider now the problem of evaluating the second-order sensitivity matrix of an arbitrary functional G (5) with respect to change of shape of free boundary S_o of structure. It was shown by Haftka and Mróz (1986) for the case of material parameters variation that the most efficient method of generating the second-order sensitivities of any functional is to use both direct and adjoint solutions of first-order sensitivity analysis. For P design parameters and M functionals this mixed approach requires 1+P +M solutions of boundary-value problems in order to get the first- and second-order sensitivities of M functionals with respect to P design parameters. In this Section the mixed approach for second-order sensitivity analysis will be applied to the case of structural shape variation.

To start our analysis, we calculate the material derivatives of (8) with respect to shape parameter b_r. Using the notation (9) and denoting the second derivatives of primary state fields by

$$u_{,pr}=\hat{u}^{pr}\ ,\ e_{,pr}=\hat{e}^{pr}\ ,\ s_{,pr}=\hat{s}^{pr}\ ,\ s_{,pr}\cdot n=\hat{T}^{pr} \tag{17}$$

it follows from (8)

$$\frac{D^2G}{Db_pDb_r} = \int\Big[\Psi,_{ss}\cdot\hat{s}^p\cdot\hat{s}^r + \Psi,_{se}\cdot(\hat{s}^p\cdot\hat{e}^r + \hat{e}^p\cdot\hat{s}^r) +$$

$$\Psi,_{su}\cdot(\hat{s}^p\cdot\hat{u}^r + \hat{u}^p\cdot\hat{s}^r) + \Psi,_{ee}\cdot\hat{e}^p\cdot\hat{e}^r + \Psi,_{eu}\cdot(\hat{e}^p\cdot\hat{u}^r + \hat{u}^p\cdot\hat{e}^r) +$$

$$\Psi,_{uu}\cdot\hat{u}^p\cdot\hat{u}^r\Big]dV + \int h,_{uu}\cdot\hat{u}^p\cdot\hat{u}^r dS_T + \int h,_{TT}\cdot\hat{T}^p\cdot\hat{T}^r dS_u +$$

$$\int(\psi,_s\cdot\hat{s}^{pr} + \Psi,_e\cdot\hat{e}^{pr} + \Psi,_u\cdot\hat{u}^{pr})dV + \int h,_u\cdot\hat{u}^{pr}dS_T + \tag{18}$$

$$\int h,_T\cdot\hat{T}^{pr}dS_u + \int\Big[(\psi,_s\cdot\hat{s}^p + \Psi,_e\cdot\hat{e}^p + \Psi,_u\cdot\hat{u}^p)v_n^r +$$

$$(\Psi,_s\cdot\hat{s}^r + \Psi,_e\cdot\hat{e}^r + \Psi,_u\cdot\hat{u}^r)v_n^p + (\Psi v_k^p),_k v_n^r\Big]dS_o$$

The fourth, fifth and sixth integrals on the right-hand side of (18) involve the second derivatives of state fields of primary structure. these integrals can be eliminated by using the solution for adjoint structure. Making use of (12) - (14), we can write

$$\int(\psi,_s\cdot\hat{s}^{pr} + \Psi,_e\cdot\hat{e}^{pr} + \Psi,_u\cdot\hat{u}^{pr})dV + \int h,_u\cdot\hat{u}^{pr}dS_T +$$

$$\int h,_T\cdot\hat{T}^{pr}dS_u = \int(e^{ai}\cdot\hat{s}^{pr} + s^{ai}\cdot\hat{e}^{pr} + f^a\cdot\hat{u}^{pr})dV + \tag{19}$$

$$\int T^{ao}\cdot\hat{u}^{pr}dS_T - \int\hat{T}^{pr}\cdot u^{ao}dS_u = \int\hat{s}^{pr}\cdot e^a - s^a\cdot\hat{e}^{pr} + f^a\cdot\hat{u}^{pr})dV +$$

$$\int T^{ao}\cdot\hat{u}^{pr}dS_T - \int\hat{T}^{pr}\cdot u^{ao}dS_u = \int T^{pr}\cdot u^a dS_o$$

In writing (19) the conditions $\hat{T}^{pr} = 0$ on S_T and $\hat{u}^{pr} = 0$ on S_u were used. These conditions follow from the fact that the shape of loaded and supported boundary portions is fixed and then the load and support conditions on these boundary portions are design insensitive. Furthermore, we should note that $T^a = 0$ on S_o for adjoint structure. The surface tractions $\hat{T}^{pr}$ on S_o can be expressed in terms of stress field of primary structure and solutions of additional boundary-value problems defined by (10). In fact, taking the material derivative with respect to design parameter b_r of both sides of load condition on S_o in (10), we get

$$\hat{T}_i^{pr} = \hat{s}_{ij}^{pr} n_j = -(s_{ij},_l^p v_l^r + s_{ij},_k^r v_k^p) n_j + s_{ij}^p v_l^r,_j n_l +$$

$$s_{ij}^r v_k^p,_j n_k + s_{ij}(v_k^p,_{lj} v_l^r n_k - v_k^p,_j v_l^r,_k n_l) +$$

$$s_{ij},_k v_k^p v_l^r,_j n_l + s_{ij},_l v_k^p,_j v_l^r n_k - \tag{20}$$

$$(s_{ij},_k v_k^p),_l v_l^r n_j - s_{ij},_k v_k^{pr} n_j + s_{ij} v_k^{pr},_j n_k$$

where v_k^{pr} denotes the local derivative $\partial v_k^p/\partial b_r$. Using now (20) in (19), after some transformations, we obtain

$$\int(\psi,_s \cdot \hat{s}^{pr} + \Psi,_e \cdot \hat{e}^{pr} + \Psi,_u \cdot \hat{u}^{pr})dV + \int h,_u \cdot \hat{u}^{pr} dS_T + \int h,_T \cdot \hat{T}^{pr} dS_u =$$

$$- \int\left[(\hat{s}^p v_n^r + \hat{s}^r v_n^p)\cdot e^a + (s\cdot e^a v_k^p),_k v_n^r + s\cdot e^a v_k^{pr} n_k\right] dS_o \tag{21}$$

and then the integrals of (18) involving the second derivatives of primary state fields are expressed in terms of primary and adjoint solutions and first-order sensitivity fields.

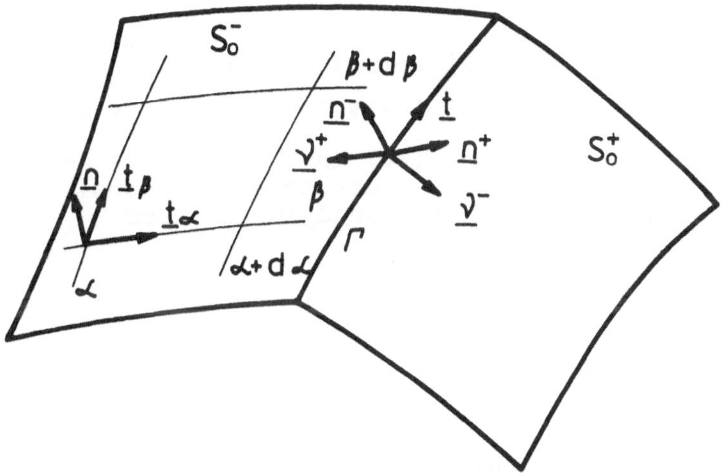

Fig. 2 Parameterization of piecewise regular surface S .

Let now the traction-free surface S_o of a structure be parameterized by an orthogonal curvilinear system (α,β) coincides with the principal curvature lines on S_o, Fig.2, where α and β are thecurvature lines parameters, so that $dS = d\alpha d\beta$. The unit tangent vectors to S_o along lines α and β are denoted by t_α and t_β and the unit normal vector $n = t_\alpha \times t_\beta$. Assuming that S_o is a piecewise regular surface, as shown in Fig. 2, denote by Γ the intersection curve between two adjacent parts of S_o and introduce the two unit vectors ν^- and ν^+ defined by

$$\nu^- = t \times n^- \quad , \quad \nu^+ = (-t) \times n^+ \tag{22}$$

where t is the unit vector tangent to Γ and n^- and n^+ are unit normal vectors to both parts of S_o along Γ. The transformation velocity vector v^p associated with shape parameter b_p can be decomposed on any part of S_o as follows

$$v_\alpha^p = v^p \cdot t_\alpha \quad , \quad v_\beta^p = v^p \cdot t_\beta \quad , \quad v_n^p = v^p \cdot n \tag{23}$$

whereas along the intersection curve Γ the following components of v^p are introduced

$$v_\nu^{p-} = v^p \cdot \nu^- \quad , \quad v_\nu^{p+} = v^p \cdot \nu^+ \tag{24}$$

Note that the pair (v_ν^{p-}, v_ν^{p+}) of the *tangential surface velocities* along Γ is, by simple geometry, uniquely defined by the pair (v_n^{p-}, v_n^{p+}) of the normal surface velocities along this line, provided that the intersection angle is not equal to π nor 0.

Substituting now (21) into (18) and performing some transformations, the components of second-order sensitivity matrix of functional G are expressed in the form

$$\frac{D^2G}{Db_pDb_r} = \int\Big[\Psi_{,ss} \cdot \hat{s}^p \cdot \hat{s}^r + \Psi_{,se} \cdot (\hat{s}^p \cdot \hat{e}^r + \hat{e}^p \cdot \hat{s}^r) +$$

$$\Psi_{,su} \cdot (\hat{s}^p \cdot \hat{u}^r + \hat{u}^p \cdot \hat{s}^r) + \Psi_{,ee} \cdot \hat{e}^p \cdot \hat{e}^r + \Psi_{,eu} \cdot (\hat{e}^p \cdot \hat{u}^r + \hat{u}^p \cdot \hat{e}^r) +$$

$$\Psi_{,uu} \cdot \hat{u}^p \cdot \hat{u}^r\Big] dV + \int h_{,uu} \cdot \hat{u}^p \cdot \hat{u}^r dS_T + \int h_{,TT} \cdot \hat{T}^p \cdot \hat{T}^r dS_u +$$

$$\int\Big\{(f^a \cdot \hat{u}^p - s^a \cdot \hat{e}^p) v_n^r + (f^a \cdot \hat{u}^r - s^a \cdot \hat{e}^r) v_n^p -$$

$$[s\cdot e^a_{,n} + s^a\cdot e_{,n} - f^a\cdot u_{,n} + 2H(\Psi - s\cdot e^a)]v^p_n v^r_n + \tag{25}$$

$$(\Psi - s\cdot e^a)(v^{pr}_k n_k + n_k v^p_{k,n} v^r_n - v^p_\alpha v^r_{n,\alpha} - v^p_\beta v^r_{n,\beta})\Big\}dS_o +$$

$$\sum\oint(\psi - s\cdot e^a)[\![v^p_\nu v^r_n]\!]d\Gamma$$

where H denotes the mean curvature of S_o and the sum of the line integrals is taken over all edges of surface S_o. $[\![v^p_\nu v^r_n]\!]$ denotes here the jump of proper components of transformation velocity fields associated with shape parameters b_p and b_r taken as the difference on both sides of intersection line Γ.

As it can be seen from (25), the second-order sensitivity of G is expressed as the sum of volume integral over structure domain, surface integrals over fixed surfaces S_u and S_T and surface and line integrals on varying piecewise regular surface S_o and it depends on derivatives of integrands of G and direct and adjoint solutions described by (10) and (12)-(14), respectively. When, in particular, the varying surface S_o is smooth, all line integrals on the right-hand side of (25) vanish. For P design parameters, the first-order sensitivity vector and second-order sensitivity matrix of an arbitrary functional G are obtained as the result of only 2+P solutions.

4. Illustrative example

In this Section we will discuss a simple example in order to illustrate the presented analysis for evaluating the first- and second-order sensitivities of an arbitrary functional. Let be given a prismatic bar with elliptical cross-section of semiaxes b_1 and b_2 subjected to torsion and bending, Fig. 3. Denoting the twisting and bending moments by M_t and M_b, respectively, the nonvanishing stress components within bar cross-section are expressed in the form

$$s_{13} = -\frac{2M_t x_2}{\pi b_1 b_2^3}, \quad s_{23} = \frac{2M_t x_1}{\pi b_1^3 b_2}, \quad s_{33} = \frac{4M_b x_2}{\pi b_1 b_2^3} \tag{26}$$

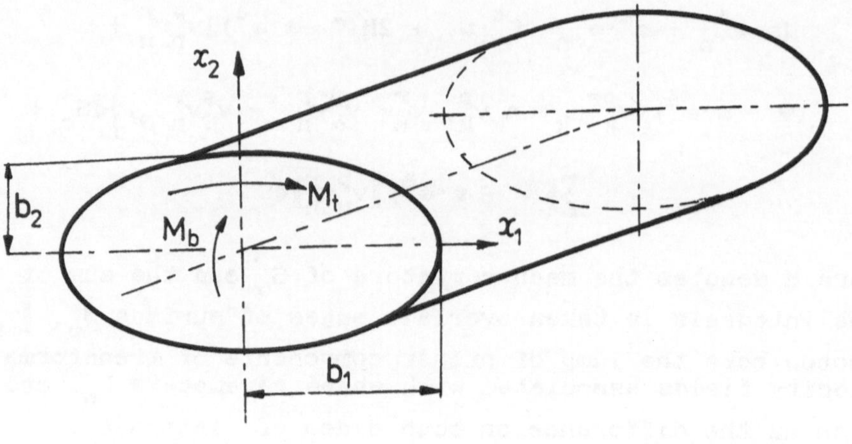

Fig. 8 Prismatic bar with elliptical cross-section.

Consider now the following functional

$$G = \int \Psi(s)\,dA = \int \left[\frac{s_M}{s_o} \right]^n dA = \int \left[\frac{\sqrt{s_{33}^2 + 3(s_{13}^2 + s_{23}^2)}}{s_o} \right]^n dA \quad (27)$$

where s_M denotes the von Mises stress and s_o is a prescribed stress level. Thus, for n tending to infinity, the functional G can be treated as the global measure of stress intensity within bar cross-section domain. Derive now the first- and second-order sensitivities of G with respect to varying semiaxes b_1 and b_2 of bar cross-section for $n = 2$ and $s_o = 1$. The transformation velocity fields associated with shape parameters b_1 and b_2 have the components

$$v_1^1 = \frac{x_1}{b_1}, \quad v_2^1 = 0, \quad v_1^2 = 0, \quad v_2^2 = \frac{x_2}{b_2} \quad (28)$$

The sensitivity solutions of boundary-value problems (10) following from direct approach can be written in the form

$$s_{13}^1 = -\frac{s_{13}}{b_1}, \quad s_{23}^1 = -\frac{3s_{23}}{b_1}, \quad s_{33}^1 = -\frac{s_{33}}{b_1}$$

$$s_{13}^2 = -\frac{3s_{13}}{b_2}, \quad s_{23}^2 = -\frac{s_{23}}{b_2}, \quad s_{33}^2 = -\frac{3s_{33}}{b_2} \quad (29)$$

The adjoint bar, in view of (12) and (13), is subjected to imposed field of initial strains

$$e^{ai}_{13} = \Psi,_{s_{13}} = 6s_{13} \quad , \quad e^{ai}_{23} = \Psi,_{s_{23}} = 6s_{23}$$
$$e^{ai}_{33} = \Psi,_{s_{33}} = 2s_{33} \tag{30}$$

with vanishing twisting and bending moments M^a_t and M^a_b. Thus, the solution for adjoint bar are given in the form

$$e^{a}_{13} = e^{ai}_{13} = 6s_{13} \quad , \quad s^a_{13} = 0$$
$$e^{a}_{23} = e^{ai}_{23} = 6s_{23} \quad , \quad s^a_{23} = 0 \tag{31}$$
$$e^{a}_{33} = e^{ai}_{33} = 2s_{33} \quad , \quad s^a_{33} = 0$$

The first-order sensitivities of functional G follow from (11) or (16), accordingly to the applied direct or adjoint approach. Using (16), we can write

$$\frac{DG}{Db_1} = \int(\psi - s\cdot e^a)v^1_n dS_o \quad , \quad \frac{DG}{Db_2} = \int(\psi - s\cdot e^a)v^2_n dS_o \tag{32}$$

Substituting now (26), (28) and (31) into (32) and integrating along outer edge of bar cross-section, the first derivatives of functional (27) with respect to b_1 and b_2 can be written in the form

$$\frac{DG}{Db_1} = -\frac{1}{\pi b_1^4 b_2^3}\left[4b_1^2M_b^2 + 3(b_1^2 + 3b_2^2)M_t^2\right]$$
$$\tag{33}$$
$$\frac{DG}{Db_2} = -\frac{3}{\pi b_1^3 b_2^4}\left[4b_1^2M_b^2 + (3b_1^2 + b_2^2)M_t^2\right]$$

Similarly, we can derive the second derivatives of (27) by using the expression (25) which is now simplified to the form

$$\frac{D^2G}{Db_pDb_r} = \int\Psi,_{ss}\cdot\hat{s}^p\cdot\hat{s}^r dA + \int\Big\{-[s\cdot e^a,_n + (\Psi - s\cdot e^a)K]v^p_n v^r_n + \tag{34}$$

$$(\Psi - s\cdot e^{a})(n_{k}v_{k}^{p}v_{n}^{r} + n_{k}v_{k}^{pr} - v_{s}^{p}v_{n,s}^{r})\Big\}dS_{o} \quad , \quad p,r=1,2$$

where K denotes the curvature of cross-sectional outer edge. Substituting the primary, direct and adjoint solutions (26), (29) and (31) as well as (28) into (34), after integration, the second derivatives of G equal

$$\frac{D^2G}{Db_1Db_1} = \frac{2}{\pi b_1^5 b_3^3}\left[4b_1^2M_b^2 + 3(b_1^2 + 6b_2^2)M_t^2\right]$$

$$\frac{D^2G}{Db_1Db_2} = \frac{3}{\pi b_1^4 b_3^4}\left[4b_1^2M_b^2 + 3(b_1^2 + b_2^2)M_t^2\right] \tag{35}$$

$$\frac{D^2G}{Db_2Db_2} = \frac{2}{\pi b_1^3 b_3^5}\left[24b_1^2M_b^2 + 3(6b_1^2 + b_2^2)M_t^2\right]$$

5. Concluding remarks

This paper supplements the results of previous works and provides a systematic variational approach to sensitivity analysis of first and second orders for structure with varying traction-free boundary. The discussed methods provide an effective tool in generating first and second variations of an arbitrary volume and/or surface integrals which can be useful in solving of optimization, redesign or identification problems.
Although the presented analysis was limited to the shape modification of free boundary part, the obtained results can be easily extended for the cases of modification of loaded and supported boundary parts.

References

Choi, K.K. and Haug, E.J.(1983). Shape design sensitivity analysis of elastic structures, J. Struct. Mech. 11,231-269.

Choi, K.K. and Seong, H.G.(1986). A domain method for shape design sensitivity analysis of built-up structures. Comp. Meth. Appl. Mech. Engng. 57, 1-15.

Dems, K. and Mróz, Z.(1984). Variational approach by means of adjoint system to structural optimization and sensitivity analysis. Int. J. Solids Struct. 20,527-552.

Dems, K. and Mróz, Z.(1985). Variational approach to first-

and second-order sensitivity analysis of elastic structures. Int. J. Num. Meth. Engng. 21, 637-661.

Dems, K. and Haftka, R.T.(1989). Two approaches to sensitivity analysis for shape variation of structures. Mech. Struct. Mach. 16, 379-400.

Dems, K.(1991). First- and second-order shape sensitivity analysis of structures. Struct. Optimization 3, 79-88.

Fujii, N.(1986). Domain optimization problems with a boundary value problem as a constraint. Proc. 4th IFAC Symp. *Control of Distributed Parameter Systems* (Edited by H.E. Rauch), pp. 5-9, Pergamon Press.

Haber, R.B.(1986). A new variational approach to structural shape design sensitivity analysis. Proc. NATO ASI *Computer Aided Optimal Design: Structural and Mechanical Systems*(Edited by C.A. Mota Soares), pp.573-587, Springer.

Haug, E.J and Arora, J.S(1978). Design sensitivity analysis of elastic mechanical systems. Comp. Meth. Appl. Mech. Engng. 15, 35-62.

Haug, E.J.(1981). Second-order design sensitivity analysis of structural systems, A.I.A.A. J. 19, 1087-1088.

Haftka, R.T.(1982). Second-order sensitivity derivatives in structural analysis. A.I.A.A. J. 20, 1765-1766.

Haftka, R.T. and Mróz, Z.(1986). First- and second-order sensitivity analysis of linear and nonlinear structures. A.I.A.A. J. 24, 1187-1192.

SHAPE SENSITIVITY ANALYSIS AND OPTIMAL DESIGN OF PLATES WITH VARYING EXTERNAL AND INTERNAL BOUNDARIES

K. DEMS
Łódź Technical University I-26
ul. Żwirki 36
90-924 Łódź
Poland

ABSTRACT. *A uniform variational formulation of sensitivity analysis for physically nonlinear plates is presented in terms of generalized stress and strains. Both the external and internal boundary shape modifications are treated within this formulation. Next the optimal design problems for stress and deflection constraints are formulated and relevant optimality conditions are derived.*

1. Introduction

The present paper is devoted to a variational formulation of sensitivity analysis and optimal design of plates subjected to flexure within small deflection and strain theory. However, a nonlinear relation is assumed between generalized stresses and strains. In optimal design problems, local or global constraints are usually set on displacements and stresses. The objective function then corresponds to a minimum of weight or cost of material of plate. In order to derive the relevant optimality conditions, explicit expressions for variations of constraint equations and objective functions in terms of the variations of design functions are to be determined. This step is called the *sensitivity analysis* and can be performed for any structure by using the *direct* or *adjoint approaches*. These two approaches have received considerable attention in the literature, cf. Choi and Haug (1983), Choi and Seong (1986), Haber(1986), Dems and Mróz (1984), Mróz (1986) and others. The general case of sensitivity analysis in the case of physical nonlinearity was discussed by Dems and Haftka (1988-89), Dems and Mróz (1987) and the case of both physically and geometrically nonlinear beams and plates was considered by Mróz *et al.* (1985) and Mróz (1987).

In this paper only the adjoint state method will be applied to derive sensitivity expressions for arbitrary va-

G. I. N. Rozvany (ed.), Optimization of Large Structural Systems, Vol. I, 493–508.

riations of plate shape design functions. The sensitivity analysis will be performed with respect to external and internal plate boundary modifications. The interfaces separating domains of different material properties, rib-stiffeners and slip or hinge lines will be considered as the internal boundaries within plate domain. The detailed analysis for discussed sensitivity expressions and relevant optimality conditions can be found in Dems and Mróz (1989, 1991) and Dems, Mróz and Szeląg (1989) and then in the present paper only the main results of these previous works will be presented.

2. Sensitivity analysis with respect to external boundary variation

Consider a plate occupying the domain A with the boundary S, Fig.1. The plate is subjected to transverse load p, whereas the generalized tractions $\bar{R}$ and displacements $\bar{w}$ are specified on boundary portion S_R and S_u, respectively. To simplify our foregoing analysis, let us assume the homogeneous form of boundary conditions along S_R and S_u. Denote the generalized stresses (i.e. bending and twisting moments within plate domain) by **M**, the associated strains (i.e curvatures and torsion) by **k** and the lateral deflection by w. The generalized stress-strain relation is given in the form

$$\mathbf{M} = \mathbf{S}(\mathbf{k}) \tag{1}$$

In the case of elasticity, **S** is generated by a potential

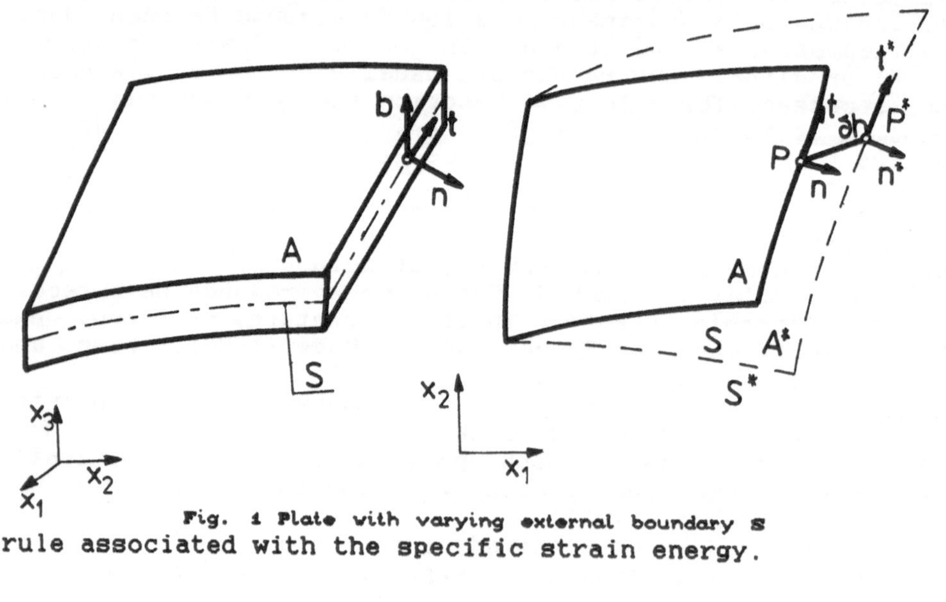

Fig. 1 Plate with varying external boundary S

rule associated with the specific strain energy.

Under applied load the plate passes from its initial configuration to a deformed one specified by the deflection field w. In addition to the *deformation process*, let us consider a *transformation process* which modifies the plate domain, $\mathbf{x}^t = \mathbf{x} + \mathbf{h}$, with imposed transformation field $\mathbf{h}(\mathbf{x})$ specified within A and along S. Obviously this transformation field modifies the shape of external boundary of a plate and affects deflection, strain and stress fields within plate domain.

Considering a simultaneous variation of transformation and state fields [cf. Dems and Mróz (1984)], any point P within plate domain, initially placed at $\mathbf{x}$, is transformed to the actual position $\mathbf{x}^*$ according to the rule

$$P \Rightarrow P^* \;:\; \mathbf{x}^* = \mathbf{x} + \delta\mathbf{h}(\mathbf{x}) \tag{2}$$

whereas the state fields for the actual configuration of plate are

$$\begin{aligned} w^*(\mathbf{x}^*) &= w(\mathbf{x}) + \mathring{\delta}w(\mathbf{x}) = w(\mathbf{x}) + \bar{\delta}w + w_{,k}\delta h_k \\ \mathbf{k}^*(\mathbf{x}^*) &= w(\mathbf{x}) + \mathring{\delta}\mathbf{k}(\mathbf{x}) = \mathbf{k}(\mathbf{x}) + \bar{\delta}\mathbf{k} + \mathbf{k}_{,k}\delta h_k \\ \mathbf{M}^*(\mathbf{x}^*) &= \mathbf{M}(\mathbf{x}) + \mathring{\delta}\mathbf{M}(\mathbf{x}) = \mathbf{M}(\mathbf{x}) + \bar{\delta}\mathbf{M} + \mathbf{M}_{,k}\delta h_k \end{aligned} \tag{3}$$

where $\mathring{\delta}$ denotes the total variation with respect to a fixedCartesian reference system and $\bar{\delta}$ is the local variation for a fixed configuration of the plate (cf. Fig. 2). Consider now any vector field $\mathbf{f}(s)$ specified along plate boundary S, whose

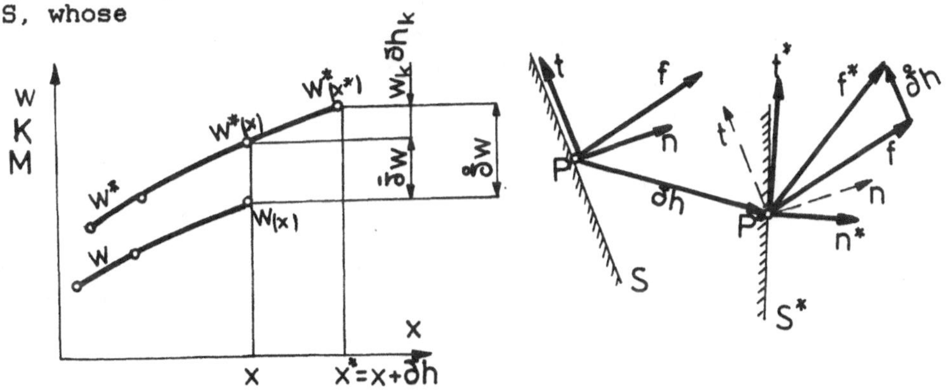

Fig. 2 Variation of state field within plate domain

components in the global fixed Cartesian coordinate system are denoted by $f_j (j=1,2,3)$ and in the local coordinate system $(\mathbf{n},\mathbf{t},\mathbf{b})$ by f_n, f_s and f_b, respectively. Following Dems and Mróz(1989), besides the total variation of $\mathbf{f}$ we can introducethe corotational variations $\delta f_n, \delta f_s, \delta f_b$ of its componentswith respect to the local reference system, which do not take into account the rotation of this system during the transformation process. The relations between the total and corotational variations of f are given in the form

$$\overset{o}{\delta f_j} = n_j \delta f_n + t_j \delta f_s + (n_j f_s - t_j f_n)(K\delta h_s + \delta h_{n,s}) \quad j=1,2$$

$$\overset{o}{\delta f_3} = \delta f_b \tag{4}$$

where K denotes the mean curvature of S and δh_n and δh_s are the normal and tangential components of transformation field $\delta \mathbf{h}$.

Consider now the following functional

$$G = \int_A \Psi(\mathbf{M},\mathbf{k},p,w)\,dA \tag{5}$$

depending on generalized stress and strain fields, transverse load and deflection within plate domain. Our goal is to determine the first variation of G with respect to variation of plate shape described by a transformation field $\delta \mathbf{h}$. Assuming Ψ to be a continuous and differentiable function of its arguments, the first variation of G equals

$$\delta G = \int(\Psi,_{\mathbf{M}}\cdot\bar{\delta}\mathbf{M}+\Psi,_{\mathbf{k}}\cdot\bar{\delta}\mathbf{k}+\Psi,_{p}\delta p+\Psi,_{w}\delta w)\,dA + \int \Psi \delta h_n dS \tag{6}$$

To eliminate local variations of state fields from (6), let us introduce an adjoint, physically linear plate of the same shape as the primary one, but subjected to the imposed fields of initial moments and curvatures specified by

$$\mathbf{M}^{ai} = \Psi,_{\mathbf{k}} \quad , \quad \mathbf{k}^{ai} = \Psi,_{\mathbf{M}} \quad \text{within A} \tag{7}$$

and loaded by

$$p^a = \Psi,_w \quad \text{within A} \tag{8}$$

Furthermore we assume that the adjoint plate is subjected to the homogeneous set of boundary conditions op the same kind as the primary plate. The stress field $\mathbf{M}^a$ within the adjoint plate is related to its strain field $\mathbf{k}^a$ by the relation

$$\mathbf{M}^a = \mathbf{D}^T\cdot(\mathbf{k}^a - \mathbf{k}^{ai}) - \mathbf{M}^{ai} \tag{9}$$

with the stiffness matrix $\mathbf{D}$ being the tangent stiffness matrix $\partial \mathbf{S}/\partial \mathbf{k}$ of the primary plate at the equilibrium point. Obviously $\mathbf{M}^a$ satisfies the equilibrium conditions for the adjoint plate and $\mathbf{k}^a$ is the associated curvature field that follows from the deflection field w^a. Using now (7)-(9) and noting that $\bar{\delta}\mathbf{M} = \mathbf{D}\cdot\bar{\delta}\mathbf{k}$, Eq.(6) can be rewritten in the form

$$\delta G = \int[(\mathbf{D}^T\cdot\mathbf{k}^{ai} + \mathbf{M}^{ai})\cdot\bar{\delta}\mathbf{k} + \Psi,_p\delta p + p^a\delta w]dA + \int\Psi\delta h_n dS =$$
$$\int(\mathbf{k}^a\cdot\bar{\delta}\mathbf{M} - \mathbf{M}^a\cdot\bar{\delta}\mathbf{k} + \Psi,_p\bar{\delta}p + p^a\bar{\delta}w)dA + \int\Psi\delta h_n dS \qquad (10)$$

Using now virtual work equations for strain and stress fields $\bar{\delta}\mathbf{k}$ and $\mathbf{M}^a$ as well as $\bar{\delta}\mathbf{M}$ and $\mathbf{k}^a$ in Eq.(10), after some transformations and integrations by parts along the plate boundary [cf. Dems and Mróz (1989)], the first variation of functional G can be expressed explicitly in terms of variations of transformation field $\mathbf{h}(\mathbf{x})$ along modified external boundary S. For homogeneous boundary conditions along edges of primary plate, δG is reduced to the form

$$\delta G = \int_A(w^a + \Psi,_p)\bar{\delta}p\,dA + \int_S(\Psi + pw^a - M_s k_s^a -$$
$$2M_{ns}k_{ns}^a + Qw^a_{,n} + M_n^a k_n)\delta h_n dS \qquad (11)$$

and is expressed in terms of integrand of (5), primary and adjoint state fields and normal component of boundary variation along S. Thus, the first variation of G is obtained via the adjoint approach as the result of solutions of only two boundary-value problems.

3. Sensitivity analysis for interface variation

Consider now a two-phase elastic plate contained in a domain A and bounded by the fixed boundary S. Assume the plate to be composed of two subdomains A_1 and A_2 separated by the interface Γ, Fig. 3. This interface can separate domains of different material properties or different plate thicknesses. Regardless of the finite jump of material properties on Γ, the deflection field, tangential curvature and torsion as well as the generalized internal tractions are continuous along interface Γ. Thus this assumption yields

$$[\![w]\!] = [\![w,_n]\!] = [\![k_s]\!] = [\![k_{sn}]\!] = 0 \quad , \quad [\![M_n]\!] = [\![Q]\!] = 0 \qquad (12)$$

where $[\![.]\!]$ denotes the jump of the enclosed quantity on Γ calculated as a difference of the respective values in the domains A_1 and A_2.

The cases when the internal tractions or generalized displacements suffer discontinuities across Γ will be treated in Sections 4 and 5.

Consider now an infinitesimal variation of plate configuration prescribed by a continuous and differentiable transformation vector field $\delta\mathbf{h}(\mathbf{x})$ satisfying the condition

$\delta h \cdot n = 0$ along S. Thus, during this transformation the external shape of a plate is fixed whereas the shape of interface Γ undergoes shape modification.

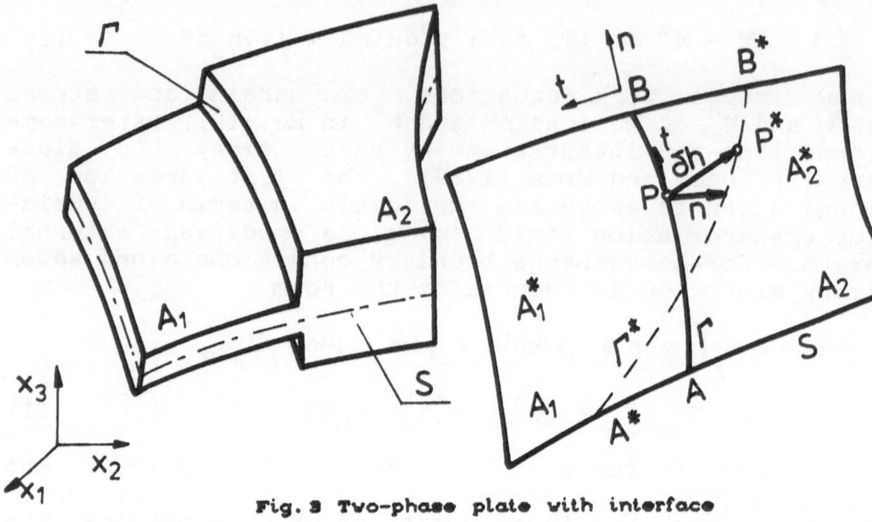

Fig. 3 Two-phase plate with interface

Similarly as in the previous Section, we introduce now an arbitrary functional G expressed in the form

$$G = G_1 + G_2 = \int_{A_1} \Psi_1(M,k,p,w)\,dA_1 + \int_{A_2} \Psi_2(M,k,p,w)\,dA_2 =$$

$$\int_{A} \Psi(M,k,p,w)\,dA \qquad (13)$$

and derive its first variation with respect to the shape variation of interface Γ. To do this, we shall utilize the results obtained in Section 2 by treating the plate domain A as two subdomains A_1 and A_2 bounded by fixed boundary portions S and varying boundary portion Γ along which the conditions (12) have to be fulfilled. Then, we introduce the adjoint plate of the same shape as the primary one that is defined by Eqs.(7)-(9). It is obvious that the adjoint solution w^a, k^a, M^a satisfies the continuity conditions along Γ, expressed in a form similar to (12). To write the expression for the first variation of (13), we apply Eq.(10) to both subdomains A_1 and A_2 of the plate domain A. Next, to eliminate the local variations of primary state fields we use the virtual work equations for state fields $\bar{\delta}k$, M^a and k^a, δM. Keeping in mind the conditions (13) for primary state fields and those similar for adjoint fields and assuming homogeneous boundary conditions along S of primary plate, the first variation of an arbitrary functional G (13) is finally expressed in the form

$$\delta G = \int_A (w^a + \Psi,_p)\bar{\delta} p dA +$$

$$\int_\Gamma ([\![\Psi]\!] + [\![p]\!] w^a - [\![M \cdot k^a]\!] + M_n [\![k_n^a]\!] + M_n^a [\![k_n]\!]) \delta h_n d\Gamma \tag{14}$$

and is expressed in terms of integrand of (13), primary and adjoint state fields and normal component of boundary variation along varying interface Γ.

4. Sensitivity analysis for stiffener shape variation

Assume the plate to be stiffened by a rib of an arbitrary shape and a cross-section symmetrical with respect to the middle surface, Fig.4. Denote the plate domain by A=A ∪A where A and A are subdomains specified by a rib intersecting the boundary S. The rib axis Γ lies within A and may have its end points on S.

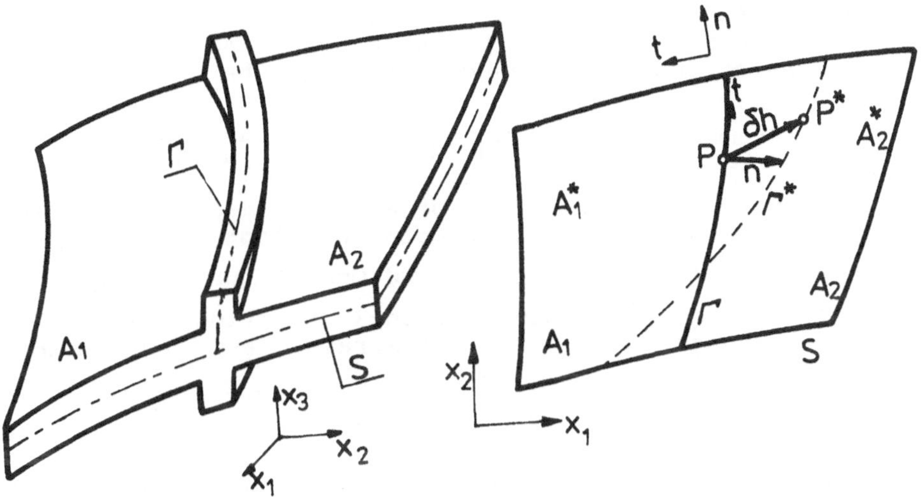

Fig. 4 Plate stiffened by a rib of varying shape

Assume now that the cross-sectional properties and the shape of rib are not specified in advance. Let the cross-sectional property and the shape of the rib be dependent on a set of material design functions $\gamma_l(x)$, $(l=1,2,..L)$, $x \in \Gamma$, and shape design functions $h_k(x)$, $(k=1,2)$, $x \in \Gamma$. Our purpose is to express explicitly the variation of an arbitrary functional in terms of variations of $\gamma_l(\mathbf{x})$ and $h_k(\mathbf{x})$, cf. Fig.4. It is also assumed that the plate domain may undergo the infinitesimal transformation $\delta\mathbf{h}(A)$ where $\delta\mathbf{h}$ is a differentiable vector field satisfying the conditions

$$\delta\mathbf{h}(A) = \delta\mathbf{h}(\Gamma) \quad \text{for } \mathbf{x} \in \Gamma \quad , \quad \delta\mathbf{h}\cdot\mathbf{n} = 0 \quad \text{for } \mathbf{x} \in S \tag{15}$$

Thus the external boundary does not undergo any normal shape transformation. On the other hand, when the rib penetrates the external plate boundary, the tangential shape transformation may only occur. The problem of optimal design of discrete stiffeners in plates was treated by Samsonov (1978). The present analysis, however, differs from that in Samsonov (1978). The optimal distribution of circular ribs in plastic plates was determined by Kozłowski and Mróz (1969). Optimization of densely stiffened plate was analyzed in Rozvany *et al.* (1982,1987) by assuming the equivalent orthotropic model.

Similarly as before, the plate is subjected to a distributed lateral pressure p. The rib within plate domain is simulated by a plane arch Γ with free or supported ends A and B, subjected to bending in **t**,**b**-plane and torsion in **n**,**b**-plane. Denote by **v**, **e**, **L** the generalized displacement, strain and stress fields specified along the rib axis, where now

$$\mathbf{v}^T = \{v\ \theta\} \quad , \quad \mathbf{e}^T = \{k_b\ k_t\} \quad , \quad \mathbf{L}^T = \{M_b\ M_t\} \tag{16}$$

Here v denotes the deflection of a rib in **b**-direction, and θ is the angle of cross-section rotation along **t**-axis, k_b and M_b denote the bending curvature and moment, whereas k_t and M_t are the torsion and cross-sectional twisting moment. The relation between the generalized strains and displacements have the form

$$k_b = \vartheta_{,s} + K\theta \quad , \quad k_t = \theta_{,s} - K\vartheta \tag{17}$$

where $\vartheta_{s} = v$, is the angle of cross-section rotation along **n**-axis.

The generalized displacements, strains and stresses within subdomains A_1 and A_1 of the plate are denoted by $\mathbf{w}^1$, $\mathbf{w}^2$, $\mathbf{k}^1$, $\mathbf{k}^2$ and $\mathbf{M}^1$, $\mathbf{M}^2$, respectively.

It is obvious, that the following kinematic relations between the generalized displacements and strains of the plate and rib hold along Γ

$$v = w^1 = w^2 \quad , \quad \vartheta = v_{,s} = w^1_{,s} = w^2_{,s} \quad , \quad \theta = -w^1_{,n} = -w^2_{,n}$$

$$k_b = w,_{ss} - Kw,_n = -k_s^1 = -k_s^2 \tag{18}$$
$$k_t = -(w,_{ns} + Kw,_s) = k_{sn}^1 = k_{sn}^2$$

The generalized stresses within a rib are related to the jump of internal forces within plate domains by the equilibrium equations of the form

$$T,_s - [\![\Omega]\!] = 0 \ , \ M_b,_s + M_t K + T = 0$$

$$M_t,_s - M_b K - [\![M_n]\!] = 0 \tag{19}$$

where $[\![M_n]\!]$ and $[\![\Omega]\!]$ denote the jumps of plate bending moment and generalized shear force along Γ, and T denotes the cross-sectional shear force within the rib.

We assume, similarly as previously, that generalized stress-strain relation within plate domain is given in the general non-linear form (1), whereas within rib domain this relation has the form

$$L = C(e,c) \tag{20}$$

Similarly as in Section 3, consider now the functional

$$G = \int \Psi(M,k,p,w)\,dA \ + \int \Phi(L,e,c)\,d\Gamma \tag{21}$$

Rewritting the functional (21) in the form

$$G = G_1 + G_2 + G_r = \int \Psi dA_1 + \int \Psi dA_2 + \int \Phi d\Gamma \tag{22}$$

its first variation equals

$$\delta G = \delta G_1 + \delta G_2 + \delta G_r \tag{23}$$

Following the analysis presented in Sections 2 and 3, we introduce now the adjoint stiffened plate of the same shape as the primary one and subjected to the same kind of boundary conditions which are specified as follows

$$\bar{w}^a = 0 \ , \ \bar{w}^a,_n = 0 \ \text{on} \ S_u \ , \ \bar{Q}^a = \ \ , \ \bar{M}^a = 0 \ \text{on} \ S_T \tag{24}$$

where S_u and S_T are the supported and loaded parts of plate outer edge, respectively. Furthermore, the adjoint plate is subject to imposed fields of initial strain and stress specified by

$$k^{ai} = \Psi,_M \ , \ M^{ai} = \Psi,_k \qquad \text{within } A = A_1 \cup A_2$$

$$e^{ai} = \Phi,_{L}\ ,\quad L^{ai} = \Phi,_{e} \qquad \text{along } \Gamma \tag{25}$$

and is loaded by the transverse pressure

$$p^{a} = \Psi,_{w} \qquad \text{within } A \tag{26}$$

The stress field M^a and L^a within domain A and along Γ are related to the strain fields k^a and e^a by the linear relation (9) and

$$L^{a} = E^{T}\cdot(e^{a} - e^{ai}) - L^{ai} \qquad \text{along } \Gamma \tag{27}$$

where $E = \partial C/\partial e$. Obviously, the fields M^a and L^a satisfy the equilibrium conditions, while k^a and e^a follow from the displacement fields w^a and v^a. The kinematic relations along Γ have the form similar to (18), while the static conditions are similar to those expressed by Eqs.(19).

Using the solutions for primary and adjoint plates we can express the variation of G explicitly in terms of material and shape variations of the plate rib. Following the analysis presented in Dems *et al.* (1989) and in Section 2, we can derive first the variations of functionals G_1 and G_2 of (22) by using (10) and next the variation of functional G defined within rib domain. This variation can be obtained in the similar way as that presented in Dems and Mróz (1987) and Dems *et al.* (1989). Thus, the first variation of functional G (21) has finally the form

$$\begin{aligned}\delta G = &\int (w^{a} + \Psi,_{p})\bar{\delta} p\, dA + \int \Big\{ [\![\Psi]\!] - \Phi K + [\![p]\!] w^{a} - [\![M_{s}]\!] k_{s}^{a} - 2[\![M_{ns}]\!] k_{ns}^{a} + \\ &[\![M_{n}^{a} k_{n}]\!] + (M_{t} k_{b}^{a} - M_{b} k_{t}^{a} + M_{t}^{a} k_{b} - M_{b}^{a} k_{t}),_{s} + T k_{t}^{a} + T^{a} k_{t} - (M_{b}^{a} k_{b} + M_{t}^{a} k_{t}) K \Big\} \delta h_{n}\, d\Gamma + \\ &\int \Big[\Phi,_{\gamma_{1}} - (e^{a} - \Phi,_{L}) \cdot \frac{\partial C}{\partial c} \Big] . \bar{\delta} c\, d\Gamma + \\ &\Big[(\Phi + M_{b}^{a} \vartheta,_{s} + M_{t}^{a} \theta,_{s}) \delta h_{s} + (M_{b} \theta^{a},_{s} - M_{t} \vartheta^{a},_{s} + M_{b}^{a} \theta,_{s} - M_{t}^{a} \vartheta,_{s}) \delta h_{n} \Big] \Big|_{A^{\Gamma}}^{B^{\Gamma}} = \\ &\delta G_{h} + \delta G_{c}\end{aligned} \tag{28}$$

and it is expressed explicitly in terms of variations of shape and material functions of a rib, as well as the solutions for primary and adjoint plates. The last term in square brackets expresses the variation of G due to shape variation at rib ends within primary plate. When the rib is closed then this term vanishes.

5. Sensitivity analysis for shape variation of hinge or slip line

In this Section, we shall consider a plate occupying the domain A with the boundary S, cf. Fig. 5.. Let us assume now that there exists within the disk domain A a line Γ, along which the displacement vector **w** can undergo some discontinuities while the tractions **R** transferred across the line are continuous. The plate is subjected to a transverse load p, whereas the generalized tractions **R** on external boundary part S_R and the generalized boundary displacements **w** on the remaining part S_w vanish due to assumption of homoge-

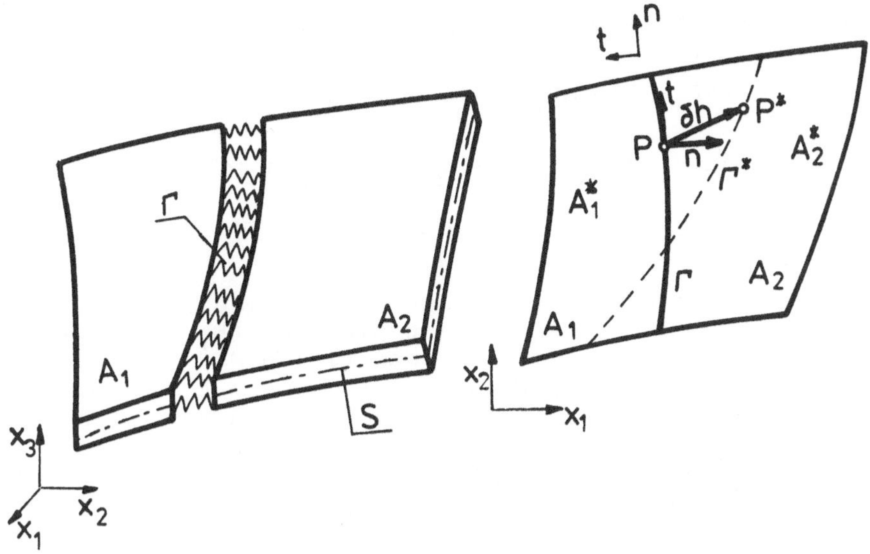

Fig. 5 Thin plate with a hinge or slip line of varying shape

neous boundary conditions. Moreover, the plate can be subjected to the imposed fields of initial strains $\mathbf{k}^i$ and stresses $\mathbf{M}^i$. The generalized stress-strain relation within plate domain is assumed in the form similar to (1), namely

$$\mathbf{M} = S(\mathbf{k}, \mathbf{k}^i, \mathbf{M}^i) \tag{29}$$

Let us now assume that the generalized displacements **w** along discontinuity line Γ within plate domain can suffer some jump in deflection w and/or its normal derivative $-w_{,n}$, which will be denoted by $\mathbf{v}(\mathbf{x}) = \mathbf{w}_2(\mathbf{x}) - \mathbf{w}_1(\mathbf{x})$, $\mathbf{x} \in \Gamma$. The surface tractions **R** along Γ are still continuous and they are related to the kinematic discontinuity vector $\mathbf{v}$ by the relation of the form

$$\mathbf{R} = \mathbf{C}(\mathbf{v},\mathbf{c}) \tag{30}$$

Assuming now that the discontinuity line Γ can undergo the shape variation satisfying conditions (15), derive the first variation of an arbitrary functional G given by

$$G = \int \Psi(\mathbf{M},\mathbf{k},p,w,\mathbf{M}^i,\mathbf{k}^i)\,dA + \int \Phi(R_\xi,v_\xi,c)\,d\Gamma \tag{31}$$

Following the analysis of Dems and Mróz(1989,1991), the first variation of (31) can be written in the form

$$\begin{aligned}\bar{\delta}G = &\int (\Psi_{,\mathbf{M}}\cdot\bar{\delta}\mathbf{M} + \Psi_{,\mathbf{k}}\cdot\bar{\delta}\mathbf{k} + \Psi_{,p}\bar{\delta}p + \Psi_{,w}\bar{\delta}w + \Psi_{,\mathbf{M}^i}\cdot\bar{\delta}\mathbf{M}^i + \\ &\Psi_{,\mathbf{k}^i}\cdot\bar{\delta}\mathbf{k}^i)\,dA - \int [\![\Psi]\!]\delta h_n d\Gamma + + \int [\Phi_{,R_\xi}(\delta R_\xi - R_{\xi,s}\delta h_s) + \\ &\Phi_{,v_\xi}(\delta v_\xi - v_{\xi,s}\delta h_s) + (\Phi\delta h_s)_{,s} - \Phi K\delta h_n]\,d\Gamma + \\ &\int \Phi_{,c}\cdot(\bar{\delta}c + \delta_n c)\,d\Gamma\end{aligned} \tag{32}$$

Similarly as in previous Sections, besides the primary plate, let us introduce an adjoint plate of the same geometry as the primary one, with imposed fields $\mathbf{M}^{ai}$, $\mathbf{k}^{ai}$ of initial stresses and strains within plate domain A and initial jump of displacements $\mathbf{v}^{ai}$ and initial continuous tractions $\mathbf{R}^{ai}$ along discontinuity line Γ. The adjoint plate is subjected to the lateral pressure p^a within A. The particular form of loading and support conditions as well as initial fields depends on the functional to be considered, namely

$$\begin{gathered}\tilde{\mathbf{w}}^a = 0 \text{ on } S_w \;, \quad \tilde{\mathbf{R}}^a = 0 \text{ on } S_R \\ \mathbf{v}^{ai} = \Phi_{,\mathbf{R}} \;, \quad \mathbf{R}^{ai} = \Phi_{,\mathbf{v}} \text{ on } \Gamma \\ \mathbf{k}^{ai} = \Psi_{,\mathbf{M}} \;, \quad \mathbf{M}^{ai} = \Psi_{,\mathbf{k}} \;, \quad p^a = \Psi_{,w} \text{ within } A\end{gathered} \tag{33}$$

The constitutive laws for the adjoint plate are assumed in the form

$$\begin{aligned}\mathbf{M}^a &= \mathbf{D}^T\cdot(\mathbf{k}^a - \mathbf{k}^{ai}) - \mathbf{M}^{ai} \quad \text{within } A\;, \\ \mathbf{R}^a &= \mathbf{E}^T\cdot(\mathbf{v}^a - \mathbf{v}^{ai}) - \mathbf{R}^{ai} \quad \text{along } \Gamma\end{aligned} \tag{34}$$

where now $\mathbf{D} = \partial\mathbf{S}/\partial\mathbf{k}$ and $\mathbf{E} = \partial\mathbf{C}/\partial\mathbf{v}$ follow from Eqs.(29)-(30) and can be regarded as the tangent stiffness matrices at the solution point. The stress field $\mathbf{M}^a$ satisfies the equilibrium and boundary conditions, whereas the adjoint total strains k^a follow from adjoint displacement field w^a.

Using (33) and (34) in (32) and applying the principle of virtual work we can eliminate terms involving the local and corotational variations of primary state fields and then the first variation of functional G can be finally expressed in the form

$$\delta G = \Big\{ \int \{ (\Psi_{,p} + w^a)\bar{\delta}p + [\Psi_{,M^i} - (k^a - k^{ai}) \cdot \frac{\partial S}{\partial M^i}] \cdot \bar{\delta}M^i + [\Psi_{,k^i} - (k^a - k^{ai}) \cdot \frac{\partial S}{\partial k^i}] \cdot \bar{\delta}k^i \} dA + \int \{ -[\![\Psi]\!] + [\![M \cdot k^a]\!] - [\![pw^a]\!] + (R \cdot v^a - \Phi)K - R_\xi v^a_{\xi,n} - R^a_\xi v_{\xi,n} + (M_{nn} v^a_{,s})_{,s} + (M^a_{nn} v_{,s})_{,s} \} \delta h_n d\Gamma - \int [\Phi_{,c} - (v^a - v^{ai}) \frac{\partial C}{\partial c}] (\bar{\delta}c + \delta_n c) d\Gamma \Big\} \tag{35}$$

Thus, the first variation of G depends explicitly on shape and material variations of discontinuity line Γ, integrands of functional G and their derivatives as well as on solutions of primary and adjoint plates.

6. Optimal shape design of plate

The typical optimal design problem involves minimization (or maximization) of any objective functional subject to the set of global or local constraints. When the global structural cost is to be minimized, the global constraints can be imposed on generalized strains, stresses or displacements. An alternative formulation would require the minimization (or maximization) of an arbitrary functional of generalized stresses, strains or displacements, which can be expressed in the form similar to (5), (13), (22) or (31). In this case, the constraint can be set on the upper bound of the structural generalized cost, that is

$$\text{min. (or max.) } G \quad \text{subject to} \quad \mathcal{K} - \mathcal{K}_o \le 0 \tag{36}$$

where $\mathcal{K}_o$ is a prescribed quantity. Introducing the Lagrange functional

$$G' = G + \lambda(\mathcal{K} - \mathcal{K}_o + \beta^2) \tag{37}$$

where λ and β denote the Lagrange multiplier and slack variable, its stationarity condition yields the following optimality conditions

$$\delta_c G = -\lambda \delta_c \mathcal{K} \quad , \quad \delta_h G = -\lambda \delta_h \mathcal{K} \tag{38}$$

with the switching and constraint conditions of the form

$$\delta\beta = 0 \quad , \quad \delta\lambda(\mathcal{K} - \mathcal{K}_\sigma + \beta^2) = 0 \tag{39}$$

Here $\delta_c G$ and $\delta_h G$ denote the variations of functional G with respect to variations of material parameters and shape of external or internal boundaries, whereas $\delta_c\mathcal{K}$ and $\delta_h\mathcal{K}$ are the variations of cost function.

The objective functional G can express, in integral form, both local and global quantities depending on generalized stresses, strains or displacements of plate. Consider, for instance, the minimization of maximal local deflection w within plate domain A. the objective functional can be represent here by

$$G = \left[\int |w|^p dA\right]^{1/p} \tag{40}$$

since for p tending to infinity, G is tending to w^{max}. Similarly, the maximum local stress component or generalized stress intensity can be obtained by considering the functional

$$G = \left[\int \Psi^p(M)\, dA\right]^{1/p} \tag{41}$$

where p is even and Ψ is assumed to be a homogeneous function of generalized stresses of order one. In fact, for $p \to \infty$, $G \to \sup.\ \Psi$, that is the functional tends to the maximum value of its integrand. Another method is to applying the penalty approach. Namely, introducing the acceptable stress intensity level Ψ_o, we can consider the functional

$$G = \int \left|\frac{\Psi(M)}{\Psi_o}\right|^p dA \tag{42}$$

For $p \to \infty$ the integrand $|\Psi/\psi_o|^p$ tends to zero for $\Psi/\Psi_o < 1$ and tends to infinity for $\psi/\psi_o > 1$. This provides a proper penalty functional which for large p takes very small values when $\psi < \psi_o$ and very large ones when $\psi > \psi_o$.

In a similar way we can convert any local quantity $\Psi(x_o)$ specified at a given point x_o of a structure into the integral form. Using the well known property of the Dirac delta function, one can write

$$\Psi(x_o) = G = \int \Psi(x)\,\delta(x - x_o)\, dA \tag{43}$$

and the the sensitivity analysis can be performed similarly as in previous Sections.

7. Concluding remarks

The present paper summarizes the results of previous works and provides a systematic variational approach to sensitivity analysis and optimal design for wide class of plates with varying external boundaries and interfaces. The analysis is limited to geometrical linear and physical nonlinear plates for which the concept of adjoint plate provides an effective tool in generating the first variation of any functional prescribed over plate domain.

References

Choi, K.K., and Haug, E.J. (1983), Shape design sensitivity analysis of elastic structures, *J. Struct. Mech.*, 11, 231-269.

Choi, K.K., and Seong, H.G. (1986), A domain method for shape design sensitivity analysis of built-up structures, *Comp. Meth. Appl. Mech. Eng.*, 57, (1), 1-15.

Dems, K., and Mróz, Z. (1984), Variational approach by means of adjoint systems to structural optimization and sensitivity analysis, *Int. J. Solids Struct.*, 20, (6), 527-552.

Dems, K, and Mróz, Z. (1987),Variational approach to sensitivity analysis and structural optimization of plane arches, *Mech. Struct. Mach.*, 15, (3), 297-321.

Dems, K. and Haftka, R.T. (1988-89), Two approaches to sensitivity analysis for shape variation of structures, *Mech. Struct. Mach.*,16, 4, 501-522
Dems, K. and Mróz, Z. (1989), Sensitivity analysis and optimal design of physically nonlinear plates, *Arch. Mech.*, 41, 4, 481-501

Dems, K. and Mróz, Z. (1991), Shape sensitivity analysis and optimal design of disks and plates with strong discontinuities of kinematic fields, *Int. J. Solids Struct.* (in print)

Dems, K., Mróz, Z. and Szeląg, D. (1989), Optimal design of rib-stiffeners in disks and plates, *Int. J. Solids Struct.*, 25, 9, 973-998Haber, R.B. (1986), A new variational approach to structural shape design sensitivity analysis, Proc. NATO ASI *Computer Aided Optimal Design: Structural and Mechanical Systems*, Ed. C.A. Mota Soares, *Springer Verlag*, 573-587.

Kozłowski, W., and Mróz, Z. (1969), Optimal design of solid plates, *Int. J. Solids Struct.*, 5, 781-794

Mróz, Z., Kamat, M.P., and Plaut, R.H. (1985), Sensitivity analysis and optimal design of non-linear beams and plates, *J. Struct. Mech.*, 13, 245-266.

Mróz, Z. (1987), Sensitivity analysis and optimal design with account for varying shape and support conditions, Proc. NATO ASI *Computer Aided Optimal design: Structural and Mechanical Systems*, Ed. C.A. Mota Soares, *Springer Verlag*, 407-439.

Rozvany, G.I.N., Olhoff, N., Cheng, K.T., and Taylor, J. (1982), On the solid plate paradox in structural optimization, *J. Struct. Mech.*, 10, 1-32.

Rozvany, G.I.N., Ong, T.G., Szeto, W.T., Sandler, R., Olhoff, N., and Bendsøe, M.P. (1987), Last-weight design of perforated elastic plates, *Int. J. Solids Struct.*,Part I., 23, (4), 521-536, Part II., 23, (4), 537-550.

Samsonov, A.M. (1978), Optimal position of an elastic thin rib on elastic plate, *Mech. Tv. Tela*, 1, 132-138.

DUAL METHODS FOR CONVEX SEPARABLE PROBLEMS

C. FLEURY
Aerospace Laboratory
University of Liege
Rue E. Solvay, 21
B-4000 Liege, Belgium

ABSTRACT. In this Lecture, dual methods for solving constrained optimization problems are presented. These methods proceed by maximizing a dual function which depends only on the Lagrangian multipliers associated with the constraints. The Lagrangian multipliers, also called dual variables, are required to remain non-negative. The dual method approach is specially useful and efficient when dealing with convex, separable problems.

In order to apply dual methods to general problems, a successful strategy consists of using convex approximation schemes. In this "Sequential Convex Programming" (SCP) approach, the primary optimization problem is replaced with a sequence of convex explicit subproblems having a simple algebraic form. Various SCP techniques have recently emerged, that have demonstrated strong potential for efficient solution of structural optimization problems. To illustrate the SCP approach, emphasis will be placed on the "Convex Linearization" (CONLIN) method, because it leads to a relatively simple dual formulation.

Next, attention is focused on convex, separable, quadratic problems for which the dual function can be written explicitly as a quadratic (but not separable) form. Efficient second order algorithms based on update formulas for the inverse Hessian matrix are presented. It is finally shown how general separable problems can be solved as a sequence of quadratic subproblems.

1. Introduction

This Lecture is focused on the use of the dual method approach to solve general constrained problems arising in structural optimization. These nonlinear programming problems can of course be solved iteratively by using primal optimization techniques such as those described in another Lecture. Each iteration begins with a complete analysis of the system behaviour in order to evaluate the objective function and constraint values along with their sensitivities to changes in the design variables (i.e. first derivatives). Most often the analysis capability is based on finite element discretization. A design iteration is concluded by employing the results of these behavioral and sensitivity analyses in a minimization algorithm which searches the n-dimensional design space for a new primal point that decreases the objective function value while remaining feasible (i.e. satisfying the constraints).

The essential difficulty in solving directly structural optimization problems via primal methods lies in the implicit character of the functions that define the problem. In other words, for each new design, these functions can only be evaluated numerically through

G. I. N. Rozvany (ed.), Optimization of Large Structural Systems, Vol. I, 509–530.

a finite element analysis. The iterative nature of the optimization process implies that many structural reanalyses must usually be accomplished before finding an acceptable solution. Those repeated analyses can lead to a prohibitive computational cost when dealing with large scale problems.

On the other hand, in the dual method approach, the constrained primal minimization problem is replaced by the maximization of a quasi-unconstrained dual function depending only on the Lagrangian multipliers associated with the constraints. These multipliers are the dual variables subject to simple non-negativity constraints. The dual method approach is well known and quite respected in the mathematical programming community [1,2]. In the context of structural optimization problems, it was initially introduced in [3], and it subsequently led to a reconciliation of optimality criteria techniques and mathematical programming methods [4]. The efficiency of the dual formulation is due to the fact that maximization is performed in the dual space, whose dimensionality is relatively low and depends on the number of active constraints at each design iteration.

It is essential to point out that duality concepts are best exploited for problems presenting special properties, including convexity and separability. Consequently, in order to effectively apply dual methods to general problems, specially devised "linearization" techniques must be used. A successful strategy consists of using convex approximation schemes. In this approach, that can be named "Sequential Convex Programming" (SCP), the primary optimization problem is replaced with a sequence of explicit approximate subproblems having a simple algebraic form. A very attractive feature of the SCP approach is that, because each subproblem is convex and separable, it can be readily solved by a dual method formulation.

Various SCP techniques have recently emerged, that have demonstrated strong potential for efficient solution of structural optimization problems. To illustrate the SCP approach, emphasis will be placed on the "Convex Linearization" (CONLIN) method, because it leads to a relatively simple dual formulation. In the convex linearization method, the initial problem is transformed into a sequence of explicit subproblems having a quite simple algebraic structure. Furthermore each subproblem is convex and separable. These properties make it attractive to solve the subproblem by using dual algorithms.

Turning to the algorithmic point of view, attention will be focused on convex, separable, quadratic problems for which the dual function can be written explicitly as a quadratic (but not separable) form. Efficient second order algorithms based on update formulas for the inverse Hessian matrix are presented. It is finally shown how general separable problems can be solved as a sequence of quadratic subproblems.

2. Primal and Dual Problems

2.1. THE LAGRANGIAN MULTIPLIER TECHNIQUE (EQUALITY CONSTRAINTS)

The origin of the dual method approach can probably be traced back to the classical Lagrangian multiplier technique for minimization problems involving equality constraints:

$$\begin{aligned} &\text{minimize} \quad f(x) \\ &\text{subject to} \quad c_j(x) = 0 \qquad j=1,m \end{aligned} \tag{1}$$

where x represents a point in the n-dimensional Euclidian space E^n, f(x) is the objective function, and $c_j(x)$ are the constraint functions (which are supposed to be linearly independent). Let us assume that this problem has at least one solution x^*. It is well known that, under rather unrestrictive conditions, there exists a vector $\lambda^* \in E^m$ of Lagrangian multipliers such that (x^*, λ^*) is a stationarity point of the Lagrangian function

$$L(x,\lambda) = f(x) - \sum \lambda_j c_j(x) \qquad (2)$$

The stationarity conditions state that:

$$\nabla_x L(x,\lambda) = 0 \qquad (3)$$

$$\nabla_\lambda L(x,\lambda) = 0 \qquad (4)$$

Therefore a stationary point is a solution of the n + m nonlinear equations:

$$\partial f(x)/\partial x_i - \sum_j \lambda_j \, \partial c_j(x)/\partial x_i = 0 \qquad i=1,n \qquad (3')$$

$$c_j(x) = 0 \qquad j=1,m \qquad (4')$$

with n + m unknowns $x_1,\ldots, x_n, \lambda_1,\ldots, \lambda_m$.

The Lagrangian multiplier technique consists of solving the system of equations (3,4) in order to obtain a solution for the constrained minimization problem (1). Obviously, this procedure increases the dimensionality of the problem. However, in some special cases, the n equations (3) can provide explicitly the variables x_i in terms of the Lagrangian multipliers λ_j, which then become the only unknowns; they are the solution of the system of nonlinear equations:

$$c_j[x(\lambda)] = 0 \qquad j=1,m$$

As an illustration, let us consider a quadratic problem with linear equality constraints:

$$\begin{aligned} &\text{minimize} \quad \tfrac{1}{2} x^T x \\ &\text{subject to} \quad C^T x = d \end{aligned} \qquad (5)$$

where C denotes the nxm matrix of the constraint gradients (which are assumed linearly independent). The Lagrangian function

$$L(x,\lambda) = \tfrac{1}{2} x^T x - \lambda^T (C^T x - d)$$

being fully explicit, the stationarity conditions lead to a closed-form solution of the problem, as shown below:

$$\nabla_x L \equiv x - C\lambda = 0 \quad ===> \quad x = C\lambda \qquad (3'')$$

$$\nabla_\lambda L \equiv C^T x - d = 0 \quad ===> \quad C^T x = d \qquad (4'')$$

Inserting (3") into (4") yields the optimal values of the Lagrangian multipliers:

$$\lambda^* = (C^TC)^{-1}d$$

from which the optimal values of the design variables can be recovered:

$$x^* = C\,\lambda^* = C(C^TC)^{-1}d$$

To introduce the concept of dual problem, it is useful to continue further on this example, and to observe that the relation $x = C\lambda$ resulting from the stationarity conditions can be interpreted as expressing the design variables x in terms of the Lagrangian multipliers λ. Now, inserting $x = C\lambda$ into the Lagrangian function yield a function which depends only on λ:

$$\ell(\lambda) = \tfrac{1}{2}\,\lambda^TC^TC\lambda - \lambda^T(C^TC\lambda - d)$$

$$= -\tfrac{1}{2}\,\lambda^TC^TC\lambda + \lambda^Td$$

This is precisely the dual function. As it will be shown later, the optimal values of the Lagrangian multipliers can be obtained by maximizing this function.

2.2. DUALITY FOR CONVEX PROBLEMS (INEQUALITY CONSTRAINTS)

In the more general case of inequality constraints, the Lagrangian multiplier technique can be generalized provided that adequate convexity conditions are stated. Although duality concepts can be used (with caution!) under weaker convexity properties, we shall simply assume that the "primal problem" to be solved,

$$\begin{aligned} &\text{minimize} \quad f(x) \\ &\text{subject to} \quad c_j(x) \geq 0 \qquad j=1,m \end{aligned} \tag{6}$$

is convex, that is, the objective function as well as the constraint functions are all convex. Then the well known Kuhn-Tucker conditions are necessary and sufficient for global optimality. These conditions can be thought of as a generalization to inequality constraints of the stationarity conditions (3,4). They state that x^* is the global minimum of the convex programming problem (6) if and only if there exist scalars $\lambda_1^*, \ldots, \lambda_m^*$ such that:

$$\partial f(x^*)/\partial x_i - \sum_j \lambda_j^*\, \partial c_j(x^*)/\partial x_i = 0 \qquad i=1,n$$

$$c_j(x^*) \geq 0 \qquad j=1,m$$

$$\lambda_j^*\, c_j(x^*) = 0 \qquad j=1,m$$

$$\lambda_j^* \geq 0 \qquad j=1,m$$

The non-negative scalars λ_j are called generalized Lagrangian multipliers or dual variables.

The first of these conditions implies that x^* is the unconstrained global minimum of the convex Lagrangian function $L(x,\lambda^*)$. Hence it is readily shown that there is a unique correspondence between the primal variables x and the dual variables λ, through the following unconstrained minimization problem, the so-called "Lagrangian problem":

$$\underset{x}{\text{minimize}}\ L(x,\lambda)$$

for which the optimality conditions are:

$$\partial L(x,\lambda)/\partial x_i = 0 \qquad i=1,n$$

Note that the minimization in the above Lagrangian problem is taken over x for any fixed λ. This permits defining the dual function, which depends solely on the dual variables λ_j:

$$\ell(\lambda) = \min_{x} L(x,\lambda)$$

Alternatively, if $x(\lambda)$ denotes the primal point minimizing the Lagrangian function for a given dual point λ, the dual function can be written:

$$\ell(\lambda) = L[x(\lambda),\lambda] = f[x(\lambda)] - \Sigma\, \lambda_j\, c_j[x(\lambda)] \tag{7}$$

It can be shown that the dual function is concave. In addition, for any feasible primal point x, and any feasible dual point λ, then:

$$\ell(\lambda) \leq f(x)$$

The "dual problem" consists of maximizing the dual function subject only to non-negativity constraints on the dual variables:

$$\text{maximize} \quad \ell(\lambda)$$

$$\text{subject to } \lambda_j \geq 0$$

It is useful to repeat here that we assume the primal problem to be convex. Under this condition, its dual is convex too and the respective solutions of both problems satisfy the same optimality conditions. As a consequence, the two problems are equivalent, in the sense that the solution to one provides a solution to the other. Furthermore their optimal values are equal:

$$\ell(\lambda^*) = f(x^*)$$

Probably the most interesting property of the dual function is that its first partial derivatives are easily available. They are given by the negative of the primal constraints:

$$\partial \ell / \partial \lambda_j = - c_j[x(\lambda)] \tag{8}$$

The optimality conditions for the quasi-unconstrained dual problem can be written:

$$\partial \ell / \partial \lambda_j \equiv c_j[x(\lambda)] = 0 \qquad \text{if } \lambda_j > 0$$

$$\partial \ell / \partial \lambda_j \equiv -c_j[x(\lambda)] < 0 \qquad \text{if } \lambda_j = 0$$

These conditions indicate that the maximum of the dual function will be attained when the constraints are satisfied as equalities for positive dual variables, and as inequalities for zero dual variables. Going a bit further, let us assume that a steepest ascent algorithm is employed to maximize the dual function, i.e.,

$$\lambda = \lambda + \alpha \nabla \ell,$$

or

$$\lambda_j = \lambda_j - \alpha c_j$$

where α denotes a step size. Thus a dual variable λ_j increases if the corresponding constraint c_j is violated, while it decreases (possibly reaching zero) if c_j is positive. From these considerations results an intuitive interpretation of the dual method approach: this approach attempts to satisfy the inequality constraints by adjusting the values of the dual variables.

Because non-negativity constraints are very simple to take into account, classical unconstrained maximization techniques can readily be adapted to solve the quasi-unconstrained dual problem. In particular, the conjugate gradient method is well suited: evaluating the dual function (7) demands the computation of the constraint values $c_j[x(\lambda)]$, so that the gradient (8) is directly obtained without additional computations.

2.3. DUALITY FOR SEPARABLE PROBLEMS

The main difficulty in the dual approach is that, at each point in the dual space, it is necessary to find the x that minimizes the Lagrangian function $L(x,\lambda)$ for fixed λ. This minimization problem in the primal space must be repeated many times, and this might well lead to a prohibitive computational cost. However, for certain problems, for example separable programming problems, this is not very cumbersome. Therefore, in addition to convexity, separability is an essential property for the dual formulation to be efficient.

A function is said to be separable if it can be written

$$c(x) = \Sigma\, c_i(x_i)$$

where each function $c_i(x_i)$ depends only on the single variable x_i. Separable functions benefit from some computationally important properties. In particular, the Hessian matrix of such a function is diagonal. Several examples of separable functions will be encountered later in this Lecture, as well as in other Lectures. Now, the constrained optimization problem (6) is a separable programming problem if each function $\{f(x), c_j(x)\}$

is itself separable. This implies that the Lagrangian function is separable too. As a result, the n-dimensional Lagrangian problem can be broken up into n one-dimensional minimization problems, and the dual function can be written as follows:

$$\ell(\lambda) = \sum_i \min_{x_i} [f_i(x_i) - \sum_j \lambda_j c_{ij}(x_i)]$$

In many cases, the single-variable minimization problem appearing above has a simple algebraic structure and it can be solved in closed form, yielding thus an explicit dual function.

As an illustration, let us reconsider the quadratic problem (5), but now with linear inequality constraints:

<u>Primal Problem</u>

$$\text{minimize} \quad \tfrac{1}{2} x^T x$$

$$\text{subject to} \quad C^T x \geq d$$

<u>Lagrangian Problem</u>

$$\text{minimize} \quad L(x,\lambda) = \tfrac{1}{2} x^T x - \lambda^T (C^T x - d)$$

To obtain the solution of this convex minimization problem, it is sufficient to state that the gradient of the Lagrangian function must vanish:

$$\nabla L = 0 \implies x(\lambda) = C\lambda$$

The dual function is obtained by inserting these fully explicit primal-dual relationships into the expression of the Lagrangian function:

$$\ell(\lambda) = \tfrac{1}{2} x^T(\lambda)\, x(\lambda) - \lambda^T [C^T x(\lambda) - d]$$

$$= \tfrac{1}{2} \lambda^T C^T C \lambda - \lambda^T (C^T C \lambda - d)$$

$$= -\tfrac{1}{2} \lambda^T C^T C \lambda + \lambda^T d$$

The gradient of the dual function is given by:

$$\nabla \ell(\lambda) = -C^T C \lambda + d$$

$$= d - C^T x(\lambda)$$

that is, by the negative of the primal constraints, in agreement with the general theory of duality.

From this latter expression, it follows that the dual Hessian matrix is constant:

$$\nabla^2 \ell(\lambda) = -C^T C$$

The dual function is therefore quadratic. It has the explicit form:

$$\ell(\lambda) = -\tfrac{1}{2}\,\lambda^T A \lambda + \lambda^T d$$

where $A = C^T C$ denotes the negative of the Hessian matrix.

In the case of equality constraints, the dual variables are unrestricted in sign. Therefore the maximum of the dual function can simply be obtained by stating that its gradient must vanish:

$$\nabla\ell(\lambda) \equiv -A\lambda + d = 0$$

leading to

$$\lambda^* = A^{-1} d$$

From this dual solution, the primal optimum is:

$$x^* = C\,\lambda^*$$

These dual and primal solutions are of course the same as those generated by the stationarity conditions of the Lagrangian function in Section 2.1.

Now, in the case of linear inequality constraints, the dual variables are required to remain non-negative, and the dual problem becomes:

<u>Dual Problem</u>

$$\text{maximize} \quad \ell(\lambda) = -\tfrac{1}{2}\,\lambda^T A \lambda + \lambda^T d$$

$$\text{subject to} \quad \lambda_j \geq 0$$

After solving this maximization problem in the dual space, the primal optimum can still be obtained as $x^* = C\,\lambda^*$

Because of the non-negativity conditions that the dual variables must fulfil, a direct solution from the stationarity conditions becomes difficult, if not impossible. For now almost two decades, the so-called "Optimality Criteria" technique has attempted to achieve such an easy direct solution, so far without any significant success. Maximizing the dual function, on the other hand, provides a quite natural and efficient alternative. An algorithm that is specially devised for that purpose is discussed in Section 4.1.

2.4. EXAMPLE: QUADRATIC PROBLEM

To illustrate the duality concepts explained above, we consider the following problem:

$$\text{minimize} \quad \tfrac{1}{2}x_1^2 + \tfrac{1}{2}x_2^2$$

$$\text{subject to} \quad x_1 + x_2 \geq 4$$

$$x_1 - x_2 \geq -4$$

The Lagrangian function for this explicit problem has the form:

$$L(x,\lambda_1,\lambda_2) = \tfrac{1}{2}\, x_1^{\,2} + \tfrac{1}{2}\, x_2^{\,2} - \lambda_1(x_1+x_2-4) - \lambda_2(x_1-x_2+4)$$

It is easily verified that the Lagrangian problem leads here to linear primal-dual relationships:

$$x_1(\lambda_1,\lambda_2) = \lambda_1 + \lambda_2$$

$$x_2(\lambda_1,\lambda_2) = \lambda_1 - \lambda_2$$

The dual function can now be obtained by substituting $x_1(\lambda_1,\lambda_2)$ and $x_2(\lambda_1,\lambda_2)$ in the definition of the Lagrangian function. This yields the following explicit dual problem:

maximize $\quad \ell(\lambda_1,\lambda_2) = -(\lambda_1 - 2)^2 - (\lambda_2 + 2)^2 + 8$

subject to $\quad \lambda_1 \geq 0$

$\qquad\qquad\quad \lambda_2 \geq 0$

As predicted by the duality theory, the first derivatives of the dual function are directly related to the primal constraints:

$$\partial\ell/\partial\lambda_1 = -2\lambda_1 + 4 \equiv 4 - x_1 - x_2$$

$$\partial\ell/\partial\lambda_2 = -2\lambda_2 - 4 \equiv -4 - x_1 + x_2$$

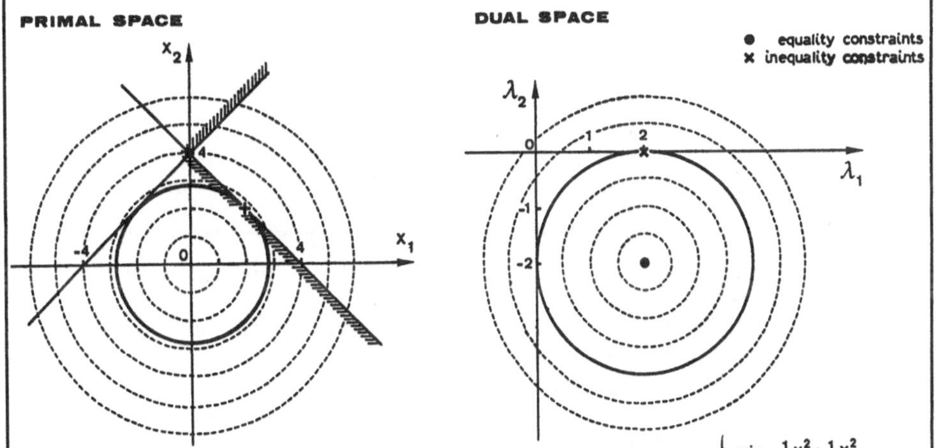

FIG. 1 EXAMPLE OF SEPARABLE QUADRATIC PROBLEM

A graphical representation of both the primal and dual spaces for this problem is shown in Fig. 1. It can be immediately concluded that the dual optimum lies at $\lambda^* = [2,0]^T$ (with

$l(\lambda^*) = 4$), corresponding to the primal solution $x^* = [2,2]^T$ (with $f(x^*)=4$). Note that the maximum value of the dual function coincides with the minimum value of the primal function. Consider now the same problem, but with equality instead of inequality constraints. The dual formulation derived above remains valid, except that the nonnegativity constraints on the dual variables must be removed. From Fig. 1, it is apparent that the unconstrained maximum of the dual function lies at $\lambda^* = [2,-2]^T$, corresponding to the primal solution $x^* = [0,4]^T$. It is worth noticing that the optimal values of the primal and dual functions are again identical ($f(x^*) = l(\lambda^*) = 8$).

The reader is encouraged to work deeply through the details of this example in order to better understand the dual approach. It is especially recommanded to resort to the matrix formulation previously presented for separable quadratic problems. In this specific case, the primal problem is described by the matrix

$$C = \begin{bmatrix} 1 & 1 \\ 1 & -1 \end{bmatrix}$$

while the dual problem is defined by the matrix

$$A = C^T C = \begin{bmatrix} 2 & 0 \\ 0 & 2 \end{bmatrix}$$

3. Sequential Convex Programming

Attention is now turned toward the use of the dual method approach to solve general structural optimization problems which do not inherently exhibit an explicit separable form. Such problems can be written in the following form:

$$\text{minimize} \quad c_0(x) \tag{9}$$

$$\text{subject to} \quad c_j(x) \leq 0 \qquad j=1,m \tag{10}$$

$$\underline{x}_i \leq x_i \leq \bar{x}_i \qquad i=1,n \tag{11}$$

where each function $c_j(x)$ depends implicitly on the design variables. In other words, for each new design, these functions can only be evaluated numerically, for example through a finite element analysis.

For reasons that will become clear later, the objective function is now denoted $c_0(x)$ instead of $f(x)$. Also, we have intentionally changed the sense of the inequality sign in the main constraints (10), often called behavior constraints. This is because they usually impose upper limits on structural response quantities such as stresses and displacements. The reader is invited to check the incidence of this change on the dual formulation presented below (expression of the Lagrangian function, of the dual gradient, etc...) It should be noted that the so-called side constraints (11) constitute a particular case of the more general constraints (10). However they are written separately in our optimization problem statement, because the dual approach described below can handle them more efficiently when considered apart from the behavior constraints.

This Section is concerned with various methods based on explicit approximation schemes, that have demonstrated a strong potential for efficient solution of design optimization problems. In these methods, the key idea is to replace the implicit problem (9-11) with a sequence of convex explicit subproblems having a simple algebraic form. Most of these approximation strategies make the assumption that the objective function and constraint functions are separable in terms of the design variables. Such an assumption is not new. During the last two decades, it has been extensively used in several research efforts dealing with structural optimization problems [4-7].

For example, it has now been well accepted that the best approaches to optimal sizing problems are those which make use of constraint linearization with respect to the reciprocal design variables. As demonstrated in Ref. [4], linearizing the behavior constraints in terms of the reciprocal variables was at the basis of both the optimality criteria approach to structural optimization [5], and the mathematical programming approach based on approximation concepts [6,7]. There is an intuitive explanation for the success of this idea, in that stresses and displacements are exact linear functions of the reciprocal sizing variables in the case of a statically determinate structure. Although there is no reason to believe that the use of reciprocal variables constitutes a universal way of solving structural optimization problems, surprisingly, this very special linearization technique has been highly successful in many cases. For shape optimal design problems, for example, there is no clear physical guideline for the selection of intermediate linearization variables. Nevertheless, the choice of reciprocal variables continues to have a beneficial effect on the convergence properties of the optimization process [8,9].

The key idea in the "Convex Linearization" method (CONLIN) [10-12] is to perform the linearization process with respect to mixed variables, either direct or reciprocal, independently for each function involved in the optimization problem. At each successive iteration point, the CONLIN method only requires evaluation of the objective and constraint functions and their first derivatives with respect to the design variables. The optimizer will then select by itself an appropriate approximation scheme on the basis of the signs of the derivatives.

The CONLIN method proceeds by linearizing each function defining the optimum design problem with respect to a properly selected mix of direct and reciprocal variables, so that a convex and separable subproblem is generated. The selection of the "intermediate" linearization variables is made on the basis of the signs of the first partial derivatives. It is easily proven that, considering any differentiable function c(x), the following linearization scheme yields a convex approximation (hence the appellation "convex linearization" suggested in Ref. [10]):

$$\tilde{c}(x) = c(x^0) + \Sigma_+ \, c_i^0(x_i - x_i^0) - \Sigma_- \, (x_i^0)^2 \, c_i^0(1/x_i - 1/x_i^0) \tag{12}$$

where c_i denote the first derivatives of c(x) with respect to the design variables, evaluated at the current design point x^0:

$$c_i^0 = \partial c(x^0)/\partial x_i$$

The symbol Σ_+ (Σ_-) means "summation over the terms for which c_i^0 is positive (negative)". One of the most interesting feature of the convex linearization scheme is that it also leads to the most conservative approximation amongst all possible combinations of mixed direct/reciprocal variables. This property was initially demonstrated in [13], where conservative approximation was employed to handle difficult buckling constraints.

The CONLIN algorithm applies this convex linearization scheme to the objective function and to all the constraint functions defining the optimization problem. It is convenient to normalize the design variables so that they become equal to unity at the current point x^0 where the problem is linearized:

$$x_i' = x_i/x_i^0 \quad ===> \quad c_i' = c_i^0 x_i^0$$

The factor $(x_i^0)^2$ disappears from (12), which becomes:

$$\tilde{c}(x') = c(x^0) + \Sigma_+ c_i'(x_i'-1) - \Sigma_- c_i'(1/x_i'-1) \tag{12'}$$

Applying the convex linearization technique to each function $c_j(x)$, and dropping the superscript $'$, the following explicit subproblem is generated:

$$\begin{aligned} &\text{minimize} \quad \Sigma_+ c_{io} x_i - \Sigma_- c_{io}/x_i - d_o \\ &\text{subject to} \quad \Sigma_+ c_{ij} x_i - \Sigma_- c_{ij}/x_i \leq d_j \qquad j=1, m \\ &\qquad \underline{x}_i \leq x_i \leq \bar{x}_i \end{aligned} \tag{13}$$

where the c_{ij} 's denote the normalized first derivatives of the objective and constraint functions evaluated at the current point x^0. Note that the constants d_j contain the zero order contributions in the Taylor series expansion:

$$d_j = \sum_i | c_{ij} | x_i^0 - c_j(x^0) \qquad j=0, m$$

Applying the dual method outlined in Section 2, the solution of the primal problem (13) can be obtained by the following "Max-Min" two-phase procedure:

$$\text{maximize} \quad \ell(\lambda) = \min_{\underline{x}_i \leq x_i \leq \bar{x}_i} L(x,\lambda) \tag{14}$$

The dual function $\ell(\lambda)$ to be maximized results from minimizing the Lagrangian function

$$L(x,\lambda) = \sum_{j=0}^{m} \lambda_j \left(\Sigma_+ c_{ij} x_i - \Sigma_- c_{ij}/x_i - d_j \right)$$

over the acceptable primal variables.

The separability of the primal problem implies that the Lagrangian function can be written as the sum of n individual functions $L_i(x_i)$, and therefore, the n-dimensional minimum problem (14) can be split into n single variable minimization problems:

$$\begin{aligned} &\text{minimize} \quad L_i(x_i) = a_i x_i + b_i/x_i \\ &\text{subject to} \quad \underline{x}_i \leq x_i \leq \bar{x}_i \end{aligned} \tag{15}$$

where a_i and b_i depend linearly on the dual variables λ_j:

$$a_i = \Sigma_+ c_{ij} \lambda_j \geq 0$$

and (16)

$$b_i = - \Sigma_- c_{ij} \lambda_j \geq 0$$

For any given set of dual variables, a_i and b_i are fixed coefficients in problem (15). These coefficients remain always non-negative in the feasible region of the dual space ($\lambda_j \geq 0$). Therefore the Lagrangian problem (15) has a unique solution, obtained by stating:

$$L_i'(x_i) \equiv a_i - b_i/x_i^2 = 0 \tag{17}$$

Since the side constraints (11) must be satisfied, it follows that:

$$x_i = (b_i/a_i)^{1/2} \quad \text{if} \quad \underline{x}_i^2 \leq b_i/a_i \leq \bar{x}_i^2 \tag{18}$$

$$x_i = \underline{x}_i \quad \text{if} \quad b_i/a_i \leq \underline{x}_i^2 \tag{19}$$

$$x_i = \bar{x}_i \quad \text{if} \quad \bar{x}_i^2 \leq b_i/a_i \tag{20}$$

Remembering that a_i and b_i depend on the dual variables (see Eq. 16), we have obtained fully explicit primal-dual relationships.

The dual problem can now be expressed in closed form:

$$\max \quad \ell(\lambda) = \sum_{j=0}^{m} \lambda_j \left[\Sigma_+ c_{ij} x_i(\lambda) - \Sigma_- c_{ij}/x_i(\lambda) - d_j \right]$$

$$\text{s.t.} \quad \lambda_j \geq 0$$

where the primal variables x_i are known explicitly in terms of the dual variables λ_j via (18-20). In order for the dual function to be bounded, the dual variables λ_j must be linked by a linear equality constraint. Following a common practice we shall simply assume that the Lagrangian multiplier λ_0 associated with the objective function is assigned a unit value.

A fundamental property of the dual function is that its first derivatives are simply given by the primal constraint values:

$$g_j \equiv \partial\ell/\partial\lambda_j = \Sigma_+ c_{ij} x_i(\lambda) - \Sigma_- c_{ij}/x_i(\lambda) - d_j$$

The dual problem can therefore be solved by applying first order algorithms based, for example, on the conjugate gradient method. Because the Hessian matrix of the dual function is readily available, second order methods can also be employed [11].

Example: linear programming problem

Although the convex linearization method does not pretend to replace the SIMPLEX algorithm, it is capable of solving efficiently a linear programming problem by transforming it into a sequence of (nonlinear) convex subproblems. To illustrate the CONLIN concepts previously described, let us consider this linear programming problem:

$$\text{minimize} \quad c_0(x) \equiv x_1 + 4x_2$$

$$\text{subject to} \quad c_1(x) \equiv x_1 - x_2 \leq 0$$

$$c_2(x) \equiv -3x_1 + 2x_2 \leq -1$$

For this problem, the function first derivatives are constant:

$$c_{10} \equiv \partial c_0/\partial x_1 = 1 \qquad c_{20} \equiv \partial c_0/\partial x_2 = 4$$

$$c_{11} \equiv \partial c_1/\partial x_1 = 1 \qquad c_{21} \equiv \partial c_1/\partial x_2 = -1$$

$$c_{12} \equiv \partial c_2/\partial x_1 = -3 \qquad c_{22} \equiv \partial c_2/\partial x_2 = 2$$

By applying the convex linearization scheme at the initial primal point $x^0 = [3,4]^T$ the following CONLIN subproblem is obtained:

$$\text{minimize} \quad x_1 + 4x_2$$

$$\text{subject to} \quad x_1 + 16/x_2 \leq 8$$

$$27/x_1 + 2x_2 \leq 17$$

Note that the linear objective function is not modified, and that only the terms corresponding to negative first derivatives are affected in the two constraint functions. Let us now write the Lagrangian function for this subproblem:

$$L(x,\lambda) = x_1+4x_2 + \lambda_1(x_1+16/x_2-8) + \lambda_2(27/x_1+2x_2-17)$$

Minimizing this function with respect to x_1 and x_2 for fixed λ_1 and λ_2, we get (see Eq. 17):

$$L_1'(x_1) \equiv 1 + \lambda_1 - 27\lambda_2/x_1^2 = 0$$

$$L_2'(x_2) \equiv 4 - 16\lambda_1/x_2^2 + 2\lambda_2 = 0$$

By solving these equations it can be concluded that:

$$x_1(\lambda_1,\lambda_2) = [27\lambda_2/(1+\lambda_1)]^{1/2}$$

$$x_2(\lambda_1,\lambda_2) = [8\lambda_1/(2+\lambda_2)]^{1/2}$$

The dual function can now be obtained by substituting $x_1(\lambda_1,\lambda_2)$ and $x_2(\lambda_1,\lambda_2)$ in the definition of the Lagrangian function. This leads to the following explicit dual problem:

$$\text{maximize} \quad \ell(\lambda_1,\lambda_2) = [27\lambda_2/(1+\lambda_1)]^{1/2} + [8\lambda_1/(2+\lambda_2)]^{1/2}$$
$$+ \lambda_1 \{[27\lambda_2/(1+\lambda_1)]^{1/2} + 16[(2+\lambda_2)/8\lambda_1]^{1/2} - 8\}$$
$$+ \lambda_2 \{27[(1+\lambda_1)/27\lambda_2]^{1/2} + 2[8\lambda_1/(2+\lambda_2)]^{1/2} - 17\}$$

$$\text{subject to} \quad \lambda_1 \geq 0$$
$$\lambda_2 \geq 0$$

It is easily verified that the first derivatives of the dual function are:

$$\partial\ell/\partial\lambda_1 = [27\lambda_2/(1+\lambda_1)]^{1/2} + 16[(2+\lambda_2)/8\lambda_1]^{1/2} - 8$$
$$\equiv x_1 + 16/x_2 - 8$$
$$\partial\ell/\partial\lambda_2 = 27[(1+\lambda_1)/27\lambda_2]^{1/2} + 2[8\lambda_1/(2+\lambda_2)]^{1/2} - 17$$
$$\equiv 27/x_1 + 2x_2 - 17$$

It is important to emphasize that the foregoing primal-dual relationships, as well as the definition of the dual function and its derivatives, become more complicated when side constraints are imposed in the problem statement. For example if the side constraints

$$4 \geq x_1 \geq 0.5$$
$$4 \geq x_2 \geq 0.5$$

are added, then the dual space is partitioned in several regions separated by second order discontinuity planes. As an exercise, the reader is invited to demonstrate that the equations of these four planes are:

$$\lambda_1 - 108\,\lambda_2 + 1 = 0 \qquad 32\,\lambda_1 - \lambda_2 - 2 = 0$$
$$16\,\lambda_1 - 27\,\lambda_2 + 16 = 0 \qquad \lambda_1 - 2\,\lambda_2 - 4 = 0$$

The definition of the dual problem changes when crossing such a discontinuity plane. The dual solution scheme must therefore take these changes into consideration. Clearly the expressions given above for the dual function and its derivatives are only valid in the region of the dual space where the primal variables are free.

After solving the dual problem, new values are obtained for the primal variables. The CONLIN scheme is then applied again at the new primal point, and an updated explicit subproblem follows, having a similar form. This convex subproblem is replaced with its dual and solved. The full process is repeated until convergence is achieved to the primal point $x^* = [1,1]^T$.

4. Second Order Dual Optimizer

This Section will concentrate on Convex Separable (CS) problems, because many general problems of practical interest can be solved as a sequence of CS subproblems via the use of convex approximation concepts. CONLIN, MMA, and the diagonal SQP method discussed in another Lecture are convincing evidence of the success of the convex approximation approach. Any CS problem can itself be broken up into a sequence of quadratic subproblems, by using second order Taylor series either in the primal space or in the dual space (or in both spaces in a primal-dual approach). Therefore the key is to solve efficiently an explicit primal problem of the following form: minimize a separable quadratic function subject to linear constraints as well as bounds on the design variables.

4.1. DUAL METHOD APPROACH FOR SEPARABLE, QUADRATIC PROBLEMS

Let us therefore consider the following separable quadratic problem, with linear inequality constraints and side constraints:

Primal Problem

minimize $\quad \frac{1}{2} \Sigma\, x_i^2$

subject to $\quad \Sigma\, c_{ij}\, x_i \geq d_j$

$$\bar{x}_i \geq x_i \geq \underline{x}_i$$

The side constraints, which impose lower and upper bounds on the design variables, can of course be viewed as linear inequality constraints. However this would increase dramatically the number of dual variables, and it is much better to treat the side constraints apart from the general linear constraints.

Lagrangian Problem

$$\min_{\bar{x}_i \geq x_i \geq \underline{x}_i} \quad L(x,\lambda) = \frac{1}{2} \sum_i x_i^2 - \sum_j \lambda_j \left(\sum_i c_{ij}\, x_i - d_j\right)$$

Because of the separability property, this n-dimensional minimization problem can be split into n one-dimensional problems:

$$\text{minimize} \quad L_i(x_i) = \frac{1}{2} x_i^2 - \left(\sum_j c_{ij}\lambda_j\right) x_i$$

$$\text{subject to} \quad \bar{x}_i \geq x_i \geq \underline{x}_i$$

Setting to zero the first derivatives of $L_i(x_i)$ yields the following primal-dual relationships:

$$x_i = \Sigma\, c_{ij}\lambda_j$$

if the side constraints are ignored. Note that, in matrix form, this relation can be written $x = C\lambda$, just as in Section 2.

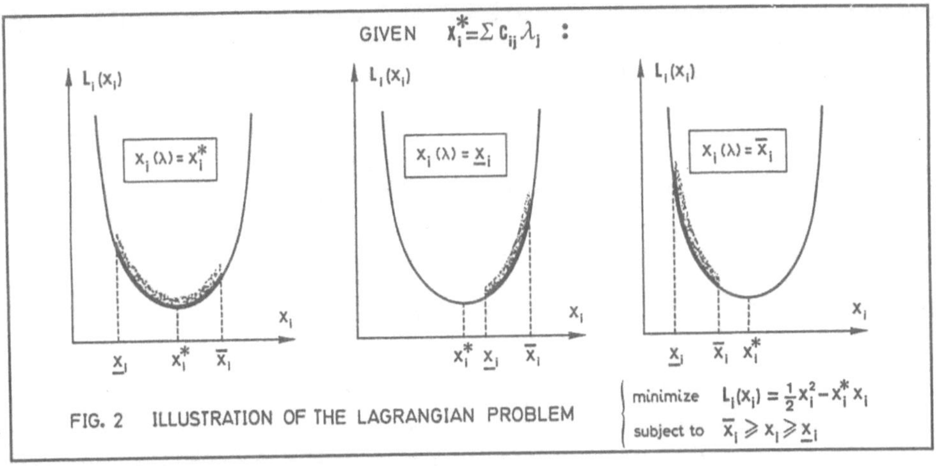

FIG. 2 ILLUSTRATION OF THE LAGRANGIAN PROBLEM

Now, to enforce satisfaction of the side constraints, we need to consider the three possible situations depicted in Fig. 2. It is clear that fully explicit primal-dual relationships $x = x(\lambda)$ are still available. At any point in the dual space, the values of the primal variables can be calculated according to the following steps:

(1) for the given λ, compute $x_i^* = \Sigma\, c_{ij}\lambda_j$

(2) if $\bar{x}_i > x_i^* > \underline{x}_i$, then $x_i(\lambda) = x_i^*$ (23)

if $x_i^* < \underline{x}_i$, then $x_i(\lambda) = \underline{x}_i$

if $x_i^* > \bar{x}_i$, then $x_i(\lambda) = \bar{x}_i$

The dual function is formed by inserting these primal-dual relationships into the Lagrangian function, which leads to the following statement of the dual problem:

<u>Dual Problem</u>

$$\text{maximize} \quad \ell(\lambda) = \tfrac{1}{2} \sum_i x_i^2(\lambda) - \sum_j \lambda_j \left(\sum_i c_{ij}\, x_i(\lambda) - d_j\right)$$

$$\text{subject to } \lambda_j \geq 0$$

The first derivatives of the dual function are simply given by the primal constraints, in agreement with the general theory of duality:

$$\partial l/\partial\lambda_j \equiv d_j - \Sigma\, c_{ij}\, x_i(\lambda)$$

The second derivatives of the dual function, yielding the Hessian matrix (denoted A as in Section 2.2), are easily available too:

$$A_{jk} \equiv \partial^2 l/\partial\lambda_j\partial\lambda_k = \partial/\partial\lambda_k[d_j - \Sigma\, c_{ij}\, x_i(\lambda)] = -\,\Sigma\, c_{ij}\, \partial x_i/\partial\lambda_k$$

But now, instead of the simple expression $A = -C^TC$ derived in Section 2.2, a slightly more complicated situation arises, because of the separate treatment of the side constraints. On one hand, for the free variables, we have:

$$\partial x_i/\partial \lambda_k = c_{ik} \qquad \text{if} \qquad \bar{x}_i > x_i > \underline{x}_i$$

On the other hand, for the fixed variables, it is obvious that:

$$\partial x_i/\partial \lambda_k = 0 \qquad \text{if} \qquad x_i = \bar{x}_i \text{ or } x_i = \underline{x}_i$$

Therefore, the Hessian matrix takes on the form:

$$A_{jk} \equiv \partial^2 l/\partial \lambda_j \partial \lambda_k = -\sum_{\text{free}} c_{ij}\, c_{ik} \tag{24}$$

where the summation is restricted to the free primal variables.

The second derivatives of the dual function are therefore discontinuous whenever a free primal variable becomes fixed, or conversely. From the primal-dual relationships (23) it is clear that the dual space is partitioned in several regions separated by second order discontinuity planes. These planes are defined by:

$$\Sigma\, c_{ij}\lambda_j = \underline{x}_i \qquad\qquad \Sigma\, c_{ij}\lambda_j = \bar{x}_i$$

4.2. UPDATE FORMULAS FOR THE INVERSE HESSIAN

The fundamental difficulty in using Newton type methods for solving the dual problem resides in the inherent discontinuities of the Hessian matrix. Those discontinuities are due to the side constraints imposed on the primal variables. Fortunately the topology of the dual space can be described in an exact mathematical way via the concept of second order discontinuity planes. Based on this concept, a very efficient and reliable algorithm can be devised to solve the dual problem.

The new algorithm works in the dual space, and it is based on the following observation: the Hessian matrix of the dual function is mathematically expressed as a sum of rank one matrices, each matrix corresponding to a free primal variable (i.e. a variable that is not currently fixed to its lower or its upper bound); Eq. (24) can indeed be rewritten in matrix form:

$$A = -\sum_{\text{free}} w_i w^T_i$$

where w_i denotes row (i) of the matrix C (i.e. $w_j = c_{ij}$).

It is therefore possible to gradually build up the inverse of the dual Hessian through a series of rank one updates. The algorithm is of course iterative in nature. At each iteration four possible situations can arise:

(1) a free primal variable becomes fixed, and consequently one rank one term must be removed from the dual Hessian;

(2) conversely a fixed primal variable may become free, and it will now contribute to the dual Hessian;

(3) a violated primal linear constraint must be activated, which means that the corresponding zero dual variable will become positive; the dimension of the dual Hessian is then increased by one;

(4) conversely a dual variable may reach the zero value (the corresponding primal constraint becomes inactive); in this case the dimension of the dual Hessian is decreased by one.

In each of these four cases, it is possible to directly compute the inverse dual Hessian via an appropriate rank one update formula (dyadic product). In fact, the dual Hessian matrix itself is not required in the algorithm, and only its inverse is actually evaluated through a series of rank one updates.

Update formula #1 (activated side constraint)

Assume that the primal variable x_i becomes fixed and let w be the corresponding row (i) of the matrix C (i.e. $w_j = c_{ij}$). Then the new Hessian is:

$$\tilde{A} = A + ww^T$$

and its inverse is updated by the formula:

$$\tilde{A}^{-1} = A^{-1} - vv^T/\gamma \qquad (25)$$

where

$$v = A^{-1}w$$

$$\gamma = w^Tv - 1$$

Update formula #2 (deactivated side constraint)

Assume now that the primal variable x_i becomes free and let again w be the corresponding row (i) of the matrix C (i.e. $w_j = c_{ij}$). Then the new Hessian is:

$$\tilde{A} = A - ww^T$$

and its inverse is updated by the same formula (25), but with

$$\gamma = w^Tv + 1$$

Update formula #3 (activated linear constraint)

Assume that the primal linear constraint c_k becomes active and let h be the corresponding column (k) of the matrix C (i.e. $h_i = c_{ik}$). Then the new Hessian is:

$$\tilde{A} = \begin{bmatrix} A & w \\ w^T & \alpha \end{bmatrix}$$

with $w_j = - \Sigma\, c_{ij}\, h_i$

$$\alpha = - \Sigma\, h_i^2$$

Its inverse is updated by the formula:

$$\tilde{A}^{-1} = \begin{bmatrix} A^{-1} - v\, v^T & \dfrac{v}{\gamma} \\ \dfrac{v^T}{\gamma} & -\dfrac{1}{\gamma} \end{bmatrix}$$

where

$$v = A^{-1} w$$

$$\gamma = w^T v - \alpha$$

<u>Update formula #4 (deactivated linear constraint)</u>

Assume now that a primal linear constraint becomes inactive. The corresponding dual variable is set to zero and it is removed from the dual subspace of positive variables. First let us interchange the last positive dual variable with the one set to zero.

Then if the old inverse Hessian was:

$$A^{-1} = \begin{bmatrix} \tilde{A}^{-1} & w \\ w^T & \alpha \end{bmatrix}$$

the new inverse Hessian is updated by the formula:

$$\tilde{A}^{-1} = A^{-1} - w w^T/\alpha$$

Preliminary results indicate that the computational cost for solving a separable quadratic problem is of the same order of magnitude as for inverting a matrix whose dimensionality is equal to the number of active linear constraints. When applied to a non quadratic problem, the method is employed iteratively at the expense of only a few matrix inversions. As an illustration of the power of this dual solver, explicit problems of the CONLIN type, with 300 design variables and 500 constraints, can be solved on a PC within a few hours.

References

[1] Lasdon, L.S. (1970) **Optimization Theory for Large Systems**, Macmillan, New York, 396-459.

[2] Lootsma, F.A. (1989) 'A comparative study of primal and dual approaches for solving separable and partially separable nonlinear optimization problems', **Structural Optimization** 1, 73-79.

[3] Fleury, C. (1979) 'Structural weight optimization by dual methods of convex programming', **International Journal for Numerical Methods in Engineering** 14, 1761-1783.

[4] Fleury, C. (1982) 'Reconciliation of mathematical programming and optimality criteria approaches to structural optimization', Chapter 10 of **Foundations of Structural Optimization: a Unified Approach** (A.J. Morris, ed.), John Wiley and Sons, 363-404.

[5] Venkayya, V.B., Khot, N.S. and Berke, L. (1973) 'Application of optimality criteria approaches to automated design of large practical structures', Proc. **Second Symp. Struct. Opt.**, AGARD-CP-123, Milan, Italy.

[6] Schmit, L.A. and Miura, H. (1976) 'Approximation concepts for efficient structural synthesis', NASA Contractor Report, NASA-CR 2552.

[7] Fleury, C. and Schmit, L.A. (1980) 'Dual methods and approximation concepts in structural synthesis', NASA Contractor Report, NASA-CR 3226.

[8] Braibant, V. and Fleury, C. (1985) 'An approximation concepts approach to shape optimal design', **Computer Methods in Applied Mechanics and Engineering** 53, 119-148.

[9] Fleury, C. (1986) 'Shape optimal design by the convex linearization method', Chapter 12 of **The Optimum Shape: Automated Structural Design** (J. Bennett and M. Botkin, eds.), Plenum Press, 297-326.

[10] Fleury, C. and Braibant, V. (1986) 'Structural optimization - a new dual method using mixed variables', **International Journal for Numerical Methods in Engineering** 23, 409-428.

[11] Fleury, C. (1989) 'CONLIN: an efficient dual optimizer based on convex approximation concepts', **Structural Optimization** 1, 81-89.

[12] Nguyen, V.H., Fleury, C. and Strodiot, J.J. (1987) 'A mathematical convergence analysis of the convex linearization method for engineering design optimization', **Engineering Optimization** 11, 195-216.

[13] Starnes, J.H. and Haftka, R.T. (1979) 'Preliminary design of composite wings for buckling, stress and displacement constraints', **Journal of Aircraft** 16, 564-570.

[14] Svanberg, K. (1987) 'Method of moving asymptotes - a new method for structural optimization', **International Journal for Numerical Methods in Engineering** 24, 359-373.

[15] Smaoui, H., Fleury, C. and Schmit, L.A. (1988) 'Advances in dual algorithms and convex approximation methods', Proc. AIAA/ASME/ASCE 29th Structures, Structural Dynamics, and Materials Conference, 1339-1347.

SEQUENTIAL CONVEX PROGRAMMING FOR STRUCTURAL OPTIMIZATION PROBLEMS

C. FLEURY
Aerospace Laboratory
University of Liege
Rue E. Solvay, 21
B-4000 Liege, Belgium

ABSTRACT. In this Lecture, several recent methods based on convex approximation schemes are discussed, that have demonstrated strong potential for efficient solution of structural optimization problems.

First, the now well established "Approximation Concepts" approach is briefly recalled for sizing as well as shape optimization problems. Next, the "Convex Linearization" method (CONLIN) is described, as well as one of its recent generalization, the "Method of Moving Asymptotes" (MMA). Both CONLIN and MMA can be interpreted as first order convex approximation methods, that attempt to estimate nonlinearity on the basis of semi-empirical rules.

Attention is next directed toward methods that use diagonal second derivatives in order to provide a sound basis for building up high quality explicit approximations of the behaviour constraints. In particular, it is shown how second order information can be effectively used without a prohibitive computational cost.

Various first and second order approaches have been successfully tested on simple problems that can be solved in closed form, on sizing optimization of trusses, and on two-dimensional shape optimal design problems. In most cases convergence is achieved within five to ten structural reanalyses.

1. Introduction

This Lecture describes recent optimization methods particularly well adapted to solve many problems arising in structural design. Mathematically the numerical optimization problem considered herein can be written in the following general form:

$$\text{minimize} \quad c_0(x) \tag{1}$$

$$\text{subject to} \quad c_j(x) \leq 0 \qquad j=1,m \tag{2}$$

Although the objective function (1) is very often a linear function of the design variables x_i (classical weight minimization with sizing variables), herein $c_0(x)$ will be assumed to be possibly a nonlinear function, so that it can represent any structural characteristic to be minimized (e.g. stress concentration), with sizing or shape design variables. The inequalities (2), often called behaviour constraints, impose limitations on structural response quantities such as stresses and displacements under static loading cases. These constraints are usually nonlinear functions, but in some situations, they might also

G. I. N. Rozvany (ed.), Optimization of Large Structural Systems, Vol. I, 531–553.

include linear functions. Of course the design variables must also be bounded from below and from above, however, these so-called side constraints are not considered explicitly in the sequel, for sake of clarity.

In this Lecture, various methods based on explicit approximation schemes are discussed, that have demonstrated strong potential for efficient solution of design optimization problems. The key idea is to replace the implicit problem (1,2) with a sequence of convex explicit subproblems having a simple algebraic form. After recalling the key role played by "reciprocal variables" in the short history of structural optimization, the "Convex Linearization" method (CONLIN) [1,2] will be described, as well as a recent generalization, the "Method of Moving Asymptotes" (MMA) [3]. In these two methods, each explicit subproblem represents a convex approximation to the primary problem, obtained through first order Taylor series expansion of the objective and constraint functions in terms of intermediate variables (e.g. direct/reciprocal variables). In addition to their theoretical and academic interest, the CONLIN and MMA methods have recently received growing attention from the industrial community. They have been recently implemented in several commercially available finite element systems.

In assessing the quality of an approximation scheme, there are a number of criteria to be considered, such as separability, convexity, conservativeness and accuracy, and a number of objectives to be pursued, namely, overall convergence of the sequence of approximate problems, convergence within a minimum number of stages, minimum effort for generating each approximate subproblem, and efficient solution of the explicit subproblems. Unfortunately, these objectives and approximation criteria often conflict with one another. In order to achieve a better compromise among them one needs more information than just the first order behaviour sensitivity derivatives. For instance, information about the way the design variables change between iterations (oscillation, steady but slow change) has been exploited in [3] as a basis for choosing the asymptotes.

Both CONLIN and MMA can be interpreted as first order convex approximation methods that attempt to estimate the curvature of the problem functions by assuming that these functions have a given separable form. This explicit separable form is governed by semi-empirical rules based on structural considerations (e.g. stress and displacement constraints are almost linear in the reciprocal variables). In other words, the Hessian matrices of these functions are assumed to be diagonal. Based on this observation, second order convex approximation strategies have begun to emerge, where exact curvature information is employed to build up the approximating subproblem.

One possible approach consists in resorting to a sequential quadratic programming strategy, restricted to the diagonal second derivatives of the Lagrangian function. For structural optimization problems involving static constraints, this latter approach can be implemented at the expense of only one additional "virtual load case" in the sensitivity analysis part of the finite element code. An alternative way of using diagonal second derivatives will be briefly described. The idea is to select intermediate linearization variables, so that the diagonal second derivatives of some combined constraint are zero. Finally, still another approach is to use the diagonal second derivatives of each function in order to define the ideal values of the moving asymptotes in the MMA method.

A very attractive feature of all these convex approximation approaches, is that they replace the primary optimization problem with a sequence of separable subproblems that can be solved efficiently by a dual method formulation. In the dual approach, discussed in another Lecture, the constrained primal minimization problem is replaced by maximizing a quasi-unconstrained dual function depending only on the Lagrangian multipliers associated with the linearized constraints. These multipliers are the dual variables subject to simple non-negativity constraints. The efficiency of the dual formulation is due to the fact that maximization is performed in the dual space, whose dimensionality is relatively low and depends on the number of active constraints at each design iteration.

These various convex approximation techniques have been successfully tested on simple problems that can be solved in closed form, on sizing optimization of trusses, and on two-dimensional shape optimal design problems.

2. First order convex approximation methods

2.1. THE KEY ROLE OF RECIPROCAL VARIABLES

The so-called approximation concepts approach to structural optimization is now widely employed to solve optimal sizing problems [4,5]. Such problems consist in minimizing the weight of thin-walled structures modeled by bar and membrane elements. Because the geometry is fixed, the design variables reduce to the transverse sizes of the structural members (i.e. bar cross-sections and membrane thicknesses). This approach consists basically of the following steps:

- a finite element analysis is performed for the initial trial design;
- from the results of the current structural analysis, an approximate optimization problem is generated (this step implies that sensitivity analysis capabilities be available in the finite element code);
- because the approximate subproblem is fully explicit, convex and separable, it can be efficiently solved by resorting to its dual formulation;
- the solution of the approximate subproblem is adopted as a new starting point in the design space and the optimization process is continued until convergence is achieved.

In the foregoing approximation concepts approach the primary optimization problem is replaced with a sequence of explicit subproblems having a simple algebraic structure. Each subproblem is generated through first order Taylor series expansion of the objective function and constraints in terms of intermediate linearization variables. For example, linearization of the constraints with respect to reciprocal variables is a well recognized technique to solve optimal sizing problems. There is an intuitive explanation for the success of this technique, in that stresses and displacements are exact linear functions of the reciprocal sizing variables in the case of a statically determinate structure. For shape optimal design problems, there is no such physical guideline for the selection of intermediate linearization variables. Nevertheless, this change of variables continues to have a highly beneficial effect on the convergence properties of the shape optimization process [6,7].

Consequently it can be argued that an efficient strategy consists in linearizing the objective function with respect to the direct variables,

$$\tilde{c}_0(x) = c_0(x^0) + \Sigma\, (\partial c_0/\partial x_i)^0(x_i - x_i^0) \quad (3)$$

while the constraint functions are linearized with respect to the reciprocal variables:

$$\tilde{c}_j(x) = c_j(x^0) - \Sigma\, (x_i^0)^2(\partial c_j/\partial x_i)^0\ (1/x_i - 1/x_i^0) \quad (4)$$

From extensive numerical experiments it can be concluded that the approximation concepts approach converges to an optimum design in usually less than ten iterations (i.e. finite element analyses). These remarkable convergence properties are generally attributed to the fact that the behaviour constraints are much less nonlinear in the space of the reciprocal variables.

More general techniques have been developed based on the idea of reciprocal variables. The Convex Linearization method (CONLIN) employs reciprocal variables in the linearization process, but it also uses direct variables, depending upon the sign of the first derivatives of the function being approximated [1]. The initial motivation for CONLIN was to further extend the range of applicability of the approximation concepts and generalized optimality criteria methods, so that these approaches could handle constraints other than upper limits on stresses or displacements, e.g. strictly linear constraints in the sizing variables. Very soon it became apparent that the CONLIN approximation scheme was well adapted to shape optimization problems [6,7].

2.2. THE CONVEX LINEARIZATION METHOD (CONLIN)

The convex linearization method (CONLIN) [1] was initially conceived as an extension to the approximation concepts approach. The key idea in the CONLIN method is to perform the linearization process with respect to mixed variables, either direct or reciprocal, independently for each function involved in the optimization problem. At each successive iteration point, the CONLIN method only requires evaluation of the objective and constraint functions and their first derivatives with respect to the design variables. The optimizer will then select by itself an appropriate approximation scheme on the basis of the signs of the derivatives. This constitutes a major improvement with respect to the regular approximation concept approach, where it is usually assumed that the objective function is linear in the direct variables (e.g., structural weight) and that the constraints can be accurately approximated as linear functions of the reciprocal variables (e.g., stresses and displacements). Furthermore, the CONLIN optimizer has an inherent tendency to generate a sequence of steadily improving feasible designs, in contrast with the previously developed approximation concepts approach using dual methods [5].

The CONLIN method proceeds by linearizing each function defining the optimum design problem with respect to a properly selected mix of direct and reciprocal variables, so that a convex and separable subproblem is generated. The selection of the "intermediate" linearization variables is made on the basis of the signs of the first partial derivatives. It is easily proven that, considering any differentiable function c(x), the following linearization scheme yields a convex approximation (hence the name "convex linearization" initially suggested in Ref. [1]):

$$\tilde{c}(x) = \Sigma_{+}(\partial c/\partial x_i)^0 x_i - \Sigma_{-}(xi^0)^2(\partial c/\partial x_i)^0/x_i + c(x^0) - \Sigma|(\partial c/\partial x_i)^0|x_i^0 \quad (5)$$

where the symbol Σ_+ (Σ_-) means "summation over positive (negative) terms". Note that the two last terms in this expression represent the contribution of the zeroth order terms in the Taylor series expansion. One of the most interesting feature of the convex linearization scheme is that it also leads to the most conservative approximation amongst all the possible combinations of mixed direct/reciprocal variables. This property was initially demonstrated in [8], where conservative approximation was employed to handle difficult buckling constraints.

Another important property, which is particularly useful in relation to the dual approach is that the explicit approximation (5) is convex. The CONLIN algorithm applies this convex linearization scheme to the objective function and to all the constraint functions. The resulting explicit approximations take on a simpler form if the design variables are normalized so that they become equal to unity at the current point x^0 where the problem is linearized. The following convex subproblem is then generated:

$$\begin{aligned} &\text{minimize} && \Sigma_{+} c_{i0}\, x_i - \Sigma\, c_{i0}/x_i - d_0 \\ &\text{subject to} && \Sigma_{-} c_{ij}\, x_i - \Sigma\, c_{ij}/x_i \leq d_j \quad j=1, m \end{aligned} \quad (6)$$

where the c_{ij} 's denote the normalized first derivatives of the objective and constraint functions evaluated at the current point x^0. Note that the constants d_j collect the zeroth order contributions in the Taylor series expansion.

In CONLIN, therefore, the linearization process is performed with respect to mixed variables, either direct or reciprocal, independently for each function involved in the optimization problem. Direct variables are used for positive first derivatives, while reciprocal variables are employed for negative first derivatives. A convex, separable subproblem is generated, that can be efficiently solved by using the dual method approach presented in another Lecture.

Because the CONLIN strategy employs conservative approximations, it has an inherent tendency to generate a sequence of steadily improving feasible designs. In addition to this desirable property, the CONLIN optimizer usually produces a nearly optimal design within less than 10 reanalyses. In fact, a rigorous mathematical study has demonstrated that the CONLIN method converges to a local optimum under rather unrestrictive assumptions [9].

2.3. THE METHOD OF MOVING ASYMPTOTES (MMA)

In practice, the CONLIN optimizer exhibits very good convergence properties when dealing with structural optimization problems. It usually converges in less than ten iterations, for sizing problems as well as for shape optimal design problems. However, in some cases, the convex approximation scheme used in CONLIN might not be appropriate, leading to inaccurate approximations, either too conservative (in which case slow convergence occurs), or not sufficiently conservative (in which case oscillations can set in). Some counter-examples illustrating this behaviour can be found in Ref. [3], where a simple and very elegant modification of the convex linearization method is proposed, called the Method of Moving Asymptotes (MMA).

In the MMA method, the intermediate linearization variables are modified such that the degree of convexity, and therefore conservativeness, of the approximation can be adjusted depending upon the problem being solved. Instead of just using direct and reciprocal variables, this method employs the intermediate variables $1/(U_i-x_i)$ and $1/(x_i-L_i)$, respectively, where U_i and L_i are user-selected parameters.

In the first order Taylor series expansion, the selection of the intermediate linearization variable is still based upon the signs of the first derivatives. The MMA approximation has the form:

$$\tilde{c}(x) = \Sigma_+ c_i/(U_i-x_i) - \Sigma_- c_i/(x_i-L_i) - d^0 \tag{7}$$

where

$$c_i = (U_i-x_i^0)^2(\partial c/\partial x_i)^0 \qquad \text{if } (\partial c/\partial x_i)^0 > 0$$

$$c_i = -(x_i^0-L_i)^2(\partial c/\partial x_i)^0 \qquad \text{if } (\partial c/\partial x_i)^0 < 0$$

The constant d^0 collects the zeroth order terms of the expansion.

As in CONLIN, the MMA method replaces the primary problem (1,2) with a sequence of convex subproblems:

$$\begin{aligned} &\text{minimize} \quad \Sigma_+ c_{i0}/(U_i-x_i) - \Sigma_- c_{i0}/(x_i-L_i) \\ &\text{subject to} \quad \Sigma_+ c_{ij}/(U_i-x_i) - \Sigma_- c_{ij}/(x_i-L_i) \leq d_j \end{aligned} \tag{8}$$

In this technique the form of each approximated function is driven by the selected values for the constants L_i and U_i, which act as "asymptotes". Taking $L_i = 0$ and $U_i = +\infty$ gives back the initial CONLIN approach; $L_i = -\infty$ and $U_i = +\infty$ leads to a sequential linear programming technique. Other reasonable values of L_i and U_i are acceptable, and these values may even be modified from one iteration to the next. That is why this method has been called the "Method of Moving Asymptotes".

2.4. OTHER CONVEX APPROXIMATIONS (POWER EXPANSIONS)

The approximation scheme used in CONLIN is powerful for conventional sizing problems with static response constraints. For a wider class of problems, for example those involving bending elements or shape design variables, as well as for problems with dynamic constraints, more elaborate approximation schemes are necessary in order to taylor the explicit approximate function to the type of nonlinearity in each problem.

In Ref. [10] the sizing of frames subject to frequency constraints is investigated using approximations of the form:

$$\tilde{c}(x) = c(x^0) + \Sigma_+ (\partial c/\partial x_i)^0(x_i-x_i^0)(x_i/x_i^0)^{p+} + \Sigma_- (\partial c/\partial x_i)^0(x_i-x_i^0)(x_i/x_i^0)^{p-} \tag{9}$$

where $p^+ \geq 0$ and $p^- < 0$ are adjustable parameters. This approximation is sometimes referred to as "hybrid power expansion". It should be pointed out that it coincides with the CONLIN approximation (5) for the values $p^+ = 0$ and $p^- = -1$, while for $p^+ = p^- = 0$, it reduces to a conventional linear Taylor series expansion in the direct variables [see Eq.

(3)]. It is shown in [10] that this approximation can be made convex over a fairly large domain around the base point x^0.

On the other hand, this idea of reciprocal variables has been extended to structural synthesis problems involving bending elements, like plates and shells [11], frames [12], etc... For example it is clear that the best linearization variable for a plate element in pure bending is the reciprocal of the cube of the thickness. For more general situations, where bending and extension effects are of the same order of magnitude, the selection of the intermediate variable is not that obvious [13]. Much research has been devoted to this topic. In particular Ref. [14] presents a systematic investigation of which intermediate variables should be used for an extensive class of structural models and behaviour constraints. A rather general change of variables is introduced, and the approximation consists of a first order Taylor series expansion in terms of the new variables. These intermediate variables are chosen in the form:

$$z_i = 1/(p-1)\; I_i^{1-p} \qquad \text{if } p \neq 1$$

$$z_i = -\ln(I_i) \qquad \text{if } p = 1$$

where $I_i = I_i(x_i)$ is a structural parameter, usually a cross-sectional property such as the moment of inertia or the area for a beam member. The exponent p is a parameter which provides a means to control the degree of curvature of the approximation. Note that the Taylor series approximations expressed in Eq. 3 (in direct variables) and in Eq. 4 (in reciprocal variables) correspond respectively to $p = 0$ and $p = 2$.

3. Second order convex approximation methods

It is important to realize that both CONLIN, MMA, and the other approximation schemes discussed in the previous Section are based on first order approximations that attempt to simulate the curvature of the objective function and constraint functions. These approximations are convex. They are also conservative, in that they tend to overestimate the true value of the approximated function.

In CONLIN the degree of conservativeness is fixed, while MMA offers much more flexibility through the moving asymptotes L and U. In other words CONLIN will provide some curvature to the approximated function, but the artificial curvature introduced by the reciprocal variables will generally be different from the true curvature.

On the other hand MMA is capable of matching the explicit curvature to the exact curvature, provided that the asymptotes are appropriately selected. Obviously, when only first derivatives are available, it is not at all straightforward to find suitable values for the asymptotes, so as to accurately represent the curvature of each function. Empirical techniques can be used to gradually update the values of L and U, depending upon the results generated at each iteration in the optimization process. For example, in Ref. [3], where the MMA method was first proposed as a generalization to the CONLIN method, it is suggested that the asymptotes be moved away from the current point x^0 if the optimization process is converging slowly. On the other hand, if the process tends to oscillate, then the asymptotes are moved closer to x^0.

The schemes that have been reported up to now to adjust the curvatures tend to be heuristic and problem dependent. In this Section, more general and rational schemes are described, based on second order sensitivity information. The methods discussed below can indeed be viewed as further generalizations of the convex approximation strategies, such as CONLIN and MMA. Because these new methods use the diagonal second derivatives of the Lagrangian function, they inherently build up the required information on the problem curvature. An important advantage of the proposed approach resides in its simplicity of use and implementation. There is no need to set up a sophisticated linearization scheme using well selected intermediate variables, nor to update control parameters in order to simulate the curvature of the constraint and objective functions. Numerical experiments on well known test problems indicate that these methods perform equally well whatever design variables are selected as base variables (direct, reciprocal, mixed, etc...).

3.1. PURE SEQUENTIAL QUADRATIC PROGRAMMING (SQP)

The approach to structural optimization which is presented in the sequel is based on a well known method in mathematical programming, first proposed and referred to by Wilson [15] as the SOLVER method, and also known as the Lagrange-Newton method [16]. Indeed, this method can be interpreted as applying Newton's method to find the stationary point of the Lagrangian function

$$L(x,\lambda) = \sum_{j=0}^{m} \lambda_j \, c_j(x) \tag{10}$$

where λ_0 is fixed to unity and $\{\lambda_j, j=1,m\}$ denote the Lagrangian multipliers associated with the constraints $c_j(x)$. This approach can also appropriately be called a Sequential Quadratic Programming (SQP) method, because it replaces the primary optimization problem (1,2) with a sequence of quadratic approximations of the form:

$$\begin{aligned} &\text{minimize} \quad \tfrac{1}{2}\,(x - x^0)^T A\,(x - x^0) + b^T (x - x^0) \\ &\text{subject to} \quad c(x^0) + C^T (x - x^0) \geq 0 \end{aligned} \tag{11}$$

where the vector b and the matrix C respectively contain the first derivatives of the objective function and those of the constraint functions, evaluated at x^0:

$$b_i = \partial c_0/\partial x_i \mid x^0$$

$$c_{ij} = \partial c_j/\partial x_i \mid x^0$$

The symmetric matrix A represents the Hessian of the Lagrangian function, and consequently, it contains information on the functions curvature:

$$A_{ik} \equiv \partial^2 L/\partial x_i \partial x_k = \sum_{j=0}^{m} \lambda_j^0 \, \partial^2 c_j/\partial x_i \partial x_k \tag{12}$$

where the λ_j^0's denote the current values of the Lagrangian multipliers associated with the constraints (λ_j^0 is zero if the constraint c_j is inactive).

The application of the method needs, as usual, a sensitivity analysis providing the first derivatives of the problem functions. In addition the A matrix needs evaluation of the second derivatives for the objective function and all the currently active constraints (i.e the constraints associated with non-zero Lagrange multipliers). In many cases the required second order sensitivity analysis is not available, or it is computationally too expensive. For that reason, the most common SQP implementations resort to a quasi-Newton approach, in which the A matrix is only an approximation to the Hessian of the Lagrangian function, gradually built up at each iteration from the first order derivatives (DFP or BFGS update formulas).

Obviously the SQP approach based upon quasi-Newton approximation will not, in general, converge as fast as a pure Newton approach using true second derivatives. The motivation of the study presented in [17], and briefly discussed below, was therefore to keep the initial formulation of the Lagrange-Newton method, which requires second order sensitivity analysis. The key idea is to take advantage of the special mathematical structure exhibited by structural optimization problems in order to generate the required second order sensitivity information with the smallest possible computational time.

The evaluation of sensitivity derivatives for static displacement and stress constraints is now well established and documented (see e.g. [18]). Differentiating the equilibrium equations of the finite element model,

$$K\,q = g \tag{13}$$

where K is the stiffness matrix, g a load vector, and q the nodal displacements, we obtain:

$$K\,\partial q/\partial x_i = \partial g/\partial x_i - (\partial K/\partial x_i)\,q \tag{14}$$

Differentiating again:

$$K\,\partial^2 q/\partial x_i\partial x_k = \partial^2 g/\partial x_i\partial x_k - (\partial^2 K/\partial x_i\partial x_k)q - (\partial K/\partial x_i)\,\partial q/\partial x_k - (\partial K/\partial x_k)\,\partial q/\partial x_i \tag{15}$$

The "reduced" second order sensitivity analysis presented in [17] consists of using the first order sensitivity equation in its direct form to get the displacement first derivatives, from which the constraint gradients can be readily evaluated. On the other hand the second order sensitivity equation is employed in its adjoint form, with the addition of only one virtual load case. It can indeed be observed that the constraint curvature term appearing in the expression (12) of the A matrix, does not require evaluation of the second derivatives for each individual constraint. Only the Lagrangian function (10) need to be considered. Therefore it is sufficient to build one virtual load case corresponding to a linear combination of the constraints with coefficients λ_j (non zero Lagrangian multipliers).

To help fix ideas, let us consider the familiar 10-bar truss example, used in this Lecture to illustrate various SCP methods. The problem consists of minimizing the weight of the truss with limits on the stresses and vertical displacements under the loading shown in Fig. 1. In this particular case the problem involves 10 design variables, so that the sensitivity analysis needs consideration of 10 pseudo-load cases, plus one virtual load case. Fig. 1 summarizes very clearly the substantial benefits that can be gained from using the foregoing approach: for only a small increase in computer time (one more right hand side in the equilibrium equations), the total number of iterations required for convergence of the optimization process is cut in half.

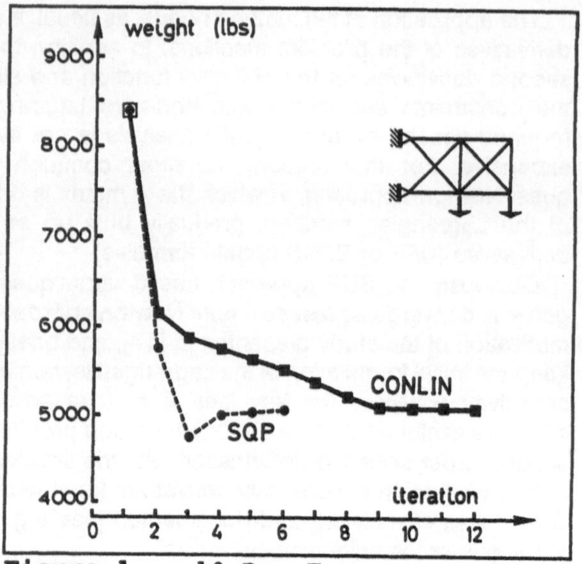

Figure 1. 10-Bar Truss Iteration History

Table 1 provides a more detailed iteration history for this problem. The "feasible weight" is obtained by multiplying the actual weight generated by the SQP algorithm, by the ratio of the most critical constraint value to its upper bound. This "scaling factor" is also tabulated. From these results it can be clearly observed that the ultimate rate of convergence is of order two: the value of the most critical constraint approaches its upper bound in a quadratic rate in the last three iterations. It is important to mention that, in a practical application, the optimization process could already be terminated after four or five iterations. However it was interesting to continue up to convergence within more than six digits accuracy, in order to demonstrate numerically the quadratic rate of convergence of the Newton-Lagrange method.

Table 1. 10-bar Truss - Iteration History with Full Hessian Matrix

Iteration	Scaling Factor	Feasible Weight
1	0.98489	8266.15
2	1.06246	6109.40
3	1.08141	5132.36
4	1.01424	5076.59
5	1.00470	5061.76
6	1.00004	5060.87
7	1.00000003	5060.85

3.2. THE DIAGONAL SQP METHOD

In the 10-bar truss example the method described above is quite efficient. However, for practical sizing problems, where the number of design variables might be large, computing and storing the nxn A matrix could become burdensome. On the other hand, for shape optimal design applications, the stiffness matrix can no longer be expressed as a simple explicit form of the design variables (e.g. positions of Bezier control nodes), and its second derivatives can be difficult to evaluate. Hence the idea to neglect the coupling between design variables, and to restrict the A matrix to its diagonal terms.

The quadratic subproblem now considered has the following separable form:

$$\text{minimize} \quad \tfrac{1}{2} \sum a_i(x_i - x_i^0)^2 + \sum b_i(x_i - x_i^0)$$
$$\text{subject to} \quad \sum c_{ij}\, x_i \geq d_j \tag{16}$$

where

$$a_i = A_{ii} + \delta_i \tag{17}$$

and A_{ii} are the diagonal elements of the Hessian matrix of the Lagrangian function (see Eq. 12). The c_{ij} and b_i have the same meaning as in problem (11), i.e. they represent the first derivatives of the constraints and of the objective function, respectively. Note that the zeroth order contributions in the linear Taylor series approximations of the constraints have been collected in the form of lower bounds d_j.

The parameters δ_i in (17) are important. They convexify the problem in the case where some of the diagonal second derivatives happen to be negative, or zero, which frequently occur in the early stages of the optimization process. In addition the control parameters δ_i act as "move limits". What has been done was simply to add to the objective function a term that represents a weighed distance with respect to x^0, i.e., the design point where the problem has been approximated:

$$\tfrac{1}{2} \sum \delta_i\, (x_i - x_i^0)^2 \tag{18}$$

One simple way to select the parameters δ_i is to ignore the linear constraints. We are left with an unconstrained problem whose solution is:

$$x_i = x_i^0 - b_i/(A_{ii} + \delta_i)$$

It is then straightforward to find δ_i such that, for example, each design variable x_i will remain within some percentage of its initial value x_i^0. In particular, in weight minimization problems (e.g. 10-bar truss), the objective function is linear, and it does not contribute to the Hessian matrix A. Therefore, at the first iteration, where the Lagrangian multipliers are usually set to zero, the A_{ii} are zero and the move limits δ_i must be given. From numerical experiments it appears that 100 % move limits are appropriate values for the first iteration. For the subsequent iterations, some of the Lagrangian multipliers become positive, and the constraint curvature starts to build up in the diagonal Hessian matrix, so that the move limits δ_i become less critical.

With this minor modification, the a_i's are positive and the quadratic subproblem (16) is strictly convex. This problem involves only linear constraints, and many algorithms are available to solve it. However, because the problem is also convex and separable, a dual approach constitutes probably the best choice.

3.3. ZERO CURVATURE APPROXIMATION

As previously mentioned reciprocal variables have played a significant role in the history of structural optimization. It is now well recognized that, for sizing problems involving bar and membrane elements, high quality approximation of the behaviour constraints can be achieved through linearization in terms of the reciprocal of the design variables (cross-sectional areas of bars; thicknesses of membrane elements). This idea has been extended to structural synthesis problems involving bending elements, like plates and shells [11], frames [12], as well as more general situations, where bending and extension effects are of the same order of magnitude [13,14].

Table 2. 10-bar Truss - Iteration History for Various Linearization Variables (1st entry: Weight; 2nd entry: Scaling Factor)

Iter.	p = 1	p = ½	p = -1	p = -½	p(i)
1	8393 0.985	8393 0.985	8393 0.985	8393 0.985	8393 0.985
2	6113 1.093	4658 1.354	6946 1.000	6608 1.030	5732 1.040
3	5544 1.181	5143 1.105	6457 0.998	6250 1.004	5805 0.992
4	5145 1.014	5458 1.009	6229 0.994	6040 0.995	5688 0.988
5	4951 1.126	5361 0.992	6044 0.995	5859 0.995	5548 0.987
6	5020 1.014	5153 1.003	5898 0.995	5713 0.995	5388 0.985
7	5033 1.007	5055 1.003	5768 0.994	5575 0.994	5193 0.998
8	5047 1.003	5059 1.002	5645 0.995	5441 0.994	5030 1.007
9	5058 1.001	5059 1.000	5526 0.995	5305 0.994	5062 1.000
10	5060 1.000	5061 1.000	5412 0.995	5187 1.000	5061 1.000
11	5061 1.000		5300 0.996	5126 1.000	
12			5202 1.000	5100 1.000	
13			5146 1.001	5084 1.000	
14			5119 1.000	5077 1.000	
15			5101 1.000		
16			5089 1.000		
17			5081 1.000		
18			5077 1.000		

Most first order methods commonly used in structural optimization are very sensitive to the choice of the intermediate linearization variables. However, because it uses limited second order information, the method described below does not suffer from this drawback. In the sequel our discussion will be focused on sizing problems, where the goal is to minimize the structural weight. The following change of variables covers most situations:

$$z_i = x_i^p$$

For example, p = +1 does not change anything and gives back the direct variables; the choice p = -1 leads to reciprocal variables; pure plate bending elements should use p = -3. Square root variables (p = +½) can also give interesting results, because they transform the linear weight objective function into a quadratic function.

The first and second derivatives of any function c[x(z)] can be obtained with respect to the new variables z_i by using the following transformations (omitting the subscript i):

$$\partial c/\partial z = 1/p\, x^{1-p}\, \partial c/\partial x$$

$$\partial^2 c/\partial z^2 = (1-p)/p^2\, x^{1-2p}\, \partial c/\partial x + 1/p^2\, x^{2(1-p)}\, \partial^2 c/\partial x^2 \tag{19}$$

Various values of the exponent p have been tried out for the classical 10-bar truss example shown in Fig. 1. The results given in Table 2 demonstrate that the method converges rapidly for any value of p. Surprisingly, direct variables (p=1), and even square root variables (p=½), lead to faster convergence than reciprocal variables (p=-1). This is probably because the objective function is represented exactly in those two cases: direct variables preserve the linear form of the weight objective function, while square root variables transform it into a quadratic separable function. However, as it might be expected, reciprocal variables yield a sequence of almost feasible designs: at each iteration, the scaling factor needed to bring back the design point on the boundary of the feasible domain, is very close to one. This behaviour is also true in the case p=-½. On the other hand, some of the designs in the p=+1 and p=+½ sequences, seriously violate the behaviour constraints.

The last column in Table 2 provides interesting results obtained with a variation of the method. The idea is to select the exponent p independently for each design variable (i), so that the second diagonal derivatives of the Lagrangian function are zero (see Eqs. 10 and 12). In other words the constraints are linearized in some space of minimum global curvature (zero curvature in the case of truly separable constraints). Because in the present case the objective function is known explicitly, the initial diagonal SQP method can be modified to further speed up convergence. Instead of solving the quadratic subproblem (16), it is indeed much better to keep the explicit nonlinear form of the weight objective function, and to solve the following linearly constrained subproblem:

$$\begin{aligned} &\text{minimize} \quad f(z) = \Sigma\, b_i\, z_i^{1/p(i)} \\ &\text{subject to} \quad \Sigma\, c_{ij}\, z_i \geq \underline{c}_j \end{aligned} \tag{20}$$

In this particular truss example, the coefficients b_i are related to the lengths of the bars. The exponents p(i) are selected so that the global constraint curvature is zero. Applying (19) to the Lagrangian function (10), and setting the resulting second derivative to zero, we obtain:

$$0 = (1-p)/p^2\, x^{1-2p}\, \partial L/\partial x + 1/p^2\, x^{2(1-p)}\, \partial^2 L/\partial x^2$$

Solving this equation for the exponent p yields, for each design variable (i):

$$p(i) = 1 + x_i^0\, (\partial^2 L/\partial x_i^2)/(\partial L/\partial x_i)$$

where the first and second derivatives are evaluated at the linearization point x^0.

The foregoing optimization strategy is interesting from a practical point of view, because it produces a sequence of nearly feasible designs while keeping the fast convergence of the diagonal SQP approach. An additional benefit gained from this modified method lies in its ability to dynamically select the "best" intermediate linearization variables for any structural optimization problem. For a statically determinate truss, the method will inherently determine that the exponent p must be equal to -1 for each design variable (i.e. automatic selection of reciprocal variables). For a statically indeterminate truss, the method will obtain different values for each p_i, most likely (but not necessarily) negative and close to -1. In the case of structural models involving in-plane as well as bending behaviour, the method would probably select values for p(i) between -1 and -3.

3.4. SECOND ORDER MMA METHOD

As shown in the previous Sections, a natural way of using second order information consists of constructing Taylor series expansions. However, from the results reported in Ref. [10], it can be argued that convex approximation schemes such as those used in the CONLIN and MMA methods, can reduce the number of explicit subproblems when compared to second order Taylor series. This suggests that convex approximation strategies, although they are based on first order information only, might have the ability to represent the exact functions even better than higher order standard methods. The drawback of generalized convex linearization methods like MMA is that they necessitate the setting of parameters with a high degree of accuracy. Hence the idea has emerged, to combine the advantages of pure SQP methods and the potential of convex approximations for high quality. This can be achieved by using second order information in order to select proper values for the tuning parameters (e.g. moving asymptotes in MMA).

The flexibility of the generalized convex approximations such as MMA is associated with adjustable parameters which control the curvature of the approximation. However the basis for selecting their appropriate values remains empirical. Indeed, the suggested approximations are all based on first order sensitivity, which does not provide any information about the curvatures. This information is either built up during the optimization process by means of empirical rules [3], or gathered from parametric studies on specific classes of problems [14].

A more rational and systematic approach based on second order information is proposed in [19]. In particular the Method of Moving Asymptotes (MMA) was implemented and successfully tested with an automated strategy for selecting the best possible values for the moving asymptotes. It is shown how restricted second order information can be used as a basis for a general and systematic rule for selecting the asymptotes of the MMA approximations. Since the approximations dealt with are separable, only diagonal terms of the second derivative matrices are needed in the process.

Distinct asymptotes for each constraint and each design variable are selected so that positive diagonal second derivatives of the constraint function and its approximate representation are identical. The explicit approximation of a function c(x) at the current design point x^0 has now the form:

$$\tilde{c}(x) = \Sigma\, a_i/(x_i-b_i) - c \tag{21}$$

where, in order to match the first and second derivatives of the exact and approximate functions:

$$a_i = (x_i^0-b_i)^2\,(\partial c/\partial x_i)^0$$

and

$$b_i = x_i^0 + 2\,(\partial c/\partial x_i)^0/(\partial^2 c/\partial x_i^2)^0 \tag{22}$$

The zero order term c is given by

$$c = \Sigma\, a_i/(x_i^0-b_i) - c(x^0)$$

To help fix ideas, the reader is invited to construct the following explicit approximations (at $x^0 = 1$) for two simple functions:

$$c(x) = x^3 \quad ===> \quad \tilde{c}(x) = 3/(2-x) - 2$$
$$c(x) = 1/x^3 \quad ===> \quad \tilde{c}(x) = 3/(4x-2) - 1/2$$

It can be seen that for the increasing cubic function, an upper asymptote is selected, while for the inverse cubic function, which is decreasing, a lower asymptote is used.

In order to preserve convexity, in case of zero or negative curvature, $(\partial^2 c/\partial x_i^2)^0$ is replaced in Eq. (22) with a small positive number. It is important to notice that this procedure permits obtaining a convex explicit approximation for a linear function. For example, at $x^0 = 0$, the increasing function $c(x) = x$ is approximated by:

$$\tilde{c}(x) = U[(U/(U-x) - 1] \ (= x/[1-x/U])$$

where U denotes a big number (upper asymptote). On the other hand, the decreasing function $c(x) = -x$ is approximated by:

$$\tilde{c}(x) = L[(L/(x-L) + 1] \ (= -x/[1-x/L])$$

where L denotes a negative number with big absolute value (lower asymptote). These two simple examples illustrate well how the MMA approximation is capable of representing linear functions. Coming back to the original MMA method (Section 2.3) it is seen that the limiting case where the asymptotes are $L_i = -\infty$ and $U_i = +\infty$ for all the approximated functions leads to a sequential linear programming technique.

Applying this second order convex approximation scheme to all the functions defining the optimization problem leads to the following explicit subproblem:

$$\begin{aligned} &\text{minimize} && \Sigma\, a_{i0}/(x_i-b_{i0}) - c_0 \\ &\text{subject to} && \Sigma\, a_{ij}/(x_i-b_{ij}) \leq c_j \qquad j=1, m \end{aligned} \tag{23}$$

It should be observed that the subscript j has been added to the asymptotes, now denoted b_{ij}. This is different from the original MMA method, where a unique pair (U_i and L_i) is associated to each primal variable, but remains the same for each approximated function. As explained in another Lecture, the explicit subproblem (23) can be efficiently solved by dual methods. However, because each asymptote depends now on both the primal variable (index i) and the constraint (index j), the Lagrangian problem becomes implicit and requires a numerical one-dimensional solution. The effort demanded by the numerical solution of the implicit Lagrangian problem, as well as the significant cost of evaluating second derivatives, represent the price one has to pay for the high quality achieved for the approximations. Nevertheless, numerical results for several test problems indicate that the approximation based on second order information requires fewer approximate problems than the original method of moving asymptotes.

4. The conlin optimizer

Based on the concepts outlined in this Lecture, as well as on the dual method approach presented in another Lecture, an advanced optimizer was developed, that has been adopted in several FEM based optimization systems. This optimizer is called CONLIN, because it was initially restricted to the pure "CONvex LINearization" method presented in Section 2.2. The CONLIN optimizer solves problems of the form:

Minimize an objective function $c_0(x) + p\,\delta^2$

subject to constraints:

$$c_j(x) \leq c_j^{max} + (\delta-1)\, c_j^{add} \qquad j=1,\ldots,m$$

$$x_i^{min} \leq x_i \leq x_i^{max} \qquad i=1,\ldots,n$$

$$1 \leq \delta \leq 2$$

where $x = (x_1,\ldots,x_n)^T$ is the vector of design variables x_i, δ represents an additional scalar variable, and p is a user-controllable weighing factor.

The additional variable δ is optional; it becomes useful only when the optimization problem does not have any solution. In that case the feasible domain, initially empty for $\delta = 1$, is opened up for $\delta > 1$, by "relaxing" the upper bounds c_j^{max}. For this reason δ is called the relaxation variable. Note that the maximum possible relaxation, corresponding to $\delta = 2$, can be controlled by the user via the weighing factor p in the objective function, as well as the increments d_j in the constraints.

At each stage of the iterative optimization process, the user must evaluate and provide the following quantities, which are employed by CONLIN to build first order explicit approximations:

- the values of the functions defining the design optimization problem (i.e. the objective function $c_0(x)$ and the constraint functions $c_j(x)$, evaluated at the current design point);
- the first derivatives of these functions:

$$c_{ij} = \partial c_j / \partial x_i \qquad i=1,2,\ldots,n \quad j=0,1,\ldots,m$$

In addition the user must decide which type of approximation will be used for each function: convex, reciprocal, or direct (linear) first Taylor series expansion.

CONLIN also contains a provision for some of the more sophisticated explicit approximations previously reviewed in this Lecture: shifted convex (MMA), diagonal quadratic or power Taylor series expansions. For these higher order separable approximations, it is necessary to input additional information related to the diagonal second derivatives of the functions (true values or some estimation).

Of course the user must also input the current values of the design variables x_i, as well as the other parameters defining the optimization problem: x_i^{min}, x_i^{max}, c_j^{max} and c_j^{add}.

4.1. TYPE OF EXPLICIT APPROXIMATIONS

The CONLIN optimizer can currently handle six types of separable explicit approximations: types 0 through 2 are first order Taylor series expansions in mixed, reciprocal or direct variables, while types 3 through 5 correspond to higher order approximations. The dual optimizers implemented in CONLIN are general, and they could easily accommodate other types of separable approximations.

Type 0: convex expansion

First order Taylor series expansion in mixed variables x_i and $1/x_i$, according to the sign of the first derivatives:

$$c(x) = \Sigma_+ c_i x_i - \Sigma_- c_i/x_i$$

Type 1: reciprocal expansion

First order Taylor series expansion in the reciprocal variables $1/x_i$:

$$c(x) = \Sigma c_i/x_i$$

Type 2: direct (linear) expansion

First order Taylor series expansion in the direct variables x_i:

$$c(x) = \Sigma c_i x_i$$

Type 3: diagonal quadratic expansion

Diagonal second order Taylor series expansion in the direct variables x_i (in addition to the first derivatives, the user must supply the diagonal second derivatives):

$$c(x) = ½ \Sigma a_i x_i^2 + \Sigma b_i x_i$$

Type 4: shifted convex expansion (MMA)

First order Taylor series expansion in the intermediate variables $1/(x_i-m_i)$ or $1/(m_i-x_i)$ according to the sign of the first derivatives, where the m_i's are user supplied "moving asymptotes" for each design variable i. The explicit approximation has the form:

$$c(x) = \Sigma_+ c_i/(m_i-x_i) - \Sigma_- c_i/(x_i-m_i)$$

Type 5: power expansion

First order Taylor series expansion in the intermediate variables $x_i^{p_i}$ where p_i is a user supplied exponent for each design variable i. The explicit approximation has the form:

$$c(x) = \Sigma c_i x_i^{p_i}$$

Note:

- for types 0 through 2, the user must only input the first derivatives of the function c(x);
- for types 3, 4 and 5, the user must provide an additional vector of n values: a_i, m_i, and p_i, respectively.

4.2. OPTIMIZATION STRATEGIES AND ALGORITHMS

Two different optimization strategies are implemented into the CONLIN program. The first strategy uses a pure convex linearization approach combined with a dual method formulation. The second strategy, based on a primal-dual formulation, is much more general, but it is less efficient for large scale problems and possibly less reliable.

Pure dual method approach

This approach is similar to the one used in previous versions of CONLIN, however it employs more efficient maximization algorithms. The explicit dual problem is replaced with a sequence of quadratic subproblems. Each subproblem is itself partially solved by either a first order or a second order maximization algorithm in the dual space. For very large scale problems, or when running on machines with small memory availability, a special out-of-core first order algorithm is also provided.

This strategy is efficient and highly reliable, but it is rather restrictive: the objective function and the constraints are ultimately approximated by convex linearization (Type 0). Therefore no equality constraints are permitted. Also only a limited number of approximation types can be accommodated. The objective function must be of type 0, while the constraints can be of types 0, 1, or 2. Higher order approximations (types 3 through 5) are absolutely not accepted.

Constraints of type 1 (reciprocal expansion) and type 2 (linear expansion) are internally replaced with a convex expansion of type 0, and a restart capability is implemented to iteratively solve the primary problem if it involves linear and/or reciprocal constraints. From these considerations, it appears that this first strategy is mostly recommended when all the problem functions are of type 0 (convex approximation).

It is important to notice that, when the user selects this strategy, he can influence the optimization results only externally, via the well known "move limit" technique. Indeed no convexification (or trust region) capabilities are provided.

Primal-dual method approach

In this second strategy, the primal explicit problem is solved by resorting to a sequential quadratic programming approach. Because all the approximating schemes used in CONLIN are separable, each quadratic subproblem will be separable too. When considering the dual of the quadratic subproblem, it appears to have the same form as in the first strategy. Therefore, similar first and second order dual algorithms can be employed to solve the sequence of subproblems. This strategy is very general and it can handle all the approximation types discussed above, as well as equality constraints.

In this primal-dual strategy, a strictly convex quadratic term, similar to (18), is added to the objective function. This term represents the distance to the current design point x0:

$$\delta\ \Sigma(x_i-x_i^0)^2$$

A large value of the constant δ leads to a small move in the design space, while a small value affects little the optimal solution to the current approximate subproblem. The constant δ can therefore be interpreted as a control parameter that acts as a "move limit" imposed on the explicit subproblem.

In summary, the derivation of a wide variety of convex approximating functions, coupled with the development of a highly efficient dual solver, can lead to a reliable general purpose optimization method.

5. Numerical examples

In addition to the 10-bar truss example that has been used as a support throughout the text, the methods using second derivatives proposed in this paper have been applied to the three problems used in [3] to assess the validity of the MMA method. Although these problems look simple and can be solved in closed form, they exhibit the typical characteristics of difficult structural optimization problems. These examples demonstrate the limitations of the methods using reciprocal variables, including CONLIN [1]. In fact, as stated in Ref. [3], when applied to these problems, the "traditional" method (i.e. of the CONLIN type) does not converge at all. The MMA method is capable of solving each problem efficiently, however the moving asymptotes L and U have to be adjusted very carefully at each iteration of the optimization process. Determining the right values for L and U might become cumbersome for larger problems. On the other hand the diagonal SQP method presented in this paper converges as fast as MMA for all the examples, without requiring any control parameters to be gradually adjusted.

5.1. CANTILEVER BEAM

The weight of a cantilever beam is to be minimized while assigning an upper limit on the tip displacement due to a given concentrated load. The beam is built up of five elements. Each beam element has a hollow square cross-section with constant thickness. The design variable associated to each element is the height (and width) of the square cross-section. This beam is simple from a structural analysis point of view, and therefore the optimization problem can be stated in closed form:

$$\text{minimize} \quad 0.0624\ (x_1 + x_2 + x_3 + x_4 + x_5)$$

$$\text{subject to} \quad 61/x_1^3 + 37/x_2^3 + 19/x_3^3 + 7/x_4^3 + 1/x_5^3 \leq 1$$

The only side constraints are that the design variables xi must remain non negative. The solution to this problem can easily be found analytically [3].

Table 3. Iteration History for Cantilever Beam
(1st entry: Weight; 2nd entry: Infeasibility)

Iter.	CONLIN	MMA (best)	Diagonal SQP (Direct)	Diagonal SQP (Reciprocal)
1	1.56 1.00	1.56 1.00	1.56 1.00	1.56 1.00
2	1.27 1.40	1.31 1.10	1.22 1.43	1.35 1.14
3	1.25 1.43	1.34 1.01	1.27 1.18	1.33 1.05
4	1.26 1.43	1.34 1.00	1.33 1.02	1.34 1.01
5	1.25 1.44		1.34 1.00	1.34 1.00
...	... ...			
13	1.26 1.42			

The diagonal SQP method was first used in the space of the direct variables, and then in the space of the reciprocal variables. The iteration history produced in both cases is given in Table 3. The initial Lagrangian multiplier is set to zero, and 100 % move limits are employed at the first iteration; otherwise the ai's would be zero in the quadratic function (18). The diagonal SQP method performs very well both in the direct space and in the reciprocal space. In the two cases, convergence is achieved in five iterations, however the constraint violation at intermediate design points is much smaller when working in the reciprocal space. It should be noted that, for this particular example, the space of zero constraint curvature would of course be the best choice. The method proposed in the preceding Section will automatically select the exponent p(i) = -3 for each design variable (i), and one single iteration is then sufficient for generating the optimum design.

For comparison the best and worst results obtained in [3] with the MMA method are also given in Table 3. The worst results correspond to using the CONLIN approach (i.e. L=0 and U=+∞ in MMA). The poor performance of CONLIN is due to the fact that the reciprocal variables do not introduce enough curvature in the approximate constraint. As a result the optimization process oscillates indefinitely between two highly infeasible designs. This behaviour can be stabilized using MMA, provided that very tight asymptotes are selected.

It should be noted that, for this particular example, the space of zero constraint curvature would of course be the best choice. The method proposed in the preceding Section will automatically select the exponent p(i) = -3 for each design variable (i), and one single iteration is then sufficient for generating the optimum design.

5.2. Two-Bar Truss

The 2-bar truss problem now considered is interesting, because it involves one sizing variable (cross-section of the bars) and one shape variable (width of the truss). The design optimization goal is to minimize the weight of the truss with upper limits on the stresses under one single load case. Again the optimization problem can be stated in closed form (see [3] for more details):

minimize $x_1 \sqrt{1 + x_2^2}$

subject to $0.124 \sqrt{1 + x_2^2}\,(8/x_1 + 1/x_1 x_2) \leq 1$

$0.124 \sqrt{1 + x_2^2}\,(8/x_1 - 1/x_1 x_2) \leq 1$

$0.2 \leq x_1 \leq 4.0 \qquad 0.1 \leq x_2 \leq 1.6$

Table 4 summarizes the iteration histories produced by the diagonal SQP method (direct and reciprocal spaces), MMA (best results presented in [3]), and CONLIN. The diagonal SQP approach performs equally well in the direct and reciprocal spaces (slightly better than MMA). Again CONLIN does not converge because it uses approximations that are not sufficiently conservative.

Table 4. Iteration History for 2-bar Truss Example
(1st entry: Weight; 2nd entry: Infeasibility)

Iter.	CONLIN	MMA (best)	Diagonal SQP (Direct)	Diagonal SQP (Reciprocal)
1	1.68 0.92	1.68 0.92	1.68 0.92	1.68 0.92
2	1.43 1.11	1.43 1.10	1.42 1.11	1.43 1.11
3	1.49 1.04	1.37 1.13	1.48 1.03	1.49 1.02
4	1.43 1.11	1.44 1.10	1.50 1.01	1.50 1.01
5	1.49 1.04	1.47 1.03	1.51 1.00	1.51 1.00
6	1.43 1.11	1.51 1.00		

5.3. Eight-Bar Truss

The eight-bar truss problem exhibits several interesting features. As shown in Table 5 the diagonal SQP approach converges rather slowly, much faster than CONLIN, but not as fast as the best results obtained by MMA [3]. In this problem there is much coupling between design variables, so that the separability assumption does not apply well. Therefore the full SQP approach leads to very good results.

In this specific example, it is possible to greatly improve the behaviour of the diagonal SQP method. Instead of positive move limits, slowing down convergence, let us introduce a negative δi in (20), in order to speed up convergence. It should be observed that the limiting case where $\delta i = -$ Aii corresponds to a sequential linear programming (SLP) approach, that is known to perform well if the solution is at a vertex in the design space. This is the case for the 8-bar truss problem. The best results generated by MMA confirm this idea: the moving asymptotes are taken larger and larger, so that in the limit, MMA behaves as a SLP approach. However, in general, it is not recommended to attempt accelerating the convergence speed by introducing negative control parameters δi. This procedure can indeed produce oscillations, or even divergence.

Table 5. Iteration History for 8-bar Trus Example

Iter	Full SQP	Diag. SQP ($\delta = 0$)	Diag. SQP ($\delta < 0$)	MMA (best)	CONLIN
1	13.05	13.05	13.05	13.05	13.05
2	11.44	11.73	11.73	12.10	11.68
3	11.25	11.62	11.62	11.67	11.66
4	11.23	11.62	11.45	11.65	11.64
5	———	11.57	11.23	11.61	11.62
6		11.52	———	11.52	11.59
7		11.47		11.42	11.57
8		11.42		11.28	11.55
9		11.37		11.23	11.53
10		11.33		———	11.52
..		...			...
15		11.23			11.43
16		———			11.42
..					...
40					11.23

REFERENCES

[1] Fleury, C. and Braibant, V. (1986) 'Structural optimization - a new dual method using mixed variables', **International Journal for Numerical Methods in Engineering** 23, 409-428.

[2] Fleury, C. (1989) 'CONLIN: an efficient dual optimizer based on convex approximation concepts', **Structural Optimization** 1, 81-89.

[3] Svanberg, K. (1987) 'Method of moving asymptotes - a new method for structural optimization', **International Journal for Numerical Methods in Engineering** 24, 359-373.

[4] Schmit, L.A. and Miura, H. (1976) 'Approximation concepts for efficient structural synthesis', NASA Contractor Report, NASA-CR 2552.

[5] Fleury, C. and Schmit, L.A. (1980) 'Dual methods and approximation concepts in structural synthesis', NASA Contractor Report, NASA-CR 3226.

[6] Braibant, V. and Fleury, C. (1985) 'An approximation concepts approach to shape optimal design', **Computer Methods in Applied Mechanics and Engineering** 53, 119-148.

[7] Fleury, C. (1986) 'Shape optimal design by the convex linearization method', Chapter 12 of **The Optimum Shape: Automated Structural Design** (J. Bennett and M. Botkin, eds.), Plenum Press, 297-326.

[8] Starnes, J.H. and Haftka, R.T. (1979) 'Preliminary design of composite wings for buckling, stress and displacement constraints', **Journal of Aircraft** 16, 564-570.

[9] Nguyen, V.H., Fleury, C. and Strodiot, J.J. (1987) 'A mathematical convergence analysis of the convex linearization method for engineering design optimization', **Engineering Optimization** 11, 195-216.

[10] Woo, T. H. (1986) 'Space Frame Optimization subject to Frequency Constraints', Proc. AIAA/ASME/ASCE/AHS 27th Structures, Structural Dynamics, and Materials Conference, 103-115.

[11] Fleury, C., Ramanathan, R.K., Salama, M. and Schmit, L.A. (1984) 'ACCESS computer program for the synthesis of large structural problems', Chapter 26 of **New Directions in Optimum Structural Design** (ATREK et al, eds), John Wiley and Sons, 541-561.

[12] Lust, R.V. and Schmit, L.A. (1986) 'Alternative approximation concepts for space frame synthesis', **AIAA Journal** 24, 1676-1684.

[13] Fleury, C. and Sander, G. (1983) 'Dual methods for optimizing finite element flexural systems' **Computer Methods in Applied Mechanics and Engineering** 37, 249-275.

[14] Prasad, B. (1984) 'Novel concepts for constraint treatments and approximations in efficient structural synthesis', **AIAA Journal** 22, 957-966.

[15] Wilson, R. B. (1963) 'A simplicial algorithm for concave programming', PhD Dissertation, Harvard University Graduate School of Business Administration.

[16] Fletcher, R. (1981) **Practical Methods of Optimization - Vol. 2: Constrained Optimization**, John Wiley & Sons.

[17] Fleury, C. (1989) 'Efficient approximation concepts using second order information', **International Journal for Numerical Methods in Engineering** 28, 2041-2058.

[18] Haftka, R.T. and Kamat, M.P. (1985) **Elements of Structural Optimization**, Martinus Nijhoff Publishers.

[19] Smaoui, H., Fleury, C. and Schmit, L.A. (1988) 'Advances in dual algorithms and convex approximation methods', Proc. AIAA/ASME/ASCE 29th Structures, Structural Dynamics, and Materials Conference, 1339-1347.

THE METHOD OF MOVING ASYMPTOTES (MMA) WITH SOME EXTENSIONS

K. SVANBERG
Optimization and Systems Theory
Royal Institute of Technology
S-10044 Stockholm, Sweden

ABSTRACT. This lecture deals with the Method of Moving Asymptotes (MMA), which is a mathematical programming method for structural optimization. In particular, some recent extensions (including "minimax" formulations) are described.

1. Introduction and background

The Method of Moving Asymptotes is a mathematical programming method which has been implemented in several large systems for structural optimization, e.g. in OPTSYS at the Aircraft division of Saab-Scania and in OASIS at ALFGAM Opt.AB.

MMA is an iterative method. In each iteration, a convex subproblem, which approximates the original problem, is generated and solved. An important role in the generation of these subproblems is played by a set of parameters which influence the "curvature" of the approximations, and also act as "asymptotes" for the subproblem. By moving these asymptotes between each iteration, the convergence of the overall process can be stabilized.

The original version of MMA was presented in [1]. In this lecture, that "pure" form of MMA is first described. Then some extensions, which are included in the implementions mentioned above, are presented. This extended version of MMA is capable of treating rather general structural optimization problems.

Consider a structural optimization problem P1 on the following form:

$$
\begin{aligned}
\underline{\text{P1}}: \quad & \text{minimize} && f_0(\mathbf{x}) \\
& \text{subject to:} && f_i(\mathbf{x}) \le \bar{f}_i\,, \quad i=1,\ldots,m \\
& && \underline{x}_j \le x_j \le \bar{x}_j\,, \quad j=1,\ldots,n
\end{aligned}
$$

G. I. N. Rozvany (ed.), Optimization of Large Structural Systems, Vol. I, 555–566.

where:

$\mathbf{x} = (x_1,\dots,x_n)^T$ is the vector of design variables, typically elemental sizes or shape variables,

$f_0(\mathbf{x})$ is the objective function, typically the structural weight,

$f_i(\mathbf{x}) \le \bar{f}_i$ are "behaviour constraints", typically limitations on stresses and displacements under different load cases,

$\underline{x}_j \le x_j \le \bar{x}_j$ are bounds ("technological constraints") on the variables.

P1 is often a difficult problem. One important reason for this is that the constraint functions are, in general, not explicitly given. Instead, it typically holds that $f_i(\mathbf{x}) = h_i(\mathbf{x},\mathbf{u})$, where h_i is a given explicit function while the vector $\mathbf{u}$ depends (implicitly) on the vector $\mathbf{x}$ by the relation: $\mathbf{K}(\mathbf{x})\mathbf{u} = \mathbf{p}$. Here, $\mathbf{K}(\mathbf{x})$ is the structural stiffness matrix (which depends on $\mathbf{x}$), $\mathbf{p}$ is a vector describing the applied load, while $\mathbf{u}$ is a vector describing the displacements of the structure.
For each new $\mathbf{x}$, the stiffness matrix $\mathbf{K}(\mathbf{x})$, which may have several thousands of rows, must be assembled before the displacement vector u is obtained as the solution of the (large) linear system $\mathbf{K}(\mathbf{x})\mathbf{u} = \mathbf{p}$. This, of course, makes each new evaluation of the constraint functions $f_i(\mathbf{x})$ expensive.

An encouraging feature of many structural optimization problems, however, is the possibility to calculate gradients of the constraint functions by a "semi-analytical" method; For a given $\mathbf{x}$, both $f_i(\mathbf{x})$ and $\nabla f_i(\mathbf{x})$ can be calculated at a cost which is not too much greater than the cost for calculating just $f_i(\mathbf{x})$. (Efficient methods for calculating gradients are described by other lectures during this meeting.)

Because of this possibility of calculating gradients, the following iterative approach is well established for solving structural optimization problems on the form P1:

STEP 0: Choose a starting solution $\mathbf{x}^{(1)}$ and let the iteration index k be equal to 1.

STEP 1: Given $\mathbf{x}^{(k)}$, calculate $f_i(\mathbf{x}^{(k)})$ and $\nabla f_i(\mathbf{x}^{(k)})$ for $i=0,1,..,m$.

STEP 2: Generate an explicit subproblem $P^{(k)}$ which approximates P1 :

$$
\begin{aligned}
\underline{P^{(k)}}:\quad & \text{minimize} \quad \tilde{f}_0^{(k)}(\mathbf{x}) \\
& \text{subject to} \quad \tilde{f}_i^{(k)}(\mathbf{x}) \le \bar{f}_i \quad , \ i=1,..,m \\
& \qquad\qquad\quad \underline{x}_j \le x_j \le \bar{x}_j \quad , \ j=1,..,n
\end{aligned}
$$

The $\tilde{f}_i^{(k)}$ are explicit functions which approximate the implicit functions f_i. The constraints $\underline{x}_j \le x_j \le \bar{x}_j$ are often replaced by more restrictive bounds on the variables, so called "move limits", to prevent $\mathbf{x}$ from going too far away from the current iteration point $\mathbf{x}^{(k)}$.

STEP 3: Solve the subproblem $P^{(k)}$, with some suitable method (dependent on how the approximating functions have been chosen). Let the optimal solution of $P^{(k)}$ be the next iteration point $\mathbf{x}^{(k+1)}$ and go to STEP 1, with k replaced by k+1.

The process is interrupted when some convergence criteria are fulfilled, or simply when the user is satisfied with the current solution $\mathbf{x}^{(k)}$.

The central step in this approach is to choose good approximating functions (i.e. STEP 2). The main information available for doing this are the calculated function values and gradients (from the current iteration as well as from the previous iterations). In addition, some important properties of the considered problem may be known. As an example, it is known that the normal stress in a truss element is monotonically decreasing as $\sigma_j = a_j + b_j/(x_j+c_j)$ if the cross section area x_j of the bar is increased while the areas of all the other bars are held fixed. This kind of information (if it is general enough) may also be useful when the approximating functions should be chosen. It is also important that the subproblem $P^{(k)}$ does not become too hard to solve. It is to prefer, e.g., that the chosen approximating functions are convex.

Several methods based on the above approach (STEP 0 - STEP 3) have been suggested. The main difference between these methods is how the approximating functions are chosen, i.e. STEP 2.

The perhaps most obvious possible method is so called "Sequential Linear Programming" (SLP) where the approximating functions are chosen as the first order Taylor expansion, i.e.:

$$\tilde{f}_i^{(k)}(\mathbf{x}) = f_i(\mathbf{x}^{(k)}) + \sum_j a_{ij}(x_j - x_j^{(k)}) \text{ , for } i = 0,1,\dots,m,$$

where $a_{ij} = \dfrac{\partial f_i}{\partial x_j}$, calculated at $\mathbf{x} = \mathbf{x}^{(k)}$.

With these $\tilde{f}_i^{(k)}(\mathbf{x})$, the subproblem $P^{(k)}$ becomes a Linear Programming (LP) problem, which may be efficiently solved by the Simplex method.

In general, SLP works well if the number of active constraints at the optimal solution $\mathbf{x}^*$ of P1, i.e. the number of constraints in P1 that are satisfied as equalities at $\mathbf{x}^*$, is equal to the number of design variables. Otherwise, the convergence to $\mathbf{x}^*$ might be very slow, obtained only through the use of decreasing move limits. It is, of course, not possible in general to know in advance how many (or which) constraints that are active at the optimal solution of P1.

For element sizing problems, Schmit (see e.g. [2]) suggested that the approximating functions for the constraints should be chosen as the first order Taylor expansion in the reciprocal element sizes $1/x_j$:

$$\tilde{f}_i^{(k)}(\mathbf{x}) = f_i(\mathbf{x}^{(k)}) + \sum_j b_{ij}(1/x_j - 1/x_j^{(k)}) \text{ , for } i = 1,\dots,m,$$

where $b_{ij} = \dfrac{\partial f_i}{\partial(1/x_j)} = -(x_j^{(k)})^2 \cdot \dfrac{\partial f_i}{\partial x_j}$, calculated at $\mathbf{x} = \mathbf{x}^{(k)}$,

while the exact objective function (assumed to be the structural weight which is a linear function of the elemental sizes) is used in $P^{(k)}$, i.e. $\tilde{f}_0^{(k)}(\mathbf{x}) = f_0(\mathbf{x})$.

This method is probably the most widely used method for element sizing problems, especially since Fleury in [3] suggested an efficient dual method for solving the subproblems. If the constraints are on nodal displacements and elemental stresses, while the design variables are elemental sizes and the objective function is the structural weight, then this method of linearization in reciprocal variables is much more reliable and efficient than SLP. The main reason for this is that nodal displacements and elemental stresses are more close to be linear in $1/x_j$ than in x_j.

A generalization of this latter method, to other structural optimization problems than just element sizing, was suggested by Fleury in [4]. Here the approximating functions are obtained by a linearization in "mixed" variables; eighter in x_j or in $1/x_j$, dependent on the signs of the derivatives at $\mathbf{x}^{(k)}$:

$$\tilde{f}_i^{(k)}(\mathbf{x}) = f_i(\mathbf{x}^{(k)}) + \sum_{j=1}^{n} \left[a_{ij}(x_j - x_j^{(k)}) + b_{ij}(1/x_j - 1/x_j^{(k)}) \right], \text{ for } i = 0,\dots,m.$$

If $\dfrac{\partial f_i}{\partial x_j} > 0$ at $\mathbf{x}^{(k)}$ then: $a_{ij} = \dfrac{\partial f_i}{\partial x_j}$ and $b_{ij} = 0$, while

if $\dfrac{\partial f_i}{\partial x_j} < 0$ at $\mathbf{x}^{(k)}$ then: $b_{ij} = -(x_j^{(k)})^2 \cdot \dfrac{\partial f_i}{\partial x_j}$ and $a_{ij} = 0$.

It is easy to check that $\tilde{f}_i^{(k)}(\mathbf{x})$ will always be a convex, first order approximation of $f_i(\mathbf{x})$.

2. Description of the "pure" MMA

MMA is a method which may be interpreded as a generalization of the above method of linearization in mixed variables. In MMA, each approximating function $\tilde{f}_i^{(k)}(\mathbf{x})$ is obtained by a linearization of $f_i(\mathbf{x})$ in variables of the type $1/(U_j - x_j)$ or $1/(x_j - L_j)$, where L_j and U_j are parameters that satisfy $L_j < x_j^{(k)} < U_j$:

$$\tilde{f}_i^{(k)}(\mathbf{x}) = \sum_{j=1}^{n} \left[\frac{p_{ij}}{U_j - x_j} + \frac{q_{ij}}{x_j - L_j} \right] + r_i \quad , \text{ for } i=0,1,..,m.$$

If $\dfrac{\partial f_i}{\partial x_j} > 0$ at $\mathbf{x}^{(k)}$ then: $p_{ij} = (U_j - x_j^{(k)})^2 \cdot \dfrac{\partial f_i}{\partial x_j}$ and $q_{ij} = 0$, while

if $\dfrac{\partial f_i}{\partial x_j} < 0$ at $\mathbf{x}^{(k)}$ then: $q_{ij} = -(x_j^{(k)} - L_j)^2 \cdot \dfrac{\partial f_i}{\partial x_j}$ and $p_{ij} = 0$.

r_i is chosen such that $\tilde{f}_i^{(k)}(\mathbf{x}^{(k)}) = f_i(\mathbf{x}^{(k)})$.

Again, it is easy to check that $\tilde{f}_i^{(k)}(\mathbf{x})$ will always be a convex, first order approximation of $f_i(\mathbf{x})$.
The values of the parameters L_j and U_j are normally changed between the iterations. Therefore, they should in fact be denoted $L_j^{(k)}$ and $U_j^{(k)}$, just as p_{ij}, q_{ij} and r_i should be denoted $p_{ij}^{(k)}$, $q_{ij}^{(k)}$ and $r_i^{(k)}$. It is straightforward to show that if one let $L_j \to -\infty$ and $U_j \to +\infty$ then the MMA-approximations become (in the limit) equal to the linear approximations used in SLP, while if one let $L_j = 0$ and $U_j \to +\infty$ then the MMA-approximations become (in the limit) equal to the above linearization in mixed variables $1/x_j$ and x_j. In MMA, however, the parameters L_j and U_j are always given finite values, which are updated at each new iteration point. This makes it possible both to stabilize and speed up the iterative scheme described above (STEP 1 - STEP 3). The "default" rule for updating the "moving asymptotes" L_j and U_j goes as follows:

* In the first two iteration, i.e. for k = 1 and k = 2, then:
  $L_j^{(k)} = x_j^{(k)} - C_0 \cdot (\bar{x}_j - \underline{x}_j)$ and $U_j^{(k)} = x_j^{(k)} + C_0 \cdot (\bar{x}_j - \underline{x}_j)$,
  where C_0 is a constant. It is often reasonable to let $C_0 = 0.5$.
* If $k \geq 3$, then for each j such that $(x_j^{(k)} - x_j^{(k-1)}) \cdot (x_j^{(k-1)} - x_j^{(k-2)}) \leq 0$, the asymptotes are moved closer to each other by the formulas:
  $L_j^{(k)} = x_j^{(k)} - C_1 \cdot (x_j^{(k-1)} - L_j^{(k-1)})$ and $U_j^{(k)} = x_j^{(k)} + C_1 \cdot (U_j^{(k-1)} - x_j^{(k-1)})$,
  where C_1 is a constant which is strictly less than 1.
  $C_1 = 0.7$ is often reasonable, but sometimes a smaller value, e.g. $C_1 = 0.5$, could be better (by making the process more stable).

* If $k \geq 3$, then for each j such that $(x_j^{(k)}-x_j^{(k-1)})\cdot(x_j^{(k-1)}-x_j^{(k-2)}) > 0$, the asymptotes are moved away from each other by the formulas: $L_j^{(k)}= x_j^{(k)}- C_2\cdot(x_j^{(k-1)}- L_j^{(k-1)})$ and $U_j^{(k)}= x_j^{(k)}+ C_2\cdot(U_j^{(k-1)}- x_j^{(k-1)})$, where C_2 is a constant which is strictly greater than 1. $C_2= 1.3$ is often reasonable, but sometimes a smaller value could be better.

C_1 and C_2 should always be chosen such that $C_1\cdot C_2 < 1$, so that one single "relaxation" of the asymptotes never fully compensates for one "tightening".

3. Extensions of MMA to more general problems

The previous section gave a brief description of the "pure" MMA.
In the rest of this lecture, some extensions of this pure form of MMA will be described, including the following:

* treatment of constraints which are <u>linear</u> in the original problem P1,
* treatment of constraints which are <u>equality</u> constraints in P1,
* <u>artificial</u> variables to avoid that the subproblem becomes infeasible,
* possibility to switch to "<u>mini-max</u>" formulations,
* first and second order <u>dual</u> methods to solve the subproblems.

The following subsets of the index set $\{1,\ldots,m\}$ of the constraints will be frequently used:

$\mathcal{L}_0$ = index set for the <u>nonlinear</u> <u>inequality</u> constraints,
$\mathcal{L}_1$ = index set for the <u>linear</u> <u>inequality</u> constraints,
$\mathcal{L}_2$ = index set for the <u>equality</u> constraints (linear or non-linear).
$\mathcal{M}_0$ = index set for the <u>ordinary</u> constraint (not mini-max).
$\mathcal{M}_1$ = index set for the "<u>mini-max</u>" constraints.

Now, consider a optimization problem P2 on the following form:

<u>P2</u>:

$$
\begin{array}{lll}
\text{minimize} & f_0(\mathbf{x}) + z & \mathbf{x} \in \mathbb{R}^n,\ z \in \mathbb{R} \\
\text{subject to:} & f_i(\mathbf{x}) \leq \bar{f}_i\ , & \text{for } i \in \mathcal{L}_0 \cap \mathcal{M}_0 \\
 & f_i(\mathbf{x}) \leq \bar{f}_i\ , & \text{for } i \in \mathcal{L}_1 \cap \mathcal{M}_0 \\
 & f_i(\mathbf{x}) = \bar{f}_i\ , & \text{for } i \in \mathcal{L}_2 \cap \mathcal{M}_0 \\
 & f_i(\mathbf{x}) - z \leq \bar{f}_i\ , & \text{for } i \in \mathcal{L}_0 \cap \mathcal{M}_1 \\
 & f_i(\mathbf{x}) - z \leq \bar{f}_i\ , & \text{for } i \in \mathcal{L}_1 \cap \mathcal{M}_1 \\
 & \underline{x}_j \leq x_j \leq \bar{x}_j\ , & \text{for } j \in \{1,\ldots,n\} \\
 & z \geq 0 &
\end{array}
$$

where $\mathcal{L}_0$, $\mathcal{L}_1$, $\mathcal{L}_2$, $\mathcal{M}_0$ and $\mathcal{M}_1$ are (the above) sets which must satisfy:

$\mathcal{L}_0 \cup \mathcal{L}_1 \cup \mathcal{L}_2 = \mathcal{M}_0 \cup \mathcal{M}_1 = \{1,\dots,m\}$, $\mathcal{L}_1 \cap \mathcal{L}_2 = \emptyset$, $\mathcal{L}_1 \cap \mathcal{L}_3 = \emptyset$, $\mathcal{L}_2 \cap \mathcal{L}_3 = \emptyset$, $\mathcal{M}_0 \cap \mathcal{M}_1 = \emptyset$ and $\mathcal{L}_2 \cap \mathcal{M}_1 = \emptyset$.

There are two important special cases of P2:

- If $\mathcal{M}_1 = \emptyset$ then the variable z disappear and P2 gets on a traditional mathematical programming form. It should be noted that P1 is obtained as a special case of P2 by letting $\mathcal{M}_1 = \mathcal{L}_1 = \mathcal{L}_2 = \emptyset$.
- If $f_0(\mathbf{x}) \equiv 0$ then the objective function, which should be minimized, may be interpreted as $z = z(\mathbf{x}) = \max_{i\in\mathcal{M}_1} \{(f_i(\mathbf{x})-\bar{f}_i)_+\}$. A possible application of this case is when the largest stress in a structure ($z(\mathbf{x}) = \max_i \{|\sigma_i(\mathbf{x})|\}$) should be minimized subject to constraints on e.g. certain displacements and the structural weight.

The subproblem $P^{(k)}$, which is generated by MMA in order to approximate P2, looks as follows: (Here, $L_j = L_j^{(k)}$ and $U_j = U_j^{(k)}$, chosen according to the the previous section.)

$\underline{P^{(k)}}$:

$$\underset{\mathbf{x},\mathbf{v},z}{\text{minimize}} \sum_{j=1}^{n} \left[\frac{p_{0j}}{U_j-x_j} + \frac{q_{0j}}{x_j-L_j} \right] + z + \frac{c}{2}\cdot(z-\bar{z})^2 + \sum_{i\in\mathcal{M}_0} (v_i^2/2 + e_i|v_i|)$$

$$\text{s.t.:} \quad \sum_{j=1}^{n} \left[\frac{p_{ij}}{U_j-x_j} + \frac{q_{ij}}{x_j-L_j} \right] - v_i \le b_i \quad \text{for } i \in \mathcal{L}_0 \cap \mathcal{M}_0$$

$$\sum_{j=1}^{n} a_{ij}x_j - v_i \le b_i \quad \text{for } i \in \mathcal{L}_1 \cap \mathcal{M}_0$$

$$\sum_{j=1}^{n} a_{ij}x_j - v_i = b_i \quad \text{for } i \in \mathcal{L}_2 \cap \mathcal{M}_0$$

$$\sum_{j=1}^{n} \left[\frac{p_{ij}}{U_j-x_j} + \frac{q_{ij}}{x_j-L_j} \right] - z \le b_i \quad \text{for } i \in \mathcal{L}_0 \cap \mathcal{M}_1$$

$$\sum_{j=1}^{n} a_{ij}x_j - z \le b_i \quad \text{for } i \in \mathcal{L}_1 \cap \mathcal{M}_1$$

$$\alpha_j \le x_j \le \beta_j \quad \text{for } j \in \{1,\dots,n\}$$

$$z \ge 0 \text{ and } v_i \ge 0 \quad \text{for } i \in \mathcal{M}_0 \cap (\mathcal{L}_0 \cup \mathcal{L}_1)$$

where:

$$p_{oj} = (U_j^{(k)}-x_j^{(k)})^2 \cdot \left[\left(\frac{\partial f_0}{\partial x_j}\right)_+ + \frac{\varepsilon_0}{U_j^{(k)}-L_j^{(k)}}\right],$$

$$q_{oj} = (x_j^{(k)}-L_j^{(k)})^2 \cdot \left[\left(\frac{\partial f_0}{\partial x_j}\right)_- + \frac{\varepsilon_0}{U_j^{(k)}-L_j^{(k)}}\right],$$

$$p_{ij} = (U_j^{(k)}-x_j^{(k)})^2 \cdot \left(\frac{\partial f_i}{\partial x_j}\right)_+ , \quad q_{ij} = (x_j^{(k)}-L_j^{(k)})^2 \cdot \left(\frac{\partial f_i}{\partial x_j}\right)_- , \quad i \ge 1,$$

$$a_{ij} = \frac{\partial f_i}{\partial x_j}, \quad b_i = \bar{f}_i - f_i(\mathbf{x}^{(k)}) + \sum_{j=1}^{n} a_{ij}x_j^{(k)} \quad \text{for } i \in \mathcal{L}_1 \cup \mathcal{L}_2 ,$$

$$b_i = \bar{f}_i - f_i(\mathbf{x}^{(k)}) + \sum_{j=1}^{n}\left(\frac{p_{ij}}{U_j^{(k)}-x_j^{(k)}} + \frac{q_{ij}}{x_j^{(k)}-L_j^{(k)}}\right) \quad \text{for } i \in \mathcal{L}_0 ,$$

$$\alpha_j = \max\{\, \underline{x}_j,\ 0.1x_j^{(k)} + 0.9L_j^{(k)}\} , \quad \beta_j = \min\{\, \bar{x}_j,\ 0.1x_j^{(k)} + 0.9U_j^{(k)}\} ,$$

$$\bar{z} = \max_{i \in \mathcal{M}_1} \{(f_i(\mathbf{x}^{(k)})-\bar{f}_i)_+\} , \quad c = 1/(2\bar{z} + 1) ,$$

ε_0 = a "reasonably small" number > 0 (e.g. 1% of an expected improvement in the objective function $f_0(\mathbf{x})$ from the start) ,

e_i = "very large" numbers, for $i \in \mathcal{M}_0$, (compared to reasonable values on the corresponding Lagrange multipliers).

By $\left(\frac{\partial f_i}{\partial x_j}\right)_+$ and $\left(\frac{\partial f_i}{\partial x_j}\right)_-$ are meant, respectively, $\max\left\{0, \frac{\partial f_i}{\partial x_j}\right\}$ and $\max\left\{0, -\frac{\partial f_i}{\partial x_j}\right\}$, where the derivatives $\frac{\partial f_i}{\partial x_j}$ are calculated at $\mathbf{x}^{(k)}$.

Some comments on this subproblem:

* Linear constraints in the original problem P2 are treated as linear constraints also in the subproblem $P^{(k)}$.
* Equality constraints in P2 are treated as linear equality constraints in $P^{(k)}$. (To make sure that $P^{(k)}$ becomes a convex problem.)
* The variables v_i are "artificial variables", which are introduced to prevent the subproblem from becoming infeasible. The variable z is the mini-max objective value. For each $\mathbf{x}$, with all $x_j \in [\alpha_j, \beta_j]$, it is always possible to choose v_i and z such that a feasible solution

$(\mathbf{x},v,z)$ of $P^{(k)}$ is obtained. This means that the subproblem $P^{(k)}$ is never infeasible. However, since the artificial variables v_i are very expensive, they will not be non-zero in the optimal solution of $P^{(k)}$ unless it is absolutely necessary.

* $p_{ij} \geq 0$ and $q_{ij} \geq 0$ for all i and j. In particular, $p_{0j} > 0$ and $q_{0j} > 0$ for all j (even if $f_0(\mathbf{x}) \equiv 0$) since ε_0 is strictly positive. Therefore, all the constraint functions in $P^{(k)}$ are convex function, while the objective function is a strictly convex function (which goes to $+\infty$ if any of the variables x_j goes to L_j or U_j). This implies (together with the previous comment on feasible solutions) that there is always a <u>unique</u> optimal solution of $P^{(k)}$.
* The objective function and the constraint functions in $P^{(k)}$ are first order approximations of the corresponding functions in P2. It should be noted that the derivatives of the objective function in $P^{(k)}$ do <u>not</u> depend on ε_0! Therefore, ε_0 does not have to be "very small". The second order derivatives of the objective functions, however, do depend on ε_0. The larger value on ε_0, the larger value on the (non-mixed) second derivatives, and the more "conservative" becomes the objective function.

4. The dual subproblem

Next, the dual problem of $P^{(k)}$ will be described. In particular, the gradient and the Hessian matrix of the dual objective function $\varphi(\mathbf{y})$ will be explicitly expressed.

The Lagrange function corresponding to $P^{(k)}$ is given by:

$$L(\mathbf{x},v,z,\mathbf{y}) = -\sum_{i=1}^{m} y_i b_i + \sum_{j=1}^{n}\left[\frac{p_j(\mathbf{y})}{U_j-x_j} + \frac{q_j(\mathbf{y})}{x_j-L_j} + a_j(\mathbf{y})x_j\right] +$$

$$+ \sum_{i\in\mathcal{M}_0}(v_i^2/2 + e_i|v_i| - y_i v_i) + \frac{c}{2}\cdot(z-\bar{z})^2 + (1-\sum_{i\in\mathcal{M}_1} y_i)\cdot z \ ,$$

where:

$\mathbf{y} = (y_1,\ldots,y_m)^T$ is the vector of dual variables (Lagrange multipliers),

$p_j(\mathbf{y}) = p_{0j} + \sum_{i\in\mathcal{L}_0} y_i p_{ij}$, $q_j(\mathbf{y}) = q_{0j} + \sum_{i\in\mathcal{L}_0} y_i q_{ij}$, and $a_j(\mathbf{y}) = \sum_{i\notin\mathcal{L}_0} y_i a_{ij}$.

The dual variables must satisfy $y_i \geq 0$ for $i \notin \mathcal{L}_2$. This will be denoted shortly as $\mathbf{y} \in \mathbf{Y}$. Then $p_j(\mathbf{y}) > 0$ and $q_j(\mathbf{y}) > 0$ for all $\mathbf{y} \in \mathbf{Y}$.

The dual objective function φ is defined as:

$$\varphi(y) = \min_{(x,v,z)} \{ L(x,v,z,y) \mid \alpha_j \le x_j \le \beta_j,\ v_i \ge 0\ i \notin \mathcal{L}_2,\ z \ge 0 \} =$$
$$= L(x(y), v(y), z(y), y).$$

For each given vector $y \in Y$, the minimizing $x_j = x_j(y)$ in the Lagrange function is, for each j, the optimal solution of the following convex one-variable problem:

$$\underline{\text{PLAG}}_j\text{: minimize}_{x_j} \quad \frac{p_j(y)}{U_j - x_j} + \frac{q_j(y)}{x_j - L_j} + a_j(y)x_j \quad \text{subject to} \quad \alpha_j \le x_j \le \beta_j .$$

If $a_j(y) = 0$ then $PLAG_j$ may be solved analytically, see [1].
If $a_j(y) \neq 0$ then it is still very easy to solve $PLAG_j$ numerically, e.g. by Newtons method.
For each given vector $y \in Y$, the minimizing $v_i = v_i(y)$ in the Lagrange function is, for each $i \in \mathcal{M}_0$, given by:

$$v_i(y) = \begin{cases} y_i - e_i & \text{if } y_i > e_i \\ 0 & \text{if } -e_i \le y_i \le e_i \\ y_i + e_i & \text{if } y_i < -e_i \end{cases}$$

This implies the following contribution to the dual objective function (for each $i \in \mathcal{M}_0$):

$$v_i^2(y)/2 + e_i|v_i(y)| - y_i v_i(y) = -v_i^2(y)/2 .$$

For each given vector $y \in Y$, the minimizing $z = z(y)$ in the Lagrange function is given by: $z(y) = (\bar{z} + \frac{1}{c}\cdot(\sum_{i \in \mathcal{M}_1} y_i - 1))_+$.

This implies the following contribution to the dual objective function: $\frac{c}{2}\cdot(z(y) - \bar{z})^2 + (1 - \sum_{i \in \mathcal{M}_1} y_i)\cdot z(y) = -\frac{c}{2}\cdot(z^2(y) - \bar{z}^2)$.

The dual objective function thus becomes:

$$\varphi(y) = -\sum_{i=1}^{m} y_i b_i + \sum_{j=1}^{n} \left[\frac{p_j(y)}{U_j - x_j(y)} + \frac{q_j(y)}{x_j(y) - L_j} + a_j(y)x_j(y) \right] -$$
$$- \sum_{i \in \mathcal{M}_0} v_i^2(y)/2 - \frac{c}{2}\cdot(z^2(y) - \bar{z}^2)$$

The dual problem is defined as:

D: maximize $\varphi(\mathbf{y})$

subject to: $\mathbf{y} \in \mathbf{Y}$ (i.e. $y_i \geq 0$ for $i \in \mathcal{L}_0 \cup \mathcal{L}_1$)

Because of the convexity of the constraint functions in $P^{(k)}$, and the strict convexity of the objective function, it is possible to prove the following: If $\mathbf{y}^*$ is an optimal solution of the dual problem D, then $(\mathbf{x}(\mathbf{y}^*), \mathbf{v}(\mathbf{y}^*), z(\mathbf{y}^*))$ is the unique global optimal solution of $P^{(k)}$.

In order to solve D, the derivatives of the dual objective function are needed. These derivatives are given by the following expressions:

If $i \in \mathcal{L}_0 \cap \mathcal{M}_0$: $$\frac{\partial\varphi}{\partial y_i} = \sum_{j=1}^{n}\left[\frac{p_{ij}}{U_j - x_j(\mathbf{y})} + \frac{q_{ij}}{x_j(\mathbf{y}) - L_j} \right] - v_i(\mathbf{y}) - b_i \ .$$

If $i \in (\mathcal{L}_1 \cap \mathcal{M}_0) \cup (\mathcal{L}_2 \cap \mathcal{M}_0)$: $$\frac{\partial\varphi}{\partial y_i} = \sum_{j=1}^{n} a_{ij}x_j(\mathbf{y}) - v_i(\mathbf{y}) - b_i \ .$$

If $i \in \mathcal{L}_0 \cap \mathcal{M}_1$: $$\frac{\partial\varphi}{\partial y_i} = \sum_{j=1}^{n}\left[\frac{p_{ij}}{U_j - x_j(\mathbf{y})} + \frac{q_{ij}}{x_j(\mathbf{y}) - L_j} \right] - z(\mathbf{y}) - b_i \ .$$

If $i \in \mathcal{L}_1 \cap \mathcal{M}_1$: $$\frac{\partial\varphi}{\partial y_i} = \sum_{j=1}^{n} a_{ij}x_j(\mathbf{y}) - z(\mathbf{y}) - b_i \ .$$

It is also possible to calculate the second order derivatives.

Let $J_f = \{\ j \mid \alpha_j < x_j(\mathbf{y}) < \beta_j\ \}$, and define, for each $j \in J_f$:

$$A_{ij} = \frac{p_{ij}}{(U_j - x_j(\mathbf{y}))^2} - \frac{q_{ij}}{(x_j(\mathbf{y}) - L_j)^2} \ \text{ if } i \in \mathcal{L}_0 \ , \quad A_{ij} = a_{ij} \ \text{ if } i \notin \mathcal{L}_0 \ ,$$

$$B_j = \frac{p_j(\mathbf{y})}{(U_j - x_j(\mathbf{y}))^3} + \frac{q_j(\mathbf{y})}{(x_j(\mathbf{y}) - L_j)^3} \ .$$

Then the second order derivatives of $\varphi(\mathbf{y})$ become:

If $k \in \mathcal{M}_0$ and $\ell \in \mathcal{M}_0$: $$\frac{\partial^2\varphi}{\partial y_k \partial y_\ell} = \frac{\partial^2}{\partial y_k \partial y_\ell}\Big(-\sum_{i \in \mathcal{M}_0} v_i^2(\mathbf{y})/2\Big) - \sum_{j \in J_f} \frac{A_{kj}A_{\ell j}}{2B_j} \ ,$$

where $$\frac{\partial^2}{\partial y_k \partial y_\ell}\Big(-\sum_{i \in \mathcal{M}_0} v_i^2(\mathbf{y})/2\Big) = \begin{cases} -1 & \text{if } k = \ell \text{ and } |y_k| > e_k \\ 0 & \text{otherwise} \end{cases}$$

If $k \in \mathcal{M}_1$ and $\ell \in \mathcal{M}_1$: $\dfrac{\partial^2\varphi}{\partial y_k \partial y_\ell} = \dfrac{\partial^2}{\partial y_k \partial y_\ell}(-cz^2(y)/2\ ) - \sum_{j \in J_f} \dfrac{A_{kj}A_{\ell j}}{2B_j}$,

where $\dfrac{\partial^2}{\partial y_k \partial y_\ell}(-cz^2(y)/2\ ) = \begin{cases} -1/c \text{ if } \sum_{i \in \mathcal{M}_1} y_i > 1 \\ 0 \text{ otherwise} \end{cases}$

If $k \in \mathcal{M}_0$ and $\ell \in \mathcal{M}_1$: $\dfrac{\partial^2\varphi}{\partial y_k \partial y_\ell} = - \sum_{j \in J_f} \dfrac{A_{kj}A_{\ell j}}{2B_j}$.

Thus, since $\varphi(y)$, $\nabla\varphi(y)$ and $\nabla^2\varphi(y)$ are straightforward to evaluate, some standard method for solving the dual problem can be applied. Eighter a first order method like the conjugate gradient method, or a second order method like Newtons method. The non-negativity constraints on some of the dual variables may be handled by an active set strategy. See e.g. [5] concerning these mentioned methods.

5. Performance of MMA

The users mentioned in section 1 have by now several years experience of MMA applied to different structural optimization problems (mostly sizing and shape optimization problems). They consider MMA to be a reliable method which in most cases generates a sequence of steadily improved solutions. In practice, the (best) solution obtained after about 10-15 iterations is usually accepted as beeing "sufficiently good", since the marginal improvement from each new iteration then has become rather small.

6. References

[1] Svanberg, K. (1987) "The method of moving asymptotes - a new method for structural optimization", International Journal for Numerical Methods in Engineering 24, 359-373.

[2] Schmit, L.A. and Farshi, B. (1974) "Some approximation concepts for structural synthesis", AIAA Journal 12, 692-699.

[3] Fleury, C. (1979) "Structural weight optimization by dual methods of convex programming", International Journal for Numerical Methods in Engineering 14, 1761-1783.

[4] Fleury, C. and Braibant, V. (1986) "Structural optimization - a new dual method using mixed variables", International Journal for Numerical Methods in Engineering 23, 409-428.

[5] Fletcher, R. (1987) "Practical Methods of Optimization", Second edition, John Wiley & Sons.

SOME SECOND ORDER METHODS FOR STRUCTURAL OPTIMIZATION

K. SVANBERG
Optimization and Systems Theory
Royal Institute of Technology
S-10044 Stockholm, Sweden

ABSTRACT. This lecture deals with second order methods for structural optimization. First the Lagrange-Newton method, also called Sequential Quadratic Programming, is summarized. Then some alternative second order methods are presented and briefly discussed.

1. Introduction and Summary

Most methods for structural optimization are "first order methods", which means that the approximations involved are of the first order, based on calculated gradients of the objective and the constraint functions. It is well known, however, that second order methods, i.e. methods based on second order derivatives (or good approximations of second order derivatives) could give much faster convergence. As an example, the Newton method usually requires much fewer iterations than the Steepest descent method, when minimizing a nonlinear function without any constraints.

The main reason that second order methods have not been very much used for structural optimization is that second order derivatives are usually considered to be too expensive to calculate. In some cases, however, the second order derivatives can be evaluated in a surprisingly efficient way. This is demonstrated in e.g. [1]-[4].

A basic tool for second order methods is the quadratic function. For unconstrained problems, the Newton method is based on sequential minimization of quadratic functions (second order Taylor expansions of the objective function). For constrained problems, the Lagrange-Newton method is based on sequential minimization of quadratic functions (second order Taylor expansions of the Lagrange function) subject to linear constraints.

Therefore, this lecture first summarizes the basic properties of quadratic programming problems. Then the Lagrange-Newton method is shortly described and interpreted as a sequential quadratic programming method. Finally, some alternative second order methods, which may be well suited for structural optimization, are discussed. In particular, some second order extensions of the Method of Moving Asymptots are presented.

G. I. N. Rozvany (ed.), Optimization of Large Structural Systems, Vol. I, 567–578.

2. Minimization of a quadratic function q(x) without any constraints

A quadratic function of n variables can always be written on the form:

$$q(\mathbf{x}) = \tfrac{1}{2}\mathbf{x}^T Q\mathbf{x} + c^T\mathbf{x} + c_0 = \tfrac{1}{2}\sum_i\sum_j q_{ij}x_ix_j + \sum_j c_jx_j + c_0$$

where $\mathbf{x} = (x_1,\dots,x_n)^T$, Q = a symmetric nxn matrix , $\mathbf{c} \in \mathbb{R}^n$ and $c_0 \in \mathbb{R}$.

The gradient $\nabla q(\mathbf{x})$ (which is a row vector) and the Hessian matrix $\nabla^2 q(\mathbf{x})$ are easily found to be: $\nabla q(\mathbf{x}) = (Q\mathbf{x} + \mathbf{c})^T$ and $\nabla^2 q(\mathbf{x}) = Q$.
Let $\mathbf{x}^* \in \mathbb{R}^n$ be given, and let $\mathbf{d} = \mathbf{x} - \mathbf{x}^*$. Then some calculations give:

$$q(\mathbf{x}) = q(\mathbf{x}^*+\mathbf{d}) = q(\mathbf{x}^*) + (Q\mathbf{x}^*+\mathbf{c})^T\mathbf{d} + \tfrac{1}{2}\mathbf{d}^TQ\mathbf{d}\ .$$

A necessary condition for $\mathbf{x}^*$ to be a minimum point of $q(\mathbf{x})$ is, clearly, that $Q\mathbf{x}^* + \mathbf{c} = \mathbf{0}$, i.e. that $\nabla q(\mathbf{x}^*) = \mathbf{0}$. Assume that this is case. Then $q(\mathbf{x}) = q(\mathbf{x}^*) + \tfrac{1}{2}\mathbf{d}^TQ\mathbf{d}$, and it immediatly follows that if Q is a <u>positive definite</u> matrix then $\mathbf{x}^*$ is a unique global minimum point of $q(\mathbf{x})$. Thus:

PROP. 1: If Q is positive definite, then the unique global minimum point of $q(\mathbf{x}) = \tfrac{1}{2}\mathbf{x}^TQ\mathbf{x} + \mathbf{c}^T\mathbf{x} + c_0$ is obtained by solving the linear equations $Q\mathbf{x} = -\mathbf{c}$.

3. Minimization of q(x) subject to linear equality constraints

Next, consider the problem of minimizing a quadratic function subject to linear equality constraints, i.e. a problem on the form:

<u>QPE:</u> minimize $q(\mathbf{x}) = \tfrac{1}{2}\mathbf{x}^TQ\mathbf{x} + \mathbf{c}^T\mathbf{x} + c_0$

subject to: $A\mathbf{x} = \mathbf{b}$

where Q is nxn and symmetric, $\mathbf{c} \in \mathbb{R}^n$, $c_0 \in \mathbb{R}$, $\mathbf{b} \in \mathbb{R}^m$ and A is a mxn matrix which is assumed to have linearly independent rows.

Let Z be an nx(n-m) matrix such that the columns of Z form a basis for the subspace $\{\ \mathbf{x} \in \mathbb{R}^n \mid A\mathbf{x} = \mathbf{0}\ \}$ = "the nullspace of A". Further, assume that $\mathbf{x}^0$ is a given feasible solution of the linear constraints, i.e. $A\mathbf{x}^0 = \mathbf{b}$. Then each feasible solution $\mathbf{x}$ of the constraints $A\mathbf{x} = \mathbf{b}$ can be written on the form $\mathbf{x} = \mathbf{x}^0 + Z\mathbf{v}$, for some vector $\mathbf{v} \in \mathbb{R}^{n-m}$, and then the above problem of minimizing $q(\mathbf{x})$ subject to $A\mathbf{x} = \mathbf{b}$ may equivalently be written as the following unconstrained quadratic problem in $\mathbf{v}$:

$$\text{minimize } q(\mathbf{x}^0+Z\mathbf{v}) = \tfrac{1}{2}\mathbf{v}^TZ^TQZ\mathbf{v} + \mathbf{v}^TZ^T(Q\mathbf{x}^0+\mathbf{c}) + q(\mathbf{x}^0)$$

Assume that the (n-m)x(n-m) matrix Z^TQZ is positive definit, i.e. that the matrix Q is positive definite on the nullspace of A.

Then the global minimum point v^* of $q(x^0+Zv)$ is obtained as the solution of the linear equations:

$$Z^TQZv = -Z^T(Qx^0+c)$$

and the global optimal solution x^* of QPE is given by: $x^* = x^0 + Zv^*$. This method of solving QPE is called the Nullspace method.

Another method is obtained from the observation that the above optimal solution x^* is the only feasible solution x for which the gradient $\nabla q(x)$ can be expressed as a linear combination of the rows of A. The direct search for such a point x leads to the following system of linear equations:

$$Qx + A^Tu = -c \quad (\Leftrightarrow \nabla q(x) = -u^TA = \text{a linear combination of rows of } A)$$

$$Ax \qquad = b \quad (\Leftrightarrow x \text{ is a feasible solution})$$

If the rows of A are linearly independent and Q is positive definite on the nullspace of A, the matrix $\begin{bmatrix} Q & A^T \\ A & 0 \end{bmatrix}$ is nonsingular, so that the system $\begin{bmatrix} Q & A^T \\ A & 0 \end{bmatrix}\begin{bmatrix} x \\ u \end{bmatrix} = \begin{bmatrix} -c \\ b \end{bmatrix}$ has a unique solution x^*, u^*.

Then x^* is the unique optimal solution of QPE.

This method of solving QPE is sometimes called the Lagrange method. (u may be interpreted as the vector of Lagrange multipliers.)

4. Minimization of q(x) subject to linear inequality constraints

Next, consider the problem of minimizing a quadratic function subject to linear inequality constraints, i.e. a problem on the form:

<u>QPI</u>: minimize $q(x) = \frac{1}{2}x^TQx + c^Tx + c_0$

subject to: $a_i^Tx \le b_i$ for $i \in I$

Sufficient conditions for a global optimum of QPI is given by the following proposition:

PROP.2: Assume that Q is positive definite. Then a given $x^* \in \mathbb{R}^n$ is a global optimum point of QPI if and only if there are real numbers u_i, $i \in I$, such that the following conditions hold:

$$Qx^* + c + \sum_{i \in I} u_i a_i = 0 \quad (\Leftrightarrow \nabla q(x) = \text{a lin.comb. of the } a_i^T)$$

and for all $i \in I$:

$a_i^Tx^* \le b_i$ (x^* is a feasible solution)

$u_i \ge 0$ (non-negative Lagrange multipliers)

$u_i(b_i - a_i^Tx^*) = 0$ ($u_i > 0$ only for active constraints)

Most methods for solving QPI are based on an "Active Set Strategy"; The method first "guesses" which of the inequality constraints that are active (satisfied with equality) at the optimal solution. Then the corresponding equality constrained problem of the type QPE is solved. It is then easy to check if the "guess" turned out to be right. (By checking the above optimality conditions.) If so, the optimal solution of QPI has been found. If not, a natural updated "guess" is made, etc. Thus, QPI is solved by solving a (hopefully rather small) number of QPE-problems.

5. The Lagrange-Newton method and Sequential Quadratic Programming

Now, consider an equality constrained nonlinear optimization problem on the form:

$$\underline{\text{PE}}: \quad \text{minimize} \quad f(\mathbf{x})$$
$$\text{subject to:} \quad \mathbf{h}(\mathbf{x}) = \mathbf{0}$$

where $\mathbf{x} \in \mathbb{R}^n$ and $\mathbf{h}(\mathbf{x}) = (h_1(\mathbf{x}),\ldots,h_m(\mathbf{x}))^T \in \mathbb{R}^m$.

$h_1,\ldots,h_m$ and f are assumed to be twice continuously differentiable functions from $\mathbb{R}^n$ to $\mathbb{R}$.

The first order optimality conditions for PE state that there exists a vector $\mathbf{u} \in \mathbb{R}^m$ (of Lagrange multipliers) which, together with the optimal solution $\mathbf{x}$, satisfies the following system of nonlinear equations:

$$\begin{aligned} \nabla f(\mathbf{x}) + \mathbf{u}^T\nabla\mathbf{h}(\mathbf{x}) &= \mathbf{0} \\ \mathbf{h}(\mathbf{x}) &= \mathbf{0} \end{aligned} \qquad \ldots(5.1)$$

If the Newton method is applied to the above (Lagrange) optimality conditions, the resulting method is called the Lagrange-Newton method.

First, a short summary of the Newton method for solving equations:

Consider a system of nonlinear equations:

$$\begin{aligned} g_1(y_1,\ldots,y_N) &= 0 \\ \vdots \quad & \quad \vdots \\ g_N(y_1,\ldots,y_N) &= 0 \end{aligned} \qquad \ldots(5.2)$$

which is written shortly as $\mathbf{g}(\mathbf{y}) = \mathbf{0}$ (where both $\mathbf{y}$ and $\mathbf{g}(\mathbf{y}) \in \mathbb{R}^N$).
The Newton method for solving (5.2) goes as follows:

Given an iteration point $\mathbf{y}^{(k)}$, the following linear equations in $\mathbf{d} \in \mathbb{R}^N$ are generated and solved:

$$\nabla\mathbf{g}(\mathbf{y}^{(k)})\mathbf{d} = -\mathbf{g}(\mathbf{y}^{(k)}) \qquad \ldots(5.3)$$

where $\nabla\mathbf{g}(\mathbf{y}^{(k)})$ is an N×N matrix with rows $\nabla g_i(\mathbf{y}^{(k)})$.

If $\nabla g(y^{(k)})$ is nonsingular, then there is an unique solution $d^{(k)}$ of these equations. The next iteration point is then given by $y^{(k+1)} = y^{(k)} + d^{(k)}$, which means that $y^{(k+1)}$ is the solution of the linearized equations: $g^{(k)}(y^{(k)}) + \nabla g^{(k)}(y^{(k)}) \cdot (y - y^{(k)}) = 0$. (Alternatively, one could let $y^{(k+1)} = y^{(k)} + \alpha_k d^{(k)}$, where α_k is given by some "linesearch", e.g. by minimizing $|g(y^{(k)} + \alpha d^{(k)})|^2$ when $\alpha \in [0,1]$.)

If Newtons method is applied on the Lagrange equations (5.1), the corresponding linear equations (5.3) become: (with $d = x - x^{(k)}$ and $w = u - u^{(k)}$, where $(x^{(k)}, u^{(k)})$ is the current iteration point):

$$\begin{bmatrix} L^{(k)} & \nabla h^{(k)T} \\ \nabla h^{(k)} & 0 \end{bmatrix} \begin{bmatrix} d \\ w \end{bmatrix} = \begin{bmatrix} -\nabla \ell^{(k)T} \\ -h^{(k)} \end{bmatrix} \qquad \ldots(5.4)$$

where: $h^{(k)} = h(x^{(k)})$, $\nabla h^{(k)} = \nabla h(x^{(k)})$,

$$\nabla \ell^{(k)} = \nabla f^{(k)} + u^{(k)T} \nabla h^{(k)} = \nabla f(x^{(k)}) + \sum_i u_i^{(k)} \nabla h_i(x^{(k)}) ,$$

$$L^{(k)} = \nabla^2 f(x^{(k)}) + \sum_i u_i^{(k)} \nabla^2 h_i(x^{(k)}).$$

By adding the vector $\begin{bmatrix} \nabla h^{(k)T} u^{(k)} \\ 0 \end{bmatrix}$ to both sides of (5.4), and by letting $u = u^{(k)} + w$, (5.4) may equivalently be written:

$$\begin{bmatrix} L^{(k)} & \nabla h^{(k)T} \\ \nabla h^{(k)} & 0 \end{bmatrix} \begin{bmatrix} d \\ u \end{bmatrix} = \begin{bmatrix} -\nabla f^{(k)T} \\ -h^{(k)} \end{bmatrix} \qquad \ldots(5.5)$$

From section 3 above, it follows that if $L^{(k)}$ (the Hessian of the Lagrangian) is positive definite on the nullspace of $\nabla h(x^{(k)})$, which is usually the case in the neighbourhood of a local minimum point, then (5.5) is equivalent to the following quadratic programming problem:

$\underline{QPE^{(k)}}$: minimize $\frac{1}{2} d^T L^{(k)} d + \nabla f^{(k)} d$

subject to: $h^{(k)} + \nabla h^{(k)} d = 0$

Thus, the Lagrange-Newton method for solving PE may be described by:

$x^{(k+1)} = x^{(k)} + d^{(k)}$, where $d^{(k)}$ is optimal for $QPE^{(k)}$, and

$u^{(k+1)} =$ the vector of Lagrange multipliers for $QPE^{(k)}$.

As mentioned above, it is possible to stabilize the method by a "line search", so that $\mathbf{x}^{(k+1)} = \mathbf{x}^{(k)} + \alpha_k \mathbf{d}^{(k)}$, where $\alpha_k \le 1$.

The interpretation of the Lagrange-Newton method as a sequential quadratic programming method (SQP method), makes it natural to generalize the method also to inequality constrained problems:

$$\underline{\text{PI}}: \quad \text{minimize} \quad f(\mathbf{x})$$
$$\text{subject to:} \quad h_i(\mathbf{x}) \le 0 \;, \; i \in I$$
$$\mathbf{x} \in \mathbb{R}^n$$

The natural generalization of the above method to this problem goes as follows, where a typical iteration is described:

Given the current iteration point $\mathbf{x}^{(k)}$ and the current estimates $u_i^{(k)}$ of the true Lagrange multipliers, the following QPI-problem is solved:

$$\underline{\text{QPI}^{(k)}}: \quad \text{minimize} \quad \tfrac{1}{2}\mathbf{d}^T L^{(k)} \mathbf{d} + \nabla f^{(k)} \mathbf{d}$$
$$\text{subject to:} \quad h_i^{(k)} + \nabla h_i^{(k)} \mathbf{d} \le 0 \;, \; i \in I$$

Let $\mathbf{d}^{(k)}$ be the optimal solution of $\text{QPI}^{(k)}$ and let λ_i be the Lagrange multipliers for the linear constraints in $\text{QPI}^{(k)}$.
Then let $\mathbf{x}^{(k+1)} = \mathbf{x}^{(k)} + \mathbf{d}^{(k)}$ and $u_i^{(k+1)} = \lambda_i$, $i \in I$.
Again, it is possible to use some kind of "line search", so that $\mathbf{x}^{(k+1)} = \mathbf{x}^{(k)} + \alpha_k \mathbf{d}^{(k)}$, where $\alpha_k \le 1$.

There are numerous different variations of the above SQP method. It is, e.g., necessary to take care of situations where the requirements on positive definiteness are not satisfied.

If it is not possible to compute the Hessian $L^{(k)}$ of the Lagrange function, then it is possible to use, instead, an approximation $S^{(k)}$ which is updated each iteration according to some quasi-Newton scheme (based on calculated gradients).

More details concerning QP and SQP methods are given in [5]-[7].

6. Sequential All Quadratic Programming (SAQP)

Consider again an inequality constrained nonlinear problem on the form:

$$\underline{\text{PI}}: \quad \text{minimize} \quad f(\mathbf{x})$$
$$\text{subject to:} \quad h_i(\mathbf{x}) \le 0 \;, \; i \in I$$
$$\mathbf{x} \in \mathbb{R}^n$$

In SQP methods, the constraint functions are linearized at the current iteration point $\mathbf{x}^{(k)}$, while the objective function is approximated by a quadratic function. The second order derivatives of the constraint functions are not neglected, however, but they are transformed to the objective function with weights given by the current estimates $u_i^{(k)}$ of the Lagrange multipliers of PI. Thus, the curvature of the constraints are transformed to the objective function.

An alternative approach would be to approximate also the constraint functions with quadratic functions. This would lead to subproblems of the following type:

$$\underline{\mathrm{AQP}^{(k)}}: \quad \text{minimize} \quad f^{(k)} + \nabla f^{(k)}\mathbf{d} + \tfrac{1}{2}\mathbf{d}^T\mathbf{F}^{(k)}\mathbf{d}$$

$$\text{subject to:} \quad h_i^{(k)} + \nabla h_i^{(k)}\mathbf{d} + \tfrac{1}{2}\mathbf{d}^T\mathbf{H}_i^{(k)}\mathbf{d} \le 0 \ , \ i \in I$$

where $\mathbf{F}^{(k)} = \nabla^2 f(\mathbf{x}^{(k)})$ and $\mathbf{H}_i^{(k)} = \nabla^2 h_i(\mathbf{x}^{(k)})$.

If $\mathbf{d}^{(k)}$ is the optimal solution of this "All Quadratic" subproblem, then the natural next iteration point would be given by: $\mathbf{x}^{(k+1)} = \mathbf{x}^{(k)} + \mathbf{d}^{(k)}$. This method may be called Sequential All Quadratic Programming (SAQP). There are two obvious drawbacks of SAQP compared to SQP:

(i) It is more difficult to solve the subproblem $\mathrm{AQP}^{(k)}$ than to solve the subproblem $\mathrm{QPI}^{(k)}$. Typically, $\mathrm{AQP}^{(k)}$ should be solved by SQP, i.e. as a sequence of QPI-problems.

(ii) It is usually more expensive to calculate the Hessian matrix of each constraint function than to calculate the Hessian matrix of just a weighted sum of constraint functions.

The possible advantage of SAQP over SQP is that the subproblem $\mathrm{AQP}^{(k)}$ might be a much better approximation (than $\mathrm{QPI}^{(k)}$) of the original problem PI. This would reduce the number of iterations needed to obtain an optimal solution of PI.

In structural optimization, the first drawback (i) may not be very crucial, because the time spent for solving the subproblem $\mathrm{AQP}^{(k)}$ is probably anyhow less than the time spent for analyzing the structure, calculating gradients of behavior constraints, etc.
The second drawback (ii) is more serious. A possible way to reduce this drawback is to use approximated Hessian, based on quasi-Newton updates. This is further discussed in section 8 below, after the introduction of two new second order methods in section 7.

7. Second order extensions of MMA

The Method of Moving Asymptotes (MMA), which was presented in [8], is a first order optimization method. In this section will be described two possible extensions of MMA to make it a second order method.
The first may be considered as a combination of MMA and SAQP, while

the second may be considered as a combination of MMA and SQP.

Consider again the above problem PI, now assumed to be a structural optimization problem written on the following form:

$$\underline{\text{PI}}: \quad \text{minimize} \quad f_0(\mathbf{x})$$
$$\text{subject to:} \quad f_i(\mathbf{x}) \le \bar{f}_i, \quad i = 1,\ldots,m$$
$$\underline{x}_j \le x_j \le \bar{x}_j, \quad j = 1,\ldots,n$$

Given the current iteration point $\mathbf{x}^{(k)}$, MMA generates and solves a subproblem on the form PMMA$^{(k)}$ below. Here, $L_j = L_j^{(k)}$ and $U_j = U_j^{(k)}$ are parameters such that $L_j^{(k)} < x_j^{(k)} < U_j^{(k)}$. Rules for chosing appropriate values on these parameters ("asymptots") are described in [8]

$$\underline{\text{PMMA}^{(k)}}: \quad \text{minimize} \quad \sum_j \left[\frac{p_{0j}}{U_j - x_j} + \frac{q_{0j}}{x_j - L_j} \right] + r_0$$
$$\text{subject to:} \quad \sum_j \left[\frac{p_{ij}}{U_j - x_j} + \frac{q_{ij}}{x_j - L_j} \right] + r_i \le \bar{f}_i, \quad i = 1,..,m$$
$$\alpha_j \le x_j \le \beta_j, \quad j = 1,\ldots,n$$

where, for $j = 1,..,n$ and $i = 0,1,..,m$:

$$p_{ij} = (U_j^{(k)} - x_j^{(k)})^2 \cdot \left[\frac{\partial f_i}{\partial x_j} \right]_+, \quad q_{ij} = (x_j^{(k)} - L_j^{(k)})^2 \cdot \left[\frac{\partial f_i}{\partial x_j} \right]_-,$$

$$\alpha_j = \max\{ \underline{x}_j,\ 0.1x_j^{(k)} + 0.9L_j^{(k)} \}, \quad \beta_j = \min\{ \bar{x}_j,\ 0.1x_j^{(k)} + 0.9U_j^{(k)} \}$$

and $$r_i = f_i(\mathbf{x}^{(k)}) - \sum_{j=1}^{n} \left[\frac{p_{ij}}{U_j - x_j^{(k)}} + \frac{q_{ij}}{x_j^{(k)} - L_j} \right].$$

The next iteration point $\mathbf{x}^{(k+1)}$ is then chosen as the optimal solution of this subproblem PMMA$^{(k)}$.

In the SAQP-MMA combination, the subproblem instead looks as follows:

$$\underline{\text{PSAM}^{(k)}}: \quad \text{minimize} \quad \sum_j g_{0j}^{(k)}(x_j) + \tfrac{1}{2}(\mathbf{x}-\mathbf{x}^{(k)})^T \mathbf{F}_0^{(k)} (\mathbf{x}-\mathbf{x}^{(k)})$$
$$\text{subject to:} \quad \sum_j g_{ij}^{(k)}(x_j) + \tfrac{1}{2}(\mathbf{x}-\mathbf{x}^{(k)})^T \mathbf{F}_i^{(k)} (\mathbf{x}-\mathbf{x}^{(k)}) \le \bar{f}_i, \quad i = 1,..,m$$
$$\alpha_j \le x_j \le \beta_j, \quad j = 1,\ldots,n$$

where, for $j = 1,..,n$ and $i = 0,1,..,m$: $\mathbf{F}_i^{(k)} = \nabla^2 f_i(\mathbf{x}^{(k)})$ and

$$g_{ij}^{(k)}(x_j) = \frac{p_{ij}}{U_j - x_j} + \frac{q_{ij}}{x_j - L_j} - c_{ij}(x_j - x_j^{(k)})^2 - d_{ij}(x_j - x_j^{(k)})^3 + r_i \ .$$

U_j, L_j, p_{ij}, q_{ij}, α_j, β_j and r_i are as above (in MMA) while c_{ij} and d_{ij} are chosen such that the second and third order derivatives of $g_{ij}^{(k)}(x_j)$ vanish at $x_j^{(k)}$. This means that:

$$c_{ij} = \frac{p_{ij}}{(U_j - x_j^{(k)})^3} + \frac{q_{ij}}{(x_j^{(k)} - L_j)^3} \quad \text{and} \quad d_{ij} = \frac{p_{ij}}{(U_j - x_j^{(k)})^4} - \frac{q_{ij}}{(x_j^{(k)} - L_j)^4} \ .$$

It may be shown that $g_{ij}^{(k)}(x_j)$ is always a convex function.

Further, $\sum_j g_{ij}^{(k)}(x_j) + \frac{1}{2}(\mathbf{x} - \mathbf{x}^{(k)})^T \mathbf{F}_i^{(k)} (\mathbf{x} - \mathbf{x}^{(k)})$ is a second order approximation of $f_i(\mathbf{x})$ around $\mathbf{x} = \mathbf{x}^{(k)}$.

This SAQP-MMA method has been tested on some truss sizing problems and on some general nonlinear programming problems from the literature. On most of the considered problem, only 3-7 iterations were needed to find an optimal solution within a very high accuracy. The price paid for this fast convergence was twofold:

(i) The subproblems PSAM$^{(k)}$ are not trivial to solve. The method used in these tests was MMA, but other methods for solving PSAM$^{(k)}$ are of corse also possible, e.g. SQP.

(ii) Second order derivatives of the (usually implicit) constraint functions have to be calculated.

As discussed above, (i) may not be very crucial for practical structural optimization problems. After all, the subproblem PSAM$^{(k)}$ is an explicit problem with functions that are easy to differentiate, both ones and twice. The time spent in the structural analysis part may anyhow be dominating.

The need for second derivatives is a more serious obstacle. Since it is usually much less expensive to evaluate and store the Hessian matrix of only the Lagrange function, than to evaluate and store the Hessian matrices of every potentially active constraint function, a combination of MMA and SQP might be more tractable than the above combination of MMA and SAQP.

In the SQP-MMA combination, the subproblem looks as follows:

<u>PSQM1</u>$^{(k)}$: minimize $\sum_j g_{0j}^{(k)}(x_j) + \frac{1}{2}(\mathbf{x} - \mathbf{x}^{(k)})^T \mathbf{L}^{(k)} (\mathbf{x} - \mathbf{x}^{(k)})$

subject to: $\sum_j g_{ij}^{(k)}(x_j) \le \bar{f}_i$, $i = 1,..,m$

$\alpha_j \le x_j \le \beta_j$, $j = 1,...,n$

Here, $L^{(k)} = \nabla^2 \ell(\mathbf{x}^{(k)})$ = the Hessian of the Lagrange function.
The functions $g_{ij}^{(k)}(x_j)$ are as above.

An alternative possibility is to let the subproblem be:

$$\underline{\text{PSQM2}^{(k)}}: \quad \text{minimize} \ \sum_j \bar{g}_j^{(k)}(x_j) + \tfrac{1}{2}(\mathbf{x}-\mathbf{x}^{(k)})^T L^{(k)} (\mathbf{x}-\mathbf{x}^{(k)})$$

$$\text{subject to:} \ \sum_j \left[\frac{p_{ij}}{U_j - x_j} + \frac{q_{ij}}{x_j - L_j} \right] + r_i \le \bar{f}_i, \quad i = 1,..,m$$

$$\alpha_j \le x_j \le \beta_j, \quad j = 1,\dots,n$$

where: $\bar{g}_j^{(k)}(x_j) = \dfrac{p_{0j}}{U_j - x_j} + \dfrac{q_{0j}}{x_j - L_j} - \bar{c}_j(x_j - x_j^{(k)})^2 - \bar{d}_j(x_j - x_j^{(k)})^3 + r_0$,

$\bar{c}_j = c_{0j} + \sum_i u_i^{(k)} c_{ij}$ and $\bar{d}_j = d_{0j} + \sum_i u_i^{(k)} d_{ij}$.

These methods have also been tested on the problems mentioned above. In general, these SQP-MMA combinations needed a few more iterations than SAQP-MMA to obtain the optimal solution with equal accuracy. On the other hand, the need for second derivatives is less in SQP-MMA. Also, it seems that the subproblems PSQM1$^{(k)}$ and PSQM2$^{(k)}$ are somewhat easier to solve (numerically) than the subproblems PSAM$^{(k)}$. The reason for this might be that the constraint functions in PSQM1$^{(k)}$ and PSQM2$^{(k)}$ are always convex (although the objective functions may be non-convex).

8. Quasi-Newton updates of the Hessians

The methods discussed in the previous section have to be modified if the second order derivatives of the constraint functions can not be provided (from the structural analysis part).

There are different suggestions on how approximated Hessian matrixes should be updated. In the recent paper [9], it was demonstrated that the symmetric rank one update (described below) could be very efficient. For SAQP-MMA, this means that the matrices $F_i^{(k)} = \nabla^2 f_i(\mathbf{x}^{(k)})$ in PSAM$^{(k)}$ are replaced by matrices $S_i^{(k)}$. Given the optimal solution $\mathbf{x}^{(k+1)}$ of this modified PSAM$^{(k)}$, new approximating matrices $S_i^{(k+1)}$ are obtained by the formula:

$$S_i^{(k+1)} = S_i^{(k)} + \frac{v_i v_i^T}{v_i^T \Delta\mathbf{x}} \qquad \dots (8.1)$$

where $\Delta\mathbf{x} = \mathbf{x}^{(k+1)} - \mathbf{x}^{(k)}$ and $v_i = \nabla f_i(\mathbf{x}^{(k+1)})^T - \nabla f_i(\mathbf{x}^{(k)})^T - S_i^{(k)} \Delta\mathbf{x}$.

If $v_i \approx 0$ or $\Delta x \approx 0$ or $|v_i^T \Delta x| < 0.01 \cdot |v_i| \cdot |\Delta x|$, then no updating of $S_i^{(k)}$ is done the current iteration, i.e. $S_i^{(k+1)} = S_i^{(k)}$. Also, if the norm of the matrix $\Delta S_i^{(k)} = \dfrac{v_i v_i^T}{v_i^T \Delta x}$ is greater than the norm of $S_i^{(k)}$, some damping may be needed: $S_i^{(k+1)} = S_i^{(k)} + \alpha \cdot \Delta S_i^{(k)}$, with $\alpha < 1$.
The first iteration, $S_i^{(0)}$ could be chosen as a diagonal matrix with diagonal elements given e.g. by $(S_i^{(0)})_{jj} = 2c_{ij}$.

In the SQP-MMA methods, it is instead $L^{(k)}$ that should be approximated by a matrix $S^{(k)}$ which is updated as follows:

$$S^{(k+1)} = S^{(k)} + \frac{vv^T}{v^T \Delta x} \qquad \ldots(8.2)$$

where $v = -S^{(k)}\Delta x + \nabla f_0(x^{(k+1)})^T - \nabla f_0(x^{(k)})^T +$

$$+ \sum_i u_i^{(k+1)} \cdot (\nabla f_i(x^{(k+1)})^T - \nabla f_i(x^{(k)})^T) \ .$$

SAQP-MMA and SQP-MMA, with the updating formulas (8.1) and (8.2), have been numerically tested on the problems mentioned above. The convergence properties were in general not as good as the convergence properties of the corresponding methods based on exact second order derivatives. In some cases, these quasi-Newton methods also behaved less stable than the pure (first-order) MMA.

However, if MMA were used during the first iterations, while the S-matrices were updated but not used , then both SAQP-MMA and SQP-MMA were able to speed up the convergence significantly during the final iterations (compared to if MMA was used all the time).

Thus, it seems to be a reasonable strategy to start with MMA and then to switch over to e.g. the quasi-Newton SQP-MMA after some iterations.

9. References

[1] Haftka R. (1982) "Second-order sensitivity derivatives in structural analysis", AIAA Journal 20, 1765-1766.

[2] Fleury C. (1988) "Efficient approximation concepts using second order information", AIAA Structures, Structural Dynamics and Materials Conference, Williamsburg, April 18-20, 1988.

[3] Ringertz U. (1988) "Newton methods for structural optimization", Report 88-19, Dep.of Aeronautical Structures and Materials, KTH, Stockholm.

[4] Svanberg K. (1989) "Optimal truss sizing based on explicit Taylor series expansions", Structural Optimization 2, 153-162.

[5] Fletcher R. (1987) "Practical Methods of Optimization", Second edition, John Wiley & Sons.

[6] Luenberger D.G. (1984) "Linear and Nonlinear Programming", Second edition, Addison-Wesley.

[7] Gill P.E., Murray W., Saunders M.A. and Wright M.H. (1989) "Constrained Nonlinear Programming", in Nemhauser, Kan and Todd (eds.) OPTIMIZATION, North-Holland.

[8] Svanberg K. (1987) "The method of moving asymptots - a new method for structural optimization", International Journal for Numerical Methods in Engineering 24, 359-373.

[9] Conn A.R., Gould N.I.M. and Toint P.L. (1991) "Convergence of quasi-Newton matrices generated by the symmetric rank one update", Mathematical Programming 50, 177-195.

LOCAL AND GLOBAL OPTIMA

K. SVANBERG
Optimization and Systems Theory
Royal Institute of Technology
S-10044 Stockholm, Sweden

ABSTRACT. This lecture deals with convexity properties of some structural optimization problems, and with the closely related question of local versus global optima.

1. Introduction and Summary

When a solution of a structural optimization problem has been obtained by some numerical method, it is often very difficult to decide whether the solution is a global optimum or just a local one. But in some cases the considered problem can be shown to possess certain convexity properties which imply that a local optimum is also a global optimum. In this lecture, the question of how to identify such nice cases is discussed.

First, some basic definitions and optimality conditions for general nonlinear programming problems (section 2) and convex programming problems (section 3) are collected. The stated propositions in sections 2 and 3 have been known for a long time. The proofs may be found in [1] and [2].

In section 4, the structural sizing problem is considered. It is noted that such problems may in general be nonconvex, with several local optima with different objective values. There is, however, a nontrivial special case of constraints which always leads to convex problems, namely upper bounds on the strain energy under different load cases. This result, which was first proved in [3], is discussed in section 4.

Another type of constraints which also leads to convex structural sizing problems is a lower bound on the lowest natural frequency. This result, which was also proved in [3], is discussed in section 5.

In section 6, some small truss sizing problems are illustrating the theoretical results from the previous sections.

G. I. N. Rozvany (ed.), Optimization of Large Structural Systems, Vol. I, 479–488.

2. Definitions and basic results for general optimization problems

Consider a nonlinear programming problem on the following form:

$$\underline{P}: \quad \text{minimize} \quad f(\mathbf{x}) \quad (\mathbf{x} \in \mathbb{R}^n) \qquad \ldots(1)$$

$$\text{subject to} \quad g_i(\mathbf{x}) \le \bar{g}_i \quad \text{for } i=1,\ldots,m \qquad \ldots(2)$$

$$\underline{x}_j \le x_j \le \bar{x}_j \quad \text{for } j=1,\ldots,n \qquad \ldots(3)$$

The constraints (3) will be written shortly as $\mathbf{x} \in X$, while the set of feasible solutions of (2) and (3) will be denoted Ω :

$\Omega = \{ \mathbf{x} \in X \mid g_i(\mathbf{x}) \le \bar{g}_i \text{ for } i=1,\ldots,m \}$ = the feasible set of P.

It is assumed that the objective function $f(\mathbf{x})$ as well as all the constraint functions $g_i(\mathbf{x})$ are twice continuously differentiable.

The gradient of $f(\mathbf{x})$ is a row vector denoted $\nabla f(\mathbf{x})$.

The Hessian matrix of $f(\mathbf{x})$ is denoted $\nabla^2 f(\mathbf{x})$.

DEF.1: $\mathbf{x}^*$ is a <u>global</u> optimum point of P if $f(\mathbf{x}^*) \le f(\mathbf{x})$ for all $\mathbf{x} \in \Omega$.

Ideally, a global optimum point is what one would like to find.
In practice, however, algorithms for numerically solving nonlinear problems of realistic sizes can not, in general, be expected to find anything better than a local optimum point:

DEF.2: $\mathbf{x}^*$ is a <u>local</u> optimum point of P if there is a real number $\varepsilon > 0$ such that $f(\mathbf{x}^*) \le f(\mathbf{x})$ for all $\mathbf{x} \in \Omega$ such that $|\mathbf{x} - \mathbf{x}^*| < \varepsilon$.

A global optimum point is also a local optimum point, but the converse is in general not true (unless P is a convex problem, see next section).

DEF.3: Assume that $\mathbf{x}^* \in \Omega$ and that there exist real numbers $\lambda_1,\ldots,\lambda_m$ such that the following conditions hold at $\mathbf{x} = \mathbf{x}^*$:

$$\text{(i)} \qquad \frac{\partial f}{\partial x_j} + \sum_{i=1}^{m} \lambda_i \cdot \frac{\partial g_i}{\partial x_j} \begin{cases} \le 0 \text{ if } x_j = \bar{x}_j \\ = 0 \text{ if } \underline{x}_j < x_j < \bar{x}_j \\ \ge 0 \text{ if } x_j = \underline{x}_j \end{cases}$$

$$\text{(ii)} \qquad \begin{cases} \lambda_i = 0 \text{ if } g_i(\mathbf{x}^*) < \bar{g}_i \\ \lambda_i \ge 0 \text{ if } g_i(\mathbf{x}^*) = \bar{g}_i \end{cases}$$

Then $\mathbf{x}^*$ is said to be a KKT-point, i.e. a point which satisfies the so called Karush-Kuhn-Tucker conditions (i) and (ii).

DEF. 4: Assume that $\mathbf{x}^* \in \Omega$. Then $I_B(\mathbf{x}^*)$ and $J_B(\mathbf{x}^*)$ are defined as:

$$I_B(\mathbf{x}^*) = \{ \, i \mid g_i(\mathbf{x}^*) = \bar{g}_i \, \}$$
$$J_B(\mathbf{x}^*) = \{ \, j \mid x_j = \underline{x}_j \quad \text{or} \quad x_j = \bar{x}_j \, \}$$

$I_B(\mathbf{x}^*)$ is the index set for the active constraints of type (2) in P, while $J_B(\mathbf{x}^*)$ is the index set for the active constraints of type (3).

Under some additional mild technical conditions (so called constraint qualifications) it may be shown that the above KKT-conditions are <u>necessary</u> conditions for $\mathbf{x}^*$ to be a local optimum point.

In general, however, the KKT-conditions are <u>not sufficient</u> conditions for a local optimum point.

Sufficient conditions are given by the following proposition:

PROP. 5: Assume that $\mathbf{x}^* \in \Omega$ and that there exist real numbers $\lambda_1,\ldots,\lambda_m$ such that the following conditions hold at $\mathbf{x} = \mathbf{x}^*$:

$$\text{(i)'} \quad \frac{\partial f}{\partial x_j} + \sum_{i=1}^{m} \lambda_i \cdot \frac{\partial g_i}{\partial x_j} \begin{cases} < 0 & \text{if } x_j = \bar{x}_j \\ = 0 & \text{if } \underline{x}_j < x_j < \bar{x}_j \\ > 0 & \text{if } x_j = \underline{x}_j \end{cases}$$

$$\text{(ii)'} \quad \begin{cases} \lambda_i = 0 & \text{if } g_i(\mathbf{x}^*) < \bar{g}_i \\ \lambda_i > 0 & \text{if } g_i(\mathbf{x}^*) = \bar{g}_i \end{cases}$$

(iii) $\mathbf{y}^T(\nabla^2 f(\mathbf{x}^*) + \sum \lambda_i \nabla^2 g_i(\mathbf{x}^*))\mathbf{y} > 0$ for all $\mathbf{y} \neq \mathbf{0}$ such that $\nabla g_i(\mathbf{x}^*)\mathbf{y} = 0$ for $i \in I_B(\mathbf{x}^*)$ and $y_j = 0$ for $j \in J_B(\mathbf{x}^*)$.

Then $\mathbf{x}^*$ is a local optimum point of P.

Some comments on PROP. 5:

(i)' and (ii)' are almost the KKT-conditions, except that there are <u>strict</u> inequalities in (i)' and (ii)'.

In words, (iii) says that the Hessian matrix of the Lagrange function is positive definite on the tangent plane of the active constraints.

In practice, to check if (i)', (ii)' and (iii) are satisfied, the following steps should be taken:
The numbers λ_i are usually provided by the numerical algorithm which solves P. If e.g. a dual method is used, then the λ_i are the optimal values of the dual variables. Given these λ_i, conditions (i)' and (ii)' are easily checked. In order to check if (iii) holds, let

$L = \nabla^2 f(\mathbf{x}^*) + \sum \lambda_i \nabla^2 g_i(\mathbf{x}^*)$ ("Hessian of the Lagrangian") and let **A** be a matrix with rows $\nabla g_i(\mathbf{x}^*)$, $i \in I_B(\mathbf{x}^*)$, and $\mathbf{e}_j^T$, $j \in J_B(\mathbf{x}^*)$.
(The rows of **A** are the gradients of the active constraints.)
Then condition (iii) may equivalently be written:

(iii)' $\mathbf{y}^T L \mathbf{y} > 0$ for all $\mathbf{y} \neq \mathbf{0}$ such that $\mathbf{A}\mathbf{y} = \mathbf{0}$.

To check if (iii)' is satisfied, let **Z** be a matrix whose columns form a basis of the null-space of **A**. **Z** may be obtained e.g. from a QR-factorization of $\mathbf{A}^T$. Now $\mathbf{A}\mathbf{y} = \mathbf{0}$ if and only if $\mathbf{y} = \mathbf{Z}\mathbf{v}$ for some vector **v**. Then (iii)' becomes equivalent to the condition that $\mathbf{v}^T \mathbf{Z}^T L \mathbf{Z} \mathbf{v} > 0$ for all vectors $\mathbf{v} \neq \mathbf{0}$, i.e. that the matrix $\mathbf{Z}^T L \mathbf{Z}$ is positive definite. To check if $\mathbf{Z}^T L \mathbf{Z}$ is positive definite, a Cholesky factorization of $\mathbf{Z}^T L \mathbf{Z}$ is performed. See e.g. [1] for more details.

3. Basic results for convex optimization problems

DEF.6: A set C in $\mathbb{R}^n$ is a convex set if $\alpha \cdot \mathbf{x} + (1-\alpha) \cdot \mathbf{y} \in C$ for every $\mathbf{x}$ and $\mathbf{y} \in C$ and every real number $\alpha \in (0,1)$.

DEF.7: A function f, defined on a convex set C, is a convex function over C if $f(\alpha \cdot \mathbf{x} + (1-\alpha) \cdot \mathbf{y}) \le \alpha f(\mathbf{x}) + (1-\alpha) f(\mathbf{y})$ for every $\mathbf{x}$ and $\mathbf{y} \in C$ and every real number $\alpha \in (0,1)$.

The most straightforward way to check whether a function is convex or not is usually to calculate its Hessian matrix:

PROP.8: A twice continuously differentiable function f is convex over a convex set C if and only if the Hessian matrix $\nabla^2 f(\mathbf{x})$ is positive semidefinite for every $\mathbf{x} \in C$.

DEF.9: The optimization problem P is said to be a convex problem if the feasible set Ω is convex and the objective function f is convex over Ω.

PROP.10: If the constraint functions g_i are convex over **X**, then the feasible set Ω is convex.

For convex problem, there are no differences between local and global optimum points! Further, the set of optimum points (if several) is a convex set:

PROP.11: If P is a convex problem and $\mathbf{x}^*$ is a local optimum point of P, then $\mathbf{x}^*$ is also a global optimum point of P.

PROP.12: If $\mathbf{x}^0$ and $\mathbf{x}^1$ are two different local optimum points of a convex problem P, then both $\mathbf{x}^0$ and $\mathbf{x}^1$ are global optimum points of P. Further, each point $\mathbf{x} = \alpha \cdot \mathbf{x}^0 + (1-\alpha) \cdot \mathbf{x}^1$, where $\alpha \in (0,1)$, is also a global optimum point of P.

For convex problems, the Karush-Kuhn-Tucker conditions are sufficient conditions for a global optimum:

PROP. 13: Assume that P is a convex problem and that $\mathbf{x}^*$ is a KKT-point. Then $\mathbf{x}^*$ is a global optimum point of P.

4. Convexity results for some structural sizing problems

In this section, the design variables are assumed to be sizing variables, e.g. cross section areas of truss elements. The objective function is assumed to be the structural weight, which is a linear function of the design variables: $f(\mathbf{x}) = w_0 + \sum w_j x_j$.

The constraint functions are assumed to be nodal displacements, or elemental stresses, written on the form $g_i(\mathbf{x}) = \mathbf{q}_i^T \mathbf{K}^{-1}(\mathbf{x})\mathbf{p}_i$, where $\mathbf{q}_i$ and $\mathbf{p}_i$ are given vectors which do not depend on $\mathbf{x}$, while $\mathbf{K}^{-1}(\mathbf{x})$ is the inverse of the structural stiffness matrix, which do depend on $\mathbf{x}$. Typically, $\mathbf{p}_i$ is an external load vector, so that $\mathbf{K}^{-1}(\mathbf{x})\mathbf{p}_i$ is the corresponding nodal displacement vector, while $\mathbf{q}_i$ is a unit vector defining the considered displacement or stress. The stiffness matrix is assumed to depend linearly on the design variables: $\mathbf{K}(\mathbf{x}) = \mathbf{K}_0 + \sum x_j \mathbf{K}_j$, where $\mathbf{K}_0, \mathbf{K}_1, \ldots, \mathbf{K}_n$ are symmetric matrices which do not depend on $\mathbf{x}$. (Even if $\mathbf{K}(\mathbf{x})$ is linear, the inverse $\mathbf{K}^{-1}(\mathbf{x})$ is nonlinear.)

It is assumed that $\mathbf{K}(\mathbf{x})$ is positive definite for all $\mathbf{x} \in \mathbf{X}$.

It should perhaps be noted that the above inverse matrix $\mathbf{K}^{-1}(\mathbf{x})$ is, of course, never actually calculated in practice! Instead, $\mathbf{u}$ is obtained as the solution of the linear equations $\mathbf{K}(\mathbf{x})\mathbf{u} = \mathbf{p}$, which are solved e.g. by a Cholesky factorization of $\mathbf{K}(\mathbf{x})$ etc. We find it convenient, however, to use the notation $g_i(\mathbf{x}) = \mathbf{q}_i^T \mathbf{K}^{-1}(\mathbf{x})\mathbf{p}_i$ when discussing convexity properties of the constraint functions.

The considered structural optimization problem may thus be written:

<u>P1</u>: minimize $f(\mathbf{x}) = w_0 + \sum w_j x_j$ ...(1)

subject to $g_i(\mathbf{x}) = \mathbf{q}_i^T \mathbf{K}^{-1}(\mathbf{x})\mathbf{p}_i \le \bar{g}_i$, $i=1,\ldots,m$ ...(2)

$\underline{x}_j \le x_j \le \bar{x}_j$, $j=1,\ldots,n$ ...(3)

In general, P1 is not a convex problem. A wellknown example of a non-convex problem of this type is the "Ten bar truss" (see section 6 below) for which there are two different local optimum points with different objective values.

There is, however, a non-trivial special case of P1 which can be shown to be convex. This special case will now be described.

Assume that $\mathbf{q}_i = \gamma_i \mathbf{p}_i$ for some positive real number γ_i.

Then $g_i(\mathbf{x}) = \mathbf{q}_i^T \mathbf{K}^{-1}(\mathbf{x})\mathbf{p}_i = \gamma_i \mathbf{p}_i^T \mathbf{K}^{-1}(\mathbf{x})\mathbf{p}_i = 2\gamma_i U_i(\mathbf{x})$, where $U_i(\mathbf{x})$ is the strain energy in the structure under the given load $\mathbf{p}_i$.

The constraint $g_i(\mathbf{x}) \le \bar{g}_i$ may then be written on the form $U_i(\mathbf{x}) \le \bar{U}_i$.

The following result was stated and proved in [3]:

PROP. 14: The strain energy $U_i(\mathbf{x}) = \frac{1}{2}\, \mathbf{p}_i^T \mathbf{K}^{-1}(\mathbf{x})\mathbf{p}_i$ is a convex function over $\mathbf{X}$.

An immediate consequence of this result is the following:

PROP. 15: If all the constraints of type (2) in P1 are upper bounds on the strain energy, under different loads, then P1 is a convex problem.

An example of a strain energy constrained problem is the "Eight bar truss" in section 6 below. This problem also illustrates the following interesting fact:

PROP. 16: If the weight of a truss structure is minimized subject to an upper bound on the strain energy under a <u>single</u> load $\mathbf{p}$ and lower bounds on the cross section areas, then there is a positive number $\bar{\sigma}$ such that, at the optimal solution $\mathbf{x}^*$, the following holds for each element, where $\sigma_j(\mathbf{x}^*)$ is the stress in the j:th element:

<u>Eighter</u> $x_j^* = \underline{x}_j$ and $|\sigma_j(\mathbf{x}^*)| \le \bar{\sigma}$

<u>or</u> $x_j^* \ge \underline{x}_j$ and $|\sigma_j(\mathbf{x}^*)| = \bar{\sigma}$

Thus, $\mathbf{x}^*$ is a so called "Fully stressed design" solution.

The connection between the strain energy constraint and Fully stressed design solutions was pointed out by C.Fleury, [4].

5. A convexity result for natural frequency constraints

The natural frequencies ω_i of a structure are obtained from the generalized eigenvalue problem: $\mathbf{K}\mathbf{y} = \omega^2 \mathbf{M}\mathbf{y}$, where $\mathbf{K}$ is the structural stiffness matrix and $\mathbf{M}$ is the structural mass matrix. As above, $\mathbf{K}$ is assumed to depend linearly on the design variables: $\mathbf{K}(\mathbf{x}) = \mathbf{K}_0 + \sum x_j \mathbf{K}_j$.

$\mathbf{M}$ is also assumed to depend linearly on the design variables: $\mathbf{M}(\mathbf{x}) = \mathbf{M}_0 + \sum x_j \mathbf{M}_j$.

For all $\mathbf{x} \in X$, both $K(\mathbf{x})$ and $M(\mathbf{x})$ are assumed to be positive definite.

To emphasize that the natural frequencies depend on $\mathbf{x}$, they are denoted $\omega_i(\mathbf{x})$.

An important example of a frequency constraint is that no natural frequency should be less than a given number. The following result concerning this type of constraint was stated and proved in [3] :

PROP. 17: Let $\Omega = \{ \mathbf{x} \in X \mid \omega_i(\mathbf{x}) \geq \underline{\omega}$ for all $i \}$ where $\underline{\omega}$ is a given positive real number. Then Ω is a convex set.

6. Some examples to illustrate the theoretical results

The "Ten Bar Truss" is a classical problem in structural optimization. The cross section areas of the different bars should be chosen such that the weight of the structure is minimized subject to constraints on the nodal displacements under a given load.

Let $\mathbf{x} = (x_1,\dots,x_{10})^T$ be the cross section areas and let $\mathbf{u} = (u_1,\dots,u_8)^T$ be the nodal displacements. Further, let:

$\mathbf{p} = (0,0,0,-444800,0,0,0,-444800)^T$ (the load vector) ,

$L_j = 9.144$ for $j = 1,4,5,6,9,10$, $L_j = \sqrt{2}\cdot 9.144$ for $j = 2,3,7,8$.

$E = 6895$, $c_j = 0.2768\cdot L_j$ for $j=1,\dots,10$, $\mathbf{c} = (c_1,\dots,c_{10})^T$,

$\gamma_j = E/L_j$ for $j = 1,4,5,6,9,10$, $\gamma_j = E/2L_j$ for $j = 2,3,7,8$,

$\mathbf{r}_1 = (1,0,0,0,0,0,0,0)^T$, $\mathbf{r}_2 = (0,0,1,-1,0,0,0,0)^T$,

$\mathbf{r}_3 = (1,1,0,0,0,0,0,0)^T$, $\mathbf{r}_4 = (0,0,1,0,0,0,0,0)^T$,

$\mathbf{r}_5 = (0,1,0,-1,0,0,0,0)^T$, $\mathbf{r}_6 = (1,0,0,0,-1,0,0,0)^T$,

$\mathbf{r}_7 = (1,-1,0,0,0,0,-1,1)^T$, $\mathbf{r}_8 = (0,0,1,1,-1,-1,0,0)^T$,

$\mathbf{r}_9 = (0,0,1,0,0,0,-1,0)^T$, $\mathbf{r}_{10} = (0,0,0,0,0,1,0,-1)^T$,

$K(\mathbf{x}) = \sum_j x_j\gamma_j \mathbf{r}_j \mathbf{r}_j^T =$ the 8×8 stiffness matrix.

Then consider the following two optimization problems P2 and P3:

<u>P2</u> :

$$\text{minimize} \quad \mathbf{c}^T\mathbf{x} \qquad \dots(1)$$

$$\text{subject to} \quad K(\mathbf{x})\mathbf{u} = \mathbf{p} \qquad \dots(2)$$

$$u_i \geq -50.8 \quad \text{for } i \in \{2,4,6,8\} \qquad \dots(3)$$

$$x_j \geq 0.645 \quad \text{for } j \in \{1,\dots,10\} \qquad \dots(4)$$

$$u_2 - u_4 \leq 22.86 \qquad \dots(5)$$

(1) is the structural weight, which should be minimized,

(2) are the equilibrium conditions for the structure,

(3) are bounds on the vertical displacements of the nodes,
(4) are lower bounds on the cross section areas,
(5) is equivalent to an upper bound on the stress in element 5.

<u>P3</u> : minimize $\mathbf{c}^T\mathbf{x}$ ...(1)

subject to $\mathbf{K}(\mathbf{x})\mathbf{u} = \mathbf{p}$ ...(2)

$u_4 + u_8 \geq -76.2$ ...(3)'

$x_j \geq 0.645$ for $j=1,\dots,10$ ...(4)

(3)' may equivalently be written on the form: $\mathbf{p}^T\mathbf{u} \leq \alpha$, which shows that (3)' is in fact an upper bound on the strain energy in the structure under the given load $\mathbf{p}$.

Consider the following three points $(\mathbf{x}^A,\mathbf{u}^A)$, $(\mathbf{x}^B,\mathbf{u}^B)$ and $(\mathbf{x}^C,\mathbf{u}^C)$:

$\mathbf{x}^A = (198.24,\ 55.10,\ 135.15,\ 154.45,\ 0.6,\ 0.6,\ 134.41,\ 0.6,\ 95.04,\ 0.6)^T$

$\mathbf{u}^A = (6.012, -18.608, -7.561, -37.227,\ 6.012, -50.800, -13.767, -50.800)^T$

$\mathbf{c}^T\mathbf{x}^A = 2302.55$

$\mathbf{x}^B = (196.90,\ 48.11,\ 135.71,\ 149.66,\ 0.6,\ 0.6,\ 138.88,\ 0.6,\ 98.20,\ 3.56)^T$

$\mathbf{u}^B = (6.071, -18.689, -7.779, -41.549,\ 4.870, -50.800, -13.794, -50.582)^T$

$\mathbf{c}^T\mathbf{x}^B = 2295.37$

$\mathbf{x}^C = (169.91,\ 120.99,\ 119.86,\ 170.71,\ 0.6,\ 0.6,\ 119.86,\ 0.6,\ 84.75,\ 0.6)^T$

$\mathbf{u}^C = (6.927, -20.782, -6.927, -20.782,\ 11.231, -51.114, -13.855, -55.418)^T$

$\mathbf{c}^T\mathbf{x}^C = 2374.99$

It is straightforward to show (by analyzing the problem with MATLAB) that both $(\mathbf{x}^A,\mathbf{u}^A)$ and $(\mathbf{x}^B,\mathbf{u}^B)$ satisfy the second order sufficient conditions for a local optimum of P2 (PROP.5). Thus, both $(\mathbf{x}^A,\mathbf{u}^A)$ and $(\mathbf{x}^B,\mathbf{u}^B)$ are local optimum points of P2. (Most people in the truss optimization community also believe that $(\mathbf{x}^B,\mathbf{u}^B)$ is a global optimum point of P2.)
It is also straightforward to show that $(\mathbf{x}^C,\mathbf{u}^C)$ is a local optimum point of P3. Moreover, since P3 is a convex problem (according to PROP.14), it follows that $(\mathbf{x}^C,\mathbf{u}^C)$ is in fact a <u>global</u> optimum point of P3.

Next, consider a three dimensional "Eight Bar Truss Pyramid" with one unsupported node and eight fixed nodes at the ground.

The cross section areas of the different bars should be chosen such that the weight of the structure is minimized subject to constraints on stresses and/or the strain energy under a given applied node at the unsupported node.

Let $\mathbf{x} = (x_1,\dots,x_8)^T$ be the cross section areas and let $\mathbf{u} = (u_1,u_2,u_3)^T$ be the displacements of the unsupported node. Further, let:

$\mathbf{p} = (40,\ 20,\ 200)^T$, $\mathbf{q} = (2/\sqrt{105},\ 1/\sqrt{105},\ 10/\sqrt{105})^T$,

$L_1 = L_2 = L_3 = L_4 = \sqrt{17}/8$, $L_5 = L_6 = L_7 = L_8 = \sqrt{18}/8$, $E = 21$,

$\alpha = 0.32917$, $\gamma_j = E/L_j^3$ and $c_j = 0.78L_j$ for $j = 1,\dots,8$,

$\mathbf{r}_1 = (\ 0.250,\ 0.250\ ,\ 0.375)^T$, $\mathbf{r}_2 = (\ 0.250, -0.250\ ,\ 0.375)^T$,

$\mathbf{r}_3 = (-0.250, -0.250\ ,\ 0.375)^T$, $\mathbf{r}_4 = (-0.250,\ 0.250\ ,\ 0.375)^T$,

$\mathbf{r}_5 = (\ 0.375,\ 0.000\ ,\ 0.375)^T$, $\mathbf{r}_6 = (\ 0.000, -0.375\ ,\ 0.375)^T$,

$\mathbf{r}_7 = (-0.375,\ 0.000\ ,\ 0.375)^T$, $\mathbf{r}_8 = (\ 0.000,\ 0.375\ ,\ 0.375)^T$.

$\mathbf{K}(\mathbf{x}) = \sum_j x_j \gamma_j \mathbf{r}_j \mathbf{r}_j^T$ = the 3×3 stiffness matrix.

Then consider the following two optimization problems P4 and P5 :

$$\underline{\text{P4}}: \quad \begin{array}{lll} \text{minimize} & \mathbf{c}^T\mathbf{x} & \dots(1) \\ \text{subject to} & \mathbf{K}(\mathbf{x})\mathbf{u} = \mathbf{p} & \dots(2) \\ & \mathbf{q}^T\mathbf{u} \le \alpha & \dots(3) \\ & x_j \ge 1.0 \quad \text{for } j = 1,..,8 & \dots(4) \end{array}$$

The constraint (3) is equivalent to an upper bound on the strain energy.

$$\underline{\text{P5}}: \quad \begin{array}{lll} \text{minimize} & \mathbf{c}^T\mathbf{x} & \dots(1) \\ \text{subject to} & \mathbf{K}(\mathbf{x})\mathbf{u} = \mathbf{p} & \dots(2) \\ & \mathbf{r}_j^T\mathbf{u} \le 10L_j^2/E \quad \text{for } j=1,..,8 & \dots(3)' \\ & x_j \ge 1.0 \quad \text{for } j=1,..,8 & \dots(4) \end{array}$$

The constraints (3)'are equivalent to upper bounds on elemental stresses. The following two points $(\mathbf{x}^D,\mathbf{u}^D)$ and $(\mathbf{x}^E,\mathbf{u}^E)$ will be referred to:

$\mathbf{x}^D = (7.18465,\ 8.84645,\ 1.00000,\ 6.78490,\ 1.0,\ 1.0,\ 1.0,\ 1.0)^T$

$\mathbf{u}^D = (0.0,\ 0.0,\ 0.337302)^T$, $\mathbf{c}^T\mathbf{x}^D = 11.22873$.

$\mathbf{x}^E = (12.96955,\ 3.06155,\ 6.78490,\ 1.00000,\ 1.0,\ 1.0,\ 1.0,\ 1.0)^T$

$\mathbf{u}^E = (0.0,\ 0.0,\ 0.337302)^T$, $\mathbf{c}^T\mathbf{x}^E = 11.22873$.

It may be shown that both $(\mathbf{x}^D,\mathbf{u}^D)$ and $(\mathbf{x}^E,\mathbf{u}^E)$ are local optimum points of P4. But P4 is a convex problem, according to PROP. 14, so it follows from PROP. 12 that both $(\mathbf{x}^D,\mathbf{u}^D)$ and $(\mathbf{x}^E,\mathbf{u}^E)$, and also each convex combination of $(\mathbf{x}^D,\mathbf{u}^D)$ and $(\mathbf{x}^E,\mathbf{u}^E)$, are in fact global optimum points of P4.

It is an interesting fact that $(\mathbf{x}^D, \mathbf{u}^D)$ and $(\mathbf{x}^E, \mathbf{u}^E)$ are local optimum points also of the stress-constrained problem P5. Moreover, each convex combination of $(\mathbf{x}^D, \mathbf{u}^D)$ and $(\mathbf{x}^E, \mathbf{u}^E)$ is also a local optimum point of P5.

Finally, to illustrate Prop. 17 concerning natural frequency constraints, consider a simple "V-formed" two bar truss.

Let x_1 and x_2 be the cross section areas of the two bars.

Then $\mathbf{K}(\mathbf{x}) = k_0 \cdot \begin{bmatrix} x_1+x_2 & x_1-x_2 \\ x_1-x_2 & x_1+x_2 \end{bmatrix}$ and $\mathbf{M}(\mathbf{x}) = m_0 \cdot \begin{bmatrix} x_1+x_2 & 0 \\ 0 & x_1+x_2 \end{bmatrix}$,

where the positive real numbers k_0 and m_0 do not depend on $\mathbf{x}$.

The natural frequencies $\omega_1(\mathbf{x})$ and $\omega_2(\mathbf{x})$ are obtained from the equation $\det[\mathbf{K}(\mathbf{x})-\omega^2\mathbf{M}(\mathbf{x})] = 0$, which leads to the two solutions:

$$\omega_1^2(\mathbf{x}) = \frac{2k_0x_1}{m_0(x_1+x_2)} \quad \text{and} \quad \omega_2^2(\mathbf{x}) = \frac{2k_0x_2}{m_0(x_1+x_2)} .$$

For simplicity, assume that $\underline{\omega}^2 = k_0/2m_0$.

Then the set $\Omega = \{ \mathbf{x} \mid \omega_i(\mathbf{x}) \geq \underline{\omega} \text{ for } i = 1,2 \}$ becomes:

$\Omega = \{ (x_1,x_2)^T \mid 3x_1-x_2 \geq 0 \text{ and } 3x_2-x_1 \geq 0 \}$, which is a convex cone.

7. References

[1] Fletcher R. (1987) "Practical Methods of Optimization", Second edition, John Wiley & Sons.

[2] Luenberger D.G. (1984) "Linear and Nonlinear Programming", Second edition, Addison-Wesley.

[3] Svanberg K., (1984) "On Local and Global Minima in Structural Optimization", in Atrek, Ragsdell, Gallagher and Zienkiewicz (eds.) New Directions in Optimum Structural Design, Wiley.

[4] Fleury C., Private communication, November 1982.

AN INTERIOR POINTS METHOD FOR NONLINEAR CONSTRAINED OPTIMIZATION

J. HERSKOVITS
Mechanical Engineering Program
COPPE / Federal University of Rio de Janeiro
Caixa Postal 68503, 21945 Rio de Janeiro, BRAZIL.

ABSTRACT. We describe a new general approach for interior points algorithms in nonlinear constrained optimization. It consists on the iterative solution, in the primal and dual variables, of Karush - Kuhn - Tucker first order optimality conditions. Based on this approach, different algorithms can be stated by taking advantage of the particular characteristics of the problem in consideration and of the order of the available information. This method is very strong and efficient, since at each iteration it only requires the solution of two linear systems with the same matrix. It is also particularly appropriated for Engineering Design Optimization, since feasible designs are obtained. We present a basic algorithm for inequality constrained problems and two of the possible particular versions. The first one is a first order algorithm and the second one uses a quasi - Newton approximation of the second derivative of the Lagrangian, in order to have superlinear asymptotic convergence. Equality constraints are introduced later.

1. Introduction

We consider the problem of minimizing a function submitted to equality and inequality constraints, when nonlinear functions are involved. Engineering Design is a natural application for this problem, since designers wants to find the best of the projects that satisfies all the requirements of feasibility. Mathematical models in physics, frequently, involve the minimization of energy and lead to numerical models including Mathematical Programming problems. In the particular case of Solid Mechanics, these models are in an advanced stage of development, so as some discretization techniques, like Finite Elements or Boundary Elements Methods.

These applications require the availability of strong numerical techniques capable of solving large size problems in an efficient way. This is also the case of an increasing number of methodologies in many areas of knowledge, such as Economics and Production Engineering. In Nonlinear Programming, these techniques are, in general, based on iterative algorithms which must be able to generate a sequence

G. I. N. Rozvany (ed.), Optimization of Large Structural Systems, Vol. I, 589–608.

converging to the required optimum.

The present conference is focused on a general approach for nonlinear constrained optimization, based on the iterative solution of Karush - Kuhn -Tucker first order optimality conditions. The equalities included in these conditions, are solved by fixed point iterates in the primal and dual variables, in a way to verify the inequalities at each iteration. Thus, the inequalities are also verified at the limit points.

When only inequality constraints are considered, the algorithms obtained by this approach, generate a sequence of interior points with decreasing values of the objective. In the general case, the nonlinear equalities are verified only at the limit. Then, an increase of the objective can be necessary to have the equalities satisfied.

The present method is simple to codify, strong and efficient. It does not involve penalty functions, active set strategies or Quadratic Programming subproblems. It merely requires to solve two linear systems with the same matrix at each iteration, and perform an inaccurate line search. It is very versatile, since different algorithms can be stated, depending on the problem under consideration and on the order of the available information.

In design optimization problems, this method produces a feasible design at each iteration. It can also be applied to several problems in solid mechanics, such as limit analysis [12] and inelastic behavior of materials. Nonlinear constitutive equations or yielding conditions can be included, with no need of linear approximations. The linear systems admit a treatment that profits from the special structure of the problem, which leads to procedures which are common in Finite Element Method.

In what follows, we make some considerations about inequality constrained problems and explain the fundamental ideas involved in the present approach. Then, a basic algorithm for inequality constrained optimization is presented. The theoretical results about global convergence are discussed in Section 5, followed by the statement of a first order and of a quasi - Newton algorithm. Finally, we introduce equality constraints and present a basic algorithm for problems including equality and inequality constraints.

2. The Inequality Constrained Problem

We consider now the inequality constrained nonlinear programming problem :

$$\underset{x}{\text{minimize}} \quad f(x)$$

$$\text{submitted to} \quad g(x) \le 0 \tag{2.1}$$

where $f \in R$ and $g \in R^m$ are smooth functions in R^n and (2.1) means $g_i(x) \le 0$; $i = 1,m$. In this section we discuss some concepts concerning this problem, that are useful to understand the basis of the present method.

Karush - Kuhn - Tucker (KKT) first order optimality conditions of problem (2.1) are expressed as follows:

$$\nabla f(x) + \sum_{i=1}^{m} \lambda_i \nabla g_i(x) = 0, \tag{2.2}$$

$$\lambda_i g_i(x) = 0; \; i = 1,m, \tag{2.3}$$

$$g_i(x) \le 0; \; i = 1,m \quad \text{and} \tag{2.4}$$

$$\lambda_i \ge 0; \; i = 1,m. \tag{2.5}$$

In the present approach, we find $x \in R^n$ and $\lambda \in R^m$ which solve the nonlinear system of equations and inequations (2.2) to (2.5).

The vectors x and λ are respectively called primal and dual variables, $\Omega \equiv \{x \in R^n / \; g(x) \le 0\}$ is the feasible set, Ω^0 its interior, $L(x,\lambda) = f(x) + \lambda^t g(x)$ the Lagrangian and $H(x,\lambda) = \nabla^2 f(x) + \sum_{i=1}^{m} \lambda_i \nabla^2 g_i(x)$ its Hessian.

If we define now $C(x) \equiv \nabla f(x)$, $A(x) \equiv [\nabla_1 g(x), \nabla_2 g(x), \ldots, \nabla_m g(x)]^t$ and a diagonal matrix $G(x) \in R^{m\text{x}m}$, such that $G_{ii}(x) = g_i(x)$, (2.2) - (2.5) become

$$C(x) + A^t(x)\lambda = 0, \tag{2.6}$$

$$G(x)\,\lambda = 0, \tag{2.7}$$

$$g(x) \le 0 \quad \text{and} \tag{2.8}$$

$$\lambda \ge 0. \tag{2.9}$$

A vector (x^*, λ^*) satisfying KKT conditions will be called a *KKT Pair* and, a *Stationary Pair*, if it only satisfies equalities (2.6) and (2.7). A vector x^* is a *KKT Point* if it exists λ^* such that (x^*, λ^*) constitutes a KKT pair and, we call it, *Stationary Point* if only (2.6) and (2.7) can be satisfied.

We make now the following definitions:

Definitions

Definition 2.1. $d \in R^n$ is a *descent direction* of a real function f(x) at $x \in R^n$ if $d^t \nabla f(x) < 0$.

Definition 2.2. $d \in R^n$ is a *feasible direction* of Ω, at $x \in \Omega$, if for some $\theta > 0$ we have $x + td \in \Omega$ for all $t \in [0,\theta]$.

Definition 2.3 A vector field d(x) defined on Ω is said to be an *uniformly feasible directions field* of Ω, if there exists $\tau > 0$ such that $x + td(x) \in \Omega$ for all $t \in [0,\tau]$. ■

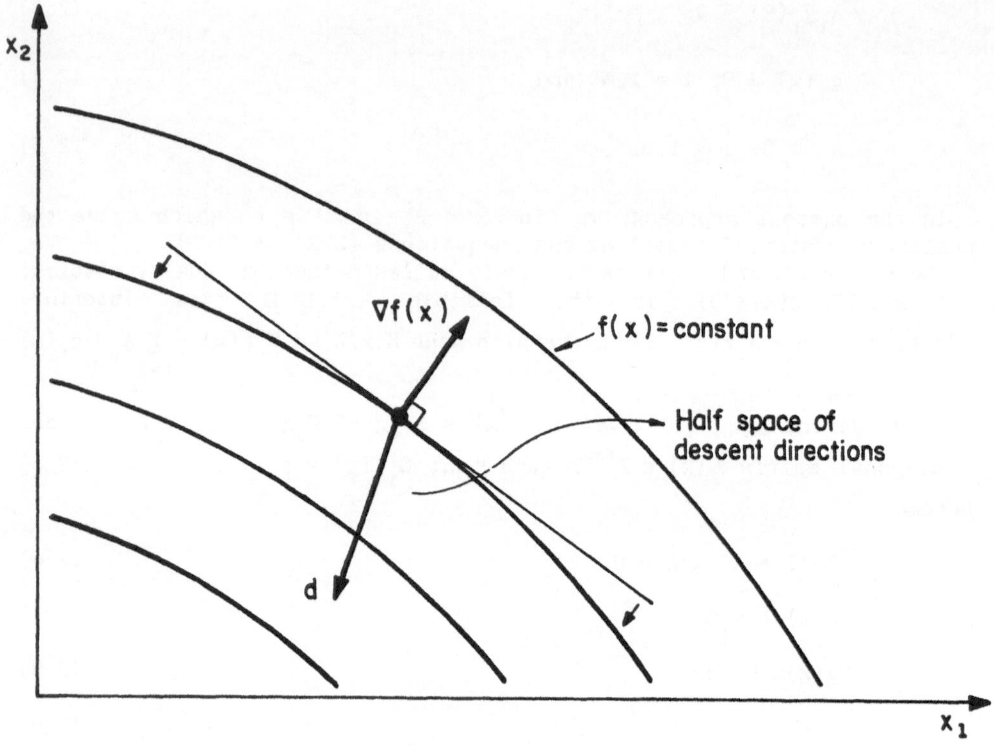

Figure 2.1

In Fig. 2.1 the isovalue contours of f(x) are represented. We can observe that f(x) decreases in any direction which makes and angle greater than 90° with ∇f(x). The set of all descent directions constitutes a half space.

The vector d in Fig. 2.2 is a feasible direction, since it supports a non zero segment [x, x + θd] contained in Ω. At an interior point, any direction is feasible. It can be proved that, at x on the boundary, d is a feasible direction if $d^t \nabla g_i(x) < 0$ for any i such that $g_i(x) = 0$. In this case, the set of feasible directions is in a cone which is orthogonal to the gradients of the active constraints.

Definition 3 introduces a condition on the vectorial field d(x), which is strongest than the simple feasibility of any element of d(x). When d(x) constitutes an uniformly feasible directions field, it supports a feasible segment [x, x + θ(x)d(x)], such that θ(x) is bounded bellow by $\tau > 0$.

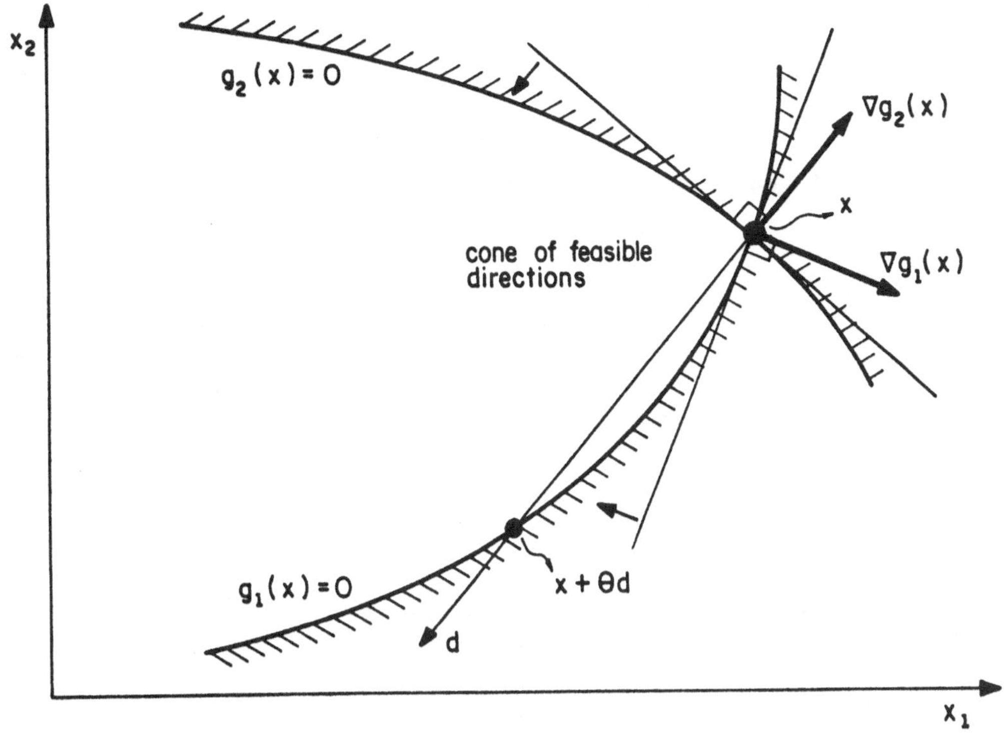

Figure 2.2

3. Foundations of the Method

The algorithms that we obtain, for a given initial interior point, generate a sequence $\{x^k\}$ of interior points with decreasing values of the objective and converging to a KKT point x^* of the problem.

At each iteration it is defined a search direction d, which is a descent direction of the objective and also a feasible direction of Ω. A line search is then performed to ensure that the new point is interior and the objective is lower.

As a consequence of the requirement of feasibility, d must actually constitute an uniformly feasible directions field. Otherwise, the step length may go to zero, forcing convergence to points which are not KKT.

In the present approach, the system of equations (2.6),(2.7) in (x,λ) is solved by means of fixed point iterates. This is done in a way to satisfy (2.8) and (2.9) at each iteration. This ensures that Karush-Kuhn-Tucker conditions are verified at the limit points.

A Newton's iteration for the solution of (2.6),(2.7) is defined by the following system:

$$\begin{bmatrix} B & A^t(x) \\ \Lambda A(x) & G(x) \end{bmatrix} \begin{bmatrix} x_0 - x \\ \lambda_0 - \lambda \end{bmatrix} = - \begin{bmatrix} C(x) + A^t(x)\lambda \\ G(x)\lambda \end{bmatrix} \tag{3.1}$$

where $B = H(x,\lambda)$, Λ is a diagonal matrix with $\Lambda_{ii} = \lambda_i$, (x,λ) is the current point and (x_0,λ_0) is a new estimate. Instead of $H(x,\lambda)$, B can be taken equal to a quasi - Newton approximation or to the identity matrix.

Then, depending on the way that $B \in R^{nxn}$ symmetric is defined, (3.1) may represent a second order, a quasi - Newton or a first order iteration. However, as a requirement for global convergence, B must be positive definite.

In what follows, we are going to introduce some modifications on iteration (3.1) in a way that, for a given interior pair (x, λ), the new estimate is interior and the objective improved.

We define a direction d_0 in the primal space as $d_0 = x_0 - x$. Then, (3.1) becomes

$$Bd_0 + A^t(x)\lambda_0 = - C(x), \tag{3.2}$$

$$\Lambda A(x)d_0 + G(x)\lambda_0 = 0. \tag{3.3}$$

As it was proved in [3], d_0 is a descent direction of f. It is easy to show that $d_0 = 0$ when x is a Stationary Point and, if d_0 is considered as a function of x, then d_0 goes to zero as x goes to a Stationary Point.

However, d_0 is not useful as a search direction, since it is not always feasible. This is due to the fact that, as any constraint goes to zero, (3.3) forces d_0 to tend to a direction tangent to the feasible set. In fact, (3.3) is equivalent to

$$\lambda_i \nabla g_i^t(x)d_0 + g_i(x)\lambda_{0i} = 0; \; i=1,m, \tag{3.4}$$

which implies that $\nabla g_i^t(x)d_0 = 0$ for i such that $g_i(x) = 0$. Thus, d_0 is tangent to the active constraints, as represented in Fig. 3.1.

Now, we define the linear system in d and $\bar{\lambda}$,

$$Bd + A^t(x)\bar{\lambda} = - C, \tag{3.5}$$

$$\Lambda A(x)d + G(x)\bar{\lambda} = - \rho\Lambda\omega, \tag{3.6}$$

obtained by adding a negative vector in the right side of (3.3),

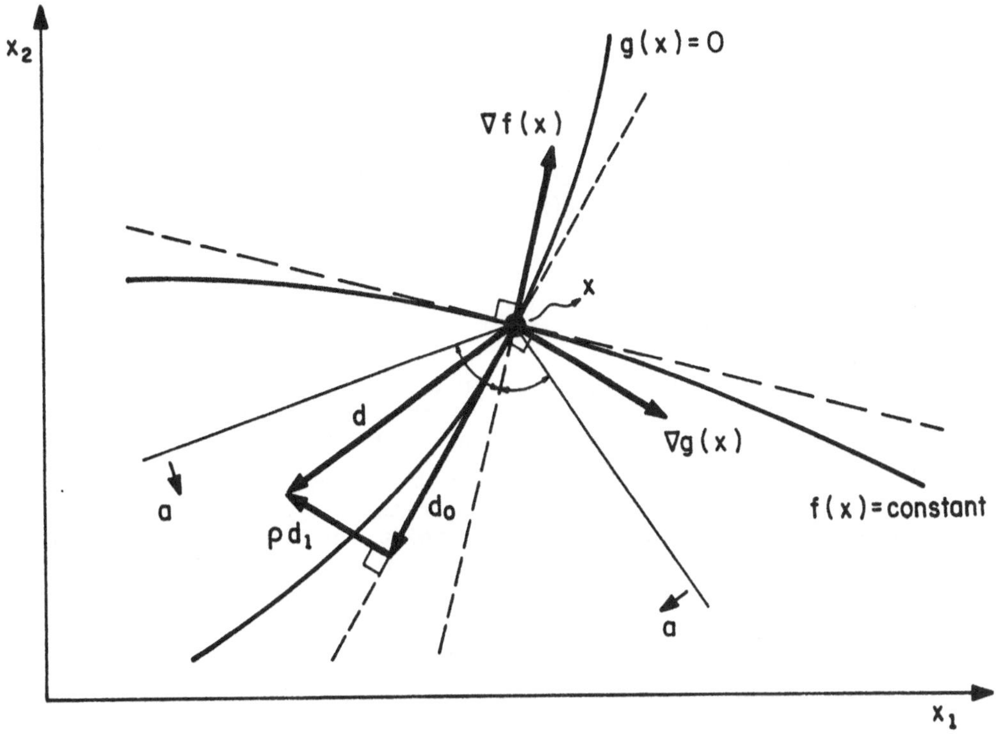

Figure 3.1

where the scalar factor ρ and $\omega \in R^m$ are positive and $\bar{\lambda}$ is the new estimate of λ. In this case, (3.6) is equivalent to

$$\lambda_i \nabla g_i^t(x)d + g_i(x)\bar{\lambda}_i = -\rho\lambda_i\omega_i; \ i=1,m, \tag{3.7}$$

and $\nabla g_i^t(x)d = -\rho\omega_i$ for the active constraints. Thus, d is a feasible direction.

We can consider that the inclusion of a negative number in the right hand of (3.4), produces the effect of deflecting d_0 in the sense of the interior of the feasible region, being the deflection of d_0 relative to the i-th constraint proportional to $\rho\omega_i$.

Finally, since the deflection of d_0 is proportional to ρ and d_0 is descent, it is possible to establish bounds on ρ which ensure that d is also a descent direction of f(x). Since $d_0^t\nabla f(x) < 0$, we obtain these bounds by imposing

$$d^t\nabla f(x) \le \alpha\, d_0^t\nabla f(x), \tag{3.8}$$

which implies $d^t\nabla f(x) < 0$.

Condition (3.8) implies that d is in a circular cone, represented by aa in Fig. 3.1, whose axis is $\nabla f(x)$. In general, the derivative of f in the direction of d will be smaller than in the direction of d_0. This is a price we must pay to obtain a feasible descent direction.

The ideas pointed above are a basis for the iterative method that we are studying. To determine a new primal point, an inaccurate line search is done in the direction of d, requiring feasibility and a satisfactory decrease of the objective. Different updating rules can be adopted to define a new positive λ.

4. A Basic Algorithm

In this section we present a basic algorithm for inequality constrained problems, that globally converges in the primal space to Karush-Kuhn-Tucker points of the problem. It is basic in the sense that several elements of the algorithm are very widely defined. This fact turns possible the implementation of different versions depending on the problem to be solved, the available information about f and g, and the desired rate of convergence.

The algorithm is stated as follows:

ALGORITHM I.

Parameters. $\alpha \in (0,1)$, $\eta \in (0,1)$, $\varphi > 0$ and $\nu \in (0,1)$.

Data. $x \in \Omega^0$, $\lambda > 0$,$B \in R^{nxn}$ symmetric and positive definite and $\omega \in R^m$ positive.

Step 1. Computation of a search direction.

(i) Compute (d_0, λ_0) by solving the linear system

$$B\, d_0 + A^t\lambda_0 = - C, \tag{4.1}$$

$\Lambda A d_0 + G \lambda_0 = 0.$ (4.2)

If $d_0 = 0$, stop.

(ii) C ompute (d_1, λ_1) by solving the linear system

$B d_1 + A^t \lambda_1 = 0,$ (4.3)

$\Lambda A d_1 + G \lambda_1 = - \Lambda \omega.$ (4.4)

(iii) If $d_1^t C > 0$, set

$\rho = \inf \{\varphi \|d_0\|^2; (\alpha-1) d_0^t C / d_1^t C\}$, or (4.5)

$\rho = \varphi \|d_0\|^2$ otherwise. (4.6)

(iv) Compute the search direction

$d = d_0 + \rho d_1$, and (4.7)

$\bar{\lambda} = \lambda_0 + \rho \lambda_1.$ (4.8)

Step 2. Line search.

Compute t, the first number of the sequence $\{1, \nu, \nu^2, \nu^3, \ldots\}$ satisfying

$f(x + td) \le f(x) + t \eta C^t d$ and (4.9)

$g_i(x + td) < 0$ if $\bar{\lambda}_i \ge 0$, or (4.10)

$g_i(x + td) \le g_i(x)$, otherwise. (4.11)

Step 3. Updates.

(i) Set

$x := x + td$

and define new values for

$\omega > 0,$

$\lambda > 0$ and

B symmetric and positive definite.

(ii) Go back to *Step 1*. ∎

In *Step 1*, the linear systems (4.1), (4.2) and (4.3), (4.4) are solved, being d and $\bar{\lambda}$ obtained as a linear combination of d_0, d_1 and λ_0, λ_1 respectively. Then, the deflection on d_0 is determined by ρd_1. Remark that, in consequence of (4.4), d_1 is a descent direction of the active constraints. Thus, d_1 points to the interior of the feasible set. In the particular case when B is taken the identity and there is a unique constraint that is saturated, d_1 is orthogonal to d, as shown in Figure 3.1.

This way of calculating d simplifies the establishment of bounds on ρ in order to ensure that (3.7) is verified. The update of ρ, stated by (4.5) and (4.6), satisfies this bounds without letting ρ to be greater than $\varphi\|d_0\|^2$. Then, ρ is at maximum of the order of $\|d_0\|^2$, which is enough for the deflected direction to constitute an Uniformly Feasible Directions Field [3].

The algorithm includes, in *Step 2*, an inexact line search based on Armijo's procedure for unconstrained optimization [8]. In addition to (4.9), which ensures a reasonable decrease of the function, conditions (4.10) and (4.11) impose to the new primal point to be interior. Moreover, (4.11) prevents saturation of constraints associated to negative dual variables, which is a requirement to have convergence to Karush - Kuhn - Tucker points.

Different algorithms can be obtained according to the way of updating λ, B and ω. In what follows we introduce some assumptions which must be satisfied by λ, B and ω to have global convergence in the primal variables to a KKT point of the problem.

Assumptions

Assumption 3.1. There exist positive numbers λ^I, λ^S and β such that $0 \le \lambda \le \lambda^S$ and $\lambda_i \ge \lambda^I$ for any i such that $g_i(x) \ge -\beta$.

Assumption 3.2. There exist positive numbers σ_1 and σ_2 such that $\sigma_1\|d^2\| \le d^tB\,d \le \sigma_2\|d^2\|$ for any $d \in R^n$.

Assumption 3.3. There exist positive numbers ω_1 and ω_2 such that $\omega_1 \le \omega \le \omega_2$. ∎

Depending on the way of updating λ, global convergence in the dual variables can also be obtained.

5. Considerations about Global Convergence

In this section, we will discuss some theoretical aspects of the study of convergence of the present method. We refer to [3], where the complete study can be found.

The following assumptions about problem (2.1) are made:

Assumptions

Assumption 5.1. There exists a real number, a , such that the set $\Omega_a \equiv \{ x \in \Omega;\ f(x) \le a \}$ is compact and has an interior Ω_a^0.

Assumption 5.2. Each $x \in \Omega_a^0$ satisfies $g(x) < 0$.

Assumption 5.3. The functions f and g are continuously differen_ tiable in Ω_a and their derivatives satisfies Lipschitz condition.

Assumption 5.4. (*Regularity Condition*) For all $x \in \Omega_a$ the vectors $\nabla g_i(x)$, for i such that $g_i(x) = 0$, are linearly independent. ∎

The main result obtained in [3] is that, if the preceding assumptions on the problem are true and the initial x belongs to Ω_a^0, then any sequence $\{x^k\}$ generated by the algorithm converges to a Karush - Kuhn - Tucker point, no matter the way of updating λ, B_k and ω, provided Assumptions 3.1 to 3.3 are satisfied. Moreover, (x^k, λ_0^k) converges to a Karush - Kuhn - Tucker pair of the problem.

Previously, it is necessary to prove that the algorithm never fails, or in other words, any of the prescribed operation can always be performed. The only possibility of failure is in the case when the solution of the linear systems (4.1), (4.2) and (4.3), (4.4) is not unique.

In Lemma 3.1 in [10], it was proved that the matrix

$$M(x,\lambda,B) \equiv \begin{bmatrix} B & A^t(x) \\ \Lambda A(x) & G(x) \end{bmatrix}$$

is non singular. In consequence, d_0, λ_0, d_1 and λ_1 obtained in *Step 1* of the algorithm are uniquely determined. Moreover, since x is in the compact Ω_a and λ and B are bounded, it follows that M is bounded out of zero and then d_0, λ_0, d_1 and λ_1 are upper bounded.

Assumption 5.2 is a requirement for the last results. This is a limitation of the present method since, even a well stated problem, may not satisfy this assumption. In particular, this happens whenever two or more active constraints are equivalent.

We consider now the special case when $d_0 = 0$ is obtained in the *Step 1* of the algorithm and, in consequence, it stops. Since all the iterates are strictly feasible, it follows from (4.1) and (4.2) that $\nabla f(x) = 0$. Thus, if the algorithm stops, x is a particular Karush - Kuhn - Tucker point.

If the algorithm never stops, an infinite sequence $\{x^k\}$ contained in the compact Ω_a is generated. Thus, it has accumulation points.

The proof of convergence is then completed by successively showing that:

i) The direction d_0 is a descent direction of f(x).

ii) The search direction d is also a descent direction of f(x) and it constitutes an Uniformly Feasible Directions Field of the problem.

iii) The step length t, defined in Step 2, has a positive lower bound.

iv) The accumulation points of the sequence $\{x^k\}$ are in fact KKT points of the problem.

It follows from (4.1), (4.2) that, if x^* is a KKT point, then $d_0 = 0$ and (x^*, λ_0) is a KKT pair. Thus, $\{x^k, \lambda_0^k\}$ converges to a KKT pair.

6. Some Particular Algorithms

Here, we present two versions of Algorithm 1. They only require first order information about the problem. The first algorithm can be considered as a gradient method, while the second one is based on a quasi - Newton approximation of $H(x,\lambda)$.

To define a particular version of Algorithm 1, it is only required to state, in Step 3, updating rules for λ, B and ω in such a way to satisfy Assumptions 3.1 to 3.3.

6.1. A FIRST ORDER ALGORITHM

Step 3 of Algorithm 1 is stated as follows:

Step 3. Updates.

Additional data. $\varepsilon > 0$.

(i) Set

$x := x + td,$

$\omega := [1, 1, \ldots, 1]$; $\omega^t \in R^m$,

$$\lambda_i := \sup\,[\lambda_{0i};\ \varepsilon \|d_0^2\|\ ];\ i = 1,m \quad \text{and} \tag{6.1}$$

$B := I.$

(ii) Go back to *Step 1*. ■

In this algorithm, ω and B are defined to be constant. Updating of λ, defined by (6.1), ensures that Assumption 3.1 is verified. Since λ_0 goes to λ^* and d_0 goes to zero as x goes to x^*, we have that λ converges to the dual variables vector λ^*. This version of the algorithm is very simple to code, which makes it stronger. It is very efficient in applications where a great accuracy in the result is not required [2,6].

6.2. A QUASI - NEWTON ALGORITHM

6.2.1. *The Quasi - Newton Approximation Matrix.* In unconstrained optimization problems by quasi - Newton method, an approximation matrix B to the second derivative of the function is built up. The formula preferred by several authors [8], for updating B is the BFGS rule

$$B := B - \frac{B\delta\delta^t B}{\delta^t B\delta} + \frac{\gamma\gamma^t}{\delta^t\gamma}, \tag{6.2}$$

where $\delta \in R^n$ is the change of the variables and $\gamma \in R^n$ is the change of the gradient of the function. It can be proved that, if the initial B is positive definite, then a new positive definite approximation matrix is obtained, provided that

$$\delta^t\gamma > 0. \tag{6.3}$$

Condition (6.3) is achieved, for an appropriated line search criterion, when the second derivative of the function is positive definite [8].

In the present algorithm, since B is an approximation of $H(x,\lambda)$, we let $\gamma \in R^n$ to be the change in x of $[C(x) + A(x)^t\lambda]$. However, $H(x,\lambda)$ is not necessarily positive definite at the solution. In order to satisfy (6.3), γ is modified, according to a procedure proposed by Powell [11].

Then, a quasi - Newton algorithm is obtained by specifying Step 3, in Algorithm 1, as follows:

Step 3. Updates.

Additional data. $\varepsilon > 0$.

(i) - Let be

$\delta = td$ and

$\gamma = [C(x + td) + A^t(x + td)\lambda] - [C(x) + A^t(x)\lambda]$.

- If

$\delta^t\gamma < 0.2\ \delta^t B\delta$,

then compute

$$\phi = \frac{0.8\ \delta^t B\delta}{\delta^t B\delta - \delta^t\gamma}$$

and set

$$\gamma := \phi\gamma + (1-\phi)B\delta. \tag{6.4}$$

- Set

$$x := x + td,$$

$$\omega := [1,\ 1,\ \ldots,\ 1]^t;\ \omega \in R^m,$$

$$\lambda_i := \sup\ [\lambda_{0i};\ \varepsilon\ \|d_0\|^2];\ i = 1,m \text{ and}$$

$$B := B - \frac{B\delta\delta^t B}{\delta^t B\delta} + \frac{\gamma\gamma^t}{\delta^t\gamma}.$$

(ii) Go back to *Step 1*. ■

The modification of γ, according to Powell's procedure, is stated in (6.4). So, it is $\delta^t\gamma \geq 0.2\ \delta^t B\delta$, which ensures that the new B is positive definite.

It can be proved that this algorithm locally converges with a superlinear rate, provided the step length is unitary near the limit point. However, to satisfy this requirement, d must satisfy Definition 2.2 for $\theta = 1$. It is clear that θ increases whenever ρ grows but, unfortunately, in some problems the upper bound on ρ may be not large enough to allow the step length to be unitary. This effect, which is similar to Maratos' effect [11], in theory can produce rates of convergence lower than superlinear. Up to now, we never detected Maratos' effect in practical applications of this algorithm. A more complete study and numerical results, can be found in [4,5].

In a quasi - Newton algorithm described in [1], Maratos' effect is avoided by looking at each iteration for a decrease of an estimate of the Lagrangian, instead of the objective function. However, global convergence is not very strong in this case, since the estimate Lagrange multipliers change at each iteration and oscillations between several accumulation points are possible.

An algorithm that also solves optimality conditions by means of fixed point iterates was presented in [10]. The authors obtained global and local superlinear convergence by applying a technique first presented by Mayne and Polak [9] in a different context. In this algorithm, a feasible descent direction d is obtained and the constraints computed at (x + d). Then, an approximate projection, $\tilde{x}$, of (x + d) on the active constraints is found. Finally, a new primal variable is determined by doing a search along a parabola which is tangent to d and contains $\tilde{x}$. In this search, there are required a decrease of the objective and feasibility of the new iterate.

7. Including Equality Constraints

Now, we extend the algorithm for inequality constrained optimization, presented in Section 4, in order to solve the Nonlinear Optimization problem

$$\begin{aligned} \underset{x}{\text{minimize}} \quad & f(x) \\ \text{submitted to} \quad & g(x) \le 0 \\ \text{and} \quad & h(x) = 0, \end{aligned} \tag{7.1}$$

where $h \in R^p$ is a smooth function in R^n.

The simplest way of including equality constraints in Algorithm 1, consists on defining a suitable penalty function of the equalities and, then, minimizing that function submitted only to the inequality constrains. Unfortunately, this approach brings up the numerical problems related to penalty functions.

In what follows, we consider KKT conditions of problem (7.1), we discuss our approach to solve them by fixed point iterates and, finally, we present an algorithm based on these approach. This algorithm requires an initial point at the interior of $\Omega \equiv \{x \in R^n / g(x) \le 0\}$, not necessarily feasible with respect to the equality constraints. It generates a sequence $\{x^k\}$ of interior points ($\{x^k\} \in \Omega^0$) which converges to a KKT point x^* of the problem. In general, the equalities are only active at the limit. Then, to have the equalities satisfied, an increase of the function can be necessary.

Karush - Kuhn - Tucker first order optimality conditions of problem (7.1) can be expressed as follows:

$$C(x) + A^t(x)\lambda + E^t(x)\mu = 0, \tag{7.2}$$

$$G(x)\ \lambda = 0, \tag{7.3}$$

$$h(x) = 0, \tag{7.4}$$

$$g(x) \le 0 \text{ and} \tag{7.5}$$

$$\lambda \ge 0, \tag{7.6}$$

where $\mu \in R^p$ is the dual variables vector corresponding to the equality constraints and $E(x) \equiv [\nabla_1 h(x), \nabla_2 h(x), \ldots, \nabla_p h(x)]^t$. Now, the Lagrangian is $L(x,\lambda,\mu) = f(x) + \lambda^t g(x) + \mu^t h(x)$, and its second derivative becomes $H(x,\lambda,\mu) = \nabla^2 f(x) + \sum_{i=1}^{m} \lambda_i \nabla^2 g_i(x) + \sum_{i=1}^{p} \mu_i \nabla^2 h_i(x)$.

A Newton's iteration for the solution of (7.2) to (7.4) is defined by the following system:

$$\begin{bmatrix} B & A^t(x) & E^t(x) \\ \Lambda A(x) & G(x) & 0 \\ E(x) & 0 & 0 \end{bmatrix} \begin{bmatrix} x_0 - x \\ \lambda_0 - \lambda \\ \mu_0 - \mu \end{bmatrix} = - \begin{bmatrix} C(x) + A^t(x)\lambda + E^t(x)\mu \\ G(x)\lambda \\ h(x) \end{bmatrix} \tag{7.7}$$

where $B = H(x,\lambda,\mu)$, (x,λ,μ) is the current point and (x_0,λ_0,μ_0) is a new estimate. Again, B can be taken equal to a quasi - Newton approximation of $H(x,\lambda,\mu)$ or to the identity matrix.

If we define $d_0 = x_0 - x$, then (7.7) becomes

$$Bd_0 + A^t(x)\lambda_0 + E^t(x)\mu_0 = - C(x), \tag{7.8}$$

$$\Lambda A(x)d_0 + G(x)\lambda_0 = 0, \tag{7.9}$$

$$E(x)d_0 = - h(x), \tag{7.10}$$

which is independent of the current value of μ. As we have done in Section 3, it can be deduced that d_0 is tangent to the active equality and inequality constraints.

Let us consider the auxiliary function

$$\phi_c(x) = f(x) + \sum_{i=1}^{p} c_i |h_i(x)|,$$

where c_i are positive constants. It can be shown [8] that, if c_i are large enough, then $\phi_c(x)$ is an Exact Penalty Function of the equality constraints. In other words, it exists a finite c such that the the minimum of ϕ_c is exactly the solution of Problem (7.1). Then, the use of ϕ_c as a penalty function is numerically very advantageous, since it does not require penalty parameters going to infinite. On the other side, ϕ_c has not derivatives at the points where some equality constraint is active. Thus, special techniques for non smooth optimization are required to minimize ϕ_c.

It can be proved [3] that d_0 obtained in (7.8) to (7.10) is a descent direction of ϕ_c, provided $c_i > |\mu_{0i}|$; $i = 1,p$. The first idea to construct an algorithm, consists on modifying (7.8) to (7.10) in order to deflect d_0 with respect to the inequality constraints. In this way, it can be obtained a search direction, that is a feasible direction of Ω and a descent direction of ϕ_c. A line search can be done, with the requirement for the new primal variable to be in the interior of Ω and to produce a decrease of ϕ_c. This approach has the important disadvantage of involving a non differentiable function in the line search.

We propose here technique which uses ϕ_c in the line search, but avoiding the non differentiable points of ϕ_c. Let be $\Delta \equiv \{x \in R^n / g(x) \le 0,\ h(x) \le 0\}$. The present algorithm, for a given initial point in Δ^0, generates a sequence $\{x^k\}$ in Δ^0, with decreasing values of ϕ_c. With this purpose, (7.8) to (7.10) is modified in order to obtain, as search direction, a feasible direction of Δ and descent direction of ϕ_c. The parameters c are updated at each iteration in order to have $c_i > |\mu_{0i}|$; $i = 1, p$. Since all the iterates are in Δ^0, the points where ϕ_c is non differentiable , are avoided in the line search.

We state the following algorithm for the solution of problem (7.1):

ALGORITHM II.

Parameters. $\alpha \in (0,1)$, $\eta \in (0,1)$, $\varphi > 0$ and $\nu \in (0,1)$.

Data. $x \in \Delta^0$, $\lambda > 0$, $B \in R^{nxn}$ symmetric and positive definite and $\omega^i \in R^m$, $\omega^e \in R^p$ and $c \in R^p$ positive.

Step 1. Computation of a search direction.

(i) Compute (d_0, λ_0, μ_0) by solving the linear system

$$B\,d_0 + A^t\lambda_0 + E^t\mu_0 = -C,$$

$$\Lambda\,A\,d_0 + G\,\lambda_0 = 0,$$

$$E\,d_0 = -h. \tag{7.11}$$

If $d_0 = 0$, stop.

(ii) Compute (d_1, λ_1, μ_1) by solving the linear system

$$B\,d_1 + A^t\lambda_1 + E^t\mu_1 = 0, \tag{7.12}$$

$$\Lambda A d_1 + G\lambda_1 = -\Lambda\,\omega^i, \tag{7.13}$$

$$E\,d_1 = -\omega^e. \tag{7.14}$$

(iii) If $c_i < -1.2\mu_{0i}$, then set $c_i = -2\,\mu_{0i}$; $i = 1,..,p$.

(iv) If $d_1^t \nabla\phi_c > 0$, set

$\rho = \inf \{\varphi \|d_0\|^2;\ (\alpha-1)\ d_0^t \nabla\phi_c / d_1^t \nabla\phi_c\}$, or

$\rho = \varphi \|d_0\|^2$ otherwise.

(v) Compute the search direction

$d = d_0 + \rho\ d_1$ and

$\bar{\lambda} = \lambda_0 + \rho\ \lambda_1$

Step 2. Line search.

Compute t, the first number of the sequence

$\{1, \nu, \nu^2, \nu^3, \ldots\}$ satisfying

$$\phi_c(x + td) \le \phi_c(x) + t\ \eta\ \nabla\phi_c^t d,$$
$$h(x + td) \le 0 \text{ and} \tag{7.15}$$
$$g_i(x + td) < 0 \text{ if } \bar{\lambda}_i \ge 0, \text{ or}$$
$$g_i(x + td) \le g_i(x), \text{ otherwise.}$$

Step 3. Updates.

(i) Set

$x := x + td$

and define new values for

$\omega^i > 0,\ \omega^e > 0,$

$\lambda > 0$ and

B symmetric and positive definite.

(ii) Go back to *Step 1.* ■

In *Step 1*, there are first obtained a direction d_0 and a direction d_1. In consequence of (7.12) to (7.14), d_1 is a descent direction both of the active inequality constraints and of the equalities. Thus, d_1 points to the interior of Δ. Updating of c ensures that d_0 is a descent direction of the resulting ϕ_c and updating of ρ, ensures that also d is a descent direction of ϕ_c. It can be deduced that c remains unchanged after a finite number of iterations, [3].

Even if the non smooth penalty function ϕ_c is involved in the line search, in consequence of (7.15), the points where ϕ_c has not derivatives are never attained.

In presence of linear equality constraints, it is simple to find an initial point verifying them. Then, it follows from (7.11) that d belongs to those constrains. Taking $\omega_i^e = 0$; for any i corresponding to the linear equality constraints, we have that they are always active. In consequence, when all the equalities are linear, no penalty function is required and a decrease of the objective is obtained at each iteration.

8. References

[1] Herskovits, J. "A two-stage feasible directions algorithm including variable metric techniques for nonlinear optimization", Research Report n° 118, INRIA, BP 105, 78153 Le Chesnay CEDEX, France, 1982.

[2] Herskovits, J. "A two-stage feasible directions algorithm for nonlinear constrained optimization", Mathematical Programming, Vol. 36, pp. 19-38, 1986.

[3] Herskovits J., " A General Approach for Interior Points Algorithms in Nonlinear Programming", Research Report, Mechanical Engineering Program, COPPE, Federal University of Rio de Janeiro, Caixa Postal 68503, 21945 Rio de Janeiro, Brazil, 1991.

[4] Herskovits, J and Asquier, "Quasi - Newton Interior Points Algorithms for Nonlinear Constrained Optimization", in DGOR - Proceedings of Operations Research 1990, Vienna, August 1990.

[5] Herskovits, J and Asquier, J., "A Superlinear Interior Points Algorithm for Engineering Design Optimization", Proceedings of the Third Air Force / NASA Symposium on Recent Advances in Multi disciplinary Analysis and Optimization, San Francisco, CA, September 1990.

[6] Herskovits, J. and Coelho, C.A.B. "An interior points algorithm for structural optimization problems", in Computer Aided Optimum Design of Structures: Recent Advances, edited by C.A. Brevia and S.Hernandez, Computational Mechanics Publications, Springer - Verlag, June 1989.

[7] Hock W. and Schittkowski, "Test Examples for Nonlinear Programming Codes", Lecture Notes in Economics and Mathematical Systems 187", Springer Verlag, Berlin, 1981.

[8] Luenberger D.G., "Linear and Nonlinear Programming", 2nd. Edition, Addison - Wesley , 1984.

[9] Mayne D.Q. and Polak E., "Feasible Directions Algorithms for Optimization Problems With Equality and Inequality Constraints", Mathematical Programming 11, pp. 67 - 80, 1976.

[10] Panier, E.R.,Tits A.L. and Herskovits J. "A QP - Free, globally convergent, locally superlinearly convergent algorithm for

inequality constrained optimization", SIAM Journal of Control and Optimization, Vol 26, pp 788-810, 1988.

[11] Powell, M.J.D. "The Convergence of Variable Metric Methods for Nonlinearly Constrained Optimization Calculations", in Nonlinear Programming 3, edited by O.L. Mangasarian, R.R. Meyer and S.M. Robinson, Academic Press, London, 1978.

[12] Zouain, N., Borges, L.A., Herskovits, J. and Feijóo, R. "An Iterative Algorithm for Limit Analysis with Nonlinear Yield Functions", Research Repport 034/90, Laboratorio Nacional de Computacion Cientifica, Rua Lauro Muller 455, 22290 Rio de Janeiro, Brazil, 1990.

A NEW MATHEMATICAL PROCEDURE FOR GLOBAL OPTIMIZATION IN NONCONVEX PROBLEMS.

Dr. S. Hernández.

Dept. of Mechanical Engineering. Zaragoza University. María de Luna 3, 50015 Zaragoza (Spain).

1. OPTIMUM DESIGN OF STRUCTURES.

The mathematical formulation of optimum design of structures with a single objective function is usually written as:

$$min\ F\ (\underline{X}) \qquad (1.a)$$

subject to

$$g_j\ (\underline{X}) \leq 0 \qquad\qquad j = 1,\ldots,m \qquad (1.b)$$

where$F\ (\underline{X})$ is the objective function, $\underline{X} = \left| x_1,\ldots, x_n \right|$ is the vector of variables and $g_j\ (\underline{X})$ the set of constraints.

Similarly, in the case of multicriteria optimization the expression (1.a) changes, and a vector $\underline{F}$ of objective functions appears

$$\underline{F} = \left| f_1,\ldots, f_K \right| \qquad (2)$$

most methods for solving multicriteria optimization propose to linearize vector $\underline{F}$ by introducing a vector $\underline{\lambda}$ and carrying out

$$F = \underline{\lambda}^T \underline{F} = \sum_{i=1}^{K} \lambda_i F_i \qquad (3)$$

G. I. N. Rozvany (ed.), Optimization of Large Structural Systems, Vol. I, 609–622.

after that the aim is to obtain Pareto minima of the problem defined by (3) and (1.2) for different combinations of vector λ . Hence, as multicriteria optimization is accomplished by solving several optimizations with a single objective function, it is possible to consider that expression (1) represents a general formulation for optimum structural design.

From the mathematical point of view it is very important to check the convexity of the problem defined by (1). If it is convex there will be only one local minimum, but if this is not the case it is possible for many local minima to exist, and the lowest of them will be the global one and the solution of the problem.

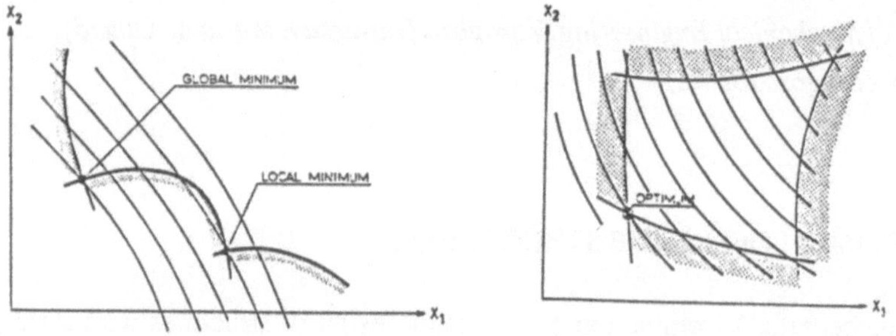

a) **Fig.1** **b)**

It is very well known that structural optimization leads in many cases to non linear problems having two very important features:

1) *Nonconvex feasible region, mainly because of the behaviour constraints.* Svanberg[1] pointed out that nonconvexity may appear in most usual practical problems. As an example it is useful to observe the results corresponding to a weight optimization of a single grillage subjected to stress constraints and solved by Moses and Onoda[2]

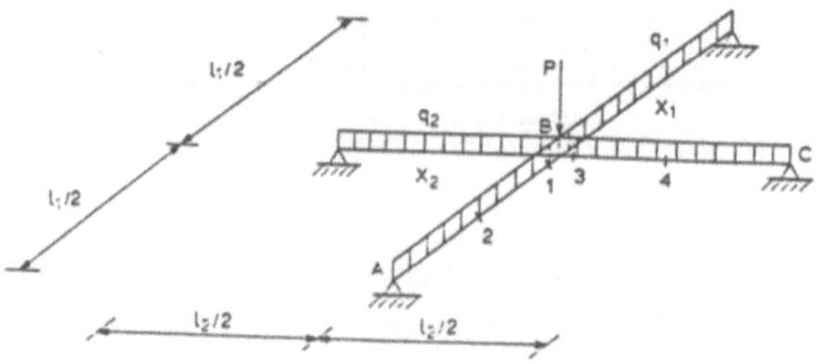

Fig.2

Feasible region for two different geometrical dimensions and loading cases are shown in Fig. 3.a and 3.b

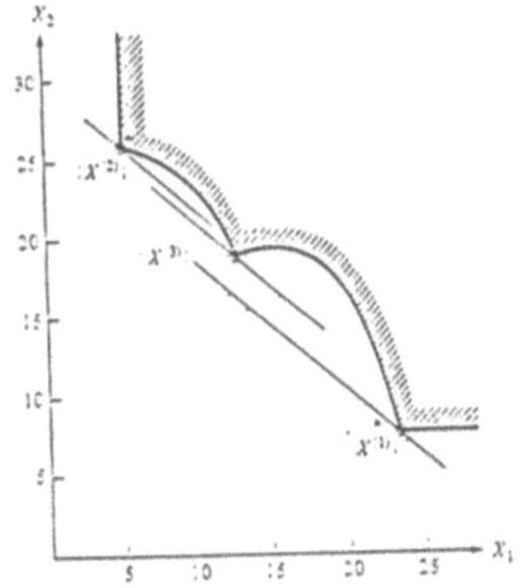

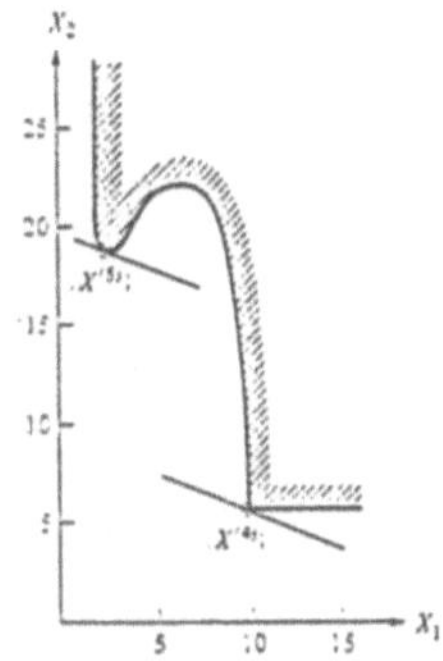

a) **Fig.3** b)

2) *Disjoints feasible regions arise when frequency limits are included in the constraints set.* Two feasible regions arising in the case of a single frame[4] undergoing a distributed load and a harmonic ground motion acting during a half period are represented in Fig. 4.

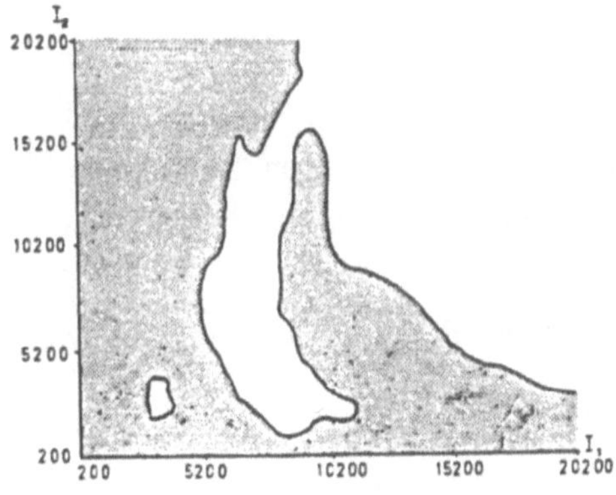

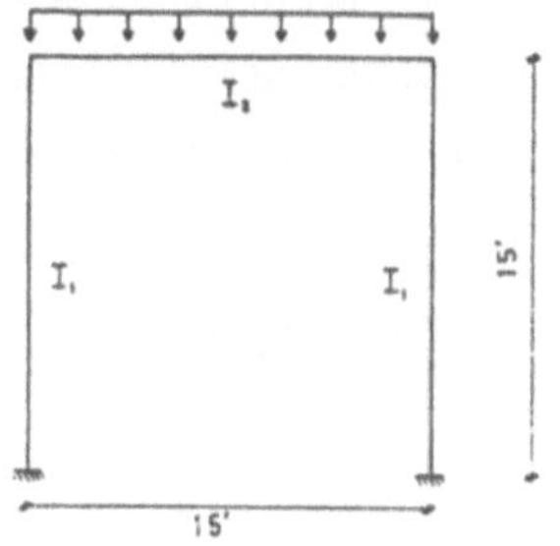

a) **Fig.4** b)

Also, optimization of the cantilever beam in Fig. 5 loaded with a torsional moment $M = M_o e^{iw_e t}$ throughout it, produces two disjoint feasible regions. In Fig. 6 two cases for different values of dynamic loading are shown.

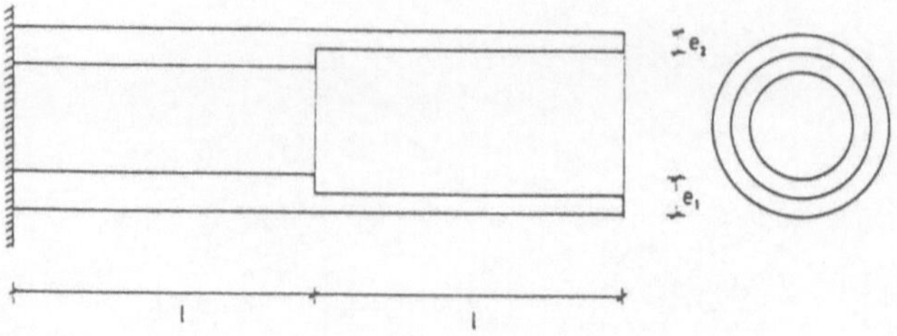

a) Fig.5 b)

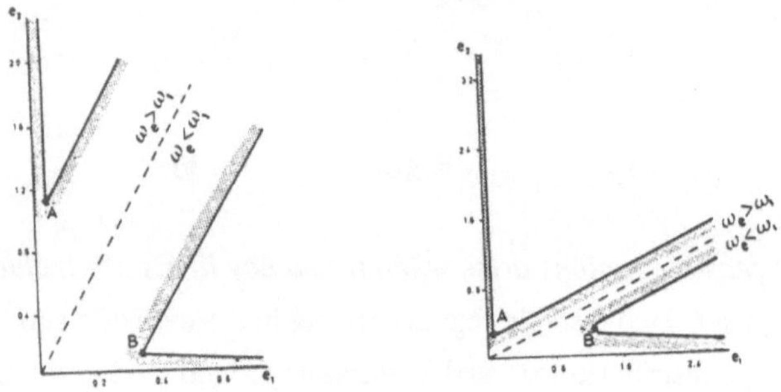

a) Fig.6 b)

Both features are of great significance when structural optimization is carried out. By using mathematical programming methods only one local minimum is guaranteed and when there are several, it depends on the initial design which will be found at the end of the process. Because of this when there are indications that many minima exist it used to be usual to begin the optimization procedure from very dissimilar starting designs to make distinct local minima easier to identify.

A greater obstacle is when the feasible region is split out into several disjoint domains, as most local optimization approaches try to find a minimum on the domain where the process started, and can not bridge to other ones. Again as before, the common way to deal with these cases was to choose several initial designs, located in different feasible regions, to start with.

2. METHODS FOR GLOBAL OPTIMIZATION

Global optimization methods try to find the global optimum over the whole feasible design region. They can be divided into deterministic and stochastic and references[5-7] describe their characteristics extensively.

Deterministic methods:

- Enumerative methods
- Cutting plane
- Branch and bound
- Bilinear programming
- Separable formulation

Most of these methods deal only with linear constraints and several asume that the objective function is quadratic. The search for the of global minimum is carried out by looking for it at many vertices of the boundary of feasible region. Due to these limitations their efficiency is very limited.

Stochastics methods:

- Multistart
- Clustering
- Multilevel

Stochastics methods aim to obtain the global optimum of the function by probabilistic techniques. Firstly, in the global phase, the set of sampling points are defined in the domain. Then a local phase is accomplished by working out with only a portion of the whole set of sampling points. The techniques above mentioned are the better known and organize searches of the global value of $F(\underline{X})$ with probability 1.

3. THE TUNNELING METHOD.

This method has been developed by Levy and Gomez[8], and Hernández[9] and works by identifying successively and in a very systematic way, the set of local minima of a function, so that as the process proceeds any new minimum produces an equal or lower value for the function $F(\underline{X})$. The process may finish if after arriving at a local minimum it is not possible to obtain another one within a reasonable computer time.

The method has two different phases:

a) *Minimization phase:* In this phase a local minimum is sought by using any numerical optimization procedure. If the process is started from point X_1^*, this phase gives a local minimum as X_1^*, appearing in Fig. 7

b) *Tunneling phase:* Departing from X_1^*, another point X_2 which produces $F(X_1^*) = F(X_2)$ and not being a local minimum is identified. After that, the procedure may proceed again from phase a) to obtain more local minima and leading finally to the global minimum X_G.

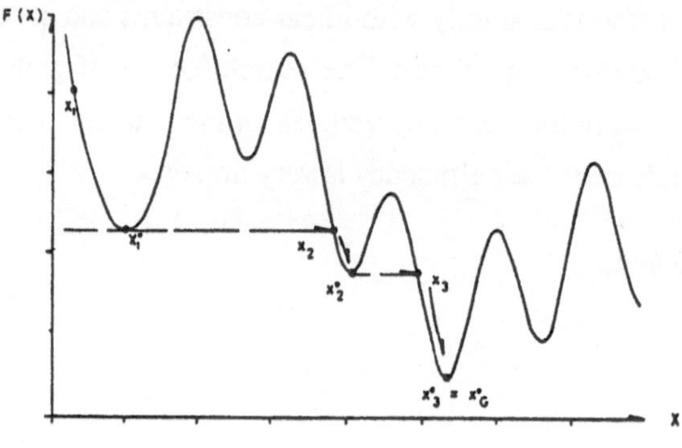

Fig.7

Phase a) does not deserve specific comments. In phase b) instead of working with the objective function $F(X)$ a tunneling function is defined as being:

$$T(X, X_i^*, \lambda_i) = \frac{F(X) - F(X_i^*)}{\left[(X - X_i^*)^T (X - X_i^*)\right]^{\lambda_i}} \quad (4)$$

The parameter λ_i in (4) is calculated on condition that the function T does not have a root at $X = X_i^*$. In a case with one variable Fig. 8-10 shows the tunneling function at a local minimum for different values of parameter λ

$$F(x) = x^6 - 16x^5 + 100x^4 - 310x^3 + 499x^2 - 394x + 140$$

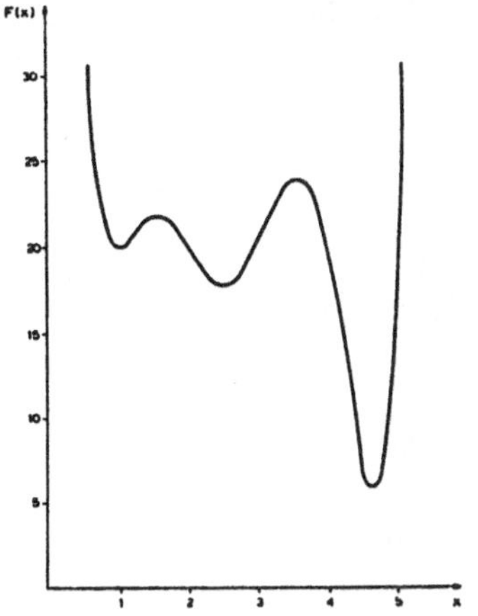

Fig.8

$$T(x,1,0.5) = \frac{F(x) - F(1)}{[(x-1)(x-1)]^{0.5}}$$

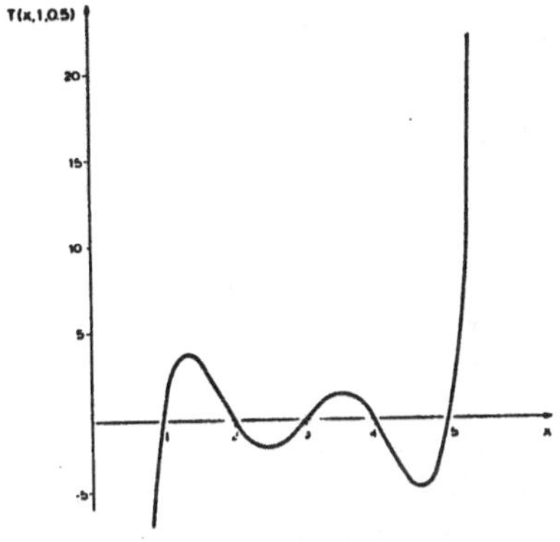

Fig.9

$$T(x,1,1.0) = \frac{F(x) - F(1)}{[(x-1)(x-1)]}$$

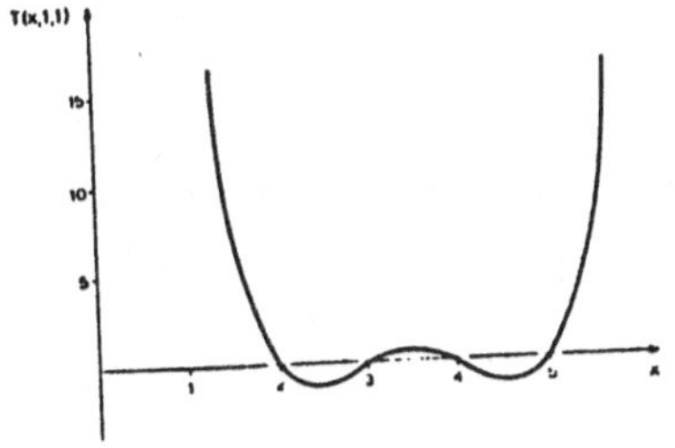

Fig.10

In most cases the parameter value should be obtained numerically and in order to do this following algorithm may be used:

Let us write $F(x) - F(x_i^*)$ as:

$$F(x) - F(x_i^*) = (x - x_i^*)^{\lambda_i} f_i(x) \tag{5}$$

the tunneling function is evaluated at points

$$x = x_i^* + e \qquad y \qquad x = x_i^* + 10e \tag{6}$$

e being a very small number such as $e = 0.0001$.

An initial value for λ is chosen.

if $\lambda \neq \lambda_i$

$$T(x_i^* + \varepsilon, x_i^*, \lambda) = \frac{F(x) - F(x_i^*)}{(x - x_i^*)^{\lambda}} = \varepsilon^{\lambda_i - \lambda} f_i(x) \tag{7}$$

$$T(x_i^* + 10\varepsilon, x_i^*, \lambda) = \frac{F(x) - F(x_i^*)}{(x - x_i^*)^{\lambda}} = (10\varepsilon)^{\lambda_i - \lambda} f_i(x) \tag{8}$$

$$\frac{T(x_i^* + 10\varepsilon, x_i^*, \lambda)}{T(x_i^* + \varepsilon, x_i^*, \lambda)} = 10^{\lambda_i - \lambda} \frac{f_i(x_i^* + 10\varepsilon)}{f_i(x_i^* + \varepsilon)} \approx 10^{\lambda_i - \lambda} \tag{9}$$

If $\lambda = \lambda_i$

$$\frac{T(x_i^* + 10\varepsilon, x_i^*, \lambda)}{T(x_i^* + \varepsilon, x_i^*, \lambda)} \cong 1 \tag{10}$$

Hence, to obtain the value of parameter λ_i it is very useful to start with a small value of λ and to increase it until convergence is obtained in accordance with expressión (10)

It may happen during the process that departing from a point $\underline{X}_i^*$ the tunneling phase will provide a point $\underline{X}_{i+1} = \underline{X}_{i+1}^*$, being also a local minimum. In this situation a new term is introduced in the tunneling function

$$T(\underline{X}, \underline{X}_i^*, \lambda_i, \underline{X}_{i+1}^*, \lambda_{i+1}) = \frac{T(\underline{X}, \underline{X}_i^*, \lambda_i)}{\left[\left(\underline{X} - \underline{X}_{i+1}^*\right)^T \left(\underline{X} - \underline{X}_{i+1}^*\right)\right]^{\lambda_{i+1}}} \tag{11}$$

and the value of parameterλ_{i+1}, is identified as previously indicated. In the case that more local minimum are found at the same objective function $F(\underline{X}_i^*)$ more terms will be added to the tunneling function.

The general scheme of the algorithm for functions with several variables is the following:

1) Start from an initial design $\underline{X}_1$ and obtain a local minimum $\underline{X}_1^*$ for any numerical optimization technique.
2) Define a search direction $\underline{S}$ and evaluate the value of parameter λ_{1S} for this direction. Then calculate a root of the tunneling function and if no root exists change the search direction. When finally a root is found go back to step 1).

Steps 1) and 2) are iteratively carried out until arriving at a local minimum $\underline{X}_i^*$ where it is not possible to obtain a root of the tunneling function. This situation requires a stopping condition: if a prearranged time in the search procedure has elapsed and no root is obtained it is asumed no more local minima exist.

This method has the advantage with respect to other existing techniques that having obtained a local minimum, it works to obtain only those corresponding to equal or lower values of the objective function,and they are found out in a systematically decreasing order. If a maximization problem is considered it performs in the opposite way.

In Fig.11 appears the function

$$F(x_1, x_2) = (4 - 2.1x_1^2 + 0.333x_1^4)x_1^2 + x_1x_2 + (4x_2^2 - 4)x_2^2 \quad (12)$$

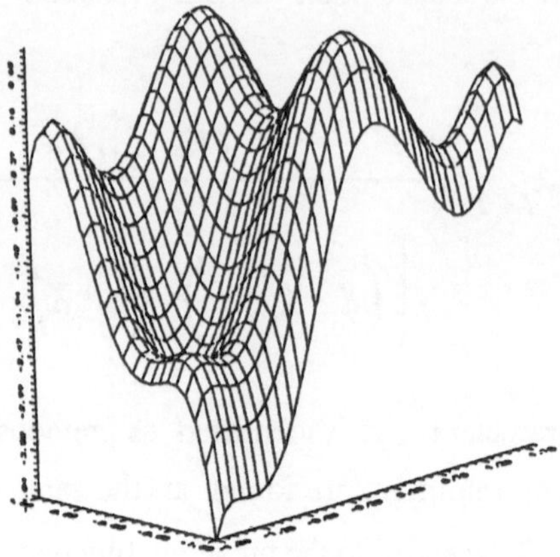

Fig.11

The results obtained along the process to identify its global maximum are:

$$\underline{X}_1^* = |\, 1.0,\ 1.0 \,| \quad (13)$$

$$\underline{X}_1^* = |\, 0.0894,\ -0.7126 \,| \qquad F(\underline{X}_1^*) = 1.0316 \quad (14)$$

$$\lambda_1 = 0.9999 \quad (15)$$

$$\underline{X}_2 = |\, -0.0078,\ 0.6178 \,| \qquad F(\underline{X}_2) = 0.9486 \quad (16)$$

$$\lambda_2 = 0.988 \quad (17)$$

$$\underline{X}_2^* = |\, -0.0894,\ 0.7124 \,| \qquad F(\underline{X}_2^*) = 1.0316 \quad (18)$$

Hence the global maximum value is $F = 1.0316$ and is placed at two points simultaneously.

Until now all examples considered were unconstrained problems. To include a set of constraints and solve constrained optimization may be carried out by combining the tunneling method with penalty functions. By doing that the problem defined in (1) is transformed into:

$$\phi(\underline{X}, r) = F(\underline{X}) + P(g_j, r) \qquad (19)$$

where P is the penalty function and r a parameter varying through the optimization procedure. The scheme of extended interior penalty function has been applied successfully[10] for this purpose. Expression (19) is writen as:

$$\phi(\underline{X}, r) = F(\underline{X}) - r\,P(g_j) \qquad (20)$$

being $P(g_j)$

$$1/g_j \qquad \text{if} \qquad g_j \leq \varepsilon \qquad (21.a)$$

$$((g_j/\varepsilon)^2 - 3(g_j/\varepsilon) + 3)/\varepsilon \qquad \text{if} \qquad g_j > \varepsilon \qquad (21.b)$$

where r, ε are parameter linked at this method.

4. EXAMPLES OF STRUCTURAL OPTIMIZATION

Two examples of optimum design of structures are included to show the performance of the method. The first one is the grillage of Fig. 12

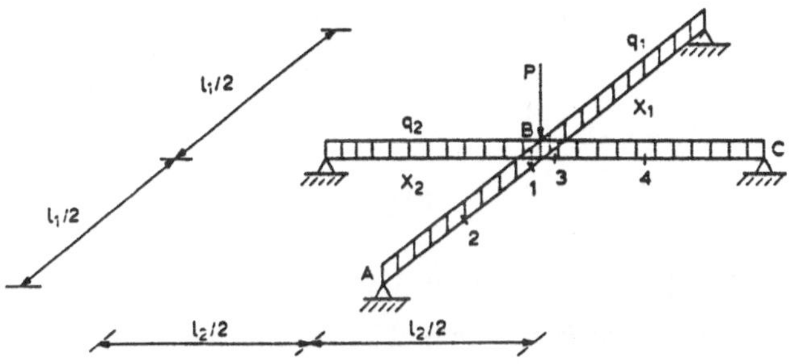

Fig.12

$$\underline{X}_1 = |\,4, 30\,| \qquad F = 9.400 \qquad (22)$$

$$\underline{X}_1^* = |\,2.247, 19.52\,| \qquad F = 6082 \qquad (23)$$

$$\underline{X}_2 = |\,9.166, 13.98\,| \qquad F = 6076 \qquad (24)$$

$$\underline{X}_2^* = |\,9.71, 6.07\,| \qquad F = 2795 \qquad (25)$$

The point $\underline{X}_2^*$ is the global minimum of the problem.

The second example is the four bar truss of Fig. 13 having two design variables and was proposed by Svanberg. It corresponds to a weight minimization subjected to the constraints

$$0.1 \leq x_i \leq 1.5 \tag{26.a}$$

$$q \leq \frac{\sqrt{0.32}}{3E} \tag{26.b}$$

x_1 = cross-sectional area of elements 1 and 2.

x_2 = cross-sectional area of elements 3 and 4.

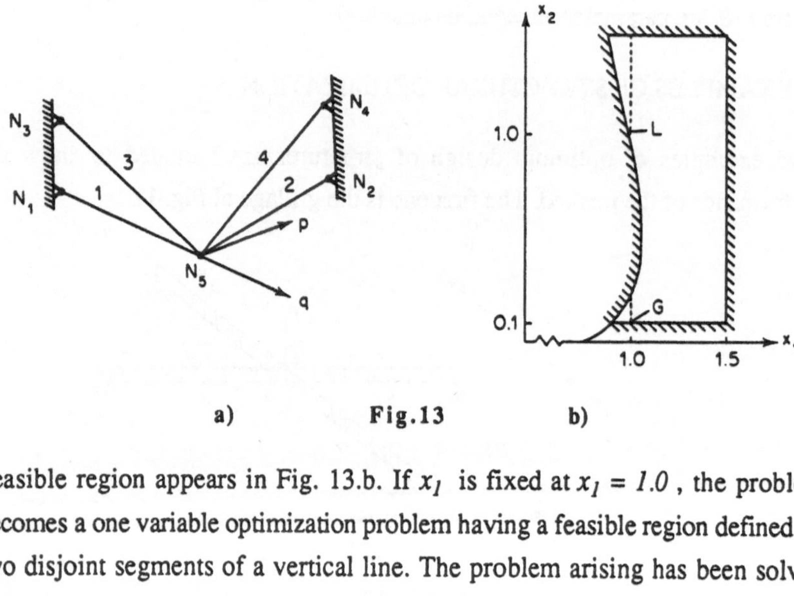

a) **Fig.13** b)

Feasible region appears in Fig. 13.b. If x_1 is fixed at $x_1 = 1.0$, the problem becomes a one variable optimization problem having a feasible region defined by two disjoint segments of a vertical line. The problem arising has been solved providing the following results

Initial design : $x_2 = 1.5$

Local minimum identified:

$x_2 = 1.025$	$F = 0.725$
$x_2 = 0.168$	$F = 0.119$
$x_2 = 0.103$	$F = 0.076$

5. CONCLUSIONS.

As structural optimization leads in many cases to nonlinear problems having nonconvex or disjoint feasible regions, more effort to develop robust global optimization methods is hardly needed.

The tunneling method has the advantage that, after obtaining a local minimum, the algorithm proceeds by finding only better minima in a very orderly way and avoiding the worst ones which are not interesting. By combining this method with penalty functions, constrained optimization problems can be solved.

Application of this approach to quite simple structural optimization problems has been very promising, but more research and checking in problems with larger number of design variables and constraints is necessary.

REFERENCES

[1] SVANBERG, K.: *On Local and Global Optimal in Structural Optimization,* in New Directions in Optimum Design, E. Atrek and alt (ed.) John Wiley, 1984.

[2] MOSES , F. and ONODA, S.: *Minimum Weight Design of Structures with Application to Elastic Grillages,* Int. Jour. Num. Meth. Eng., vol.1, pp-311-331, 1969.

[3] CASSIS, J.H.: *Optimum Design of Structures Subjected to Dynamic Loads,* UCLA-ENG-7451, UCLA School of Engineering and Applied Science, (Ca.), June 1974.

[4] JOHNSON, E.H., RIZZI, P., ASHLEY, H., and SEGENREICH, S.A.: *Optimization of Continuons One-Dimensional Structures under Steady Harmonic Excitation,* AIAA Journal, Vol. 14, Nº 2, pp. 1690-1698, Dec 1976.

[5] PARDALOS, P.M. and ROSEN, J.B.: *Constrained Global Optimization: Algorithms and Applications,* Springer Verlag, 1987.

[6] TORN, A. and ZILINSKAS, A. : *Global Optimization,* Springer Verlag, 1987.

[7] RATSCHEK, H. and RUKNE, I.: *New Computer methods for Global Optimization,* John Wiley, 1988.

[8] LEVY, A.V. and GOMEZ, S. : *The Tunneling Method Applied to Global Optimization,* in Numerical Optimization 1984, SIAM Conference, Boggs, P.T., Byrd, R.H. and Shnabel, R.B. (ed.), 1984.

[9] HERNANDEZ, S.: *Mathematical Theory of Structural Optimization*, in Advances in Design Optimization, Adeli, H. (ed.) Chapman & Hall, (to be published).

[10] HERNANDEZ, S.: *El método de tunelización. Un método de optimización global en problemas no convexos en la ingeniería.* IX Congreso Nacional de Ingeniería Mecánica. Zaragoza, 1991 (in Spanish).

SUBJECT INDEX